COMPREHENSIVE CHEMICAL KINETICS

COMPREHENSIVE

CHEMICAL KINETICS

EDITED BY

C. H. BAMFORD

M.A., Ph.D., Sc.D. (Cantab.), F.R.I.C., F.R.S.
Campbell-Brown Professor of Industrial Chemistry,
University of Liverpool

AND

C. F. H. TIPPER

Ph.D. (Bristol), D.Sc. (Edinburgh)
Senior Lecturer in Physical Chemistry,
University of Liverpool

VOLUME 15

NON-RADICAL POLYMERISATION

ELSEVIER SCIENTIFIC PUBLISHING COMPANY
AMSTERDAM – OXFORD – NEW YORK
1976

ELSEVIER SCIENTIFIC PUBLISHING COMPANY
335 JAN VAN GALENSTRAAT
P.O. BOX 1270, AMSTERDAM, THE NETHERLANDS

AMERICAN ELSEVIER PUBLISHING COMPANY, INC.
52 VANDERBILT AVENUE
NEW YORK, NEW YORK 10017

ISBN 0-444-41252-2

WITH 144 ILLUSTRATIONS AND 116 TABLES

PRINTED IN THE NETHERLANDS

Contributors to Volume 15

C. H. BAMFORD
The Donnan Laboratories,
The University,
Liverpool, England

H. BLOCK
The Donnan Laboratories,
The University,
Liverpool, England

S. BYWATER
Division of Chemistry,
National Research Council of Canada,
Ottawa, Canada

W. COOPER
Dunlop Ltd., Research Centre,
Birmingham, England

F. DOBINSON
Monsanto Company,
Pensacola, Fla.,U.S.A.

M. P. DREYFUSS
Corporate Research,
The B.F. Goodrich Co.,
Brecksville, Ohio, U.S.A.

P. DREYFUSS
Institute of Polymer Science,
The University of Akron,
Akron, Ohio, U.S.A.

A. LEDWITH
The Donnan Laboratories,
The University,
Liverpool, England

J. H. SAUNDERS
Monsanto Company,
Pensacola, Fla.,U.S.A.

J. ŠEBENDA
Institute of Macromolecular Chemistry,
Czechoslovakian Academy of Sciences,
Prague, Czechoslovakia

D. C. SHERRINGTON
Department of Pure and Applied Chemistry,
The University of Strathclyde,
Glasgow, Scotland

O. VOGL
Polymer Science and Engineering Department,
University of Massachusetts,
Amherst, Mass., U.S.A.

Preface

Section 5 is concerned with polymerization reactions in the liquid and vapour phases, including Zeigler–Natta polymerizations, emulsion, suspension and precipitation polymerizations, but not solid phase polymerizations which are to be dealt with in a later section. It seemed reasonable to include polymer degradation reactions of all types, despite the fact that most of these involve solid polymers. Calculation of molecular weight distributions and averages is included, but there is no discussion of the experimental techniques for determining molecular weights. The microstructure of polymer chains is considered when relevant to the kinetics. Rate parameters of individual elementary steps, as well as of overall processes are given when available.

Volume 15 deals with those polymerization processes which do not involve free radicals as intermediates. Chapters 1 and 2 cover homogeneous anionic and cationic polymerization, respectively, and Chapter 3 polymerizations initiated by Zeigler–Natta and related organometallic catalysts. Chapters 4, 5 and 6 deal with the polymerization of cyclic ethers and sulphides, of aldehydes and of lactams, respectively. Finally, in Chapter 7 polycondensation reactions, and in Chapter 8 the polymerization of N-carboxy-α-amino acid anhydrides, are discussed.

C. H. Bamford
C. F. H. Tipper

Liverpool
July, 1975

Contents

Chapter 3 (W. Cooper)

Chapter 4 (P. Dreyfuss and M. P. Dreyfuss)

Polymerization of cyclic ethers and sulphides 259

Chapter 1

Anionic Polymerization of Olefins

S. BYWATER

1. Introduction

Anionic polymerizations comprise those systems in which the growing polymer chain has a terminal reactive carbon atom with a partial or full negative charge. It follows that the most effective initiators will be compounds of the electropositive Group IA or IIA elements. Although it has been known for many years that the alkali metals or their alkyls would initiate vinyl polymerization [1—4], systematic kinetic studies have taken place only in the last dozen years or so. Information on polymerization initiated by the alkaline earth metals is still only qualitative in nature. Early work with the alkali metals and alkyls served principally to define the conditions required to polymerize various olefins. Ziegler [5], in 1936, summarized the results of his experiments and was able to show (largely from product analysis) that, for instance, in the polymerization of butadiene the mechanism could be regarded as a "continuing metal—organic synthesis"

$$R\text{—Metal} + C_4H_6 \rightarrow R\,.\,C_4H_6\,.\,\text{Metal} \rightarrow R\,.\,(C_4H_6)_2\,.\,\text{Metal}\ldots$$

He was aware that termination reactions were of little importance, since successive additions of monomer could be made to polymerize after the complete reaction of the first added portion, even after the reaction mixture had been allowed to stand for some time. The relative efficiency of a group of initiators was also described. In the polymerization of butadiene, for instance, the deep red colour of cumylpotassium disappeared almost immediately, whereas with triphenylmethylsodium, the colour persisted during the polymerization process. With metallic sodium initiation, polymerization was shown to proceed at both ends of the polymer chain via an α,ω-disodium adduct of the diene. It can be said, therefore, that by 1936 the essential characteristics of the process had been described, but interest in this topic was limited for many years afterwards. A few papers appeared in the literature. Beaman [6], in 1948, described the polymerization of methacrylonitrile and methylmethacrylate by sodium and triphenylmethylsodium in liquid ammonia, and similar experiments with styrene were described in 1949 [7, 8] using sodium or potassium amide as initiator.

Higginson and Wooding [9], in 1952, reported what is probably the first kinetic analysis of an anionic polymerization system, that of styrene initiated by potassium amide in liquid ammonia. They found the rate of polymerization to be proportional to $[styrene]^2 [KNH_2]^{1/2}$ and suggested a mechanism which involved initiation by the amide anion, propagation (which must involve free anions [10]) and chain limitation by proton transfer from ammonia to the chain-end anion. The kinetic analysis involved the classical stationary state method which would predict a second order dependence on the styrene concentration as it is involved in both initiation and propagation steps. The square root dependence on initiator concentration results if the real initiator (NH_2^-) is present in small amounts in equilibrium with a large excess of undissociated KNH_2. The dissociation constant was determined to be about 1×10^{-4} which justifies the assumption made. This system is, of course, quite different to those in inert solvents, described by Ziegler where the stationary state method is completely inapplicable. The important chain transfer reaction ($\overline{DP} \sim 12$) in liquid ammonia, together with the negligible consumption of KNH_2, produces a system where a quasi-stationary concentration of ions would occur. In the "metal—organic synthesis" systems of Ziegler, the concentration of active polymer chains would increase during the initiation step until all the initiator was reacted, at which point a true constant concentration of propagating polymer chains would exist (equal to the initial initiator concentration). Subsequent monomer consumption would then be expected to be a first order process.

Wooding and Higginson [11] also attempted a correlation with the base strength of the initiator, of the ability of various sodium salts to initiate polymerization of acrylonitrile, methylmethacrylate, styrene and butadiene. They were able to show that even relatively weak bases such as sodium methoxide would polymerize acrylonitrile; methylmethacrylate required stronger bases such as fluorenyl- or indenylsodium, and styrene or butadiene required very strong bases of the type triphenylmethylsodium or sodamide or even metal alkyls.

The monomers studied in these early investigations are characteristic of the wider range now known to polymerize anionically. These involve vinyl monomers having substituent groups which will stabilize negative charge when the unit is incorporated at the growing chain end. They may be either non-polar monomers such as the dienes and styrene and its derivatives or polar monomers such as nitroethylene, acrylonitrile, acrylates, acrylamides, vinyl ketones and vinyl chloride and their derivatives. In general, detailed kinetic studies are possible only with the first group of monomers, where polymerizations are relatively free of unwanted side reactions. With the second group, attack of the initiator or the growing polymer chain-end on the polar group almost always leads to a complex reaction mechanism which is difficult to interpret.

The almost explosive growth of the study of polymerization initiated by organo-metallic compounds dates from the years 1954—1956. The discovery of the industrially important Ziegler—Natta complex catalysts [12, 13] encouraged research into organo-metallic chemistry generally. At about the same time, research into the simpler anionic systems was stimulated by the discovery that the lithium metal or lithium alkyl initiated polymerization of isoprene could yield a highly *cis*-1,4 poly-isoprene [14]. In the same year, Szwarc [15] described a powerful system for kinetic study of anionic polymerization; initiators such as sodium naphthalene were used in cyclic ether solvents such as tetrahydrofuran. These systems have the advantage that the initiator is soluble, easy to prepare in reasonable purity, and generally initiates polymerization rapidly. The solvents are superior to diethyl ether in terms of solubility of the polymer formed, and the higher dielectric constant enables some measurable ionization to occur, with the possibility of measurement of free anion and ion-pair contributions to the rate. In contrast to reactions in liquid ammonia, proton transfer from the solvent cannot occur and the major side reaction, ring cleavage, is usually slow enough not to affect the rate measurements. Szwarc was able to show with this system that initiation occurred by electron transfer from the aromatic ion (pair) to styrene monomer to form a radical anion which immediately (or after prior addition of monomer) dimerized to form an α,ω-dianion (pair) propagating from both ends. Termination reactions were not important and the final DP of the polymer was simply determined by the ratio [monomer]$_0$/0.5 [catalyst]$_0$.

More detailed kinetic studies of anionic polymerization are developments of the pioneer work of Ziegler in hydrocarbon solvents and of Szwarc in cyclic ether solvents. These are the so-called "living polymer" systems [15] carried out under conditions where reactions other than chain initiation and propagation can be neglected. Both reaction steps can, in principle, be studied independently. The method of approach is necessarily different from that used for free radical or cationic polymerizations. Chain initiation can be measured from the rate of disappearance of initiator or spectroscopically from the rate of formation of carbanionic centres. Once all the initiator is consumed, the rate of monomer consumption simply measures the propagation rate. If the initiation rate is much faster than the propagation rate, all chains are formed essentially simultaneously and apart from a small statistical factor, grow at equal rates producing a polymer of very narrow (Poisson-type) molecular weight distribution.

The ability to measure the propagation rate in solvents of widely differing dielectric constant and dipole moment, has provided much information on the active species present. In hydrocarbon solvents both the ion-pair and its aggregates are present, viz.

$$(P^-M^+)_n \rightleftharpoons nP^-M^+$$

In polar solvents, self association is usually of no importance but the ion-pairs become solvated and also dissociate to a small extent to free ions, viz.

$$P^-M^+ \rightleftharpoons (P^-M^+).S \rightleftharpoons P^- + M^+.S$$

All these entities differ in their reactivity towards monomer and the study of anionic polymerization is, in large measure, the effort to determine the concentration and reactivity of each type of ionic species.

2. Some general experimental considerations

The initiators and active polymer chains involved in anionic polymerization react instantly with water, carbon dioxide and oxygen. Polymerizations must, therefore, be carried out in high-vacuum systems or under a blanket of carefully purified inert gas. The former method is, in general, more satisfactory. At concentrations of organo-metallic agent much less than millimolar, the presence of reactive impurities on the walls of the vessel produces an important source of error, for a measurable part of the initiator will be destroyed. Initiator concentrations determined as the amount originally added are then too high. This effect can be circumvented by measurement of the true organo-metallic concentration during or after the polymerization. The determination can be achieved spectroscopically, by titration with *tert*-butanol or butylchloride at the end of reaction, or from the degree of polymerization of the polymer formed. Even so, a source of error still remains, for the inorganic salts produced by initiator destruction may, in some cases, have an effect on the rate of polymerization. The most satisfactory solution is to wash all reaction vessels and ancillary glassware in vacuum immediately before use with a solution of initiator or similar reactive organo-metallic reagent. If this is non-volatile, the solution can be drained from the reaction flask, which can then be rinsed by solvent refluxed from the solution. Finally the solvent used for the polymerization can also be distilled from the same solution.

It follows from this discussion that all solvents and monomers used must be carefully purified. Hydrocarbons should be stirred over sulphuric acid for many days and ethers refluxed over sodium—potassium alloy or sodium fluorenone before fractionation. Traces of unsaturated materials in aliphatic hydrocarbons can be removed by silica gel. After fractionation, a preliminary drying over calcium hydride can be followed by storage over sodium—potassium alloy for ethers, or a treatment with butyllithium or similar non-volatile reactive organometallic reagent for hydrocarbons. Monomers cannot be treated quite so drastically, but fractionation followed by a pre-polymerization in vacuum over butyllithium appears to be reasonably effective. The polymerization can be prevented from becoming explosive by breaking a fragile bulb containing polymethylmethacrylate into the vessel when the mixture becomes viscous. The organo-metallic centres are destroyed by the ester groups on the polymer, producing no volatile products. The monomer can be distilled into a storage vessel from this mixture. Use of techniques similar to those described above can be shown to be suitable for the study of polymerizations at initiator concentrations as low as 2×10^{-5} M. More detailed descriptions of the type of apparatus required can be obtained from several references [16—20].

One of the most useful properties of polymerization systems involving styrene or the dienes is the appearance in the solutions of colour associated with the actual active centres. For example, "living" polystyrene solutions in tetrahydrofuran show a strong red colouration which disappears immediately on the introduction of a trace of oxygen, water or carbon dioxide. The source of this colour is a strong absorption band ($\epsilon \sim 10^4$) in the near-ultraviolet region of the spectrum. The positions of the maximum and intensity of absorption are not very sensitive to changes in solvent or counter-ion (Fig. 1) [λ_{max} polystyryl anion 328—346 mμ, ϵ_{max} 1.2—1.4 $\times$ 10^4; λ_{max} polyisoprenyl anion 270—315

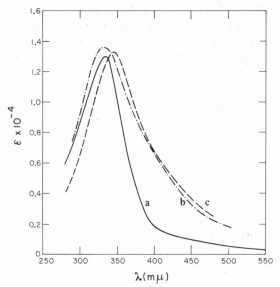

Fig. 1. Absorption spectra of polystyryl anions. (a) polystyryllithium in benzene; (b) polystyrylpotassium in benzene; (c) polystyrylcesium in tetrahydropyran.

mμ, ϵ_{max} ~8 $\times$ 10^3; λ_{max} polybutadienyl anion 275—312 mμ, ϵ_{max} ~8 $\times$ 10^3]. The occurrence of this absorption band is connected with the presence of a charge-delocalized anion, which is the actual active centre for polymerization, viz.

NMR measurements on such species confirm that there is extensive charge delocalization particularly with styryl anions. Molecular orbital theory suggests that for such odd-alternant systems there will be an electronic transition from the highest occupied level to the next higher one and that this will have the same energy in cation or anion. This prediction is borne out in practice, at least for styryl ions, although the rapid isomerization of the cations made this at first difficult to confirm [21, 22]. The energy levels will be perturbed by the physical surroundings (counter-ion, association of ion-pairs) but this effect does not seem to be large.

It is clear that the measurement of optical density at the appropriate wavelength is a valuable tool for routine monitoring of the real concentration of active centres. This is particularly valuable since accurate measurement of initiator concentrations is a time-consuming process when working with completely sealed systems. In addition, this technique gives clear warning if side reactions occur. Slow reaction of the active centres with solvent or impurity is recognized by a decrease in absorption. Isomerization reactions produce an intensity decrease at the normal wavelengths and an increase in absorption elsewhere, invariably at longer wavelengths. Both types of side-reaction can be observed with polystyryl sodium in tetrahydrofuran at room temperature [23—25]. In this case isomerization was shown [26] to occur via the elimination of sodium hydride and the ultimate formation of the ion

viz.

$$\sim\!CH_2-\underset{\underset{Ph}{|}}{CH^-}Na^+ \longrightarrow \sim\!CH=\underset{\underset{Ph}{|}}{CH} + NaH$$

$$\sim\!CH_2-\underset{\underset{Ph}{|}}{CH^-}Na^+ + \underset{\underset{Ph}{|}}{CH}=CH\!\sim \longrightarrow$$

$$\sim\!CH_2-\underset{\underset{Ph}{|}}{CH_2} + Na^+\underset{\underset{Ph}{|}}{CH}\cdots\bar{C}H\cdots\underset{\underset{Ph}{|}}{C}\!\sim$$

The final product has increased charge delocalization and is unreactive in polymerization. Isomerization in this case is fortunately slower than polymerization although important over several hours. In non-polar solvents, such as benzene, isomerization reactions also occur, but these require many days at room temperature (Fig. 2) and are completely unimportant in the polymerization process. They are important at elevated temperatures [27].

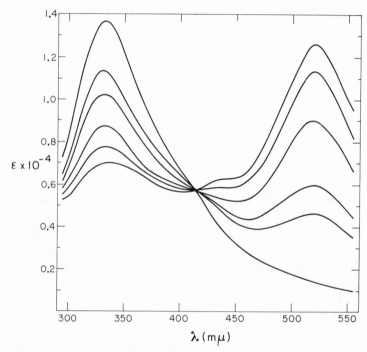

Fig. 2. Changes in absorption spectrum of polystyrylpotassium in benzene at room temperature. Time in order of increasing absorption at 520 mμ, 1, 21, 45, 117, 291, 460 h.

Isomerization reactions with diene monomers are very rapid in tetrahydrofuran at room temperature [28, 29]. Within a few minutes, extensive isomerization has occurred and it is difficult to measure the rate of monomer addition to the normal anions. These processes are much slower at lower temperatures [30].

In solvents of moderate dielectric constant (e.g. THF, $D \sim 7.6$ at room temperature) the polymerization systems show measurable conductance. It is clear that the concentration of free ions cannot be large, since conventional inorganic salts are not highly dissociated in such solvents. Figure 3 shows the variation of equivalent conductance with concentration for polystyrylsodium (two ended, $Na^+CH(Ph)CH_2 . (CHPhCH_2)_{50}$-$CH_2CH(Ph)^-Na^+$) in THF at 20°C. Λ varies as $[C]^{-0.5}$ as expected if the degree of dissociation to free ions is low. A dissociation constant of

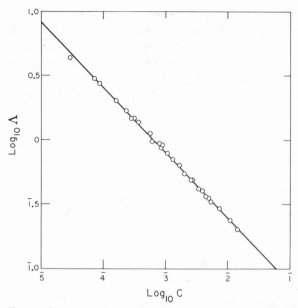

Fig. 3. Conductance of polystyrylsodium in tetrahydrofuran at 20°C. (Λ) equivalent conductance; (C) total ionic concentration [24].

$1.5-1.6 \times 10^{-7}$ mole l^{-1} can be determined from the results [24] which corresponds to about 1% free ions at total concentrations about milli-molar. In general, dissociation constants in the range 10^{-5} to 10^{-11} are found in cyclic ether solvents ($D > 5$). In no case is the concentration of free carbanions more than $1-2\%$ of the total ionic concentration but their presence is extremely important.

Another valuable technique for study of the physical form of the active centres is viscometry. It is reasonably simple to add a vacuum viscometer to the sealed polymerization vessel and to measure the viscosity of the solution after polymerization is complete but while active centres still exist [31]. These values can be compared with the viscosity after destruction of the centres by traces of alcohol. As the solution viscosity is sensitive to the apparent total molecular weight of the polymer any association of polymer molecules via their active ends will be readily apparent. This technique was used [32] in very early experiments to show that in hydrocarbon solvents extensive association of polymer chains often occurs and has an important effect on the polymerization kinetics. Light scattering techniques can be employed in a similar manner [33] enabling the apparent molecular weight to be directly determined.

The above-mentioned techniques have had an important effect on elucidation of kinetics and mechanism of anionic polymerization. They depend, naturally, on the fact that there is essentially no termination in the systems, and that the concentration of active centres is relatively large. Measurements made after all monomer has been consumed apply to a system which is identical to that during the polymerization except for the DP of the polymer. The rapid elucidation of mechanisms depends in no minor manner on this advantage which is not available with free radical or cationic systems.

3. Polymerization of styrene and dienes in hydrocarbon solvents

3.1. THE INITIATION STEP

Studies of the mechanism of chain initiation in hydrocarbon solvents are restricted to experiments with lithium alkyls. No alkyls or aryls of the other alkali metals are sufficiently soluble in these solvents. The solubility is connected with a high degree of association in solution to form an "inverted micelle" with the lithium atoms inside the structure, surrounded by the alkyl chains. This apparently serves to solubilize the aggregate. More detailed descriptions of the physical properties of lithium alkyls are outside the scope of this review and can be found elsewhere [34]. The degree of association is normally four or six for simple alkyls in either benzene or cyclohexane ($n = 6$; EtLi, n-BuLi; $n = 4$; $tert$-BuLi, $sec.$-BuLi, iso-PrLi) [35, 36]. The determinations of association number are usually made cryoscopically and are made at concentrations rather higher than used in polymerization studies. There is, however, no general trend with concentration and in some cases measurements have been made at concentrations as low as 10^{-3} M with no evidence of dissociation. Only in one or two systems [35] where $n = 4$ in the dilute region does the degree of association rise to six at high concentrations. Steric requirements are obviously an important factor in determining the degree of association. The only compounds reported having an appreciably lower n value are menthyllithium [37] and benzyllithium [38] where $n = 2$.

The association phenomenon must have some influence on the rate of reaction of lithium alkyls. Indeed, early experiments showed that the overall rate of polymerization in hydrocarbon solvents is insensitive to the concentration of alkyl used [39, 40]. This led to the suggestion that the reactive entities were the dissociation products of the aggregates. More detailed studies require the measurement of the initiation rate itself, which is quite slow under many conditions [17, 41]. Figure 4 shows results at a rather low monomer/initiator ratio. While this exaggerates the slowness of the initiation step, even at higher ratios with many lithium alkyls appreciable amounts of monomer are consumed before chain initiation is complete. The figure also shows that in benzene as solvent, no induction period is present and an initial rate of chain initiation can be readily measured. It is necessary to use initial rates because as soon as appreciable amounts of polymer-containing lithium species are formed by the initiation process, these tend to form mixed aggregates with the

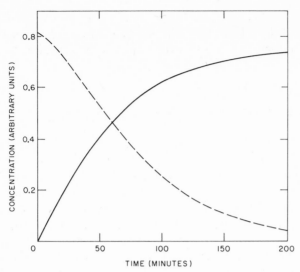

Fig. 4. Concentration of styrene monomer (- - -) and polystyryllithium (———) after mixing styrene (1.4×10^{-2} M) with n-butyllithium (1.1×10^{-3} M) in benzene at 30°C.

initiator [32, 42] and thereby alter the effective reactivity of the lithium alkyl.

Figure 5 shows the rate of reaction of n- and $sec.$-butyllithium with styrene in benzene as measured spectroscopically from initial rates of

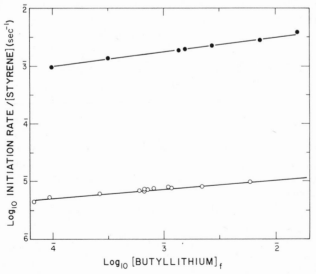

Fig. 5. The initial rate of reaction with styrene of $sec.$-butyllithium, (●) and n-butyllithium, (○) in benzene at 30°C as a function of formal concentration of butyllithium [17, 36].

formation of active centres, $Bu(CH_2-CHPh)_n Li$, in benzene [17, 36] at 30°C, viz.

$$BuLi + PhCH=CH_2 \rightarrow Bu . CH_2-CH(Ph)Li$$

There is a large difference in initiation rates between the two initiators, but in both cases the reaction order in lithium alkyl is fractional, whereas the dependence on monomer concentration is, as expected, of the first order. The lines drawn have slopes of 1/4 (*sec.*-BuLi) and 1/6 (*n*-BuLi). There seems to be a clear relationship between association number (n) and reaction order and simple mechanisms can be suggested [17] (although not, of course, proved) of the type,

$$(BuLi)_n \rightleftharpoons nBuLi \qquad K_1 = \frac{[BuLi]^n}{[(BuLi)_n]} \qquad (1)$$

$$BuLi + M \rightarrow Bu . M . Li \qquad k_i \qquad (2)$$

$$R_i = k_i K_1^{1/n} [(BuLi)_n]^{1/n} [M]^\star$$

To predict the required orders, the equilibrium must be rapidly established and the concentration of free butyllithium very small and its reactivity with monomer not large enough to disturb the equilibrium. The dissociation process can proceed via intermediate degrees of association, provided that the major species present is the *n*-fold associate and that intermediate species are also unreactive. Similar results have been found for isoprene [33, 36], diphenylethylene and fluorene [43—45] in benzene. The results are summarized in Table 1.

TABLE I

Kinetic parameters for the reaction of lithium alkyls with styrene, isoprene and 1,1-diphenylethylene in benzene at 30°C

Initiator	Monomer	$k_i(K_1/n)^{1/n}$	$E_{apparent}$ (kcal mole^{-1})	Ref.
n-BuLi	Styrene	$2.3_3 \times 10^{-5}$	18.0	17
sec.-BuLi	Styrene	9.8×10^{-3}		36
n-BuLi	Diphenylethylene	$4.1_7 \times 10^{-5}$	16.4 (15.8)	43
tert.-BuLi	Diphenylethylene	$2.4_7 \times 10^{-3}$	15.5	44
EtLi	Diphenylethylene	2.4×10^{-5}	17.3	45
sec.-BuLi	Isoprene	2.8×10^{-3}		36
tert.-BuLi	Isoprene	$3.5_6 \times 10^{-3}$		46

★ If the concentration of butyllithium added is defined in terms of formula weight $[BuLi]_f = n[(BuLi)_n]$, then

$$R_i = k_i(K_1/n)^{1/n} [BuLi]_f^{1/n} [M]$$

The general validity of such schemes has been questioned [47] on the basis that the rate of dissociation of the aggregated alkyl would not be fast enough to produce the experimentally observed initiation rates. The dissociation process is postulated to proceed via dimer and it is suggested that the activation energy for dimer dissociation would be as high as 37 kcal mole^{-1}, which would cause a very low rate of production of butyllithium monomer. No concrete information on this dissociation energy is, of course, available and the validity of the criticism depends on the plausibility of this estimated dissociation energy obtained by extrapolation from data on the dissociation of other dimers in solution. A major weakness of the argument is that it is too general and takes no account of solvent effects. The dissociative mechanism, if true, can only be expected to hold under a small range of conditions.

Experiments in cyclohexane and hexane show that benzene cannot be regarded as an inert solvent and that the initiation mechanism differs between aliphatic and aromatic solvents. Studies on intermolecular exchange [38, 48—50], which presumably reflects the primary dissociation process, also indicate differences between the two types of solvent as well as between different lithium alkyls. Recent experiments with *tert.*-butyllithium [46], which has the lowest intermolecular exchange rate of the alkyls, indicate a more complex behaviour in the initiation reaction in benzene. This may be connected with a lower rate of dissociation of this alkyl.

The dissociation mechanism certainly cannot hold in aliphatic hydrocarbon solvents. In cyclohexane or hexane the initiation rate curve is sigmoidal in shape [33, 51—53] (Fig. 6). This effect should not be confused with the same form found if the overall monomer consumption is measured; this occurs even in benzene (Fig. 4) and is a result of competing initiation and propagation steps followed by the inevitable depletion of reactants. The sigmoidal form of the initiation reaction process is not always immediately apparent since the initial slow initiation period gets shorter as the initiation rate increases (initial and maximum rates approach each other as the initiator concentration is increased [51]). The inaccuracy of gas chromatographic analysis and presence of inorganic lithium salts also serve to make it less obvious in some cases. The overall rates are markedly lower than in benzene solution, but still show pronounced differences between different lithium alkyls [54]. The apparent orders in initiator are much closer to unity (range 0.7—1.4) as determined from the maximum rate of initiation. No exact meaning can be attributed to the maximum rate, however, for the process is too complex, involving as it does at this point, the presence of mixed aggregates containing polymer species.

Only in one case (*sec.*-BuLi plus isoprene in hexane) has an attempt been made to measure the true initial rate at very low conversions [51]. In this case, the most favourable for experimental measurements, it was

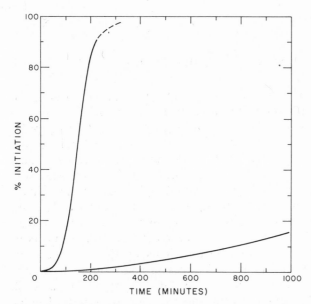

Fig. 6. The reaction of butyllithium with isoprene in hexane at 30°C. Upper curve, *sec.*-butyllithium (7.9×10^{-5} M); lower curve, *n*-butyllithium (1.0×10^{-4} M). Isoprene concentration 0.04M.

found that the initial rate of initiation is first order in monomer and initiator (Fig. 7) and very much lower than in benzene. This suggests that the initiation process in this case involves direct attack of a monomer molecule on the *sec.*-BuLi aggregates to incorporate a polyisoprenyl-

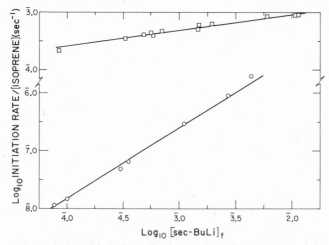

Fig. 7. The initial rate of reaction of *sec.*-butyllithium with isoprene as a function of formal concentration of butyllithium: (○) Hexane; (□) benzene; temperature 30°C. [36, 51].

14

lithium (DP = 1) molecule, viz.

$$(sec.\text{-BuLi})_4 + M \rightarrow [(sec.\text{-BuLi})_3(sec.\text{-Bu}-M-Li)_1] \tag{3}$$

It seems probable that the same process occurs with other initiators in aliphatic solvents. The subsequent increase in initiation rate can only be attributed to the increased reactivity of aggregates containing poly-isoprenyllithium [36, 51]. The rate curve can be fitted up to 3% conversion by assuming that the first mixed aggregate formed reacts twelve times faster with monomer than do simple *sec.*-BuLi aggregates. It is clear that a complete analysis of the overall rate of initiation in aliphatic hydrocarbon solvents is impossible, because, as initiation proceeds, even more complex mixtures of aggregates are formed of the type [(*sec.*-BuLi)$_m$ (polyisoprenyl-Li)$_n$] each of differing reactivity and with a constantly changing concentration distribution. This at first produces an increasing initiation rate, until finally both initiator and monomer concentrations are depleted and the rate necessarily falls, eventually to zero. It is conceivable that under some conditions the rate of intermolecular exchange between various aggregates in solution could have some influence also on the reaction path. What information is available [49], however, suggests that aggregates of *n*- and *sec.*-butyllithium exchange rapidly with polyisoprenyllithium aggregates and an equilibrium distribution would always be present. Once again, however, *tert.*-butyllithium behaves anomalously showing a much slower equilibration in its interchange with other alkyls and a non-statistical distribution in its mixtures with polyisoprenyllithium.

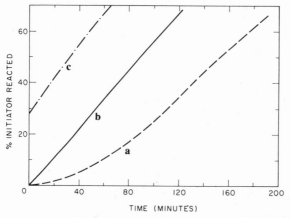

Fig. 8. The reaction of *sec.*-butyllithium with isoprene in hexane at 30°C. (a) normal conditions [*sec.*-BuLi] = 1.1 x 10^{-3} M, [isoprene] = 5 x 10^{-2} M; (b) in the presence of 2 x 10^{-3} M lithium *tert.*-butoxide; (c) in the presence of sufficient polyisoprenyl-lithium to simulate 30% conversion at time zero [51].

It can be confirmed that it is the polyisoprenyllithium formed in the initiation step which provides the autocatalytic mechanism [51]. If isoprene and *sec.*-butyllithium are mixed in presence of polyisoprenyllithium (the latter having been formed earlier in a separate reaction), then the initial rate of reaction of the *sec.*-butyllithium and isoprene is much faster than normal and no induction period occurs. The actual rate is then roughly that expected had the polyisoprenyllithium been formed normally as part of the initiation step (Fig. 8). These results also imply rapid mixing of *sec.*-BuLi and polyisoprenyllithium aggregates because if exchange was not rapid, the added polymer species could have little effect on the initiation process.

The slow initial phase in aliphatic solvents can also be eliminated artificially by the addition of lithium alkoxides to the reaction mixture [51, 52, 55, 56] as shown in Fig. 8. This provides one possible method to increase the overall initiation rate in solvents such as hexane or cyclohexane, the combination *sec.*-butyllithium plus *n*-butoxylithium being particularly efficient with an enhanced initiation rate at all conversions [56]. Other combinations of alkoxide and alkyl give an increased initial rate* but no development of autocatalysis so that the overall rate is actually diminished (e.g. *n*-butyllithium plus *tert.*-butoxylithium [56]). The situation is obviously too complex to analyse at finite conversions involving as it does, three types of lithium compound all of which may be involved in cross-association. It can be seen at least qualitatively that the overall rate may be accelerated, depressed or unchanged depending on the reactivity of butyllithium in a particular environment compared with that in which it finds itself in a normal polymerization (mixed with polyisoprenyllithium). Generally it can be said that, in the initial stage, the reactivity of butyllithium is enhanced by complexing with lithium butoxide.

A knowledge of the action of lithium alkoxides in these systems is important because they are always present to some extent, formed by fortuitous catalyst destruction by traces of oxygen. The other likely impurity is lithium hydroxide produced from traces of water in the system. It apparently decreases the initiation rate [53] but primarily [55] because it reacts with organo-lithium compounds

$$RLi + LiOH \rightarrow RH + Li_2O \tag{4}$$

Lithium oxide precipitates out of solution and has little effect on the polymerization, but the overall effect is to reduce the real initiator concentration, and that of polymeric lithium compounds.

---

* Care is necessary in comparison of results of different authors. In some cases the rates reported are maximum rates as the slow initial period was obscured by experimental scatter.

In summary, a comparison of initiation reactions in benzene with those in cyclohexane or hexane reveals large differences in behaviour. The rates of initiation and kinetic orders in initiator concentration are very different, particularly in the initial stages which represent the only time at which a simple alkyl—olefin reaction occurs. Even after the auto-acceleration phase in aliphatic solvents the rates are well below those observed in benzene. Only with menthyllithium, which is much less highly aggregated in solution ($n = 2$), do these generalizations appear to fail. Even in cyclohexane, no induction period is observed and the rates of initiation although lower than in toluene are of comparable magnitude [57]. The lower degree of association together with a bulky hydrocarbon group could severely limit the formation of mixed aggregates. The extremely fast rates observed suggest that association number is a very important factor influencing reactivity. This is also suggested from the comparative rates of initiation of n- and $sec.$-butyllithium.

In benzene, the initiation rate decreases with time mainly attributable to decrease in reagent concentrations. In fact in the styrene—benzene—n-butyllithium system the monomer consumption curve can be fitted satisfactorily [17, 58] using kinetic constants derived for the initiation and propagation steps. The former rate coefficient is obtained under conditions where little polystyryllithium is present and the latter co-efficient is obtained under conditions where no initiator is present. Yet they fit the reaction curve reasonably well over intermediate conversions where mixed aggregates can occur. Viscosity measurements [46] show that polystyryllithium does form mixed aggregates with n-butyllithium in benzene, but their effect on the polymerization rate appears negligible. Similar results are observed for isoprene polymerization in benzene although no quantitative measurements were made in this case. One must suppose that for a dissociative process the effect of complexing the initiator with the growing chains is not large and perhaps results in a small *decrease* of initiator reactivity. This would be much harder to detect than an acceleration of rate.

3.2 THE PROPAGATION STEP

Once chain initiation is complete, the monomer consumption rate is determined only by the chain propagation step. With the less efficient lithium alkyl initiators in hexane or cyclohexane, rather large amounts of monomer are needed to complete chain initiation. The appearance of a first order decay in monomer concentration, invariably obtained in these experiments, is not a very sensitive indication of the complete absence of initiator. Analysis of trial samples for hydrolysis products of lithium alkyls or spectroscopic determination that the polymer anion concentration has reached a plateau are preferable. A "seeding technique" is often used [32, 59] where the real initiator is a pre-formed active polymer of low molecular weight. It cannot always be assumed that this does not still contain residual lithium alkyl, except with the very reactive initiators, but it undoubtedly reduces the time required for complete initiation compared with use of the lithium alkyl itself.

The propagation rate can also be determined for the other alkali metals. In this case it is preferable to produce a polymer of low DP by the reaction of a monomer solution with an alkali metal film. The oligomeric active polymer, which is soluble in hydrocarbons, can be filtered-off and used as initiator. The simple alkyls or aryls of the higher alkali metals are almost insoluble in these solvents and are not easy to produce in a high state of purity.

Figures 9 and 10 show logarithmic plots of $(1/M)(dM/dt)$ versus formal concentration of $R(M)_n Li$ for the monomers styrene, isoprene and

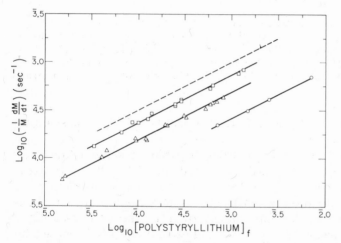

Fig. 9. The rate of chain propagation in the lithium alkyl initiated polymerization of styrene as a function of concentration of growing chains. The lines drawn have a slope of 0.50 corresponding to a reaction order of one half. (○) Cyclohexane 30°C [62]; (△) benzene 30°C [17]; (□) cyclohexane 40°C [61]. The broken line indicates the rates expected in benzene at 40°C from the activation energy determined in ref. 17.

butadiene. In all cases the reaction order is fractional in lithium compound in benzene, cyclohexane or hexane. The one half order for styrene can be directly compared with the association number which is two [32, 52, 60], as measured under actual polymerization conditions. A simple dissociative mechanism of the type,

$$(R . M_n . Li)_2 \rightleftharpoons 2 R . M_n Li \qquad K_2 \qquad (5)$$

$$R . M_n . Li + M \rightarrow RM_{n+1} Li \qquad k_p \qquad (6)$$

$$(1/M)(dM/dt) = k_p K_2^{1/2} [(R . M_n Li)_2]^{1/2}$$

18

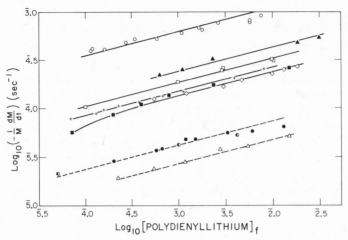

Fig. 10. The rate of chain propagation in the lithium alkyl initiated polymerization of isoprene (——) and butadiene (- - -) as a function of concentration of growing chains. The lines drawn have a slope of 0.25. (○) Cyclohexane 40°C [61]; (▲) benzene 30°C [64];(□) cyclohexane 30°C [64]; (+) cyclohexane 30°C [33]; (■) heptane 30°C [65]; (◇) hexane 30°C [64]; (●) cyclohexane 30°C [52]; (◖) heptane 30°C [63]; (△) hexane 30°C [64].

with the same conditions noted for the dissociative initiation mechanism, would be hard to refute.

With butadiene and isoprene, the orders in lithium alkenyl are near one quarter (or perhaps even one sixth for butadiene [52, 63]). Data exists where the reported order is nearer one half [59, 66], characterized by rates which are close to those given in Fig. 10 at initiator concentrations near 10^{-2} M, but which become much lower as the concentration is decreased. In one case, later work has indicated that this is caused by increased initiator destruction at low concentrations and it seems reasonable to suppose that this explanation is valid for the whole group of experiments. Comparison of kinetic order with degree of association is hindered by the fact that there is no agreement as to the association number of polybutadienyl or polyisoprenyllithium, it being variously described as two [32, 60] or near four [33, 61]. The association phenomenon, however, undoubtedly plays a role in the observed kinetics.

For all three monomers, the observed rate is not a simple measure of the rate of the propagation step (6). Attempts have been made to measure the actual rate coefficient, k_p, either by measurement of the dissociation constant of the aggregates, or by decreasing the active polymer concentration to the point at which association no longer exists. Concentrated solution viscosities (which are extremely sensitive to the apparent molecular weight, $\eta \sim M^{3.4}$) have been used [67—69] to determine association numbers for polyisoprenyllithium in hexane and polystyryllithium in benzene. The major species present was found to be the dimer in both cases. Dissociation constants were determined to be 5.4×10^{-7} and 1.6×10^{-6} at $30°C$ leading to k_p values of 5 and 17 l mole^{-1} sec^{-1}. The difficulties of accurate determination of K_2 in these systems are very large being dependent on accurate determination of small deviations from an association number of two [70]. The values of rate coefficient obtained should therefore be considered more as estimates, with an error which would tend to make them minimum values.

The propagation coefficient for polyisoprenyllithium in heptane has been estimated [65] by working at concentrations as low as 5×10^{-7}M. The reaction order was found to increase below 10^{-4}M (see Fig. 10) and became first in polyisoprenyllithium at about 5×10^{-6}M. Assuming this corresponds to complete dissociation of the aggregates, a k_p value of 0.65 l mole^{-1} sec^{-1} at $20°C$ is determined. Unfortunately, the apparent reaction order continuously increases and approaches two at concentrations about 5×10^{-7}M, which raises serious doubts about this method of estimation of k_p. It is more likely that at least part of the fall-off of rates below 10^{-4}M concentrations was caused by the presence of lithium salts produced by reactions with trace moisture. In general no techniques described in the literature are adequate below concentrations of about 2×10^{-5}M when dealing with these extremely sensitive systems [70].

With o-methoxystyrene, the presence of the methoxy group appears to reduce the tendency to self-association of active centres [71]. With n-butyllithium in toluene, chain initiation is rapid, as would be the case if the ether group was added as THF or dioxane. In the propagation reaction, the order in active centres increases from 0.5 at 10^{-2}M to 0.7 at 5×10^{-4}M. If the dissociative mechanism is assumed to hold, both K_2 and k_p can be evaluated from the data. Values obtained are 10^{-3} mole l^{-1} and 0.83 l mole^{-1} sec^{-1}, respectively, at $20°C$. The propagation rate coefficient is comparable with values obtained for styrene polymerization in benzene in the presence of small amounts of added cyclic ethers.

The other alkali metals have been less extensively studied. The propagation rates of polystyrylsodium, -potassium, -rubidium and -cesium have been measured in benzene and cyclohexane [72, 73]. The sodium compound still shows half order kinetics in active centre concentration and is presumably associated to dimers. The rates for the rubidium and cesium compounds are directly proportional to the concentrations of the active chains which are presumably unassociated in solution. Absolute k_p values can be determined from the propagation rate in this case. Polystyrylpotassium shows intermediate behaviour (Fig. 11), the reaction order being close to unity at a concentration of the potassium compound near 5×10^{-5}M and close to one half at concentrations around 10^{-3}M. It could be shown by viscosity measurements that association was absent in the low concentration range. In this system both K_2 and k_p can be measured. The results are summarized in Table 2. The half order reactions show a large increase in $k_p K_2^{1/2}$ between lithium and potassium which

TABLE 2
Kinetic parameters for the propagation step in hydrocarbon solvents
(a) Complex rate coefficients; active species predominantly in associated form, in $(l\ mole^{-1})^{1/n}\ sec^{-1}$ units.

Monomer	Counter-ion	Solvent[a]	Temp. ($^{\circ}$C)	$k_p(K/n)^{1/n}$	E_{app} (kcal mole^{-1})	Ref.
Isoprene	Lithium	BEN	30	$1.3_8 \times 10^{-3}$		64
Isoprene	Lithium	CYH	40	$3.3_1 \times 10^{-3}$		61
Isoprene	Lithium	CYH	30	$1.0_5 \times 10^{-3}$		64
Isoprene	Lithium	CYH	30	$0.8_3 \times 10^{-3}$		33
Isoprene	Lithium	HEP	30	$0.7_8 \times 10^{-3}$		65
Isoprene	Lithium	HEX	30	$0.7_5 \times 10^{-3}$	19.5	64, 46
Butadiene	Lithium	CYH	30	0.23×10^{-3}		52
Butadiene	Lithium	HEX	30	0.15×10^{-3}		64
Styrene	Lithium	CYH	40	2.4×10^{-2}		61
Styrene	Lithium	BEN	30	$1.5_5 \times 10^{-2}$	14.3	17
Styrene	Sodium	BEN	30	$\sim 1.7 \times 10^{-1}$		72
Styrene	Potassium	BEN	30	1.8	11.0	72

(b) Absolute rate coefficients in $l\ mole^{-1}\ sec^{-1}$ units

Monomer	Counter-ion	Solvent[a]	Temp. ($^{\circ}$C)	$k_p^{\pm}$	E_p	Ref.
Styrene	Li^+	BEN	25	15.5	3.8	69
Styrene	Li^+2THF	BEN	25	0.9	9.2	74
Styrene	Li^+2DIOX	BEN	25	0.4		75
Styrene	K^+	BEN	25	47	7.2	72
Styrene	Rb^+	BEN	25	24		72
Styrene	Cs^+	BEN	25	18	7.3	72
Styrene	K^+	CYH	40	30		73
Styrene	Rb^+	CYH	40	22.5		73
Styrene	Cs^+	CYH	40	19		73
o-Methoxy-styrene	Li^+	TOL	20	0.8	12	71

[a] BEN, benzene; CYH, cyclohexane; HEP, heptane; HEX, hexane; TOL, toluene.

must be caused mostly by increases in the dissociation constant of the dimers. The true propagation rate coefficients, k_p, where measurable, show a much smaller change with alkali metal. The value for lithium determined as described earlier seems out of line with the other values, in terms of ΔS, and ΔH for both the activation process and for the monomer—dimer equilibrium. It is interesting that, generally, the reactivity of alkali metal compounds is usually described as $Cs > K > Na > Li$; the present results suggest that in many cases this may not be a true measure of reactivity, but caused by the decreased tendency in the compounds of higher alkali metals to associate to inactive forms. The degree of association follows that expected for dipole—dipole interaction, with a separation determined by the relative sizes of alkali metal cations.

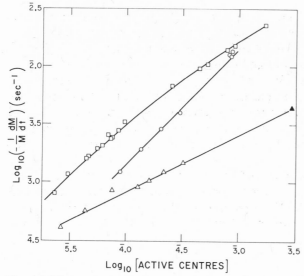

Fig. 11. The rate of chain propagation for polystyrylsodium (△), -potassium (□), and -cesium (○), as functions of growing chain concentration. Solvent, benzene 30°C [72].

One other system has been studied in which there is evidence of lack of association of active centres at high dilutions [76]; the propagation rate of polybutadienyl potassium in cumene at −30°C. At concentrations about 10^{-4} M the rate appears to be first order in potassium compound with a rate coefficient of about 7×10^{-3} l mole^{-1} sec^{-1}.

3.3 EFFECT OF POLAR MATERIALS ON CHAIN PROPAGATION

Lithium butoxides increase the rate of reaction of lithium alkyls with olefins in cyclohexane or hexane but decrease it in benzene. The propagation rate is, however, decreased in both types of solvent [77, 78] according to information presently available. In fact, as far as is known, butoxides reduce rates where the mechanism has been suggested to be dissociative and increase them in the other cases. More data are still required to confirm that this always happens. The experiments with polystyryllithium [77] show that the polymer dimers in solution are not dissociated by lithium *tert.*-butoxide as would be expected if mixed aggregates of the type (PstLi . BuOLi$_n$) were formed. In this case, at least, the rate effect appears to be caused by addition of butoxide to the polystyryllithium dimers. The reaction still shows half order characteristics, and the rate depression is almost complete at a 1:1 ratio of butoxide to polymer chains. The major species present in solution would seem to be (PstLi . BuOLi)$_2$ at this point. Similar results have been obtained with polyisoprenyllithium in cyclohexane [78]. The nature of the butoxide (*n*-, *sec.*- or *tert.*-) seems to be relatively unimportant. It was also shown in this latter case that butyllithium when present also depresses the propagation rate. Ethyllithium has the same effect [79], as does lithium hydroxide, although in the latter case it is due to destruction of active polymer chains [55]. These experiments go far to explain the differences in reported propagation rates. Incomplete initiation, or initiator destruction by moisture or oxygen will all lead to low rates. In general it is usually safe to assume that the highest rates reported are most nearly correct; only traces of polar solvents present will serve to give too high a rate.

The effect of small amounts of cyclic ethers on the rates of lithium catalyzed reactions can be more pronounced, especially in styrene polymerization (Fig. 12). The observed propagation rate first increases rapidly, passes through a maximum, and then decreases for both dioxane

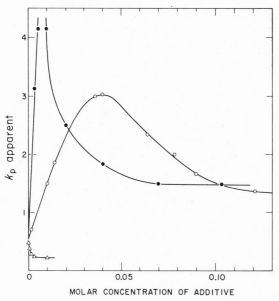

Fig. 12. Apparent k_p (l mole^{-1} sec^{-1}) for the propagation step of polystyryllithium in benzene as a function of the molar concentration of added: (●) tetrahydrofuran; (○) dioxane; (△) lithium-*tert.*-butoxide 30°C. Concentration of active chains ~1 x 10^{-3} M [74, 75, 77].

and tetrahydrofuran. Despite the large differences in dipole moment and basicity, both solvents are good solvating agents for cations. At concentrations higher than those corresponding to the maximum in the curves, the polystyryllithium dimers are completely dissociated as can be demonstrated by viscosity methods. The propagation rate is then first order in polystyryllithium. The form of the curves has been explained [74] in terms of the formation of two new species, a monoetherate and a dietherate according to the scheme

$$(PstLi)_2 \rightleftharpoons 2\ PstLi \tag{7}$$

$$PstLi + E \rightleftharpoons PstLi:E \tag{8}$$

$$PstLi:E + E \rightleftharpoons PstLi:2E \tag{9}$$

where E denotes ether. The polymerization active species are considered to be the unsolvated ion-pair and the two etherates. To produce the maximum the dietherate must be less reactive than either of the other two species. The kinetic analysis agrees with the above scheme but yields complex rate coefficients except for the propagation coefficient of the dietherate. This is found to be 0.5 l mole^{-1} sec^{-1} for dioxane and 1.2 l mole^{-1} sec^{-1} for tetrahydrofuran (30°C), much lower than any reasonable estimate of the rate coefficient of the unsolvated species. The propagation coefficient of the dioxanate is virtually identical to that observed for pure dioxane solvent; that for pure tetrahydrofuran is much higher, but in this case other active species are present.

With isoprene, the effect of tetrahydrofuran is much smaller and no maximum is observed at low tetrahydrofuran concentrations [33, 68]. With 0.01M THF in cyclohexane (which corresponds to a maximum rate for styrene) the rate of polymerization is barely changed. At 0.1M THF the order in polyisoprenyllithium has increased from one quarter to one half and complete aggregate dissociation has probably not been achieved. Evidence for specific solvation is thus difficult to obtain since so much THF would have to be added that the nature of the medium is drastically changed. This suggests that the self-association of polyisoprenyllithium must be much stronger than for polystyryllithium, for there is no reason to doubt that THF does in fact strongly solvate all systems having a lithium counter-ion.

4. Polymerization of styrene and dienes in polar solvents

4.1 THE INITIATION STEP

Information on the initiation reaction is restricted to the solvent tetrahydrofuran, in conjunction with the monomers α-methylstyrene and 1,1,-diphenylethylene which can only form oligomers, at least at room temperature. Initiation may once again be produced by metal alkyls or aryls, by direct contact of the olefin with an alkali metal film, or by use of the addition products of alkali metals to condensed aromatic ring compounds.

The class of initiators [80], such as sodium naphthalene, are produced by the reaction of an alkali metal with polycyclic hydrocarbons. Their stability depends on the solvent used [81, 82] and the electron affinity of the hydrocarbon. Solvents with good cation solvating capabilities such as tetrahydrofuran and dimethoxyethane are necessary for their formation. The fact that these compounds have an unpaired electron was discovered by early ESR measurements [83]. The unpaired electron is accommodated in the lowest energy unoccupied π-orbital and the compounds are described as radical-anions, e.g. $N^{\cdot -}$.

These compounds can produce polymerization by electron transfer to monomers [15, 84]★

$$N^{\cdot -} + M \rightleftharpoons M^{\cdot -} + N \tag{10}$$

Alkali metals themselves may function as electron (and cation) donors [85] providing an alternative source of $M^{\cdot -}$ when in contact with monomer.

The rate coefficients for electron transfers of this type, reaction (10), are known to be high, of the order of $10^7 - 10^9$ 1 mole^{-1} sec^{-1}. The position of the equilibrium will depend on the electron affinities of the two olefins. The electron affinity of naphthalene, for instance, is higher than that of styrene and considerable amounts of $N^{\cdot -}$ should be present with equivalent quantities of the two hydrocarbons. The situation is more favourable for formation of styrene$^-$ if sodium biphenyl is used. The equilibrium will be grossly perturbed, however, by subsequent reaction of the $M^{\cdot -}$ species so that none is detectable by ESR measurements even with equimolar amounts of naphthalene$^-$ and styrene [86]. This follows because the $M^{\cdot -}$ species can dimerize

$$2 M^{\cdot -} \rightarrow {}^-M.M^- \tag{11}$$

or react with monomer

$$M^{\cdot -} + M \rightarrow {}^\cdot M.M^- \tag{12}$$

to form species which possess real separated radical and ionic ends. Reaction (11) is possible for styrene but not for naphthalene for in the latter case there is a large loss in delocalization energy in formation of the covalent bond. Whether reaction (11) or (12) predominates will depend on reaction conditions. For diphenylethylene the values of k_{11} and k_{12} have been determined to be $1-2 \times 10^6$ and ~800 1 mole^{-1} sec^{-1} respectively [87]. If these results are typical, reaction (12) may predominate at high monomer : initiator ratios commonly used in polymerization, despite its rather unfavourable rate coefficient. The net result of all these reactions

---

★ The species are written without counter-ions for simplicity. It is known that these compounds exist in ether solvents mainly as ion-pairs. The relative reactivity of free anions and ion-pairs in these systems is not known.

will be an a,ω-dicarbanion propagating from both ends, formed directly in reaction (11) or from the product of reaction (12). The latter species has the following possibilities for reaction

$$e^- + {}^{\bullet}M \cdot M^- \rightarrow {}^-M \cdot M^- \tag{13}$$

$$2\,{}^{\bullet}M \cdot M^- \rightarrow {}^-M \cdot M \cdot M \cdot M^- \tag{14}$$

Reaction (14) should be somewhat faster than reaction (11) as the charge repulsion effect which causes reaction (11) to be rather slower than normal will not be so effective.

Evidence for the fate of ${}^{\bullet}M \cdot M^-$ can be obtained from product analysis in such initiation systems but unfortunately in both cases studied, the results are inconclusive. On mixing a radical anion with monomer, a rather high concentration of radicals will be produced, so recombination reactions (11) or (14) should normally be very effective leading to anionic polymerization only. Alkali metal initiation might, because of slower initiation, give evidence for a radical reaction. In just one system has it been claimed that both radical and anionic polymerization occurred from the respective ends of ${}^{\bullet}M \cdot M^-$. This is the lithium metal initiated copolymerization of styrene and methylmethacrylate [85]. Appreciable amounts of styrene are incorporated into the polymer, whereas lithium alkyl initiation produces almost pure polymethylmethacrylate. The radical end should give a random copolymer and the anionic chain end, homopolymethylmethacrylate. There is, however, no evidence for the presence of randomly copolymerized units in the polymer [88] and it is suggested that the styrene is incorporated by preferential initiation at the metal surface [89]. There remains, therefore, no evidence for appreciable amounts of radical polymerization in electron-transfer initiated polymerizations although, of course, the possibility that it is important in some special case cannot be entirely excluded.

Attempts have been made in the sodium metal initiated polymerization of α-methylstyrene [90] to assess the importance of steps (12) and (14). Rapidly stirred, finely divided, emulsions of sodium, especially in presence of a polycyclic hydrocarbon carrier, give good yields of 2,5-dimethyl-2,5-diphenyladipic acid after carbonation [91] as expected from reactions (11) or (13). Milder conditions involving a sodium mirror and no stirring gave mostly a tetramer [90, 92] whose structure was suggested to be head—head, tail—tail, viz.

$$\underset{\underset{\text{Ph}}{|}}{\overset{\overset{\text{CH}_3}{|}}{\text{Na}^{+-}\text{C}}}-\text{CH}_2\text{CH}_2-\underset{\underset{\text{Ph}}{|}}{\overset{\overset{\text{CH}_3}{|}}{\text{C}}}-\underset{\underset{\text{Ph}}{|}}{\overset{\overset{\text{CH}_3}{|}}{\text{C}}}-\text{CH}_2-\text{CH}_2-\underset{\underset{\text{Ph}}{|}}{\overset{\overset{\text{CH}_3}{|}}{\text{C}^-}}\,\text{Na}^+$$

as expected from dimerization of

$$\overset{\overset{\displaystyle CH_3}{|}}{\underset{\underset{\displaystyle Ph}{|}}{\cdot C}}-CH_2CH_2-\overset{\overset{\displaystyle CH_3}{|}}{\underset{\underset{\displaystyle Ph}{|}}{C^-}}$$

in step (14), of the thermodynamically favoured product of step (12). NMR characterization of the protonated tetramer [93, 94] has failed, however to confirm the head, head—tail, tail structure and suggests that it has the normal structure,

$$H.\overset{\overset{\displaystyle CH_3}{|}}{\underset{\underset{\displaystyle Ph}{|}}{C}}-CH_2-\overset{\overset{\displaystyle CH_3}{|}}{\underset{\underset{\displaystyle Ph,}{|}}{C}}-CH_2CH_2-\overset{\overset{\displaystyle CH_3}{|}}{\underset{\underset{\displaystyle Ph}{|}}{C}}-CH_2-\overset{\overset{\displaystyle CH_3}{|}}{\underset{\underset{\displaystyle Ph}{|}}{C}}.H$$

produced by addition of two molecules of α-methylstyrene to the dimer di-anion

$$Na^+ \; ^-\overset{\overset{\displaystyle CH_3}{|}}{\underset{\underset{\displaystyle Ph}{|}}{C}}-CH_2-CH_2-\overset{\overset{\displaystyle CH_3}{|}}{\underset{\underset{\displaystyle Ph}{|}}{C^-}} \; Na^+$$

It is clear that it is difficult to obtain evidence of the exact sequence of reactions following electron transfer, and that their relative importance is likely to be a function of reaction conditions.

Compounds such as sodium naphthalene can initiate polymerization by mechanisms involving addition. This is known to occur for ethylene oxide [95], viz.

Half the naphthalene is recovered and half remains bonded to polymer chains. A similar mechanism probably occurs for cyclic siloxanes [96]. Similar addition mechanisms have been suggested for vinyl monomers e.g. methylmethacrylate with some (but not all) aromatic radical anions [97], and even for α-methylstyrene initiated by sodium diphenylacetylene [98]. There remains a possibility [99] that the dihydro-aromatic groups observed in the polymer, in these cases, were not incorporated in the initiation step.

With lithium alkyl or aryl initiation in tetrahydrofuran, information is available only on the addition reaction to 1,1-diphenylethylene. Preliminary assessment of the rates showed a wide variation in initiator efficiency [100]. Direct comparison of initiation rates is difficult, however, since the orders in initiator vary between one quarter and unity [101] (Fig. 13) although the order in diphenylethylene is, as expected, unity. The reaction order in lithium compound is 0.27 ± 0.03 (methyl-Li); 0.34 ± 0.1 (vinyl-Li); 0.66 ± 0.04 (phenyl-Li); 1.1 ± 0.2 (benzyl-Li).

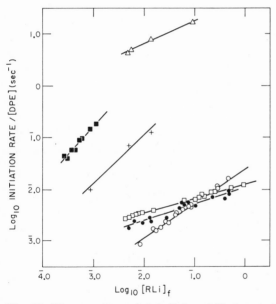

Fig. 13. The reaction of 1,1-diphenylethylene with various lithium compounds in tetrahydrofuran. Variation of rate with formal concentration of lithium alkyl or aryl. (△) n-Butyllithium; (■) benzyllithium; (+) allyllithium; (□) methyllithium; (●) vinyllithium; (○) phenyllithium. Solvent tetrahydrofuran [101].

The situation is reminiscent of that observed with hydrocarbon solutions and suggests association of some of these compounds even in tetrahydrofuran. Colligative measurements [102] confirm that this is so. Methyllithium is tetrameric in both diethyl ether and THF and n-butyllithium is tetrameric in ether. The latter compound reacts too rapidly with THF to measure its degree of association and similarly an accurate reaction order cannot be obtained. Earlier reports that n-butyllithium exists as a solvated dimer in ether are not conclusive [103]. Benzyllithium is unassociated which parallels the behaviour of the polymeric analogue, polystyryllithium. The apparent association numbers *decrease* at high concentrations, which can be explained by complex formation between the aggregates and tetrahydrofuran. Thus the methyllithium data are well fitted on the assumption of the existence of $(MeLi)_4 8THF$. For phenyllithium the results suggest the dimer is the major entity present but with partial dissociation, both the dimer and monomer present being solvated.

There is again an obvious correlation between reaction order and degree of association. The one quarter order for methyllithium could be explained satisfactorily by the usual assumption that the reactive species is methyllithium monomer in equilibrium with its tetramer. For phenyllithium a partial dissociation would lead to an order between one half and unity, as observed, if the dissociated product only were active. Other schemes involving some reactivity of both species would be equally plausible [101]. The first order behaviour with benzyllithium would require that the major reactive species is the ion-pair and not the free benzyl anion which must be present in small concentration.

4.2 THE PROPAGATION STEP

The initiation step is normally fast in polar solvents and an initiator-free "living polymer" of low molecular weight can be produced for study of the propagation reaction. The propagation step may proceed at both ends of the polymer chain (initiation by alkali metals, sodium naphthalene, or sodium biphenyl) or at a single chain end (initiation by lithium alkyls or cumyl salts of the alkali metals). The concentration of active centres is either twice the number of polymer chains present or equal to their number respectively. In either case the rates are normalized to the concentration of bound alkali metal present, described variously as concentration of active centres, "living ends" or sometimes polystyryl-lithium, potassium, etc. Much of the elucidation of reaction mechanism has occurred with styrene as monomer which will now be used to illustrate the principles involved. The solvents commonly used are dioxane ($D = 2.25$), oxepane ($D = 5.06$), tetrahydropyran ($D = 5.61$), 2-methyl-tetrahydrofuran ($D = 6.24$), tetrahydrofuran ($D = 7.39$) or dimethoxy-ethane ($D = 7.20$) where D denotes the dielectric constant at $25°C$.

Figure 14 illustrates the effect of solvent on the propagation rate of polystyrylsodium at $25{-}30°C$. The ordinate represents the apparent propagation coefficient ($k_{app} = R_p / [\text{monomer}] [\text{active centres}]$) i.e. assuming a simple bimolecular reaction

$$\cdots \overline{(n)} \mathrm{Metal}^+ + \mathrm{St.} \longrightarrow \cdots \overline{(n+1)} \mathrm{Metal}^+ \tag{15}$$

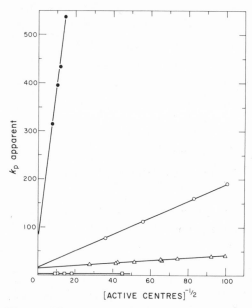

Fig. 14. The propagation step in the polymerization of styrene (counter-ion, sodium). Variation of k_{app} ($l \, \mathrm{mole}^{-1} \, \mathrm{sec}^{-1}$) with inverse square root of concentration of active centres. Solvents: (●) tetrahydrofuran 25°C [16]; (○) tetrahydropyran 30°C [106]; (△) oxepane 30°C [105]; (□) dioxane 25°C [104].

It will be noticed that a wide range of rates can be obtained, the rates in dimethoxyethane being so high that they cannot be accommodated on the scale of the figure. Except for dioxane as solvent, the reactions are not first order in active centres, although they are in fact always first order in monomer concentration. The cause of this deviation [107, 108] is the presence and reactivity of both ion-pairs and free anions in the solutions, viz.

$$\cdots \overline{(n)} \mathrm{Metal}^+ \rightleftharpoons \cdots \overline{(n)} + \mathrm{Metal}^+$$

$$+\mathrm{M} \downarrow k_p^{\pm} \qquad\qquad +\mathrm{M} \downarrow k_p^{-} \tag{16}$$

As indicated earlier, the concentration of free anions even in tetra-hydrofuran is only $\sim$1 % of the total at millimolar concentration so they must be of very high reactivity to influence the propagation rate. The increase in k_{app} with D at moderately low concentrations of active centres is largely caused by the increased concentration of free anions. Only in dioxane is the observed rate coefficient that of the ion-pair since the extent of dissociation is completely negligible in this solvent. Generally then, the propagation rate is given by

$$R_p = [M](k_p^-[\text{anions}] + k_p^\pm[\text{ion-pairs}])$$

At equilibrium

$$K_d = [\text{cations}][\text{anions}]/[\text{ion-pairs}]$$

then

$$R_p = [M](k_p^- K_d^{1/2}[\text{ion-pairs}]^{1/2} + k_p^\pm[\text{ion-pairs}])$$
$$= [M][\text{ion-pairs}](k_p^\pm + k_p^- K_d^{1/2}/[\text{ion-pairs}]^{1/2})$$

As dissociation is negligible, [ion-pairs] $\approx$ [active centres] i.e. essentially all the bound alkali metal is present as ion-pairs. Thus,

$$k_{app} = k_p^\pm + k_p^- K_d^{1/2}/[\text{active centres}]^{1/2}$$

The intercept in Fig. 14 gives, therefore, the ion-pair rate coefficient and the slopes of the lines yield $k_p^- K_d^{1/2}$. Conductance measurements can be used to determine K_d and hence a value for k_p^- is obtained. The validity of these concepts can be checked by carrying out the polymerization in the presence of sodium tetraphenylboride. This salt dissociates to a much greater extent than polystyrylsodium and its presence suppresses the ionization of the latter by a common ion effect. Under appropriate conditions a simple first order reaction in active centres can be observed in its presence with a rate equal to that measured by extrapolation to infinite concentration of active centres.

The dissociation constant of polystyrylsodium varies from 7×10^{-12} (oxepane, 30°C) to 1.5×10^{-7} (THF, 25°C), but the k_p^- evaluated in the two solvents using these constants varies little, being 1.2×10^5 l mole^{-1} sec^{-1} in oxepane and either 0.65×10^5 or 1.3×10^5 l mole^{-1} sec^{-1} in tetrahydrofuran. The difference is within the limits of experimental error and an examination of published data in other solvents suggests no meaningful trend (Table 3). The A and E factors are quite normal and are comparable with those obtained for free radical polymerization. The ion-pair rate coefficient, $k_p^\pm$, as obtained in the presence of sufficient tetraphenylboride or by extrapolation as in Fig. 14, varies more systematically with solvent. Values reported for polystyrylsodium are (in l mole^{-1} sec^{-1} at 25°C) 3.5 [104], 5.5 [114] (dioxane); 10.5 [105] (oxepane); <5 [109], ~15 [106, 115—117] (tetrahydropyran); 80 [16], 120 [118], 220 [119] (tetrahydrofuran); 3600 [112], 4700 [120] (dimethoxyethane).

TABLE 3

Absolute rate coefficients for propagation via free anions
(k_p^- at 25°C (l mole^{-1} sec^{-1}) in the polymerization of styrene.)

Solvent[a]	$10^{-5} k_p^-$	$10^{-8} A_p$	E_p	Ref.
OXP	1.0 (30°C)			105
THP	0.6	2.0	4.8	109
THP	1.4 (30°C)			106
THP	1.2	0.63	3.8	110
THF	0.65	11.0	5.9	111
THF	1.3	1.6	4.1	110
DME	0.4			112
2-Methyl THF	0.7			113

[a] OXP, oxepane; THP, tetrahydropyran; THF, tetrahydrofuran; DME, dimethoxyethane; DIOX, dioxane.

The ion-pair coefficients appear to increase with dielectric constant. Reactions in which charge separation is involved in the transition state are favoured by increased dielectric constant [121] but the variation seems rather large, especially for the last two solvents, which should have similar rates on this hypothesis. The effects appear to be more specific as shown in Table 4, in which are collected data for the other alkali metal counter-ions. It shows a spread of rate coefficients which can be as high as a factor of two between experiments carried out in different laboratories.

TABLE 4

Absolute rate coefficients for propagation via ion-pairs
($k_p^\pm$ at 25°C (l mole^{-1} sec^{-1}) in the polymerization of styrene.)

Solvent[a]	Counter-ion				Ref.
	Li^+	Na^+	K^+	Cs^+	
DME		3600		$\sim$170	112
DME		4700 (20°C)			120
THF	160	80	$\sim$60	22	16
THF		220			119
THF		170			118
2-Methyl THF	12	11	7.5	22	109
2-Methyl THF	25	23	17	23	113
THP	$<$5	$<$5			109
THP	19.5 (30°C)	17 (30°C)	30 (30°C)		106
THP		14.8			116
THP		14	73	53	115
THP	10	13	61	64	117
DIOX	0.94	3.4	19.8	24.5	104
DIOX		6.5	28	15 $\pm$ 5	118
DIOX		5			114

[a] Solvents: abbreviations as in Table 3.

References pp. 60—65

Nevertheless trends can be discerned. In dioxane and tetrahydropyran there is a tendency for the rate coefficient to increase from Li^+ to Cs^+. This is the behaviour expected [104] if the factor influencing the rates is the increased energy required to separate anion and counter-ion, as the latter becomes smaller (increased coulombic interaction energy). The variation of $k_p^{\pm}$, in fact, is not large over all counter-ions in dioxane, tetrahydropyran, and 2-methyltetrahydrofuran, and for cesium is almost non-existent even including tetrahydrofuran and benzene if the rather high values in tetrahydropyran are excluded. In tetrahydrofuran and dimethoxyethane, the trend is reversed, the largest values of $k_p^{\pm}$ being observed with the smaller counter-ions. The ion-pair dissociation constants in tetrahydrofuran also increase going from Cs^+ to Li^+ [122] whereas in tetrahydropyran differences are much smaller [106]. The mobilities of the free cations in tetrahydrofuran are lower for the smaller cations [122]. Evidently in tetrahydrofuran (and dimethoxyethane) the smaller free cations undergo very strong specific solvation by these two solvents. It has been suggested that specific solvation also occurs in the ion-pairs [111, 116], so as to produce besides the normal ions in contact (contact-pair) a fraction in the form of solvent-separated pairs. There are two sets of evidence for the existence of two distinct forms of ion-pair in this type of system.

Much of the information is obtained from studies of the related fluorenyl compounds. Examination of the ultraviolet absorption spectra of fluorenyllithium or sodium in solvent mixtures shows changes in absorption spectra as a function of solvent composition [123]. This is illustrated in Fig. 15 for fluorenyllithium in toluene—tetrahydrofuran mixtures. The band at 374 mμ increases and that at 349 mμ decreases as the concentration of tetrahydrofuran increases. The presence of an isosbestic point indicates we are dealing with two different related species. Fluorenylsodium shows the same behaviour in tetrahydrofuran alone as the temperature is lowered and it is suggested that the band at longer wavelength corresponds to solvent-separated ion-pairs and that at shorter wavelength to contact-pairs [123]. The longer wavelength band cannot correspond to the free anion for this is not present in the required concentration. Examination of various fluorenyl salts in different solvents in this manner, leads to estimates of the fraction of solvent-separated ion-pairs. Fluorenyllithium is present, even at 25°C, in tetrahydrofuran predominantly as separated pairs but the sodium compound, under these conditions, exists mainly as contact pairs. Solution in an even better cation-solvating solvent such as dimethoxyethane, or cooling the tetrahydrofuran solution to −70°C is required to convert fluorenylsodium to solvent-separated pairs.

Conductance measurements [124] in tetrahydrofuran confirm these results. The ion-pair dissociation process should be quite exothermic due to strong solvation of the free cation. More extensive ion-pair solvation, equivalent to partial dissociation, would make the dissociation less

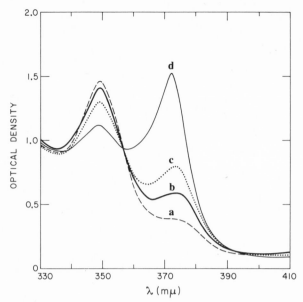

Fig. 15. Absorption spectra of fluorenyllithium in toluene—tetrahydrofuran mixtures. Volume percentages of THF: (a) 43%; (b) 50%; (c) 63%; (d) 100%.

exothermic. The process is illustrated diagrammatically as

$$F^-Na^+ \rightleftharpoons F^- \quad Na^+ \rightleftharpoons F^- + \quad Na^+ \tag{17}$$

contact pair   solvent-separated   free ions
pair

The numbers of solvent molecules are not to be regarded as exact but the variation with [THF] suggests an extra two molecules are involved in the separated pair. Some solvation is likely to be present in the contact-pair. Evidence for such solvation was presented earlier for polystyryllithium involving two molecules of tetrahydrofuran. It must be regarded as being peripheral. With fluorenyllithium the heat of dissociation to free ions is quite low and of the order of -1 to -2 kcal mole^{-1}. There is a small change of ΔH with temperature as expected because of the change in D. With fluorenylsodium, in contrast, ΔH varies markedly with temperature, being ~-8 kcal mole^{-1} around room temperature decreasing to ~-1 kcal mole^{-1} at $-70°C$ (Fig. 16). This corresponds to the spectroscopically observed change to predominantly solvent-separated ion-pairs at low

34

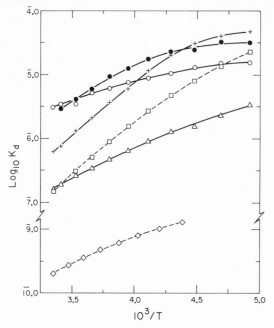

Fig. 16. Temperature dependence of ion-pair dissociation constants of fluorenyl (———) and polystyryl (- - -) salts of the alkali metals. (+) Fluorenylsodium in tetrahydrofuran [125]; (●) fluorenylpotassium in dimethoxyethane [125]; (○) fluorenyllithium in tetrahydrofuran [125]; (△) fluorenylpotassium in tetrahydrofuran [125]; (□) polystyrylsodium in tetrahydrofuran [111]; (◇) polystyrylsodium in tetrahydropyran [109].

temperatures. Fluorenylpotassium shows the same large temperature coefficient of ΔH in dimethoxyethane [125], requiring an even more powerful cation solvating solvent and low temperature to convert it to separated-pairs. Other compounds such as dimethylsulphoxide, hexamethylphosphoramide or the polyglycoldimethylethers ($CH_3O-(CH_2CH_2O)_n CH_3$) are even more efficient producers of solvent-separated ion-pairs [126] being effective at very low concentrations in a medium such as dioxane or tetrahydropyran. The position of the equilibrium between the two types of ion-pairs as a function of added solvating agent gives some idea of the numbers of molecules of the latter involved in the process.

The fluorenyl salts seem to be particularly susceptible to the formation of solvent separated pairs. Their formation is favoured by the presence of a large charge-delocalized anion plus a small compact cation which can be strongly solvated (or presumably vice-versa). The solvating agent should be highly polar, preferably small in size, or with the ability to offer multiple coordination as with the polyglycol-dimethylethers. With polystyryl salts, the formation of solvent-separated ion-pairs is less extensive. The absorption spectra are not particularly conclusive because the absorption band is rather broad and could conceal spectral changes. The temperature dependence of the dissociation constant of polystyrylsodium to free ions gives a heat of dissociation of ~ -8 kcal mole^{-1} in tetrahydrofuran at 25°C, decreasing to -5 to -6 kcal mole^{-1} at -70°C [111] (Fig. 16). At room temperature it exists predominantly as contact-pairs, but an assumption of a heat of dissociation of -1 to -2 kcal mole^{-1} for the solvent separated pair would correspond to about 30% of such species at -70°C. Although there is no evidence at room temperature for appreciable amounts of solvent-separated pairs, even much less than 1% of them would be important if their reactivity were comparable more to the free anion than the contact-pair.

There is, in fact, other evidence that this is the case as is shown in Fig. 17. The activation energy for the propagation reaction ($k_p^{\pm}$) for polystyrylsodium is quite normal in solvents such as dioxane ($A \sim 5 \times 10^7$,

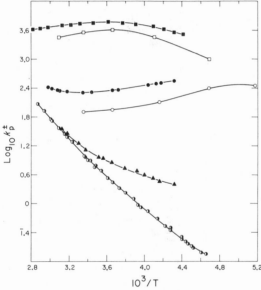

Fig. 17. Variation of ion-pair propagation coefficients, $k_p^{\pm}$, with temperature in styrene polymerization (counter-ion, sodium). Solvents: (◑) oxepane [105]; (▲) tetrahydropyran [116]; (○) tetrahydrofuran [111]; (●) tetrahydrofuran [110]; (□) dimethoxyethane [112]; (■) dimethoxyethane [110]. More data are now available over a wider range of experimental conditions, see ref. 233.

$E = 9.5$ kcal mole^{-1}) [114, 118], oxepane ($A = 1.6 \times 10^8$, $E = 9.8$ kcal mole^{-1}) [105] and in tetrahydropyran ($A = 1.1 \times 10^6$, $E = 6.7$ kcal mole^{-1}) [115] ($A = 8 \times 10^6$, $E = 7.9$ kcal mole^{-1}) [116] around room temperature although some decrease is observed with the latter two solvents at lower temperatures. For tetrahydrofuran and dimethoxyethane, the temperature dependence of $k_p^{\pm}$ is quite abnormal [110, 112], as it corresponds to a negative value of E_p at about 25°C. This is, of course, impossible for a simple single reaction step and an explanation must involve the reaction as proceeding in more than one step or via two species of different reactivity and whose proportion varies with temperature. The latter explanation is the most plausible in view of the above evidence. Under these conditions [111] with an equilibrium between contact (1) and solvent-separated (2) ion-pairs defined by $K_s = (1-\gamma)/\gamma$ where γ is the fraction of (1),

$$k_p^{\pm} = \gamma k_1^{\pm} + (1 - \gamma)k_2^{\pm} = k_1^{\pm}/1 + K_s + K_s k_2^{\pm}/(1 + K_s)$$

where $k_1^{\pm}$ and $k_2^{\pm}$ are the propagation coefficients of the two types of ion-pair. If the proportion of solvent-separated pairs is relatively small and $k_2^{\pm} \gg k_1^{\pm}$, so that the second term predominates in the above equation, an apparent negative activation energy may be observed. This requires that the change from contact to solvent separated pairs is exothermic with an enthalpy change (ΔH_s) numerically greater than E_2. Over a very wide temperature range the observed activation energy will vary from E_1 (high temperature) to E_2 (low temperature) and the rate coefficient when plotted against $1/T$ will show two inflection points in the intermediate range [116, 127]. A large fraction of this possible range has been observed in tetrahydrofuran. The Arrhenius parameters derived [128] are $A_1 = 6.3 \times 10^7$, $E_1 = 8.6$ kcal mole^{-1}; $A_2 = 2.0 \times 10^8$, $E_2 = 4.7$ kcal mole^{-1}, $\Delta H_s = -6.5$ kcal mole^{-1}. With dimethoxyethane as solvent, the maximum is displaced to the left of that in tetrahydrofuran as expected if it is more efficient in producing separated pairs. An even more efficient way of increasing the proportion of solvent-separated ion-pairs is to add polyglycol-dimethylethers to a solution of polystyrylsodium in tetrahydropyran [129]. These compounds promote the ionization of polystyrylsodium but also increase the value of $k_p^{\pm}$ when present only in millimolar concentrations.

In summary, the wide range of propagation rates in different solvents and with different counter-ions can be rationalized in terms of the presence of three reactive species

$$\text{Pst}^- \text{Metal}^+ \rightleftharpoons \text{Pst}^- \| \text{Metal}^+ \rightleftharpoons \text{Pst}^- + \text{Metal}^+ \qquad (18)$$

| contact pair | solvent separated pair | free ions |

The ease of formation of solvent-separated pairs increases in the two series $\text{Cs}^+ < \text{Rb}^+ < \text{K}^+ < \text{Na}^+ < \text{Li}^+$ and dioxane $<$ oxepane $<$ tetrahydropyran $<$ tetrahydrofuran $<$ DME $<$ polyglycol ethers. In solvents such as dioxane only the contact pairs contribute to the observed propagation rate, the concentration of other forms being completely negligible. With solvents of higher dielectric constant (tetrahydropyran, oxepane) the overall propagation rates increase, primarily because of increased ionization to form very small but significant quantities of the highly reactive free anion. The ion-pair rate changes only modestly. In tetrahydrofuran or dimethoxyethane and particularly with counter-ions lithium and sodium, specific solvation of ion-pair and free cation produce abnormally high ionization and small amounts of reactive solvent-separated ion-pairs even at room temperature. The combination produces extremely high rates, both overall and when limited to the ion-pair rate. With the larger counter-ions (e.g. Cs^+) lack of even solvation of the free cation except in dimethoxyethane and the inability to form solvent-separated pairs keeps ionization at a relatively low level and the ion pairs in the form of relatively unreactive contact pairs. In the intermediate region (2-methyl-tetrahydrofuran, tetrahydropyran) the ion pair rate coefficient may change only slightly with counter-ion. At room temperature there may be the conversion of very small amounts of ion-pairs to the reactive solvent-separated form if the counter-ion is small. This tends to increase $k_p^{\pm}$; it can be balanced by a lower reactivity of the contact-pairs having smaller counter-ions as observed in dioxane. As the temperature is decreased however, more solvent separated ion-pairs may form with Li^+ or Na^+ and the activation energy appears to decrease even in oxepane as shown in Fig. 16.

The formulation of two types of ion-pair is an attractive hypothesis which has been used for other systems [130] to explain differences in reactivity. The polymerization of styrene-type monomers in ether solvents, all of which solvate small cations efficiently, seems to be a particularly favourable case for the formation of thermodynamically distinct species. Situations can be visualized, however, in which two distinct species do not exist but only a more gradual change in properties of the ion-pair occurs as the solvent properties are changed. These possibilities, together with the factors influencing solvent-separated ion-pair formation, are discussed elsewhere [131, 132]. In the present case some of the temperature variation of rate coefficient could be explained in terms of better solvation of the transition state by the more basic ethers, a factor which will increase at lower temperatures [111]. This could produce a decrease in activation energy, particularly at low temperatures. It would, however, be difficult to explain the whole of the $k_p^{\pm}$ versus $1/T$ curve in tetrahydrofuran with its double inflection by this hypothesis and the independent spectroscopic and conductimetric evidence lends confidence to the whole scheme.

Some information can be obtained about the *rate coefficients* of ion-pair dissociation from measurements of molecular weight distribution. If the lifetime in any of the ionic states is short compared with its rate of reaction with monomer, the molecular weight distribution will be the simple narrow Poisson type expected for rapid initiation, propagation via a single species and no termination. A slower ion-pair dissociation process will cause a broadening of the distribution from which the dissociation rate can be measured [133]. This broadening has been observed [134, 135] in the polymerization of polystyrylsodium in tetrahydrofuran. Only in the presence of sodium tetraphenylboride did the distribution approach the expected Poisson distribution. From this data the rate coefficient for dissociation of the ion-pair was found to be 75 sec^{-1} at 25°C and the association rate coefficient, 1.5 x 10^9 l mole^{-1} sec^{-1}. The results imply that the equilibration between the two types of ion-pair is rapid under the experimental conditions. By working with tetrahydropyran solvent at different temperatures, however, some broadening of the distribution was observed even with sodium tetraphenylboride present which enabled estimates to be made of the rate of interconversion of the two types of ion-pairs [110]*.

Detailed kinetic studies on other monomers than styrene are not numerous. The data are assembled in Table 5. α-Methylstyrene has received the most extensive study with data available for several counter-ions and three solvents [20, 118, 136, 142]. The $k_p^{\pm}$ values in the table are those extrapolated to 25°C, the actual temperature range studied was lower owing to the unfavourable equilibrium between monomer and polymer at room temperature. Some inaccuracy may result from this extrapolation. In dioxane [142] the activation energy decreases from 13 to 8 kcal mole^{-1} between Na$^+$ and Rb$^+$, evidently the characteristics of a contact-ion-pair. In tetrahydropyran [20] the $k_p^{\pm}$ value of the lithium compound is much higher than the others, presumably caused by a contribution from solvent-separated pairs. This is confirmed by the fact

TABLE 5

Absolute rate coefficients for propagation via ion-pairs
($k_p^{\pm}$ (l mole^{-1} sec^{-1}) for compounds other than styrene at 25°C)

Solvent[a]	Monomer	Counter-ion					Ref.
		Li$^+$	Na$^+$	K$^+$	Rb$^+$	Cs$^+$	
DIOX	α-Methylstyrene		0.016	0.098	0.062		118
THP	α-Methylstyrene	2.6	0.047	0.246	0.351	0.261	20
THF	α-Methylstyrene		0.32	0.25			136
2-Methyl-THF	o-Methylstyrene		1				137
2-Methyl-THF	p-Methylstyrene		5				137
THF	p-Methyoxy-styrene		40	40		15	138
DIOX	2-Vinylpyridine		3—4 x 10^3				139
THP	2-Vinylpyridine		~4.5 x 10^3				139
THF	2-Vinylpyridine		2 x 10^3			1.3 x 10^3	139
Ether	Isoprene	0.03 (20°C)					140
THF	Isoprene	0.20 (30°C)					141

[a] Solvents: abbreviations as in Table 3.

* See ref. 234 for a full compilation of interconversion rates.

that the observed activation energy is close to zero ($+9°$ to $-12.5°C$). Conductance measurements on the model oligomer, 2-phenyl-pentyllithium, show a decrease in exothermicity of ion-pair dissociation from -6.5 kcal mole^{-1} at $25°C$ to -1.0 kcal mole^{-1} at $-50°C$ [143], again indicative of extensive solvent-separated pair formation at low temperatures. The situation for the sodium compound in tetrahydropyran is less sure. The activation energy is positive at 5.5 kcal mole^{-1}, but less than with the higher alkali metal counter-ions, and the pre-exponential factor is very low (6.3×10^2); there may be some small contribution from separated pairs. The other alkali metals show approximately normal behaviour for contact pairs. In tetrahydrofuran [136] the sodium compound shows strong evidence for appreciable concentrations of solvent-separated pairs from spectroscopic studies [144] and from the fact that the propagation coefficient goes through a maximum at ~ -30 to $-40°C$ (cf. styrene $-75°C$) and E_p is negative above this temperature. The potassium compound gives only evidence for one species with a constant activation energy of 5.2 kcal mole^{-1} between -60 and $+10°C$; the proportion of solvent separated pairs must be negligible.

The K_d values both in tetrahydrofuran and tetrahydropyran are of the same order of magnitude as those of polystyrene compounds at $25°C$. Only the lithium compound of α-methylstyrene has an appreciably higher K_d than that of styrene. The free anion rate coefficient, 830 l mole^{-1} sec^{-1} (extrapolated, $25°C$; $A = 1.5 \times 10^8$, $E = 7.2$ kcal mole^{-1}) [145] is smaller than for styrene, as are the ion-pair coefficients, but the ratio between the two is roughly the same. A major factor producing low rates appears to be a higher activation energy than is found for styrene polymerization.

2-Vinylpyridine is an interesting case where *intra*-molecular solvation of the ion-pair can occur. This is indicated by an ion-pair dissociation constant about one hundredth of that of the analogous styrene derivative. The dissociation constant of the polystyrene derivatives is higher than expected for a reasonable ion-pair separation in solvents such as tetrahydrofuran, because of the stabilization of the dissociated cation by solvation. Stabilization of the ion-pair by internal $Na^+ \cdots N$ interaction will reduce this driving force. In contrast, $k_p^{\pm}$ is much higher than for styrene homopolymerization. This has been explained in terms of a more reactive "looser" ion-pair with increased dipole moment caused by increase of the $-C^- Na^+$ distance by intramolecular solvation [139]. An important factor in the high homopropagation coefficient, however, must be the high reactivity of vinyl pyridine monomer. Copolymerization experiments [146] indicate that active vinylpyridine polymers add styrene only very slowly, whereas the active polystyrene polymers add to vinylpyridine at an extremely fast rate. The variation of $k_p^{\pm}$ with solvent is much lower than with styrene. Since solvent-separated ion-pair formation is less likely in presence of intramolecular solvation, this is not surprising. It does indicate, perhaps, that given a constant form of ion-pair the effect of dielectric constant as supposed in classical theories of reactivity of ions and ion-pairs may be quite small.

The data on isoprene suggest a similar interpretation [140, 141]. There is little difference in the $k_p^\pm$ values in Table 5 between diethylether and tetrahydrofuran, and the difference would be even smaller if the results were corrected to the same temperature. It would seem that with this monomer solvent-separated ion-pairs are not easily formed even in the favourable case of Li^+ counter-ions. The charge delocalization is restricted to three carbon atoms in the anion giving a less diffuse ion. These results must suggest the possibility that the apparent importance of solvent-separated ion-pairs in anionic polymerization is to some extent caused by the fact that most detailed studies have been made largely on styrene and its derivatives.

5. Polymerization of polar monomers

A relatively large number of vinyl monomers possessing electron-withdrawing groups can be polymerized by organo-lithium or magnesium compounds. The polymerizations seldom show the characteristics of the "living polymer" systems described in earlier sections. The initiator, besides adding to the double bond of the monomer, can react with its polar groups to produce inactive products. Some growing polymer chains can be formed, but even they are susceptible to side-reactions.

The chlorine-containing monomers, vinyl chloride [147, 148], vinylidene chloride [149] and chloroprene [150] can be polymerized at room temperature by organo-lithium and magnesium compounds [151, 152]. The polymerization ceases before all the monomer is consumed. Kinetic analysis has been attempted allowing for an exponential decay in the concentration of active centres, together with two competing reactions of monomer with initiator, one leading to active chains, the other to initiator deactivation. The formulation of kinetic schemes seems premature in the absence of more detailed knowledge of the reaction process. Metallation of vinyl chloride by lithium alkyls can occur [153]. Free radicals have been detected in the reactions of n-butyllithium and sodium naphthalene with alkyl or aryl halides [154, 155]. The reaction of *tert.*--butylmagnesium chloride with vinyl chloride has been indicated to produce high polymer by a free radical mechanism and oligomers anionically [156]. Similar doubts as to the nature of the reaction must exist for the polymerization initated by lithium alkyls, for it is reported that the presence of air has little effect on the polymerization of vinyl chloride and that tetrahydrofuran completely inhibits polymerization [147].

Qualitative information is available on several other vinyl monomers including acrylonitrile [157—159], methacrylonitrile [160], *tert.*-

butylvinylketone [161] and methylisopropenyl ketone [162]. In these cases polymerization is effective at temperatures as low as $-70°C$, which makes the possibility of radical reactions less likely. The polymerizations still show, however, the same limited conversions and low initiator efficiency as with the chlorine containing compounds. For acrylonitrile some of the reactions leading to initiator loss have been described [163, 164]. Only the acrylate monomers show some of the characteristics of "living polymers" [165–167]. Polymerization of acrylates by organolithium or sodium compounds goes to completion at low temperatures, and addition of more monomer, even after the solutions have been standing some time, results in its polymerization. The number of active chains producing high polymer is, however, less than the quantity of added initiator and the molecular weight distributions, at least with toluene as solvent, are very broad [168, 169]. As initiation is known to be very rapid in these systems [167, 169] some complications must occur in the propagation step.

The progress of the reaction with methylmethacrylate depends somewhat on initiator, temperature and solvent. Investigations have been carried out using fluorenyllithium [167, 168, 170], phenylmagnesium bromide [171, 172], butyllithium [173] and 1,1-diphenylhexyllithium [174] in toluene solution with or without the presence of ethers. Product analysis shows that two basic reactions occur with the monomer both with magnesium compounds [171, 175] and with butyllithium [176], viz.

It is not certain if reaction (20) goes to completion. Reaction (19) leads to propagation whereas (20) does not. The relative importance of the two steps will depend on initiator, solvent and temperature, but insufficient evidence is available to discuss variations systematically. Multiple attack on the monomer is possible at high initiator : monomer ratios but is not likely to be important under polymerization conditions. Metallation of the monomer has been suggested to occur with butyllithium. It seems unlikely to occur in this system and the butane detected in the hydrolysis products of the reaction can come from other sources. Reaction (20) is not the only source of initiator loss as indicated by fractionation of the polymer. Much of the initiator only leads to polymer of very low molecular weight [168, 171, 173, 174] so few of the products of step (19) lead to high polymer. In the phenylmagnesium bromide initiated polymerization [171] appreciable amounts of methyl 5-benzyl-4-oxo-1,3,5-trimethylcyclohexane-1,3-dicarboxylate, viz.

were found, based on the presence of infrared absorption in low molecular weight products at 1712 cm^{-1} as well as the normal 1740 cm^{-1} position in carboxylates. This compound is produced from carbanion attack on the polymer carbonyl group three units back in an analogous manner to reaction (20). Evidence of such a trimer was also found in the fluorenyllithium initiated polymerization [168]. The lowest fraction isolated by counter-current extraction (equivalent to ~25% of added initiator) had a molecular weight of 484, and strong absorption at 1712 cm^{-1}. Other fractions also contained this absorption which decreased as molecular weight increased.

Experiments on termination with tritiated acetic acid [168] showed that all except the fraction of lowest molecular weight would add a proton with at least 80% efficiency. It seems reasonable to suppose that these chains would also be reactive towards monomer, but the shorter chains apparently do not propagate at all or if they do so it is with low efficiency. This leads to the suggestion [168] that the shorter chains are "pseudo-terminated" by intra-molecular complexing, e.g.

Such chains might have a small probability of propagating by dissociation of the complex, but could also lead to the cyclic ketone above in a true termination reaction, or in the reaction with tritiated acetic acid. In the latter case, tritium would be found in the product. To explain the experimental results, the tendency to intra-molecular cyclization must decrease rapidly as the polymer chain becomes longer. The original suggestion that the decrease could be correlated with a different microstructure of oligomeric products, cannot be confirmed [170], and a major difficulty of the theory remains.

Whatever the explanation, there is no doubt that polymer of molecular weight 500—800 is formed extremely rapidly at the start of polymerization and can be isolated from the final product. With fluorenyllithium [168], (toluene—ether, —60°C) a first order disappearance of monomer is observed, which extrapolates at zero time, not to the original added monomer concentration but to a concentration corresponding to the immediate loss of three molecules of monomer per initiator molecule. With 1,1-diphenylhexyllithium [174] (toluene, —30°C) this extrapolation corresponds to the rapid addition to the initiator of about five monomer units. In this case termination at various times and isolation of precipitant-soluble material confirms that polymer of molecular weight ~830 is formed rapidly and does not change appreciably in amount throughout the polymerization. With butyllithium [173] (toluene, —30°C) the course of reaction is more complex in the initial stages but eventually a steady concentration of active centres is probably formed as the reaction settles down to first order decay in monomer. A second addition of monomer at the end of the reaction then produces a first order disappearance of monomer immediately. The two first order rate coefficients are identical. Evidently products are produced with butyllithium which disturb the reaction, and until these are removed a steady concentration of active centres is not achieved.

Some indications of what may be happening are gained from analysis of the acetic acid terminated products at various reaction times (Fig. 18). With butyllithium initiation a large amount of methanol (based on initiator) can be isolated at very short reaction times, whereas in the diphenylhexyllithium initiated process only a slower and smaller methanol formation is observed. Now this must correspond to lithium methoxide formed either directly in step (20), or in the termination of product (20), or from the cyclization reaction, or from termination of active chains [177] in reaction (21),

$$\underset{\underset{\text{COOCH}_3}{|}}{\overset{\overset{\text{CH}_3}{|}}{\sim\sim\text{C}^-\text{Li}^+}} \xrightarrow{\;+\text{M}\;} \underset{\underset{\text{COOCH}_3}{|}}{\overset{\overset{\text{CH}_3}{|}\quad\;\overset{\text{CH}_3}{|}}{\sim\sim\text{C}-\text{CO}-\text{C}=\text{CH}_2}} + \text{CH}_3\text{O}^-\text{Li}^+ \qquad (21)$$

Attack on polymer carbonyl groups in the same manner is also possible. With both initiators large amounts of precipitant soluble polymer are formed rapidly, so if the immediate formation of terminated cyclic

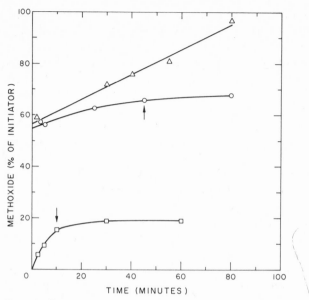

Fig. 18. Percentage of initiator appearing as methoxide ion in the polymerization of 0.125M methylmethacrylate in toluene. ($\triangle$) Initiator 7.6×10^{-3}M butyllithium at $10°C$; ($\bigcirc$) initiator 7.6×10^{-3}M butyllithium at $-30°C$; ($\square$) initiator 3.2×10^{-3}M diphenylhexyllithium at $-30°C$. Arrows indicate time for 70% conversion of monomer to polymer [174].

product were responsible for the methoxide formation, the two would be expected to show similar behaviour. The most reasonable explanation for the different results is that butyllithium largely attacks monomer via reaction (20) to produce immediately substantial amounts of methoxide, but diphenylhexyllithium attacks via route (19). The origin of the slower, longer term methoxide generation is more difficult to assign. Some contribution from reaction (21) is possible which is probably important at temperatures above $0°C$. At the lower temperatures normally used in polymerization it is less important, but it would be difficult to detect a small contribution. The amount of methoxide finally formed is, however, larger than the estimated concentration of growing chains of high molecular weight with diphenylhexyllithium. A slow cyclization of "pseudo-terminated" short chains seems to be the only other possible source. With butyllithium initiation, alkoxides are formed in large excess over active polymer chains. Their presence is known to affect the reaction [178] and in addition if reaction (20) proceeds to completion, an alkyl vinyl ketone is produced which also could disturb the polymerization process [162].

Kinetic analysis of the polymerization of methylmethacrylate must be made with the foregoing complications in mind. The analysis depends on the assumption that a constant fraction of the initiator rapidly forms growing chains and that side products do not affect the subsequent propagation step. For the fluorenyllithium initiated polymerization in toluene—diethylether (2—20% ether) or in toluene—tetrahydrofuran (0.1—15% THF) at —60°C evidence was presented that ~17% or 34% of the initiator formed active chains respectively [170]. At high ether concentration, the propagation process is first order in monomer and initiator, but tends to second order in initiator as the ether concentration is reduced. It was suggested that the species active in propagation changes from an ion-pair to an ion-pair dimer as the ether concentration becomes low. In media containing tetrahydrofuran the overall reaction is apparently half order in fluorenyllithium, a result which can be explained if the predominant reactive species is the free anion in labile equilibrium with a large excess of ion-pairs (see Section 4.2). The observed propagation rate coefficient although determined from an internal first order disappearance of monomer is, however, dependent on initial monomer concentration. This suggests that the real cause of the complex kinetic pattern is the assumption that a constant fraction of the initiator forms growing polymer chains. A careful examination of the data for tetrahydrofuran mixtures shows that the initiator efficiency (as determined from $\overline{M}_n$ of precipitable polymer) often decreased as the initiator concentration increased. This would provide an alternative explanation for deviations from first order behaviour, one which would not require the unlikely assumption of predominantly free ion propagation under conditions where ionization should be completely negligible. The data reported at high diethylether concentrations better support the assumption of a constant fraction of initiator producing chains. It is perhaps significant that these are the conditions in which simple reaction orders are observed. The assumption appears to fail at low ether concentrations and once again the reaction orders require an implausible mechanism involving associated species. There is no evidence that the methacrylate ion-pairs self associate in toluene [179] (unlike styrene), presumably because intramolecular solvation by the ester group makes this unnecessary in non-polar media.

The propagation reaction in the 1,1-diphenylhexyllithium initiated polymerization (toluene, —30°C) shows simple first order kinetics in monomer and initiator over a limited range of concentrations [174]. The precipitant-soluble (and presumably inactive) polymer chains always form, in this case, a constant fraction of the added initiator. These represent about 80% of the potential polymer chains. $k_p^{\pm}$ values evaluated on the basis that the active chain concentration is always 20% of that of the added initiator are shown in Fig. 19. The results correspond to an activation energy of 5.0 kcal mole^{-1} together with a pre-exponential factor of ~10^5. Extrapolation to 25°C leads to a value of $k_p^{\pm}$ of ~20 l mole^{-1} sec^{-1}. The values are reasonable for an ion-pair process, but may represent minimum values since it is not sure that all high polymer chains are active. Figure 19 also includes a point at —60°C determined from data

46

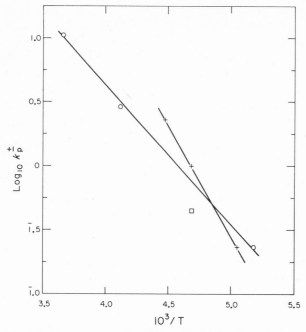

Fig. 19. Temperature dependence of $k_p^\pm$ for methylmethacrylate polymerization in toluene. ($\circ$) Initiated by diphenylhexyllithium [174]; (+) initiated by butyl-lithiumdiethylzinc [180]; ($\square$) initiated by fluorenyllithium [170] (0.1% THF present).

on fluorenyllithium initiated polymerization (calculated from estimates of the concentration of active chains determined from $\overline{M}_n$ of the precipitable polymer). It is in reasonable agreement with the other results in toluene. The presence of 0.1% THF in this case should not have much influence on the rates in methacrylate polymerization, although higher concentrations do produce noticeable acceleration of rate.

The complex of n-butyllithium with diethylzinc has also been studied as a polymerization initiator [180] for methylmethacrylate. In pure toluene the results obtained with this system were complex and not readily interpretable, but in presence of 1—10% tetrahydrofuran a simpler behaviour was reported. Monomer disappears by a first order process, and the observed first order coefficients can be expressed in terms of initiator, monomer and tetrahydrofuran concentrations in the following way

$$R_p = \{k_1 + k'[\text{THF}]\}\, f[\text{I}]_0[\text{M}]_0$$

where f is the fraction of initiator leading to polymer chains of high molecular weight. $[\text{I}]_0$ and $[\text{M}]_0$ are the initial concentrations of initiator and monomer. Measurements of $\overline{M}_n$ of precipitable polymer lead to an f value of 0.34 for all experiments. The results can be explained in terms of two types of active centre, a normal ion-pair and one solvated by one molecule of tetrahydrofuran [180], if it is assumed that the former is always in excess. Values of k_1 obtained by extrapolation of the data to zero tetrahydrofuran concentration should give $k_p^{\pm}$ in pure toluene. The values obtained are plotted in Fig. 19. The rate coefficient is close to that obtained for simple lithium—aryl initiators at $-75°C$ but increases more rapidly with temperature ($E = 7.5$ kcal mole^{-1}). Some differences would be expected if the counter-ion is LiZnEt$_2$ as the authors suggest rather than Li^{+}. The most important effect of complexing the butyllithium has been to decrease drastically its side reactions. Use of n-butyllithium alone leads to so few active polymer chains that a kinetic analysis is impossible [173].

Only one kinetic study exists on initiation of methacrylate polymerization by a sodium compound. The initiator was the disodium oligomer ("tetramer") of α-methylstyrene and polymerization was investigated at $25°C$ in toluene in presence of 0.05—0.2 mole fraction of tetrahydrofuran [181]. An internal first order disappearance of monomer was observed, the first order coefficient being directly proportional to active chain and tetrahydrofuran concentrations. The rate coefficients evaluated, e.g. $k_p^{\pm} = 3.1—13 \times 10^{-2}$ l mole^{-1} sec^{-1} at various tetrahydrofuran concentrations, are much lower than those for lithium initiators. They were, however, evaluated using a methyl iodide titration technique to estimate the active chain concentration. In view of the reactivity of tritiated acetic acid with many short chains which are clearly not active in chain propagation, there must be suspicion of similar behaviour with methyl iodide. If this happens, the active chain concentration would be over-estimated and the derived $k_p^{\pm}$ value would be too low. Unfortunately no molecular weights of the precipitable polymer were determined, so that it is impossible to check on active chain concentration using this alternative method.

A few kinetic experiments have been reported on initiation of methylmethacrylate by Grignard reagents [172, 182]. The polymerization process is obviously complex. At constant initiator and monomer concentrations the propagation rates show a maximum at about $-25°C$. A simple first order consumption of monomer is not observed, the apparent rate coefficient decreasing slowly at longer times. No simple dependence of initial rates on monomer concentration is observed. The results suggest that the initiator efficiency depends strongly on reaction conditions. Some estimates were made of this parameter using $^{14}C_2H_5MgBr$ as initiator. From the radioactivity of the precipitable polymer and assuming one initiator residue per polymer chain, efficiencies between 2% and 10% could be obtained. Estimation of $k_p^{\pm}$ from this type of experiment leads to values $\sim 2 \times 10^{-2}$ l mole^{-1} sec^{-1} with only a small change between -40 and $0°C$.

Further progress in the establishment of more detailed kinetic analysis of acrylate polymerization will depend on better estimates of the true

References pp. 60—65

concentration of active chains. For a given alkali metal, initiators must be chosen which show the least tendency to undergo side reactions, as established by analysis of reaction products, otherwise the side products are liable to influence the reaction. More data are required using sodium and the other alkali metal compounds together with experiments in polar solvents★. It seems unlikely, however, that as detailed a knowledge of the reactivity of different types of ionic species as is known for styrene will ever be possible in the anionic polymerization of polar monomers.

6. Stereospecificity in the propagation step

The use of organo-lithium or magnesium compounds in non-polar solvents often leads to a high degree of stereospecificity in the propagation step. The standard method for the preparation of highly isotactic acrylates involves the use of phenylmagnesium bromide in toluene solution. The metal halide component is important, the chlorides producing noticeably less regular polymer [183]. The presence of appreciable amounts of ether also decreases the specificity of the reaction.

Organo-lithium compounds under similar conditions also give isotactic polymer with acrylate monomers, the steric purity of the product depending on the particular initiator. The variations seem to be connected with the influence of products from the side reactions [174]. This behaviour is specific to the lithium—hydrocarbon systems; sodium or potassium alkyls even in toluene produce polymer of mixed microstructure [184]. Fairly highly syndiotactic polymers can be prepared with diethylmagnesium or with lithium alkyls in strongly solvating solvents such as tetrahydrofuran at low temperatures. In this case, the stereoregulating effect is not so strong and the microstructure of the polymer resembles that obtained from a free radical initiated polymerization at low temperature. This is the behaviour expected if the structure is determined by steric repulsion of bulky groups on monomer and chain end. Intermediate conditions, such as the presence of small amounts of ether or changing the counter-ion to Na^+, K^+, or Cs^+ produces mixed structures, some of which appear to contain reasonably long sequence lengths of iso- and syndiotactic units [170, 184].

The diene monomers give predominantly 1,4-polymers in hydrocarbon solvents if polymerized using lithium-based initiation. Isoprene, under these conditions, gives a predominantly *cis*-1,4 polymer but with butadiene the proportions of *cis*- and *trans*-1,4 are fairly evenly distributed. Once again this phenomenon is characteristic of lithium compounds; sodium- and potassium-based initiation gives mixed structures even in hydrocarbon solvents. Polymerization in polar solvents such as tetrahydrofuran leads to largely 3,4-polyisoprene or 1,2-butadiene with

★ Results are now available for THF with the counter-ions Na^+ and Cs^+. It appears that side reactions do not play an important role in this solvent at least at low temperatures [235].

not much dependence of structure on counter-ion [70]. Even small amounts of tetrahydrofuran in a hydrocarbon solvent causes the structure of the polymer to change from 1,4 to 3,4 (or 1,2).

Styrene only forms an isotactic polymer under heterogeneous reaction conditions. This is the case with sodium or potassium alkyls [185, 186]. It has been reported that the lithium alkyls produce isotactic polystyrene under homogeneous reaction conditions, but later investigation showed that the presence of lithium hydroxide or oxide is necessary [187].

Theories of the mechanism of stereoregulation have been mainly concerned with the lithium-based initiators (and to some extent on those involving magnesium compounds). It would be too complex a problem to produce a general explanation of the observed microstructures with all the counter-ions in different solvents. With lithium compounds, the mechanism has been related to their strong tendency to associate, which suggests that the carbon—lithium bond is able to interact with other polar groups or electron-rich groupings such as double bonds. The mechanisms suggested for stereoregular polymerization in hydrocarbon solvents normally involve the formation of a six-membered cyclic transition state. In the case of isoprene [188—190] this involves coordination of the C—Li bond with the incoming monomer in the cis-form. The configuration of each unit is thus supposed to be determined in the reaction step leading to its incorporation in the chain and implicitly cannot change in the time interval before the addition of the next monomer unit. The fact that butadiene forms a mixed cis-trans structure weakens the argument for such a mechanism, for although alternative cyclic transition states can be devised to produce trans insertion, the attractive simplicity and uniqueness of the cis complex is thereby lost. The mechanism is associated with the hypothesis that the C—Li bond is covalent but ionizes in ether solvents to give ionic structures which favour addition at the 3-position rather than the 1-position of the chain end [190]. It is usually proposed that the compounds of the higher alkali metals are predominantly ionic under all conditions.

In the polymerization of acrylates, interaction of lithium with carboxyl groups of incoming monomer and the penultimate unit of the polymer chain is the dominant feature of theories of stereoregulation [168, 191]. The stable form of the chain end is proposed to involve secondary bonding of the lithium to the penultimate carboxyl group thus holding it in a fixed configuration, viz.

The incoming monomer unit would then be forced, either because of steric interactions, or by the interaction of its carboxyl group with lithium at the chain-end, to add in a specific manner to re-form the same loose ring structure present initially. One variant of this mechanism [192] involves a covalently bonded six membered ring formed by enolization of the active chain end followed by alkoxide ion attack on the penultimate carboxyl group. In polar solvents, or in the presence of moderate amounts of them, competition for solvation of the counter-ion would be produced and the intramolecular solvation producing the stereospecificity would be reduced in effectiveness as the ether concentration is increased. Replacing the lithium counter-ion with sodium or other alkali metal would be expected to reduce the interaction Metal$^+$ $\cdots$ O=C— and hence weaken the stereo-regulating forces. In polar solvents solvent-separated ion-pairs and even free anions could exist, eventually both probably leading to syndiotactic polymer by the usual mechanism suggested for free radical polymerization. In intermediate cases, it is possible to conceive of a slow equilibrium between two types of ion-pairs each tending to add monomer in a different way and leading to block structures. Unfortunately, our knowledge of the ionic species present in acrylate polymerization is not at the advanced stage that it is for styrene polymerization and such mechanisms must be speculative.

The above mechanisms for *cis*-1,4 polymerization of isoprene or isotactic polymerization of acrylates assume that the configuration of each unit is fixed at the moment of reaction and that no racemization occurs between additions of monomer molecules. Little evidence for the validity of the mechanisms was available when suggested. Recently it has been possible to obtain information, from NMR studies, on the reaction path. This evidence is of two types and depends on the polymerization of stereospecifically deuterated monomers to determine the mode of approach of monomer molecules and on direct observations of NMR spectra of the terminal monomer unit in the polymer.

If an acrylic monomer is prepared with specific deuteration in the methylene group, R'OOC—C=C—D, it is possible to determine from the configuration of the —CHD— groups incorporated into the polymer, the relative positions of the two ester groups on monomer and chain end in the reaction approach [193, 194]. If the two groups are aligned, then the polymer will have deuterium atoms on the same side (in the planar zig-zag conformation) as its ester groups. A polymer which is isotactic with respect to the tertiary carbon and also to the —CHD group is formed, named *threo*-di-isotactic. This is found to be the case with organo-lithium initiated polymerization of acrylates in toluene [179]. As the most reasonable explanation for such alignment is lithium interaction with carboxyl groups of monomer and chain end, one feature of the proposed mechanisms is confirmed.

In the presence of small amounts of ethers, where the polymer remains predominantly isotactic in configuration, the amount of *threo*-CHD— groups decreases as the ether concentration is increased. Eventually the polymer CHD groups have the reversed (erythro) configuration. Approach with ester groups on chain end and monomer opposed to each other is being observed, as if repulsion between bulky groups was more important in the reaction approach. In this case the CHD groups observed are still in meso (isotactic like) diads which implies that subsequent to reaction, rotation has occurred around the terminal bond [179]. Otherwise a racemic (syndiotactic like) relationship would have been formed between the end two ester groups in the chain. This confirms that interaction of the lithium ion with chain-end terminal and penultimate carbonyl groups is strong enough to invert the configuration of the chain end carbanion if monomer approaches in the "wrong" direction. Larger concentrations of ethers, of course, destroy this interaction by extensive solvation of Li$^+$ and the configuration at the tertiary carbon atoms remains syndiotactic like.

The situation in the phenylmagnesiumbromide initiated polymer-izations is more difficult to explain. In the virtual absence of diethylether the polymer formed is mostly *erythro*-di-isotactic (opposed ester presen-tation) but becomes *threo*-di-isotactic in the presence of moderate amounts of ether [179]. The reason for this is not clear. However, with both initiators there is evidence that the configuration of each unit is not entirely determined at the point of reaction, but reorganization to a more stable form of the active chain end is possible. The relative thermo-dynamic stabilities of the different conformations of the organo-metallic terminal unit also play a role.

In the case of isoprene, low molecular weight model compounds (dimer, trimer, etc.) can be prepared in benzene to produce oligomeric analogues of the polyisoprenyllithium active in polymerization [195]. The NMR spectra of such oligomers show that *cis* and *trans* forms of the lithium bearing terminal unit occur, and that one predominates at room temperature [195, 196]. It is probably the *cis* form, although this is difficult to establish without doubt★. Except for the one unit chain, the *cis—trans* ratio varies reversibly with temperature. Transfer to tetrahydro-furan-rich mixtures at low temperatures shows that isomerization occurs when the solution is warmed to −40°C, probably to the *trans* form. This is the stable form in such solvents at all temperatures. The NMR spectra are basically the same in both hydrocarbon and ether solvents. Only the resonance due to the proton on the γ-carbon is shifted upfield in polar

---

★ Later experiments have established that it is, in fact, probably *trans* [236].

solvents. This corresponds to a greater fraction of the negative charge being on this carbon atom. In benzene, most of the charge is on the terminal $-CH_2$ group (α), although the γ-shift in benzene corresponds to some charge on this carbon. In tetrahydrofuran mixtures the terminal unit can be described as a charge-delocalized allylic anion,

$$
(\sim\!\!\sim\!\!\sim CH_2 - CH \cdots \overset{\displaystyle \overset{CH_3}{|}}{C} \cdots \bar{C}H_2)
$$

$$
Li^+
$$

$$
\delta \quad \gamma \quad \beta \quad \alpha
$$

Hindered rotation around terminal and penultimate C—C bonds can be observed in the temperature range —80 to —40°C. No evidence of two separate and distinct 3,4 and 1,4 structures in either solvent can be detected.

These observations can be correlated with polymer microstructure. Where the charge is largely on carbon-α, the polymer microstructure is 1,4 and as charge is redistributed to the γ-carbon, the polymer structure becomes 3,4 even though in all cases most of the negative charge may still be on carbon-α. Rather than simple charge redistribution on moving from non-polar to polar solvents other authors [197] have suggested two different forms in dynamic equilibrium. The covalent 1,4 form (*cis* and *trans*) is then supposed to be the predominant species in hydrocarbon solvents and the allylic form in polar solvents. The *cis*-1,4 structure of the polymer in hydrocarbon solvents seems to be connected with the presence of predominantly one stable isomer of the active end, i.e. the thermodynamic stability of the *cis* and *trans* forms is important. Now the species responsible for the NMR spectrum in hydrocarbon solvents is of course the associated form of polyisoprenyllithium. It is possible, therefore, that the *cis—trans* ratio in the polymer is determined by packing preferences of *cis* and *trans* ends in the aggregates [198]. An important check on the validity of this approach would be a determination of the configurational preference of polybutadienyllithium in hydrocarbon solvents. If the *cis—trans* distribution of the lithium-bearing unit were fairly random, as it is known to be in the final polymer, then this would provide confirmatory evidence of thermodynamic rather than kinetic control. Preliminary evidence [199] on the structure of the 1:1 addition product of *sec.*-butyllithium with butadiene, suggests a 3:1 preference of *trans* over *cis*. There are, however, some doubts if this result would be typical of active polymers of higher molecular weight★. The microstructure of polybutadienes of very low molecular weight is notably different from that of the high polymer. This is a situation which does not occur with isoprene polymers. It is premature, therefore, to make a final decision on the validity of this method of steric control.

★ The *trans* preference seems to be maintained at higher DP [237].

7. Copolymerization

In the study of anionic copolymerization it is possible to use two types of approach. The first method is the use of the classical copolymer composition equations developed for free radical polymerization. The second is unique to anionic polymerization and depends on the fact that for "living" systems it is possible to prepare an active polymer of one monomer and to study its reaction with the second monomer. The initial rate of disappearance of one type of active end, or the appearance of the other type (usually determined spectroscopically) or the rate of monomer consumption gives directly the reactivity of polymer-1 with monomer-2. It is in principle possible to compare the two methods to see if additional complications occur when both monomers are present together.

First of all, it is necessary to investigate the applicability of standard copolymerization theory to anionic polymerization. In the copolymerization of two monomers, the rates of incorporation into the polymer are given by eqn. (22).

$$-d[M_1]/dt = k_{11}[P_1^*][M_1] + k_{21}[P_2^*][M_1] \qquad (22a)$$

$$-d[M_2]/dt = k_{22}[P_2^*][M_2] + k_{12}[P_1^*][M_2] \qquad (22b)$$

Here $[P_x^*]$ is the concentration of growing centres ending in monomer x and k_{xy} is the absolute rate coefficient of reaction of P_x^* with monomer y. Two difficulties arise in anionic polymerization. In hydrocarbon solvents with lithium and sodium based initiators, $[P_x^*]$ is not the total concentration of polymer units ending in unit x but, due to self-association phenomena, only that part in an active form. The reactivity ratios determined are, however, unaffected by the association phenomena. As each ratio refers to a common active centre, the effective concentration of active species is reduced equally to both monomers. In polar solvents such as tetrahydrofuran, this difficulty does not arise, but there will be two types of each reactive centre P_x^*, one an anion and the other an ion-pair. Application of eqn. (22) will give apparent rate coefficients as discussed in Section 4 if total concentrations of P_x^* are used. Reactivities can change with concentration if defined on this basis.

The copolymer composition equation [200]

$$d[M_1]/d[M_2] = [M_1](r_1[M_1] + [M_2])/[M_2](r_2[M_2] + [M_1]) \qquad (23)$$

where $r_1 = k_{11}/k_{12}$ and $r_2 = k_{22}/k_{21}$, is usually used at low conversion in free radical initiated polymerization to determine values of r_1 and r_2 from the initial copolymer composition. Except under special conditions ($r_1 = r_2 = 1$), the composition will change with conversion as one monomer is preferentially depleted. Often the composition is extrapolated to zero time to avoid this difficulty. This is a potentially dangerous procedure with anionic systems, for it is equivalent to extrapolation to low molecular weight and the effect of preferential initiation to one monomer may be important. In addition, eqn. (23) depends on the equilibrium of the two cross-propagation steps, viz.

$$k_{21}[P_2^*][M_1] = k_{12}[P_1^*][M_2] \tag{24}$$

This is an accurate assumption in free radical polymerizations where individual active centres are of very short life, and the steady state concentrations rapidly established. Equilibration may not be so rapid in anionic polymerization.

This problem has been considered [201] for the simplified case where the concentrations of each monomer are held constant. If there is preferential initiation to one monomer, which is then the reactive monomer in the propagation step, equilibration is rapidly achieved. If the reverse occurs, i.e. one monomer is about ten times more reactive to the initiator but about ten times less reactive in chain propagation, eqn. (23) will be correct when the polymer DP has reached one hundred. If both reactivity ratios are much larger than unity (i.e. there is a tendency for block copolymers to form) then the copolymer composition equation will fail to give the correct reactivity ratios until very long chains are formed. The actual situation will be worse than is indicated by these calculations as the feed composition will in fact normally change and unless an integrated form of eqn. (23) is used, the errors will be larger. There is insufficient data available to assess the errors due to these causes; generally the copolymer composition is measured at 5—20% conversion and the derived reactivity ratios appear to be at least of the correct order of magnitude when compared with the values determined from individual rate coefficients.

Early investigations of ionic copolymerization [202] led to the conclusion that the product $r_1 r_2$ is approximately unity. This result will be produced exactly if $k_{11}/k_{12} = k_{21}/k_{22}$, as expected if the competition between two monomers for reaction with an ionic centre is independent of the nature of that centre [202]. The copolymer composition equation then becomes

$$d[M_1]/d[M_2] = r_1[M_1]/[M_2] \tag{25}$$

A special case arises if $k_{11} = k_{21}$ and $k_{12} = k_{22}$, i.e. the absolute reactivities of the two types of chain-end are identical. In that case,

$$\frac{d[M_1]}{d[M_2]} = \frac{k_{11}[M_1]}{k_{22}[M_2]} \tag{26}$$

Equation (26) is the "ideal" copolymer composition equation suggested [203] early in the development of copolymerization theory but which had to be abandoned in favour of eqn. (23) as a general description of radical copolymerization. Only in this particular case are the rates of incorporation of each monomer proportional to their homopolymerization rates. It was shown that the reactivity of a series of monomers in stannic chloride initiated copolymerization followed the same order as their homopolymerization rates [202] and so eqn. (26) could be at least qualitatively correct for carbonium-ion polymerizations and possibly for reactions carried by carbanions. This, in fact, does not seem to be correct for anionic polymerizations since the reactivities of the ion-paired species at least, differ greatly. The methylmethacrylate ion-pair will, for instance, not add to styrene monomer, whereas the polystyryl ion-pair adds rapidly to methylmethacrylate [204]. This is a general phenomenon; no reaction will occur if the ion-pair is on a monomer unit which has an appreciably higher electron affinity than that of the reacting monomer. The additions are thus extremely selective, more so than in radical copolymerization. There is no evidence that eqn. (26) holds and the approximate agreement with eqn. (25) results from other causes indicated below.

The copolymerization of styrene and the dienes has received extensive study and illustrates the principles involved [205—208]. In hydrocarbon solvents with lithium based initiators at equimolar feed ratio, the initial copolymer composition is mainly diene plus 5—10% of styrene. The initial copolymerization rate is roughly that of the homopolymerization of the diene, which is lower than that of styrene. Similar results are observed in copolymerization of isoprene and butadiene. The initial polymer contains mostly butadiene although this is the slower polymerizing monomer. The value of $r_1 r_2$ is close to unity, so eqn. (25) holds at least approximately. Preferential absorption around the active chain-ends* of the monomer which adds most rapidly was suggested to be the reason for this behaviour [205]. There is, of course, no reason why this should be necessary since unless eqn. (26) holds, for which there is no evidence, there is no simple relationship between copolymerization and homopolymerization rates. The real reason for this behaviour is found in the extremely high rate of one of the cross-propagation steps [209]. This point was confirmed for the styrene—butadiene case [210, 211] and for styrene—isoprene [61]; chain-ends terminating in the styryl anion-pair add the two dienes very rapidly. Under these conditions very little styrene can be incorporated in the polymer at normal feed ratios. The effect of excess butadiene monomer on the rate of addition of styrene monomer to "living" polybutadiene has also been checked [211]. No effect was noticed; if

---

* Absorption or complexing of monomer has been suggested to occur even in homopolymerization. It is, in principle, unlikely, for anionic chain-ends add monomer rapidly, and no convincing evidence has ever been presented for its occurrence.

preferential absorption of butadiene was expected round the chain-end, the rate of addition of styrene should be depressed. It has also been pointed out that in this type of copolymerization characterized by one monomer being much more reactive to both types of chain-end ($r_1 \gg 1$, $r_2 \ll 1$ where subscript 1 refers to the diene) eqn. (23) can be approximated by [211].

$$\frac{d[M_1]}{d[M_2]} = r_1 \frac{[M_1]}{[M_2]} + 1 \qquad (27)$$

It will be experimentally difficult to distinguish between eqns. (25) and (27) when r_1 is large and the true value of the product $r_1 r_2$ may be anywhere between 1, eqn. (25), and ≈ 0, eqn. (27).

An examination of reported reactivity ratios (Table 6) shows that the behaviour $r_1 \gg 1$, $r_2 \ll 1$ or vice versa is a common feature of anionic copolymerization. Only in copolymerizations involving the monomers 1,1-diphenylethylene and stilbene, which cannot homopolymerize, do we find $r_1 < 1$, $r_2 < 1$ [212—215], and hence the alternating tendency so characteristic of many free radical initiated copolymerizations. Normally one monomer is much more reactive to either type of active centre in the order acrylonitrile > methylmethacrylate > styrene > butadiene > isoprene. This is the order of electron affinities of the monomers as measured polarographically in polar solvents [216, 217]. In other words, the reactivity correlates well with the overall thermodynamic stability of the product. Variations of reactivity ratio occur with different solvents and counter-ions but the gross order is predictable.

TABLE 6

Reactivity ratios in anionic copolymerization

(Determined from the copolymer composition equation except for starred values which were determined from measurements of the four individual rate coefficients.)

M_1	M_2	Solvent[a]	Counter-ion	Temp. (°C)	r_1	r_2	Ref.
Styrene	Isoprene	BEN	Li	30	0.14	7.0	207
Styrene	Isoprene	TOL	Li	27	0.25	9.5	218
Styrene*	Isoprene	CYH	Li	40	0.046	16.6	61
Styrene	Isoprene	THF	Li	27	9	0.1	218
Styrene	Isoprene	THF	Li	−35	40	~0	218
Styrene*	Isoprene	THF	Na	25	~30		221
Styrene	Butadiene	TOL	Li	25	0.1	12.5	218
Styrene	Butadiene	BEN	Li	30—50	0.05	20	206
Styrene*	Butadiene	BEN	Li	29	0.06	3	210
Styrene*	Butadiene	CYH	Li	40	<0.04	26	211

TABLE 6—*continued*

M_1	M_2	Solvent[a]	Counter-ion	Temp. (°C)	r_1	r_2	Ref.
Styrene	Butadiene	HEP	Li	30	~0	7	208
Styrene	Butadiene	THF	Li	−35	8	0.2	218
Styrene	Butadiene	Ether	Li	30	0.1	1.8	219
Styrene*	Butadiene	THF	Na	25	~55		221
Isoprene	Butadiene	HEX	Li	50	0.5	3.4	205
Isoprene	Butadiene	BEN	Li	40	0.5	3.6	220
Styrene*	Divinyl-benzene	BEN	Li	30	0.094	10.0	230
Styrene	p-Methyl-styrene	TOL	Li	0	2.5	0.26	222
Styrene	p-Methyl-styrene	BEN	Li	30	2.5	0.4	223
Styrene	p-Methyl-styrene	THF	Li	0	1.3	0.9	222
Styrene	p-Methyl-styrene	THF	Na	0	2	0.4	222
Styrene*	p-Methyl-styrene	THF	Na	25	5.3	0.18	224, 146
Styrene	p-Methoxy-styrene	TOL	Li	0	11	0.05	222
Styrene	p-Methoxy-styrene	THF	Li	0	3	0.23	222
Styrene	p-Methoxy-styrene	THF	Na	0	4	0.13	222
Styrene*	p-Methoxy-styrene	THF	Na	25	19	0.05	146
Styrene*	α-Methyl-styrene	THF	Na	25	35	0.03	224
Styrene	α-Methyl-styrene	THF	Na	25	17	0.015	225
Styrene	1,1-Diphenyl-ethylene	THF	Li, Na	30	0.1	~0	213
Styrene*	1,1-Diphenyl-ethylene	THF	Na	25	0.3, 0.15[b]	~0	212
Styrene	1,1-Diphenyl-ethylene	BEN	Li	30	0.7	~0	213
Styrene*	1,1-Diphenyl-ethylene	BEN	Li	30	0.7	~0	226
Isoprene	1,1-Diphenyl-ethylene	THF	Li, Na, K	0	0.11	~0	214
Methyl-methacrylate	Acrylo-nitrile	BULK	Li	−8	0.4	7	227
Methyl-methacrylate	Acrylo-nitrile	BULK	Na	−12	0.14	5	227
Methyl-methacrylate	Acrylo-nitrile	TOL	Mg	−78	0.03	5	228

References pp. 60—65

TABLE 6—*continued*

M₁	M₂	Solvent[a]	Counter-ion	Temp. (°C)	r_1	r_2	Ref.
Methyl-methacrylate	Acrylo-nitrile	NH₃	Na	−50	0.25	8	202
Methyl-methacrylate	Acrylo-nitrile	THF	Mg	−80	0.3	10	227
Methyl-methacrylate	Methacrylo-nitrile	NH₃	Na	−50	0.7	5	229
Methyl-methacrylate	Methacrylo-nitrile	Ether/TOL	Mg	−78	2.0	0.3	228
Acrylonitrile	Methacrylo-nitrile	TOL	Mg	−50	0.9	0.5	228
Acrylonitrile	Methacrylo-nitrile	NH₃	Na	−50	1.0	0.05	228

[a] Solvents: abbreviations as in Tables 2 and 3.
[b] Ion-pair and free anion respectively.

Only in the lithium catalysed copolymerization of styrene with the dienes in hydrocarbons do the monomers show an unexpected order of reactivity, an anomaly which disappears for polar solvents. The preference for the diene in the copolymer vanishes even in hydrocarbon solvents if sodium is used as initiator [231]. With the dienes, however, an added complication exists which makes simple experiments on electron affinity measured in solvents such as dioxane—water a poor guide to reactivity. They can react in more than one way to give a 3,4 (or 1,2) structure or alternatively a 1,4 structure. There appears to be good correlation between the amount of styrene in the copolymer and the percentage of 1,4 structure in a polyisoprene homopolymerized with the same initiator and solvent [232]. Under conditions in which isoprene polymerizes in the 1,4 form it seems to be a more reactive monomer than styrene, but if the product is the 3,4 structure it is less reactive. Once again the final state of the reaction appears to play an important role, which suggests that the transition state in anionic polymerization has a large contribution from the final state. It is interesting that in the copolymerization of isoprene and butadiene, the latter is always the more reactive monomer. The microstructures of both are similarly affected by reaction conditions and perturbations caused by changing polymer structure disappear.

As noted earlier, reactivity ratios measured with polar solvents should be determined for both the free anions and the ion-pairs. This has not in general been carried out and the reactivity ratios represent effective values often determined at about millimolar concentrations of active centres.

This is true for all the values reported in Table 6 with one exception. Only for solvents such as diethylether and dioxane, where the free anion contribution to the rate is negligible, do they represent true values corresponding to a single species, in this case the ion-pair. If the reactivity ratios are determined from measurements of the rates of the four individual reaction steps, then it is possible to analyze the results, if each is carried out at a series of concentrations, in the manner described in Section 4.2, for k_{xy}^- and $k_{xy}^\pm$. If the reactivity ratios alone are required it is not necessary to measure the ion-pair dissociation constants because these cancel in the determination of r_1 and r_2, e.g.

$$k_{11}^{\text{apparent}} = k_{11}^\pm + k_{11}^- K_{d(1)}^{1/2} \,[\text{active centres}]_1^{-1/2}$$

$$k_{12}^{\text{apparent}} = k_{12}^\pm + k_{12}^- K_{d(1)}^{1/2} \,[\text{active centres}]_1^{-1/2}$$

and similarly for k_{22} and k_{21}.

Thus by plotting k_{xy}^{apparent} for the four possible steps as a function of $[\text{active centres}]_x^{-1/2}$, the intercepts give the four values $k_{11}^\pm$, $k_{12}^\pm$, $k_{22}^\pm$, $k_{21}^\pm$ and hence $r_1^\pm$ and $r_2^\pm$. The ratio of the slopes of the graphs of k_{11}^{apparent} and k_{12}^{apparent} gives r_1^- directly. r_2^- can be determined in the same way. Figure 20 illustrates the method for styrene, p-methylstyrene copolymerization in tetrahydrofuran [146]. The data were obtained

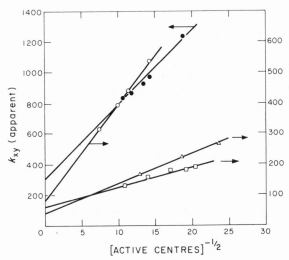

Fig. 20. Variation of the apparent propagation coefficients with reciprocal square root of concentration of active centres in the copolymerization of styrene (1) with p-methylstyrene (2). Solvent: tetrahydrofuran. ($\bigcirc$) k_{11}; ($\bullet$) k_{21}; ($\triangle$) k_{22}; ($\square$) k_{12} [146].

before the importance of the two types of reactive centre was fully realized and a wider range of concentrations would be necessary for accurate determinations. Nevertheless, qualitatively the results indicate some differences in reactivity ratios between ions and ion-pairs. $r_1^{\pm} \sim 1$ and $r_2^{\pm} \sim 0.13$ can be determined from the intercepts. $r_1^- = \text{slope}_{11}/\text{slope}_{12} \sim 6$ and $r_2^- = \text{slope}_{22}/\text{slope}_{21} \sim 0.2$. The values in Table 6 (5.3, 0.18) determined as effective values at active centre concentrations $\sim 2{-}3\text{mM}$ are essentially those of the free anions. This is the result expected for styrene and its derivatives in this concentration range. Most of the reaction is carried by free anions. It is probable that all the reactivity ratios in Table 6 determined in tetrahydrofuran are close to the free anion values but firm conclusions would require much more detailed study. There is obviously much scope for further investigations in this field, particularly on the rates of individual reaction steps rather than on reactivity ratio measurements from copolymer composition.

REFERENCES

1 F. E. Matthews and E. H. Strange, Brit. Pat., 24,790 (1910).
2 C. Harries, U.S. Pat. 1,058,056 (1913).
3 K. Ziegler, F. Crössmann, H. Kleiner and O. Schäfer, Justus Liebigs Ann. Chem., 473 (1929) 1.
4 K. Ziegler, F. Dersch and H. Wollthan, Justus Liebigs Ann. Chem., 511 (1934) 13.
5 K. Ziegler, Angew. Chem., 49 (1936) 499.
6 R. G. Beaman, J. Amer. Chem. Soc., 70 (1948) 3115.
7 J. J. Sanderson and C. R. Hauser, J. Amer. Chem. Soc., 71 (1949) 1595.
8 M. G. Evans, W. C. E. Higginson and N. S. Wooding, Recl. Trav. Chim. Pays-Bas., 68 (1949) 1069.
9 W. C. E. Higginson and N. S. Wooding, J. Chem. Soc., (1952) 760.
10 M. Szwarc and J. Smid, Progr. React. Kinet., 2 (1964) 219.
11 N. S. Wooding and W. C. E. Higginson, J. Chem. Soc., (1952) 774.
12 K. Ziegler, E. Holzkamp, H. Breil and H. Martin, Angew Chem., 67 (1955) 541.
13 G. Natta, J. Polym. Sci., 16 (1955) 143.
14 F. W. Stavely and coworkers, Ind. Eng. Chem., 48 (1956) 778.
15 M. Szwarc, Nature (London), 178 (1956) 1168.
16 D. N. Bhattacharyya, C. L. Lee, J. Smid and M. Szwarc, J. Phys. Chem., 69 (1965) 612.
17 D. J. Worsfold and S. Bywater, Can. J. Chem., 38 (1960) 1891.
18 L. Fetters, J. Res. Nat. Bur. Stand., Sect. A, 70 (1966) 421.
19 M. Szwarc, Carbanions, Living Polymers and Electron Transfer Processes, Interscience, New York, 1968, Chap. IV.
20 F. S. Dainton, K. M. Hui and K. J. Ivin, Eur. Polym. J., 5 (1969) 387.
21 S. Bywater and D. J. Worsfold, Can. J. Chem., 44 (1966) 1671.
22 G. Olah, C. V. Pittman, R. Waack and M. Doran, J. Amer. Chem. Soc., 88 (1966) 1488.
23 M. Levy, M. Szwarc, S. Bywater and D. J. Worsfold, Polymer, 1 (1960) 515.
24 D. J. Worsfold and S. Bywater, J. Chem. Soc., (1960) 5234.
25 J. Barriac, M. Fontanille and P. Sigwalt, Bull. Soc. Chim. Fr., (1969) 3487.
26 G. Spach, M. Levy and M. Szwarc, J. Chem. Soc., (1962) 355.
27 S. Bywater and D. J. Worsfold, J. Polym. Sci., 58 (1962) 571.
28 S. Bywater, A. F. Johnson and D. J. Worsfold, Can. J. Chem., 42 (1964) 1255.

29 D. K. Polyakov, Y. L. Spirin, A. R. Gantmakher and S. S. Medvedev, *Dokl. Akad. Nauk SSSR, 150* (1963) 1051.
30 A. Gourdenne and P. Sigwalt, *Eur. Polym. J., 3* (1967) 481.
31 H. Brody, D. H. Richards and M. Szwarc, *Chem. Ind. (London)*, (1958) 1473.
32 M. Morton, E. E. Bostick and R. Livigni, *Rubber Plastics Age, 42* (1961) 397.
33 D. J. Worsfold and S. Bywater, *Can. J. Chem., 42* (1964) 2884.
34 T. L. Brown, *Advan. Organometal. Chem., 3* (1965) 365.
35 H. L. Lewis and T. L. Brown, *J. Amer. Chem. Soc., 92* (1970) 4664.
36 S. Bywater and D. J. Worsfold, *J. Organometal. Chem., 10* (1967) 1.
37 W. H. Glaze and C. H. Freeman, *J. Amer. Chem. Soc., 91* (1969) 7198.
38 T. L. Brown, *Accounts Chem. Res., 1* (1968) 23.
39 K. F. O'Driscoll and A. V. Tobolsky, *J. Polym. Sci., 35* (1959) 259.
40 F. J. Welch, *J. Amer. Chem. Soc., 81* (1959) 1345.
41 E. N. Kropacheva, B. A. Dolgoplosk and E. M. Kuznetsova, *Dokl. Akad. Nauk SSSR, 130* (1960) 1253.
42 M. Morton, R. A. Pett and L. J. Fetters, *Macromolecules, 3* (1970) 333.
43 A. G. Evans and D. B. George, *J. Chem. Soc.*, (1961) 4653.
44 R. A. H. Cassling, A. G. Evans and N. H. Rees, *J. Chem. Soc., B* (1966) 519.
45 A. G. Evans, C. R. Gore and N. H. Rees, *J. Chem. Soc.*, (1965) 5110.
46 J. E. L. Roovers and S. Bywater, Unpublished work.
47 T. L. Brown, *J. Organometal. Chem., 5* (1966) 191.
48 G. E. Hartwell and T. L. Brown, *J. Amer. Chem. Soc., 88* (1966) 4625.
49 F. Schué and S. Bywater, *Macromolecules, 2* (1969) 458.
50 M. Y. Darensbourg, B. Y. Kimura, G. E. Hartwell and T. L. Brown, *J. Amer. Chem. Soc., 92* (1970) 1236.
51 J. E. L. Roovers and S. Bywater, *Macromolecules, 1* (1968) 328.
52 A. F. Johnson and D. J. Worsfold, *J. Polym. Sci., Part A, 3* (1965) 449.
53 A. Guyot and J. Vialle, *J. Polym. Sci., Part B, 6* (1968) 403.
54 H. L. Hsieh, *J. Polym. Sci., Part A, 3* (1965) 163.
55 H. L. Hsieh, *J. Polym. Sci., Part A1, 8* (1970) 533.
56 A. Guyot and J. Vialle, *J. Macromol. Sci. Chem., 4* (1970) 79.
57 C. M. Selman and H. L. Hsieh, *J. Polym. Sci., Part B, 9* (1971) 219.
58 R. C. P. Cubbon and D. Margerison, *Polymer, 6* (1965) 102.
59 M. Morton, E. E. Bostick, R. A. Livigni and L. J. Fetters, *J. Polym. Sci., Part A, 1* (1963) 1735.
60 M. Morton, L. J. Fetters, R. A. Pett and J. F. Meier, *Macromolecules, 3* (1970) 327.
61 D. J. Worsfold, *J. Polym. Sci., Part A1, 5* (1967) 2783.
62 H. L. Hsieh, *J. Polym. Sci., Part A, 3* (1965) 173.
63 Yu. L. Spirin, A. R. Gantmakher and S. S. Medvedev, *Dokl. Akad. Nauk SSSR, 146* (1962) 368.
64 M. Morton, R. A. Pett and J. F. Fellers, Preprints I.U.P.A.C. Macromolecular Symposium Tokyo, 1 (1966) 69. J. F. Fellers, Thesis, University of Akron, (1967).
65 H. Sinn, *Angew. Chem., 75* (1963) 805.
66 D. N. Cramond, P. S. Lowry and J. R. Urwin, *Eur. Polym. J., 2* (1966) 107.
67 M. Morton, L. J. Fetters and E. E. Bostick, *J. Polym. Sci., Part C, 1* (1963) 311.
68 M. Morton and L. J. Fetters, *J. Polym. Sci., Part A, 2* (1964) 3311.
69 M. Morton, *Polymer Preprints, 5* (1964) 1092.
70 S. Bywater, *Fortschr. Hochpolym.-Forsch., 4* (1965) 66.
71 J. Geerts, M. Van Beylen and G. Smets, *J. Polym. Sci., Part A1, 7* (1969) 2859.
72 J. E. L. Roovers and S. Bywater, *Trans. Faraday Soc., 62* (1966) 701.
73 J. E. L. Roovers and S. Bywater, *Can. J. Chem., 46* (1968) 2711.
74 S. Bywater and D. J. Worsfold, *Can. J. Chem., 40* (1962) 1564.

62

75 I. A. Alexander and S. Bywater, *J. Polym. Sci., Part A1*, 6 (1968) 3407.
76 R. V. Basova and A. R. Gantmakher, *Dokl. Akad. Nauk SSSR*, 169 (1966) 368.
77 J. E. L. Roovers and S. Bywater, *Trans. Faraday Soc.*, 62 (1966) 1876.
78 A. Guyot and J. Vialle, *J. Macromol. Sci. Chem.*, 4 (1970) 107.
79 B. François, V. Sinn and J. Parrod, *J. Polym. Sci., Part C*, 4 (1964) 375.
80 N. D. Scott, U.S. Pat., 2,181,771 (1939).
81 N. D. Scott, J. F. Walker and V. L. Hansley, *J. Amer. Chem. Soc.*, 58 (1936) 2442.
82 A. I. Shatenstein and E. S. Petrov, *Russ. Chem. Rev.*, 36 (1967) 100.
83 D. Lipkin, D. E. Paul, J. Townsend and S. I. Weissman, *Science*, 117 (1953) 534.
84 M. Szwarc, M. Levy and R. Milkovich, *J. Amer. Chem. Soc.*, 78 (1956) 2656.
85 K. F. O'Driscoll and A. V. Tobolsky, *J. Polym. Sci.*, 31 (1958) 123.
86 M. Levy and M. Szwarc, *J. Amer. Chem. Soc.*, 82 (1960) 521.
87 J. Jagur-Grodzinski and M. Szwarc, *Proc. Roy. Soc., Ser. A.*, 288 (1965) 212.
88 C. G. Overberger and N. Yamamoto, *J. Polym. Sci., Part B*, 3 (1965) 569.
89 C. G. Overberger and N. Yamamoto, *J. Polym. Sci., Part A1*, 4 (1966) 3101.
90 C. L. Lee, J. Smid and M. Szwarc, *J. Phys. Chem.*, 66 (1962) 904.
91 C. E. Frank, J. R. Leebrick, L. F. Moormeier, J. A. Scheben and O. Homberg, *J. Org. Chem.*, 26 (1961) 307.
92 A Vrancken, J. Smid and M. Szwarc. *Trans. Faraday Soc.*, 58 (1962) 2036.
93 R. L. Williams and D. H. Richards, *Chem. Commun.*, (1967) 414.
94 D. H. Richards and N. F. Scilly, *Chem. Commun.*, (1968) 1641.
95 D. H. Richards and M. Szwarc, *Trans. Faraday Soc.*, 55 (1959) 1644.
96 M. Morton, A. Rembaum and E. E. Bostick, *J. Polym. Sci.*, 32 (1958) 530.
97 A. V. Tobolsky and D. B. Hartley, *J. Amer. Chem. Soc.*, 84 (1962) 1391.
98 A. V. Tobolsky, A. Rembaum and A. Eisenberg, *J. Polym. Sci.*, 45 (1960) 347.
99 F. Bahsteter, J. Smid and M. Szwarc, *J. Amer. Chem. Soc.*, 85 (1963) 3909.
100 R. Waack and M. A. Doran, *J. Org. Chem.*, 32 (1967) 3395.
101 R. Waack and M. A. Doran, *J. Amer. Chem. Soc.*, 91 (1969) 2456.
102 P. West and R. Waack, *J. Amer. Chem. Soc.*, 89 (1967) 4395.
103 R. Waack and P. West, *J. Organometal. Chem.*, 5 (1966) 188.
104 D. N. Bhattacharyya, J. Smid and M. Szwarc, *J. Phys. Chem.*, 69 (1965) 624.
105 G. Löhr and S. Bywater, *Can. J. Chem.*, 48 (1970) 2031.
106 A. Parry, J. E. L. Roovers and S. Bywater, *Macromolecules*, 3 (1970) 355.
107 D. N. Bhattacharyya, C. L. Lee, J. Smid and M. Szwarc, *Polymer*, 5 (1964) 54.
108 H. Hostalka, R. V. Figini and G. V. Schulz, *Makromol. Chem.*, 71 (1964) 198.
109 M. Van Beylen, M. Fischer, J. Smid and M. Szwarc, *Macromolecules*, 2 (1969) 575.
110 G. V. Schulz, L. L. Böhm, M. Chmeliř, G. Löhr and B. J. Schmitt, *Kinetics and Mechanics of Polyreactions*, Akadémiai Kiado, Budapest, 1971, p. 223.
111 T. Shimomura, K. J. Tölle, J. Smid and M. Szwarc, *J. Amer. Chem. Soc.*, 89 (1967) 796.
112 T. Shimomura, J. Smid and M. Szwarc, *J. Amer. Chem. Soc.*, 89 (1967) 5743.
113 H. Hiroha, K. Takaya, M. Nakayama and N. Ise, *Trans. Faraday Soc.*, 66 (1970) 3163.
114 G. Allen, G. Gee and C. Stretch, *J. Polym. Sci.*, 48 (1960) 189.
115 F. S. Dainton, K. J. Ivin and R. LaFlair, *Eur. Polym. J.*, 5 (1969) 379.
116 L. Böhm, W. K. R. Barnikol and G. V. Schulz, *Makromol. Chem.*, 110 (1967) 222.
117 H. Hirohara, M. Nakayama, K. Takaya and N. Ise, *Trans. Faraday Soc.*, 66 (1970) 1165.
118 F. S. Dainton, G. E. East, G. A. Harpell, N. R. Hurworth, K. J. Ivin, R. T. LaFlair, R. H. Pallen and K. M. Hui, *Makromol. Chem.*, 89 (1965) 257.

119 H. Hostalka and G. V. Schulz, *Z. Phys. Chem. (Frankfurt)*, *45* (1965) 286.
120 G. Löhr and G. V. Schulz, *Makromol. Chem.*, *117* (1968) 283.
121 K. J. Laidler and H. Eyring, *Ann. N.Y. Acad. Sci.*, *39* (1940) 303.
122 D. N. Bhattacharyya, C. J. Lee, J. Smid and M. Szwarc, *J. Phys. Chem.*, *69* (1965) 608.
123 T. E. Hogen-Esch and J. Smid, *J. Amer. Chem. Soc.*, *88* (1966) 307.
124 T. E. Hogen-Esch and J. Smid, *J. Amer. Chem. Soc.*, *88* (1966) 318.
125 T. Ellingsen and J. Smid, *J. Phys. Chem.*, *73* (1969) 2712.
126 L. L. Chan, K. H. Wong and J. Smid, *J. Amer. Chem. Soc.*, *92* (1970) 1955.
127 M. Szwarc, *Carbanions, Living Polymers and Electron Transfer Processes*, Interscience, New York, 1968, Chap. 7.
128 B. J. Schmitt and G. V. Schulz, *Makromol. Chem.*, *142* (1971) 325.
129 M. Shinohara, J. Smid and M. Szwarc, *J. Amer. Chem. Soc.*, *90* (1968) 2175.
130 S. Winstein, E. Clippinger, A. H. Fainberg and G. C. Robinson, *J. Amer. Chem. Soc.*, *76* (1954) 2597.
131 L. L. Chan and J. Smid, *J. Amer. Chem. Soc.*, *90* (1968) 4654.
132 P. Chang, R. V. Slates and M. Szwarc, *J. Phys. Chem.*, *70* (1966) 3180.
133 R. V. Figini, *Makromol. Chem.*, *71* (1964) 193.
134 G. Löhr and G. V. Schulz, *Makromol. Chem.*, *77* (1964) 240.
135 R. V. Figini, G. Löhr and G. V. Schulz, *J. Polym. Sci., Part B*, *3* (1965) 985.
136 K. M. Hui and T. L. Ng, *J. Polym. Sci., Part A1*, *7* (1969) 3101.
137 M. Nakayama, H. Hirohara, K. Takaya and N. Ise, *J. Polym. Sci., Part A1*, *8* (1970) 3653.
138 K. Takaya, H. Hirohara, M. Nakayama and N. Ise, *Trans. Faraday Soc.*, *67* (1971) 119.
139 M. Fisher and M. Szwarc, *Macromolecules*, *3* (1970) 23.
140 H. Sinn and F. Bandermann, *Makromol. Chem.*, *62* (1963) 134.
141 S. Bywater and D. J. Worsfold, *Can. J. Chem.*, *45* (1967) 1821.
142 F. S. Dainton, G. A. Harpell and K. J. Ivin, *Eur. Polym. J.*, *5* (1969) 395.
143 J. Comyn, F. S. Dainton and K. J. Ivin, *Eur. Polym. J.*, *6* (1970) 319.
144 J. Comyn and K. J. Ivin, *Eur. Polym. J.*, *5* (1969) 587.
145 J. Comyn, F. S. Dainton, G. A. Harpell, K. M. Hui and K. J. Ivin, *J. Polym. Sci., Part B*, *5* (1967) 965.
146 M. Shima, D. N. Bhattacharyya, J. Smid and M. Szwarc, *J. Amer. Chem. Soc.*, *85* (1963) 1306.
147 J. Furukawa, T. Tsuruta, Y. Fujita and A. Kawasaki, *J. Chem. Soc. Jap., Ind. Chem. Sect.*, *63* (1960) 645.
148 A. Guyot and P. Q. Tho, *J. Polym. Sci., Part C*, *4* (1964) 299.
149 A. Konishi, *Bull. Chem. Soc. Jap.*, *35* (1962) 197.
150 B. L. Erussalimsky, I. G. Krasnoselskaya and V. V. Masurek, *Vysokomol. Soedin. Z.*, *6* (1964) 1294.
151 B. L. Erussalimsky, I. G. Krasnoselskaya and I. V. Kulevskaya, *Russ. Chem. Rev.*, *37* (1968) 874.
152 B. L. Erussalimsky, V. V. Masurek, I. G. Krasnoselskaya and V. Ghassan-Zade, *J. Polym. Sci., Part C*, *23* (1968) 215.
153 V. Jisova, M. Kolinsky and D. Lim, *J. Polym. Sci., Part A1*, *8* (1970) 1525.
154 G. A. Russell and D. W. Lamson, *J. Amer. Chem. Soc.*, *91* (1969) 3967.
155 J. F. Garst, R. H. Cox, J. T. Barbas, R. D. Roberts, J. I. Morris and R. C. Morrison, *J. Amer. Chem. Soc.*, *92* (1970) 5761.
156 A. Guyot and J. Mordini, *C.R. Acad. Sci., Ser. C*, *271* (1970) 576.
157 M. Frankel, A. Ottolenghi, M. Albeck and A. Zilkha, *J. Chem. Soc.*, (1959) 3858.
158 M. L. Miller, *J. Polym. Sci.*, *56* (1962) 203.
159 B. L. Erussalimsky, I. Kulevskaya and V. Masurek, *J. Polym. Sci., Part C*, *16* (1967) 1355.

64

160 C. G. Overberger, E. M. Pearce and N. Mayes, *J. Polym. Sci., 34* (1959) 109.
161 C. G. Overberger and A. H. Schiller, *J. Polym. Sci., Part C, 1* (1963) 325.
162 S. Bywater, Unpublished Data.
163 N. Kawabata and T. Tsuruta, *Makromol. Chem., 98* (1966) 262.
164 B. L. Erussalimsky, I. G. Krasnoselskaya, V. N. Krasulina, A. V. Novoselara and E. V. Zashtsherinsky, *Eur. Polym. J., 6* (1970) 1391.
165 F. Wenger, *Chem. Ind. (London)*, (1959) 1094.
166 R. K. Graham, D. L. Dunkelberger and E. S. Cohn, *J. Polym. Sci., 42* (1960) 501.
167 D. L. Glusker, E. Stiles and B. Yoncoskie, *J. Polym. Sci., 49* (1961) 297.
168 D. L. Glusker, I. Lysloff and E. Stiles, *J. Polym. Sci., 49* (1961) 315.
169 B. J. Cottam, D. M. Wiles and S. Bywater, *Can. J. Chem., 41* (1963) 1905.
170 D. L. Glusker, R. A. Galluccio and R. A. Evans, *J. Amer. Chem. Soc., 86* (1964) 187.
171 W. E. Goode, F. H. Owens and W. L. Myers, *J. Polym. Sci., 47* (1960) 75.
172 P. E. M. Allen and A. G. Moody, *Makromol. Chem., 81* (1965) 234.
173 D. M. Wiles and S. Bywater, *Polymer, 3* (1962) 175.
174 D. M. Wiles and S. Bywater, *Trans. Faraday Soc., 61* (1965) 150.
175 F. H. Owens, W. L. Myers and F. E. Zimmerman, *J. Org. Chem., 26* (1961) 2288.
176 N. Kawabata and T. Tsuruta, *Makromol. Chem., 86* (1965) 231.
177 H. Schreiber, *Makromol. Chem., 36* (1959) 86.
178 D. M. Wiles and S. Bywater, *J. Phys. Chem., 68* (1964) 1983.
179 W. Fowells, C. Schuerch, F. A. Bovey and F. P. Hood, *J. Amer. Chem. Soc., 89* (1967) 1396.
180 G. L'Abbé and G. Smets, *J. Polym. Sci., Part A1, 5* (1967) 1359.
181 P. E. M. Allen, D. O. Jordan and M. A. Naim, *Trans. Faraday Soc., 63* (1967) 234.
182 P. E. M. Allen and A. G. Moody, *Makromol. Chem., 83* (1965) 220.
183 W. E. Goode, F. H. Owens, R. P. Fellmann, W. H. Snyder and J. E. Moore, *J. Polym. Sci., 46* (1960) 317.
184 D. Braun, M. Herner, U. Johnson and W. Kern, *Makromol. Chem., 51* (1962) 15.
185 W. Kern, D. Braun and M. Herner, *Makromol. Chem., 28* (1958) 66.
186 D. Braun, M. Herner and W. Kern, *J. Polym. Sci., 51* (1961) s2.
187 D. J. Worsfold and S. Bywater, *Makromol. Chem., 65* (1963) 245.
188 R. S. Stearns and L. E. Forman, *J. Polym. Sci., 41* (1959) 381.
189 C. E. H. Bawn, *Rubber Plastics Age*, (1961) 267.
190 M. Szwarc, *J. Polym. Sci., 40* (1959) 583.
191 C. E. H. Bawn and A. Ledwith, *Quart. Rev., Chem. Soc., 16* (1962) 361.
192 D. J. Cram and K. R. Kopecky, *J. Amer. Chem. Soc., 81* (1959) 2748.
193 C. Schuerch, W. Fowells, A. Yamada, F. A. Bovey, F. P. Hood and E. W. Anderson, *J. Amer. Chem. Soc., 86* (1964) 4481.
194 T. Yoshino, J. Komiyama and M. Shinomiya, *J. Amer. Chem. Soc., 86* (1964) 4482.
195 F. Schué, D. J. Worsfold and S. Bywater, *J. Polym. Sci., Part B, 7* (1969) 821.
196 M. Morton and R. D. Sanderson, *Amer. Chem. Soc., Div. Org. Coatings Plast. Chem., 30* (1970) 227.
197 M. Morton, R. D. Sanderson and R. Sakata, *J. Polym. Sci., Part B, 9* (1971) 61.
198 F. Schué, D. J. Worsfold and S. Bywater, *Macromolecules, 3* (1970) 509.
199 W. H. Glaze and P. C. Jones, *Chem. Commun.*, (1969) 1434.
200 F. R. Mayo and F. M. Lewis, *J. Amer. Chem. Soc., 66* (1944) 1594.
201 T. Fueno and J. Furukawa, *J. Polym. Sci., Part A, 2* (1964) 3681.
202 Y. Landler, *J. Polym. Sci., 8* (1952) 63.
203 F. T. Wall, *J. Amer. Chem. Soc., 63* (1941) 1862.
204 R. K. Graham, D. L. Dunkelberger and W. E. Goode, *J. Amer. Chem. Soc., 82* (1960) 400.

205 G. V. Rakova and A. A. Korotkov, *Dokl. Akad. Nauk SSSR*, *119* (1958) 982.
206 A. A. Korotkov and N. N. Chesnokova, *Vysokomol. Soedin. Z.*, *2* (1960) 365.
207 A. A. Korotkov and G. V. Rakova, *Vysokomol, Soedin. Z.*, *3* (1961) 1482.
208 I. Kuntz, *J. Polym. Sci.*, *54* (1961) 569.
209 K. F. O'Driscoll, *J. Polym. Sci.*, *57* (1962) 721.
210 M. Morton and F. R. Ells, *J. Polym. Sci.*, *61* (1962) 25.
211 A. F. Johnson and D. J. Worsfold, *Makromol. Chem.*, *85* (1965) 273.
212 E. Ureta, J. Smid and M. Szwarc, *J. Polym. Sci., Part A1*, *4* (1966) 2219.
213 H. Yuki, J. Hotta, Y. Okamato and S. Murahaski, *Bull. Chem. Soc. Jap.*, *40* (1967) 2659.
214 H. Yuki and Y. Okamato, *Bull. Chem. Soc. Jap.*, *42* (1969) 1644.
215 H. Yuki, Y. Okamato, K. Tsubota and K. Kosai, *Polym. J.*, *1* (1970) 147.
216 P. Sigwalt, *Bull. Soc. Chim. Fr.*, (1964) 423.
217 T. Fueno, K. Morokuma and J. Furukawa, *Bull. Chem. Soc. Jap.*, *32* (1959) 1003.
218 Yu. L. Spirin, A. A. Arest-Yakubovich, D. K. Polyakov, A. R. Gantmakher and S. S. Medvedev, *J. Polym. Sci.*, *58* (1962) 1181.
219 A. A. Korotkov, S. P. Mitsengendler and K. M. Aleyev, *Vysokomol. Soedin. Z.*, *2* (1960) 1811.
220 J. Furukawa, T. Saegusa and K. Irako, *J. Chem. Soc. Jap., Ind. Chem. Sect.*, *65* (1962) 2079.
221 M. Shima, J. Smid and M. Szwarc, *J. Polym. Sci., Part B*, *2* (1964) 735.
222 A. V. Tobolsky and R. J. Boudreau, *J. Polym. Sci.*, *51* (1961) s53.
223 K. O'Driscoll and R. Patsiga, *J. Polym. Sci., Part A*, *3* (1965) 1037.
224 J. Smid and M. Szwarc, *J. Polym. Sci.*, *61* (1962) 31.
225 J. Tölle, P. Wittmer and H. Gerrens, *Makromol. Chem.*, *113* (1968) 23.
226 Z. Laita and M. Szwarc, *Macromolecules*, *2* (1969) 412.
227 N. F. Zutty and F. J. Welch, *J. Polym. Sci.*, *43* (1960) 445.
228 F. Dawans and G. Smets, *Makromol. Chem.*, *59* (1963) 163.
229 F. C. Foster, *J. Amer. Chem. Soc.*, *72* (1950) 1370.
230 D. J. Worsfold, *Macromolecules*, *3* (1970) 514.
231 D. J. Kelley and A. V. Tobolsky, *J. Amer. Chem. Soc.*, *81* (1959) 1597.
232 A. V. Tobolsky and C. E. Rogers, *J. Polym. Sci.*, *38* (1959) 205.
233 B. J. Schmitt and G. V. Schulz, *Eur. Polym. J.*, *11* (1975) 119.
234 L. L. Böhm, G. Löhr and G. V. Schulz, *Ber. Bunsenges. Phys. Chem.*, *78* (1974) 1064.
235 G. Löhr and G. V. Schulz, *Eur. Polym. J.*, *10* (1974) 121.
236 S. Brownstein, S. Bywater and D. J. Worsfold, *Macromolecules*, *6* (1973) 715.
237 S. Bywater, D. J. Worsfold and G. Hollingworth, *Macromolecules*, *5* (1972) 389.

Chapter 2

Kinetics of Homogeneous Cationic Polymerization

A. LEDWITH and D. C. SHERRINGTON

1. Introduction

Cationic polymerization is the conversion of low molecular weight monomeric molecules into high molecular weight polymeric ones, via a mechanism involving stepwise growth of a carbonium (R_3C^+), carboxonium ($R\overset{+}{O}=CR_2$), oxonium (R_3O^+), sulphonium (R_3S^+) or immonium ($R_2\overset{+}{N}=CR_2'$) ion. Both vinyl and cyclic monomers are susceptible to polymerization by such intermediates, e.g.

$$n\,CH_2{=}CHY \longrightarrow \text{\tiny\char"223} C^+ \longrightarrow \text{\Large(}CH_2{-}CHY\text{\Large)}_n$$

With olefinic monomers the substituent, Y, is often electron donating e.g. $-OR$, $-NR_2$ $-Ar$, and therefore capable of stabilizing a positive charge formally residing on the α carbon atom, e.g.

If Y were electron withdrawing, e.g. $-Cl$, $-CN$, it would exert a substantial destabilizing effect on any such positive charge, and in fact these monomers are not susceptible to polymerization by a cationic mechanism. As far as cyclic monomers are concerned the propagating intermediate is also stabilized, since most of the positive charge resides on the basic heteroatom, e.g.

The overall driving force for the cationic polymerization of olefinic monomers is the same as that in free radical processes, i.e., the reduction in free energy associated with the loss of unsaturation in forming polymeric chains. This fall in free energy arises from a large negative enthalpy change, which more than compensates for the accompanying unfavourable entropy change. With cyclic monomers the thermodynamic driving force arises from the relief of ring strain in forming linear polymeric chains, although here this is a smaller enthalpy change (negative) and entropy change (positive). The thermodynamics of cationic polymerization have recently been reviewed [1].

Except in very special circumstances (Section 6.4) electroneutrality is maintained in cationic polymerizations by the presence of a negatively charged counter-ion or gegenion, X^-. This species has no analogue in free radical polymerizations, and indeed much of the early work on cationic (and anionic) reactions was carried out with almost total disregard for the effect of counter-ion. In fact it turns out as we shall see that the counter-ion, and its physical relationship with the growing cation, is of vital importance in the interpretation of kinetic data derived from these systems. The present authors take the view that, while results obtained from polymerizations where such relationships are unknown are qualitatively useful, any quantitative data obtained must be treated with caution.

Kinetic and mechanistic aspects of cationic polymerizations were first comprehensively reviewed in a book edited by Plesch [2]. Since then other leading workers in the field have contributed additional surveys [3—9]. It is the intention in the present survey to analyse the limited reliable kinetic data now available*, and not to undertake a gross review of old and new literature. To this end an outline of the relatively complex nature of cationic systems will first be presented to justify the subsequent detailed consideration of a few important systems.

2. Complex nature of cationic systems

The occurrence of positively charged intermediates particularly in vinyl polymerizations has still to be shown directly. For example the growing cation in the polymerization of styrene initiated by protonic acids has still to be detected spectroscopically (or by other methods), despite considerable interest in this particular system [10—22]. There is, however, a large amount of circumstantial evidence for the occurrence of cations, both in vinyl and ring opening polymerizations, and the literature has been reviewed recently [9]. Particularly convincing evidence has been produced by the use of additives and impurities, though copolymerization and conductance data, as well as solvent effects, have also been compelling.

Carbonium ions in particular are much more reactive than correspond-

---

* Since this review was completed, a substantial body of relevant information (presented at the International Symposium on Cationic Polymerization, Rouen, 1973) has been published in collected form, *Makromol. Chem.*, 175 (1974) 1017—1328.

ing radicals (or anions), and it is not surprising that they have eluded direct detection in polymerization processes. In addition to the possibility of recombining with attendant counter-ions, they may undergo isomerization to form more stable structures, so that the resulting polymer can possess a structure differing from that of the monomer unit from which it was formed [23]. The general difficulty of generating carbonium ions in a quantitative manner, coupled with the ease of their destruction, or diversion, often precludes the establishment of a stationary state concentration of active centres. Even when such a state is achieved, equilibrium concentrations are often very small, and this again leads to problems in the identification of the intermediates. In contrast to other types of polymerization, most cationic systems, particularly those involving vinyl monomers, are typified by complex initiation equilibria, rapid propagation reactions, and a chain breaking process dominated by facile monomer transfer rather than true termination reactions. The complexities introduced by catalyst fragments from use of excess initiator, can be devastating, and the question of whether a co-catalyst is necessary or not has still to be resolved in many systems. Recently, Kennedy [24] has made a number of suggestions which help to remove some of the ambiguities, but, even so, there is no doubt that a substantial proportion of the inconsistencies which arise in cationic polymerization are derived directly from complex and ill-defined initiation procedures. In addition to these inherent problems, many of which may not be resolvable, there are side reactions arising from the presence of impurities. Very often these are not detected until, perhaps fortuitously, some more rigorous purification technique is used. Phenomena and data which had previously been assigned as intrinsic to the system, must then be re-examined to allow for the possibility of a direct dependence on the impurities originally present.

All of these factors contribute to making almost every individual cationic polymerization unique. A change in the initiator system can introduce a completely different set of so-called pre-initiation equilibria, each with its own equilibrium constant. Merely changing the initiator concentration can disturb the equilibrium between the growing positive centre and its counter-ion; while alteration of the latter can have an effect almost as dramatic as altering the monomer species. Solvent and temperature also are important variables, and can produce not only a change in reaction rates, but also in polymer molecular weights and molecular weight distributions. Clearly then before any kinetic experiment can produce well defined results, all of these parameters must be carefully monitored and controlled and meticulous control of purity must be maintained at all times.

3. Methods of initiation

All cationic polymerizations must involve firstly the generation of a positively charged species, and an accompanying counter-ion, viz.

$$AB \rightarrow A^+B^-$$

In the case of chemical initiation, as opposed to initiation by ionizing radiation, this process can be carried out in situ in the polymerization mixture [25, 26] or, in special cases, it may be undertaken externally by producing a stable salt, capable of initiating polymerization at some subsequent time [27]. In the former case the equilibria involved must be included in the kinetic scheme, and contribute to the overall rate of polymerization. Kennedy [24] has suggested that initiation by Lewis acids ($TiCl_4$, $SnCl_4$, BF_3, $SbCl_5$, $AlCl_3$, AlR_3, etc.) falls into two categories depending on the type of monomer. Where the latter possess an allylic hydrogen atom, e.g. isobutene or α-methylstyrene, then a "self initiation" process is possible via abstraction of a hydride ion, e.g.

$$CH_2=C\begin{smallmatrix}CH_3\\CH_3\end{smallmatrix} + SnCl_4 \;\overset{K}{\rightleftharpoons}\; CH_2=C\begin{smallmatrix}\overset{+}{C}H_2\\CH_3\end{smallmatrix} + SnCl_4H^-$$

If, however, the monomer has no such reactive hydrogen atom, some protogenic (e.g. H_2O or HCl) or catiogenic (e.g. RCl) co-initiator is required to provide a source of protons or carbonium ions respectively, e.g.

$$AlCl_3 + RCl \;\overset{K'}{\rightleftharpoons}\; R^+AlCl_4^-$$

$$BF_3 + H_2O \;\overset{K''}{\rightleftharpoons}\; H^+BF_3OH^-$$

In either case a knowledge of the respective equilibrium constants (or forward rate coefficients) is necessary for a complete kinetic analysis. Generally speaking such data is not available though considerable advances are being made towards this end [25, 28]. Where it is possible to generate a reactive positive charge and isolate it as a stable salt then considerable simplification of the polymerization technique and the kinetic analysis may result, e.g.

$$Ph_3CCl + SbCl_5 \;\rightleftharpoons\; Ph_3C^+SbCl_6^-$$

This approach has been used with some success over the last few years [26, 28, 29] both with vinyl monomers [27, 29, 30] and cyclic monomer systems [31, 32]. Similarly pre-formed oxonium and acylium ion salts have been isolated and used, subsequently, to initiate polymerization, e.g.

$$(CH_3CO)_2O + HClO_4 \;\rightleftharpoons\; CH_3CO^+ClO_4^- + CH_3COOH$$

$$R_2O + RCl + SbCl_5 \;\rightleftharpoons\; R_3O^+SbCl_6^-$$

These salts are most efficient in the polymerization of cyclic monomers [33, 34] though some use in vinyl systems has been reported [34]. Protonic acids, e.g., $HClO_4$ (and H_2SO_4) also fall within the definition of pre-formed initiators. As already indicated a considerable amount of work has been carried out on the polymerization of styrene [10—22] by these acids, where an additional question concerning the role of covalent perchlorate ester species is raised. This problem is dealt with in more detail later (see Section 6.2).

One final, though extremely important, method of generating reactive cations is by the use of ionizing radiation. Though irradiation with γ-rays from a ^{60}Co source was known to initiate free radical, liquid state, polymerizations [36], it was not until 1957 that the polymerization of isobutene at —78°C was shown unequivocally to be a cationic process [37]. Presumably on irradiation an electron is ejected from a suitable liquid monomer with the generation of a radical cation [38] which can then propagate.

$$CH_2{=}CHR \xrightarrow{k_i} \overset{\cdot}{C}H_2{-}\overset{+}{C}HR + e^-$$
$$\text{γ-ray}$$

Such a process constitutes the "initiation reaction" of a cationic polymerization, all pre-initiation phenomena being absent. This situation can be regarded as a model for the reaction which must follow the charge formation process of any chemical initiation, and marks the point at which some generalizations can be introduced into cationic reaction mechanisms, e.g.

$$A^+B^- + \text{monomer} \xrightarrow{k_i} \text{monomer}^+X^-$$

4. Kinetic schemes

A limited number of attempts have been made to set up a general mechanistic scheme describing cationic systems in terms of fundamental reactions, in a similar manner to that used in free radical polymerizations, and to derive generally applicable kinetic equations [3—4]. Because of the individuality of each cationic system, however, this approach has met with little success, and there has been a greater tendency towards treating each polymerization in isolation for detailed kinetic analysis. It is possible, however, to postulate at least token schemes which can be used as a guide. After the pre-initiation equilibria, polymerization can be considered in terms of classical initiation, propagation, transfer and termination reactions, i.e. for vinyl monomers

$$A^+B^- + CH_2{=}CHR \xrightarrow{k_i} A CH_2{-}\overset{+}{C}HR, B^- \quad \text{initiation}$$

$$\text{\textasciitilde\textasciitilde}CH_2\overset{+}{C}HR, B^- + CH_2{=}CHR \xrightarrow{k_p} \text{\textasciitilde\textasciitilde}CH_2\overset{+}{C}HR, B^- \quad \text{propagation}$$

$$\text{\textasciitilde\textasciitilde}CH_2\overset{+}{C}HR, B^- + CH_2{=}CHR \xrightarrow{k_{trM}} \text{\textasciitilde\textasciitilde}CH{=}CHR + CH_3\overset{+}{C}HR, B^-$$

$$\text{monomer transfer}$$

$$\text{\textasciitilde\textasciitilde}CH_2\overset{+}{C}HR, B^- \xrightarrow{k_t} \text{dead polymer (spontaneous termination)}$$

and for cyclic monomers

$$A^+B^- + \overset{\frown}{X} \xrightarrow{k_i} A{-}\overset{+}{\overset{\frown}{X}}B^- \quad \text{initiation}$$

$$A{-}\overset{+}{\overset{\frown}{X}}B^- + X \xrightarrow{k_p} \text{\textasciitilde\textasciitilde}\overset{+}{\overset{\frown}{X}}B^- \quad \text{propagation}$$

$$\text{\textasciitilde\textasciitilde}\overset{+}{\overset{\frown}{X}}B^- \xrightarrow{k_t} \text{dead polymer (spontaneous termination)}$$

In vinyl polymerizations there are few, if any, true termination processes in the absence of impurities, and the main chain limiting steps are transfer reactions. Indeed transfer to monomer can be so facile that it is possible for the rate coefficient for monomer transfer, k_{trM}, to be only one order of magnitude below that for propagation, k_p [27]. Thus in a pure system, even at 100% conversion of monomer to polymer, many cationic olefinic polymerizations can still be regarded as "living", in the sense that little or no destruction of active centres has occurred although the reactive charges reside on a low molecular weight species at the end of polymerization, unlike the case of some anionic reactions where both transfer and termination processes can be absent producing high molecular weight "living" polymeric species [39]. As far as cyclic monomers are concerned there is, in general, little facility for simple proton transfer to monomer, and in the absence of excess catalyst fragments and also any reaction of the growing centre with its counter-ion or dead polymer, the production of a "living" polymer becomes possible. Indeed such "living" species have been prepared and characterized [40, 41].

Cationic polymerizations are normally studied by observing the overall rate of formation of polymer or, more conveniently, loss of monomer (M), $-d[M]/dt$, as a function of monomer and initiator (A^+B^-) concentrations, i.e.,

$$-\frac{d[M]}{dt} = k_{obs}[M]^a[A^+B^-]^b$$

where, k_{obs} is the observed overall rate coefficient. As with many kinetic experiments, attempts are made to determine $-d[M]/dt$ at low conversion, so that initial values of [M] and [A$^+$B$^-$] may be used to determine k_{obs}, and the orders of reaction "a" and "b", by conventional methods. Where A$^+$B$^-$ is generated in situ from a known catalyst/co-catalyst combination, then the equation must be modified to include these reactants, i.e.

$$-\frac{d[M]}{dt} = k'_{obs}[M]^a[cat]^b[cocat]^c$$

where k'_{obs} now includes the equilibrium constant relating catalyst, co-catalyst and A$^+$B$^-$.

If the monomer is involved in the pre-initiation equilibrium [24] then the exponent "a" is usually equal to two. However, if in addition, the monomer contributes to the solvation and stabilization of the active centre, its order may be even higher. Similar abnormally high dependences have also been observed with respect to the other components and serve to highlight the complexities which can arise in ionic reactions. In comparatively high dielectric constant solvents (e.g., methylene chloride), where considerable solvation by the solvent itself is possible these difficulties are minimized and this enables the overall rate coefficient for the polymerization to be estimated, but does not allow computation of elementary coefficients, i.e., k_i, k_p, k_t etc. In a limited number of systems polymerization is characterized by a fall in reaction rate to zero before 100% conversion. Some intrinsic termination reaction apparently consumes active centres, leaving a proportion of monomer unreacted [11, 42]. In these circumstances where no stationary state concentration of active centres is set up, a term, $\exp(-k_t t)$ must be included in the rate expression (where k_t is the rate coefficient governing the particular termination reaction responsible, and t is the reaction time). On the assumption that initiation is very rapid [11], it is possible to determine separate values of k_p and k_t (see Section 6.2). There is, however, no general method whereby separate rate coefficients can be determined as in free radical polymerization, although, with the cationic systems if a steady state concentration of active centres is assumed, the average degree of polymerization of the polymer formed is determined by the ratio of the rate of chain growth to the sum of the rates of all reactions resulting in the formation of dead polymer chains, i.e.

$$\overline{D.P.} = \frac{k_p[M]}{k_t + k_{trM}[M] + \sum_i k_{trY}[Y]_i}$$

where [Y] = impurity transfer agent, and k_t is the rate coefficient for spontaneous termination. This equation can be further embellished [43], although the exercise is of little practical value. In by far the majority of cationic polymerizations transfer to monomer dominates all other chain limiting processes, reducing the above equation to a simple ratio, $\overline{D.P} = k_p/k_{trM}$, and as pointed out by Plesch [44], once the rate coefficient for propagation, k_p, is known, the other rate coefficients can usually be calculated from an overall rate coefficient, k_{obs}, and/or molecular weight data.

Probably the most fruitful approach to the estimation of k_p has been the simplest. Polymerization reactions are chosen in which initiation can be assumed to be speedy and efficient with the generation of a fixed number of active centres, $[P_n^+]_0$. True termination is assumed to be absent, so that the rate expression takes the form

$$-d[M]/dt = k[M][P_n^+]_0$$

If initiation is sufficiently rapid that it can be considered as complete prior to propagation, then the observed rate coefficient, k, will approximate to the rate coefficient for propagation, k_p. Thus considering data at time $t = 0$ yields the equation

$$\left(-\frac{d[M]}{dt}\right)_0 = k_p[M]_0[P_n^+]_0$$

Even here, however, some assumptions as to the relative values of initiation rate coefficients are required. The only other problem remaining is the estimation of $[P_n^+]_0$. One method of defining this at the outset is to introduce a fixed amount of a pre-formed ionic initiator, $[I]_0$, a technique which has been used in both vinyl and cyclic polymerizations as mentioned earlier. Under these circumstances $[P_n^+]_0 = [I]_0$. Another powerful method finding increasing application is the premature termination of a polymerization by addition of a basic impurity [45, 46] which attacks, and adds directly to, the active centres. The terminal units formed are then counted in some convenient way e.g. spectrophotometrically. Detailed application of both of these approaches will be discussed later. Both methods have the additional advantage of being able to cope, with minor modification, with the additional complexities associated with ion pair equilibria. These complexities arise mainly because ionic species can exist in several physical forms in association with their counter-ions, each form having a quite different reactivity. Thus instead of there being one set of fundamental rate reactions, each individual reaction can itself be regarded as being composed of a number of processes involving a different ionic species. The propagation reaction, for example, can be represented as

$$\text{\textasciitilde}M^+ + X^- + M \xrightarrow{\ k_p^+\ } \text{\textasciitilde}M^+ + X^-$$

$$\text{\textasciitilde}M^+X^- + M \xrightarrow{\ k_p^\pm\ } \text{\textasciitilde}M^+X^-$$

$$\text{\textasciitilde}M^+//X^- + M \xrightarrow{\ k_p^{+\ -}\ } \text{\textasciitilde}M^+//X^-$$

where free ions, contact ion pairs and solvent separated ion pairs ($\text{\textasciitilde}M^+//X^-$) all make a contribution, each with its accompanying propagation rate coefficient; the other fundamental processes can be represented in the same manner.

5. Free ion—ion pair equilibria

Winstein et al. [47] have shown conclusively that many organic ionic materials (R^+X^-) in a suitable solvent can exist in at least three forms, a contact (or intimate) ion pair, a solvent separated ion pair, and free solvated ions, e.g.

$$R^+X^- \rightleftharpoons R^+//X^- \xrightleftharpoons{\ K_d\ } R^+ + X^-$$

| contact ion pair | solvent separated ion pair | free solvated ions |

The value of K_d, the ion pair dissociation constant, varies according to the nature of R^+ and X^-, the solvating and dissociating ability of the solvent employed and the temperature. Dennison and Ramsey [48] have treated quantitatively the effect of solvent and have shown the following relationship to hold:

$$-\ln K_d = Z^2 /a\epsilon kT$$

where Z is charge on ions; a, sum of van der Waal's ionic radii; ϵ, dielectric constant of the solvent; k, Boltzmann constant; T, temperature ($^\circ K$). The treatment, however, is oversimplified to some extent in that it requires ions to be present only as very close pairs or completely free. Winstein was able to show that the rates at which these various species react with a given substrate can be substantially different and, in general, a free cation would be expected to react more quickly than any of its corresponding ion pairs, because of its higher effective charge density. The fundamental source of reactivity of a cation can be regarded as its ability to polarize susceptible molecules, and clearly the larger the charge density, the greater the polarizing power. It has been argued that, under special circumstances, the ion pair might be more reactive employing a so-called "push—pull" mechanism. In this model the geometry of the transition state is such that the presence of the counterion aids the formation of a charged product, e.g.

$$R^+X^- + S \longrightarrow \left[\begin{array}{c} X^- \\ R\text{------}S \\ \delta+ \quad \delta+ \end{array} \right] \longrightarrow RS^+X^-$$

Thus in ionic polymerizations generally, in addition to the difficulty of deciding the chemical nature of the species involved in initiation, propagation etc., there is also the problem of the physical nature of each in relationship to its counter-ion.

Fortunately there is available a simple conductance technique for monitoring ion pair—free ion equilibria, which, while being unable to give information about different ion pairs, readily produces data on the proportion of free ions to all ion pairs. The equilibrium is treated as approximating to the dissociation of a weak electrolyte, e.g. acetic acid in water,

$$R^+X^- \xrightleftharpoons{K_d} R^+ + X^-$$

all ion
pairs

For such equilibria Ostwald's Dilution Law represents, fairly accurately, the behaviour of free and paired species, viz.

$$\alpha = \frac{\lambda}{\lambda_0} \quad \text{and} \quad \frac{1}{\lambda} = \frac{1}{\lambda_0} + \frac{c\lambda}{K_d(\lambda_0)^2}$$

where λ is equivalent conductance of a solution of concentration, c; λ_0, equivalent conductance at infinite dilution; α, degree of dissociation.

Fuoss and Accascina [49] have shown, however, that for ionic concentrations as low as 10^{-4}M, Ostwald's equation is not exact because of its neglect of long range interionic attraction upon the conductance and activities of ions. Maintaining the concept of the "sphere in continuum" model, in which ions are regarded as hard spheres immersed in a continuous medium, Fuoss corrected the equation from first principles and derived the relationship

$$\frac{F(z)}{\lambda} = \frac{1}{\lambda_0} + \frac{cf_\pm^2}{F(z)K_d\lambda_0^2}$$

where, $\quad F(z) = (4/3)\cos^2(1/3)\cos^{-1}\left(-\frac{3\sqrt{3z}}{2}\right)$

$$z = \text{constant}\sqrt{\lambda c}\,(\lambda_0)^{-3/2}$$

$f_\pm$ = mean activity coefficient.

From measurements of λ at various values of c in a given solvent, and at a given temperature, and application of this expression, it is possible to obtain data for K_d, i.e. in a static system it is possible to determine the proportion of free ions and ion pairs.

If the complications which arise in the generation of a charged initiator species (A^+B^-) are neglected, the process of initiation and propagation can be summarized, showing both ion pair (all types) and free ion contributions, as

$$A^+B^- + M \xrightarrow{k_i^\pm} AM^+B^- \xrightarrow[M]{k_p^\pm} \sim\!\!\sim M^+B^- \xrightarrow{M} \text{etc.}$$

$$\Big\updownarrow K_1 \qquad\qquad \Big\updownarrow K_2 \qquad\qquad \Big\updownarrow K_3$$

$$A^+ + B^- + M \xrightarrow{k_i^+} AM^+ + B^- \xrightarrow{k_p^+} \sim\!\!\sim M^+ + B^- \xrightarrow{M} \text{etc.}$$

where $k_i^\pm$ = rate coefficient for initiation by ion pairs

$\qquad k_i^+$ = rate coefficient for initiation by free ions

$\qquad k_p^\pm$ = rate coefficient for propagation by ion pairs

$\qquad k_p^+$ = rate coefficient for propagation by free ions

$\qquad K_1$ = ion pair dissociation constant for initiator

$K_2 \approx K_3$ = ion pair dissociation constant of propagating species.

Recently, Higashimura [7] has reviewed the data on elementary rate coefficients (k_i, k_p, k_{tM} and k_t) in cationic polymerization of vinyl monomers. Information available on initiation and termination reactions is extremely limited, and virtually no attempt [50] has been made to elucidate, either qualitatively or quantitatively the role of free ions and ion pairs in these processes. Numerical data on the separate contributions to propagation by free ions and ion pairs is slowly becoming available, though in a less ordered fashion than in the case of anionic systems. It seems likely that the most fruitful approach to the problem of absolute reactivity, in initiation processes at least, will be an examination of reactions of non-polymerizable monomer models, where electronic factors may be maintained close to those of polymerizable monomers, but where steric factors prevent complication due to propagation.

As far as the propagation reaction is concerned, the rate of propagation, R_p, can be regarded, for simplicity, as being composed of a free ion and a single ion pair contribution, the latter combining the effect of the various types of ion pair present. The observed propagation rate coefficient, $k_{p_{obs}}$, will then be given by

$$k_{p_{obs}} = \alpha k_p^+ + (1 - \alpha)k_p^{\pm}$$

where α is the degree of dissociation of ion pairs into free ions.

It is evident, therefore, that in any kinetic experiment which is to be of quantitative value, first of all data on the rate of propagation alone, as opposed to an overall rate of polymerization, must be ascertained. Secondly the proportion of free ions and paired species (i.e. α or K_d) must be known and finally, in order to determine the two unknowns k_p^+ and $k_p^{\pm}$, either their approximate relative values must be known, or the proportion of each species must be altered by some arbitrary procedure (e.g. changing the total concentration, or use of "common ion" effect). Details of this type of treatment will be given later for specific examples. If the system under investigation can be adjusted so that only one type of active centre is present, i.e. all free ions, or all ion pairs, then clearly the manipulation becomes greatly simplified. Unfortunately because of the occurrence of excessive transfer, and therefore lack of true "living" cationic polymers in vinyl systems, it is impossible to determine the ion pair dissociation constant (and hence α) of the polymeric growing ion pairs directly, though recently a fairly successful study of living polymeric oxonium ion pairs has been carried out [46]. At the present time it is necessary to derive estimates for α using K_d data on more stable ion pair systems under the same conditions as those used in the polymerizations [46, 51, 34]. Such methods undoubtedly involve some degree of error, though it has been shown that a substantial variation in the structure of ion pair components produces relatively small changes in the dissociation constant, providing the ions concerned are fairly large and diffuse [51].

6. Polymerization of vinyl monomers

6.1 MOLECULAR IODINE AS AN INITIATOR

Molecular iodine was shown to be an effective initiator of the polymerization of active olefins (e.g. alkyl vinyl ethers) as long ago as 1878 [52]. The mechanism operative is now accepted as being a cationic one, and a number of kinetic investigations have been carried out from time to time [53—58]. In addition parallel ultraviolet/visible spectrophotometric investigations have been pursued [59, 60] in an attempt to identify some of the intermediates involved. These claim to show the presence of π-complexes between iodine and monomer molecules, $I_2 \leftarrow \overset{CH_2}{\underset{CHR}{\parallel}}$, presenting additional complications as regards the inter-

pretation of kinetic data. More recent work [61] has shown that the ultra-violet absorption bands in alkyl vinyl ether/I_2 systems may not in fact be due to π-complexes but to appropriately substituted 1,2-di-iodo-

ethanes, e.g. $ICH_2-\overset{I}{\underset{OR}{\overset{|}{\underset{|}{C}}}}H$. Much of the previous quantitative data was

obtained under the presumption that such π-complexes did contribute, and without regard for ion pair equilibria. Nevertheless, the work of Okamura et al. on the initiation of p-methoxystyrene [62, 63], alkyl vinyl ethers [64, 65], p-methylstyrene [65] and p-chlorostyrene [65] was the first detailed attempt to compute rate coefficients of each elementary reaction in cationic polymerization, with particular emphasis on the propagation coefficient, k_p. The procedure used, therefore, is worthy of closer examination, and the quantitative results obtained will be reassessed later in this chapter.

It was postulated that free iodine could react with a monomer molecule to form a π-complex, or an active centre capable of propagating, i.e.,

$$I_2 + M \xrightarrow{\ K\ } I_2M \qquad \text{complexation}$$
$$n\,I_2 + M \xrightarrow{\ k_i\ } M^+ \qquad \text{initiation}$$

Two additional assumptions were made, firstly the complex was inactive and could not initiate polymerization, secondly termination was a simple unimolecular process. The propagation and termination steps could therefore be written as

$$\text{\tiny\textasciitilde\textasciitilde}M^+ + M \xrightarrow{\ k_p\ } \text{\tiny\textasciitilde\textasciitilde}MM^+$$

$$\text{\tiny\textasciitilde\textasciitilde}MM^+ \xrightarrow{\ k_t\ } \text{dead polymer}$$

Alternative schemes of initiation were considered only in a qualitative manner.

From these reactions it was demonstrated that

$$\frac{[I_2]_{TOTAL}}{[X]} = 1 + \frac{k_i\,[M]\,[X]^{n-1}}{k_t\,(1 + K[M])^n}$$

where, in the early stages in the polymerization

$$[X] = [I_2]_{TOTAL} - [M^+] = [I_2]_{FREE} + [MI_2]$$

Now the rate of polymerization is given by

$$R_p = k_p [M^+] [M]$$

and when $n = 1$

$$R_p = \frac{k_p (k_i/k_t)[M]^2 [I_2]_{TOTAL}}{1 + K[M] + (k_i/k_t)[M]}$$

A dilatometric experimental method showed that there was no induction period and true stationary state polymerizations were set up immediately. For p-methoxystyrene the experimental rate laws took the form

$$R_p \ \alpha[M][I_2]^{2.3} \ \text{(in carbon tetrachloride)}$$

$$R_p \ \alpha[M][I_2]^{1.3} \ \text{(in ethylene dichloride)}$$

and, for isobutyl vinyl ether,

$$R_p \ \alpha[M][I_2] \ \text{(in ethylene dichloride)}$$

$$R_p \ \alpha[M][I_2]^2 \ \text{(in } n\text{-hexane)}$$

In both cases the $[I_2]$ exponent was larger for polymerizations in the solvent of lower dielectric constant where the counter-ion, (deliberately omitted in the kinetic schemes), was, therefore, envisaged as tri-iodide ion, I_3^-, corresponding to $n = 2$. In ethylene dichloride $n = 1$ and the gegenion was assumed to be unassociated, i.e. I^-. Visible spectrophotometry was used to measure $[X]$, $(= [I_2]_{FREE} + [I_2 M])$ and, from a plot of $[I_2]_{TOTAL}/[X]$ versus $[M][X]^{n-1}/(1 + K[M])^n$, the ratio k_i/k_t was obtained. Data for K were obtained independently by ultraviolet spectrophotometry using well characterized methods [66, 67]. Thus having determined k_i/k_t, substitution of the experimental data, R_p, $[M]$ and $[I_2]_{TOTAL}$, into the final rate expression enabled values of k_p to be calculated. Iodine is also conveniently estimated by iodometric titration

TABLE 1
Rate coefficients for propagation at 30°C, with molecular iodine as initiator [62-65].

Monomer	Solvent	$k_p(l \ mole^{-1} \ sec^{-1})$
p-methoxystyrene	$CH_2 CH_2 Cl_2$	16.0
p-methoxystyrene	CCl_4	0.13
isobutyl vinyl ether	$CH_2 CH_2 Cl_2$	6.5
isobutyl vinyl ether	n-hexane	0.045
p-methylstyrene	$CH_2 CH_2 Cl_2$	0.09
p-methylstyrene	CCl_4	$\sim 5 \times 10^{-4}$

using sodium thiosulphate, and parallel experiments using this technique confirmed the results from spectrophotometry. The data obtained are shown in Table 1.

Variation of k_p with solvent polarity was confirmed in the case of p-methoxystyrene by the use of $ClCH_2CH_2Cl/CCl_4$ mixtures [65]. The relevance of all these data in the context of the possible ionic equilibria involved in propagation will be discussed in Section 6.5. Similar attempts to obtain rate coefficients for propagation with styrene using metal halides, e.g. $SnCl_4$, as initiators have not been so successful, and considerable difficulty in obtaining reproducibility has been experienced [7].

6.2 POLYMERIZATION BY PROTONIC ACIDS — PSEUDOCATIONIC POLYMERIZATION

For a protonic acid to initiate cationic polymerization the corresponding anion must be non-nucleophilic to avoid simple addition across the carbon—carbon double bond with formation of a covalent bond to the counter-ion, i.e.

$$HX + CH_2=CHR \rightarrow CH_3CHRX$$

It is not surprising to find therefore that acids such as perchloric and sulphuric are more efficient initiators than, say, simple hydrogen halides. As already pointed out, the polymerization of styrene by these acids has been widely studied [10—22, 68]. In the case of sulphuric acid in ethylene dichloride, there is an initial fast reaction which stops abruptly before all the monomer is consumed [11]. Pepper et al. [11] have analysed this non-stationary state polymerization employing the scheme

Initiation $CH_2=CHPh + H_2SO_4 \xrightarrow{k_i} CH_3-\overset{+}{C}HPh\cdots HSO_4^-$

Propagation $CH_3\overset{+}{C}Ph\cdots HSO_4^- + M \xrightarrow{k_p} P_n^+\cdots HSO_4^-$

Transfer

(a) Spontaneous

$\sim\sim CH_2\overset{+}{C}HPh\cdots HSO_4^- \xrightarrow{k_{tr}} \sim\sim CH=CHPh + H_2SO_4$

(b) to monomer

$\sim\sim CH_2CHPhCH_2\overset{+}{C}HPh\cdots HSO_4^- + M \xrightarrow{k_{trM}}$

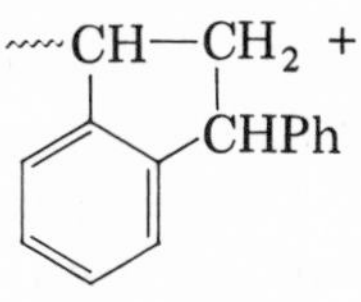

$$\sim\sim CH-CH_2 + M^+ \cdots HSO_4^-$$
$$\underset{CHPh}{|}$$

Termination

$$\text{~~CH}_2\overset{+}{\text{C}}\text{HPh} \cdots \text{HSO}_4^- \xrightarrow{\ k_t\ } \text{~~CH}_2\text{CHPhOSO}_3\text{H}$$

Assuming $k_i \gg k_p$ then the initial concentration of growing chains is equal to the concentration of the initiator $[I]_0$, and if the two transfer reactions re-initiate new chains instantaneously then it can be shown that

$$X_t = \ln([M]_0/[M]_t) = (k_p/k_t)[I]_0 [1-\exp(-k_t t)]$$

and

$$X_\infty = \ln([M]_0/[M]_\infty) = (k_p/k_t)[I]_0$$

Experimental data substituted in the second equation produces a value for the ratio k_p/k_t. This value is substituted in the first equation, and the absolute value of k_t calculated such that the equation fits the observed time/conversion data. Once k_t is determined k_p follows from the known ratio. At 25°C in ethylene dichloride [11] (effective dielectric constant, 9.72) k_p has the value 7.6 l mole^{-1} sec^{-1}.

In the polymerization of styrene initiated by anhydrous perchloric acid the rate of initiation is assumed to be very fast by analogy with initiation by sulphuric acid, but, in contrast to the latter system, true termination does not seem to occur, and the main molecular weight limiting processes are transfer reactions. Thus to a good approximation the total concentration of active centres is always equal to the initial concentration of perchloric acid. In ethylene dichloride and methylene dichloride solutions, the reaction proceeds to high conversions according to the experimental rate law

$$-\frac{d[M]}{dt} = k[HClO_4]_0[M]$$

Since initiation is rapid the rate coefficient, k, can be regarded as k_p directly. Unfortunately, however, a complex situation has arisen because of additional uncertainty as to the nature of active centres, over and above the question of ion pairing. There appears to be agreement among the various groups of workers that a covalent ester species participates in an equilibrium also involving the propagating species, i.e.,

$$\text{~~}\overset{+}{\text{C}}\text{HPh} + \text{ClO}_4^- \rightleftharpoons \text{~~}\overset{+}{\text{C}}\text{HPh, ClO}_4^- \rightleftharpoons \text{~~CHPhOClO}_3$$
$$\text{covalent ester}$$

The question of whether such a species is dormant, or can itself propagate by "pseudocationic" propagation, has come to dominate discussion of this particular system. In the temperature range −90 to +30°C the experimental facts are not in dispute but there is disagreement in the interpretation. There appear to be three stages in the propagation process [44]. Stage I occurs at low temperatures and consists of an extremely rapid reaction for which conductance experiments have confirmed the presence of ionic species, possibly including free oligostyryl cations. As the temperature is raised the contribution from Stage I diminishes and is followed by a much slower Stage II, (Activation energy, $\sim 8-12$ kcal mole^{-1}). Within this stage, ionic conductance is low and the simple experimental rate law observed is

$$-\frac{d[M]}{dt} = k_2[HClO_4]_0[M]$$

When Stage II is almost complete the hitherto colourless solution suddenly turns yellow, there is a simultaneous increase in conductance, and the residual monomer is consumed extremely rapidly (Stage III). The yellow colour has been shown to be due to the 3-aralkyl-1-phenylindanyl cation, and Stage III (and I) is generally regarded as involving propagation by a very small concentration of oligostyryl cations in equilibrium with the indanyl species, i.e.,

$$CH_3(\underset{\underset{Ph}{|}}{CH}-CH_2)_n-\underset{\underset{Ph}{|}}{CH}-CH_2-\overset{+}{C}HPh$$

Ethanol and water have little effect on Stage II, though addition of common ion, e.g. Li$^+$ClO$_4^-$, produces a fall in the value of k_2 to a limiting value (k_2^0) at high salt concentrations. A careful re-examination of this stage has also shown that after the first half-life, the reaction accelerates steadily until Stage III is reached. Two questions arise concerning Stage II. Firstly, does propagation take place via a small concentration of free ions or a larger concentration of some much less reactive species? Secondly, if species of low reactivity are involved, are these ion pairs or covalent ester groups (pseudocationic species?). Plesch [44] has acknowledged that the common ion studies indicate an ionic species in Stage II, and takes the view that there is a small free ion contribution to propagation, which increases with conversion, with the remaining contribution from the propagating ester. Using approximated, but quantitative, electrochemical arguments based on generally accepted values for a free ion propagation coefficient, k_p^+, and an ion pair dissociation constant, K_d, for the growing species, Plesch has concluded that ion pairs can play no significant role in Stage II. Pepper et al., [13—15] however, take the view that ion pairs are responsible, and that the fraction of oligostyryl species present as ester units is dormant and these do not themselves propagate. The rate coefficient, k_2, has been shown to have a value of ~17.0 l mole^{-1} sec^{-1} at 25°C in ethylene dichloride (ϵ = 9.72, 95% solvent, 5% monomer), and falls to 1.2 x 10^{-3} l mole^{-1} sec^{-1} at 20°C in carbon tetrachloride.

More recently, independent work has been carried out by Hamann et al. [22], using NMR and gas—liquid chromatographic techniques and carbon tetrachloride and carbon tetrachloride/methylene dichloride mixtures as solvents. At ambient temperatures in the former solvent the polymerization was found to be only first order overall, i.e.

$$-\frac{d[M]}{dt} = k_1 [HClO_4]_0$$

in contrast to the results of Pepper and Reilly [13]. In experiments where the proportion of methylene chloride reached as high as 35%, a change towards the more familiar second order relationship was found, viz.

$$-\frac{d[M]}{dt} = k_2 [HClO_4]_0 [M]$$

It appears that in Pepper's work carried out at a lower temperature, the percentage conversion was too small to distinguish accurately between these two reaction orders, and conversion of the apparent second order rate coefficient, k_2, to the corresponding first order coefficient, k_1, (by including a typical value for [M]), yields a value of ~5.2 x 10^{-4} sec^{-1} at 20°C comparing favourably with k_1 = 1.1 x 10^{-3} sec^{-1} at 30°C from the work of Hamann et al. [22]. The lack of dependence on [M] in the carbon tetrachloride reaction means that the rate determining step cannot be attack of an active centre on a free monomer molecule in solution, as is usual in more polar solvents [13]. Rather it must be the rearrangement of a molecular complex which already has one or more monomer molecules built into it, essentially as a separate "phase" from the solution. This type of model has been used before [69], and in the present case the active centre, whatever it is, can be regarded as being solvated by monomer molecules in preference to carbon tetrachloride. Gandini and Plesch [19], in particular, have previously suggested this kind of association, with involvement of up to four monomer molecules. The kinetic scheme and analysis proposed by Hamann et al. [22] is

$$\mathrm{M}_n\,\mathrm{HClO}_4 + x\mathrm{M} \underset{\text{fast}}{\overset{K}{\rightleftharpoons}} \mathrm{M}_n\,\mathrm{HClO}_4\,x\mathrm{M}$$

active centre   associated active centre

$$\mathrm{M}_n\,\mathrm{HClO}_4\,x\mathrm{M} \xrightarrow[k_1]{\text{slow}} \mathrm{M}_{n+1}\,\mathrm{HClO}_4\,(x-1)\mathrm{M}$$

Hence,

$$-\frac{d[\mathrm{M}]}{dt} = k_1[\mathrm{M}_n\,\mathrm{HClO}_4\,x\mathrm{M}]$$

and

$$K = \frac{[\mathrm{M}_n\mathrm{HClO}_4\,x\mathrm{M}]}{[\mathrm{M}_n\mathrm{HClO}_4]\,[\mathrm{M}]^x}$$

Also, from a mass balance equation

$$[\mathrm{HClO}_4]_0 = [\mathrm{M}_n\mathrm{HClO}_4] + [\mathrm{M}_n\,\mathrm{HClO}_4\,x\mathrm{M}]$$

Thus

$$-\frac{d[\mathrm{M}]}{dt} = \frac{k_1\,K[\mathrm{HClO}_4]_0[\mathrm{M}]^x}{1 + K[\mathrm{M}]^x}$$

When K is large, this reduces to the observed overall first order equation found with carbon tetrachloride as solvent, i.e.

$$-\frac{d[\mathrm{M}]}{dt} = k_1[\mathrm{HClO}_4]_0$$

As the solvent is changed to include increasing proportions of methylene dichloride, the dielectric constant increases and so does the solvating power of the medium. The active centres therefore become solvated by the solvent in preference to monomer molecules i.e. $x \to 1$ and $K \to 0$. In this case the slow step becomes the collision and reaction of the active centre with a monomer molecule in solution as proposed by Pepper and Reilly [13]. Hamann et al. have also speculated further as to the nature of the active centre. NMR spectra showed no evidence for the presence of perchlorate esters; this, though not conclusive, is somewhat disappointing for the proponents of pseudocationic propagation. They have also re-examined Pepper's data for k_2 as a function of the dielectric constant, ϵ, of the solvent. Previously Pepper and Reilly [13] had speculated that the linear relationship between log k_2 and the function $(\epsilon - 1)/(2\epsilon + 1)$ indicated that the active species was dipolar in character (ion pairs?). Hamann et al. applied the Laidler Eyring theory [70] of dielectric effects in bimolecular reactions in more detail, and have concluded that Pepper's results are more consistent with a transition state for propagation being *more* polar than the initial active centre. This implies that the latter must be a π-complex, such as that proposed earlier by Bywater and Worsfold [71], or an ester, not an ion pair, though the transition state may involve a fully developed ion pair structure.

Consideration of the effect of pressure on the polymerization of styrene initiated by $HClO_4$ [72] appears to confirm the occurrence of a highly polar transition state for propagation. To what extent these arguments are valid depends upon the nature of the data for k_2. As Plesch [44] has pointed out the observed values for k_2 could well be composite, and involve contributions from more than one type of species, i.e. $k_2 = xk'_p + yk''_p + xk'''_p$. Thus, before the above criteria can be applied in a strict and rigid sense, it is essential that the numerical data being manipulated must refer to one type of active centre in isolation. The nature of the data for k_2 will be discussed later in Section 6.5.

6.3 STABLE CARBONIUM ION SALTS AS INITIATORS

A number of years ago triphenylmethyl cation, Ph_3C^+, formed in situ by dissociation of triphenylmethyl chloride, was shown [73] to initiate the polymerization of 2-ethylhexyl vinyl ether in *m*-cresol solvent. More recently certain stable carbonium ion salts, notably hexachloroantimonate ($SbCl_6^-$) salts of cycloheptatrienyl (tropylium, $C_7H_7^+$) and triphenylmethyl cations have been shown [74, 50] to be very efficient initiators of the cationic polymerization of many reactive monomers [27, 29, 75]. Since the discovery of the effectiveness of the $SbCl_6^-$ salt, triphenylmethyl salts with different anions have also been used [76—78]. The most detailed kinetic studies using these initiators have been carried out on alkyl vinyl ethers [27, 30] and *N*-vinylcarbazole [39] in homogeneous solution in methylene chloride.

In the case of *N*-vinylcarbazole at 0 and −25°C, initiation has been shown to be the direct addition of the carbonium ion ($C_7H_7^+$) to monomer, a process which appears to take place virtually instantaneously. Under these circumstances the concentration of active centres is assumed equal to the initial salt concentration, $[I]_0$. Significant termination appears to be absent during kinetic lifetimes, (up to 100% conversion) and polymer molecular weights are limited only by transfer reactions. The experimental rate law can be expressed as

$$-\frac{d[M]}{dt} = k[C]_0[M]$$

and from the initial slopes of conversion/time curves the coefficient k can be evaluated. Since initiation is almost instantaneous k represents the rate coefficient for propagation, k_p, directly. Reaction rates were found to be extremely large and an adiabatic calorimetric technique [79] was required for their determination. The data obtained are shown in Table 2.

TABLE 2
Polymerization of N-vinylcarbazole by tropylium salts in methylene chloride solution

Counter-ion	Temperature ($^\circ$C)	$10^{-5}k_p$ (1 mole^{-1} sec^{-1})	Activation Energy (kcal mole^{-1})
ClO_4^-	0	2.2	6.5
ClO_4^-	-25	0.66	6.5
$SbCl_6^-$	0	4.6	5.7
$SbCl_6^-$	-25	1.6	5.7

In the case of alkyl vinyl ethers [27, 30] reaction rates were again high and the same experimental technique was used. However, the initiation reaction did not appear to be as fast as that in the polymerization of N-vinylcarbazole, and the conversion/time data showed evidence of an initial acceleration to a maximum rate of polymerization, particularly in runs carried out at -25°C. The mechanism of initiation was assumed to involve direct addition of the initiating carbonium ion to the double bond of the monomer, in the light of related evidence from similar reaction in the presence of strong nucleophiles [80, 81]. At 0°C there was also an indication of a contribution from a termination reaction. Polymer yields were always in excess of 75%, however, and the termination process was neglected in the kinetic analysis. The simple scheme envisaged is

$$C^+ + M \xrightarrow{k_i} CM^+$$

$$CM^+ + M \xrightarrow{k_p} CM_2^+ \xrightarrow{k_p} CM_n^+$$

from which

$$\frac{d[CM^+]}{dt} = k_i[C^+][M] = -\frac{d[C^+]}{dt}$$

$$-\frac{d[M]}{dt} = k_i[C^+][M] + k_p[CM^+][M]$$

Also from a mass balance

$$[C^+] + [CM^+] = [C]_0,$$

the initial salt concentration.

Appropriate manipulation of these equations yields the expression

$$\frac{d^2 \ln[M]}{dt^2} + k_i \frac{[M] d \ln[M]}{dt} + k_i k_p [C]_0 [M] = 0$$

Thus if $\ln[M]$ is plotted against t, $d^2 \ln[M]/dt^2$ vanishes at the point of maximum slope, so that k_p can be evaluated from the relationship

$$-\frac{d \ln[M]}{dt} = k_p [C]_0.$$

The data obtained by this treatment are shown in Table 3.

TABLE 3
Polymerization of alkyl vinyl ethers by stable carbonium ion salts in methylene chloride

Alkyl vinyl ether	Initiator	Temperature ($^\circ$C)	$10^{-3} k_p$ (l mole^{-1} sec^{-1})
isobutyl	$C_7H_7{}^+SbCl_6^-$	0	6.8
isobutyl	$C_7H_7{}^+SbCl_6^-$	−25	2.0
isobutyl	$Ph_3C^+SbCl_6^-$	0	4.0
isobutyl	$Ph_3C^+SbCl_6^-$	−25	1.5
isobutyl	$Ph_3C^+BF_4^-$	0	2.8
methyl	$C_7H_7{}^+SbCl_6^-$	0	0.14
methyl	$C_7H_7{}^+SbCl_6^-$	11.8	0.4
tert-butyl	$C_7H_7{}^+SbCl_6^-$	0	3.5
tert-butyl	$C_7H_7{}^+SbCl_6^-$	11.8	3.8

In both of these studies initiator concentrations were in the range $10^{-4}-10^{-6}$ M, and conductance experiments [51] have shown that at the temperatures employed in the polymerizations the initiating salts are, for all practical purposes, completely dissociated into free ions. A study [51] of a large variety of stable salts of cations of widely differing structure with the same anion ($SbCl_6^-$), showed that values for the ion pair dissociation constants in methylene chloride did not vary by more than one order of magnitude at a given temperature. Thus dissociation constants for the propagating polymeric ion pairs would not be expected to differ substantially from those of the initiators. The kinetic data obtained are assumed, therefore, to refer to the reaction of free propagating cation with monomer, k_p^+. This fact is confirmed by the lack of dependence of the rate coefficients, within experimental error, on both the counter-ion, and the carbonium ion used as initiator. Significantly the results from both N-vinylcarbazole and vinyl ethers are many orders of magnitude higher than any other data reported in the literature for chemically initiated cationic polymerizations, supporting the argument that, in these systems at least, the reactivity of free cations has at last been exposed semi-quantitatively.

6.4 POLYMERIZATION BY γ-RAY IRRADIATION

As mentioned previously, irradiation of suitably reactive bulk (liquid) monomers with γ-rays from a ^{60}Co source can initiate cationic polymerization [37], i.e.

$$CH_2{=}CHR \xrightarrow{\ k_i\ } \overset{\bullet}{C}H_2{-}\overset{+}{C}HR + e^-$$

$$\nearrow \text{γ-ray}$$

An electron is ejected from the monomer and becomes solvated by the latter or is captured by an impurity in the system, forming an anion, Y^-. The monomeric radical cation formed is then free to propagate via a radical mechanism at one end, and a cationic one at the other although the radical ends would soon terminate by combination or disproportionation. Using this technique extremely low concentrations of active centres can be achieved and the cations formed can be regarded as being completely free from any counter-ion (solvated electron or Y^-). It is now accepted that free cations propagate many orders of magnitude more rapidly than free radicals (Section 6.5), and so many cationic propagation steps will take place for each radical addition. Thus even if the radical ends are never destroyed for example by dimerization, their contribution to polymerization will be undetectably small. With those monomers which are polymerizable by anionic mechanisms, the contribution to polymer formation by radical processes also appears to be minimal [89]. Polymerization by γ-ray irradiation does not involve problems of initiation equilibria and reactions with resulting initiator fragments, nor is there any necessity to consider contributions from aggregated species. The method, therefore, allows the propagation of free cations to be investigated, and has proved to be an invaluable yardstick for more conventional chemically initiated systems.

A number of monomers has already been investigated, e.g., cyclopentadiene [86], styrene [89, 93], α-methylstyrene [87, 89, 95], formaldehyde [94], alkyl vinyl ethers [85, 89], isobutene [84], nitroethylene [96] and also the cyclic monomer cyclohexene oxide [97]. For most of these reliable quantitative data is now available. Because the number of active centres formed is small all of these systems are particularly susceptible to traces of impurities, especially water and spurious basic materials. Indeed much of the early data [82, 83] from γ-ray initiation was confused because of the use of relatively "wet" monomers. Furthermore the intrinsic lifetime of a free cation is limited because the counter-ion in the system, often a solvated electron, is a reactive species, readily capable of terminating polymerization by charge recombination or neutralization. Hence, unlike many chemically initiated systems which have stable non-nucleophilic counter-ions, radiation systems always have an inherent termination reaction, as well as the possibility of termination with adventitious impurities. Taylor and Williams [84] have argued that if the lifetime of a free cation (τ), as determined by charge neutralization experiments, is to be significant, the concentration of reactive impurities, [X], must be very low. In fact k_{tx} [X] must be less than $1/\tau$, where k_{tx} is the rate coefficient for reaction with X. Since τ is typically $\sim 10^{-3}$ sec or longer at those rates of practical interest [83], [X] must be reduced below 10^{-7} M if the diffusion controlled value, in mobile liquids, of 10^{10} l mole^{-1} sec^{-1} is adopted for k_{tx}. Such an impurity level demands the use of high vacuum apparatus approaching the sophistication of so-called ultra-high vacuum techniques [84].

Rates of radiation induced polymerizations are normally determined by dilatometric [85] or gravimetric [84] experiments. Some of the first quantitative results from cyclopentadiene [86] and α-methylstyrene [87] were obtained by competitive kinetic methods, based on the retarding effect of ammonia and amines. This approach tends to yield maximum values for R_p^+. More recently, however, a procedure combining stationary state kinetic and conductance measurements has been described [88, 89], and further refined [85]. Because the ions generated by γ-ray irradiation have a transient existence, the kinetic treatment leads to expressions which are very similar to those derived for homogeneous free radical polymerizations [90]. A simplified version of the kinetic scheme is as follows:

$$ M \xrightarrow{\gamma\text{-ray}} \cdot M^+ + e^- \quad \text{Rate, } R_i = IG_i/100 $$

where e^- = solvated electron or trapped electron (Y^-),

$\qquad I$ = dose rate (eV cm^{-3} sec^{-1}),

$\qquad G_i$ = 100 eV yield of free ions (excluding geminate recombinations)

$$ \cdot M^+ + M \xrightarrow{k_i} M_2^+ $$
$$ M_2^+ + M \xrightarrow{k_p} M_n^+ $$
$$ M_n^+ + M \xrightarrow{k_{tr}} M_n + M^+ $$
$$ M_n^+ + Y^- \xrightarrow{k_t} \text{products.} $$

Assuming a stationary state concentration of free ions,

$$ R_i = k_t [M_n^+] [Y^-] $$
$$ \therefore \ [M_n^+] = (R_i/k_t)^{1/2} = [Y^-] $$

By definition, τ, the mean lifetime in the stationary state is given by

$$\tau = \frac{[M_n^+]}{R_i}$$

and the rate of polymerization, R_p, by

$$R_p = k_p [M] [M_n^+]$$

or

$$R_p = k_p [M] \tau R_i$$

or

$$R_p = k_p [M] (R_i/k_t)^{1/2}$$

It also follows that, $R_p \alpha I^{1/2}$, for a free ion mechanism and also

$$G(-m) = k_p [M] \tau G_i$$

where $G(-m)$ is the number of monomer molecules consumed per 100 eV.

More complicated schemes which cater for the presence of finite amounts of impurity have also been deduced [85, 89]. The dilatometric or gravimetric data provides a quantitative measure of k_p. R_i is obtained from the conductance measurements, or providing that $R_p \alpha I^{1/2}$, R_i can be assumed to equal 0.1, since this yield of free ions is typical for many hydrocarbons [91]. In the case of insignificant impurity termination k_p can be calculated by assigning k_t a diffusion controlled value per charge recombination [84], i.e.,

$$k_t = 4\pi r_c (D^+ + D^-)$$

where $\qquad r_c = e^2 / \epsilon k T$, for singly charged ions, with

$\qquad\qquad e$ = electron charge

$\qquad\qquad \epsilon$ = dielectric constant

$\qquad\qquad k$ = Boltzmann constant

$\qquad\qquad T$ = absolute temperature

and D^+, D^- = diffusion constants for recombining cation and anion.

D^- is likely to be much larger than D^+ since the cation is a polymeric species, hence $k_t \approx 4\pi r_c D^-$. A reasonable value for D^- is 4.5×10^{-5} cm^2 sec^{-1} at 25°C, the value for the hydrated electron [92], yielding a corresponding value for k_t of $10^{11}-10^{12}$ at 0°C and in a solvent of dielectric constant 2. An alternative approach, particularly when additional impurities are present, is to estimate τ or τ', from the conductance measurements, where τ' is the modified lifetime in the stationary state, as a result of contributions to termination from processes other than simple charge recombination. In this case the rate equation is modified so that

$$R_p = k_p [M] \tau' R_i$$

or

$$\frac{R_p}{\tau'} \alpha I$$

and the rate coefficient for propagation, k_p, can then be calculated [85]. A number of liquid monomers have now been studied by these techniques and a summary of the propagation rate coefficients obtained is shown in Table 4.

TABLE 4
Kinetic data from γ-ray polymerizations

Monomer	Solvent	Temperature ($^\circ$C)	k_p (l mole^{-1} sec^{-1})	Activation energy (kcal mole^{-1})	Ref.
styrene	bulk	15	3.5×10^6	0	89
α-methyl styrene	bulk	0	4×10^6	—	89
α-methyl styrene	bulk	30	3×10^6	0	87
isobutyl vinyl ether	bulk	30	3×10^5	6.6	89
isobutyl vinyl ether	bulk	0	3.8×10^4	9.6	85
isobutyl vinyl ether	bulk	25	1.2×10^5	9.6	85
isobutyl vinyl ether	bulk	50	6.0×10^5	9.6	85
isobutyl vinyl ether	bulk	42.5	1×10^6	—	98
cyclopentadiene	bulk	-78	6×10^8	<2	86
isobutene	bulk	0	1.5×10^8	0	84

The most startling feature of these results is the magnitude of k_p. Once again these values are many orders of magnitude in excess of most of those determined from chemically initiated polymerizations. Even accepting that some of these estimations are upper limits, e.g. cyclopentadiene, the differences are enormous when a comparison is made with previous information.

6.5 A COMPARISON OF RATE PROPAGATION DATA

6.5.1 Relevance of numerical values

Earlier on in this chapter it was pointed out, in some detail, that over the last few years the dominant feature in studies of cationic polymerization has been the attempts to elucidate quantitatively the role of free ions and ion pairs, particularly in the propagation reaction. Sufficient data are now available to be able to draw a number of definite conclusions. Probably the most instructive picture can be obtained from a consideration of representative propagation rate coefficients for alkyl vinyl ethers (particularly isobutyl vinyl ether) which have been very widely studied (Table 5).

TABLE 5

Propagation data for isobutyl vinyl ether

Initiator	Solvent	Temperature ($^\circ$C)	k_p (l mole^{-1} sec^{-1})	Ref.
I_2	$CH_2CH_2Cl_2$	30	6.5	64, 65
I_2	CCl_4	30	0.045	64, 65
$C_7H_7^+SbCl_6^-$	CH_2Cl_2	0	6.8×10^3	27
$Ph_3C^+SbCl_6^-$	CH_2Cl_2	0	4.0×10^3	27
$Ph_3C^+BF_4^-$	CH_2Cl_2	0	2.8×10^3	27
γ-ray	bulk	0	3.8×10^4	85
γ-ray	bulk	30	3×10^5—10^6	89,98

There seems little doubt that in radiation induced polymerizations the reactive entity is a free cation (vinyl ethers are not susceptible to free radical or anionic polymerization). The dielectric constant of bulk isobutyl vinyl ether is low (<4) and very little solvation of cations is likely. Under these circumstances, therefore, the charge density of the active centre is likely to be a maximum and hence, also, the bimolecular rate coefficient for reaction with monomer. These data can, therefore, be regarded as a measure of the reactivity of a non-solvated or "naked" free ion and bear out the high reactivity predicted some years ago [110, 111]. The experimental results from initiation by stable carbonium ion salts are approximately one order of magnitude lower than those from γ-ray studies, but nevertheless still represent extremely high reactivity. In the latter work the dielectric constant of the solvent is much higher (CH_2Cl_2, $\epsilon \sim 10$, 0°C) and considerable solvation of the active centre must be anticipated. As a result the charge density of the free cation will be reduced, and hence the lower value of k_p represents the reactivity of a solvated free ion rather than a "naked" one. Confirmation of the apparent free ion nature of these polymerizations is afforded by the data on the ion pair dissociation constant, K_d, of the salts used for initiation, and, more importantly, the invariance, within experimental error, of k_p with the counter-ion used ($SbCl_6^-$ or BF_4^-). Overall effects of solvent polarity will be considered shortly in more detail.

In comparison to these two sets of data, the results from polymerizations initiated by molecular iodine present a dramatic contrast, and one which is characteristic of many monomers. The rate coefficient for propagation is at least three orders of magnitude below that for a solvated free ion, and at least four below that for a "naked" free ion. Ethylene dichloride has very similar solvating properties to methylene chloride, and hence it would be expected that at least a proportion of the active centres would be in the form of free ions (solvated). However, in these systems the counter-ions are likely to be I^- and I_3^- and the ion pair dissociation constants for the propagating species in methylene chloride may be much smaller than those predicted [27, 61] when $SbCl_6^-$ is the counter-ion, and the proportion of free ions could be insignificant. Under such circumstances the data from initiation by iodine could be regarded as an estimate of the rate coefficient for ion pair propagation, although this is by no means certain. For instance, in a solvent of much lower dielectric constant, CCl_4, the rate coefficient appears to be approximately two orders of magnitude less, an observation difficult to explain if ion pairs only were reactive in $ClCH_2CH_2Cl$. Conceivably this variation arises as a result of contributions from solvent separated ion pair species as indicated by studies [99] of added salts on the polymerization of alkyl vinyl ethers in methylene chloride and benzene initiated by iodine. Considerable rate accelerations have been observed and attributed to a forced charge in the counter-ion, the effect understandably being greater in solvents of lower dielectric constant. Similar observations [77] have been reported for polymerization of alkyl vinyl ethers by triphenylmethyl salts in low polarity solvents. These observations tend to confirm that the major propagating species are ion pairs in the iodine systems, but it is at least a possibility that the data represent statistical averages of free ion and ion pair contributions.

A similar though less comprehensive comparison is possible for the polymerization of styrene. Here much more data is available from chemically initiated polymerizations, but unfortunately none of these is conclusive with respect to the nature of the active centre (Table 6).

TABLE 6
Propagation data for styrene

Initiator	Solvent	Temperature ($^\circ$C)	k_p ($l\,mole^{-1}\,sec^{-1}$)	Ref.
$HClO_4$	$ClCH_2CH_2Cl$	25	17.0	13
$HClO_4$	CCl_4	20	1.2×10^{-3}	13
H_2SO_4	$ClCH_2CH_2Cl$	25	7.6	11
$SnCl_4$	$ClCH_2CH_2Cl$	30	0.42	100
I_2	$ClCH_2CH_2Cl$	30	3.5×10^{-3}	101
γ-ray	bulk	15	3.5×10^6	89

As with isobutyl vinyl ether the result from γ-ray irradiation must represent the reactivity of "naked" free ions and this k_p is at least 10^5

times larger than any other value available to date. Once again interpretation of the remaining data is speculative. Certainly such values cannot be representative of a propagating free ion because the k_p values are simply numerically too low. Also considerable variation occurs with the initiator system used, a good indication of substantial contributions from species other than freely dissociated ones. On the other hand, the variation of k_p with solvent in the perchloric acid work, is the reverse of what might be expected if ion pairs only were responsible for propagation, and indicates the presence of at least two species of differing reactivity in equilibrium with each other. However, as already discussed this system is particularly complex, and there is little to be gained from further analysis of existing data which are better regarded as composite in nature. A similar query must remain on the other values for k_p obtained from initiation by iodine and $SnCl_4$. In these studies again a solvent of relatively high solvating power has been used, and it is difficult to eliminate completely the idea of a substantial contribution from dissociated species in spite of the values for k_p being so low. In these cases k_p has probably been inappropriately determined, and is in fact a composite rate coefficient involving initiation and termination factors.

The most recent work on styrene [102] has been carried out in benzene solution with initiation by boron trifluoride etherate, $BF_3O(C_2H_5)_2$. An accurate estimate of the number of active centres, $[P_n^+]$, was achieved by a method similar to that used in Saegusa et al. [45] in ring opening polymerizations (see Section 7.1). Reactions were prematurely terminated by addition of 2-bromothiophene, and the bromine content of the polymers produced estimated by radio activation analysis. Application of the simple relationship (which assumes rapid initiation)

$$R_p = k_p [M] [P_n^+]$$

enabled k_p to be calculated. Water was evaluated as a co-catalyst, but in general k_p remained fairly constant and assumed a value of 0.25 l mole^{-1} sec^{-1} at 30°C. In benzene there is little possibility of a contribution from dissociated species, and so this data must represent the reactivity of ion pairs.

Experiments similar to those made with alkyl vinyl ethers, involving the addition of stable salts, e.g. $Bu_4N^+ ClO_4^-$ to polymerizing mixtures of styrene initiated by, for example, stannic chloride and boron trifluoride have been carried out in methylene chloride solutions [103]. The results show some effect of counter-ion but again are quantitatively inconclusive because of the polar solvent employed. It is difficult to know the nature of the active centre before the addition of salt, and so any change produced can be interpreted only in a very qualitative way although some ion pairing must take place.

6.5.2 *Solvent effects*

The effects of solvents on ionic polymerizations have long been misinterpreted, largely because of two factors. Firstly the failure to isolate the process of propagation from the other elementary reactions which comprise polymerization, and secondly the failure to compare isolated free ion and ion pair data. It is worthwhile, therefore, to examine these simple considerations from a general point of view.

In propagation, where the active centre is a fully developed free ion or ion pair, as opposed to a covalent species, the transition state will usually involve some degree of charge dispersion. Solvents of high polarity will not stabilize the transition state as much as the initial state. Such solvents, therefore, increase the activation barrier and reduce the rate of propagation.

Initiation of ionic polymerization normally requires a charge formation process, i.e.

$$AB \rightleftharpoons A^+B^-$$

Such a reaction must involve a transition state which is more polar than the reactants. Solvents of high polarity will, therefore, stabilize the transition state more than the initial one, with a consequent reduction in the activation barrier and an acceleration of the rate of initiation. Solvents of lower solvating ability will have the opposite effect, i.e.

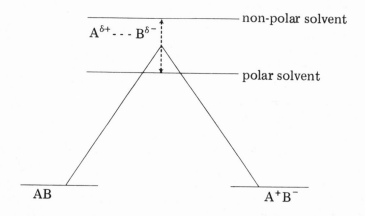

The effect on ion pair propagation might be expected to be less than that on free ion reactivities, because the effective charges are smaller anyway.

Coupled with these considerations is the effect of solvent on the ion pair dissociation constant, which governs the proportion of each species present in a system at any instant. Obviously the more polar the medium the more highly dissociated an ionic species and vice versa. Since free ions react more rapidly than ion pairs, increasing the proportion of these increases the rate of a reaction and its apparent rate coefficient. Thus, if k_p increases with increasing dielectric constant of the solvent, then either both free ions and ion pairs are present and the proportion of the former is merely increased, or k_p is not a true propagation coefficient but embodies a contribution from initiation. The converse is also true. If a rate coefficient for propagation is determined, and it represents contributions from one ionic species alone (free ion *or* ion pair), then increasing the solvent polarity should decrease its value.

The polymerization of p-methoxystyrene (and to a lesser extent isobutyl vinyl ether and p-methylstyrene (Table 1)), initiated by iodine has been studied as a function of the dielectric constant of the solvent. The information derived is shown in Table 7. Clearly as the solvent polarity

TABLE 7

Effect of the polarity of solvent on k_p for p-methoxystyrene
(Iodine as initiator at $30°C$)[65]

Solvent	k_p ($1 \, mole^{-1} \, sec^{-1}$)
$CH_2Cl_2 : CCl_4$	
4:0	17
3:1	1.8
2:2	0.31
1:3	0.28
0:4	0.12
$CHCl_3$	0.14

falls so does the apparent value of k_p. Using the general arguments presented before, this can be interpreted in two ways. Firstly both free ions and ion pairs may be contributing to the propagation, and the proportion of the latter merely increases as CCl_4 is added (or solvent separated ion pairs and contact pairs, and the latter increase as CCl_4 is added). Secondly the data may be seriously affected by a kinetic term from the initiation processes, i.e. the decrease in k_p as CCl_4 is added may reflect merely a decrease in the rate of *initiation*. A similar trend has been reported for the polymerization of styrene by perchloric acid [13], where k_p is reduced by a factor of 1.5×10^4 on changing from ethylene dichloride to carbon tetrachloride. The interpretation in this case is more difficult because of the possible contribution of a covalent species.

There has not yet been a comprehensive study of solvent effects although isobutyl vinyl ether has been polymerized in bulk and in methylene chloride solution. In both cases free ion propagation data is reported. This should, therefore, be unaffected by ion pair/free ion considerations. In bulk [85] k_p has a value of $\sim 4 \times 10^4 \, 1 \, mole^{-1} \, sec^{-1}$ at $0°C$, while in methylene chloride at the same temperature k_p assumes a value [27] $\sim 6 \times 10^3$ 1 mole^{-1} sec^{-1}. Thus on increasing the dielectric constant from 4 to 10, the rate coefficient is decreased by a factor of about 10. This is consistent with similar data from reactions involving small molecules [104] and confirms the free ion nature of both of these systems.

6.5.3 Energetics of propagation reactions

The energetics of any propagation reaction can in the first instance be described by a simple Arrhenius relationship

$$k_p = A_p \, \exp\left(\frac{E_{act}}{RT}\right)$$

where A_p = pre-exponential factor

E_{act} = energy (enthalpy) of activation

In general activation energies for propagation in cationic systems are low ($<$10 kcal mole^{-1}), and in the reactions which involve only free ions, an approach to zero activation energy can be observed [84]. Indeed on occasions apparently negative activation energies have been described [1], illustrating the caution required in interpretation of this parameter. On reducing the temperature of a polymerization, the dielectric constant of the solvent increases, and hence the proportion of dissociated ions increases. Since the latter react much more quickly than ion pairs this increases the observed rate coefficient, k_p. In favourable circumstances this increase can more than offset the decrease due to simple Arrhenius behaviour. As a result the overall activation energy may be negative. For some considerable time the fast reactions observed, even at low temperatures, in many cationic polymerizations were attributed solely to a relatively large concentration of active centres compared with free radical systems, and not to any intrinsic high reactivity. The emergence of propagation data many orders of magnitude higher than those for corresponding free radical reactions now provides an additional and more correct explanation.

In the free radical polymerization of styrene [105] at 25°C, k_p has a value $\sim$40 1 mole^{-1} sec^{-1} and E_{act} = 6.3 kcal mole^{-1}. This compares with k_p for free cationic growth of $\sim 3.5 \times 10^6$ 1 mole^{-1} sec^{-1} at 15°C and $E_{act} \simeq 0$. It is worthwhile to consider why the styryl cation reacts about 10^5 times faster with its monomer than does the styryl radical. A similar comparison is possible in the case of N-vinylcarbazole for which the reported [106] free radical propagation coefficient is 6.0 1 mole^{-1} sec^{-1} at 10°C with E_{act} = 6.9 kcal mole^{-1}. This compares with a value [29] of 4.0×10^5 1 mole^{-1} sec^{-1} at 0°C with E_{act} = 6.0 kcal mole^{-1} for propagation of the corresponding free cation. Again a difference of $\sim 10^5$

is apparent. Chapiro [107] has studied the radiation induced polymerization of "wet" styrene where only free radicals contribute to polymerization. From analogies between the kinetic treatments of free ion and free radical polymerizations [89] and neglecting impurity termination, it follows that (see Section 6.4)

$$\frac{k_{p\ \text{cationic}}}{k_{p\ \text{radical}}} = \frac{R_{p\ \text{(cationic)}}}{R_{p\ \text{(radical)}}} \left[\frac{k_{t\ \text{(cationic)}} G_{i\ \text{(radical)}}}{k_{t\ \text{(radical)}} G_{i\ \text{(ionic)}}} \right]^{1/2}$$

Substitution of the appropriate rate values yields a ratio of about 10^5, extremely close to that obtained from chemically initiated polymerization.

Studies of the anionic polymerization of various monomers [39] have yielded data for both free ion growth and ion pair propagation. Indeed compared with the situation in cationic systems, data from anionic polymerizations are rather elegant. Swzarc et al. [108] have deduced a value of 6.5×10^4 l mole^{-1} sec^{-1} for the free anionic propagation coefficient of styrene in tetrahydrofuran at 25°C, with [109] E_{act} = 6 kcal mole^{-1}. Corresponding ion pair growth coefficients fall in the range 10—100 l mole^{-1} sec^{-1} depending on the counter-ion. The free anionic data is extremely important because simple theoretical considerations would predict free anionic propagation to require rather more activation than free cationic growth, i.e. $k_p^+ > k_p^-$. The former process requires the use of antibonding orbitals of the monomer in the transition state, since all bonding orbitals on both the growing carbanion and the monomer being attached are already filled. On the other hand, a propagating free cation has vacant bonding orbitals favouring formation of the transition state.

Comparison of the data for free cationic propagation of isobutene, styrene, isobutyl vinyl ether and N-vinylcarbazole is also useful. Elementary electronic arguments would predict a reactivity sequence as follows:

$$\underset{\underset{\text{CH}_3}{|}}{\overset{\overset{\text{CH}_3}{|}}{\sim\!\!\sim\!\!\text{CH}_2\!-\!\overset{+}{\text{C}}}} > \underset{\underset{\text{Ph}}{|}}{\sim\!\!\sim\!\!\text{CH}_2\!-\!\overset{+}{\text{CH}}} > \underset{\underset{\text{OR}}{|}}{\sim\!\!\sim\!\!\text{CH}_2\!-\!\overset{+}{\text{CH}}} > \sim\!\!\sim\!\!\text{CH}_2\!-\!\overset{+}{\text{CH}}$$

in keeping with the idea that cations from isobutene and styrene would be considerably less stabilized by their substitutents, than would those derived from isobutyl vinyl ether and N-vinylcarbazole. The high reactivity of N-vinylcarbazole probably arises because the nitrogen lone pair electrons, which could normally stabilize the adjacent positive charge, comprise part of a 14 π-electron aromatic system (cf. anthracene).

Observed values of rate coefficients do not depend simply on energy (enthalpy) terms but are also controlled by entropy effects, reflected in the pre-exponential factor, A_p.

Higashimura [7] has used a crude statistical thermodynamic approach to compare pre-exponential factors for free ion, free radical and ion pair propagation reactions. According to Eyring's theory the rate coefficient for propagation has the form

$$k_p = \left(\frac{kT}{h}\right) \exp\left(\frac{\Delta S_p^{\ddagger}}{R}\right) \exp\left(\frac{-\Delta E_p^{\ddagger}}{RT}\right)$$

k = Boltzmann constant

T = absolute temperature

h = Planck's constant

$\Delta E_p^{\ddagger}$ is the enthalpy of activation and corresponds to E_{act} in the Arrhenius equation

$\Delta S_p^{\ddagger}$ is the entropy of activation ($A_p \propto \exp \Delta S_p^{\ddagger}$)

The entropy term can be expressed in terms of partition functions as follows,

$$\Delta S_p^{\ddagger} = R \ln(f_T/f_I) + RT\partial \ln(f_T/f_I)/\partial T$$

where, f_T is the partition function for the transition state and f_I that for the initial state. The second term in this expression can be neglected in comparison to the first, and in addition, since other parts of the growing chain are common to both, f_T and f_I can be assigned to the monomer only for the transition, f_T (M), and initial, f_I (M), states. Thus

$$\Delta S_p^{\ddagger} = R\ln \{f_T(M)/f_I(M)\}$$

The initial state of the monomer can be regarded as the same irrespective of the mechanism of propagation and hence

$$\frac{A_{p\,(radical)}}{A_{p\,(cationic)}} = \frac{f_T(M)_{(radical)}}{f_T(M)_{(cationic)}}$$

Vibrational and rotational modes of adding monomer in the transition state for free radical and free cationic growth will be virtually the same, and are shown diagrammatically as (I). However, for ion pair propagation the presence of counter-ion allows only vibrational movements, illustrated as (II).

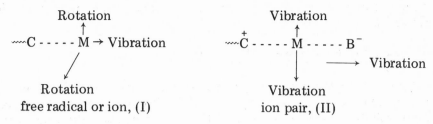

free radical or ion, (I)   ion pair, (II)

Thus the ratio,

$$\frac{A_{p\,(\text{free radical})}}{A_{p\,(\text{free ion})}} = \frac{f_{rot}^2 f_{vib}}{f_{rot}^2 f_{vib}} = 1$$

i.e. the pre-exponential factor for free ion and radical growth should be approximately the same. However, the ratio of free ion (or free radical) to ion pair factors is quite different, i.e.

$$\frac{A_{p\,(\text{free ion})}}{A_{p\,(\text{ion pair})}} = \frac{f_{rot}^2 f_{vib}}{f_{vib}^3} \sim \frac{f_{rot}^2}{f_{vib}}$$

Appropriate substitution of the partition functions yields the ratio

$$\frac{A_{p\,(\text{free ion or radical})}}{A_{p\,(\text{ion pair})}} = \frac{8\pi^2 IkT}{h^2} \Big/ \left(\frac{kT}{h}\right)^2 \prod_{i=1}^{2} \left(\frac{1}{\nu_i}\right)$$
$$\simeq 10^3$$

where I, the moment of inertia, and ν_i the vibrational frequency of the monomer, are taken as 10^{-38} gm cm^{-2} and $\simeq 10^{13}$ sec^{-1} respectively. This expression serves merely to quantify the qualitative observation that the presence of a counter-ion in a transition state must restrict the amount of freedom enjoyed by the inserting monomer molecule. Thus the entropy change due to monomer immobilization in going from initial to transition state must be more negative for the ion pair insertion than for either the free ion or free radical addition. Hence $A_{p\,(\text{ion pair})}$ must be less than $A_{p\,(\text{free ion or radical})}$. These criteria have been applied to the polymerization of styrene initiated by iodine [101], where the ratio $A_{p\,(\text{radical})}/A_p\,(I_2)$ was found to be in the range 10^4-10^6, indicating, the authors suggest, an ion pair mechanism. However, great care must be exercised in this type of correlation since a number of inconsistencies still exist. In the polymerization of styrene by perchloric acid for example, the A_p value is of the same order as that for free radical propagation, indicating a free ion mechanism. However, the known absolute values of k_p along with other experimental observations show this to be impossible. Pepper and Reilly [13] have suggested that the entropy of activation must include a term accounting for changes in solvation, $\Delta S_p^{\ddagger}$ (solv), as well as monomer immobilization, $\Delta S_p^{\ddagger}$ (immob), in going from initial to transition states, i.e.

$$\Delta S_p^{\ddagger} = \Delta S_p^{\ddagger}(\text{immob}) + \Delta S_p^{\ddagger}(\text{solv})$$

The first term is a negative quantity, and for free cations approximates to the value for free radical propagation ($-28\,\text{eu}$). Generally the solvation term will be positive, because in the transition state the charge distribution is larger, and hence the degree of solvation lower than that of the compact ionic reactant. Eyring's theory shows $A_p \propto \exp(\Delta S_p^{\ddagger})$, so that any positive contribution to $\Delta S_p^{\ddagger}$ will increase the value of A_p. This solvation component may be expected to be smaller in ion pair growth than in the free ion reaction, since smaller differences in solvation are likely to occur.

Application of these arguments to those data which are unequivocally free ion values does seem to work, however. For example the radiation induced polymerizations produce pre-exponential factors of 10^7-10^9 l mole^{-1} sec^{-1}, slightly larger than the accepted range for the free radical polymerization of styrene. Similarly the work on N-vinylcarbazole [29] produces $A_p \simeq 10^{10}$ l mole^{-1} sec^{-1}, somewhat larger than the free radical value [106], and presumably affected by solvation contributions. The data for isobutyl vinyl ether [31] indicate a value for A_p of $\simeq 10^8-10^9$, consistent with the idea that free cations are the main contributors.

7. Polymerization of cyclic monomers

A large number of cyclic monomers has been shown to polymerize by cationic mechanisms, and of these by far the majority are heterocyclic compounds containing oxygen, sulphur or nitrogen [112]. The kinetics and mechanisms of the reactions have been extensively reviewed [5], and some of the more practical aspects of the polymers produced have also been discussed [113]. In the present Section we shall deal with those polymerizations whose kinetics have been examined in detail. Of particular interest are 1,4 epoxides, notably tetrahydrofuran, and 1,3 episulphides, thietans. Some reference will also be made to polymerization of 1,3 epoxides, and the cyclic formals, 1,3 dioxolane and 1,3 dioxepan. Formaldehyde will also be considered briefly because this molecule can be regarded as the smallest possible oxygen heterocycle, giving a polymer which is structurally analogous to those from epoxides, i.e.

where $x = 1$ for formaldehyde

1,2 epoxides will not be included since the homopolymerization of these monomers e.g. ethylene oxide, is best carried out by anionic means [114].

Detailed analysis of the kinetics of polymerization of cyclic monomers has only relatively recently reached the sophistication of analogous vinyl systems. Despite this an ever increasing volume of relevant data is emerging, and, indeed many reactions are proving to be more amenable to investigation in an absolute manner, than the more extensively studied vinyl polymerizations. This situation has arisen partly as a result of the inherent lower reactivity of these monomers, and also from the fact that, in general, the polymerizations are less susceptible to impurities and side reactions. Even monomer transfer is absent in most of these systems, and a situation approaching that found in anionic vinyl polymerization prevails.

Generalized methods of initiating the polymerization of these monomers have recently been reviewed in detail [9], and were also mentioned briefly earlier in this Chapter. As with vinyl monomers initiation can be efficient and rapid, with the production of a fixed number of active centres. Propagation appears to be much slower, however, and rates of polymerization are comparable to those in free radical addition polymerizations. Techniques such as dilatometry, spectrophotometry etc. are therefore convenient for kinetic investigation of this type of cationic reaction.

7.1 POLYMERIZATION OF TETRAHYDROFURAN, THF

The polymerization of this 1,4 epoxide has been widely studied [115] and the polymer produced, polytetramethylene oxide or polytetrahydrofuran, has found commercial application. The most characteristic feature of the polymerization is the occurrence of a ceiling temperature within the range of normal laboratory experiments, $\sim 85°C$. In any polymerization the propagation reaction is normally shown as a reaction going 100% to completion although in fact this reaction, and the other elementary steps, are really chemical equilibria. Some reactant, usually monomer in the case of propagation, is present at the end of the reaction although the value of the

$$\text{equilibrium constant} \left(= \frac{\text{rate coefficient for propagation}}{\text{rate coefficient for depropagation}} \right)$$

is generally so large that the amount of monomer present at equilibrium is undetectably small. The polymerization of tetrahydrofuran is unusual, though by no means unique, in that an approach to a measurable equilibrium between monomer and polymer is attained under normal laboratory temperatures. In other words, for every temperature there corresponds a finite equilibrium monomer concentration, $[M]_e$, that is thermodynamically determined. Furthermore, there is a critical temperature or ceiling temperature, T_c, above which no significant polymerization occurs. This temperature has been determined [41] by measuring the equilibrium yield of polymer at various temperatures, and making a small extrapolation of the curve obtained to zero conversion. From plots of this type, a value of $T_c = 85 \pm 2°C$ has been obtained [41, 116]. The equilibrium monomer concentration $[M]_e$, at any temperature, T, is related to the heat of polymerization, ΔH_p, and the entropy changes, ΔS_p, per mole according to the equation [117]

$$\ln[M]_e = \frac{1}{T}\frac{\Delta H_p}{R} - \frac{\Delta S_p}{R}$$

A number of determinations [118] of these parameters have been made and good agreement achieved; $\Delta H_p = -(4-5)$ kcal mole^{-1}, $\Delta S_p = -(17-20)$ cal deg^{-1} mole^{-1}.

In many instances the polymerization of tetrahydrofuran seems to proceed without termination, and "living" polymers, in which transfer reactions are also absent, have been prepared [40, 41]. A simple kinetic scheme suffices therefore to describe the situation, i.e.

Here k_i, k_p and k_d are the rate coefficients for initiation, propagation and depropagation respectively, and, for convenience only, the counter-ion is shown as accompanying the reactive cation. For such a scheme Vofsi and Tobolsky [119] have expressed the rate of polymerization, $-d[M]/dt$, as

$$-\frac{d[M]}{dt} = k_i[I]([M] - [M]_e) + k_p([I]_0 - [I])([M] - [M]_e)$$

where $[I]_0$ is the total initiator concentration, $[I]$ is that proportion of $[I]_0$ which after time, t, has still not been consumed in forming active centres $(= [I]_0 - [I])$ and $[M]$ is the monomer concentration at time, t. If initiation is rapid and virtually complete prior to propagation then $[I]$ will be negligibly small, and the number of active centres will correspond almost instantaneously to $[I]_0$. Providing there are no termination processes this number of reactive species will remain constant throughout the polymerization, and the following simple rate law applies:

$$-\frac{d[M]}{dt} = k_p [I]_0 ([M] - [M]_e)$$

from which integration yields:

$$k_p t = 2.303 \log [([M]_0 - [M]_e)/([M] - [M]_e)]/[I]_0$$

where $[M]_0$ is the initial monomer concentration. Experimentally it is found that this equation fits the kinetics of tetrahydrofuran polymerizations with most catalysts, where initiation is rapid. As with vinyl systems, it is the quantitative interpretation of $[I]_0$ which provides the difficulty. Where complex initiator/co-initiator systems are used $[I]_0$ must be determined by some indirect method; however, if a preformed salt is used, e.g. $R_3O^+X^-$, then $[I]_0$ may be taken simply as the concentration of initiator used.

Preformed triethyloxonium tetrafluoroborate, $Et_3O^+BF_4^-$, has been used by Vofsi and Tobolsky [119] as an initiator at $0°C$ in ethylene dichloride solvent. Using ^{14}C labelled catalyst, it was possible to show that the initial initiator concentration corresponded to the concentration of active centres formed. When only 15% of the monomer had been consumed, the initiator conversion was better than 90%. The rate at which the initiator disappeared followed the experimental law

$$-\frac{d[I]}{dt} = k_i [I] ([M] - [M]_e)$$

whilst the rate of monomer consumption was consistent with the simplified rate law mentioned earlier. The rate coefficient for propagation, k_p, was found to have a value 4.8×10^{-3} l $mole^{-1}$ sec^{-1} and that for initiation, k_i, about an order of magnitude larger.

More recently Saegusa et al. [120, 45] have developed a technique for the determination of the concentration of active centres, $[P_n^+]$, by terminating the polymerization with the sodium salt of phenol, Na^+OPh^-, and estimating the PhO-groups in the polymer spectrophotometrically. A closely related method has been used by Jaacks et al. [121] to estimate the concentration of tertiary oxonium ions in the polymerization of 1,3 dioxolan (Section 7.3). Saegusa has shown that the chromophore "polymer—OPh" has an absorption maximum, $\lambda_{max} = 272$ nm and an extinction coefficient, $\epsilon, = 1.96 \times 10^3$ l $mole^{-1}$ cm^{-1} in methylene chloride. Consumption of monomer as a function of time was followed by a gravimetric method and the results interpreted [122] according to the kinetic scheme.

For an active centre concentration $[P_n^+]$ at any time, t, the net loss of monomer is given by

$$-\frac{d[M]}{dt} = k_p[P_n^+][M] - k_d[P_n^+]$$

at equilibrium, $-d[M]/dt = 0$ and hence,

$$k_p[P_n^+][M]_e = k_d[P_n^+]$$

i.e.

$$k_d = k_p[M]_e$$

Thus substituting for k_d yields the expression, similar to that derived by Tobolsky,

$$-\frac{d[M]}{dt} = k_p[P_n^+]([M] - [M]_e)$$

and integrating this between times, t_1 and t_2 gives

$$\ln\left\{\frac{[M]_{t_1} - [M]_e}{[M]_{t_2} - [M]_e}\right\} = k_p \int_{t_1}^{t_2}[P_n^+]\,dt$$

This therefore, allows evaluation of k_p without making any assumptions concerning the rate of initiation or the absence of termination. The scheme does however assume the formation of long polymer chains so that the consumption of monomer in initiation can be ignored. Saegusa's results indicate a slight fall off in $[P_n^+]$ with time, from which a termination coefficient, k_t, was calculated using the expression for spontaneous termination, i.e.

$$-\frac{d[P_n^+]}{dt} = k_t[P_n^+]$$

Application of these techniques and interpretations to the polymerization of THF by $Et_3O^+BF_4^-$ in methylene chloride at $0°C$ has yielded a value for the rate coefficient for propagation, k_p, of $\sim 3.5 \times 10^{-3}$ l mole^{-1} sec^{-1}, which agrees closely with the results obtained by Tobolsky. The corresponding termination coefficient, k_t, is $\sim 2 \times 10^{-5}$ sec^{-1}. When $Et_3O^+AlCl_4^-$ was used as the initiator excessive termination was shown to occur, probably involving dissociation of the very unstable counter-ion, and the formation of a halogen end group in the polymer.

Simple Lewis acid cationic catalysts, e.g. BF_3, $SnCl_4$, AlR_3, PF_5, $SbCl_5$ etc., have been extensively used in the study of the polymerization of THF [118]. With these initiators the problem of estimating the concentration of propagating species becomes acute, and is not made any easier by the use of so-called promoters such as propene oxide,

$$CH_2-CHCH_3$$
$$\diagdown O \diagup$$

and epichlorohydrin

$$CH_2-CHCH_2Cl,$$
$$\diagdown O \diagup$$

used in catalytic proportions. Their function is to accelerate the initiation process, which with some of these catalysts is quite slow, although the precise mechanism by which they enhance the rate of initiation is not really understood. It seems likely that they react more rapidly with the initiating species than does THF itself, and that the oxonium ion so formed then initiates the polymerization of THF, e.g.

$$R^+X^- + O\overset{R_2\diagup R_1}{\triangleleft} \longrightarrow R-\overset{+}{O}\overset{R_2\diagup R_1}{\triangleleft} \quad X^-$$

$$THF \downarrow k_i$$

$$ROCR_1R_2CH_2-\overset{+}{O}\Box \quad X^-$$

Unfortunately this simple picture does not explain all the experimental observations. For example, propene oxide will accelerate the rate of polymerization of bulk THF even when added several hours after the start of polymerization [123] indicative of the presence of a dormant species in the reaction. Mechanistic details of these co-initiation phenomena have been discussed in more detail elsewhere [124, 50].

Some of the Lewis acid systems that have been studied [125—127] provide data for the propagation reaction in THF polymerization which do not depart too far from those derived from the better defined initiators, and which are amenable to kinetic analysis. Here again the "termination" method of Saegusa et al. [128, 129] has proved particularly invaluable, and some of the results from these studies are shown in Table 8 which is given at the end of this Section.

Triphenylmethyl, tropylium, xanthylium, acylium and diazonium salts have also been used successfully in the study of the polymerization of THF [50]. In fact the true equilibrium conversions were first demonstrated using p-chlorophenyl diazonium hexafluorophosphate [41, 115], and at the same time the polymerizations were shown to be "living". The lower equilibrium yields [50] observed when other counter-ions are employed, e.g. $SbCl_6^-$, appear to arise because of a termination mechanism associated with the anions. Kinetic studies [50] of the polymerization of bulk THF with $Ph_3C^+SbCl_6^-$ have established the apparent rate law

$$- \frac{d[M]}{dt} = k_p [Ph_3C^+SbCl_6^-] [M]$$

and at $50°C$, $k_p = 1.4 \times 10^{-2}$ l mole^{-1} sec^{-1}. The dissociation of this salt in methylene chloride has been studied [51] quantitatively and the ion pair dissociation constant, K_d, has a value 1.9×10^{-4} l mole^{-1} at $25°C$. Using the Denison and Ramsey equation [48] (see Section 5) and appropriate values for the dielectric constants of methylene chloride and THF, it can be shown that K_d in THF is at least one order of magnitude lower than the value in methylene chloride. Thus at the initiator concentrations used in the kinetic work ($\sim 10^{-3}$ M) the triphenylmethyl salt exists predominantly as ion pairs in THF. Providing the dissociation constant for the propagating species is no larger than that for the initiator salt, then the value of k_p reported refers to the reaction of oxonium ion pairs and represents the first attempt to elucidate independent reactivities for different ionic species in ring opening polymerizations. Recent work on $Et_3O^+X^-$ [34] and $Et_3S^+X^-$ salts [130], and also on living polytetrahydrofuran ion pairs [46], indicates that the assumption about the dissociation constant of the growing ion pairs is valid.

Ledwith and his co-workers [50] have also studied the mechanism of initiation by triphenylmethyl salts in some detail, by following the decay of the characteristic visible absorption maximum of the initiating cation. The kinetic equation which has been established for both methylene chloride and THF solvents is

$$- \frac{d[Ph_3C^+]}{dt} = k_2 [Ph_3C^+SbCl_6^-] [THF]$$

In the former medium k_2 has a value of $\sim 6 \times 10^{-3}$ l mole^{-1} sec^{-1} ([THF] = 0.4 M) while in pure THF the value is $\sim 21 \times 10^{-3}$ l mole^{-1} sec^{-1}, both being extrapolated values at $25°C$. The concentrations of salt employed were such that in methylene chloride largely free ions were likely to be present, while in THF ion pairs would predominate. Hence it seems that in these circumstances the free triphenylmethyl ion is less reactive than the corresponding ion pair. As we shall see shortly, this result is not too remarkable in the light of more recent data on the propagation reaction, though a direct comparison may be misleading since the mechanism by which each of these ionic species initiate polymerization of THF may in fact differ somewhat [50].

The most enlightening results reported to date on the polymerization of THF come from recent work by Sangster and Worsfold [46], who have succeeded in evaluating separate rate coefficients for growing free ions and ion pairs. Furthermore, the data they have obtained seems very reasonable when compared with results from other cyclic monomers (Section 7.2). THF was polymerized in methylene chloride at $-0.5°C$ using $Et_3O^+BF_4^-$ as initiator and initial rates of polymerization (<5% conversion) obtained dilatometrically. Within this range little or no destruction of active centres was observed and the concentration of initiator used was estimated by the method of Saegusa et al. [45]. In addition, conductance measurements on the initiator salt and low molecular weight living polymer in methylene chloride yielded values of ion pair dissociation constants. $Bu_4N^+BF_4^-$ was used in some of the kinetic experiments to disturb the free ion—ion pair equilibria by a common ion effect. The rate of polymerization is given by Vofsi and Tobolsky's relationship [119]

$$-\frac{d[M]}{dt} = k_p[P_n^+]([M] - [M]_e)$$

where k_p is the overall rate coefficient for propagation, and $[P_n^+]$ is the total concentration of active centres. This was assumed to equal the initial concentration of catalyst salt used, an assumption shown to be valid providing only low conversions to polymer are considered. The rate coefficient, k_p, was evaluated from the gradient of the tangent to the time/conversion curves at time zero, and, where both free ions and ion pairs contribute in propagation, k_p can be shown [39] to take the form

$$k_{p\ overall} = \alpha k_p^+ + (1-\alpha)k_p^\pm$$

or

$$k_{p\ overall} = k_p^\pm + (k_p^+ - k_p^\pm)K_d^{1/2}[P_n^+]^{-1/2}$$

where $k_p^\pm$ = rate coefficient for ion pair growth

k_p^+ = rate coefficient for free ion growth

K_d = dissociation constant of the growing ion pair

α = degree of dissociation of the growing ion pair

For data obtained in the presence of a common ion salt, e.g. $Bu_4N^+BF_4^-$, this relationship becomes

$$k_{p \text{ overall}} = k_p^{\pm} + (k_p^+ - k_p^{\pm})K_d/[BF_4^-]$$

where $[BF_4^-]$ is calculated from the ion pair dissociation constant of the added salt. A plot of the observed propagation coefficient, $k_{p \text{ overall}}$, versus $[P_n^+]^{-1/2}$, from normal kinetic runs, yields a straight line intercept, $k_p^{\pm}$, and gradient $(k_p^+ - k_p^{\pm})K_d^{1/2}$, and, from the common-ion experiments, a plot of $k_{p \text{ overall}}$ versus $[BF_4^-]^{-1}$ yields a similar straight line, now depressed slightly, with intercept $k_p^{\pm}$ and gradient, $(k_p^+ - k_p^{\pm})K_d$. From the graphical treatments independent data for $k_p^{\pm}$, k_p^+ and K_d in methylene chloride at $-0.5°C$ were obtained, and are, respectively 1.40×10^{-3} l mole^{-1} sec^{-1}, 1.0×10^{-2} l mole^{-1} sec^{-1} and 3.7×10^{-6} l mole^{-1}. The value of K_d determined in this way was in good agreement with data calculated directly from conductance measurements on the initiator salt, 5.4×10^{-6} l mole^{-1}, and on the low molecular weight living polymeric oxonium ion salt, 4.4×10^{-6} l mole^{-1}. Such agreement lends considerable credibility to the suggestion that two different ionic species are active in this system. Interestingly the reactivity of the free polymeric oxonium ion appears to be only about a factor of 7 greater than that of the ion pair, in contrast to the situation prevailing in cationic vinyl polymerization, a point which will be developed later. A comparison of data for the polymerization of THF is given in Table 8.

7.2 POLYMERIZATION OF THIETANS

The study of the polymerization of thietan and its substituted derivatives has been undertaken relatively recently [131—134]. Detailed kinetic experiments have been reported by Goethals et al. [33, 130, 135—137] and, as we shall see, the results parallel closely those found in the polymerization of THF. Most of the work has been carried out using 3,3-dimethylthietan,

in methylene chloride solution with initiation by $Et_3O^+BF_4^-$. Early results [33] indicated a rapid initiation process thought to be the generation of a cyclic tertiary sulphonium ion, which subsequently propagates by ring opening, viz.

TABLE 8

Rate coefficients for the propagation reaction in THF polymerizations

Initiator	Co-initiator[a]	Solvent	Temperature (°C)	$k_p \times 10^3$ (l mole^{-1} sec^{-1})	Ref.
$BF_3 \cdot Et_2O$	ECH	Et_2O	0	3.7	125
$BF_3 \cdot Et_2O$	ECH	Et_2O	20	16.6	125
$AlEt_3 \cdot H_2O^b$	ECH	Bulk	0	8—12	127
$AlEt_3$	ECH	Bulk	0	5.6	127
BF_3	ECH	Bulk	0	4.5	128
BF_3	ECH	CH_2Cl_2	0	4.1	128
$SnCl_4$	ECH	Bulk	0	6	128
$AlEtCl_2$	ECH	Bulk	0	7.8	128
$Et_3O^+BF_4^-$	none	$CH_2CH_2Cl_2$	0	4.8	119
$Et_3O^+BF_4^-$	none	CH_2Cl_2	0	3.5	122
$Ph_3C^+SbCl_6^-$	none	Bulk	50	14^c	50
$Et_3O^+BF_4^-$	none	CH_2Cl_2	−0.5	$+10.0^d$	46
$Et_3O^+BF_4^-$	none	CH_2Cl_2	−0.5	$\pm 1.4^d$	46

a ECH = Epichlorohydrin and [ECH] $\simeq$ [Initiator].
b ratio $AlEt_3/H_2O$ in range 10/1 to 1/1.
c probably an ion pair value.
d separate free ion, +, and ion pair, ±, data.

$$Et_3O^+BF_4^- + (CH_3)_2\!-\!\!\diamondsuit\!\!-\!S \xrightarrow[\text{fast}]{k_i} (CH_3)_2\!-\!\!\diamondsuit\!\!-\!\overset{+}{S}\!-\!Et + Et_2O$$
$$BF_4^-$$

$$k_p \Big\downarrow (CH_3)_2\!-\!\!\diamondsuit\!\!-\!S$$

$$EtS\!-\!CH_2\!-\!\underset{\underset{CH_3}{|}}{\overset{\overset{CH_3}{|}}{C}}\!-\!CH_2\!-\!\overset{+}{S}\!\!\diamondsuit\!(CH_3)_2$$
$$BF_4^-$$

. Unlike THF this system has an inherent termination reaction since the polymer yields are always less than 100%. When triethyl sulphonium tetrafluoroborate, $Et_3S^+BF_4^-$, was found to be inactive as an initiator it was concluded that termination involved reaction of a backbone sulphur atom with an active centre, to form a stable tertiary sulphonium ion, viz.

$$\sim\!\!\overset{+}{S}\!\!\diamondsuit\!\!-\!\underset{\underset{CH_2}{|}}{\overset{\overset{CH_2}{|}}{}} + \underset{\underset{CH_2}{|}}{\overset{\overset{CH_2}{|}}{S}} \xrightarrow{k_t} \sim\!\!SCH_2\underset{\underset{CH_3}{|}}{\overset{\overset{CH_3}{|}}{C}}\!-\!CH_2\!-\!\underset{\underset{CH_2}{|}}{\overset{\overset{CH_2}{|}}{\overset{+}{S}}} \quad BF_4^-$$

Molecular weight data was also in agreement with this proposed termination mechanism.

It is known that the basicity of cyclic sulphides is generally less than that of linear sulphides [133, 138] depending upon substitution, and confirmation of these suggestions was achieved when 1-ethyl-3,3-dimethyl trimethylene sulphonium tetrafluoroborate,

$$Et\!-\!\overset{+}{S}\!\!\underset{CH_2}{\overset{CH_2}{<}}\!\!C\!\!\underset{CH_3}{\overset{CH_3}{<}}$$
$$BF_4^-$$

was shown [136] to function as an effective initiator. Kinetic data obtained using this salt were identical to those obtained for $Et_3O^+BF_4^-$. Goethals and Drijvers [33] have proposed the following kinetic analysis for this non-stationary state polymerization. The rate of polymerization, R_p, is given by

$$R_p = -\frac{d[M]}{dt} = k_p[P_n^+][M]$$

and the rate of termination, R_t, by

$$R_t = -\frac{d[P_n^+]}{dt} = k_t[P_n^+]\underbrace{([M]_0 - [M])}$$

concentration of monomeric
units in polymer chains

or

$$-\frac{d[P_n^+]}{[P_n^+]} = k_t([M]_0 - [M])\,dt$$

from which integration yields

$$-\ln[P_n^+] = k_t \int ([M]_0 - [M])\,dt + K$$

When $t = 0$, $[M] = [M]_0$ and hence, $K = -\ln[P_n^+]_0$. However, if initiation is instantaneous then $[P_n^+] = [I]_0$, the initial concentration of initiator salt. Thus the equation above becomes

$$[P_n^+] = [I]_0 \exp\left[-k_t \int ([M]_0 - [M])\,dt\right]$$

or

$$[P_n^+] = [I]_0 \exp(-k_t A)$$

where A is the area under the curve of a plot of $([M_0] - [M])/t$ up to time t. Combining this expression with that for the rate of polymerization gives the final relationship

$$R_p = k_p[M][I]_0 \exp(-k_t A)$$

or

$$\ln\left\{\frac{R_p}{[M][I]_0}\right\} = \ln k_p - k_t A$$

Polymerizations were followed by a nuclear magnetic resonance technique and R_p was obtained from the slope of the tangent at various points on the time/conversion curves. A plot of $\ln(R_p/[M][I]_0)$ against A yields a straight line of slope, k_t, and ordinate intercept, $\ln k_p$. A similar analysis has been used by Penczek and Kubisa [139, 140] in the polymerization of 3,3-bis(chloromethyl)oxetan (Section 7.3) and a summary of the results obtained by Goethals and Drijvers is shown in Table 9.

From these results entropies and enthalpies of activation for propagation and termination were deduced, i.e. $\Delta H_p^{\ddagger} = 12.5$ kcal mole^{-1}, $\Delta S_p^{\ddagger} = -26$ cal deg^{-1} mole^{-1}, $\Delta H_t^{\ddagger} = 7.4$ kcal mole^{-1}, $\Delta S_t^{\ddagger} = -49$ cal deg^{-1} mole^{-1}. Since the activation enthalpy is greater for propagation than for

TABLE 9

Polymerization of 3,3-dimethylthietan in CH_2Cl_2 initiated by $Et_3O^+BF_4^-([M]_0 = 2.0$ M, $[Et_3O^+BF_4^-] = 0.02$ M)

Temp. ($^\circ$C)	$10^3 k_p$ (l mole^{-1} sec^{-1})	$10^4 k_t$ (l mole^{-1} sec^{-1})
0	1.15	1.33
10	3.17	2.67
20	6.50	3.33
30	12.5	6.17

termination it might be expected that $k_p < k_t$. However, this is not the case. The high negative entropy of activation for the termination reaction has the effect of making $k_p > k_t$ over the whole temperature range. The relative values of the two activation enthalpies are not surprising, however, in view of the difference in basicities between the sulphur atom in the cyclic monomer and the linear sulphide function of the polymer segment. Subsequent studies [136, 130] have shown, as we shall now see, that this data is composite in nature, and includes both free ion and ion pair terms. Since the proportion of these species can vary as the temperature is changed, and their reactivities are different (though not excessively so), the absolute values of enthalpy and entropy parameters must be regarded only as general guides.

As mentioned already when a cyclic tertiary sulphonium ion salt [136] was used as an initiator then almost identical results were obtained. However, it was observed that the value of k_p fell slightly as the initial catalyst concentration was increased. In fact careful analysis of the data for initiation by $Et_3O^+BF_4^-$ indicates a similar trend. This variation of k_p is an indication that more than one ionic species contributes to propagation, and Drijvers and Goethals [130] have analysed in this way data from polymerizations using ethyltetramethylene sulphonium tetrafluoroborate,

A simplified procedure for obtaining the overall value of k_p was adopted. Since at the beginning of polymerization the concentration of linear sulphide residues (i.e. polymer) is negligible, termination can be neglected and the initial rate of polymerization, R_0, can be expressed as

$$R_0 = k[P_n^+]_0[M]_0$$

which, making the same assumptions as before, becomes

$$R_0 = k_p [I]_0 [M]_0$$

As in the polymerization of THF, k_p takes the form

$$k_p = \alpha k_p^+ + (1 - \alpha)k_p^\pm$$

where k_p^+ and $k_p^\pm$, as before, are the separate rate coefficients for free ion and ion/pair growth and α is the degree of dissociation. Rewriting this as,

$$k_p = k_p^\pm + (k_p^+ - k_p^\pm)\alpha$$

and substituting in the rate equation gives the final form

$$R_0/[I]_0 [M]_0 = k_p^\pm + (k_p^+ - k_p^\pm)\alpha$$

A plot of $R_0/[I]_0 [M]_0$ versus α yields a straight line of intercept $k_p^\pm$ and gradient $(k_p^+ - k_p^\pm)$. Experimentally R_0 was obtained, as previously, from conversion/time data using an NMR technique and the degree of dissociation, α, obtained by assuming that the growing ion pair had a dissociation equilibrium similar to that of the initiating ion pair. The dissociation constant for the latter was obtained from separate conductance experiments on the initiator salt dissolved in the appropriate solvent. In the polymerizations a large excess of sulphide functions (monomer and polymer) and diethyl ether, arising because the initiator is generated in situ from $Et_3O^+BF_4^-$ and tetrahydrothiophene, are present as well as the solvent. Conductance measurements were also made, therefore, in the presence of these species, but no significant variation was found. Studies were carried out with three solvents, methylene chloride, nitrobenzene and benzene. In the latter solvent no ionic conductance was detectable and so the kinetic analysis above could not be applied. With nitrobenzene as solvent, $(k_p^+ - k_p^\pm)$ is practically zero and the appropriate plot yielded a straight line parallel to the abscissa. In order to get a better evaluation, therefore, a modified plot $R_0/[I]_0 [M]_0\alpha$ versus $k_p^+ + k_p^\pm(1 - \alpha)/\alpha$ was made and the results are shown in Tables 10 and 11.

TABLE 10

Dissociation constants (K_d) of model compounds in various solvents at 20°C

Sulphonium salt	K_d (mole l^{-1})	
	CH_2Cl_2	$C_6H_5NO_2$
$(C_2H_5)_3S^+BF_4^-$	3.6×10^{-5}	1.35×10^{-2}
CH_2-CH_2 $\quad\quad\quad S^+-C_2H_5 BF_4^-$ CH_2-CH_2	5.6×10^{-5}	1.65×10^{-2}
Average values used in analysis	4.6×10^{-5}	1.5×10^{-2}

The results were checked by use of a common-ion procedure once again similar to that used by Sangster and Worsfold [46] in the polymerization of THF, and by workers in the field of anionic polymerizations [39]. In the case of methylene chloride as solvent, addition of a common salt depressed the overall value of k_p indicating $k_p^{\pm} < k_p^{+}$, whereas in the case of nitrobenzene no effect was observed, so that $k_p^{\pm} \simeq k_p^{+}$. Appropriate analysis of rate data obtained in the presence of common ion yielded values for k_p^{+} and $k_p^{\pm}$ in both solvents, very similar to those already calculated (Table 11).

TABLE 11
Rate coefficients for propagation with 3,3'-dimethylthietan at 20°C

Solvent	$10^3 \, k_p^{+}$ (l mole^{-1} sec^{-1})	$10^3 \, k_p^{\pm}$ (l mole^{-1} sec^{-1})
CH_2Cl_2	66.6	1.66
$C_6H_5NO_2$	3.16	3.32

Once again it would seem that the difference in reactivity between propagating free ions and ion pairs in the polymerization of cyclic monomers is considerably smaller (or zero) than the differences in ionic vinyl propagation, a point discussed further in Section 7.4. The lower reactivity of the free cyclic sulphonium ion in nitrobenzene compared to that in methylene chloride is consistent with the general concepts of solvent effects in ionic reactions (Section 6.5). Propagation requires charge dispersion, hence the higher the polarity of the medium, the greater the activation barrier produced, and the smaller the rate coefficient. Alternatively, in nitrobenzene the free active centre can be regarded as more solvated than in methylene chloride with a consequent reduction in charge density and reactivity. Goethals and Drijvers have suggested that the lower rates of polymerization obtained in benzene can be attributed to a very low concentration of free ions, undetectable by normal conductance techniques. It would seem more likely, however, that, in a solvent of such a low dielectric constant, a more plausible explanation would be that the only significant species present are ion pairs. With this assumption the analysis gives a value of $k_p^{\pm}$ in benzene of $\sim 1.5 \times 10^{-3}$ l mole^{-1} sec^{-1} very similar to the other two ion pair values.

Most recently a rather similar study has been carried out on the polymerization of propene sulphide [137]. This system, however, is kinetically more complicated because of the formation of a relatively stable twelve membered ring sulphonium ion, which can apparently slowly re-initiate a second stage polymerization. A comparative study of various substituted thietans has been reported and confirms that the rates and the equilibria of reactions between sulphides and sulphonium salts depend upon the difference in basicity of the reacting and product sulphides.

7.3 POLYMERIZATION OF OXETANS, CYCLIC FORMALS AND FORMALDEHYDE

Polymerization of 3,3-bis(chloromethyl)oxetan,

by $(\text{i-}C_4H_9)_3\text{Al/H}_2\text{O}$ in chlorobenzene has been studied in detail over the temperature range 55—95°C by Penczek et al. [139—141] and the rate of consumption of monomer estimated by a vacuum dilatometric technique. The ring strain in oxetans is much higher than in THF, and as a result the polymer—monomer equilibrium is much less pronounced, and high yields of polymer are readily obtained. Conversion/time data has an overall "S" shaped appearance indicating a non-stationary state polymerization with slow initiation, fast propagation and slow-degradative chain transfer to polymer (effectively termination). This latter process has been postulated before in the polymerization of such cyclic ethers. For example Rose [142] has proposed intramolecular chain transfer to polymer to account for the formation of low molecular weight cyclic oligomers in the polymerization of oxetan and 3,3'-dimethyl oxetan. Such a reaction is equivalent to the termination process suggested more recently in the polymerization of propene sulphide [137].

Penczek's kinetic scheme is

$$\text{AlR}_3/\text{H}_2\text{O} + \text{M} \xrightarrow{k_i} \text{P}_1^+$$

$$\text{P}_1^+ + \text{M} \xrightarrow{k_p} \text{P}_n^+$$

$$\text{P}_i^+ + \text{P}_j \xrightarrow{k_t} \text{P}_i\text{P}_j \ (\text{inactive})$$

If $k_p \gg k_i$, i.e. if initiation is virtually instantaneous, then this scheme is identical to that used by Goethals and Drijvers [33] for the polymerization of thietans (Section 7.2), and a similar analysis of experimental data is possible. However, if initiation is slow, as it is in this case, and compares with the rate of propagation, then the rate of change of the concentration of active centres takes a modified form, i.e.

$$-\frac{\text{d}[\text{P}_n^+]}{\text{d}t} = k_i[\text{Catalyst}][\text{M}] + k_t[\text{P}_n^+]([\text{M}]_0 - [\text{M}])$$

which now includes a term accounting for the generation of active centres over a finite period of time. Manipulation of this equation to yield $[\text{P}_n^+]$, with substitution into the usual equation defining the rate of polymerization

$$-\frac{d[M]}{dt} = k_p [P_n^+] [M]$$

produces the complex relationship [139]

$$-\frac{d \ln[M]}{dt} = k_p [I]_0 \left[1 - \exp\left(-k_i \int_{[M]_0}^{[M]_t} [M]_{dt} \right) \right]$$

$$-k_t \left\{ [M]_0 \ln \left\{ \frac{[M]_0}{[M]} \right\} - ([M]_0 - [M]) \right\}$$

Once again it is assumed that polymer chains are long so that the consumption of monomer in the initiation step can be neglected. As in the polymerization of thietans an estimation of $[I]_0$ is required in this analysis, and with the particular catalyst system in question (AlR_3/H_2O) this is considerably more difficult. Careful scrutiny of the effect of varying the absolute concentrations and relative proportions of these components, has shown that, on average, ten molecules of water are required for the formation of one active centre at the beginning of the polymerization, i.e., $[I]_0 \simeq 0.1 [H_2O]_{total}$. Furthermore at the outset of polymerization, when the amount of polymer is negligible, termination can be neglected, and hence the kinetic equation collapses to the more manageable form

$$\log \left[\frac{d \ln[M]}{[I]_0 dt} + k_p \right] = \log k_p - \frac{k_i}{2.303} \int_{[M]_0}^{[M]} [M] \, dt$$

A plot of the left hand logarithmic function versus $\int_{[M]_0}^{[M]} [M] \, dt$, from appropriate conversion/time data, should therefore be a straight line if the correct value for k_p is chosen. The latter is obtained by a trial and error process, and k_i is calculated from the gradient of the "best" straight line. At 70°C the mean value of k_p was found to be 8.5 l mole^{-1} sec^{-1} and $k_1 = 1.6 \times 10^{-3}$ l mole^{-1} sec^{-1}, with activation energies (using similar data at 85°C) 6.0 kcal mole^{-1} and 8.5 kcal mole^{-1} for propagation and initiation, respectively. This method of computation was found to be valid only up to ~20% conversion, and for temperatures above 60°C. The dielectric constant of chlorobenzene is very low, and so it seems highly improbable that freely dissociated ions occur to any great extent in this system. The data derived therefore, reflect the reactivity of propagating ion pairs, and, even taking account of the high temperature employed, this seems to be high compared with the results of other ring opening polymerizations. Data from a parallel study in a solvent of high dielectric constant would be extremely interesting.

Probably the first reference to the polymerization of cyclic acetals (formals) was by Hill and Carothers [143]. Since then a number of groups of research workers have shown interest in these monomers, particularly in 1,3-dioxolan(I). 1,3-dioxepan(II) has been studied to a much lesser extent [144, 145], while 1,3 dioxan(III) does not appear to polymerize, but merely forms crystalline dimer and trimer.

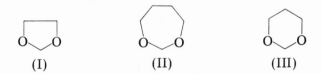

(I)   (II)   (III)

1,3-dioxolan is readily polymerized by conventional cationic initiators [151], e.g. $HClO_4$ [146] and $SbCl_5$ [149], and also by stable oxonium ion [147, 148] and carbonium ion salts [149, 150]. Both 1,3 dioxolan and 1,3 dioxepan exhibit ceiling temperature phenomena similar to that observed in the polymerization of THF. Plesch and Westerman [146] report a ceiling temperature of +1°C (1 M solution) and an entropy and enthalpy of polymerization in methylene chloride solution of -18.6 cal deg^{-1} $mole^{-1}$ and -5.2 kcal $mole^{-1}$, respectively, for 1,3 dioxolan. The corresponding data for dioxepan [144] is (+27°C, 1 M solution) -11.5 cal deg^{-1} $mole^{-1}$ and -3.6 kcal $mole^{-1}$, respectively, again in methylene chloride solution.

Under conditions where polymer does form it is still by no means clear what is the polymerization mechanism [153], nor is it entirely understood whether cyclic, as opposed to linear, macromolecules are formed. Yamashita et al. [148] have proposed that the active centre is a linear carboxonium ion, i.e.

$$\sim\!\!OCH_2CH_2OCH_2OCH_2CH_2\overset{+}{O}CH_2\,X^-$$

$$\updownarrow$$

$$\sim\!\!OCH_2CH_2OCH_2OCH_2CH_2\overset{+}{O}\!=\!CH_2\,X^-$$

while monomer containing species [146, 152] have also been suggested, e.g.,

In fact, Plesch and Westerman [144, 146] have gone further and argued that under dry conditions, propagation may involve successive ring expansion to produce cyclic polymers, viz.

Furthermore, in order to account for the fast transfer to polymer which can occur within this system, Penczek [154] has suggested yet another possibility for the structure of the active centre, a tertiary cyclic or linear oxonium ion, i.e.

More recently Jaacks et al. [155] have established that a "backbiting" mechanism seems to be the most probable interpretation of ring formation in the polymer, and hence the idea of a macrocyclic active centre appears unlikely. Since great confusion still remains in this area it is not surprising that unambiguous kinetic data have yet to be reported. It has been established [156, 147] that initiation using $Et_3O^+BF_4^-$ is not the simple and quantitative alkylation proposed by Yamashita et al. [148], though initiation by the corresponding SbF_6^- salt does seem to be more straightforward. Side reactions, involving hydride ion abstraction from the monomer, occur with the formation of a dioxolenium salt

$$Et_3O^+BF_4^- + \underset{\underset{H_2}{\overset{|}{C}}}{\overset{\ulcorner\qquad\urcorner}{O\diagdown\diagup O}} \longrightarrow Et_2O + EtH \cdot \underset{\underset{H}{\overset{|}{C}}\, BF_4^-}{\overset{\ulcorner\qquad\urcorner}{O\diagdown\overset{+}{}\diagup O}}$$

which if capable of initiating polymerization does so only with difficulty. Similar reactions when triphenylmethyl salts are used have been reported [149], and in both of these systems the concentration of active centres produced, cannot be equated directly to the concentration of initiator salt. Data for the rate coefficient for propagation, k_p, reported by Yamashita et al. [148] depends directly on this assumption and hence must be regarded as erroneous. These workers have applied the kinetic expression for equilibrium polymerization to their system, i.e.,

$$-\frac{d[M]}{dt} = k_p [P_n^+]_0 ([M] - [M]_e)$$

and report a value for k_p of 4.3×10^{-2} l mole^{-1} sec^{-1} at 30°C in methylene chloride. Since the value of $[P_n^+]$ used was in fact $[I]_0$, the initiator concentration, and independent evidence suggests $[P_n^+]$ is much less than $[I]_0$, the real value for k_p must be somewhat larger than that quoted above. Similarly the work of Berman et al. [157] requires re-interpretation.

Plesch and Westerman [144, 146] have used vacuum dilatometry to study the kinetics of polymerization of 1,3 dioxolan in methylene chloride initiated by $Et_3O^+BF_4^-$ at 25°C. In agreement with Yamashita's results, this system showed a considerable induction period (~2 hours), and eventually gave a conversion/time relationship with considerable "S" shaped character, i.e., a slow initiation reaction giving way to a faster propagation process. Depolymerization of the polymer by heating to 40°C and cooling back to 25°C, enabled the reaction to be followed many times. No induction period other than the initial one was observed. Conductance experiments showed that the number of ions present initially falls to a minimum during polymerization, and hence the ions present initially are not responsible for the initiation of polymerization. On the contrary, the initiating species are formed only very slowly and correspond merely to a fraction of the salt originally introduced. From the linear portion of the conversion curves a first order rate coefficient, k_1, was calculated, and found to be a function of the initial concentration of catalyst with the ratio $k_1/[I]_0 \approx 0.17$ l mole^{-1} sec^{-1} at 25°C. When perchloric acid was used as initiator no induction period was observed and $k_1/[I]_0 \approx 21$ l mole^{-1} sec^{-1} at 25°C. With $Et_3O^+SbF_6^-$ also no induction period was observed and the ratio $k_1/[I]_0$ is of the same order as that for perchloric acid. More recently Plesch [44] has tentatively estimated a value for k_p of 10 l mole^{-1} sec^{-1} and 3×10^3 l mole^{-1} sec^{-1} for 1,3 dioxolan and 1,3 dioxepan, respectively, in methylene chloride at 0°C using perchloric acid as catalyst. Bearing in mind the dielectric constant of the solvent, and the concentration of initiator used, he has suggested that this data is a composite representation of the reactivities of free ions and ion pairs in these systems.

Formaldehyde is yet another monomer which readily polymerizes by a cationic mechanism, and as pointed out earlier it can be regarded as an oxygen heterocycle. The polymer obtained is polyoxymethylene and the system exhibits the usual characteristic equilibrium conversion. By far the most important kinetic studies have been carried out by Jaacks et al. [121, 158, 159], who have developed a novel technique [121] for the estimation of the concentration of active centres in their polymerizations. Reactions are terminated by the use of an excess of a sodium alkoxide, the polymer obtained is purified, then subjected to acid hydrolysis [13] to yield the corresponding free alcohol which is then estimated by a quantitative gas chromatographic technique. A wide variety of initiator systems and reaction solvents has been investigated [158], but the most detailed work has been carried out on stannic chloride and perchloric acid at −78°C in diethyl ether solvent [159]. Initiation is generally found to be rapid and essentially complete prior to propagation. At these low temperatures significant yields of polymer are obtained, and in fact propagation appears to be inhibited by diffusion of monomer to the active centres occluded in the precipitated polymer. No indication of genuine kinetic termination has been found. Conversion data have been analysed using the simple expression

$$-\frac{d[M]}{dt} = k_p [P_n^+] [M]$$

which applies when initiation is fast, and true termination absent. $[P_n^+]$ was established as already outlined. As polymerization proceeds the effective concentration of active centres decreases, presumably because of inaccessibility in the heterogeneous network. An initial value of the rate coefficient for propagation, k_p, has been estimated as ~0.50 l mole^{-1} sec^{-1} at −78°C in diethyl ether using stannic chloride as initiator, while under the same conditions with perchloric acid a value of ~0.66 l mole^{-1} sec^{-1} was calculated. Polymerization rates appear to be of the same order in toluene, though the rate increases in the sequence tetrahydrofuran > nitroethylene > methylene chloride > acetone indicating some increase in k_p as solvent polarity rises, e.g., the value in methylene chloride is 2.0 l mole^{-1} sec^{-1}. A tentative explanation might be that in the solvents of lower dielectric constant predominantly ion pairs are responsible for polymerization, whereas in the more polar media a larger contribution from free ions is observed. In the absence of dissociation data a more detailed interpretation cannot be made, it does seem, however, that any variation in reactivity between different ionic species may once again be less than in the case of vinyl propagation.

The closely related monomer, *sym*-trioxan, is a crystalline trimer of formaldehyde, and again polymerizable by cationic means [151, 160]. However, considerable induction periods are observed in its solution polymerization, and the interpretation of kinetic data is additionally complicated by the formation of equilibrium amounts of monomeric formaldehyde as well as polyoxymethylene.

7.4 SUMMARY OF PROPAGATION DATA AND COMPARISON WITH VINYL MONOMERS

A summary of all the important rate data on the polymerization of tetrahydrofuran has already been given in Table 8, including data [46] for the separate reactivities of growing free ions and ion pairs. The remaining data on cyclic monomers are summarized in Table 12.

Examination of the data in these tables shows that where values of k_p referring unequivocally to free ion growth are available, i.e. k_p^+ then the reactivity is many orders of magnitude less than that for free ions propagating in cationic (or anionic) vinyl systems, e.g. k_p^+ for THF in CH_2Cl_2 at $0°C$ is 1.0×10^{-2} l mole^{-1} sec^{-1} and for dimethylthietan in CH_2Cl_2 at $20°C$ is 6.7×10^{-2} l mole^{-1} sec^{-1}, whereas k_p^+ for N-vinyl-carbazole in CH_2Cl_2 at $0°C$ is $\sim 3 \times 10^5$ l mole^{-1} sec^{-1} and for isobutyl vinyl ether in CH_2Cl_2 at $0°C$ $\sim 4 \times 10^3$ l mole^{-1} sec^{-1}. The only exception to this appears to be the case of dioxepan where the composite rate coefficient may be as high as 3×10^3 l mole^{-1} sec^{-1} in CH_2Cl_2, with the possibility that k_p^+ may be even higher. The other outstanding result is that ion pair reactivities on the whole are not significantly lower than free ion values, and may even be equal to or possibly larger than these under special circumstances. This certainly seems to be the case in the initiation reaction of the polymerization of THF by triphenylmethyl salts [50].

It is not really surprising to find that the freely dissociated active centres in cyclic monomer polymerizations are less reactive than free cations (or anions) derived from vinyl monomers. The former ions have positive charge residing substantially on a relatively basic atom, e.g. O or S. Such ions are considerably more stable than the carbonium ions in vinyl systems, even taking account of stabilization of the latter by substituents. In addition propagation in ring opening polymerization involves essentially attack by monomer on the growing chain at a carbon atom in a position α to the positively charged atom, e.g.

References pp. 127—131

TABLE 12
Kinetic data for cyclic monomers

Monomer	Initiator	Solvent	Temp. (°C)	k_p(l mole^{-1} sec^{-1})[b]	Ref.
3,3'-dimethylthietan	$Et_3O^+BF_4^-$	CH_2Cl_2	0	1.15×10^{-3}	33
3,3'-dimethylthietan	$Et_3O^+BF_4^-$	CH_2Cl_2	10	3.17×10^{-3}	33
3,3'-dimethylthietan	$Et_3O^+BF_4^-$	CH_2Cl_2	20	6.50×10^{-3}	33
3,3'-dimethylthietan	$Et_3O^+BF_4^-$	CH_2Cl_2	30	12.5×10^{-3}	33
3,3'-dimethylthietan	$(CH_2)_4^+SEtBF_4^-$	CH_2Cl_2	20	$66.6 \times 10^{-3} +, 1.66 \times 10^{-3} \pm$	130
3,3'-dimethylthietan	$(CH_2)_4^+SEtBF_4^-$	$C_6H_5NO_2$	20	$3.16 \times 10^{-3} +, 3.32 \times 10^{-3} \pm$	130
3,3'-dimethylthietan	$(CH_2)_4^+SEtBF_4^-$	C_6H_6	20	$1.5 \times 10^{-3} \pm$	130
BCMO[a]	$(C_4H_9)Al/H_2O$	C_6H_5Cl	70	$8.5 \pm$	139
Dioxolan	$HClO_4$	CH_2Cl_2	0	10	44
Dioxepan	$HClO_4$	CH_2Cl_2	0	3×10^3	44
Dioxepan	$SnCl_4$	Et_2O	-78	$0.50 \pm$	159
Formaldehyde	$HClO_4$	Et_2O	-78	$0.66 \pm$	159
Formaldehyde	$SnCl_4$	CH_2Cl_2	-78	2.00	158

[a] BCMO = 3,3-bis (chloromethyl)oxetan.
[b] +, free ion data; ±, ion pair data; others composite or unknown.

whereas in vinyl propagation the attack takes place directly on the positively charged atom, e.g.

It would seem that in the case of cyclic formals 3,3-bis(chloromethyl) oxetan and formaldehyde the value of k_p^+ is likely to lie somewhere between the extremes represented by say THF and N-vinylcarbazole. This is reasonable for the oxetan and formaldehyde where the driving force for polymerization is likely to be higher than for THF and hence the reactivity of derived cations larger. In the oxetan the ring strain is higher than in THF, and in formaldehyde there is formal unsaturation. As far as 1,3-dioxolan is concerned, the ring strain is similar to that in THF but reactivity in propagation is likely to be higher because of the more unstable electronic structures of formals as compared with ethers, i.e.

$$R-O-CH_2-OR' \text{ versus } R-O-R'$$

The question of why the growing free ion from dioxepan should be even more reactive is difficult to answer, and must await the results of further studies.

The similarity in the reactivities of free ions and corresponding ion pairs derived from the same cyclic monomer is more intriguing. Whereas ion pair reactivities are about 10^3 times smaller than corresponding free ion values in the anionic polymerization of vinyl monomers [39], and probably of the same relative proportions in cationic systems, the difference in cationic ring opening polymerizations is considerably less. For polymerization of THF in methylene chloride the factor is only ~ 7, and for polymerization of 3,3-dimethylthietan, ~ 40 in methylene chloride and ~ 1 in nitrobenzene. Because the overall reactivity in cyclic monomer reactions is lower than for olefinic polymerizations, it might be expected that difference between free ion and ion pair reactivities, within one system, would also be less. However, this does not seem to be the whole answer. Plesch [44] has pointed out that in the polymerization of cyclic ethers and thietans (and presumably, therefore, other cyclic monomers) the propagating cation is likely to be highly solvated by monomer molecules, e.g.

Although the dielectric constant of these monomers is smaller than that for say methylene chloride, the ability to solvate cations is high. The ion pair dissociation constant of $Et_3O^+PF_6^-$ in methylene chloride at $0°C$ is reported to increase by a factor of ~2 when traces of diethyl ether are added in spite of it being a less polar solvent [34]. This has been interpreted in terms of specific solvation of the cation by ether molecules with consequent reduction in the effective charge density of the positive species and hence in the coulombic force favouring ion pairing. Comparable solvation by monomer molecules of the active centres in ring opening polymerization would, therefore, reduce the charge density of the growing cation and hence contribute to the lower reactivity of these free cations relative to those derived from vinyl monomers. It can be argued also that the reduction in charge density, and hence reactivity, brought about by the pairing of the solvated cation with an anion may be relatively insignificant, when compared to the effect of ion pairing on the reactivity of the free ions in vinyl propagation. It must be noted, however, that in the anionic polymerization of ethylene oxide initiated by potassium alkoxides in hexamethylphosphoramide [161] the difference in reactivity between the growing free ion and ion pair was also found to be small, though complications in the kinetic behaviour make the data somewhat less reliable. Here the active centres are anions (paired or free) and specific solvation by the monomer is almost certainly absent.

One final piece of experimental evidence is also of relevance, i.e. the data reported on the radiation induced polymerization of 1,2 cyclohexene oxide [97]. Under very dry conditions the mechanism appears to be a free cationic one and apparently the ratio of the rate coefficient for termination to that for propagation, k_t/k_p, is ~2.4. If termination is assumed to be diffusion controlled, then k_p would be of the same order $(10^9-10^{10}$ l mole^{-1} sec$^{-1})$. The authors have pointed out that this is merely a rough estimate and represents an upper limit. However, even if this data is in error by 3 or 4 orders of magnitude, such a measure of reactivity is considerably higher than any reported from chemical initiation of similar monomers.

REFERENCES

1 H. Sawada, *J. Macromol. Sci. Rev. Macromol. Chem.*, *C7* (1972), 161.
2 P. H. Plesch (Ed.), *The Chemistry of Cationic Polymerization.* Pergamon Press, New York, 1963.
3 P. H. Plesch, in J. C. Robb and F. W. Peaker (Eds.), *Progress in High Polymers*, Vol. 2, 1968, p. 137.
4 Z. Zlamal, in G. E. Ham (Ed.), *Kinetics and Mechanism of Polymerization*, Vol. I, Pt. II., Marcel Dekker, New York, 1969, 231.
5 K. C. Frisch and R. L. Reegen (Eds.), *Kinetics and Mechanism of Polymerization*, Vol. 2, Marcel Dekker, New York, 1969.
6 A. Ledwith and C. Fitzsimmons, in J. P. Kennedy and E. Tornquist (Eds.), *Polymer Chemistry of Synthetic Elastomers*, Pt. I, Wiley—Interscience, New York, 1968, p. 377.
7 T. Higashimura, in T. Tsuruta and K. F. O'Driscoll (Eds.), *Structure and Mechanism in Vinyl Polymerization*, Marcel Dekker, New York, 1969, p. 313.
8 A. Tsukamoto and O. Vogl, in A. D. Jenkins (Ed.), *Progress in Polymer Science*, Vol. 3, Pergamon, 1971, p. 199.
9 A. Ledwith and D. C. Sherrington, in A. D. Jenkins and A. Ledwith (Eds.), *Reactivity Mechanism and Structure in Polymer Chemistry*, Wiley, London, 1974, p. 244.
10 D. O. Jordan and F. E. Treloar, *J. Chem. Soc.*, (1961) 729 and 734.
11 D. C. Pepper, R. E. Burton, M. J. Hayes, A. Albert and D. H. Jenkinson, *Proc. Roy. Soc. Ser. A*, *263* (1961) 58, 63, 75, 82.
12 D. C. Pepper and P. J. Reilly, *J. Polym. Sci.*, *58* (1962) 639.
13 D. C. Pepper and P. J. Reilly, *Proc. Roy. Soc. Ser. A*, *291* (1966) 41.
14 L. E. Darcy, W. P. Millrine and D. C. Pepper, *Chem. Commun.*, (1968) 1441.
15 B. MacCarthy, W. P. Millrine and D. C. Pepper, *Chem. Commun.*, (1968) 1442.
16 A. Gandini and P. H. Plesch, *Proc. Chem. Soc., London*, (1964) 240.
17 A. Gandini and P. H. Plesch, *J. Polym. Sci., Part B*, *3* (1965) 127.
18 A. Gandini and P. H. Plesch, *J. Chem. Soc.*, (1965) 4826.
19 A. Gandini and P. H. Plesch, *Eur. Polym. J.*, *4* (1968) 55.
20 V. Bertoli and P. H. Plesch, *J. Chem. Soc., B*, (1968) 1500.
21 K. Ikeda, T. Higashimura and S. Okamura, *Chem. High Polym. (Japan)*, *26* (1969) 364.
22 S. D. Hamann, A. J. Murphy, D. M. Solomon and R. I. Willing, *J. Macromol. Sci., Chem. A6* (1972) 771.
23 J. P. Kennedy, *Encyclopedia of Polymer Science and Technology*, Vol. 7, Wiley, 1967, p. 754.
24 J. P. Kennedy, *J. Macromol. Sci., Chem., A6* (1972) 329.
25 J. P. Kennedy, *J. Macromol. Sci., Chem., A3* (1969) 861, 885.
26 C. G. Overberger, R. J. Ehring and R. A. Marcus, *J. Amer. Chem. Soc.*, *80* (1958) 2456.
27 C. E. H. Bawn, C. Fitzsimmons, A. Ledwith, J. Penfold, D. C. Sherrington and J. A. Weightman, *Polymer*, *12* (1971) 119.
28 A. Priola, S. Cesca and G. Ferraris, *Makromol. Chem.*, *160* (1972) 41.
29 P. M. Bowyer, A. Ledwith and D. C. Sherrington, *Polymer*, *12* (1971) 509.
30 A. Ledwith, E. Lockett and D. C. Sherrington, *Polymer*, *16* (1975) 31.
31 C. E. H. Bawn, R. M. Bell, C. Fitzsimmons and A. Ledwith, *Polymer*, *6* (1965) 661.
32 I. Kuntz, *J. Polym. Sci. Part B*, *4* (1966) 427.
33 W. J. Goethals and W. Drijvers, *Makromol. Chem.*, *136* (1970) 73.
34 F. R. Jones and P. H. Plesch, *Chem. Commun.*, (1970) 1018.
35 T. Masuda and T. Higashimura, *Polym. Lett.*, *9* (1971) 783.

128

36 A. Chapiro and P. Wahl, *C. R. Acad. Sci.*, *238* (1954) 1803.
37 W. H. Davidson, S. H. Pinner and R. Worrall, *Chem. Ind.* (*London*), *38* (1957) 1274.
38 Y. Tabata, in G. E. Ham (Ed.), *Kinetics and Mechanism of Polymerization*, Vol. 1, *Vinyl Polymerization* Pt. II, Marcel Dekker, New York, 1969, p. 305.
39 M. Szwarc, *Carbanions, Living Polymers and Electron Transfer Processes*, Interscience, New York, 1968.
40 M. P. Dreyfuss and P. Dreyfuss, *Polymer*, *6* (1965) 93.
41 M. P. Dreyfuss and P. Dreyfuss, *J. Polym. Sci. Part A*, *4* (1966) 2179.
42 G. J. Blake and D. D. Eley, *J. Chem. Soc.*, (1965) 7412.
43 Z. Zlámal and A. Kazda, *J. Polym. Sci. Part A*, *1* (1966) 1783.
44 P. H. Plesch, *Advan. Polymer Sci.*, *8* (1971) 137.
45 T. Saegusa, S. Matsumoto and Y. Hashimoto, *Polym. J.*, *1* (1970) 31.
46 J. M. Sangster and D. J. Worsfold, *Polym. Prepr., Amer. Chem. Soc., Div. Polym. Chem.*, *13* (1972) 72.
47 S. Winstein, P. E. Klinedinst Jr. and G. C. Robinson, *J. Amer. Chem. Soc.*, *83* (1961) 885.
48 J. T. Dennison and J. B. Ramsey, *J. Amer. Chem. Soc.*, *77* (1955) 2615.
49 R. M. Fuoss and F. Accascina, *Electrolyte Conductance*, Interscience, New York, 1959.
50 A. Ledwith, *Advan. Chem. Ser.*, *91* (1969) 317.
51 P. M. Bowyer, A. Ledwith and D. C. Sherrington, *J. Chem. Soc. B*, (1971) 1511.
52 J. Wislicenius, *Ann.*, *192* (1878) 106.
53 D. D. Eley and A. W. Richards, *Trans. Faraday Soc.*, *45* (1949) 425.
54 D. D. Eley and J. Saunders, *J. Chem. Soc.*, (1952) 4162.
55 A. G. Evans, P. M. S. Jones and J. M. Thomas, *J. Chem. Soc.*, (1957) 105.
56 D. Giusti and F. Andruzzi, Symp. on Macromol. Chem., Prague, (1965), Paper P. 492.
57 N. Kanoh, T. Higashimura and S. Okamura, *Makromol. Chem.*, *56* (1962) 65.
58 D. D. Eley, F. L. Isack and C. M. Rochester, *J. Chem. Soc., A*, (1968) 872, 1651.
59 D. D. Eley and A. Seabrooke, *J. Chem. Soc.*, (1964) 2231.
60 T. Higashimura, N. Kanoh, Y. Yonezawa, K. Fukui and S. Okamura, *J. Chem. Soc. Jap, Pure Chem. Sect.*, *81* (1960) 550.
61 A. Ledwith and D. C. Sherrington, *Polymer*, *12* (1971) 344.
62 S. Okamura, N. Kanoh and T. Higashimura, *Makromol. Chem.* *47* (1961) 19.
63 N. Kanoh, K. Ikeda, A. Gotoh, T. Higashimura and S. Okamura, *Makromol. Chem.*, *86* (1965) 200.
64 S. Okamura, N. Kanoh and T. Higashimura, *Makromol. Chem.*, *47* (1961) 35.
65 N. Kanoh, K. Ikeda, A. Gotoh, T. Higashimura and S. Okamura, *Makromol. Chem.*, *86* (1965) 200.
66 H. A. Benesi and J. H. Hildebrand, *J. Amer. Chem. Soc.*, *71* (1949) 2703.
67 L. J. Andrews and R. M. Keefer, *J. Amer. Chem. Soc.*, *74* (1952) 458.
68 S. Bywater and D. J. Worsfold, *Can. J. Chem.*, *44* (1966) 1671.
69 C. M. Fontana and G. A. Kidder, *J. Amer. Chem. Soc.*, *70* (1948) 3745.
70 K. J. Laidler and H. Eyring, *Ann. N.Y. Acad. Sci.*, *39* (1940) 303.
71 S. Bywater and D. J. Worsfold, *Can. J. Chem.*, *44* (1966) 1671.
72 A. A. Zharov, A. A. Berlin and N. S. Enikolopyan, *J. Polym. Sci. Part C*, *16* (1967) 2313.
73 D. D. Eley and A. W. Richards, *Trans. Faraday Soc.*, *45* (1949) 436.
74 A. Ledwith, *J. Appl. Chem.*, *17* (1967) 344.
75 A. D. Eckard, A. Ledwith and D. C. Sherrington, *Polymer*, *12* (1971) 444.
76 C. Aso, T. Kunitake, Y. Matsuguma and Y. Imaizumi, *J. Polym. Sci. Part A1*, *6* (1968) 3049.
77 T. Kunitake, Y. Matsuguma and C. Aso, *Polym. J.*, *2* (1971) 345.

78 Y. Matsuguma and T. Kunitake, *Polym. J.*, *2* (1971) 353.
79 R. Biddulph and P. H. Plesch, *Chem. Ind. (London)*, (1959) 1482.
80 D. N. Kursanov, M. E. Vol'pin and I. S. Akhrem, *Dokl. Akad. Nauk. SSSR, 120* (1958) 531.
81 M. E. Vol'pin, I. S. Akhrem and D. N. Kursanov, *Zh. Obshch. Khim.*, *30* (1960) 159.
82 F. Williams in P. Ausloos (Ed.), *Fundamental Processes in Radiation Chemistry*, Interscience, New York, 1968, p. 515.
83 T. H. Bates, J. V. F. Best and F. Williams, *Trans. Faraday Soc.*, *58* (1962) 192.
84 R. B. Taylor and F. Williams, *J. Amer. Chem. Soc.*, *91* (1969) 3728.
85 K. Hayashi, K. Hayashi and S. Okamura, *J. Polym. Sci. Part A1*, *9* (1971) 2305.
86 M. A. Bonin, W. R. Busler and F. Williams, *J. Amer. Chem. Soc.*, *87* (1965) 199.
87 E. Hubmann, R. B. Taylor and F. Williams, *Trans. Faraday Soc.*, *62* (1966) 88.
88 K. Hayashi, Y. Yamazawa, T. Takagaki, F. Williams, K. Hayashi and S. Okamura, *Trans. Faraday Soc.*, *63* (1967) 1489.
89 F. Williams, K. Hayashi, K. Ueno, K. Hayashi and S. Okamura, *Trans. Faraday Soc.*, *63* (1967) 1501.
90 C. H. Bamford, W. G. Barb, A. D. Jenkins and P. F. Onyon, *Kinetics of Vinyl Polymerization by Radical Mechanisms*, Butterworths, London, 1958, pp. 12, 29, 38.
91 W. F. Schmidt and A. O. Allen, *J. Phys. Chem.*, *72* (1968) 3730.
92 M. S. Matheson, *Advan. Chem. Ser.*, *50* (1965) 45.
93 R. C. Potter and D. J. Metz, *J. Polym. Sci. Part A1*, *4* (1966) 2295.
94 K. Boehlke and V. Jaacks, *Makromol. Chem.*, *142* (1971) 189; W. Kern, E. Eberius and V. Jaacks, *Makromol. Chem.*, *141* (1971) 63.
95 D. J. Metz, *Advan. Chem. Ser.*, *66* (1967) 170.
96 H. Yamaoka, F. Williams and K. Hayashi, *Trans. Faraday Soc.*, *63* (1967) 376.
97 D. Cordischi, A. Mele and R. Rufo, *Trans. Faraday Soc.*, *64* (1968) 2794.
98 K. Ueno, K. Hayashi and S. Okamura, *J. Macromol. Sci. Chem.*, *A2* (1968) 209.
99 T. Masuda, Y. Miki and T. Higashimura, *Polym. J.*, *3* (1972) 724.
100 F. S. Dainton, in P. H. Plesch (Ed.), *Cationic Polymerization and Related Complexes*, Heffer, Cambridge, 1953, p. 148.
101 N. Kanoh, T. Higashimura and S. Okamura, *Makromol. Chem.*, *56* (1962) 65.
102 T. Higashimura, H. Kusano, T. Masuda and S. Okamura, *Polym. Lett.*, *9* (1971) 463.
103 T. Masuda and T. Higashimura, *J. Polym. Sci. Part A1*, *9* (1971) 1563.
104 G. W. Cowell, A. Ledwith, A. C. White and H. J. Woods, *J. Chem. Soc. B*, (1970) 227.
105 G. M. Burnett, *Trans. Faraday Soc.*, *46* (1950) 772.
106 J. Hughes and A. M. North, *Trans. Faraday Soc.*, *62* (1966) 1866.
107 A. Chapiro, *Radiation Chemistry of Polymeric Systems*, Interscience, New York, 1962, pp. 138, 159.
108 M. Szwarc, D. N. Bhattacharyya, C. L. Lee and J. Smid, *J. Phys. Chem.*, *69* (1965) 612.
109 M. Szwarc, T. Shimomura, K. J. Tolle and J. Smid, *J. Amer. Chem. Soc.*, *89* (1967) 796.
110 A. G. Evans and M. Polanyi, *Nature (London)*, *152* (1943) 738; *J. Chem. Soc.*, (1947) 252.
111 F. S. Dainton, *Sci. Proc. Roy. Dublin Soc.*, *25* (1951) 148.
112 J. P. Kennedy in C. E. M. Bawn (Ed.), *M.T.P. Internat. Review of Science, Macromolecular Science*, Vol. I, Butterworth, London, 1973, p. 49.
113 J. P. Kennedy and E. Tornquist (Eds.), *Polymer Chemistry of Synthetic Elastomers, Pt. I*, Wiley—Interscience, New York, 1968.
114 Y. Ishi and S. Sakai, in K. C. Frisch and R. L. Reegen (Eds.), *Kinetics and Mechanism of Polymerization*, Vol. 2, Marcel Dekker, New York, 1969, p. 13.

130

115 P. Dreyfuss and M. P. Dreyfuss, *Advan. Polym. Sci.*, *4* (1967) 528.
116 D. Sims, *J. Chem. Soc.*, (1964) 864.
117 F. S. Dainton and K. Ivin, *Quart. Rev.*, 12 (1958) 61.
118 P. Dreyfuss and M. P. Dreyfuss, in K. C. Frisch and R. L. Reegen (Eds.), *Kinetics and Mechanism of Polymerization*, Vol. 2, Marcel Dekker, New York, 1969, p. 111.
119 D. Vofsi and A. V. Tobolsky, *J. Polym. Sci. Part A*, *3* (1965) 3261.
120 T. Saegusa and S. Matsumoto, *J. Polym. Sci. Part A1*, *6* (1968) 1559.
121 V. Jaacks, K. Boehlke and E. Eberius, *Makromol. Chem.*, *118* (1968) 354.
122 T. Saegusa and S. Matsumoto, *J. Macromol. Sci. Part A*, *4* (1970) 873.
123 I. Kuntz and M. T. Melchior, *J. Polym. Sci. Part A1*, *7* (1969) 1959.
124 T. Saegusa, S. Matsumoto and T. Ueshima, *Makromol. Chem.*, 105 (1967) 132.
125 B. A. Rozenberg, O. M. Chekhuta, E. B. Lyudig, A. R. Gantmakher and S. S. Medvedev, *Vysokomol. Soedin. Zh.*, *6* (1964) 2030; *Polym. Sci. USSR, 6* (1964) 2246.
126 H. Imai, T. Saegusa, S. Matsumoto, T. Tadasa and J. Furukawa, *Makromol. Chem.*, *102* (1967) 222.
127 T. Saegusa, H. Imai and S. Matsumoto, *J. Polym. Sci. Part A1*, 6 (1968) 459.
128 T. Saegusa and S. Matsumoto, *Macromols.*, *1* (1968) 442.
129 T. Saegusa, S. Miyaji and S. Matsumoto, *Macromols.*, *1* (1968) 478.
130 W. Drijvers and E. J. Goethals, Paper presented at the IUPAC Symposium on Macromolecular Chemistry, Boston, July 1971, p. 663.
131 V. S. Foldi and W. Sweeney, *Makromol. Chem.*, *72* (1964) 208.
132 G. L. Brode, A.C.S. *Polym. Prepr.*, *Amer. Chem. Soc., Div. Polym. Chem.*, *6* (1965) 626.
133 J. K. Stille and J. A. Empen, *J. Polym. Sci. Part A1*, *5* (1967) 273.
134 C. C. Price and E. A. Blair, *J. Polym. Sci. Part A1*, *5* (1967) 171.
135 E. J. Goethals and E. Du Prez, *J. Polym. Sci. Part A1*, *4* (1966) 2893.
136 W. Drijvers and E. J. Goethals, *Makromol. Chem.*, *148* (1971) 311.
137 E. J. Goethals, A.C.S. *Polym. Prepr.*, *Amer. Chem. Soc., Div. Polym. Chem.*, *13* (1972) 51.
138 E. Lippert and H. Prigge, *Justus Liebigs Ann. Chem.*, *659* (1962) 81.
139 S. Penczek and P. Kubisa, *Makromol. Chem.*, *130* (1969) 186.
140 S. Penczek and P. Kubisa, Symposium on Macromolecular Chemistry, Budapest, Paper 2/12, 1969.
141 P. Kubisa, J. Brzezinski and S. Penczek, *Makromol. Chem.*, *100* (1967) 286.
142 J. B. Rose, *J. Chem. Soc.*, (1956) 542.
143 J. W. Hill and W. H. Carothers *J. Amer. Chem. Soc.*, *57* (1935) 925.
144 P. H. Plesch and P. H. Westerman, *Polymer*, *10* (1969) 105.
145 Y. Yamashita, M. Okada, K. Suyama and H. Kasahara, *Makromol. Chem.*, *114* (1968) 146.
146 P. H. Plesch and P. H. Westerman, *J. Polym. Sci. Part C*, *16* (1968) 3837.
147 F. R. Jones and P. H. Plesch, *Chem. Commun.*, (1969) 1230.
148 Y. Yamashita, M. Okada and H. Kasahara, *Makromol. Chem.*, *117* (1968) 256.
149 P. Kubisa and S. Penczek, *Makromol. Chem.*, *144* (1971) 169.
150 S. Slomkowski and S. Penczek, *Chem. Commun.*, (1970) 1347.
151 J. Furukawa and K. Tada, in K. C. Frisch and R. L. Reegen (Eds.), *Kinetics and Mechanism of Polymerization*, Vol. 2, Marcel Dekker, New York, 1969, p. 159.
152 V. Jaacks, K. Boehlke and E. Eberius, *Makromol. Chem.*, 118 (1968) 354.
153 K. Boehlke, P. Weyland and V. Jaacks, to be published.
154 S. Penczek, *Makromol. Chem.*, *134* (1970) 299.
155 T. Helen, D. Schlotterbeck and V. Jaacks, to be published.
156 F. R. Jones and P. H. Plesch, *Chem. Commun.*, (1969) 1231.

157 E. L. Berman, E. V. Ludvig, B. A. Ponomorienko and S. S. Medviediev, *Vysokomol. Soedin, Ser. A, 11* (1969) 200.

158 W. Kern, E. Eberius and V. Jaacks, *Makromol. Chem., 141* (1971) 63.

159 K. Boehlke and V. Jaacks, *Makromol. Chem., 142* (1971) 189.

160 R. Mateva, C. Konstantinov and V. Kabaivanov, *J. Polym. Sci. Part A1, 8* (1970) 3563.

161 J. E. Figueruelo and D. J. Worsfold, *Eur. Polym. J., 4* (1968) 439.

Chapter 3

Kinetics of Polymerization Initiated by Ziegler—Natta and Related Catalysts

W. COOPER

1. Introduction

The origins of the discovery that ethylene was polymerized by catalysts containing transition metals go back to the early 1940s [1], and in the period 1950—1955 supported transition metal oxide catalysts were developed which would give linear, high molecular weight, high density polyethylene [2]. However, it was the discovery about 1954 by Ziegler [3] that combinations of aluminium alkyls and titanium tetrachloride would polymerize ethylene to linear high molecular weight polymer at low temperatures and pressures which stimulated the great scientific and industrial developments in this field. Ziegler's discovery arose during a systematic study of the effects of transition metals and their compounds on the addition of ethylene to aluminium trialkyls and it was established relatively quickly that many catalyst combinations were effective for the polymerization of ethylene and α-olefins from organometal compounds from Groups I to IV and compounds of the transition metals from Groups IV to VI. In addition to simple olefins, styrene and substituted styrenes, and conjugated dienes such as butadiene and isoprene were polymerized, and for the latter, catalysts based on Group VIII metal compounds were also found to be effective [4]. In Ziegler catalysts the most widely employed organo-metallic compounds are the alkyls, alkyl halides and alkyl hydrides of aluminium, but the organic derivatives of the alkali metals, magnesium, beryllium, zinc, tin and lead have frequently been used. Typical of the transition metal compounds are the higher valency halides and oxyhalides of titanium, vanadium and chromium.

The catalysts are usually prepared in hydrocarbon solvents, essentially in the absence of air or moisture and are mixtures of ill-defined composition. In many instances dark-coloured precipitates are formed of variable stoichiometry containing complexes of the organo-metal compound with the transition metal in a lower valence state. Natta [5] showed that pure lower valence transition metal compounds, such as titanium or vanadium trichloride, when treated with organo-metal compounds were effective catalysts, and were particularly suitable for the preparation of crystalline high melting point polyolefins. The close identity of these two classes of catalyst has led to their description as

References pp. 249—257

Ziegler—Natta catalysts. The metal oxide catalysts are, in fact, of a similar character to the Ziegler—Natta types and a more generally descriptive term which covers all of these is coordination catalysts. This derives from the view that an essential step in the polymerization is coordination of the olefin with the transition metal.

For some time it was thought that a catalyst surface played an important role in the polymerization and to some extent this view remains. It was soon found, however, that, by choice of groupings attached to the transition metal, catalysts which were completely soluble in hydrocarbon solvents were obtained. The first examples of these were bimetallic complexes of titanium and aluminium solubilized by π-cyclopentadienyl groups attached to the titanium [6], but later single-species organo-transition metal compounds (e.g. π-allyl nickel halides) were found to be polymerization catalysts [7]. The soluble catalysts have been particularly useful for studies on polymerization mechanism, but they tend to be more suited for the polymerization of diene monomers and not to give tactic polymers from the α-olefins. In general, hydrocarbon monomers are the most suitable, but others, such as the vinyl ethers, can also be polymerized. Monomers containing groupings such as carbonyl and hydroxyl, which would react with the catalyst, may not polymerize efficiently, although there have been several reports of the polymerization of acrylonitrile and methyl methacrylate by coordination catalysts.

The importance of these catalysts stems from the fact that they are the only ones known which will give high molecular weight polymers from olefins such as propene and butene-1, and with these and other monomers the polymers have specific molecular configurations [8]. (It has been appreciated for some years that monomer units could be disposed in a polymer chain in various stereochemical arrangements [9], but before the discovery of coordination catalysts there was little expectation, with the possible exception of the dienes, that the polymer structure could be varied or controlled by the choice of initiator.)

Ziegler—Natta catalysts exhibit great variations in catalyst stability and activity, and polymer yield varies greatly with different monomers. Some catalysts are relatively unstable but lifetimes can vary from minutes to months, and are dependent on the preparative conditions. Catalyst stoichiometry, the presence or absence of monomer during its preparation and the type of solvent, can have major effects; loss of activity is usually ascribed to decomposition of some organo-transition metal complex. Organo-transition metal compounds vary greatly in stability, but are frequently stabilized by complexation with other metal compounds and by adsorption on insoluble metal halides. The nature of the groupings attached to the metal may also have a pronounced influence on stability. Methyl titanium trichloride is relatively stable at room temperatures, but substitution of more halogen atoms by methyl groups or the substitution of methyl by higher alkyl groups greatly reduces the stability. The presence of hydrogen atoms α to the metal atom in the organic grouping facilitate decomposition to form the metal hydride and olefin [10]. Aryl derivatives are more stable than alkyl compounds and stability is further enhanced by the presence of alkoxyl groups [11].

As the first transition series is ascended the metal—carbon bond becomes weaker and the alkyls of iron, cobalt and nickel are stable at ambient temperatures only when coordinated with strong π donors such as 1,1 bipyridyl [12]. Organic groupings which form delocalized bonds with the metal are more stable and many π complexes or π-allylic compounds of transition metals are known. The metal—cyclopentadienyl bond is too stable for these compounds to initiate polymerization, but some π-allyl compounds will polymerize monomers such as butadiene where a comparable structure will be retained in the propagating chain.

Kinetic studies on polymerizations using coordination catalysts present considerable difficulties owing to their complexity and variability, to the incomplete understanding of the reaction mechanisms, and to the problems in handling highly reactive and in many cases heterogeneous systems. Over the past fifteen years there have been many kinetic investigations, but a substantial proportion of them have had as their goal rather limited objectives. They have been concerned with distinguishing between alternative proposals on mechanism and in relatively few are sufficient data, from which meaningful kinetics can be assembled, presented. The sensitivity of the catalyst to air, water and other oxygen-containing compounds necessitates high standards of purity in the catalyst components, monomers and solvents and in the technique employed. In many investigations the precautions taken have been inadequate. It is essential for a proper appreciation of the kinetics data that there should be some knowledge of the catalyst structures and polymerization mechanism. Unfortunately, this is still very imperfect and there are many gaps to be filled.

2. Catalyst structure

2.1 NATTA CATALYSTS

The most extensively studied catalysts are those from the purple alpha form of titanium trichloride, in which a suspension of relatively large crystals (ca. 10 μm) of the halide in a hydrocarbon is treated with an organo-metal compound, e.g., AlEt$_3$, AlEt$_2$Cl, BeEt$_2$. The crystals possess a laminar structure of alternate layers of chlorine and titanium atoms (Fig. 1a) [13], and electron photomicrographs show that polymer growth starts at the edges or growth spirals of crystals [14, 15], (Figs. 2 and 3). As polymerization proceeds diaggregation of the relatively coarse crystals occurs, and the small portions of crystal containing active sites become encapsulated within small spheres of polymer [16].

136

(a)

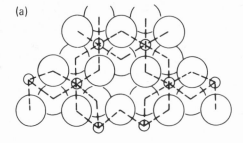

(b)

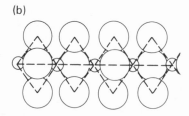

Fig. 1. (a) α-TiCl₃ (laminar); (b) β-TiCl₃ (fibrillar).

Fig. 2. Electromicrograph of polypropene on α-TiCl₃ showing polymer growth at steps on crystal surface [14].

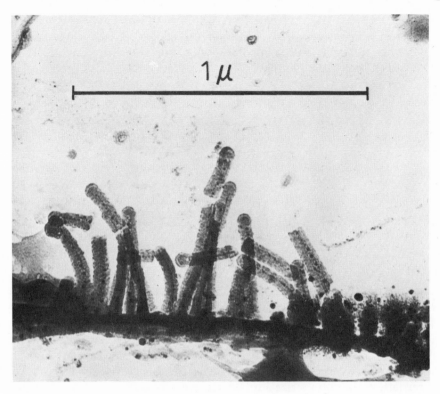

Fig. 3. Electromicrograph of polypropene on edges of α-TiCl$_3$, showing outward growth of polymer from reactive sites [15].

The activity of the catalyst depends on the conditions of preparation and purity of the titanium trichloride. In high activity catalysts the initial crystal aggregates are broken down completely and most of the metal atoms have been claimed to become involved in the polymerization [17]. The formation of the active sites at the crystal edges is considered to involve complexation or alkylation by the organo-metal compounds [18], viz.

$$TiCl_3 + AlR_3 \rightleftarrows TiCl_3 . AlR_3 \rightleftarrows RTiCl_2AlR_2Cl$$

$$RTiCl_2 . AlR_2Cl + AlR_3 \rightleftarrows RTiCl_2 . AlR_3 + AlR_2Cl$$

In addition to reaction of metal alkyl with the lateral edges of the titanium trichloride crystal, reaction also occurs at the main faces although this does not initiate polymerization. A study of the stoichiometry and mechanism of the TiCl$_3$/AlMe$_3$ reaction indicates that a complex of the structure TiCl$_3$—TiCl$_2$AlMe$_2$ is formed on the 001 face, in which the titanium and chlorine atoms maintain their original positions in the crystal lattice [19], but these presumably do not have vacancies for coordination and insertion of monomer molecules. It is not yet settled whether the catalytic entity involves only the transition metal (I) or whether it is a bimetallic complex (II or III), viz.

I   II   III

Analysis of the catalysts shows the formation of stable complexes from the metal alkyl with the subhalide, and the dependence of polymerization rate on metal alkyl concentration is consistent with the formation of a complex as the catalytic entity. There is also evidence for the displacement of one compound by another at the crystal surface. Competition between $AlEt_2Cl$ and $AlEtCl_2$ for the $TiCl_3$ surface is shown by retardation of polymerization on adding the $AlEtCl_2$ and activation when it is removed [20]. Addition of titanium tetrachloride to the inactive complex $TiCl_3/AlRCl_2$ gives a catalyst, probably with the structure $RTiCl_3 . TiCl_3$, which is reduced in activity by an increase in the concentration of $AlRCl_2$ which competes with the $RTiCl_3$ for the $TiCl_3$ surface [21].

It is not essential for the complex to contain a second metal such as aluminium in addition to the transition metal because methyl titanium trichloride adsorbed on α-$TiCl_3$ is a catalyst for olefin polymerization [22]. Another catalyst of comparable structure but very different activity is from $TiCl_2$ and $MeTiCl_3$, probably containing $TiCl_3 . MeTiCl_2$ as the active complex [14]. These, and other catalysts which do not contain a second metal alkyl, lend support to the view that the active site need consist only of an alkylated transition metal compound. The two theories are not mutually exclusive, however. The essential feature is that the Ti—C bond at which addition occurs should be activated for monomer insertion and it is conceivable that this could be achieved either by coordination of monomer at a vacant site in an alkylated $TiCl_3$ lattice or at a bimetallic complex with an organo-metal compound. Catalyst activity and polymer stereoregularity are affected by the structure of the organo-metal compound, which has often been held to be evidence for the involvement of the metal alkyl in the active complex, but this could result from changes in the degree of alkylation or reduction of the titanium trichloride rather than from differences resulting from coordination of a different organometal compound.

The sites for coordination of monomer (indicated by the open dotted lines in (I) to (III)) have been the subject of much speculation in regard to the mechanism of stereoregulation [23, 24]. It has also been suggested that the active site may contain two vacancies [25]. In the above, catalyst complexes have been represented as bridged structures but they have also been considered to be ionic complexes [26], e.g.

$$ClTiR^+AlRCl_3^-$$

In the heterogeneous catalysts this would be no more than a formal representation of the distribution of electronic charge in the active sites, but with soluble catalysts from $(\pi\text{-}C_5H_5)_2TiCl_2/AlMe_2Cl$ there is evidence from electrodialysis studies that the active species possesses a positive charge and probably has the structure $(\pi\text{-}C_5H_5)_2Ti^+Me$ [27]. This type of structure is consistent with certain features of butadiene polymerization by soluble nickel and cobalt catalysts [28].

2.2 ZIEGLER CATALYSTS

Heterogeneous catalysts produced in situ by reduction of transition metal compounds in their higher valency states (e.g. $TiCl_4$, VCl_4, $VOCl_3$, CrO_2Cl_2) are of variable composition. The initial reaction is halogen—alkyl interchange followed by decomposition of the complex to give the subhalide and a mixture of alkanes and alkenes. The extent of reduction depends on the structure of the organometal compound and on the molar ratio of metal alkyl to transition metal compound. Most of the published information relates to the reduction of titanium and vanadium halides. There are discrepancies in the published data but they are in agreement in that rapid reduction occurs to give a mixture of lower valency compounds of the transition metal [29], this being in marked contrast with the behaviour of the insoluble subhalides, which undergo little reduction [30] except under special circumstances [31].

Finely divided $TiCl_3$ reacts to a limited extent with aluminium alkyls, as judged by the extraction of chloride ion into the solution. With a sample of high surface area (145.5 m^2 g^{-1}) reaction was essentially complete after 3—4 h and increased with increase in temperature: ca 15% at 10°C and 50% at 70°C [273]. The kinetics of reaction followed the relationship

$$d[Cl^-]/dt = k\{[Cl^-] - [Cl^-]_\infty\}$$

where $[Cl^-]_\infty$ is the final concentration of chloride ion in the $TiCl_3$ [274].

Titanium tetrachloride and aluminium triethyl form a hydrocarbon soluble complex at low temperatures which decomposes at $-30°C$ to give the trichloride as a major product [32]. Complexes containing tetravalent titanium stabilized by adsorption on titanium trichloride apparently persist in catalysts prepared at Al/Ti ratios below 1.0 [33], but at higher ratios there are some Ti(II) sites present in the catalyst [34]. Analysis shows that at Al/Ti ratios above 1.0 the solid precipitate contains divalent titanium or even lower valency states of the metal [35]. Reduction of $TiCl_4$ with $AlEt_2Cl$ is less rapid and extensive than with $AlEt_3$ and even at high Al/Ti ratios [36] reduction does not proceed much below the trivalent state. Aluminium alkyl dihalides are still less reactive and reduction to $TiCl_3$ is slow and incomplete except at high Al/Ti ratios or elevated temperatures [37].

Titanium trichloride prepared by reduction of the tetrachloride by metal alkyls is the brown (β) form with a different lattice structure from α-$TiCl_3$ (Fig. 1b). As prepared it is of high surface area [38] (100—500 m^2 g^{-1}), and gives catalysts of high activity for ethylene polymerization [21]. Propene polymers prepared with β-$TiCl_3$ are of lower tacticity than those obtained with α-$TiCl_3$ [40] and, with isoprene, complexes of β-$TiCl_3$ with $AlEt_3$ or $AlEt_2Cl$ give high cis 1,4 polymer instead of the high trans 1,4 polymer [41]. β-Titanium trichloride is transformed into the γ form* with some reduction in surface area by heating the solid at $250°C$ [42] or at $170°C$ in the presence of suitable hydrocarbon solvents such as n-decane [43].

There is little information on the reduction of other transition metal halides other than of $VOCl_3$ and VCl_4 with $AlEt_3$ or $AlEt_2Cl$. The intermediate alkyl vanadium complexes are very unstable and precipitates are obtained, even at $-78°C$, containing mainly trivalent vanadium. With $AlEt_3$ prolonged reaction times or high Al/V ratios result in some reduction to the divalent state [44]. In Table 1 are given data on the composition of Ziegler catalysts from titanium and vanadium halides.

2.3 METAL OXIDE SUPPORTED CATALYSTS

These catalysts consist of the oxides of vanadium, chromium, molybdenum or tungsten deposited by co-precipitation or by impregnation on a silica or silica—alumina support. The catalysts require an activation process before use, and for those containing chromic oxide this consists of

---

* The γ-form of $TiCl_3$ is purple in colour and in common with α-$TiCl_3$ has a layer structure of titanium and chlorine atoms. The α-form has the close packed hexagonal arrangement and the gamma form the close packed cubic structure. Another form (δ-$TiCl_3$) produced by milling α-$TiCl_3$ with $AlCl_3$, has a mixed $\alpha + \gamma$ lattice structure and is highly active in polymerization [39]. The γ- and δ-forms resemble the α-form in polymerization behaviour, but differ in activity.

TABLE 1

Reduction of transition metal halides by organometal compounds

	Ratio of reactants Al/Ti	Cl/Ti	Composition of precipitate Al/Ti	Et/Ti	Ref.
Titanium tetrachloride					
(a) AlEt$_3$	1.2	3.08	0.07	—	89
	2.5	2.15	0.30	—	
	5	1.65	0.31	—	
	1[a]	3.17	0.27	0.37	35
	2	2.78	0.58	1.19	
	10	1.51	0.55	1.21	
(b) AlEt$_2$Cl	1.5[a]	3.20	0.14	0.34	35
	4.5	3.21	0.20	0.38	
	7.5	3.14	0.29	0.26	
	15	3.33	0.28	0.33	
			Average valence state of Ti in precipitate		
	1[a]		3.12		35
	3		2.99		
	5		2.53		
(c) AlEtCl$_2$		% Ti reduced			
	1.00[b]	64.5	3.36		37
	2.01	79.5	3.22		
	4.15	76.5	3.24		

	Ratio of reactants Al/V	Composition of precipitates Al/V	Cl/V	Ref.
Vanadium tetrachloride				
(a) AlEt$_3$	1.0[c]	0.46	3.0	
	2.0	0.48	2.8	
	8.0	0.62	2.6	90
(b) AlR$_2$Cl	2 (R = Et)[c]	0.63	3.5	90
	8	0.50	3.6	
		Extent of reduction[d]	Reaction time	
	5 (R = Et)	1.3—5.1%V^{4+} 94.9—98.7%V^{3+}	300 min	44
	4 (R = i-Bu)	99.8%V^{3+} 0.2%V^{2+}	16 h	
Vanadium oxytrichloride				
(a) AlEt$_3$	1[c]	0.82	2.2	
	2	0.77	2.1	90
	8	0.94	2.0	

TABLE 1—*continued*

	Ratio of reactants Al/V	Composition of precipitates Al/V	Cl/V	Ref.
(b) $AlEt_2Cl$	1^c	0.82	3.2	
	2	0.90	3.2	90
	8	0.93	3.0	
Vanadium trichloride				
(a) AlR_3	2 $(R = Et)^e$	60 min at $60°C$, $2.1\%V^{2+}$		30
		60 min at $100°C$, $5.78\%V^{2+}$		
	0.5 $(R = Et)$	$98.9\%V^{3+}$ $1.1\%V^{2+}$ $\}$ 2.5 min		44
	2.0 $(R = i\text{-}Bu)^f$	$31.6\%V^{3+}$ $69.4\%V^{2+}$ $\}$ 2.6 min		
		$19.4\%V^{2+}$ $80.6\%V^{2+}$ $\}$ 120 min.		
(b) $AlEt_2Cl$	4	180 min at $90°C$, $0\%V^{2+}$		30
		17 hrs at $-78°C$, $10\%V^{2+}$		44
Vanadyl acetylacetonate	$VO(acac)_2$			
$AlEt_2Cl$		V^{III}	V^{IIg}	
	20	75	25	139
	50	33	67	
	100	6	94	
	300	6	94	

[a] Temp., $20°C$; solvent—petroleum, bp $100-120°C$; [Al] = 0.0125—0.25 mole l^{-1}.
[b] Temp., $35°C$; no solvent.
[c] Temp., $15°C$.
[d] Solvent, toluene; temp., $-78°C$; [V] = 0.019 mole l^{-1}.
[e] Solvent, n-heptane; [Al] = 0.025 mole l^{-1}.
[f] Conditions as [d], VCl_3 obtained from VCl_4 and $Al(i\text{-}Bu)_2Cl$ at $-78°C$ (Al/V = 4) and probably of high surface area.
[g] Calc. from ESR signal; solvent, toluene; temp., $20°C$.

oxidation in air at $600-800°C$. Polymerization of ethylene is concurrent with reduction of the chromium and the catalyst becomes blue in colour [45]. Other agents such as carbon monoxide also reduce the metal to give active catalysts [46]. Catalysts fully reduced to Cr^{3+} are ineffective and it is likely that the active centre contains chromium in two valence states, at one of which the monomer is coordinated*. The formation of an organo-chromium compound as the propagating entity is reasonable and in accord with observations on other Ziegler type systems, but the mechanisms of the initiation and propagation reactions are unknown. The efficiency of the catalyst depends on the transition metal content (passing through a maximum with increasing chromium content), the activation

* Cr^{II} has been reported in catalysts reduced by ethylene or carbon monoxide [161]. Elsewhere mixed $Cr^{III}-Cr^{VI}$ entities have been proposed [144], and also Cr^{V} [232].

process and the nature of the support. Catalysts based on molybdenum or tungsten oxides are not reduced by ethylene and a reducing co-catalyst, e.g., hydrogen, sodium or sodium hydride, is necessary [47]. An analogous catalytic entity to that in the chromium based catalysts is probably involved [48]. These supported oxide catalysts are usually employed at higher temperatures and pressures than Ziegler—Natta catalysts and are for practical purposes restricted to polymers and copolymers of ethylene.

2.4 SOLUBLE CATALYSTS

Many soluble catalysts are known which will polymerize ethylene and butadiene. High activity soluble catalysts are employed commercially for diene polymerization but most soluble types are inefficient for olefin polymerization. A few are crystalline and of known structure such as blue $(\pi\text{-}C_5H_5)_2TiCl \cdot AlEt_2Cl$ [49] and red $[(\pi\text{-}C_5H_5)_2TiAlEt_2]_2$ [50]. The complex $(\pi\text{-}C_5H_5)_2TiCl_2 \cdot AlEt_2Cl$ polymerizes ethylene rapidly but decomposes quickly to the much less active blue trivalent titanium complex. Soluble catalysts are obtained from titanium alkoxides or acetyl acetonates with aluminium trialkyls and these polymerize ethylene and butadiene. Several active species have been identified, dependent on the temperature of formation and the Al/Ti ratio. Reduction to the trivalent state is slow and incomplete and maximum activity for ethylene polymerization occurs at about 25% reduction to Ti^{III} [51].

The structure of the catalyst has been studied using ESR which indicates that at low Al/Ti ratios structural units of the type IV are present, while at higher ratios a second ethyl group may be attached to titanium(V) [52, 53] viz.

$$EtTi(OR)_2 \cdot 2AlEt_3$$

IV   V

Maximum activity for butadiene polymerization occurs at Al/Ti $\simeq$ 6, significantly higher than that observed for ethylene, and suggestive of a different catalytic entity.

From the $Ti(acac)_3/AlEt_3$ reaction products the species VI and VII

VI   VII

have been tentatively identified [54]. The higher level of alkylation is observed at higher temperatures and for optimum activity high Al/Ti ratios are required.

The soluble catalyst from $VO(acac)_2/AlEt_2Cl$ contains V^{II} and V^{III} species (Table 1) (Al/V = 20—300) both of which are active for ethylene polymerization [139].

Soluble chromium catalysts are obtained from $Cr(acac)_3$ with $AlEt_3$ [55], $AlEt_2Cl$ and $AlEtCl_2$ [139]. As with other catalysts ligand exchange takes place and from $Cr(acac)_3/AlEt_3$ the complex VIII

$$Cr[(acac)AlEt_2]_3$$

VIII

has been isolated [56]. ESR studies on $Cr(acac)_3/AlEt_2Cl$ at high Al/Cr (100—300) ratios indicates that the chromium is present in a high spin Cr(II) complex [139].

Soluble cobalt and nickel catalysts for conjugated diene polymerization are usually prepared in the presence of monomer with the formation of a π-allylic structure as a relatively stable intermediate, but the nature of attachment of other ligands to the active site is not known. Aluminium halides and cobalt halides react to form complexes of the structure (IX) [57]

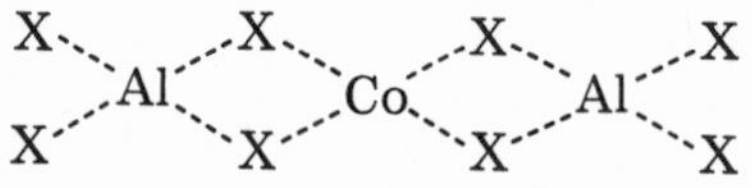

$$X = Cl, Br, I$$

IX

which polymerize butadiene to high cis 1,4-polymer [58]. Reaction with monomer probably transforms these to allylic complexes and by analogy the aluminium alkyl halide/cobalt salt catalysts are possibly represented by structures of the type X and XI.

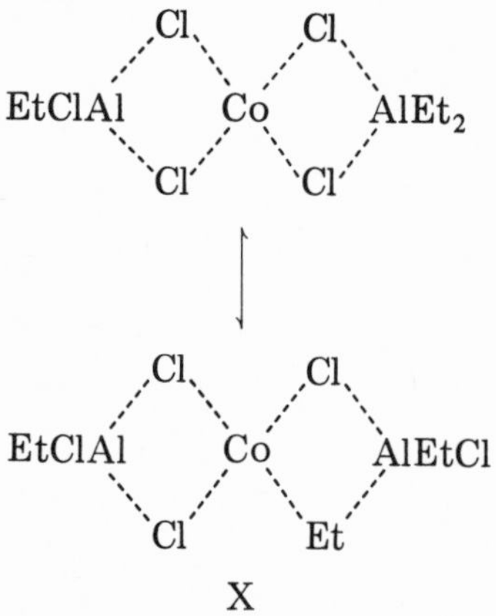

X

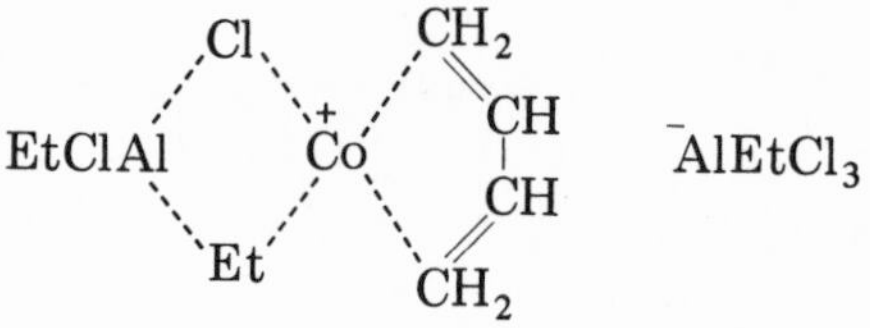

XI

2.5 OTHER CATALYSTS

In addition to the binary catalysts from transition metal compounds and metal alkyls there are an increasing number which are clearly of the same general type but which have very different structures. Several of these are crystalline in character, and have been subjected to an activation process which gives rise to lattice defects and catalytic activity. Thus, nickel and cobalt chlorides, which untreated are not catalysts, lose chlorine on irradiation and become active for the polymerization of butadiene to high *cis* 1,4-polymer [59]. Titanium dichloride, likewise not a catalyst, is transformed into an active catalyst (the activity of which is proportional to the Ti^{II} content) for the polymerization of ethylene [60]. In these the active sites evidently react with monomer to form organo-transition metal compounds which coordinate further monomer and initiate polymerization.

Soluble single species catalysts are also known, such as the bis(π-allyl nickel halides) [7]. These can be prepared separately or in situ by reacting bis-allyl nickel (which is an ologomerization catalyst for butadiene but does not give high molecular weight polymer) with an equimolar quantity of nickel halide, and thus bears some resemblance to the catalysts from titanium subhalides and alkyl titanium halides. It is of interest to note that the active species is the monomeric form of the initiator* as π-complex with butadiene (XII) [61].

These catalysts are of relative low activity, possibly as a result of small extents of dissociation to the active species. The addition of Lewis acids such as $AlCl_3$ increases the activity greatly, probably by making all the transition metal available in an active form. These compounds are ionic complexes typically containing the cation (π-$C_3H_5Ni^+ \cdot C_6H_6$) [7]. Bis-(π-allyl nickel trifluoroacetate) is a high activity catalyst for the polymerization of butadiene, here also because complete dissociation occurs in the presence of monomer to give the active catalyst [63]. This type of catalyst shows interesting behaviour in that the microstructure of the polymer is markedly influenced by the presence of other ligands. Thus, in saturated hydrocarbons the *cis* 1,4-polymer is produced whereas in benzene solvent "equibinary" polymers containing alternating units of *cis*

---

* Dependent on the halogen *cis* or *trans* 1,4-polybutadiene is obtained. This may reflect differences in the mode of coordination of butadiene.

1,4- and *trans* 1,4-structure are obtained. In the presence of triphenyl-phosphite or alcohols a different complex is produced which produces high *trans* 1,4-polybutadiene.

Combinations of π-allyl nickel chloride and bis(π-allyl) nickel produce macrocyclic oligomers from butadiene containing from four to above eight monomer units [138]; in the absence of the chloride dimers and trimers only are produced. This is of interest in that termination occurs by ring closure rather than by hydrogen transfer.

Tris-allyl chromium gives poly 1,2-butadiene [62] and the active entity involves two chromium atoms [63]. Propagation presumably involves monomer addition to the allyl grouping without undue disturbance of the electronic distribution around the metal but there is no information on the manner of monomer coordination or of the activation process which results in polymerization.

Titanium and zirconium tetrabenzyl and the mixed metal—benzyl halides are soluble in hydrocarbon solvents and will polymerize ethylene and α-olefins, the latter to stereo-specific polymers [64]. The structures of the true initiators are not known but they are unlikely to be the simple organo-metal compounds. Catalysts of higher activity are obtained when they are used in combination with aluminium alkyls. It is of interest to note that titanium tetra(dimethyl amide) reacts with acrylonitrile to form an active species, which then forms high molecular weight polymer by coordination polymerization [65].

3. Initiation

From considerations of catalyst structure it is apparent that complex formation, interchange reactions and valence state changes may occur prior to the growth of a polymer chain, and in coordination polymer-ization the term initiation tends to be used loosely to connote all these reactions. When the active organo-transition metal compound is pre-formed and contains an initiating group of comparable structure to that of the propagating chain the rate of initiation will be the same as propaga-tion; this situation occurs with π-crotyl nickel iodide in butadiene polymerization [61], and is presumably so also for preformed catalysts containing alkyl groups in olefin polymerization. When the catalysts are prepared in situ, initiation will not be kinetically distinguishable if alkylation of the transition metal is fast compared with subsequent propagation, but catalysts containing less reactive initiating groups than the propagating chains will show an acceleration stage. This is observed in olefin polymerization by catalysts containing transition metal—hydride bonds. With heterogeneous catalysts a slow acceleration in rate is also observed but this is the result of changes in the physical structure of the catalyst giving rise to an increasing number of propagating centres. With weak alkylating agents the number of active centres may increase slowly with time, irrespective of changes in surface area, with a corresponding dependence of initial polymerization rate on the time of prereaction of the catalyst components. This is the case for propene polymerization with α-TiCl$_3$/AlEt$_2$Cl or AlEtCl$_2$ and with the latter, activity is found only after prolonged reaction [66].

Induction periods and an accelerating stage in the polymerization may result from the presence of impurities which are slowly removed from the system by reaction with the catalyst components. In butadiene polymerization by the soluble catalyst from nickel salicylate/BF$_3$Et$_2$O/LiBu there is a marked induction from traces of 1,2 butadiene (below 100 p.p.m.) in the monomer [67]. In the absence of 1,2 butadiene polymerization starts immediately. An induction period has been found with the similar catalyst, nickel naphthenate/BF$_3$Et$_2$O/AlEt$_3$ [68], but the origins of this were not identified.

4. Propagation and stereo regulation

In coordination polymerization it is generally accepted that the monomer forms a π-complex with the transition metal prior to insertion into the growing chain. In general these complexes are insufficiently stable to be isolated although complexes of allene [69] and butadiene [70] have been reported. With allene the complex was formed prior to polymerization with soluble nickel catalysts, and *cis* coordinated butadiene forms part of the cobalt complex, $CoC_{12}H_{19}$, which is a dimerization catalyst.

Cossee [24] has suggested that the complexed monomer adds by a flow of electrons in the transition state, with transfer of the growing chain to the coordination position previously occupied by monomer, thus leaving a vacancy for coordination with the next monomer molecule, XIII, viz.

XIII

It has been established from the polymerization of 1-d_1-deuteropropene that the double bond opens in the *cis* direction [71], in conformity with the proposed mechanism. A repetition of this process with the monomer molecules approaching the active centre in the same direction would give the isotactic polymer while alternate approach to mirror image positions would give the syndiotactic form.

In heterogeneous catalysts from α-TiCl₃ the geometry of these sites will be determined by chloride ion vacancies at crystal surfaces and edges, and there are many speculations on the nature of these sites in relation to the mechanism of stereo-regulation [25]. Conjugated dienes can assume single or two point coordination to the active centre and as butadiene normally exists in the *s-trans* conformation [72] the formation of *cis* 1,4-polymer is considered to result from two point attachment of monomer in the *s-cis* form; single point attachment would produce the *trans* 1,4 or 1,2 polymers. Catalysts, such as VCl₃/AlEt₃, which will only polymerize conjugated dienes to *trans* 1,4-polymers, will not polymerize a monomer which is sterically constrained in the *s-cis* conformation, e.g., 2-*t*-Bu butadiene [73].

These mechanistic proposals leave uncertain the precise location of the growing chain at the catalyst centre. Early attempts to solve the problem were based on the identification in the polymer of characteristic groupings attached to various parts of the catalyst [74]. It is apparent, however, that the more reactive groupings will be the first to initiate polymerization and alkyl or aryl interchange reactions can result in their involvement in the initiation step, irrespective of the part of the catalyst complex to which they were originally combined. If they were firmly bound and unlikely to undergo exchange reactions (e.g. π-cyclopentadienyl groupings) they would also be unlikely to initiate polymerization. The currently held view is that the propagating chain is attached to the transition metal, but the evidence for this is indirect — namely the existence of catalysts containing only transition metal compounds, the dependence of monomer reactivity ratios in olefin copolymerization on the structure of the transition metal compound but not on the structure of the organo-metal compound [75], and from studies with optically active catalysts [76].

5. Transfer

It is usual for the molecular weight of polymer prepared with a coordination catalyst to reach a constant value with polymerization time or conversion, which indicates the existence of chain terminating steps. Since the polymerization rate may either remain constant or decline with increase in conversion the reactions limiting molecular weight are either chain transfer or termination with deactivation of the catalyst centre. Three types of transfer reaction have been found.

5.1 TRANSFER WITH MONOMER

In this reaction a hydrogen atom is transferred from the propagating chain to the monomer, with the formation of a stable molecule and a new growing chain, XIV, viz.

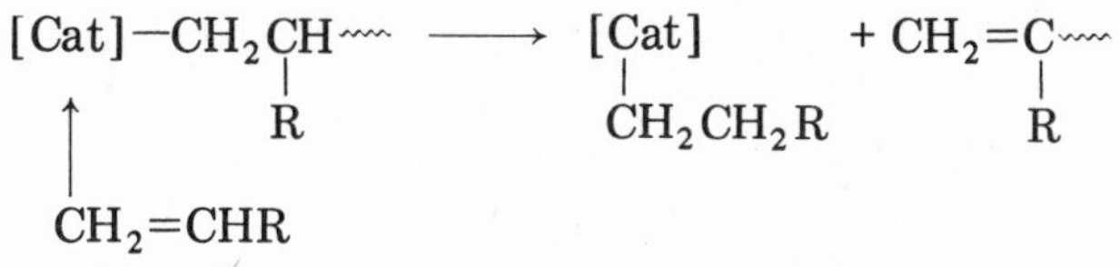

$$[Cat]-CH_2CH\sim \longrightarrow [Cat] \quad + CH_2=C\sim$$

$$\uparrow \qquad | \qquad \qquad | \qquad \qquad |$$
$$\qquad R \qquad CH_2CH_2R \qquad R$$
$$|$$
$$CH_2=CHR$$

XIV

This mechanism is in agreement with the presence of terminal vinylidene groups in polypropene [77], and the dependence of molecular weight on monomer concentration.

Molecular weights of polymers from styrenes deuterated in the side chain are the same as those of polystyrene prepared under the same conditions [78], and this is true also of polymers from ethylene and deuteroethylene [79]. If hydride ion transfer were rate determining an isotope effect would be expected with higher molecular weights in the deuterated polymers. The rate determining step would therefore appear to be coordination of monomer followed either by rapid transfer or insertion into the polymer chain.

5.2 TRANSFER WITH METAL—ALKYL

For most catalysts it is usual to employ an excess of metal alkyl above that which is required to alkylate the transition metal, and it is well established that this excess reduces the molecular weights of the polymers formed. The transfer reaction is formally alkyl interchange, XV, viz.

$$[Cat] - P_n + AlR_3 \rightarrow [Cat]-R + P_n \cdot AlR_2$$

XV

well established for low molecular weight organo-metal compounds [80]. The transfer reaction depends both on the structure of the transition metal and on the organo-metal compound.

5.3 TRANSFER WITH TRANSITION METAL COMPOUND

The occurrence of this reaction has been deduced [81] from the effects of the concentration of transition metal compound on polymer molecular weight, and the mechanism is not understood. It is conceivable that the chains are transferred to transition metal atoms without vacant sites for coordination of monomer and which do not then propagate or that the chain is eliminated with deactivation but exposes a fresh transition metal atom for alkylation and subsequent propagation.

5.4 TRANSFER WITH HYDROGEN

Relatively few other compounds transfer growing polymer chains with coordination catalysts, but molecular hydrogen and its isotopes are effective. The transfer mechanism is

$$[Cat]\,CH_2CHXP_n + H_2 \rightarrow [Cat]H + CH_3CHXP_n$$

XVI

In agreement with this mechanism terminal groupings (methyl from polyethylene and isopropyl from polypropene) increase in concentration. When deuterium is used the end groups are CH_2D [34] and with tritium two atoms per molecule have been found; one from initiation by the metal hydride bond and the other on chain transfer [82].

Spontaneous deactivation to give a stable molecule will also produce a hydride derivative of the catalyst, XVII, viz.

$$[Cat]\,.\,CH_2{-}\underset{\underset{R}{|}}{CH}{-}P_n \rightarrow [Cat]H + CH_2 = \underset{\underset{R}{|}}{C}{-}P_n$$

XVII

After reaction with the monomer to form a new propagating chain the position is formally the same as transfer with monomer. However, the two mechanisms can be distinguished kinetically if realkylation of the catalyst is slow compared with propagation. There is no direct evidence for this reaction although it is well established that the relatively stable alkyls of magnesium and aluminium form metal hydride bonds on decomposition at elevated temperatures [83]. The existence of spontaneous termination has been deduced from a consideration of the kinetics, and by analogy with the effects of hydrogen on the polymerization.

6. Termination

A true termination reaction with destruction of the catalyst site occurs usually with the metal reduced to a lower valency state*. From simple organo-metallic compounds the decomposition products result from combination and disproportionation of the organic groupings [37]. The reaction would appear to be scission of the transition metal—carbon bond,

---

* With some systems it is possible to reactivate the catalytic site (p. 166). The rate of polymerization is maintained with the formation of a large number of polymer molecules per transition metal atom [234]. In such cases termination becomes a transfer reaction.

but since free radicals do not appear to be formed [84] it is conceivably a concerted bimolecular reaction [85].

In the polymerization of ethylene by $(\pi\text{-}C_5H_5)_2TiCl_2/AlMe_2Cl$ [111] and of butadiene by $Co(acac)_3/AlEt_2Cl/H_2O$ [87] there is evidence for bimolecular termination. The conclusions on ethylene polymerization have been questioned, however, and it has been proposed that intra-molecular decomposition of the catalyst complex occurs via ionic inter-mediates [91]. Smith and Zelmer [275] have examined several catalyst systems for ethylene polymerization and with the assumption that the rate at any time is proportional to the active site concentration ($[C_t^*]$), second order catalyst decay was deduced, since $\{1 - [C_t^*]\}/[C_t^*]$ was linear with time. This evidence, of course, does not distinguish between chemical deactivation and physical occlusion of sites. In conjugated diene polymerization by Group VIII metal catalysts ·the unsaturated polymer chain stabilizes the active centre and the copolymerization of a mono-olefin which converts the growing chain from a π to a σ bonded structure is followed by a catalyst decomposition, with a reduction in rate and polymer molecular weight [88].

7. Kinetics

7.1 GENERAL FEATURES

Ziegler—Natta polymerization kinetics are, as indicated in the intro-duction, difficult to study experimentally, particularly those based on insoluble transition metal compounds. They are also complicated by the interaction of concurrent chemical reactions and by physical processes which may have either accelerating or retarding influences on the polymerization.

These will be apparent from the discussion on polymerization mecha-nism but it is perhaps useful to summarize them, with the observation that few if any of the reported investigations have taken them all into account.

(a) Surface activity. Some of the active sites on crystal edges and other lattice defects will be poisoned, some reversibly, others irreversibly. The initial concentration of sites will depend on the composition of the catalyst, i.e., the incorporation of defects during its preparation, and on the size of the crystallites.

(b) New site formation. Break up of crystallites as a result of the growth of polymer chains will expose fresh sites. Stirring breaks up agglomerations of polymer coated catalyst particles and, if sufficiently vigorous, may result in cleavage of crystals.

(c) Multiplicity of catalyst species. More than one type of initiating species may be formed both in homogeneous and heterogeneous catalysts, but in the latter identification and determination are rendered more difficult since they cannot be studied directly by physical techniques such as ESR and NMR.

(d) Catalyst decay. Catalyst sites can undergo decomposition with loss of activity and in some cases the activity can be restored, e.g., by oxidation. There may be great differences in catalyst stability even with those of allied structure obtained from the same reactants.

(e) Mass transfer of monomer. Solution of gaseous monomers may become rate controlling and will be affected by the speed and efficiency of stirring.

(f) Encapsulation of catalyst sites. Propagation, metal—alkyl transfer and monomer transfer may become diffusion controlled when the catalyst particles become encapsulated with insoluble polymer.

7.1.1 Polymerization rate

The concentration of active centres $([C^*])^*$ and polymerization rate (R_p) are given by

$$[C^*] = f[T]^a [A]^b \text{ and } R_p = k_p [C^*] [M]_0^{c\dagger} \tag{1}$$

---

* $[C^*]$ is the concentration of active centres under steady state conditions. The effects of initiation, transfer and termination are considered elsewhere.

$\dagger$ With heterogeneous catalysts where concentrations of active species result from adsorption equilibria of components on to the surface the rate is more properly expressed in the form

$$R_p = k'_p \theta_M \theta_A S \tag{1'}$$

where θ_M and θ_A are the fractions of catalyst surface covered by monomer and metal alkyl and S is the surface area of the transition metal compound [41]. However, most polymerizations are first order in monomer hence θ_M will be proportional to $[M]$, S clearly will be related to $[T]$ at the steady state and as indicated above at a fixed ratio of A/T the effect of the metal alkyl can be included in the rate constant. It is thus convenient to express the rate constants using molar concentrations as in eqn. (1). In general calculated reaction rate coefficients are based on molar concentrations as this facilitates comparisons between heterogeneous and homogeneous systems.

As no completely satisfactory representation of the formation and nature of active species in heterogeneous catalysts has yet been devised this may be an over-simplification. Clearly the first order dependence of rate on monomer concentration is indicative of comparable solvent and monomer adsorption, even with what might be considered more strongly adsorbed monomers, such as butadiene, in comparison with mono-olefins or aliphatic hydrocarbons. The role of the more strongly adsorbed metal alkyl is more difficult to assess. The proportion of active alkylated transition metal atom sites will obviously increase with increase in $[A]$ up to a limiting value.

If alkylation of the surface were irreversible and catalysts were preformed (i.e. no concurrent initiation and propagation) the polymerization rate should rise to a sharp maximum and then decline as occupancy of the surface by monomer and solvent is reduced. However, in view of the relatively low activity of most heterogeneous catalysts the maximum rate would be expected at much lower levels of metal alkyl (i.e.

Al/Ti $\ll$ 1) than are observed. Broad maxima at Al/Ti $\gg$ 1 are more consistent with reversible site formation or occupation, but with the qualification that only a fraction of the available sites (θ'_A) are catalytically active. The reaction rate can then be written $R_p = k'_p[M]\theta'_A S$.

The influence of increasing $[A]/[T]$ is represented in a simplified form in Fig. 3a(a) for values of K_A and K (the dissociation constant of dimeric metal alkyl) reasonably close to those reported in the literature. At all values of $[A]$ reactions will be first order in $[M]$, and rates at a particular monomer concentration will increase with increase in alkyl concentration to a maximum defined by the ratio of θ'_A/θ_A and then decline. The height of the maximum and the value of $[A]$ at which it occurs will obviously fall as θ'_A/θ_A decreases. Fig. 3a(b) shows the dependence of polymerization rate on metal alkyl concentration for $R_p \propto \theta_A$ and $R_p \propto \theta'_A$. The curves are clearly of a similar shape although since $\theta'_A < \theta_A$ the absolute rates would be different.

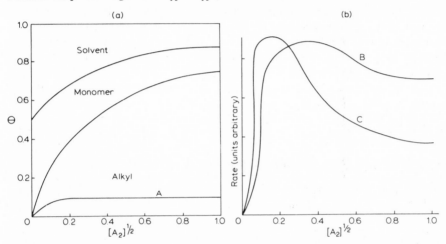

Fig. 3a. (a) Effect of metal alkyl concentration on catalyst surface coverage. $[S]$ = $[M]$; $K_S \simeq K_M \ll K_A$; $\theta_S = \theta_M$; $K_A = 10^2$ $K = 10^{-3}$; Curve A, Alkyl surface coverage (θ'_A) rising to a maximum of 0.1. (b) Effect of metal alkyl concentration on polymerization rate. Curve B, $R_p \alpha\, \theta_M \theta_A$; curve C, $R_p \alpha\, \theta_M \theta'_A$.

Quantitative data on surface coverage of catalyst are sparse, but Burfield and Tait [300] quote a value of θ_M of 0.216 for a monomer concentration of 2.00 mole/l in the polymerization of 4-methylpentene-1 by VCl_3/AlR_3. Proportionality between $[M]$ and θ_M would lead to a value of θ_M = 0.85 for pure monomer. Burfield, McKenzie and Tait [299] have given a value of θ_A equal to 0.125 for rather similar experimental conditions (the same monomer concentration but a higher metal alkyl concentration.) It is undoubtedly an oversimplification to assume that the solvent, benzene, and monomer can replace one another in proportion to concentrations on the catalyst surface without affecting the adsorption of the metal alkyl. Nevertheless, the sum of θ_M and θ_A is relatively close to unity and gives confidence in the correctness of expressing reaction velocity constants in second order units.

The position when monomer added in gaseous form is not in equilibrium with the solvent has been considered by Boocock and Haward [252]. They derive the relationship

$$\frac{1}{R_p} = \frac{(k_d + k_p)}{k_d k_p[T][M]_0} + \frac{1}{k_c X[M]_0}$$

[T], [A] and [M] are the concentrations of transition metal compound, metal alkyl and monomer and the coefficients a, b and c will depend on the polymerization mechanism; f is the efficiency of utilization of the transition metal compound.

The coefficient, a, will be unity unless the active centre is formed from the transition metal complex in an aggregative or dissociative step, and, in general, only a fraction of the transition metal will participate in the polymerization. At a fixed ratio of [A]/[T] the concentration of metal alkyl need not be considered, its effect on [C*] being accounted for in the composite rate coefficient. When the [A]/[T] ratio is increased the rate usually increases either to a steady value or to a maximum and then declines, and, dependent on the range of values of [A]/[T] and the type of catalyst, b, will be positive, zero or negative. As one or two monomer molecules may coordinate with the catalyst the exponent, c, will have a value between 0 and 2, dependent on the interaction between monomer and catalyst and the rate of propagation.

If active sites are produced by reversible reaction of the metal alkyl with the transition metal compound, viz.

$$T + A \; \underset{}{\overset{K_A}{\rightleftharpoons}} \; C*$$

their concentration at equilibrium is given by

$$[C*] = \frac{fK_A[A][T]}{1 + K_A[A]} \quad (\text{provided } [A] \gg f[T]) \tag{2}$$

(a) If monomer does not coordinate with the catalyst complex the rate of polymerization will be given by

$$R_p = k_p[C*][M] = \frac{fk_pK_A[A][T][M]}{1 + K_A[A]} \tag{3}$$

As the concentration of metal alkyl is increased the rate will rise to the maximum value ($fk_p[T][M]$), and at constant [M] and [T] there will be

---

k_c and k_d are the coefficients for transfer of gaseous monomer to the solvent and diffusion of monomer across the liquid film at the particle surface, $[M]_0$ is the saturation concentration of monomer and X the gas/liquid area.
For $k_d \gg k_p$

$$\frac{1}{R_p} = \frac{1}{k_p[T][M]_0} + \frac{1}{k_cX[M]_0}$$

This predicts that at fixed [T] as X increases, i.e. with improved agitation, the rate will rise to a maximum value, and there will be a linear relationship between $1/R_p$ and $1/[T]$. It has been assumed that monomer is in equilibrium with the solution and in practice the problem is restricted almost entirely to ethylene polymerization.

a linear relationship between $[A]/R_p$ and $[A]$. As the lower aluminium alkyls are dimeric in solution but coordinate on the catalyst surface in the monomeric form the concentration term will be $[A]^{1/2}$.

(b) If the monomer is coordinated prior to addition the rate is not necessarily directly proportional to the concentration in solution. The reaction scheme may be written

$$M_n C^* + M \underset{k_{-1}}{\overset{k_1}{\rightleftharpoons}} M_n C^* M \xrightarrow{k_p'} M_{n+1} C^*$$

$$(K_M)$$

The total concentration of active centres, eqn. (2), will be the sum of uncoordinated and coordinated sites, viz.

$$[C^*] = [\Sigma M_r C^*] + [\Sigma M_r C^* M]$$

and $R_p = k_p' [\Sigma M_r C^* M]$ is given by

$$\frac{k_p' k_1 [C^*] [M]}{k_1 [M] + (k_{-1} + k_p')}, \tag{4}$$

(i) When the complex between monomer and catalyst is weak ($k_1 \ll k_{-1}$) the polymerization is first order in monomer and

$$R_p \simeq \frac{k_1 k_p' [C^*] [M]}{(k_{-1} + k_p')} \tag{5*}$$

(ii) If the monomer is strongly coordinated $k_1 \gg k_{-1}$ but is not rate controlling in respect of propagation, i.e., $k_1 \gg k_p'$, the rate will be independent of monomer concentration ($R_p \simeq k_p' [C^*]$).

Second order dependence of rate on monomer will result from insertion of weakly coordinated monomer molecules in pairs or if the coordination of a second monomer molecule facilitates the insertion of a weakly complexed molecule, i.e.

$$M_n C^* + M \rightleftharpoons M_n C^* M + M \rightleftharpoons M_n C^* M_2 \rightarrow M_{n+2} C^*$$

or

$$M_n C^* + M \rightleftharpoons M_n C^* M + M \rightarrow M_{n+1} C^* M$$

In the second case if the transition metal—olefin complex is stable the polymerization will be first order in monomer [25].

---

* The composite term in (5) equates to k_p in eqn. (1).

References pp. 249—257

If monomer and metal alkyl are adsorbed at the transition metal to produce the growing chains, an excess of one or other of the reagents will occupy all the sites. Hence with increasing $[A]/[T]$ the polymerization rate should rise to a maximum and then decline. The concentration of active centres is given by

$$[C^*] = \frac{fK_A K_M [A][M][T]}{(1 + K_A[A] + K_M[M])^2} \simeq \frac{fK_A K_M [A][M][T]}{(1 + K_A[A])^2} \text{ for } K_A \gg K_M \quad (6)$$

The maximum in rate with increase in $[A]$ at constant $[M]$ and $[T]$ occurs at

$$[A]_{max} = \frac{K_M[M] + 1}{K_A}$$

with

$$[C^*]_{max} = \frac{f[M][T]}{4(K_M[M] + 1)} \quad (7)$$

If the catalyst decomposes at a significant rate during the time of the polymerization there will be a fall in the concentration of active sites and a decrease in polymerization rate with time. The relationships between the yield of polymer $[P]'$ or conversion $[P] = [M]_0 - [M]/[M]_0$) for first and second order decomposition (k_d) with either constant or declining monomer concentration, are

Catalyst decompositon reaction	Constant M	Declining M (Reaction first order in monomer)
None	$[P]' = [M]_0 k_p [C^*]t$	$-\ln(1-[P]) = k_p[C^*]t$
First order	$[P]' = [M]_0 \dfrac{k_p}{k_d} [C^*]_0(1-e^{-k_d t})$	$-\ln(1-[P]) = \dfrac{k_p}{k_d} [C^*]_0(1-e^{-k_d t})$
Second order	$[P]' = [M]_0 \dfrac{k_p}{k_d} \ln(1+[C^*]_0 k_d t)$	$-\ln(1-[P] = \dfrac{k_p}{k_d} \ln(1+[C^*]_0 k_d t)$

Appropriate graphical plots can be employed to distinguish between the different cases [92].

Results from published work appear to fit most of the cases considered. With Natta type catalysts from metal subhalides and aluminium alkyls the rate rises to a maximum value with increase in $[A]/[T]$ ratio consistent with the relationship in eqn. (1) (Fig. 4). Ingberman et al. [271] have shown the data to be equally well represented by a Langmuir type

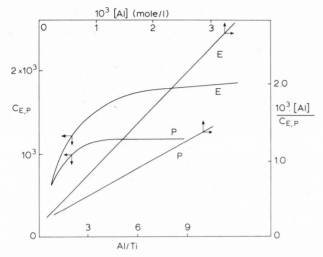

Fig. 4. Effect of catalyst composition on polymerization of olefins [96]. E, ethylene; P, propene; $TiCl_3/AlR_3$, $[Ti] = 2.6 \times 10^{-3}$ mole l^{-1} E: 450 torr; temp., $50°C$: P: 355 mm; temp, $56°C$. $C_{E,P}$, ml of monomer polymerized.

adsorption or by the Freundlich approximation, $R_p = a[A]^{1/n}$ (Fig. 4a). The reversibility of the reaction does not appear to have been established unequivocally, however, and this point is considered later in the discussion on the initiation reaction.

The data available do not clearly distinguish between proposed schemes. Thus, Keii [269] has tested the equations (2) and (6) in their linear forms $[A]/R_p$ versus $[A]$ and $\{[A]/R_p\}^{1/2}$ versus $[A]$, for

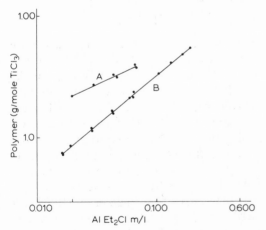

Fig. 4a. Polymer yield versus $AlEt_2Cl$ concentration for the polymerization of propene by α/γ $TiCl_3/Al(i\text{-}Bu)_3$; Al/Ti = 1/2 (curve A), and α/γ $TiCl_3/AlEt_2Cl$; Al/Ti = 0.6/1 (curve B). Solvent n-heptane, $40°C$; $TiCl_3 = 10 \pm 0.2 \times 10^{-3}$ mole; time 2.5 h [271].

published data on olefin polymerization [93, 236], and has shown them to be equally good at low concentrations of A, and for there to be deviations with both at high concentrations of A.

Ziegler systems from halides of the transition metal in its highest valence state frequently exhibit rates which pass through maxima, in some cases sharply (Fig. 5), and then fall as the ratio of transition metal to metal alkyl is decreased. It is unlikely, however, that this results from equilibria giving rise to eqns. (6) and (7) and the explanation is more probably that the polymerization rate reflects the activity of the catalyst sites produced by the particular stoichiometry. Table 1 shows the dependence of catalyst composition on molar ratio of reagents.

The position with Natta type catalysts, which in general give simpler and more stable catalytic entities, may be different. Thus, although many

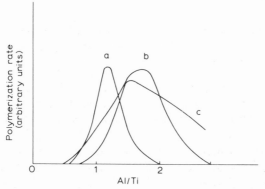

Fig. 5. Effect of catalyst composition on polymerization rate. (a) Isoprene, $TiCl_4/Al(i\text{-}Bu)_3$ [41]; (b) butene-1, $TiCl_4/Al(i\text{-}Bu)_3$ [98]; (c) ethylene, $TiCl_4/LiAlR_4$ [97].

authors have observed the polymerization rate to rise to a steady value with increase of A/T, Vesely [93] for propene using $\alpha\text{-}TiCl_3/AlEt_3$ found the rate to pass through a maximum, and, with use of eqn. (7), K_M and K_A were calculated to be 0.163 and 21.2 l mole^{-1} respectively (in n-heptane at 50°C). Soga and Keii [94] reported $K_A = 140$ l mole^{-1} for $AlEt_3$ and 40 l mole^{-1} for $ZnEt_2$ under the same conditions of temperature and solvent (n-heptane at 43.5°C). From the temperature coefficients of K_A, heats of absorption, (ΔH_a) of -12 and -3 kcal mole^{-1} were obtained for $AlEt_3$ and $ZnEt_2$ while Vesely et al. [95] reported the significantly higher value of $\Delta H_a = -35$ kcal mole^{-1} ($\Delta G = 2$ kcal mole^{-1}) for $AlEt_3$.

For 4-methylpentene-1 with $VCl_3/Al(i\text{-}Bu)_3$ McKenzie et al. [236] also observed a maximum in rate, at Al/V = 2.2—2.3. This alone does not rule out the possibility of sites of lower activity being produced at higher Al/V ratios, but in parallel experiments where the concentration of aluminium alkyl in solution remained constant the rate per unit concentration of vanadium increased to a steady value with increase in Al/V ratio, indicating competition between monomer and metal alkyl for adsorption on the catalyst surface. With this particular system the effects on rate become noticeable at a ratio of M/A below about 40. Surface coverage of the catalyst by monomer was estimated to fall from 0.216 at $[A]$ = 33.7×10^{-3} mole l^{-1} to 0.079 at $[A]$ = 558×10^{-3} mole l^{-1} ($[T]$ and $[M]$ constant at 18.5×10^{-3} mole l^{-1} and 2.0 mole l^{-1}) and K_M was estimated to be 0.164 ± 0.02 l mole^{-1} at 30°C. Values of K_A reported are 5.12 ± 0.20 l mole^{-1} for Al(i-Bu)$_3$ and 224 ± 10 l mole^{-1} for AlEt$_3$ at 30°C [237]. For Al(i-Bu)$_3$, ΔH_a = 5.2 kcal mole^{-1}. (The high value for AlEt$_3$ compared with Vesely's value of 21.2 l mole^{-1} is accounted for by the use in the latter of metal alkyl dimer concentrations. When allowance is made for the low dissociation constant of (AlEt$_3$)$_2$ ($K_d^{1/2}$ = 0.035), the results are in agreement*.)

With most systems polymerization rates have been found to be first order in both transition metal compound and monomer. The few instances reported where polymerizations are second order in monomer cannot be regarded as satisfactorily established as involving two monomer molecules in the propagation reaction. Apparent second order kinetics can, for example, arise as a result of chain transfer with slow reinitiation or from termination reactions. With one system [63] the rate has been found to be independent of monomer concentration.

Catalyst stabilities vary greatly. Those from preformed subhalides tend to be more stable and some of them remain active when stored for many weeks. Ziegler catalysts contain species of variable stability and rates decline rapidly with time, unless an ageing period is given to the catalyst.

7.1.2 Polymer molecular weight

Average molecular weight is the other property from which rate coefficients can be evaluated provided there is adequate information on the mechanism. In coordination polymerization it is usual for the molecular weight to rise to a limiting value (Fig. 6) indicative of transfer reaction†.

In a system in which the active centres have been preformed the number average degree of polymerization will be given by

$$\bar{P}_n = \int_0^t R_p \, dt / \{[C^*] + \int_0^t \Sigma R_{tr} \, dt\} \tag{8}$$

---

* Monomer—dimer equilibrium constants and heats of dissociation of AlEt$_3$ and AlMe$_3$ are reported by Robb et al. [253].

† In the absence of transfer reaction

$$d\bar{P}_n/dt = k_p[M]$$

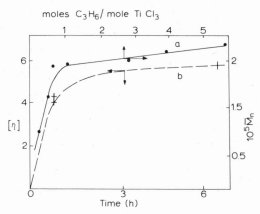

Fig. 6. Molecular weight dependence of polypropene on conversion. (a) $TiCl_3/AlEt_2Cl$ [20]; temp., 60°C. (b) $TiCl_3/AlEt_3$ [81]; temp., 15°C; Al/Ti = 1.

As the transfer reactions are the result of interactions of growing chains with catalyst components or with monomer

$$\Sigma R_{tr} = [C^*]\{k_{tr_A}[A]^i + k_{tr_T}[T]^i + k_{tr_M}[M]\}$$

k_{tr_A}, k_{tr_T} and k_{tr_M} are the transfer coefficients for reaction with metal alkyl, transition metal and monomer respectively.

The exponents, i, will be unity unless there is dissociation of the catalyst components involved in the transfer reactions. When the catalyst is stable two simple cases can be identified.

(a) Constant monomer concentration.

$$\int_0^t R_p dt = k_p[C^*][M]_0 t \quad \text{and} \quad \int_0^t \Sigma R_{tr} dt = \Sigma R_{tr} t$$

and thus

$$1/\bar{P}_n = \frac{1}{k_p[M]_0 t} + \frac{k_{tr_M}}{k_p} + \frac{k_{tr_A}[A]^i}{k_p[M]_0} + \frac{k_{tr_T}[T]^i}{k_p[M]_0} \tag{9}$$

The slope of $1/\bar{P}_n$ versus $1/t$ gives k_p and the intercept the sum of the transfer reactions. These are individually resolved by measuring molecular weights as functions of the concentration of each component. The change in slope of $1/\bar{P}_n$ versus $1/t$ with temperature gives the activation energy for propagation, E_p.

(b) Falling monomer concentration and with transfer of growing chains by monomer. Here it is convenient to express the results in terms of conversion, [P], when

$$\int_0^t R_p \, dt = [M]_0 - [M] = [P][M]_0$$

$$\int_0^t k_{tr_M}[C^*][M] = k_{tr_M}([M]_0 - [M])$$

$$1/\bar{P}_n = (k_{tr_M}/k_p) + ([C^*]/[P][M]_0) \tag{10}$$

From a plot of $1/\bar{P}_n$ versus $1/[P]$, $[C^*]$ can be calculated and, in conjunction with rate data, the rate coefficients. Systems in which in addition there are transfer reactions dependent on time only or where there is catalyst decay give more complex relationships not amenable to simple graphical presentations.

More complex situations have been treated analytically, such as reversible deactivation of initially active catalyst either by dimerization $(2C^* \rightleftharpoons C_2)$ or bimolecular reaction $(C^* + C^* \rightleftharpoons 2C)$ [99]. Approach to equilibrium concentration of active centres would be accompanied by a fall in rate to a steady value (assuming constant monomer concentration and a stable catalyst) and a rise in molecular weight with time either to a maximum value or to a steady rate of increase dependent on the presence or absence of transfer reactions. The effect on average molecular weight of transfer reactions in which the catalyst entities possess two active centres has been calculated [100]. Although some ionic catalysts may behave in this way there is no evidence to indicate that these mechanisms apply to any known coordination catalyst.

7.1.3 Catalyst association—dissociation

Certain catalysts, such as the π-allyl nickel halides, which are effective for butadiene, exist in dimeric form and are converted into active species by dissociation. Partial dissociation followed by addition of monomer to the active catalyst fragments would result in the rate of polymerization being proportional to $[C_2]^{1/2}[M]$, where C_2 represents the dimeric catalyst species.

Some complexes dissociate fully in solution (e.g., π-allyl nickel trifluoroacetate) whence [63] $R_p \propto [C_2][M]$. Donor solvents (triphenylphosphite, nitrobenzene) can also coordinate with the metal and influence catalyst dissociation with effects on rate and polymer molecular weight; the structure of the polymer may also be affected [173].

There is evidence to show that diene monomer is involved in the association—dissociation equilibria [61], and there are several possibilities with effects on reaction rates and orders in monomer and catalyst. The following have been considered by Harrod and Wallace [250].

(a) Dissociation of catalyst followed by rapid equilibrium coordination of catalyst fragments with monomer, viz.

$$C_2 \rightleftharpoons 2C; C + M \rightleftharpoons CM$$

(i) Catalyst mainly as C_2, when $R_p \propto [C_2]^{1/2}[M]$
(ii) Catalyst mainly as CM, when $R_p \propto [C_2][M]^0$
(b) Pre-coordination of monomer with catalyst, dissociation of the adduct and rapid equilibrium coordination of a second monomer unit to one of the fragments, viz.

$$C_2 + M \rightleftharpoons C_2M; C_2M \rightleftharpoons CM + C; C + M \rightleftharpoons CM$$

(i) With catalyst mainly as C_2, then $R_p \propto [C_2]^{1/2}[M]$
(ii) With catalyst mainly as C_2M and $C_2M + M \rightleftharpoons 2CM$
If weakly dissociated by monomer then $R_p \propto [C_2]^{1/2}[M]^{1/2}$
If fully dissociated by monomer, then $R_p \propto [C_2][M]^0$
(c) Pre-coordination of two monomer molecules with catalyst, followed by dissociation to the active species, viz.

$$C_2 + M \rightleftharpoons C_2M; C_2M + M \rightleftharpoons C_2M_2; C_2M_2 \rightleftharpoons 2CM$$

(i) Catalyst mainly as C_2 and coordinated dimer fully dissociated to CM; $R_p \propto [C_2]^{1/2}[M]$
(ii) Catalyst mainly as C_2M; $R_p \propto [C_2]^{1/2}[M]^{1/2}$
(iii) Catalyst mainly as C_2M_2; $R_p \propto [C_2]^{1/2}[M]^0$

With suitable values for the equilibrium constants the order in monomer will fall with increase in its concentration.

With these catalysts molecular weights would, in the absence of chain termination reactions, increase directly with conversion, with Poisson distributions. Monomer transfer, which has been observed with the π-allyl nickel halide catalysts, would lead to lower values and a broadening of the distribution.

The degree of polymerization can be expressed in the form

$$\frac{1}{\bar{P}_n} = \frac{2[C_2]}{\text{conversion}} + \frac{k_{tr_M}}{k_{tr_M} + k_p}$$

7.1.4 Retardation and inhibition of polymerization

There is no adequate kinetic treatment for this aspect of coordination polymerization although many compounds have a major influence on the reactions. Some have a retarding effect, e.g., water, alcohols, carbon dioxide [101], while others accelerate polymerization. Examples of the latter are hexachlorocyclopentadiene [102], ethyl trichloroacetate [103], oxygen [104], and phosphorus trichloride [105].

Donor molecules such as thioethers and tertiary amines, may accelerate or retard the rate [106, 107], and there may be changes in the microstructure of the polymers [63, 108].

Those additives which are chemically transformed by the catalyst have effects which depend on the order of addition to the system. For example, water in small amounts reacts with aluminium alkyls to form aloxanes, $(R_2 AlO)_2$, which are effective catalysts, but if added first to α-TiCl$_3$ it results in an induction period proportional to the amount added followed by acceleration up to the normal rate as the coordinated water is removed from the catalyst sites. A detailed examination of the kinetics of the effect of water has been carried out by Kissin et al. [109]. The rate of removal of adsorbed water follows the law

$$-d[H_2O]/dt = K[A][H_2O]$$

giving rise to the following variation of rate with time:

$$R'_p = (R'_p)_0 (1 - \exp K[A]t)$$

$(R'_p)_0$ is the rate without added water; R'_p and $(R'_p)_0$ are the reduced rates at unit monomer and transition metal concentrations. Taking into account the surface area of the catalyst it was estimated that one molecule of water blocks a site of $\sim 35 \text{Å}^2$.

Methanol has an effect which is more complex. There is an initial retardation resulting from sites on the surface being blocked, giving rise to proportionality between $((R'_p)_0 - R'_p)/(R'_p)_0$ and $[D]/[T]$ (Fig. 7), [D]

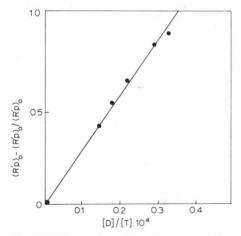

Fig. 7. Effect of methanol concentration on polymerization of propene by α-TiCl$_3$/AlEt$_3$ (initial rates) [109].

is the concentration of methanol and $(R_p')_0$ and R_p' refer to the initial rates. As the alcohol is removed it produces a more active co-catalyst and the rate then rises to a higher value than the control (Fig. 8). With the assumption that one alcohol molecule blocks one site the area of an active centre is calculated to be ~35Å^2, in agreement with the findings obtained with water.

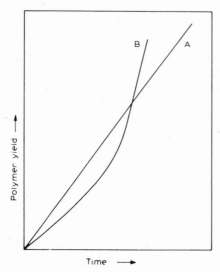

Fig. 8. Polymerization of propene by α-TiCl$_3$/AlEt$_3$ in the presence (curve B) and absence (curve A) of methanol [109]. A, Al/Ti = 2.88; B, Al/Ti/MeOH = 3.48/1/2.6 $\times$ 10^{-4}.

Vesely [93] has expressed the effects of a number of additives in terms of competitive adsorption at the catalyst site by the insertion of a term K_D [D] in the denominator of eqn. (6). Values of K_D quoted for propene polymerization by α-TiCl$_3$/AlEt$_3$ are given in Table 2. However, if the additives are removed by chemical reaction the use of an equilibrium constant as a measure of their ability to coordinate with the catalytic centre would appear to be a considerable over-simplification.

TABLE 2
Values of K_D in the polymerization
of propene by α-TiCl$_3$/AlEt$_3$ [93].

Substance	K_D
COS	2.4 x 10^3
CS$_2$	1.7 x 10^3
Me$_2$S	4.1 x 10^2
H$_2$O	1.3 x 10^2
H$_2$S	50

Soga et al. [272] have studied the effect of $SeOCl_2$ on propene polymerization by $TiCl_3/AlEt_3$. The effect is to inhibit polymerization completely at $SeOCl_2/AlEt_3$ ratios above $\frac{1}{2}$. It was concluded that a non-catalytic complex is formed which is adsorbed on to the catalyst surface. From the change in rate with $[SeOCl_2]$, K_D was calculated to range from 4.5—6.5 to $3.2 \cdot 10^3$ l mole^{-1} for temperatures 31 to 61°C. Polymer molecular weights are unaffected and the observed increase in isotacticity was the result of a reduction in $AlEt_3$ concentration.

Burfield and Tait [300] have studied the effect of triethylamine on the $VCl_3/Al(i\text{-}Bu)_3$ catalyzed polymerization of 4-methylpentene. Rates passed through maxima at ratios of $NEt_3/Al(i\text{-}Bu)_3$ ca. 0.5—0.7 (Al/V = 2/1) and K_D was calculated to be 360 l mole^{-1} at 30°C, obtained from the relationship

$$1/R_p = \frac{1 + K_M[M] + K_D[D]_F}{k_p[C^*]K_M[M]}$$

($[D]_F$ is the concentration of free, uncomplexed donor = $[D]_0 - [A]$. It was assumed that $k_p[C^*]$ remained constant throughout, hence the slope and intercept of the line relating $1/R_p$ versus D are related to K_D by

$$K_D = \frac{\text{Slope}}{\text{Intercept}} \{1 + K_M[M]\}$$

The higher adsorptive power of NEt_3 compared with $Al(i\text{-}Bu)_3$ or even the more strongly adsorbed $AlEt_3$ (see p. 159) is apparent. The active centre concentrations (see pp. 172—179) in the presence of triethylamine were comparable (ca. 4×10^{-4} mole/mole VCl_3) to that in the absence of donor. Activation by NEt_3 is thought to result by removal of adsorbed aluminium alkyl compounds which give lower activity catalytic complexes on the VCl_3 surface.

Equilibria in heterogeneous systems is complicated by the fact that species produced in the polymerization, particularly the initiation reaction (see p. 137) participate in the reactions. Thus $AlEtCl_2$ is well established as an interfering substance in the polymerization of propene by $TiCl_3/AlR_2Cl$ [20]. Here reaction rates follow the law

$$(R_p)/(R_p)_0 = 1/(1 + k[I] + [I_0])$$

$(R_p)_0$ is the maximum rate without inhibitor (I = $AlEtCl_2$). R_p is the rate in the presence of added inhibitor [I] or generated in preparation of the catalyst $[I_0]$. Complexing agents used to remove $AlEtCl_2$, e.g. KEt_2AlCl_2, ($AlEtCl_2 + KEt_2AlCl_2 \rightarrow AlEt_2Cl + KEtAlCl_3$), restore the rate to the maximum value and rates become much less dependent on conversion.

Burfield and Tait [300] have reassessed the data of Caunt [20] for $AlEtCl_2$ on the α-$TiCl_3$/$AlEt_2Cl$ system and of Boor [106] on the effect of $N(n\text{-Bu})_3$ on the polymerization of propene by $TiCl_3$/$ZnEt_2$ and show that a Langmuir type adsorption curve $1/R_p = A + B[D]$ applies. ([D] represents the concentration of $N(n\text{-Bu})_3$ or $AlEtCl_2$). The peak in the rate curve for the effect of i-Pr_2O concentration on the polymerization of isoprene by VCl_3/$AlEt_3$ [107] was attributed to changes in concentration of active sites since k_p was not significantly different in the presence of the donor and values for [C*] were much increased — clearly very different from the VCl_3/AlR_3 catalyst for 4-methylpentene-1. These considerations confirm the view that competitive adsorption at the heterogeneous catalyst surface dominates the polymerization kinetics and explains why reactions are influenced by order of addition of catalyst, monomer and additive. Declining polymerization rates, so characteristic of hetero-geneous catalysts, can be interpreted as by Caunt [20], solely in terms of equilibria between complexed species of differing activity. In cases where there are true termination reactions — particularly the catalysts prepared with more strongly adsorbed and active reducing agents such as $AlEt_3$ — the results are more difficult to interpret.

The effects of oxygen are not fully understood. Reaction with an aluminium alkyl will convert some of the metal—alkyl bonds to alkoxide, with a possible change in co-catalyst activity, but a greater effect could result from a higher valence state in the transition metal and it has been suggested that the maintenance of high catalytic activity in titanium and vanadium catalysts for ethylene polymerization results from oxidation of inactive lower valence metal compounds [104]. Hexachlorocyclopenta-diene likewise may operate by oxidizing divalent vanadium compounds to active trivalent complexes [102]. Keii [269] has reported data of Doi et al. for the effect of oxygen on propene polymerization by $TiCl_3$/$AlEt_3$ which indicates that oxygen and $AlEt_3$ react with the stoichiometry $A + \frac{3}{2}O_2 = X$; X is considered to be $Al_2(OEt)_3Et_3$. From the dependence of rate on oxygen concentration the adsorption constant of the alkoxide was estimated to be $K_D = 59{-}26 \times 10^3$ l $mole^{-1}$ over the temperature range 30 to 50°C.

Unreactive compounds which can coordinate reversibly with the catalytic complex in competition with monomer, such as aromatic hydrocarbons or pyridine and other tertiary amines, may accelerate the rate in small amounts [106] but in high concentrations have a retarding effect. These effects have been interpreted in terms of site blocking and activation and, in the case of heterogeneous catalysts, by increasing the number of active sites. For a soluble catalyst the scheme

$$C*M + D \underset{}{\overset{K_D}{\rightleftharpoons}} CD + M$$

$$C*M \overset{k_p}{\longrightarrow} P_1C*$$

$$P_1C^* + M(\text{or } D) \xrightarrow{\text{fast}} P_1C^*M \text{ (or } P_1CD) \text{ etc.}$$

leads to

$$R_p = k_p \Sigma [C^*M] = \frac{k_p f[T][M]^2}{[M] + K_D[D]}$$

i.e.

$$\frac{(R_p)_0}{R_p} = \frac{[M] + K_D[D]}{[M]}$$

Data for K_D for a number of aromatic hydrocarbons in the polymerization of butadiene by a soluble cobalt catalyst are given in Table 3 [110].

TABLE 3
Values of K_D for aromatic hydrocarbons in the polymerization of butadiene by $Co(acac)_3/Al_2Et_3Cl_3$ (K_D butadiene = 1.0) [110]

Donor	K_D
Benzene	6.7×10^{-3}
Toluene	0.41
p-Xylene	2.27
o-Xylene	2.46
m-Xylene	6.11
1,2,4-Trimethyl benzene	9.8
1,3,5-Trimethyl benzene	12.2
1,2,4,5-Tetramethyl benzene	105
Pentamethyl benzene	292
Hexamethyl benzene	2798

For a further comment see note added in proof (C), page 248.

7.1.5 Retardation resulting from encapsulation of catalyst by insoluble polymer

The growth of insoluble polymer around catalyst particles forms a barrier through which monomer must diffuse and, if this is slow in comparison with reaction, a gradient of monomer concentration will be established with a progressive fall in rate to a steady value.

Allen and Gill [154] considered the effect of this on change of polymerization rate with time for two models distinguished by either a steady concentration of monomer in the polymer layer or by a falling concentration of monomer through the polymer coating, viz.

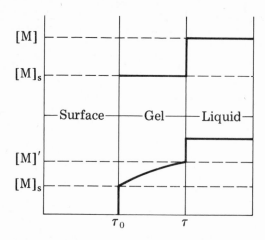

With the first model the rate declines from the normal unobstructed rate to a steady rate when the particles are completely covered, viz.

$$R_p = (R_p)_0 + \frac{[P]}{[P]_s} \{(R_p)_s - (R_p)_0\}$$

$(R_p)_0$ and $(R_p)_s$ are the initial and steady rates, $[P]$ the conversion and $[P]_s$ the conversion at the steady rate. The significant assumption is that gel merely encapsulates the particle and its effect is independent of the thickness of the layer.

With the diffusion controlled model it was deduced that the rate of polymerization per particle was given by

$$R'_p = k'_p [M]_s r_0^2; \quad [M]_s = \frac{rD[M]'}{(D + k'_p r_0^3)r - k'_p r_0^4}$$

where k'_p is a propagation rate coefficient, r_0 and r are the radii of the catalyst particles and encapsulated particles, D the diffusion constant and $[M]_s$ and $[M]'$ the concentrations of monomer at the catalyst and encapsulated particle surfaces. This predicts a rate rising to a constant value as the radius of the capsule increases, i.e. with conversion, but not in finite time.

Iguchi [266] considered a mechanism in which growing chains are occluded, and thus become non-propagating, by a process kinetically analogous to the interaction of crystallites growing from the polymer melt and following Avrami kinetics. (It was assumed that chemical transfer and termination reactions do not occur.) The polymerization rate is given by the concentrations of monomer and active centres at any time, i.e. $R_p = k_p [C^*]_t [M]_t$. If the decay of catalytic activity follows the relationship

$$[C^*]_t = [C^*]_0 \left\{ 1 - \frac{[M]_0 - [M]}{[M]_0} \right\}$$

the polymerization will become second order in monomer

$$\frac{[M]_0}{[M]_t} = k_p [C^*]_0 t + 1$$

Number and weight average molecular weights were also calculated. The former is the total monomer polymerized divided by the active site concentration, while the latter, a more complex function, becomes equal to $2\overline{M}_n$ at complete conversion.

7.2 POLYMERIZATION DATA

7.2.1 Initiation stages

There is little information on rates of initiation in coordination polymerization. With the soluble catalyst $(\pi\text{-}C_5 H_5)_2 TiCl_2 - AlMe_2 Cl$ for ethylene, Chien [111] reported $k_i = 4.99 \times 10^{-3}$ l mole^{-1} s^{-1} at 15°C (E_i = 15.5 kcal mole^{-1}). This was determined from the rate of incorporation of C_{14} labelled $AlMe_2 Cl$ into the polymer. In most systems examined the initiating centres have been preformed and reaction starts immediately. This is the case with Natta type catalysts from transition metal subhalides with the alkyls of aluminium, beryllium and zinc. With some catalysts the rate rises rapidly to a steady value but with others there is a rapid rise to a maximum followed by a decline to a constant rate, the latter being

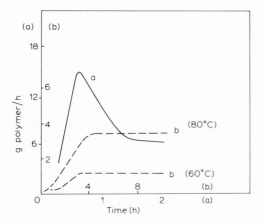

Fig. 9. Propene polymerization by aluminium triethyl and vanadium halides [30]. (a) $VCl_3/AlEt_3$; $AlEt_3$ = 0.019 mole l^{-1}, Al/V = 3; propene pressure, 1 atm; temp., 60°C. (b) $VCl_3/AlEt_2 Cl$; $AlEt_2 Cl$ = 0.064 mole l^{-1}; Al/V = 4; [M] = 1.785 mole l^{-1}; n-heptane solvent.

References pp. 249—257

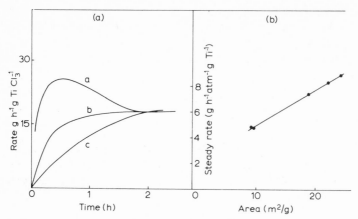

Fig. 10. Polymerization of propene by α-TiCl$_3$/AlEt$_3$ [81]. Al/Ti = 4 to 7.5; temp., 70°C; $p_{C_3H_6}$ = 1450 torr. (a) Finely ground, (b) lightly ground, (c) unground.

proportional to the surface area of the catalyst. Mechanical grinding of the transition metal subhalide greatly accentuates the early acceleration stage.

The different patterns of behaviour are shown in the polymerization of propene by VCl$_3$/AlEt$_2$Cl and by VCl$_3$/AlEt$_3$ (Fig. 9). In Fig. 10(a) is shown the polymerization of propene by α-TiCl$_3$/AlEt$_3$ and Fig. 10(b) illustrates the dependence of the steady rate on surface area. Keii [112, 269] has studied the phenomena in the early stages of the polymerization of propene by TiCl$_3$/AlEt$_3$. Acceleration of polymerization prior to attainment of the steady rate (Fig. 10a, curve c) has been found to follow the equation

$$\frac{d(R_p)}{dt} = k\{(R_p)_s - (R_p)\}$$

i.e.

$$(R_p)_t = (R_p)_s (1 - \exp(-kt))$$

where $(R_p)_t$ is the rate after time t up to the steady value $(R_p)_s$.

The observation that the time to 75% of the steady rate is inversely proportional to $(R_p)_s$ [270] leads to the time dependence of R_p in the early stages of reaction, when $t \ll 1/k$, as $(R_p)_t \simeq (R_p)_s kt \simeq (R_p)_s^2 t$. Since the steady rate is proportional to [M] it follows that the acceleration stage is second order in monomer. The activation energy for this stage was found to be 20 kcal mole^{-1}.

If allowance is made for concurrent first order catalyst decay during build-up of catalyst sites

$$(R_p)_t = \frac{k(R_p)_s}{k - k_d} \{\exp(-k_d t) - \exp(-kt)\}$$

where k_d is the catalyst decay constant.

In the case where acceleration is followed by a declining rate from powdered $TiCl_3$ (Fig. 10a, curve a) the reaction is first or second order in monomer, dependent on the order of addition of monomer and metal alkyl to the $TiCl_3$, viz.

$$R_p = k[M]^2 \frac{K[A]t}{1 + K[A]} \quad \text{and} \quad R_p = K'[M][A]t \text{ (propene first)}$$

The activation energy for the acceleration stage was 17 kcal mole^{-1}.

The explanation for the dependence of the kinetics on order of addition was that of competitive equilibrium in the formation of active centres from metal alkyl, $TiCl_3$ and monomer. The build-up of polymerization is much faster than in those systems which rise to a steady rate, and the maximum rate is at least double the steady rate. The formation of the active but short-lived species in the example shown in Fig. 10a has not been associated with a large change in surface area since the steady rate is unchanged. However, more intensive grinding gives rise to a greater increase in the initial acceleration and to higher steady rates.

The dependence of rate on time in the declining stage of the polymerization follows the relationship

$$\frac{(R_p)_t - (R_p)_s}{(R_p)_0 - (R_p)_s} = \exp(-k't)$$

where $(R_p)_0$ is a constant which represents the maximum rate in the acceleration period. The differential form of this equation

$$\frac{-d(R_p)}{dt} = k'\{(R_p) - (R_p)_s\}$$

is thus of a similar form to that of the acceleration stage.

As k' has been found to be insensitive to the type of $TiCl_3$ used and to the experimental conditions, Keii [269] concluded the origins of the rate changes to be physical rather than chemical. The same type of rate equation has also been found for ethylene polymerization by $TiCl_3/AlEt_3$ [269]. However, with isoprene using the heterogeneous $VCl_3/AlEt_3$ system a rather different relationship was found in which increase in the active centre concentrations was proportional to conversion rather than to time [108].

The kinetic features of the early stages in the polymerization with this catalyst appear to be firmly established, and quantitative agreement between different groups of workers are in good agreement, considering the difficulties and differences in experimental conditions. Interpretation of the kinetics in terms of mechanism is still speculative, particularly the explanation for the decline in rate to a steady value.

There appear to be significant differences in the changes in polymerization kinetics with time with other catalyst systems. Buls and Higgins [265] found a change in the kinetics of propene polymerization by $TiCl_3/AlEt_2Cl$ with conversion, from first order in monomer to second order after a period of about 0.5 h. From the dependence of the slopes of the second order plots on monomer concentration the reaction is really first order and a physical explanation was proposed for this change in kinetics and the fall in polymerization rate, namely that proposed by Iguchi [266] for occlusion of growing chains by polymer.

The activation energy for polymerization was found to be 14 ± 1 kcal mole^{-1}. The initial rate was related to the catalyst composition by

$$R_p = k_1 [T][A]^{1/2} \exp(-k_2[T][A]^{1/2})$$

The rate was increased by the presence of hydrogen. A rather complex polymerization mechanism is proposed in which the propagation step is regarded as an alkyl transfer reaction, followed by realkylation with monomer of the metal hydride so formed.

A rapid declining rate is observed with unaged catalysts from $TiCl_4$ or VCl_4 with aluminium alkyls with ethylene [113], propene [30] or ethylene-propene [114] and here also unstable catalytic entities are produced; no correlations of steady rates with catalyst composition or surface area have been made.

Unfortunately, there is no information on the concentrations of active centres in the early stages of reaction with these types of catalyst. Indeed, the question of active centre concentrations in heterogeneous catalysts is beset with difficulties, estimates vary greatly and the methods are sometimes of dubious reliability. Before quoting published values some account of the methods used may be desirable.

7.2.2 Active centre concentrations

The first attempt at such an estimation involved determining the amount of metal alkyl adsorbed on the transition metal halide and the fraction of this which became combined with the polymer. Clearly this can only give a rough indication of the actual concentration of active centres.

Table 4 gives data which indicates that about 1—10% of α-TiCl$_3$ becomes complexed with AlEt$_3$ or AlEt$_2$Cl. Not all of this necessarily initiates polymerization, since it may be in non-active positions on the main faces of the subhalide crystals. Natta [81] observed that when metal alkyl was adsorbed at high temperatures (100°C) it all appeared in the polymer, but at lower temperatures only a relatively small fraction became combined with polymer.

TABLE 4

Adsorption of aluminium alkyls on α-TiCl$_3$

Aluminium Compound	Adsorption time and temperature	Solvent	Conc. of ethyl groups (mole per mole TiCl$_3$)
AlEt$_3$	3 h at $-18°$C	n-Heptane	0.045[a]
AlEt$_3$	30 min at 20°C	Benzene	0.0482[a]
AlEt$_2$Cl	30 min at 70°C	n-Heptane	0.003[a]
AlEt$_2$Cl	Overnight at ambient temp.	n-Heptane	0.091[b] 0.11

[a] C14 labelled organo-aluminium compound; determination of ethane after acid treatment [81].
[b] Reacted with ^{131}I$_2$ and assay of labelled ethyl iodide [86].

There is some disagreement as to how firmly metal alkyls are adsorbed on titanium trichloride. Keii et al. [112] state that AlEt$_3$ is removed by washing with hydrocarbon solvent with loss of polymerization activity which is restored by the addition of further AlEt$_3$. Natta [81] found that washed catalyst contained substantial amounts of adsorbed aluminium alkyl and retained its polymerization activity. Chien [86] likewise concluded that treatment of TiCl$_3$ with AlEt$_2$Cl resulted in surface alkylation. With the precipitates from TiCl$_4$ and aluminium alkyls there are similar discrepancies. Most data (Table 1) show the presence of substantial amounts of adsorbed alkyl, but Saltman et al. [41] and Tepenitsyna et al. [225] both reported that washing with hydrocarbon solvent removed adsorbed aluminium alkyl and polymerization activity*.

The explanation may be concerned with the experimental methods. Thus the formation of the surface complex TiCl$_3$. AlEt$_3$ may be reversible, but if conditions are such that desorption of AlEt$_2$Cl has occurred in the presence of excess metal alkyl to give EtTiCl$_2$. AlEt$_3$, activity would not be removed by washing off weakly adsorbed AlEt$_3$.

Direct measurements of active centre concentrations have been made by terminating the polymerization with some reagent such as iodine [86], carbon dioxide [115] or an alcohol [116], labelled with an isotope which introduces radioactivity into the polymer; e.g.

$$P_n^- M^+ + {}^{131}I_2 \rightarrow P_n {}^{131}I + M^+ I^-$$
$$+ {}^{14}CO_2 \rightarrow P_n {}^{14}COO^- M^+$$
$$+ RO^3 H \rightarrow P_n {}^3 H + M^+ OR^-$$

With tritiated alcohols there is an isotope effect resulting from differential reactivity of normal and labelled alcohols with active chain ends.

* See note added in proof (B), page 247.

References pp. 249—257

With methanol using $TiCl_4/AlR_2H$ Feldman and Perry [116] obtained a value of 3.7 for ethylene polymerization, while with $TiCl_3$ based systems for propene values of 1.3 [126] and 1.56 [121] have been obtained. n-Butanol gave isotope factors of 2.6 [176] for ethylene, 1.63—2.36 [213] and 1.0 [132] for propene and 1.37 [213] for butene-1, all with $TiCl_3$ based catalysts. Burfield and Tait [238] have reported 3.20 ± 0.15 for MeO^3H for VCl_3/AlR_3 with 4-methylpentene-1.

There is, in addition, the possibility of calculating active centre concentrations from measurements of polymerization rate and polymer molecular weight.

The termination reactions appear to be quantitative and specific for the reaction of saturated polymer molecules attached to aluminium and titanium [116], but applied to diene polymerizations the method is less satisfactory mainly because of the greater stability of allylic carbon—transition metal bonds. Polybutadiene has been labelled by terminating with tritiated methanol with the $Cr(acac)_3/AlEt_3$ catalyst [55], and similarly polyisoprene prepared with $VCl_3/AlEt_3$ [107]. Polybutadiene prepared with $TiI_4/Al(i$-$Bu)_3$ has been labelled using $^{14}CO_2$ [115].

Catalysts from Group VIII metals have given unsatisfactory results. In the polymerization of butadiene with soluble cobalt catalysts tritium is not incorporated when dry active methanol is employed [115], although it is combined when it has not been specially dried [117, 118]. Alkoxyl groups have been found when using dry alcohol [115, 119] but the reaction is apparently slow and not suited to quantitative work [119]. Side reactions result in the incorporation of tritium into the polymer other than by termination of active chains [118], probably from the addition of hydrogen chloride produced by reaction of the alcohol with the aluminium diethyl chloride [108]. Complexes of nickel, rhodium and ruthenium will polymerize butadiene in alcohol solution [7, 120], and with these it has not been possible to determine active site concentrations directly.

In systems where termination is quantitative and there are no significant side reactions it is clear that the measurements will involve all polymer molecules containing reactive metal—carbon bonds, including non-propagating chains attached to aluminium after transfer. No entirely satisfactory correction procedure has been found, but the usual method is to extrapolate the linear region of the increase in metal—carbon bond concentration with time to zero time. Provided the transfer reactions are constant in rate the concentration of metal bonds [N] is given by

$$[N] = \frac{\int R_p\, dt}{\bar{P}_n} = [C^*] + \Sigma k_{tr}[C^*][X]^i t$$

In Fig. 11 are shown experimental curves for changes in active site concentrations with time. The problems involved in accurate estimates of

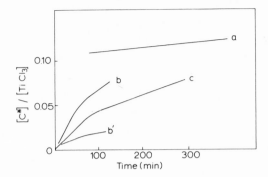

Fig. 11. Active centres in propene polymerization by $TiCl_3/AlEt_2Cl$. (a) ^{14}C, Chien [86]. (b) BuO^3H (b')I_2, Schnecko and Kern [213]. (c) MeO^3H, Coover and Guillet [121].

zero time concentrations are shown particularly by the curves in which there are two regions of linear increase in metal—polymer bonds. It has been suggested [121] that the metal alkyl which participates in transfer later in the polymerization is retarded by diffusion through a layer of polymer, and therefore that extrapolation from the second linear region gives high estimates of active centre concentrations.

Errors in the opposite sense will occur if there is an induction period, e.g., from the presence of an inhibitor which is removed by reacting with the catalyst, from slow initiation or disaggregation of a crystalline solid, and low results will be obtained [20]. A procedure to meet this problem is to extrapolate the yield of polymer [P] against active site concentration since, if polymerization rate and monomer concentration remain constant,

$$[N] = [C^*] + \frac{\Sigma k_{tr}[X]^i[P]}{k_p M}$$

An example of this extrapolation procedure is shown in Fig. 12. An alternative procedure, which emphasizes the build-up of growing centres in systems where the monomer concentration falls, is a plot of C* versus log (100-% conversion) (Fig. 13) [238].

An assessment of the methods for determining active centre concentrations has been made by Schnecko and Kern [213], and they conclude that radioactive assay is more reliable than indirect kinetic methods. Propene and butene-1 polymerized by $TiCl_3/AlEt_2Cl$ on termination with iodine or BuO^3H gave comparable extrapolated values of ca. 0.5% for catalyst efficiency. This is lower than earlier estimates but higher than the very low values (0.1%) obtained by Coover and Guillet [121], although the shapes of the curves obtained by the two groups were very similar.

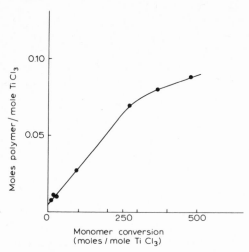

Fig. 12. Increase in active polypropene concentration with conversion using α-TiCl$_3$/AlEt$_2$Cl catalyst [20].

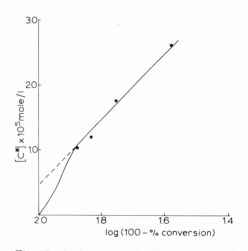

Fig. 13. Active centres in the polymerization of 4-methylpentene-1 by VCl$_3$/Al(i-Bu)$_3$ [238]. [VCl$_3$] = 18.5 x 10^{-3} mole l^{-1}, [M] = 2.0 mole l^{-1}, 50°C.

Soluble catalysts are of interest in that all the transition metal is available for alkylation. The concentration of active species will be represented by eqn. (2) only if the extent of reaction is small ($K \ll 1$), such that the change in concentration of metal alkyl is small. Otherwise the approximate equation gives spuriously high estimates of [C*] at low A/T ratios, and the exact equation

$$K_A = [C^*]/\{[T]_0 - [C^*]\}\{[A]_0 - [C^*]\}$$

must be used. The necessity for high Al/Ti ratios for maximum catalytic activity with systems such as $Ti(OR)_4/AlEt_3$ [51, 146], $Ti(acac)_3/AlEt_2Cl$ [145] or $(\pi-C_5H_5)_2 TiEtCl/AlEtCl_2$ [216], indicates that K_A may not always be large and the rate of transformation of the initial complex into the active species may be slow. It is possible that more than one reaction is involved, viz.

$$T + A \xrightleftharpoons[K_A]{} C_1^*; \quad C_1^* + A \xrightleftharpoons[K_{A'}]{} C_2^* \text{ etc.}$$

The polymerization rate will then depend on the relative proportions and activities of C_1^*, C_2^*, etc. (which will be determined by A/T and K_A, $K_{A'}$ etc., and the propagation coefficients.) Thus, it has been found with $(\pi-C_5H_5)_2 TiEtCl/AlEtCl_2$ that the rate of ethylene polymerization accelerates with time as active species are formed and with increase in Al/Ti ratios the rate rises to a steady value at constant Ti concentration and $[C^*] \simeq [Ti]_0$, and passes through a maximum at constant metal alkyl halide concentration.

The system $(\pi-C_5H_5)_2 TiCl_2/AlMe_2Cl$ is complicated by relatively rapid decomposition of the active Ti^{IV} complex but initiation is slow compared with propagation [111]. In the related systems $(\pi-C_5H_4R)_2$-$TiCl_2/AlEt_2Cl$ (R = H, Me, Et) a linear relationship has been found between (conversion)$^{1/2}$ and time for periods up to 10 min at 5°C [217], in accord with slow initiation, since

$$P = \int R_p \, dt = k_p(1 -- e^{-k_i t}) [M] [C^*]_0 \simeq k_i k_p t^2 [M] [C^*]_0/2$$

Alkyl substitution in the cyclopentadiene ring reduces the rate of alkylation and in consequence the rate of increase in polymerization rate.

This catalyst [111] gives several polymer molecules per titanium atom,[*] and there are similar findings for $Cr(acac)_3/AlEt_3$ for butadiene and $Cr(acac)_3/AlEt_2Cl$ or $AlEtCl_2$ for ethylene [139]. It appears reasonable to assume, therefore, that all the transition metal participates in the reaction. With $Cr(acac)_3/AlEt_2Cl$ the average number of polymer molecules approximates to one per Cr atom which additionally would indicate the absence of a transfer reaction. This point needs to be confirmed, however, with separate measurements of active centre concentrations and chain transfer, since two soluble nickel catalysts [68, 124] have been reported to have low catalyst efficiencies.

The results of a number of determinations of active centre concentrations are given in Table 5. With the exception of the soluble catalyst for ethylene, the results for which are not corrected for the known transfer reaction, and the surprisingly high values for the $Cr(acac)_3/AlEt_3$ sys-

---

[*] At 0°C the system $(\pi-C_5H_5)_2 TiEtCl/AlEtCl_2$ was found to give one polymer molecule per Ti atom [239].

TABLE 5

Efficiencies of coordination catalysts

Catalyst	Monomer	Method	[C*]/[T]	Ref.
$TiCl_4/Ali\text{-}Bu_2H$	Ethylene	MeO^3H	0.155 (unaged) 0.03—0.003 (aged at 69°C)	116
$\gamma TiCl_3/AlEt_2Cl$	Ethylene	$R_p/\bar{P}_n$	0.015	133b
$TiCl_3/AlEt_3$	Ethylene	^{14}C	0.035—0.059	14
$TiCl_3/AlEt_2Cl$	Ethylene	BuO^3H	0.0069	214
		BuO^3H	0.0088	176
$(\pi\text{-}C_5H_5)_2TiCl_2/AlMe_2Cl$	Ethylene	^{14}C	$0.56—7.2^a$	111
$CrO_3\text{-}SiO_2$	Ethylene	MeO^3H	0.005	125
$Zr(CH_2Ph)_4/Al_2O_3$	Ethylene	$^{14}CO_2$	0.007—0.009	320
$Zr(CH_2Ph)_4/SiO_2$	Ethylene	$^{14}CO_2$	0.009—0.044	320
$Ti(CH_2Ph)_4/Al_2O_3$	Ethylene	$^{14}CO_2$	0.002—0.004	320
$\alpha\text{-}TiCl_3/AlEt_3$	Propene	^{14}C	0.003—0.011	81
		MeO^3H	0.003	121
		BuO^3H	0.038^b 0.068^c	132
$TiCl_3/AlEt_2Cl$	Propene	$\bar{P}_n$	0.02	213
		I_2	0.003	213
		BuO^3H	0.053	213
		MeO^3H	0.001	121
		^{14}C	0.03—0.10	86
		MeO^3H	0.014—0.09	126
		$^{131}I_2$	0.004	20
		$R_p/\bar{P}_n$	0.012—0.014	123
		$R_p/\bar{P}_n$	0.057 0.019^d	147
$\alpha/\gamma TiCl_3/AlEt_2Cl$	Propene	$R_p/\bar{P}_n$	0.007—0.018	271
$\alpha\text{-}TiCl_3/AlEtCl_2/HMP^e$	Propene	MeO^3H	0.0002	121
VCl_3/AlR_3	Propene	$R_p/\bar{P}_n$	0.008	182
$TiCl_3/AlEt_2Cl$	Butene-1	I_2	0.001	213
	Butene-1	BuO^3H	0.0035	213
	Butene-1	BuO^3H	0.0044	176
$VCl_3/AlEt_3$	Isoprene	MeO^3H	0.006	107
$VCl_3/AlEt_3/iPr_2O$	Isoprene	MeO^3H	0.005—0.10	107
$Cr(acac)_3/AlEt_3$	Butadiene	MeO^3H	4—6	55
$Ni(naphthenate)_2/BF_3 \cdot Et_2O/AlEt_3$	Butadiene	$R_p/\bar{P}_n$	0.02—0.07	68
$TiI_4/Al(i\text{-}Bu)_3/iPr_2O$	Butadiene	$R_p/\bar{P}_n$	0.33	122

TABLE 5—*continued*

Catalyst	Monomer	Method	[C*]/[T]	Ref.
Ti(On-Bu)$_4$/AlEt$_3$	Butadiene	ESR	0.03	124
		$R_p/\bar{P}_n$	0.001	146
RhCl$_3$	Butadiene	$R_p/\bar{P}_n$	$\simeq 2$	180
VCl$_3$/AlR$_3$	4-Methyl pentene-1	MeO^{3}H	$2.3-6.1 \times 10^{-4}$	238

[a] Not corrected for chain transfer. From R_p and $\bar{M}_n$ on the related $(\pi C_5 H_5)_2$ TiEtCl/EtAlCl$_2$ system 0.7—1.4 chains per Ti were found [235], which indicate no chain transfer.
[b] TiCl$_4$ reduced by hydrogen.
[c] TiCl$_4$ reduced by organo-aluminium compounds.
[d] TiCl$_3$ contained 4.7% Al.
[e] Hexamethylphosphoramide.

tem — with butadiene — most of them are low and for the heterogeneous catalysts are of the right order for the involvement in the polymerization of accessible transition metal atoms at the edges of the crystals. The lower end of the range (ca. 0.1%) would correspond to TiCl$_3$ crystals of ca. 1 μm in size while the higher estimate (ca. 10%) would be equivalent to reaction at the edges of a crystal of ca. 100 Å across. Electron micrographs show the coarse particles of α-TiCl$_3$ (20—40 μm) to be aggregates of plate-like lamella ranging from 100—1000 Å in size; an estimate of the plate thickness is 175 Å [127]. Surface areas from gas adsorption [95] range from 5—100 m^2 g^{-1}, and clearly relate to the finer particles. Spherical particles of polypropene ca. 0.4 μm in diameter and containing 0.46% of titanium have been detected on the surface of α-TiCl$_3$ crystals. From this the basic crystal has been calculated to have an average volume of 1.8×10^9 Å^3. This is equivalent to a square plate ca. 800 Å across and 200 Å thick with 1—2% of the titanium in active centres [16].

The fraction of TiCl$_3$ involved in polymerization depends on its method of preparation and the product prepared by reduction of titanium tetrachloride with aluminium shows greater reactivity than that from reduction with hydrogen [128], probably due to the presence of co-crystallized AlCl$_3$. It has been shown that ball-milling TiCl$_3$ with AlCl$_3$ increases polymerization activity, up to a maximum at a ratio of Ti/Al of 4/1 [129]. Electron micrographs and measurements of polymer yields and molecular weights suggest that TiCl$_3$/AlCl$_3$ crystals are porous and that up to half the transition metal atoms initiate polymerization [17]. (The yields of polypropene ranged from 0.16 to 0.53 mole per mole TiCl$_3$, and termination experiments with tritiated methanol indicated an efficiency of 0.4 mole of active centres per mole TiCl$_3$.) The relative

References pp. 249—257

TABLE 6

Propagation rate coefficients

Monomer	Catalyst	Temp. (°C)	k_p (l mole^{-1} sec^{-1})	E_p (kcal mole^{-1})	Ref.
Ethylene	$(\pi\text{-}C_5H_5)_2TiEtCl/AlEtCl_2$	0	6	—	202
	$(\pi\text{-}C_5H_5)_2TiCl_2/AlMe_2Cl$	45	39	12.2	111
	$TiCl_4/Al(i\text{-}Bu)_2H$	39.4	52(± 2.7)	11.8	116
	$TiCl_3/AlEt_2Cl$	60	78	—	214
	$\gamma\text{-}TiCl_3/AlEt_2Cl$	40	76	—	133
	$CrO_3/SiO_2\text{—}Al_2O_3, SiO_2$	75	~1000, 720	4.2—5.4	45, 46 125
	$(\pi\text{-}C_5H_5)_2TiEtCl/AlEtCl_2$	10	15	5.8	276
	$(\pi\text{-}C_5H_5)_2TiEtCl/AlEtCl_2$	6.4	20	8.7	277
	$(EtO)_3TiCl/AlEt_2Cl$	−20	6×10^{-2}	9.2	280
	$Zr(CH_2Ph)_4/Al_2O_3$	80	21.8×10^3	—	279
	$Zr(CH_2Ph)_4/Al_2O_3$	75	$1.4\text{—}2.6 \times 10^3$	—	320
	$Zr(CH_2Ph)_4/SiO_2$	75	$1.0\text{—}2.4 \times 10^2$	—	320
	$Ti(CH_2Ph)_4/SiO_2$	75	$7.3\text{—}15.7 \times 10^2$	—	320
	$(\pi\text{-}C_5H_5)_2TiC_6H_{13}Cl/AlEt_2Cl$	10	47.7	—	319
Propene	$\alpha\text{-}TiCl_3/AlEt_3$	70	2.8^a	10	81
	$\alpha\text{-}TiCl_3/AlEt_3$	70	48	—	121
	$\alpha\text{-}TiCl_3/AlEt_2Cl$	50	0.4	13.0	86
	$\alpha\text{-}TiCl_3/AlEt_2Cl$	70	40	—	121
	$\alpha\text{-}TiCl_3/AlEt_2Cl$	70	5.6	—	123
	$\alpha\text{-}TiCl_3/AlEt_2Cl$	50	1.2 ± 0.2	—	132
	$\alpha\text{-}TiCl_3/AlEt_2Cl$	50	4.2^b	—	126
	$\alpha\text{-}TiCl_3/AlEtCl_2/HMP$	70	50	—	121
	$TiCl_3/AlEt_2Cl$	60	$6.2, 7.5^c$	—	147
		60	18	—	213
	$VCl_3/Al(i\text{-}Bu)_3$	40	9	12.9	182
Butene-1	$TiCl_3/AlEt_2Cl$	60	7.3	—	213
	$VCl_3/AlEt_3$	—	—	2.3	128
Butadiene	$Ni(naphthenate)_2/ BF_3OEt_2/AlEt_3$	40	2—6	—	68
	$Cr(acac)_3/AlEt_3$	30	0.0042	15.0	55
	$Ti(On\text{-}Bu)_4/AlEt_3$	—	0.7	—	146
	$TiI_4/Al(i\text{-}Bu)_3$	40	4.2	12.05 ± 1.21	318
Isoprene	$VCl_3/AlEt_3$	50	1.4	—	107
Methyl methacrylate	$Cr(\pi\text{-}C_3H_5)_3$	30	4.4×10^{-3}	11.1	229
	$Cr(\pi\text{-}MeC_3H_4)_3$	30	40×10^{-3}	9.0	229

^{a} Assuming 1 % catalyst efficiency; E_p refers to steady state polymerization.
^{b} Estimated assuming solubility of propene in hexane at 50°C and 1 atm. = 0.43 mole l^{-1}.
^{c} Higher value for $TiCl_3$ containing 4.7% Al.

proportions of polymer molecules attached to aluminium and titanium were not determined, however, nor were details given of the correction for metal—alkyl transfer in the labelling experiments. Therefore, while it would appear that these catalysts are of high efficiency the precise value is not completely certain.

7.2.3 Propagation rates

From the active centre concentrations and polymerization rates propagation rate coefficients can be calculated. Values taken or calculated from the literature are given in Table 6. The marked differences in the values for propene reflect the disparate estimates of catalyst efficiencies. Data on comparable systems are not available but the larger values for propene consequent on low catalyst efficiencies seem high when compared with the values found for ethylene. Comparisons of the relative activities of β-TiCl$_3$ and α-TiCl$_3$ with aluminium alkyls so far as rates of polymerization are concerned, have shown the former to be from half [121] to five to six [21] times as active as the latter, the rather wide disparity possibly reflecting differences in surface area of the catalyst samples. However, ethylene polymerizes from about 20 to 100 times as fast as propene with β-TiCl$_3$ type systems [21, 134], and while the growth of ethylene from sites which are not accessible to propene may account for some of the difference (as has been suggested for the VCl$_4$/Al(C$_6$H$_{13}$)$_3$ system [135]) it would seem that k_p will be much greater for ethylene than for propene.

7.2.4 Transfer

(a) Metal—alkyl. Transfer of growing chains can occur with excess metal—alkyl present in the system and the molecular weight will fall with increase in metal—alkyl/transition metal ratio. If the metal—alkyl is dimeric in solution and it is the monomeric form which is involved in the transfer reaction, e.g.

$$Al_2R_6 \underset{K}{\rightleftharpoons} 2AlR_3$$

$$AlR_3 + P_nTiX_n \cdot AlR_2Cl \xrightarrow{k_{tr_A}} RTiX_nAlR_2Cl + AlR_2P_n$$

the rate (R_{tr_A}) will be given by

$$R_{tr_A} = K^{1/2}k_{tr_A}[A]^{1/2}[C^*]$$

and the fall in molecular weight will be linear with $[A]^{1/2}$. This has been found to be the case for propene with α-TiCl$_3$/AlEt$_3$ (Fig. 13a) [81]. Transfer by AlEt$_2$Cl is slower than with AlEt$_3$ and Chien [86] and Bier

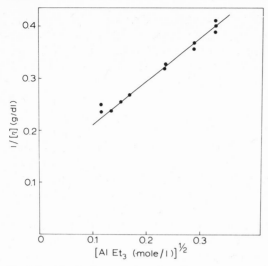

Fig. 13a. Dependence of molecular weight of polypropene on AlEt$_3$ concentration [81]. AlEt$_3$/TiCl$_3$ = 3/1; temp., 70°C; [M] = 0.41 mole l^{-1}.

TABLE 7

Polymerization of propene with α-TiCl$_3$/ZnEt$_2$ [150b]

Zn/Ti	[η]
1.4	2.3
2.1	2.2
3.0	1.7
4.4	1.3

Temp., 50°C; solvent, heptane; [Zn] + [Ti] = 9 x 10^{-2} mole l^{-1}; conversions (20 h), 45—64%.

[136] have reported the average molecular weights to increase with time of polymerization for many hours (Fig. 14). This is not in agreement with the values for R_{tr}/R_p in Table 8, but these refer to the early stages of the reactions and conditions may change to give longer-lived chains later in the polymerization. The transfer reaction has been found by Coover and Guillet [121] to depend also on the monomer concentration, viz.

Propene pressure (p.s.i.g.)	[M] (mole l^{-1})	R_{trA} (mole^{-1} sec^{-1})	R_{trA}/[M] (sec^{-1})
10	0.45	3.24	6.9
40	1.3	9.25	6.7

It was suggested that the monomer was involved in the transfer reactions, but an alternative interpretation [237] is that the effect results from slow chain initiation at low monomer concentrations.

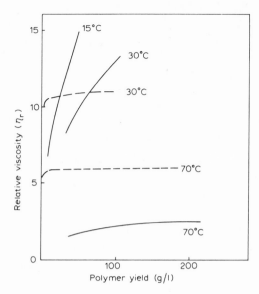

Fig. 14. Increase in polypropene solution viscosity with catalyst from titanium trichloride and organo-aluminium compounds [136]. (———), $TiCl_3/AlEt_2Cl$, (- - - - - -), $TiCl_3/AlEt_3$.

Differences in the rate of transfer have been observed between beryllium and aluminium alkyls in butene-1 polymerization with $TiCl_3$ catalysts [128], the former giving the higher values. Aluminium alkyls differ among themselves, the rates diminishing for VCl_3/AlR_3 with 4-methyl pentene-1 as monomer in the order $Et > i$-Bu $> n$-Bu $> n$-Hex. The values expressed as (mole VCl_3 l^{-1} min^{-1}) x 10^6 were respectively 17.2, 3.24, 1.53 and 0.87 [238]. Zinc alkyls participate readily in transfer reactions with the α-$TiCl_3/AlEt_3$ catalyst and reduce the molecular weight of the polymer [150a], and in Table 7 is shown the effect of zinc diethyl concentration on the solution viscosity of polypropene prepared with the α-$TiCl_3/ZnEt_2$ catalyst [150b].

The rates of the transfer reactions have been obtained from the slopes of the increase in metal—carbon concentration with reciprocal time, and the very few data from the literature are given in Table 8. In some cases only ratios are available. The results are not easily compared in that R_{tr} depends on the catalyst concentration, and since different mechanisms are assumed the rate coefficients may be in different units.

(b) Transition metal. As indicated in the discussion on polymerization mechanism, little is known of chain transfer reactions involving the transition metal and few kinetic data are available. Natta [81] has reported a half order dependence on $TiCl_3$ concentration for the molecular weight dependence of polypropene prepared with $TiCl_3/AlEt_3$, the reaction being represented as involving a monomer molecule, while Chien [86] obtained first order dependence on $TiCl_3$ concentration. The only

TABLE 8
Transfer reactions (units: mole, l, s)

Monomer	Catalyst	Temp. (°C)	k_{tr_A}	k_{tr_T}	k_{tr_M}	k_{tr_A}/k_p	k_{tr_M}/k_p	R_{tr_A}/R_p	Ref.
Ethylene	γTiCl$_3$/AlEt$_2$Cl	40	—	—	10^{-2}	—	1.8×10^{-4}	—	133b
	(EtO)$_3$TiCl/AlEt$_2$Cl	−20	—	—	1.3×10^{-2e}	—	—	—	280
Propene	TiCl$_3$/AlEt$_3$ (Al/Ti = 1/1)	70	—	—	—	—	—	1.15×10^{-3}	121
	TiCl$_3$/AlEt$_2$Cl(Al/Ti = 1/1)	70	—	—	—	—	—	0.58×10^{-3}	121
	TiCl$_3$/AlEt$_2$Cl(Al/Ti = 2/1)	50	6.54×10^{-5a}	9.54×10^{-4b}	—	1.54×10^{-4}	—	0.31×10^{-4}	86
	TiCl$_3$/AlEt$_2$Cl(Al/Ti = 2/1)	60	—	—	—	—	—	2.3×10^{-4}	20
	TiCl$_3$/AlEtCl$_2$/HMP (Al/Ti/HMP = 1/1/0.6)	70	—	—	—	—	—	0.70×10^{-3}	121
4-Me-pen-tene-1	VCl$_3$/Al(i-Bu)$_3$	60	—	—	—	1.4×10^{-3c}	2.4×10^{-3c}	—	195
	VCl$_3$/Al(iBu)$_3$	30	—	—	—	2.2×10^{-5}	0.82×10^{-3}	—	237, 245
Butadiene	Cr(acac)$_3$/AlEt$_3$ (Al/Cr = 11.6)	30	—	9.2×10^{-3}	—	—	—	—	55
	Ti(On-Bu)$_3$/AlEt$_3$ Ni(naphthenate)$_2$/	—	—	—	2×10^{-5}	—	3×10^{-5}	—	146
	BF$_3$. Et$_2$O/AlEt$_3$ (Ni/B/Al = 1/7.2/6.5)	40	—	—	2—8×10^{-4}	—	$\sim 10^{-4}$	—	68
	(π-C$_5$H$_5$NiI)$_2$	50	—	—	—	—	2×10^{-2}	—	61
	TiI$_4$/Al(i-Bu)$_3$	40	0.79^d	—	—	0.188	—	—	318
Methyl methacrylate	Cr(π-2MeC$_3$H$_4$)$_3$	0	—	—	2.34×10^{-5}	—	3.25×10^{-3}	—	229, 230

a Units l$^{1/2}$ mole$^{-1/2}$ sec^{-1}; E_{tr_A} = 18 kcal mole^{-1}.
b E_{tr_T} = 7.6 kcal mole^{-1}.
c From viscosity measurements using $\overline{M}_w/\overline{M}_n$ = 10.
d E_{tr_A} = 12.36 ± 1.10.
e E_{tr} = 6.7 kcal mole^{-1}.

other report on transfer by transition metal compound is for butadiene polymerized by $Cr(acac)_3/AlEt_3$ [55].

(c) Monomer. A few values are available for the monomer transfer reaction in diene polymerization, but they are insufficient for any clear conclusions to be drawn. Monomer transfer occurs with propene using the α-TiCl$_3$/AlEt$_3$ catalyst [86]. Monomer transfer is a minor reaction with α-TiCl$_3$/AlEt$_2$Cl for propene [136], in view of the fact that most polymer chains are attached either to transition metal or aluminium [17].

(d) Spontaneous termination followed by reinitiation. There are few kinetic studies on this reaction, but there is one value reported [133] for ethylene at 40°C, using γ-TiCl$_3$/AlEt$_2$Cl, of ca. 3×10^{-4} l mole^{-1} sec^{-1}. This was obtained with the assumption that the rate of the subsequent initiation reaction is the same as succeeding growth steps.

In Section 7.1 the effects of chain-transfer on polymerization rate were not considered. There is, however, evidence from the dependence of rate on monomer concentration to show that initiation reactions in olefin polymerization are slower than propagation. If the rate equation

$$R_p = k_p f[T][M]$$

holds, $[M]/R_p$ would be independent of monomer concentration. Experimentally it has been found that this is not so for propene with TiCl$_3$/AlEt$_3$ [277] or ethylene with TiCl$_4$/Al(i-Bu)$_2$H [228], where rates follow the relationship

$$[M]/R_p = A + B/[M]$$

Similar observations have been made for 4-methylpentene-1 with VCl$_3$/Al(i-Bu)$_3$ [236].

If the effect of the transfer reactions and subsequent re-initiation of growing chains on monomer consumption and active centre concentration are taken into account the polymerization rate is given by [227]

$$R_p = \frac{f[T]\{k_p[M] + k_{t_l} + k_{tr_A}[A]^{1/2} + k_{tr_M}[M]\}}{1 + k_{t_l}/k_i[M] + k_{tr_A}[A]^{1/2}/k_i'[M] + k_{tr_M}/k_i''}$$

k_{t_l}, k_{tr_A} and k_{tr_M} are the rate coefficients of the internal transfer reaction and those with metal alkyl and monomer respectively, and k_i, k_i', k_i'' the corresponding rate coefficients for initiation. This gives rise to the experimental form of the rate equation if one or more of the coefficients k_i, k_i' or k_i'' are small compared with k_p. It is apparent that if they were comparable to k_p, as only the first term in the numerator is large compared with the others,

$$R_p \simeq \frac{k_p f[T][M]}{1 + 1/\bar{P}_n}$$

which approximates to $k_p f[T][M]$, since the smallest value of $\bar{P}_n$ is found to range from 10^2 to 10^3.

Natta et al. [227] have suggested that reinitiation is slow from ethyl groups in the $TiCl_3/AlEt_3$ catalyst with zinc diethyl as transfer agent for propene, ($k_p/k_i' = 60-120$) and that $k_p/k_i > k_p/k_i'$, while Schindler [228] identified the slow reinitiation as the addition of monomer to the catalyst hydride bond.[*] The latter author has also shown that the rate of polymerization of ethylene is slower when it is mixed, at the same partial pressure with an inert gas such as nitrogen. The experimental results are consistent with occupation of some of the sites on the catalyst by nitrogen and this concept, in addition to slow re-initiation from spontaneously transferred chains, has been introduced into the kinetic treatment. The scheme proposed by Schindler (in slightly changed notation) is

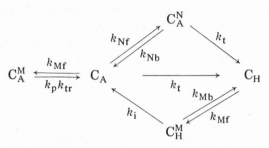

The subscripts A and H correspond to sites with alkyl (polymer) and hydride groupings and the superscripts M and N to adsorbed monomer and nitrogen. The coefficients k_{Mf}, k_{Mb}, k_{Nf}, k_{Nb} refer to adsorption and desorption of monomer and nitrogen respectively while k_t, k_i, k_{tr} and k_p are the rate coefficients of termination, re-initiation, transfer and propagation. This leads to a rate relationship (G, constant total no. of active sites)

$$[M]/(R_p)_{N_2} = [M]/(R_p)_0 + B_N[N]_2/[M]$$

$$B_N = \frac{k_{Nf}}{k_{Mf}}[k_t(k_i + k_{Mb})(k_p + k_{tr})/(k_t + k_{Nb})k_i k_{Mf} k_p G]$$

---

[*] Hydrogen can increase the polymerization rate of olefins with certain catalysts [175]. With butene-1 using $3TiCl_3/AlCl_3/Al(i-Bu)_3$ as catalyst the rate was increased by hydrogen by a factor of about two. The effect is not fully explained, but either the hydride sites produced by transfer are more active than those in the original catalyst, which must be of lower activity than those of the propagating chains to be kinetically limiting, or there has been an increase in the number of active centres present as a consequence of changes in the crystalline structure. Hydrogen was found to increase the polymer yield in the polymerization of 4-methyl pentene-1 with γ-$TiCl_3/AlEt_2Cl$ [242], with the usual reduction in polymer molecular weight. The explanation proposed for the increase in rate was that the normal termination reaction produces a terminal olefinic grouping which remains attached to the catalytic centre, whereas with hydrogen chain transfer gives a saturated chain end which does not compete with monomer for the catalyst site. From the yield of the polymer with time a second order catalyst decomposition reaction was proposed.

In this treatment $k_{\mathrm{M\,b}}$ has been assumed to be small compared with k_p or k_{tr} and it is implicit that re-initiation following monomer transfer is comparable in rate with propagation. Similar equations have been derived for the participation of hydrogen and deuterium in the reaction, except that these not only compete with monomer for adsorption on the catalyst but participate in transfer reactions.

(e) Hydrogen. The effects of hydrogen on the rate of polymerization of propene and polymer molecular weight are given by the relationships [137]

$$(R_p)_{\mathrm{H}} = (R_p)_0 - f[\mathrm{H_2}]^{1/2}$$

and

$$\bar{P}_n = \frac{1}{A + B[\mathrm{H_2}]^{1/2}}$$

The constants A and B were found to be 2.15×10^{-6} and 15.5×10^{-6} for propene and 1.54×10^{-6} and 3.11×10^{-6} for ethylene with α-$\mathrm{TiCl_3}/\mathrm{AlEt_3}$ at 75°C. The results obtained were:

$p_{\mathrm{H_2}}$ (atm)	0	0.2	0.4	0.5	0.8	1.0
Rate (g polymer per g $\mathrm{TiCl_3}$ per h)	15	11	9.5	9	6.7	6
$\bar{M}$w $\times 10^{-3}$	480	110	—	72	—	60

For a transfer reaction with molecular hydrogen the changes in rate and molecular weight should be dependent on the hydrogen pressure, viz.

$$\{1/(R_p)_{\mathrm{H}} - 1/(R_p)_0\} = \{1/(P_n)_{\mathrm{H}} - 1/(P_n)_0\} \simeq \frac{k_{tr}[\mathrm{H_2}]}{k_i[\mathrm{M}]}$$

To explain the half power relationships in hydrogen concentration Natta et al. [137] have suggested that the hydrogen dissociates reversibly under the influence of the catalyst. A rate coefficient of ~ 3 l mole^{-1} s^{-1} has been obtained for the system γ-$\mathrm{TiCl_3}/\mathrm{AlEt_2\,Cl}$ and ethylene [133], approximately one-thirtieth of the propagation rate coefficient.

No other groups of compound have clearly defined chain transfer activity. Ethyl trichloroacetate reduces the molecular weight of ethylene/propene copolymers using $\mathrm{VOCl_3}/\mathrm{Al_2\,Et_3\,Cl_3}$ [108] but as the rate of polymerization is also increased the effect may be caused by an increase in the number of active centres. α-Olefins and allenes reduce the molecular weights of cis polybutadiene obtained with soluble cobalt catalysts [139], but in this case it is not clear whether transfer or termination processes are involved.

7.2.5 Lifetimes of growing chains

The growth times for individual polymer molecules in coordination polymerization are generally much longer than in free-radical initiated polymerization, as demonstrated by the increase in average molecular weight of polymer in the early stages of reaction. With the exception of ethylene, propagation rate coefficients are of the order of one-tenth to one-hundredth of those for most free radical propagation reactions.

To calculate the average lifetime ($\bar{\tau}$) in systems where the catalyst species are unstable it is necessary to know at any specific time the polymerization rate, the active site concentration and the number average degree of polymerization of the polymer being formed, since

$$\bar{\tau} = \bar{P}_n [C^*]/R_p$$

When the catalyst centres are stable and the monomer concentration is kept constant during the polymerization, lifetimes can be calculated from the steady molecular weight and the slope of $1/\bar{P}_n$ versus $1/t$

$$\bar{\tau} = \bar{P}_n \, d(1/P_n)/d(1/t) = \bar{P}_n/k_p[M]_0$$

(The lifetime may also be expressed in terms of the transfer reaction rates using eqn. (8)).

Values in the literature relate to olefin polymerization with heterogeneous titanium based catalysts where the necessary information has been acquired, and the data are given in Table 9. It would not appear to

TABLE 9

Chain lifetimes in coordinated polymerization of olefins

Catalyst	Temp. (°C)	$\bar{\tau}$ (min)	Ref.
Ethylene			
$TiCl_4/Al(i\text{-}Bu)_2H$	—	4 (40 min)	116
$\gamma\text{-}TiCl_3/AlEt_2Cl$	40	30	133b
Propene			
$\alpha\text{-}TiCl_3/AlEt_3$	70	4—7	81
$\alpha\text{-}TiCl_3/AlEt_2Cl$	50	410	86
	60	12	20
	30	226	123
	50	42.6	
	70	22—28	
	90	12	
$TiCl_3/AlEt_2Cl$	60	10.7—37	213
4-Methylpentene-1			
$VCl_3/Al(i\text{-}Bu)_3$	30	14.8	245
Butene-1			
$TiCl_3/AlEt_2Cl$	60	2.5— 4.8	213

present much difficulty, however, to evaluate chain lifetimes for homo-geneous catalysts with olefins and dienes. Thus, if it is assumed that the polymerization rate of ethylene by $Cr(acac)_3/AlEtCl_2/(Al/Cr = 300)$ [139] remains constant and that each chromium atom initiates one polymer chain the average lifetime after a polymerization time of 248 min is about 70 min.

The precise experimental conditions for the measurements of chain lifetimes of polyethylene with the $TiCl_4/Al(i\text{-}Bu_2)H$ catalyst are not explicitly stated, but there is clear evidence for a steady increase in lifetime with polymerization time. For an average lifetime of 4 min after 40 min polymerization time, the instantaneous values were 4 min after 18 min polymerization and 10 min after 40 min polymerization. As the concentration of active centres remains almost steady after a sharp initial fall, the increase cannot be accounted for wholly by changes in the mono-mer/active sites ratio. The explanation may lie in a reduced rate of chain transfer with increase in conversion, as has been found for propene with $\alpha\text{-}TiCl_3/AlEt_2Cl$ [121]. In accord with this view average chain lifetimes of polypropene have been calculated to increase with conversion [123].

For propene with $\alpha\text{-}TiCl_3/AlEt_2Cl$ both Bier [136] and Chien [86] deduced very long growth periods, in contrast with the calculations of Caunt [20] and Tanaka and Morikawa [123]. The discrepancy between the widely varying estimates was ascribed by the latter authors to inadequate allowance for the breadth of the molecular weight distribution when viscosity average molecular weights were employed to calculate numbers of polymer molecules formed. Chain lifetimes will clearly be reduced by the presence of chain transfer agents in proportion to the reduction in molecular weight. The presence of 7% (v/v) of hydrogen in propene reduced the average lifetime by a factor of 4 [123].

7.2.6 Termination

Two measurements of the bimolecular termination rate coefficient in the polymerization of ethylene by Ti^{IV} complexes are available. For $(\pi\text{-}C_5H_5)_2TiCl_2/AlMe_2Cl$ a value of $0.5 \text{ l mole}^{-1} \text{ sec}^{-1}$ at $0°C$ has been reported [111] and for $(\pi\text{-}C_5H_5)_2TiEtCl/AlEtCl_2$, $0.6 \text{ l mole}^{-1} \text{ sec}^{-1}$ at $0°C$ [202]. The rate coefficient of decomposition of the latter complex in the absence of ethylene was much lower being $5 \times 10^{-3} \text{ l mole}^{-1} \text{ sec}^{-1}$ at $20°C$ [202].

7.2.7 Molecular weight distribution

The diversity of initiation termination and transfer reactions and the changes in concentrations of types and activities of catalyst species during polymerization have so far precluded a theoretical treatment of molecular weight distributions. With some of the soluble catalysts, e.g. $Zr(CH_2C_6\text{-}$

$H_5)_4$ for styrene [278], $Ti(On-Bu)_4/AlEt_3$ for butadiene (125) or $(\pi-C_4H_7\underline{Ni}I)_2$ for butadiene [61] molecular weight distributions are fairly narrow $(\bar{M}_w/\bar{M}_n = 1-2)$, where the propagating species have long lifetimes and their concentrations remain reasonably constant throughout the polymerization. The "soluble" catalysts based on vanadium compounds likewise give relatively narrow distributions with ethylene or ethylene/propene $(\bar{M}_w/\bar{M}_n \sim 2)$. Polymers prepared with heterogeneous catalysts

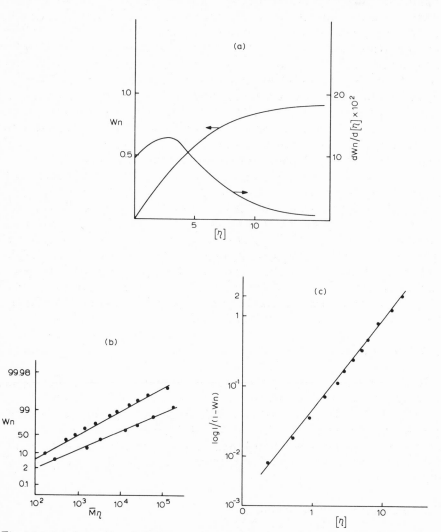

Fig. 14a. (a) Integral and differential distribution curves for polyethylene prepared with catalyst $TiCl_4/TiCl_3/C_3H_7Al(OC_8H_{17})_2 = 5/1/4$; $\bar{M}_n = 59,000$ $\bar{M}_w/\bar{M}_n = 5.1$ $[\eta]$ = 5.35 dl g^{-1} [305b]. (b) Log-normal distributions for polyethylenes prepared with (a) $TiCl_4/Al(i-Bu)_3$ and (b) $TiCl_4/Al(i-Bu)_2Cl$ catalysts [305a]. (c) Exponential distribution curve for polyethylene prepared with catalyst as Fig. 14a [305b].

generally have broad molecular weight distributions (ca. $\bar{M}_w/\bar{M}_n$ 5—20) Fig. 14a, which have been explained qualitatively as resulting from chain transfer and termination processes and from occlusion of active sites by precipitated polymer. This last cannot be a general feature, however, since under conditions such that polymer is not precipitated from solution, broad molecular weight distributions are still obtained with heterogeneous catalysts.

Experimentally, linear log-normal distributions are observed for poly-ethylenes and polypropenes prepared using Ziegler—Natta catalysts (Fig. 14a—b). It has been shown that the main features of the distribution can either be explained by variable catalyst site activity or by termination of a propagating chain by a desorption process from the catalytic surface proportional to molecular weight. Such experimental evidence as is available favours the variable site hypothesis [302].

The broadness of the distribution[*] is given by the parameter γ, obtained from the slope ($\tan \theta$) of the log—normal plot where $\tan \theta = \gamma \sqrt{}/2$. The polydispersity ratio ($\bar{M}_w/\bar{M}_n = \exp(\gamma^2/2)$ no and x are constants in the distribution function not determinable from the log-normal line. (Special cases of the general distribution function have also been employed, e.g., the Lansing and Kraemer relationship where x = 0 [304] and the Wesslau relationship [305], where x = —1). Fractionation data can also be adequately represented by other exponential distributions such as that of Tung [306] $W(n) \, dn = abn^{b-1} \exp(-an^b) \, dn$. ($W(n) = 1 - \exp(-an^b)$). Fig. 14a(c).

8. Individual monomers

There are relatively few detailed kinetic investigations on coordination polymerization from which general conclusions can be drawn, but there have been many studies which although limited in scope give indications of overall monomer and catalyst reactivity. There seems to be merit, therefore, in reviewing the general polymerization behaviour of some of the more important monomers with coordination catalysts. Rate data are frequently given, albeit sometimes in a form which is difficult to transform into standard units,[†] but useful information on the dependence

---

[*] The generalized distribution function for the nth polymer species is given by [303]

$$W(n) \, dn = An^x \exp \left\{ -\left(\frac{1}{\gamma} \ln \frac{n}{n_0} \right)^2 \right\} \, dn.$$

[†] Attempts have been made to extract or calculate relevant data in the tables, even though information provided in some cases is ambiguous or incomplete (particularly in regard to monomer concentration).

of polymer molecular weights on experimental conditions is more rarely encountered.

8.1 ETHYLENE

Ethylene is polymerized by heterogeneous and some soluble types of catalyst. Most are effective at ambient temperatures and atmospheric pressure but the supported Group VI oxide catalysts are more generally used at temperatures in the range 180—270°C and at pressures of 10—70 kg cm^{-2}. In Table 10 are given some data on polymerizations with a variety of catalyst systems. The values reflect the high reactivity of ethylene and, so far as the titanium halide catalysts are concerned, they are in fair agreement with one another if allowance is made for the probability of quite large differences in catalyst efficiency.

Reaction orders in monomer and transition metal compound are normal with most catalysts, but reactions second order in monomer have been reported for $TiCl_4/ZnBu_2$ [140] and $Al/TiCl_4$ [141]. No data on the stabilities of the active sites are available so it is not possible to decide whether these observations imply the involvement of two monomer units in the propagation step. It is to be noted that, with catalysts from zinc alkyls with α-$TiCl_3$, the rates decline rapidly with time at constant monomer concentration [94], while the latter system is complex in that active species are produced in a multistage reaction sequence (indicated by an induction period and an acceleration stage) and the operating conditions are severe (150°C and 10—70 kg cm^{-2} ethylene pressure).

A characteristic feature of $TiCl_4/AlEt_3$ or $AlEt_2Cl$ catalysts is a rapid rise in rate to a maximum followed by a fall to a steady value [113, 134]. The maximum and the steady values are dependent on the Al/Ti ratio (Fig. 15). The instability in the early stages of reaction is conceivably due to the presence of Ti^{IV} complexes adsorbed on β-$TiCl_3$, and at least a proportion of these may be removed by ageing the catalyst. The steady rate tends to increase with ageing time with titanium catalysts from aluminium alkyl halides and may be associated with the formation of stable trivalent titanium species. The complex α-$TiCl_3$—$AlEtCl_2$ has very weak catalytic activity, reflecting the weak alkylating power of $AlEtCl_2$, but it becomes highly active with the addition of a strong donor such as hexamethylphosphoramide [121]. The mode of action of the donor is open to speculation, but may be associated with the formation of some $AlEt_2Cl$ from the $AlEtCl_2$ by alkyl—halogen interchange [142]*.

The system $TiCl_4/\beta$-$TiCl_3/AlRCl_2$ is also active [38] and the activity increases with ageing time; with suitable compositions (typically $Ti^{IV}/$

---

* A complex $2AlEtCl_2 . \frac{1}{2}HMP$ is, however, produced from $AlEt_2Cl$ and $AlCl_3HMP$ [142b], showing that very little $AlEt_2Cl$ can be formed on adding hexamethylphosphoramide to $TiCl_3/AlEtCl_2$ catalyst.

TABLE 10
Polymerization of ethylene

Catalyst	Reaction order		Polym. rate coefficient, $k_p f$ (l mole⁻¹ sec⁻¹)	E_ϕ (kcal mole⁻¹)	Ref.
	Monomer	Transition Metal			
$TiCl_4/AlR_3$	1.0	1.0	23 (100°C)[a]	—	97
$TiCl_4/Al(i\text{-}Bu)_3$	—	—	~1 (30°C)[b]	10.4	131
$TiCl_4/AlEt_3$ (Al/Ti = 1.0)	1.0	1.0	0.23 (37°C)	—	149
$TiCl_4/AlEt_3$ (Al/Ti = 3)	—	—	0.43 (70°C)[c]	6.2	134
$TiCl_4/Al(i\text{-}Bu)_2Cl$(Al/Ti = 0.69	—	—	0.18		
0.85	—	—	0.34 30°C[j]	—	113
1.83	—	—	0.55—1.1		
$TiCl_3/AlEt_3$	1.0	—	1.0 (40°C)[k]	6.1	269
$\gamma\text{-}TiCl_3/AlEt_2Cl$	1.0	1.0	1.2 (40°C)	17.3[d]	133
$TiCl_4/Al(i\text{-}Bu)_2H$(Al/Ti = 1/2)	—	—	5.2 (67°C)[e]	11.8	116
$Ti(OBu)_4/AlEt_3$(Al/Ti = 3)	1.0[f]	1.0	0.13 (35°C)	13	51
$(\pi\text{-}C_5H_5)_2TiCl_2/AlMe_2Cl$	1.0	1.0	ca. 28 (15°C)	11.8	111
$VCl_4/Al(Hex)_3$(Al/V = 2.5)	1.0	1.0	12.5 (25°C)	6.6	193
$VO(acac)_2/AlEt_2Cl$(Al/V = 50)	—	—	30 (20°C)[g]	—	139
$CrO_3\text{—}SiO_2$ (2.5% Cr)	1.0	1.0	3.4 (75°C)[i]	10	125
$Cr(acac)_3/AlEt_2Cl$(Al/Cr = 300)	—	—	24 (20°C)[h]	—	139

[a] Initial rate.
[b] Estimated from Fig. 1 assuming [M] $\simeq$ 0.1 mole l⁻¹.
[c] Steady rate.
[d] Over range 30—40°C.
[e] Catalyst aged at 69°C; catalyst efficiency taken as 0.03.
[f] Over restricted range of monomer pressures. From Fig. 7 in ref. 51 assuming [M] = 0.1 mole l⁻¹.
[g] Calc. from Fig. 6 in ref. 139[*], assuming [M] = 0.065 mole l⁻¹.
[h] Calc. from Fig. 8 in ref. 139[*], assuming [M] = 0.065 mole l⁻¹.
[i] Based on values of k_p and estimate of [C*].
[j] [M] estimated to be 0.42 mole l⁻¹ at 5 atm.
[k] Calculated from Fig. 5.10 in ref. 269 assuming [M] = 0.06 mole l⁻¹ at 400 torr and 40°C.

[*] The monomer concentration was that reported in an earlier investigation by the same authors [202], and is low compared with literature values for ethylene solubility in hydrocarbon solvents. Thus, using the relationship reported by Meshkova et al. [134] the solubility in n-heptane at 1 atm. and 20°C is calculated to be 0.14 mole l⁻¹. The value in m-xylene from the relationship quoted by Nakamura et al. [239] is 0.13 mole l⁻¹ at 20°C and 1 atm. pressure and 0.09 mole l⁻¹ in benzene at 50°C and 1 atm. pressure [240].

Ti[II] /Al = 1/3/1) high conversion rates are attained: 100—200 g polymer per g $TiCl_3$ per h at 60°C and 1 atm ethylene pressure. Rates are proportional to ethylene pressure and to $(TiCl_4)^{0.5}$, and molecular weights, which are controlled by transfer reactions, rise to a constant value with increase in conversion; estimates of k_p/k_{tr} ranged from 4.7 to 7.6 x 10³.

Titanium dichloride gives catalysts with aluminium alkyls but alone is inactive for ethylene polymerization unless mechanically ground. Grinding

194

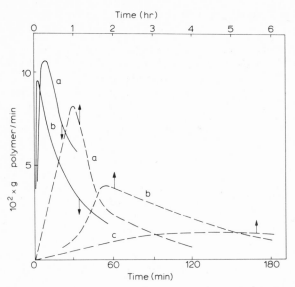

Fig. 15. Ethylene polymerization by $TiCl_4/AlEt_3$ and $TiCl_4/Al(i\text{-}Bu)_2Cl$ [113, 134].

		Al/Ti	Temp. (°C)
——$TiCl_4/AlEt_3$	a	1.7	−1
	b	2.86	−1
- - - - -$TiCl_4/Al(i\text{-}Bu)_2Cl$	a	3.3	30
	b	1.9	30
	c	1.0	30

produces significant catalytic activity accompanied by a change in chemical composition [60]. The activity, however, correlates with the Ti^{II} content, and an empirical relationship, $(R_p)_g = (R_p)_0 - C[Ti^0]$, was found which relates the rate of polymerization $(R_p)_g$ (per m^2 of surface area) after grinding to the hypothetical rate $(R_p)_0$ for pure active $TiCl_2$ and the concentration of metallic titanium, $[Ti^0]$, formed. C is a constant. Typically, $(R_p)_0 = 14\,g\,m^{-2}\,h^{-1}$ and $C = 0.556\,g\,m^{-2}\,h^{-1}$ per mole Ti^0. Polymerization temperatures and pressures required at 145°C and 30 kg cm^{-2} and cyclohexane solution are considerably greater than for conventional Natta catalysts, with rates up to 120 g per g catalyst per h reported.

Titanium trichloride alone is not a catalyst,* but when subjected to γ-radiation in hydrocarbon solvent it becomes active, presumably as the

---

* It does, however, give weakly active catalysts without metal alkyls by the addition of electron donors, e.g., tertiary amines [106].

result of the formation of an alkyl titanium halide which is then adsorbed on the crystalline $TiCl_3$ [143].

Boocock and Haward [252] have examined the polymerization of ethylene by α- and γ-$TiCl_3$/$AlEt_3$ under conditions such that mass transfer of monomer to the active centres was limiting, with the following observations:

(a) A linear relationship between $1/R_p$ and $1/[T]$ was obtained, in agreement with theory.

(b) The rate rose to a limiting value as the speed of stirring was increased.

(c) There was a reduction in the rate of ethylene absorption when it was diluted with nitrogen to a much greater extent than would have been calculated from the ethylene concentration. This is probably due to differences in average composition of gas bubbles in the solvent.

(d) With a mixture of ethylene and hydrogen there was a fall in polymer molecular weight with increase in concentration of $TiCl_3$, attributed to a change in the ethylene/hydrogen ratio at the particle surface as a result of mass transfer limiting the concentration of ethylene.

Swingler [307] has also investigated the γ-$TiCl_3$/$AlEt_2Cl$ system for ethylene polymerization. In this investigation, contrary to the earlier work [133], the initial stages of the polymerization show considerable complexity. The rate falls sharply in the first stage of the polymerization and then slowly accelerates to a maximum value at temperatures below 70°C. At higher temperatures the rate declines slowly from the maximum rate, due to a deactivation process which is second order in active centres. At sufficiently high Al/Ti ratios the maximum rate declines, consistent with competition between metal alkyl and monomer for the catalyst surface. (Table 10A.)

TABLE 10A
Effect of Al/Ti ratio on polymerization of ethylene by
γ-$TiCl_3$/$AlEt_2Cl$ at 70°C

Al/Ti	$(k_pf)_{max}$ (l mole^{-1} sec^{-1})
8.3	7.1 (est)
24	8.00
47	6.02
90	5.56

These rate coefficients are in substantial agreement with other data for ethylene polymerization with the γ-$TiCl_3$/$AlEt_2Cl$ system [133] (Table 10) suggesting comparable activity in the catalyst. Ageing the catalyst prior to polymerization gives increased activity, e.g. 90 mins at 80°C increases subsequent polymerization activity at 40°C by a factor of about three.

Activation energies were somewhat dependent on catalyst treatment and on the various stages of reaction, but in general were of the order of 9 kcal mole^{-1}.

Aluminium triethyl as co-catalyst gives, also in agreement with earlier data, more active but less stable catalysts, which become less active on ageing prior to polymerization. Kinetic treatment is more difficult because of the changes in rate with time and conditions of experiment (e.g. pretreatment of catalyst with reagents, co-catalyst concentration), but activation energies for both the initial and maximum rates were considerably higher, at 17—18 kcal mole^{-1} than those for the AlEt$_2$Cl systems.

A comparison of the γ form of TiCl$_3$ with the α and β forms gave the following polymerization rates at 80°C.

Type of TiCl$_3$	Rate coefficient (l mole^{-1} sec^{-1}) Cocatalyst	
	AlEt$_2$Cl^a	AlEt$_3$
α	1.5	8.15 — 9.51
β	7.72 — 8.90	29.6 — 33.0
γ	9.55 — 17.4	43.0 — 43.2

^{a} Activation energies for polymerization comparable in the range 7.1—9.8 kcal mole^{-1}.

Activated catalysts prepared by reduction of titanium tetrachloride by organo-metal compounds under conditions such that metal halides are incorporated into the titanium trichloride lattice are of considerable commercial importance. When organo-aluminium compounds are employed as reducing agent it is usual for variable proportions of aluminium chloride to be present in the precipitate (Table 1) with enhancement of activity, and some of the more active types have the approximate structure TiCl$_3$ 0.3AlCl$_3$. Metals other than aluminium can also be employed to give catalysts of enhanced activity. Thus, Haward, Roper and Fletcher [296] have demonstrated high activity in catalysts containing magnesium halides, and polymer yields of over 10^6 grams/gram of TiCl$_3$ have been reported [298]. A catalyst prepared at 2/1 molar proportions of (Mg(C$_{12}$H$_{23}$)$_2$/MgC$_{12}$H$_{23}$Br) to TiCl$_4$ had the empirical composition TiCl$_{2.88}$. 1.47(MgCl$_{1.60}$Br$_{0.04}$) showing the concentration of magnesium halide to exceed that of the titanium halide.

Table 10B shows relative polymerization rates for several catalysts prepared under different conditions. At progressively lower concentrations of catalyst (ca. 5 x 10^{-6} mole l^{-1}) and operating so that mass transfer of monomer was not a limiting factor much higher rate coefficients were observed (> 10^3 l mole^{-1} sec^{-1} at 80°C). Maximum reaction rates were determined for a number of titanium concentrations and propagation rates calculated using the equation on p. 174. In this way $k_p f$ was

TABLE 10B

Effect of reducing agent on the activity of catalysts for the polymerization of ethylene

Alkyl used for reduction of $TiCl_4$ [A]/[Ti] = 1/1	Co-catalyst [A]/[Ti] = 10/1	Relative polymerization rate	$[\eta]$ polymer
$AlEt_2Cl$[a]	$AlEt_3$	1[b]	1.86
LiBu	$Al(C_{11}H_{23})_3$	2.0	2.15
$CdBu_2$	$Al(C_8H_{17})_3$	1.85	2.55
MgBuBr	$Al(C_{11}H_{23})_3$	5.8	4.50

[a] Precipitate converted into γ form by heating at 160°C for 2 h.

[b] 376 g polymer g $TiCl_3^{-1}$ h^{-1} bar^{-1} (C_2H_4) at 80°C. This would correspond to a rate coefficient of ca. 15 l mole^{-1} sec^{-1} assuming [M] = 0.04 mole l^{-1} at 80°C and 1 bar [297].

estimated to be $3.4 \pm 0.1 \times 10^6$ (dm^3 C_2H_4) (mole $TiCl_3$)$^{-1}$ h^{-1} at 50°C and for a monomer pressure of one bar [297]. This is equivalent to 350 l mole^{-1} sec^{-1} in the units used in this review, taking the monomer concentration to be 0.12 mole l^{-1}. Overall activation energy for polymerization was 18.4 kcal mole^{-1} similar to that observed by Berger and Grieveson [133].

The large value for the rate coefficient compared with those given in Table 10 raises the point as to whether it results from very high activity of catalytic sites, from a high proportion of active centres, or from a combination of both effects. The efficiency of the magnesium activated catalysts (obtained from $R_p t/\bar{P}_n$ vs. t) gives a value of $f \simeq 0.5$ [297] compared with less than 0.01 for most Ti based catalysts. Thus when allowance is made for the inactive centres in the catalyst a value of k_p ca. 0.7×10^3 l mole^{-1} sec^{-1} is obtained. However, values of k_p for the other titanium systems listed in Table 6 are below 100 l mole^{-1} sec^{-1}, even when allowance has been made for their low levels of efficiency (f = 0.007—0.006), which suggests much greater activity in the modified catalyst sites. It is, of course, conceivable that the earlier and lower estimates of k_p have resulted either from over estimates of catalyst efficiency (cf. the work of Coover et al. on propene polymerization [121]) or from reduced polymerization rates consequent on mass transfer limitations on the ingress of monomer. Evidence to support this comes from measurements by Boucher, Parsons and Haward [298] on a conventional $TiCl_3/AlEt_2Cl$ catalyst, which gives a value of $k_p f = 27$ l mole^{-1} sec^{-1} at 50°C, $(2.6 \times 10^4$ (dm^3 C_2H_4) (mole $TiCl_3$)$^{-1}$ h^{-1} bar^{-1}), considerably greater than most of the values in Table 10.

In view of this uncertainty no definite conclusion can be drawn as to the activity of the sites in magnesium activated titanium catalysts. On balance it may be reasonably presumed that there is a positive effect, but doubts remain as to whether the earlier values are not underestimates by a considerable margin.

There is no internal anomaly, of course, in the data reported in ref. 298. For the active catalyst an efficiency of ca. 60% was obtained (from a plot of $[P]/\bar{P}_n$ vs. t) whereas for the conventional unmodified catalyst a value of $0.44 \pm 0.5\%$ was found. Polymerization rates were 3.4×10^6 and 2.6×10^4 dm^3 C$_2$H$_4$ (mole TiCl$_3$)$^{-1}$ h^{-1} respectively. The ratio $(3.4 \times 10^6)/(2.6 \times 10^4) \cdot 0.44/60$ is sufficiently close to unity as to indicate comparable activities of the sites in modified and unmodified catalysts. The high value of k_p calculated from polymerization rates with both of these catalysts, compared with the values quoted in Table 10, remains unexplained unless earlier values are low, as suggested, as a consequence of restricted transfer of monomer or from overestimates of active centres in prior publications.

The deposited metal oxide catalysts exhibit complex behaviour and most of the information available relates to the commercially important CrO$_3$ catalyst supported on silica or silica—alumina. The nature of the support and the concentration of chromium affect the catalyst activity and, apparently, the propagation rate coefficient [46]. The activity passes through a maximum with increase in chromium content on the support. Maximum activity is found with chromium levels between 1—2.5% [46], but in terms of active chromium concentration between 0.003 and 0.01% [45]. The catalyst activity depends also on the conditions of activation and pretreatment with reducing agents such as carbon monoxide, ammonia and hydrogen cyanide. Hogan [45] from relative changes in polymerization activity with chromium concentration and from the inhibiting effect of triethylamine estimated active site concentrations in the region of 2.5×10^{-5} mole per g catalyst for a 1% Cr catalyst activated at 800°C in dry air.

The polymerization of ethylene from the gas phase by CrO$_3$/SiO$_2$ has been studied [280, 281]. A linear relationship was observed between the rate of polymerization of ethylene and the capacity of the catalyst for adsorption of carbon monoxide — this latter increasing with time of pretreatment at 300°C.

The nature of the catalyst sites was given particular attention. Reduction of the catalyst, from the amount of carbon dioxide produced, indicated an average valence state of the chromium of 4.5 ± 0.2 comparable to that observed in reduction by ethylene. The conclusion was that the active sites contained chromium in a valence state above Cr^{III} — possibly Cr^V. However, in view of a relationship between polymerization activity and Cr^{II} concentration [161] for extensively reduced catalysts, other catalytic species may exist.

Carbon monoxide is less strongly adsorbed than is ethylene but acts as a catalyst poison. This suggests that there are sites of variable polymerization activity and that carbon monoxide is most strongly adsorbed on the most active ones.

Ermanov et al. [46], using CH$_3$O^{3}H for active centre measurements, reported values of 0.5 — 0.9 x 10^{-6} mole per g catalyst for a 2.5% Cr catalyst deposited on silica—alumina (97/3) activated at 400°C and 710°C in air and reduced at 300°C with CO. These catalysts were of lower activity, however, and in conjunction with actual rates the magnitudes of the propagation rate coefficients for both are calculated to be similar at ca. 10^3 l mole^{-1} sec^{-1}. Deuterium in the ethylene stream gave polymers of lower molecular weight containing CH_2D groupings, which was interpreted as resulting from addition of hydride ions from the catalytic sites rather than hydride transfer directly from the terminating chain [45].

Other chromium catalysts for ethylene polymerization employ chromocene [246] and bis(triphenylsilyl) chromate [247] deposited on silica—alumina. The catalyst support is essential for high activity at moderate ethylene pressures (200—600 p.s.i.). The former catalyst is activated further by organo-aluminium compounds. Polymerization rates are proportional to ethylene pressure and molecular weight is lowered by raising the temperature or with hydrogen (0.1—0.5 mole fraction) in the monomer feed; wide molecular weight distributions were observed.

A typical catalyst containing 2.5% of chromium in n-hexane (20 mg l^{-1}) activated by aluminium diethyl ethoxide (0.3 mg l^{-1}) gave a yield of 140 g polymer per h at 88—91°C and a monomer pressure of 300 p.s.i.g. The chromocene based catalysts are also of high activity, this being dependent on the conditions of catalyst preparation. Dehydration at 670°C gave the highest yield of polymer. Detailed kinetics were not carried out, but yields of 130—1670 g polymer per mmole Cr per 100 p.s.i. ethylene per h at 60°C were reported.

In ethylene—propene copolymerization the former monomer is greatly favoured and a value for r_1 of 72 was found. Hydrogen is particularly active as a chain transfer agent for this catalyst, a value of $k_{tr,H_2}/k_{tr,M} = 3.8$ x 10^3 being quoted, some ten times greater than that for a conventional Ziegler system [133b]. The active species in both these systems was ascribed to a low valence Cr^{II} complex.

In a subsequent publication Karol et al. [254] have given average values of the transfer coefficients with monomer and hydrogen for the chromacene catalyst, viz.

$$C_M = k_{tr_M}/k_p = 1.42 \times 10^{-4}; \quad C_H = k_{tr_{H_2}}/k_p = 0.465$$

The polymerization rate was proportional to the concentration of Cr$^{(II)}$, which accounted for 90% of all the chromium found in the catalyst, but whether the whole of the chromium present is active was not determined. The activation energy for polymerization was 10.1 kcal mole^{-1} over the range 30—56°C. Replacement of hydrogen by deuterium had little effect on polymer yield but molecular weights were somewhat lower.

The behaviour of supported oxide catalysts based on molybdenum, tungsten and vanadium resembles to some extent the chromium catalysts except in the means of activation, but there are significant differences such as the effects of the type of catalyst support [144]. Rates generally are lower than with chromium, although they are difficult to compare since there are such wide variations in activation procedure and operating conditions. There has been one kinetic study of the $Mo_2O_3/Al_2O_3/Na$ system [47]. (The fully oxidized catalyst was reduced with hydrogen at 555°C before adding the sodium activator.) The rates of polymerization over the temperature and pressure ranges 200—275°C and 44—70 kg cm^{-2}, were proportional to monomer and catalyst concentrations, but tended to be low (0.4 to 2.5 g polymer per g catalyst per h). The rate of polymerization passed through a maximum at 235°C with increase in temperature, the activation energies in the two branches below and above the maximum being 14 kcal mole^{-1} and —18 kcal mole^{-1} respectively. This was considered to be the result of a reduction at the higher temperatures in the effective concentration of ethylene at the catalyst surface. Propagation rate coefficients with supported oxide catalysts from Group V and VI metals decreased in the order Cr > Mo > W and Cr > V [46].

Ethylene has been polymerized by the heterogeneous catalyst $Zr(CH_2Ph)_4/Al_2O_3$ [279]. Using the kinetic scheme proposed for styrene, in which the monomer coordination and insertion steps for initiation and propagation are distinguished, the propagation rate coefficient is given by

$$k_p(=K_2k_2) = 21.8 \times 10^3 \text{ l mole}^{-1}\text{sec}^{-1}$$

at 80°C. This is the highest which has been recorded for ethylene polymerization. High rates (R_p = 2.7 x 10^{-4} mole l^{-1} s^{-1} at 80°C for [M] = 0.055 mole l^{-1} and [C]$_0$ 4.5 x 10^{-5} mole l^{-1} are obtained even though the proportion of active catalyst is small [C*]/[C]$_0$ = 0.005.

Hydrogen is an effective chain transfer agent,

$$k_{tr_{H_2}}/k_p = 1.25 \times 10^{-2},$$

about 10^3 times that of the spontaneous termination coefficient.

Zakharov et al. [320] using labelled carbon dioxide as terminating agent obtained comparable figures for the efficiencies of supported catalyst for zirconium and titanium tetrabenzyls (Table 5). The catalyst from $Zr(CH_2Ph)_4/Al_2O_3$ was considerably more active than either the corresponding one on a silica support or $Ti(CH_2Ph)_4/Al_2O_3$. The supports were of high surface activity (250 m^2 g^{-1}). Propagation rate coefficients calculated for these catalysts are very high (Table 6) although lower than the value reported above. Carbon monoxide was found to inhibit the polymerization quantitatively but about ten times the amount compared with carbon dioxide was required.

Soluble catalysts for ethylene polymerization from Group IV to VI transition metal compounds contain the metal in π-allylic compounds or binary complexes of soluble salts with aluminium alkyl compounds. Titanium (IV) complexes from $(\pi\text{-}C_5H_5)_2TiCl_2/AlR_2Cl$ [111] and $(\pi\text{-}C_5H_5)_2TiRCl/AlRCl_2$ [202, 216] polymerize ethylene rapidly but activity is quickly lost as the Ti^{III} complex is formed (Fig. 16). Other soluble titanium complexes from $Ti(OR)_4/AlEt_3$ [51] and $Ti(acac)_3/AlEt_2Cl$ [145] polymerize ethylene, but are of lower activity and contain the metal wholly or partly in the trivalent state.

The soluble $VO(acac)_2/AlEt_2Cl$ catalyst [139] is active over a wide range of catalyst compositions, with a maximum at $Al/V = 50$. An increase in the Al/V ratio from 50 to 500 results in a fall in rate from about 2×10^{-4} mole l^{-1} sec^{-1} to 0.9×10^{-4} mole l^{-1} sec^{-1} at $20°C$ for $[V] = 10^{-4}$ mole l^{-1} and ethylene 700 torr in toluene.

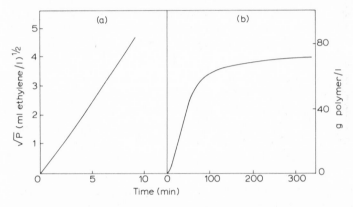

Fig. 16. Polymerization of ethylene using soluble catalysts. (a) $(\pi\text{-}C_5H_5)_2TiCl_2/AlEt_2Cl$; $Al/Ti = 5/2$, $[Ti] = 10^{-2}$ mole l^{-1}. Temp., $5°C$; solvent, benzene [217]. (b) $(\pi\text{-}C_5H_5)_2TiCl_2/AlMe_2Cl$; $AlTi = 5/2$, $[Ti] = 2 \times 10^{-2}$ mole l^{-1}. Temp., $15°C$; solvent, toluene [111].

Chromium catalysts $Cr(acac)_3/AlEt_2Cl$ or $AlEtCl_2$ [139] give high molecular weight polymers and are non-terminating. Rates are comparable to those obtained with the soluble titanium(IV) and vanadium catalysts, but the chromium catalysts would appear to be more stable. Over-all rates are in fact of a similar magnitude to those reported for the deposited chromium catalysts. In the latter the fraction of active metal is small (Table 5, p. 178) [125] and the propagation rate coefficients are therefore much greater (k_p 700—1000 l mole^{-1} sec^{-1} at $75°C$ compared with ca. 20 l mole^{-1} sec^{-1} at $20°C$ for the $Cr(acac)_3/AlEt_2Cl$ catalyst).

Detailed kinetic measurements have not been carried out on the soluble allylic compounds from Group IV, V and VI transition metals but rates depend on the transition metal, the presence and number of halogen atoms, and the structure of the allylic grouping [231]. Chromium tris(2-methallyl) was found to be among the more active initiators. Halogen substitution for allyl groups reduced the activity of chromium but raised the activity of corresponding zirconium complexes. Methyl substitution in the allylic grouping increased polymerization rate while phenyl substitution had the opposite effect. Molecular weights were reduced by hydrogen.

Further studies on the soluble catalysts show them to be more complex than was initially thought. The active species, of which there may be several, clearly depend on the Al/Ti ratio and it has been suggested that they are in equilibrium. Reichert and Meyer [276] have shown the rate of ethylene polymerization by $(\pi\text{-}C_5H_5)_2\text{TiEtCl/AlEtCl}_2$ to increase after an initial steady rate to a maximum, and then decline linearly with time as polyethylene is precipitated. The rate is increased by the addition of small amounts of water.

The propagation rate coefficient (15 1 mole^{-1} sec^{-1} at 10°C) in the initial homogeneous stage is close to other reported values (202, 235, 266) in Table 6. The activation energy is in the range 5—8 kcal mole^{-1}. Oligomers are produced in the first few seconds of the polymerization ($\bar{P}_n$ = 4—12) and the velocity constant falls; the rate of insertion of the first ethylene unit was estimated, by extrapolation, to be ca. 30 1 mole^{-1} sec^{-1} at 10°C. The overall velocity constant ($(R_p)_0/[\text{M}][\text{Ti}]$) depends on the Al/Ti ratio and passes through a maximum.

In Table 11A are shown the differences between soluble titanium catalysts of allied composition. Borisova et al. [277] obtained a comparable value for the velocity constant (Table 6) with the catalyst $(\pi\text{-}C_5H_5)_2\text{TiEtCl/AlEtCl}_2$ at Al/Ti = 1/1 in benzene or heptane/benzene (98/2). No induction period was observed and the rate diminished continuously with time — this being accompanied by reduction of Ti^{IV} to Ti^{III}, in agreement with earlier observations. In this work measurements were carried out while polymer was precipitating from the solution.

A linear relationship was obtained between the reciprocal of the yield of polymer and time in accord with a second order disproportionation reaction but the slopes of these lines were inversely proportional to [Ti]$_0$.

Excess AlEtCl$_2$ inhibits the polymerization and its effect attributed to an exchange reaction to form an inactive complex

$$(\pi C_5H_5)_2\text{TiP}_n\text{Cl}\,.\,\text{AlEtCl}_2 + \text{AlEtCl}_2$$
$$\rightarrow (\pi C_5H_5)_2\text{TiCl}_2\,.\,\text{AlEtCl}_2 + \text{AlP}_n\text{EtCl}$$

This complex could be realkylated to form the active species.

Studies of reactor kinetics in Ziegler—Natta type polymerization have received little attention but Schnell and Fink [319] have recently published results on the polymerization of ethylene by $(\pi\text{-}C_5H_5)_2\text{TiEtCl/}$ AlEtCl$_2$ in a continuous plug flow reactor. The formation of oligomers in the early stages of reaction (0.095—1.74 sec) was followed using gel chromatography on the outflow, as a function of Al/Ti ratio and the nature of $R(C_2$ to $C_6)$ in the catalyst. The propagation rate coefficient calculated for $R = C_6H_{13}$ was 47.7 l mole^{-1} sec^{-1} at 10°C. This result is in fair agreement with other values for this type of catalyst.

Waters and Mortimer [235] with catalysts $(\pi\text{-}C_5H_5)_2\text{TiRCl}/\text{AlR}'\text{Cl}_2$ found polymerization rates to be unaffected by the structure of the alkyl groups, for ethyl and higher homologues. The rate in toluene was lower (by a factor of 4 at Al/Ti = 1.0) than in dichloromethane, which was attributed to involvement of the aromatic solvent in the polymerization, presumably in the monomer coordination step. Rates increased with ethylene pressure and with Al/Ti ratio, and, as the latter was increased, the time before which a decay in rate occurred decreased; maximum rates with 0.01 mole l^{-1} of titanium complex, 760 torr ethylene pressure and 0°C, increased from 8 to 120 x 10^{-4} mole l^{-1} sec^{-1} with increase in Al/Ti ratio from 1.0 to 4.0. Assuming ethylene solubility of approximately 0.15 mole l^{-1} these would correspond to apparent rate coefficients of 0.5 to 7 l mole^{-1} sec^{-1}, not greatly different from other quoted values (Table 10).

With catalysts of the structure $(\pi\text{-}C_5H_5)_2\text{TiRCl}/\text{EtAlCl}_2$ with R = Ph, Me the polymerization exhibited an acceleration state [243] indicating slow initiation, and thus differed from catalysts containing ethyl and higher alkyl groups. This observation of slow initiation at a CH$_3$ transition metal bond confirms an earlier report [111].

Olivé and Olivé [291] have studied transfer reactions in ethylene polymerization by $(\text{EtO})_3\text{TiCl}/\text{EtAlCl}_2$. Polymerization reaches a maximum value at Al/Ti = 7 and is first order in monomer. As polymer molecular weight is independent of catalyst and monomer concentrations it is concluded that transfer involves monomer and that spontaneous termination is absent.

The polymerization rate at -20°C and a monomer pressure in toluene of 12 atms ([M] estimated to be 3.4 mole l^{-1}) was 4.9 x 10^{-3} mole l^{-1} sec^{-1} for [Ti] = 2 x 10^{-2} mole l^{-1} and $\overline{M}_n = 153$. With the assumption of $[C^*] = [\text{Ti}]_0$ at Al/Ti $\geqslant 7$ and the relationships

$$R_p = (k_p + k_{\text{trM}})[C^*][M]; \quad \overline{P}_n = k_p/k_{\text{trM}} + 1$$

k_p and $k_{\text{tr,M}}$ were calculated (Tables 6 and 7).

The transition state complex suggested for the transfer reaction was

This structure was considered to be not inconsistent with the low values for $E_{\text{tr,M}}$ and $A_{\text{tr,M}}$ in comparison with other β hydride elimination reactions such as

$$R_2AlCH_2CH_2R^1 \rightarrow R_2AlH + R^1CH=CH_2$$

A limited kinetic investigation has been carried out on promotors for vanadium catalysts ($VO(Ot\text{-}Bu)_3/Al_2Et_3Cl_3$) in ethylene polymerization [234]. It was shown that esters of trichloroacetic acid, added continuously during the polymerization, reactivated the catalyst and permitted polymerization to be carried out at $120°C$. Under these circumstances over 250 polymer chains were produced per vanadium atom and the polymers had $\overline{M}_w/\overline{M}_n$ ratios close to 2.0, which would be anticipated for a single catalytic entity.

8.2 PROPENE

Most kinetic data on propene polymerization have been obtained with the heterogeneous catalysts from $\alpha\text{-}TiCl_3$ and VCl_3 which, after the early stages of reaction, give steady rates. The catalysts from $TiCl_4$ or VCl_4

TABLE 11

Propene polymerization by coordination catalysts

Catalyst	Reaction order		Polym. rate coefficient k_{pf} ($l\ mole^{-1}\ sec^{-1}$)	E (kcal $mole^{-1}$)	Ref.
	Monomer	Catalyst			
$TiCl_4/AlEt_3(Al/Ti = 3)$	—	—	$4.6 \times 10^{-3}(70°C)$	—	137
$\alpha\text{-}TiCl_3/AlEt_3(Al/Ti = 3)$	1.0	1.0	$2.8 \times 10^{-2}\ (70°C)$	10	81
$\alpha\text{-}TiCl_3/AlEt_3(Al/Ti = 3.3/1)$	—	—	$2.5 \times 10^{-2}\ (70°C)$	—	151
$\alpha\text{-}TiCl_3/AlEt_3$	—	—	$14.4 \times 10^{-2}\ (70°C)$	—	121
$\alpha/\gamma\text{-}TiCl_3/AlEt_2Cl(Al/Ti = 3.6)$	1.0	1.0	$2.1 \times 10^{-2}\ (40°C)^e$	12	271
$\alpha\text{-}TiCl_3/AlEt_2Cl$	—	—	$5.6 \times 10^{-2}\ (70°C)$	—	123
$\alpha\text{-}TiCl_3/AlEt_2Cl$	—	—	$4 \times 10^{-2}\ (70°C)$	—	121
$\alpha\text{-}TiCl_3/AlEt_2Cl$	—	—	$5.8 \times 10^{-2}\ (50°C)^a$	—	20
$\alpha\text{-}TiCl_3/AlEt_2Cl$	1.0	—	$6.2 \times 10^{-2}\ (50°C)$	—	126
$TiCl_3/AlEt_2Cl$	—	—	$9.5 \times 10^{-2}\ (60°C)$	—	213
$\alpha\text{-}TiCl_3/AlEt_2Cl$	1.0	1.0	$3.5 \times 10^{-2}\ (50°C)$	—	86
$TiCl_3/AlEt_2Cl(Al/Ti = 7—10)$	—	—	$3.5 \times 10^{-2}\ (60°C)$	—	147
	—	—	$14.3 \times 10^{-2}\ (60°C)^b$	—	147
$TiCl_3/AlEtCl_2/HMP$ (Ti/Al/HMP = 1/1/0.6)	—	—	$10^{-2}\ (70°C)$	—	121
$VCl_3/AlEt_3(Al/V = 3/1)$	—	—	$6.5 \times 10^{-2}\ (60°C)^c$	—	30
$VCl_3/AlEt_2Cl(Al/V = 4/1)$	—	—	$1.4 \times 10^{-3}\ (60°C)^d$	11	30
$VCl_4/Al(Hex)_3(Al/V = 2.5)$	1.0	1.0	$6.7 \times 10^{-3}\ (25°C)$	6.6	193

[a] From Fig. 1 in ref. 20, curve O, [M] assumed to be 0.4 mole l^{-1}.

[b] $TiCl_3$ produced by reduction of $TiCl_4$ with aluminium, and containing 4.7% Al.

[c] From Fig. 4 in ref. 30.

[d] From Fig. 3 in ref. 30, steady rate; [M] assumed to be 0.4 mole l^{-1}.

[e] Calculated from data in Table II of ref. 271. [M] assumed to be 0.75 mole l^{-1} at 8 psig and $40°C$ in n-heptane.

TABLE 11A
Soluble titanium catalysts for ethylene polymerization

Catalyst	Induction period	$(R_p \max \times 10^4)$
$(\pi\text{-}C_5H_5)_2 TiCl_2/AlEtCl_2$	none	6
$(\pi\text{-}C_5H_5)_2 TiCl_2/AlEt_2Cl$	yes	3
$(\pi\text{-}C_5H_5)_2 TiEtCl/AlEtCl_2$	none	50
$(\pi\text{-}C_5H_5)_2 TiEtCl/AlEt_2Cl$	yes	0.5

$M = 0.11$ mole l^{-1} $Ti = 10^{-2}$ mole l^{-1} $Al/Ti = 1.9$ Benzene $15°C$

with aluminium alkyls contain unstable species which can be eliminated by ageing before addition of monomer. They are finely dispersed, but not truly soluble, and are much more effective for the polymerization of ethylene than of propene. In Table 11 are summarized data for a number of catalysts.

TABLE 12
Comparison of co-catalysts in propene polymerization [148].

Co-catalyst	Polymerization rate (l per g $TiCl_3$ per min)	
	(a)	(b)
$AlEt_3$	7.5	10.2
$BeEt_2$	16	27

Surface area of $TiCl_3$ (a) 5.6 $m^2 g^{-1}$ (b) 9.2 $m^2 g^{-1}$ (Be or Al/Ti = 3/1).

TABLE 13
Effect of catalyst structure on yield of atactic polypropene [152]

Catalyst		% Atactic polymer[a]
$TiCl_3$	$ZnEt_2$	60—70
	$MgEt_2$	15—22
	$AlEt_3$	15—20
	$BeEt_2$	4—6
	$AlEt_2Cl$	6—9
	$AlEtCl_2 . D^b$	1—5
VCl_3	$AlEt_3$	27
$TiCl_4$	$AlEt_3$	52

[a] Values depend on experimental conditions but are not greatly affected by polymerization temperature.
There are small differences between α- γ- and δ-forms of $TiCl_3$.
[b] D = pyridine or quaternary ammonium halides.

There are significant differences in rate and polymer tacticity, dependent on the nature of the transition metal compound and organo-metal compound (Tables 12 and 13).

Solvent effects have not been greatly studied but Ingberman et al. [271] observed increases in rates of polymerization of propene by α/γ $TiCl_3$/$AlEt_2Cl$ in the order n heptane $<$ toluene $<$ chlorobenzene $<$ o-dichlorobenzene. These data were interpreted in terms of the ability of the solvent to desorb impurities from the catalyst surface. However, polymerizations were conducted at constant propene pressure and individual monomer concentrations were not reported. In agreement with other reports the polymerization rate was reduced by the presence of $AlEtCl_2$.

Guttman and Guillet [310] examined the polymerization of propene from the gas phase by single crystals of α-$TiCl_3$, activated by reaction with aluminium trimethyl vapour, at 25°C. It was assumed that each fibril (Fig. 3) represented an active site and from the propene concentration in the gas phase and the rate of growth of the fibrils k_p was estimated to be 2.5×10^6 l mole^{-1} sec^{-1}. This estimate is clearly tentative (the density of the fibrils is taken to be the same as the solid polymer and it cannot be certain that only one site is involved per fibril), and to be compared with the values in Table 6 it implies that a similar relationship exists between adsorption of monomer at the surface in solution or in the gas phase. Even making allowance for error the quoted value for k_p is so much higher than any other reported that a detailed kinetic study of gas phase polymerization of olefins by coordination catalysts would be of great value.

A limited kinetic investigation of vapour phase polymerization of propene has been reported by Grigor'ev et al. [321]. Samples of $TiCl_3$ prepared by reducing $TiCl_4$ by aluminium and with surface areas of $15.7-27.2$ m^2 g^{-1} were used. (They contained ca. 4% of Al). Maximum rates occurred at Al/Ti 0.6 to 1.1. It is not possible to apply a monomer concentration term to the data for comparison with results in solution, but at a propene pressure of 1 atm. rates were high at ca. $3-5 \times 10^{-2}$ moles/mole Ti/sec. α-$TiCl_3$ with no combined aluminium but of comparable surface area had about half the activity of the aluminium-containing samples.

The relatively steady polymerization rate observed with catalysts from $TiCl_3$ plus $AlEt_3$ or $AlEt_2Cl$ (after initial stages of the reaction) is in contrast with the behaviour of $ZnEt_2$ as co-catalyst, where the rate declines rapidly and continuously with time [94]. With the latter the rate has been found to fit the empirical relationship

$$R_p = B p_{C_3H_6} (1 - f e^{-kt})$$

B is a constant which depends on the $ZnEt_2$ concentration, f and k are constants defining the decay of polymerization rate, and $p_{C_3H_6}$ is the propene pressure. The fall in activity could result from the decomposition of relatively unstable catalytic species or the formation of less active sites by displacement of the metal alkyl by a different co-catalyst (e.g. ZnEtCl). The activation energy for the polymerization (6.5 kcal mole^{-1}) is considerably lower than for $AlEt_3$ (10—13 kcal mole^{-1}) or $BeEt_2$ [148] (16.2 kcal mole^{-1}).

A recent study of mass transfer controlled kinetics of propene polymerization by $TiCl_3/AlEt_3$ catalyst has shown the monomer absorption rate to be proportional to the speed of stirring and to fall with increase in conversion [289].

The steady polymerization rate R_p obeyed the relationship

$$(R_p)_s = [\beta k_\infty G/\beta + k_\infty G][M]_0;$$

β is the mass transfer coefficient, G the weight of $TiCl_3$ and k_∞ the stationary polymerization rate constant, which is the same form as found by Boocock and Haward [252].

In the declining rate period, when the effect of mass transfer of monomer has been allowed for in the over-all polymerization rate, it is shown that the polymerization rate coefficient is independent of the amount of polymer produced (i.e. speed of stirring or concentration of catalyst). The fall in polymerization rate with time can thus be due to changes in activity or active site concentrations but there is no physical barrier from polymer on the catalyst retarding the access of monomer. Calculations show that mass transfer of monomer is not limiting in the early stages of polymerization but as polymer accumulates it has a progressively larger effect and can reduce the rate by up to 50%.

These observations do not, however, preclude a catalyst occlusion phenomenon in other systems — as suggested for styrene polymerization by Allan and Gill [154], but it does suggest that these latter results would be worthy of reinvestigation, preferably with a stable catalyst.

There is little information on the mechanism or kinetics of formation of the amorphous polymer produced concurrently with the crystalline polymer. In general, polymerization rates and molecular weights are lower, but there is no clear relationship between rate of formation of amorphous polymer and catalyst composition. Catalysts from $TiCl_4$, $VOCl_3$ or VCl_4 which tend to produce colloidally dispersed or appreciably soluble catalysts give higher amounts of amorphous polymer and in some instances little or no crystalline material is produced. There is a tendency with most catalysts for the amount of amorphous polymer to increase with increase in metal—alkyl concentration.

The combination, $VCl_4/AlEt_2Cl(Al/V = 2—20)$, at low temperatures (−78°C) gives the syndiotactic form of polypropene [130]; the rate is proportional to vanadium concentration. Addition of $AlEt_3$ to the catalyst changes the structure of the polymer to the isotactic form.

8.3 STYRENE

There have been a substantial number of kinetic studies on the polymerization of styrene with coordination catalysts. Most give a mixture of isotactic and amorphous polymer, but there is little information on how this depends on catalyst structure and other reaction variables. Polymerizations are relatively slow and in most instances rates decline with time. This in part may be due to the decomposition of relatively unstable catalyst centres, but in several cases it is the result of encapsulation of the catalyst with polymer through which monomer must diffuse. Polystyrene precipitates from aliphatic hydrocarbon solvents when monomer concentrations are below $2-3$ mole l^{-1}, and at higher concentrations or in benzene solution, when polymer precipitation does not occur, the rate increases [153]. The kinetics of polymerization in which diffusion of monomer to the active centres is a determining factor has been considered by Allen and Gill [154]. In most investigations over-all rate coefficients, activation energies and reaction orders in monomer and catalyst have been determined (Table 14). As reactions may have orders higher than unity in catalyst and monomer the over-all rate coefficients do not permit direct comparison of polymerization rates.

Catalyst efficiencies are likely to be variable, but there is only one report on this, based on yields and molecular weights of crystalline polystyrene with $AlEt_3/TiCl_4$ catalyst [155]. The values ranged from 0.028 to 0.003, and the propagation rate coefficient for an average value of 0.01 was calculated to be 0.17 l mole^{-1} sec^{-1} at 70°C.

The experimental data, in general, appears rather unsatisfactory and reproducibility is probably not of a high order. Also as polymerizations fall in rate with time, complete reliance cannot be placed on the rate coefficients given. The reaction orders given in Table 14 in some cases only hold over a restricted range of monomer and catalyst concentrations and there are numerous peculiarities which are not readily explained, more particularly changes in reaction order with relatively minor differences in catalyst composition. Even with the $TiCl_3/AlEt_3$ catalyst, which is less complex than most, Otto and Parravano [156] observed a maximum in rate at Al/Ti = 0.6 followed by a fall in rate at higher ratios (proportional to $AlEt_3^{-0.5}$) with the order 1.5 in monomer. Burnett and Tait [153], however, reported rates which were independent of Al/Ti ratio over a wide range and were either first or second order in monomer dependent on concentration.

The catalyst from $PhTi(O(i-Pr))_3/Ti(O(i-Pr))_3$ is a 1/1 complex which decomposes to give free radicals which then initiate polymerization. The rate is proportional to the half power of the titanium complex and to the 1.5 power of the monomer concentration (Fig. 17).

For most of the other systems the question of free radical, anionic or cationic propagation reactions has not been considered, although as with

TABLE 14

Styrene polymerization by coordination catalysts

Catalysts	Reaction order		Polymerization rate coefficient ($k_p f$)	E (kcal mole^{-1})	Ref.
	Transition metal	Monomer			
α-TiCl$_3$/AlEt$_3$	1.0	1.0^a	3.8×10^{-5} (60°C)	9.4	153
α-TiCl$_3$/AlEt$_2$Cl	1.0	2.0^b	2.3×10^{-5} (60°C)		
α-TiCl$_3$/AlEt$_2$Br	1.0		2.1×10^{-5} (60°C)		
TiCl$_3$/AlEt$_3$(Al/Ti = 1.5/1)	1.0	1.0	1.5×10^{-4} (75°C)	11.0	158
TiCl$_3$/AlEt$_3$(Al/Ti = 3/1)	1.0	1.5	$\sim 10^{-3}$ (60°C)^c	8.1	156
TiCl$_4$/AlEt$_3$(Al/Ti = 3/1)			2.9×10^{-4} (60°C)		78a
TiCl$_4$/AlEt$_3$(Al/Ti = 3/1)	1.0	1.0	1.7×10^{-3} (70°C)	10	155
TiR$_4$/R$'$MgBr R = OBu, Cl; R$'$ = Bu, Ph		1.0		$\sim$20	159
TiCl$_4$/LiBu(Li/Ti = 1.5/1)	1.0	1.0^d	2.3×10^{-2} (30°C)	2.1	160
(π-C$_5$H$_5$)$_2$TiCl$_2$/AlEt$_2$Cl (Al/Ti = 18.2/1)	1.0		0.97×10^{-2} (26°C)		162
PhTi(Oi-Pr)$_3$/Ti(OiPr)$_3$	0.5	1.5			157
VOCl$_3$/AlEt$_3$ (Al/V = 2/1)	1.0	2.0	4.8×10^{-5} (40°C)	7.4	163
VOCl$_3$/Al(i-Bu)$_3$(Al/V = 3/1)	2.0	2.0	9.6×10^{-1} (40°C)	11.3	163
VCl$_4$/AlEt$_3$(Al/V = 2/1)	1.0	1.0	e	5.5	164
VCl$_4$/Al(i-Bu)$_3$(Al/V = 3/1)	0	1.0	e	5.3	164
VCl$_3$/AlEt$_3$		1.0			156
Cr(acac)$_3$/AlEt$_2$Br(Al/Cr = 3/1)	1.0	1.0	3.38×10^{-4} (30°C)	11.0	165
Cr(acac)$_3$/AlEt$_3$	0.5	1.0	1.42×10^{-5} (30°C)^f	10.5	165
ZrCl$_4$/AlEt$_3$ (Al/Zr = 1.5/1)	1.0	2.0^g	3.2×10^{-3} (40°C)	10.9	166
ZrCl$_3$/AlEt$_3$(Al/Zr = 2/1)	2.0	2.0^h	2×10^{-3} (40°C)	6.5	166

^{a} M > 3.6 mole l^{-1}.

^{b} M < 3.6 mole l^{-1}.

^{c} Est. from Fig. 3, ref. 156.

^{d} M < 2.1 mole l^{-1}.

^{e} Quoted values for these two systems are 2.69×10^{-8} and 3.5×10^{-8} l mole^{-1} sec^{-1} at 40°C, which do not appear to correlate with the data in Figs. 6—8 of ref. 164.

^{f} Units given in l mole^{-1} sec^{-1}, inconsistent with reaction order.

^{g} Up to catalyst concentration of 1.4×10^{-2} mole l^{-1}.

^{h} Up to catalyst concentration of 1.6×10^{-2} mole l^{-1}.

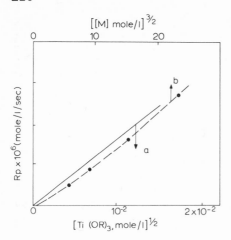

Fig. 17. Polymerization of styrene by $PhTi(Oi-Pr)_3/Ti(O-iPr)_3$ [157]. $PhTi(Oi-Pr)_3/Ti(O-iPr)_3 > 1$; temp., 18.7°C; solvent, benzene. (a) $[M] = 6.20$ mole l^{-1}. (b) $[Ti(OR)_3] = 10^{-4}$ mole l^{-1}. (Rates taken from Fig. 10 in ref. 157 at $[M]$ 8.71, 6.20, 4.35 and 2.32 mole l^{-1}.)

other monomers which can propagate by more than one type of active species this must always be a possibility. Indeed, with the more polar monomers it has been suggested that the reaction mechanism may be influenced by the nature of the solvent as well as by the catalyst [215]. In the case of styrene the formation of isotactic polymer, the dependence of rate on transition metal/metal alkyl ratio and the reduction in molecular weight with added zinc diethyl, support the view that the reactions are coordinated polymerizations. Unlike the α-olefins, styrene will polymerize rapidly with the free Lewis acids, but this should not arise at the catalyst ratios usually employed. Activation energies, although variable, are higher than those found with cationic catalysts.

The polymerization of styrene by $Zr(CH_2Ph)_4$ has been studied [278]. Only one of the four benzyl groupings participates in the polymerization and each metal centre initiates one chain (C14 labelled benzyl groups and molecular weight determinations gave values of 1.4—1.6 but the high value resulted apparently from alkylation of the polymer by $Zr(CH_2Ph)_4$).

Transfer reactions were not observed but termination, spontaneous β-hydrogen abstraction or involving monomer, was reported. Polymerization rates were low, e.g. $R_p = 4 \times 10^{-6}$ mole l^{-1} sec^{-1} at 30°C for $[C]_0 = 3 \times 10^{-2}$ mole l^{-1} and $[M]_0 = 5$ mole l^{-1} in PhMe. The rate follows the relationship

$$R_p = k_2 K_2 [C^*][M]$$

In this k_2 and K_2 are the monomer insertion reaction rate and the equilibrium constant for monomer coordination, arising from the kinetic scheme

$$C^* + M \rightleftharpoons CM^* \quad K_i$$
$$CM^* \rightarrow P_1^* \quad k_i$$
$$P_n^* + M \rightleftharpoons P_n M^* \quad K_2$$
$$P_n^* M \rightarrow P_{n+1}^* \quad k_2$$
$$P_n^* \rightarrow P_n + C^1 \quad k_t$$
$$P_n^* + M \rightarrow P_n C \quad k_t^1$$

$k_2 K_2$ is thus equivalent to k_p in systems where the coordination and insertion steps are not differentiated.

The proportion of catalyst consumed at conversions in the region of 2.6—4.8% ranged from 0.008 to 0.012 using C^{14} labelled catalyst, while the fraction of growing chains determined by quenching with tritiated water ranged from 0.001—0.002. Most of the chains attached to transition metal were inactive and the estimated fraction of propagating chains was 1/4300. Values for the rate coefficients are (units, mole, l, sec).

	$30°C$	$40°C$	$E(kcal\ mole^{-1})^a$
$k_i K_i$	3×10^{-8}	6.7×10^{-8}	15.2
$k_2 K_2$	0.07	0.2	20
k_t	1.7×10^{-4}	4×10^{-4}	16
k_t^1	0.44×10^{-4}	2×10^{-4}	25
$[C]^*/[C]_0$	0.23×10^{-3}	0.2×10^{-3}	—

for $[M]_0 = 3$ mole l^{-1} $[C]_0 = 3 \times 10^{-2}$ mole l^{-1} toluene.

^{a} Over-all reaction 16.6 kcal mole^{-1}.

K_i and K_2 were considered to have values in the range 10^{-3} mole^{-1}. Whence $k_2 \simeq 70$ sec^{-1}. Values for $\bar{M}_w/\bar{M}_n$ were close to 2 and average molecular weights were independent of catalyst concentration. $\bar{P}_n$ was related to monomer concentration by the relationship $1/\bar{P}_n = 1/A[M]_0 + B$. For the analogous heterogeneous system $Zr(CH_2Ph)_4/Al_2O_3$ the following values were obtained [279].

$k_2 K_2$ $2.1 - 2.2$ at $30°C$

$[C^*]/[C]_0$ $0.4 - 0.6 \times 10^{-3}$

8.4 BUTADIENE

Butadiene polymerizes with heterogeneous and homogeneous catalysts and there have been several detailed kinetic investigations. Features of special interest are the marked dependence of polymer microstructure on catalyst composition and the existence of soluble Group VIII element catalysts, specific for the polymerization of conjugated dienes. A summary of kinetic data on butadiene polymerization is given in Table 15.

TABLE 15

Polymerization of butadiene with coordination catalysts

Catalyst	Order		Polymerization rate coefficient (1 mole^{-1} sec^{-1})	E (kcal mole^{-1})	Ref.
	Monomer	Initiator			
Ti(On-Bu)$_4$/AlEt$_3$ (Al/Ti = 5.85)	1.0	1.0	0.8 x 10^{-4}(5°C) 0.5 x 10^{-3}(25°C)^a	11.5	146
TiCl$_4$/AlHCl$_2$. Et$_2$O/AlI$_3$ (1/5.5/1.3)	1.0	1.0	0.29(0°C)^b	8.5	177
(π-C$_4$H$_7$NiI)$_2$	1.0	0.5	2.08 x 10^{-4}M$^{-0.5}$ (50°C)	14.5	61
Co(acac)$_3$/AlEt$_2$Cl/ H$_2$O(Al/H$_2$O = 50)	1.0	1.0	1.6 (−2°C)^c	11.9	168
Co(naphthenate)$_2$/ AlEt$_2$Cl/H$_2$O (Al/H$_2$O = 10)	—	—	3.6 (25°C)	—	107
Co(acac)$_3$/Al$_2$Cl$_3$Et$_3$	1.0	1.0	129—136 (41°C)	5.3	110
Ni(naphthenate)$_2$/ BF$_3$. Et$_2$O/AlEt$_3$ (Ni/B/Al = 1/7.3/6.5)	1.0	1.0	0.24 (40°C)^d	ca. 5	68
(π-C$_3$H$_5$)Ni . CF$_3$COO)$_2$ (benzene solvent)	1.0	1.0	1.8 x 10^{-3}(40°C)^e	13	173
Cr(acac)$_3$/AlEt$_3$ (Al/Cr = 11.6/1)	1.0	1.0	2 x 10^{-2}(30°C)	20	55

^{a} Calc. on Ti from Figs. 6 and 7, ref. 146.
^{b} Calc. on Ti.
^{c} Calc. from Fig. 1, ref. 168.
^{d} Calc. from Fig. 4 in ref. 68.
^{e} Taken from Figs. 5 and 6, ref. 173.

8.4.1 Cobalt catalysts

Active catalysts for butadiene polymerization are obtained from aluminium alkyl halides and soluble Co^{II} and Co^{III} salts and complexes. The structure of the organic grouping attached to the cobalt is not important, but compounds most widely employed are acetylacetonates and carboxylic acid salts such as the octoate and naphthenate. The activity of the catalyst and structure of the polymer are affected by the groupings in the complex. Catalysts from aluminium trialkyls and cobalt salts other than halides are relatively unstable and give syndiotactic 1,2-polybutadiene. If halogens are present, e.g., from CoCl$_2$ or CoBr$_2$, more stable catalysts are produced by exchange reactions and these give high cis-1,4-polybutadiene. There is no kinetic information on these systems but those from aluminium alkyl halides, which give cis-1,4-polymer, have been studied in detail.

(a) $Co(acac)_3/AlEt_2Cl/H_2O$. This catalyst is soluble in benzene and heptane and high yields of high molecular weight polymers are obtained at temperatures ranging from —40—50°C. Combinations of cobalt salts with aluminium diethyl chloride do not polymerize butadiene unless an activator is present. Water is usually employed but oxygen, aluminium chloride and methanol have a similar activating effect [167]. The amount of water is not critical. The optimum is about 10 mole % on the aluminium diethylchloride but some activity remains with 100 mole % addition, this corresponding to the conversion of aluminium dialkyl halide to the aloxane $(EtClAl)_2O$. Rates are first order in monomer and cobalt

TABLE 16

Effect of polymerization variables on molecular weight of polybutadiene from catalyst $Co(acac)_3/AlEt_2Cl/H_2O$ [168].

[M]	$\bar{P}_v$	Temp. (°C)	$\bar{P}_v$	Al/Co	$\bar{P}_v$	% conversion	$\bar{P}_v$	[Co] × 10^4	$\bar{P}_v$
0.5	140	—40	980	800	940	5	650	0.83	3810
1.0	410	—2	780	500	1040	30	1000	1.50	2040
2.0	780	25	260	200	730	56	1140	2.50	920
3.0	940	50	220	100	870	82	1650	5.00	750

Except for the variable (temperature, conversion, etc.), experimental conditions were [M] = 2.0 mole l^{-1}, [Co] = 2.5 × 10^{-4} mole l^{-1}. Conversion, 20%. Al/Co = 400/l, Al/H$_2$O = 50. Temperature, —2°C. Solvent, toluene.

concentration and independent of Al/Co ratio over the range 80—400 [168], but decline rapidly with time above —20°C consistent with second order deactivation of active centres. The stability of catalysts is very dependent on composition and those from $CoCl_2$, $AlCl_3$ and $AlEt_3$ retain polymerization activity even when stored for many months [169].

The dependence of molecular weight of polymers on monomer concentration, temperature, Al/Co ratio and conversion are shown in Table 16. As molecular weights are lower than would be calculated from polymer and catalyst concentrations and polymer molecular weights, there is a transfer reaction but its nature is not yet established. Similar findings for the relationship between polymerization variables and molecular weight were found by Zgonnik et al. [170].

The $CoCl_2/AlEt_2Cl/H_2O$ catalyst is heterogeneous and rate measurements have not been reported, but, similar to the soluble catalysts, molecular weights increase in proportion to monomer concentration, and are independent of Al/Co ratio over the range 1—10/1 [167].

214

There is a continuous increase in molecular weight as the ratio H_2O/Al is increased, the rate passing through maxima at H_2O/Al in the range 0.2—0.5. At low conversion and constant monomer concentration the polymerization rate [292]

$$R_p \propto k[H_2O]^{1/2}[CoCl_2][AlEt_2Cl]^{1/2}$$

(b) $Co(acac)_3/Al_2Cl_3Et_3$. Catalysts from aluminium ethyl sesqui-chloride do not require the presence of water to initiate polymerization (Fig. 18). Rates are independent of Al/Co ratio and proportional to [Co] and to [M] in aliphatic hydrocarbon solvents [110]. From the over-all rate coefficient (Table 15) this catalyst would appear to be more active than that from $AlEt_2Cl/H_2O$. Aromatic solvents coordinate with the catalyst and those which are stronger donors than butadiene reduce the rate of polymerization (Table 3, p. 167) [110].

(c) $CoX_2/AlCl_3 : X = Br$, Cl. These complexes, of structure IX (p. 144), are soluble in benzene and cyclohexane and polymerize butadiene to high molecular weight 1,4 polymer. The polymerizations are sensitive to trace impurities which result in irreproducibility and may produce low molecular weight or crosslinked polymer [58]. Some of these difficulties arise from cationic activity in the catalyst (possibly from excess alu-minium halide), and polymerizations are faster and polymers of higher molecular weight are obtained with the addition of thiophene [171]. Thiophene forms part of the catalyst and becomes incorporated into the polymer and it has been suggested that the active species is an alkylated cobalt halide—aluminium chloride—thiophene complex [172]. Molecular weight was considered to be controlled by breakdown of uncomplexed (inactive) organo-cobalt halide into polymer and cobalt halide. A reduc-tion in the concentration of the former by complexation with thiophene would thus explain the increase in molecular weight with increase in thiophene/cobalt ratio (Table 17).

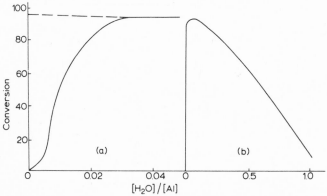

Fig. 18. Effect of water on polymerization of butadiene by soluble cobalt catalysts [167]. Al/Co = 500/1; time, 19 h; temp., 5°C. (———), $AlEt_2Cl$; (------), $Al_2Cl_3Et_3$. (a) Low water concentration. (b) High water concentration.

TABLE 17

Effect of thiophene concentration on molecular weight of polybutadiene from catalyst $CoCl_2$/$AlCl_3$/thiophene [172] [Co] = 0.3×10^{-3} mole l^{-1}; $AlCl_3$. Thiophene = 0.34×10^{-3} mole l^{-1}; [M] = 1.5 mole l^{-1}

Thiophene/[Co]	$\overline{M}_n \times 10^{-3}$
0	16
0.5	225
1.0	421
1.5	622

(d) $CoCl_2$/py_2/$AliBu_2$ Cl. Medvedev [283] has reported on the effect of water on the polymerization of butadiene by $CoCl_2$/py_2/$AliBu_2$ Cl. Either below or above an optimum water concentration (5.2×10^{-4} mole l^{-1} independent of $[Co]_0$ or $[Al]_0$ but proportional to $[M]_0$) polymerization rates declined. Addition of further water to a system containing less than the optimum concentration caused an increase in polymerization rate but to a lower value than the steady maximum. It was suggested, therefore, that water participates both in initiation and termination steps. The polymerization rate at optimum water concentration obeys the relationship

$$(R_p)_s = (k_i k_p / k_t)[H_2O]_0 [Co]_0 [M]_0$$

k_i, k_p and k_t are the rates of initiation, propagation and termination.

Typical rate data for Al/Co = 400, $[M]_0$ = 1.56 mole l^{-1} at 21°C in benzene solution are:

$[Co]_0 \times 10^5$	$(R_p)_s$ mole l^{-1} sec$^{-1} \times 10^4$
2.5	6.0
2.0	4.70
1.2	3.12
0.8	1.56
0.5	1.17

Molecular weights were reduced by chain transfer with excess AlR_2 Cl and a linear relationship was observed between $1/\overline{M}_\eta$ and $1/[Al]_0^2$

$[Al]_0 \times 10^3$ mole l^{-1}	Al/Co	$\overline{M}_\eta \times 10^6$
1.68	210	1.66
3.28	410	1.14
6.4	803	0.50

References pp. 249—257

Monomer was involved in the transfer process, possibly as a π complex with $k_{tr}/k_p = 4.7 \times 10^{-4}$ [M] mole l^{-1}.

8.4.2 Nickel catalysts

Binary or ternary catalyst systems from nickel compounds with Group I—III metal—alkyls have many features in common with those from cobalt and it may be inferred that a similar type of catalytic entity is involved. The composition for optimum activity may be different, however, and in the soluble catalyst Ni(naphthenate)$_2$/BF$_3$. Et$_2$O/AlEt$_3$ (Ni/B/Al = 1/7.3/6.5) [68] the ratio of transition metal to aluminium is much higher than in cobalt systems. Rates were proportional to [M] and [Ni], molecular weight was limited by transfer with monomer and catalyst efficiency was relatively low (Table 5, p. 178). With the system AlEt$_3$/Ni(Oct)$_2$/BF$_3$—Et$_2$O (17/1/15) the molecular weight rose with increase in [M]/[Ni] ratio and 3—9 chains were produced per nickel atom. It was observed that as molecular weight increased so the *cis* content of the polymer increased — from ca. 50% up to ca. 90% [292].

The dimeric allylic nickel halide complexes are simpler in that no problems of reaction with organo-aluminium compounds are involved. They are of low activity and give polymer of low molecular weight owing to a rapid monomer transfer reaction [61]. The active centre is formed by dissociation of the complex in the presence of monomer and since the rate is proportional to $[Ni]_0^{1/2}$ and [M] the extent of dissociation is small and the complex with monomer of low stability.

More recently it has been shown that in the polymerization with π-crotylnickel iodide the order in monomer falls from a value close to unity at [M] below 0.5 mole l^{-1} to below 0.5 at [M] > 4 mole l^{-1}. These observations have been interpreted in terms of scheme (c) on p. 162, namely coordination of two monomer molecules with the catalyst and with most of the catalyst existing in the complex (inactive) state. The molecular weights of the polymers are double those calculated from the kinetic scheme put forward [61] and this is attributed to coupling of "live" polymer chains on termination [251]. Molecular weight distributions are binodal consistent with slow propagation and transfer.

Lewis acids increase the activity of bis(π-allyl nickel halides) and the effect is evidently to convert them into monomeric species [63], since rates become first order in nickel and the activation energy for polymerization is lower at 6 kcal mole^{-1}.

The kinetics of butadiene polymerized by bis(π-allyl nickel trifluoroacetate) has been studied by Teyssié et al. [173]. Equilibrium constants for the formation of the complex (with aromatic compounds which give the equibinary polymer), calculated from cryoscopic measurements or polymer micro-structure, are 6—7 and 26—35 respectively for benzene and nitrobenzene. The higher concentration of active centres compared with the π-allyl nickel halides is shown by polymerization rates being faster by a factor of 10—100 (Table 15, p. 212) [63, 173]. The addition of nitrobenzene gives faster rates and the rate increases with increase in conversion. In the first stages of the reaction an oligomer is formed, the molecular weight being controlled by chain transfer with monomer. High molecular weight polymer is produced later, and after the monomer has been consumed the concentration of low molecular weight polymer falls. Typically, the oligomer has a molecular weight of 900—1000 while the high polymer, which appears at conversions above 10%, has a molecular weight [173] in excess of 10^5.

The activity of these complexes is influenced markedly by the inductive effect of the acid. In π-allyl nickel haloacetates and complexes of 2,6,10-dodecatriene-1,12-diyl nickel with haloacetic acids the rate of polymerization increases with the increase in acid strength [233] (Fig. 19).

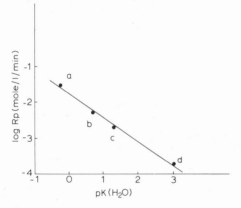

Fig. 19. Polymerization of butadiene by $C_{12}H_{18}Ni/R$. COOH [233]. [Ni]/ [RCOOH] = 1/1; [M] = 3.4 mole l^{-1}; [Ni] = 1.4 $\times$ 10^{-3} mole l^{-1}. Solvent, n-heptane; temp., $55°C$. R = (a) CF_3 (b) CCl_3 (c) $CHCl_2$ (d) CH_2Cl

The polymerization rate depends on the nature of the π-allylic grouping attached to the initial catalyst and there does not appear to be an acceleration stage corresponding to a slow initiation step followed by a faster propagation reaction [248], which would have been expected. Values for k_p (1 mole^{-1} sec^{-1} at $30°C$) and activation energies (kcal mole^{-1}) for π-allyl, π-methallyl and π-crotyl nickel trifluoroacetates are 1.2 $\times$ 10^{-3} and 12.2, 1.66 $\times$ 10^{-4} and 10.7 and 4.3 $\times$ 10^{-5} and 10.2, respectively.

There are also differences in polymer microstructure. Similar considerations on rate and structure apply to the effects of different solvents. It appears therefore that the behaviour of the catalyst is determined throughout the course of the polymerization by the structure of the initial catalytic entity, which in turn depends upon structure and conditions of preparation and use. In the reaction of these π-allylic complexes with butadiene the rate decreases in the order allyl > methallyl > crotyl, and the *syn* form of the new allylic complex produced is the more stable [255]. The configuration of the allylic complex is *syn* irrespective of whether the *cis* or *trans* 1,4 polymer is produced [256, 257].

8.4.3 *Titanium catalysts*

(a) $TiCl_4/Al(i\text{-}Bu)_3$. With this heterogeneous catalyst the rate was first order in titanium and the order in monomer was 1.0 for Al/Ti ratios 1 to 1.6, and 2.0 for Al/Ti ratios > 2.0 (Fig. 20) [174]. Over the region where first order kinetics were observed the rate increased linearly with Al/Ti ratio, while maximum activity was at Al/Ti = 2.7. The existence of a polymerization in which two monomer molecules are coordinated in the propagation step is of interest. It is unfortunate, therefore, that other features of the polymerization were not studied so as to put this deduction on a firm basis, and in particular the question of catalyst

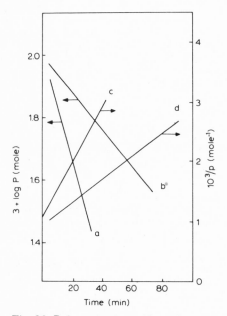

Fig. 20. Polymerization of butadiene by $TiCl_4/Al(i\text{-}Bu)_3$ [174].

[TiCl$_4$] (mole l^{-1})	Al/Ti
(a) 14.2×10^{-3}	1.43
(b) 4.1×10^{-3}	1.43
(c) 6.6×10^{-3}	2.0
(d) 3.3×10^{-3}	2.0
Solvent, heptane; temp., 25°C.	

stability was not considered. With catalysts from other Ti^{IV} halides there is evidence for a rapid fall in activity with time [99, 122], and an excess of aluminium alkyl is known to accelerate the decomposition of organo-titanium compounds [49]. Moreover, although the polymer microstructure changes as the Al/Ti ratio is increased the effect is progressive without any dramatic change which might be anticipated for a different polymerization mechanism. If there was a significant reduction in catalyst concentrations during the time of the experiment, this would have the effect of increasing the apparent order in monomer. Indeed, by choice of propagation and catalyst decomposition (first or second order rate coefficients), polymerizations which are actually first order in monomer will accurately follow second order kinetics over limited ranges of conversion.

(b) $TiI_4/Al(i\text{-}Bu)_3/i\text{-}Pr_2O$. Titanium tetraiodide is a sparingly soluble compound but with the co-catalyst forms a dark coloured complex, apparently homogeneous in hydrocarbon solvents, which gives polybutadiene of ca. 95% *cis* 1,4 content [122]. It is active over a wide range of Al/Ti ratios (2.8—25.9) but, after a fast initial reaction, polymerizations cease at low conversions (30—60%) and rates become almost independent of catalyst concentration with $[Ti] > 0.5 \times 10^{-3}$ mole l^{-1}, and catalytic activity declines with ageing. Polymer molecular weights rise to a maximum with increase in conversion, increase with monomer concentration and fall with increase in catalyst concentration. It is suggested that initiation is rapid and that chain transfer with excess metal alkyl occurs.

A detailed kinetic investigation of butadiene polymerization using $TiI_4/Al(i\text{-}Bu)_3$ has been reported by Loo and Hsu [318]. Rates were first order in monomer and catalyst concentration. Maximum catalyst activity was observed at Al/Ti = 1.5, whereas Henderson [122] with comparable catalysts containing diisopropylether found comparable activity over a wide range of composition and at Al/Ti ratios above 2. This difference indicates, perhaps, the diminished reducing power of metal alkyl-donor complexes. Catalyst activity was found to fall during polymerization but the data were considered to be more consistent with first rather than second order decay. Chain transfer with metal alkyl was the major transfer reaction and monomer transfer shown to be negligible.

Molecular weight distributions of the polymers ranged from $\overline{M}_w/\overline{M}_n = 2$—2.5 and were relatively symmetrical as wt. vs. log $\overline{M}$ plots. The authors give the following relationships for the molecular weights at low fractional conversions (x).

$$\overline{M}_n = m_0[M_0][k_p x/\{k_p[C_0] - k_{tr_A}[A]\ln(1-x)\}]$$

$$\overline{M}_w = \left[\frac{2m_0 k_p[M]_0}{x}\right][k_p[C_0]/\{k_{tr_A}[A] - k_p[C_0]\}]$$

$$[1/2k_p[C_0]\{1-(1-x)^2\} - \{1/k_p[C_0] + k_{tr_A}[A]\}$$

$$\{1-(1-x)(1 + k_{tr_A}[A]/k_p)\}]$$

(m_0 is the molecular weight of the monomer).

Using a curve fitting procedure from the experimental data propagation and alkyl transfer reaction rates and activation energies were calculated. Values for k_p and $k_{tr,A}$ at $40°C$ are included in Tables 6 and 8. The catalyst efficiency varied from 0.016 to 0.10 with most of the data in the range 0.01 to 0.04.

(c) $TiCl_2I_2/Al(i\text{-}Bu)_3$. This catalyst is also soluble and the polymerization rate falls rapidly with time to a low constant value, with an energy of activation [99] of 12.5 ± 0.4 kcal mole^{-1}. Polymer yields and average molecular weights indicate that the molar concentration of polymer remains constant throughout the polymerization. It was considered that initiation was fast, that transfer and termination do not occur, and that the fall in rate was the result of reversible association of the active centres to form inactive dimeric species*. An analysis of the data for two mechanisms, $2C* \rightleftharpoons C_2$ and $2C* \rightleftharpoons 2C'$ ($C*$, C_2 and C' represent active centres and the two forms of inactive species) was carried out to calculate rates and equilibrium constants. Active centre concentrations were not determined nor was the possibility of alternative explanations for the kinetic behaviour, such as the existence of two types of active catalyst, one unstable and the other stable, considered [99].

(d) $Ti(O(n\text{-}Bu))_4/AlEt_3$ [146]. Butadiene is polymerized to polymer of high vinyl content by this soluble catalyst. The maximum rate is observed at Al/Ti = 6, but the concentration of active Ti^{III} complex only forms a small fraction of the total catalyst. Activity is lost on ageing the catalyst, although growing polymer chains are non-terminating. Rates are first order in titanium concentration and (at low concentrations) in monomer. Molecular weights rise to a constant value with increase in conversion, possibly the result of transfer with monomer. The polymers are, however, of narrow molecular weight distribution.

As discussed on p. 143 there are a number of species in this catalyst dependent on the temperature of reaction and molar ratios of the reactants, and only a small proportion of the titanium atoms are active in polymerization. A characteristic ESR signal appears at $g = 1.983$ in the presence of butadiene which has been identified with the growing chains [124], and Table 17a shows its small proportion. An analogous species ($g = 1.978$) is obtained from the soluble $Ti(acac)_3/AlEt_3$ system but at much higher Al/Ti ratios [54].

(e) $TiCl_4/AlI_3/AlHCl_2 . Et_2O$. With this rather complex system reduction of the titanium to the trivalent state occurs, with ligand exchange to give attachment of iodide groups to titanium [177]. In consequence high

---

* This mechanism could conceivably explain the fall in activity of the $TiI_4/Al(i\text{-}Bu)_3$ catalyst.

TABLE 17a

Active centre concentrations in butadiene polymerization by $Ti(OBu)_4/AlEt_3$ and $Ti(acac)_3/AlEt_3$ [124, 154]

Catalyst	Al/Ti	Polymer conc. $\times 10^5$ mole l^{-1}	$Ti^{III} \times 10^5$ mole l^{-1}	$\dfrac{Ti^{III}}{[Ti]_0}$
$Ti(OBu)_4$	1.9	0	0	—
$-AlEt_3$	3.4	2.8	3.8	1.2×10^{-3a}
	3.9	4.5	4.5	1.6×10^{-3}
	4.8	6.9	9.2	3.4×10^{-3}
$Ti(acac)_3$	3.0	—	0	—
$-AlEt_3$	8.0	—	0.18	—
	58	—	14	—
	108	120	17	2.4×10^{-2}

a These values were calculated from the data in ref. 124, Table 1. Elsewhere in the paper the fraction of active titanium is stated to be 0.03.

cis 1,4-polybutadiene is formed. Activity is greatly influenced by the ratios of the three components and a kinetic investigation has been made for the composition $TiCl_4/AlHCl_2 . Et_2O/AlI_3 = 1/5.5/1.3$ (Table 15, p. 212).

8.4.4 Chromium catalysts

(a) $Cr(acac)_3/AlEt_3$. This catalyst is soluble and gives 1,2 polybutadiene. It is of low activity, notwithstanding high concentrations of metal alkyl bonds, and rates and molecular weights are low [55]. Chain transfer with excess metal-alkyl occurs and, as the rate also falls with increase in Al/Cr ratio, $AlEt_3$ retards the reaction, possibly by complexing with and deactivating the catalytic species.

(b) $Cr(\pi\text{-allyl})_3$. Chromium tris-allyl is a catalyst for butadiene polymerization without added organo-metal compound and like the other chromium based catalysts gives poly 1,2-butadiene. Details of the kinetics have not been given but the reaction is second order in catalyst and zero-order in monomer [63]. This suggests that a relatively stable complex is formed between the dimeric catalyst species and the monomer.

Dolgoplosk et al. [301] have reported kinetic data on the polymerization of dienes using π-allylic (π-allyl and π-crotyl) chromium compounds in solution and deposited on a silica support, and on the copolymerization of butadiene and isoprene using the supported π-allyl catalyst and a $CrO_3/SiO_2-Al_2O_3$ catalyst. Full details of the kinetics, including the units employed, were not given, but the following features are appropriate to this review:

Monomer	Catalyst[a]	Kinetics	Polymer
Butadiene	$Cr(\pi\text{-R})_3$	Second order in [Cr] $E = 19$ kcal mole^{-1}	MW 8—9×10^3 80% 1,2 structure
Isoprene	$Cr(\pi\text{-R})_2\,OSiO<$	First order in [Cr]. Second order in [M] $E = 9.8$ kcal mole^{-1}	MW 45—558×10^3 rising with conversion 94—97% *trans* 1,4 structure ($\overline{M}_v/\overline{M}_n \simeq 1.0$)
Isoprene	$CrO_3/$- SiO_2—Al_2O_3	—	(97—99% *trans* 1,4) Molecular weight increases linearly with conversion

[a] R, allyl, crotyl.

The essentially similar behaviour of the supported catalysts from $Cr(\pi$-allyl$)_3$ or CrO_3 suggests a similar catalytic entity. In contrast with the soluble chromium catalysts the polymers are predominantly 1,4 structure and the reactions are essentially non-terminating.

8.4.5 Rhodium catalysts

Butadiene is polymerized by rhodium compounds in aqueous or alcoholic solution [178]. It is generally accepted that the active species is a π-allyl rhodium complex of low valency [28, 179] which is not rapidly terminated by reaction with water or alcohol. No clear kinetic pattern was observed in the earlier papers but a recent investigation [180] has shown the rate and molecular weight data to be accommodated by a scheme involving monomer transfer and physical immobilization of the active centres in precipitated polymer. In the initial stages the polymerization is first order in rhodium and, at constant monomer concentration, is (pseudo) zero order ($E = 14.8$ kcal mole^{-1}). This is followed by a declining rate which is almost independent of temperature. Molecular weights rise slowly to a maximum value with time (ca. 4000 after 22 h at 70°C).

From the linear relationship between $1/\overline{P}_n$ and $1/\{[M]_0 - [M]\}$ all the rhodium initiates polymerization and $[C^*]/[Rh] = 2$ (Fig. 21); hence the slow rate must be the result of a slow propagation reaction. The existence of a monomer transfer reaction is shown by a linear relationship between $1/\overline{P}_n$ and $1/[M]_0$. The effect of emulsifiers, known to be important in accelerating the rate and rather specific in behaviour [181], were considered to stabilize the active centres against termination.

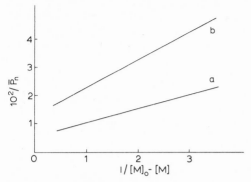

Fig. 21. Molecular weight dependence on conversion for butadiene polymerization by RhCl$_3$ [180]. [RhCl$_3$] = 5 x 10^{-3} mole l^{-1}; sodium dodecyl sulphonate, 6 x 10^{-2} mole l^{-1}; [M] = 5.0 mole l^{-1}. (a) 60°C; (b) 90°C.

8.5 OTHER MONOMERS

8.5.1 Butene-1

This olefin is less reactive than propene and much less reactive than ethylene, and certain catalysts which polymerize ethylene readily will not polymerize butene-1, e.g., AlR$_2$Cl/VCl$_4$/anisole. The monomer has been polymerized with TiCl$_4$/AlR$_3$ [98] and with TiCl$_3$/AlR$_3$ and VCl$_3$/AlR$_3$ [183]. The first of these gave results characteristic of mixed catalytic entities of widely varying stability, viz. a fast initial reaction which was first order in monomer, followed by a much slower polymerization. The reaction order for the initial rate in TiCl$_4$ was high (2—3), which cannot be reconciled with any known mechanism.

With the sub-halide catalysts the most interesting finding is that (from the slopes of $1/\bar{M}_v$ versus $1/t$) the propagation rate coefficient is not

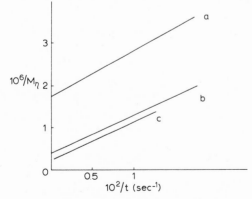

Fig. 22. Molecular weight of polybutene-1 [128]. Temp., 0°C; solvent, toluene. Catalyst (a) δ-TiCl$_3$/BeEt$_2$ (b) δ-TiCl$_3$/AlEt$_2$Cl (c) δ-TiCl$_3$/AlEt$_2$I.

References pp. 249—257

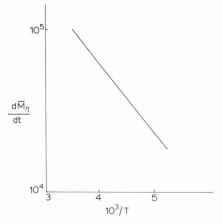

Fig. 23. Effect of temperature on change of molecular weight of polybutene-1 using VCl$_3$/AlEt$_3$ catalyst [128].

affected by the structure of the metal-alkyl (AlEt$_2$Cl, AlEt$_2$I, Al(i-Bu)$_2$Cl, BeEt$_2$; Fig. 22), which supports the view that the metal in the co-catalyst does not form an essential part of the catalytic species. The metal alkyls show significant differences in the rates of the transfer reactions and are greater for BeEt$_2$ than for aluminium alkyls or aluminium alkyl halides. The apparent propagation rate is greater with VCl$_3$ than with TiCl$_3$. From the change in slopes of $1/\overline{M}_v$ vs. $1/t$ with temperature (Fig. 23) E_p was found to have the low value, 2.3 kcal mole^{-1}. Butene-1 polymerizes at about one-quarter the rate of propene with the soluble VCl$_4$/AlR$_3$ catalyst [184], and about half the rate with heterogeneous TiCl$_3$/AlEt$_2$Cl [213].

8.5.2 Hexene-1

This olefin polymerizes with TiCl$_4$/Al(i-Bu)$_3$ (Al/Ti = 1/2) to form low molecular weight polymers [185]. Rates are first order in monomer concentration and from the initial values the apparent propagation rate coefficient is ca. 6 x 10^{-2} l mole^{-1} sec^{-1} at 50°C, the activation energy being 9.5 kcal mole^{-1}. This is very similar to the rates observed with propene and butene-1, and suggests that k_p has a comparable magnitude.

8.5.3 4-Methylpentene-1

The polymerization of 4-methylpentene-1 by VCl$_3$/AlR$_3$ in benzene and heptane solution has been studied by McKenzie et al. [236]. With Al(i-Bu)$_3$ the rate was found to be proportional to VCl$_3$ concentration and to reach a maximum at Al/V = 2.0. Polymerization was first order in monomer above [M] = 1.0 mole l^{-1}; below this concentration at 30°C the rate was second order. As all the data followed the relationship [M]/R$_p$ =

$A + B/[M]$ the polymerization would appear to be proceeding from two active sites. Overall activation energies were 13.7 and 16.6 kcal mole^{-1}, respectively, with benzene and heptane solvents, the former giving the faster rates.

For a concentration of VCl_3 = 18.9 x 10^{-3} mole l^{-1}, Al/V = 2 and [M] = 2.0 mole l^{-1} the rate in benzene at 30°C was 5.56 x 10^{-3} mole l^{-1} min^{-1}, corresponding to an apparent rate coefficient of 2.45 x 10^{-3} l mole^{-1} sec^{-1}, substantially slower than the reaction of propene. From active centre concentrations and rate data the propagation rate coefficient was calculated [238]. This was reported as

$$R_p = 3.0 \pm 0.5 \times 10^3 \ (\text{min}^{-1}) \ \theta_m \ [C*] \ \text{at } 30°C$$

and independent of the structure of the metal alkyl. Expressed in the form employed in Table 6 this is equivalent to 5.4 l mole^{-1} sec^{-1}, which is of the same order as that observed for propene with the same catalyst.

Average lifetimes were calculated from molecular weights and polymerization rates (Table 9, p. 188) [244]; lifetimes were independent of VCl_3 concentration but fell with increase in Al/V ratio above 2/1, because of transfer with metal alkyl. At a fixed ratio of Al/V = 2/1 the molecular weight was dependent on the nature of the aluminium alkyl employed, being much lower with $AlMe_3$ than with the higher alkyls.

Molecular weights fell at temperatures above 40°C, and from their dependence on monomer concentration the rate of the transfer reaction with monomer was determined (Table 7, p. 182).

The polymerization of 4-methyl pentene-1 has also been studied using γ-TiCl$_3$/AlEt$_2$Cl as catalyst [242]. At 54°C there was a steady decline in polymerization rate with time (a tenfold decrease in 3 h) attributed to first order decay of the catalyst, and no sign of the establishment of a steady rate was observed, even after prolonged reaction times. This behaviour is in contrast to propene polymerization where TiCl$_3$/AlR$_2$Cl catalysts give steady rates over long periods. Estimates of rate coefficients are thus hardly meaningful but an indication of the order of activity can be obtained from the yield of 28.8 g polymer after 1 h at 54°C from 550 ml monomer and 0.5 x 10^{-3} mole TiCl$_3$ (Al/Ti = 3/1). This corresponds to an average rate coefficient of 3—4 x 10^{-2} l mole^{-1} sec^{-1}, which is comparable to the values observed for propene with TiCl$_3$ based catalysts (Table 11, p. 204) and faster than the rates obtained with 4MP1 using VCl$_3$ [236].

8.5.4 Other olefins

There are no detailed kinetic measurements on olefins other than those dealt with above, but there have been several investigations on the effects of olefin structure on polymerizability. For the most part these have consisted of single determinations of polymer yield under standard conditions, from which it is not possible to calculate rate coefficients, nor is it possible to compare one group of measurements with another as the experimental conditions have not been the same. Boor [186] has collated much of the published work and with some additional experimental data has drawn the following conclusions:

(a) There is little change in polymerization rate with linear α-olefins as the chain length is increased from pentene-1 to decene-1.

(b) With branched olefins the rate falls as the substituent is nearer to the double bond but even when the branch point is not in close proximity the rate is lower than for linear olefins.

(c) Disubstitution at the carbon atom adjacent to the double bond prevents polymerization, and rates are very low when the substituents are on the next carbon atom.

(d) Two methyl groups on one carbon atom have a greater effect on rate than a closed ring substituent.

The magnitude of the effects occasioned by changes in olefin structure are shown by the data in Table 18.

TABLE 18
Effect on monomer structure on polymerization rate [186]

Monomer	Arbitrary rate coefficient	
	(a)[a]	(b)[a]
Heptene-1	69	617
4-Methylhexene-1	1.38	109
3-Methylpentene-1	0.84	43
3-Ethylpentene-1	0.052	1.3
3,3-Dimethylbutene-1	0	0

[a] Catalysts; (a) $VCl_3/AlEt_3$, (b) γ-$TiCl_3/Al(i\text{-}Bu)_3$.
3-Methylbutene has been observed to polymerize at ca. 25 g polymer g $TiCl_3^{-1}$ h^{-1} at 60°C in bulk with $TiCl_3/Al(i\text{-}Bu)_3$ [316]. This corresponds to $k_p f = 1.7 \times 10^{-3}$ l mole^{-1} sec^{-1}, much lower than values reported for linear α-olefins.

8.5.5 Isoprene

This monomer has been studied using the $VCl_3/AlEt_3$ catalyst, which gives high *trans* 1,4 polymer [107]; the reaction is accelerated by the addition of di-isopropylether. There is an acceleration of rate in the early stages to a steady value (R_m) which fits the relationship

$$P_t = \log \left\{ \frac{R_m - R_t}{R_m - R_0} \right\}$$

R_0 is the initial rate, R_t the rate and P_t the yield of polymer after time t.

This equation is consistent with the formation of new active centres ($N*$) according to

$$dN*/dP = k(N_m^* - N*)$$

where N_m^* is the maximum concentration of active centres. The overall rate coefficient is 3.7×10^{-3} l mole^{-1} s^{-1} at 50°C and $k_p = 1.4$ l mole^{-1} sec^{-1}. Although the rate is increased by the addition of di-isopropylether to the catalyst the propagation rate coefficient is not increased, so the effect is that of increasing the number of active centres. There is a transfer reaction between growing chains and aluminium alkyl and a termination reaction which reduces catalytic activity at temperatures above 60°C [108].

8.5.6 1,3-Pentadiene

This monomer is usually obtained as a mixture of the *cis* and *trans* isomers both of which have been polymerized with coordination type catalysts. Polymerization of the *cis* form is considered to be preceded by isomerization, since those catalysts which do not isomerize the *cis* monomer (e.g. cobalt salt—organo aluminium halide) selectively polymerize the *trans* isomer. A kinetic study of the polymerization of *cis* 1,3-pentadiene using Ti(OBu-n)$_4$/AlEt$_3$ (Al/Ti = 1.3—6) as catalyst has been published [267]. This gives a polymer containing ca. 73% *cis* 1,4; 15—16% *trans* 1,4 and 11—12% 3,4 microstructure.

The kinetic relationship was expressed in the form

$$M_c \xrightarrow{k_I} M_T \xrightarrow{k_p} P$$

where M_c, M_T and k_I refer to the *cis* and *trans* forms of the monomer and the isomerization rate coefficient respectively, and k_p the propagation coefficient for conversion to polymer.

If initiation and termination processes are ignored

$$\frac{d[M_c]}{dt} = -k_I[M_c]; \quad \frac{d[P]}{dt} = k_p[M_T]$$

$$[M_T] = \frac{k_I}{k_p - k_I}[M_c]_0 \{\exp(-k_It) - \exp(k_pt)\}$$

whence, since $[M_c]_0 = [M_c] + [M_T] + [P]$

$$\frac{[P]}{[M_c]_0} = 1 - \frac{1}{k_p - k_I}\{k_p \exp(-k_It) - k_I \exp(-k_pt)\}$$

$[M_c]_0$, $[M_c]$, $[M_T]$ and $[P]$ refer respectively to the concentration of *cis* monomer initially and at any time, the concentration of *trans* monomer and the moles of monomer polymerized.

For Al/Ti = 4, good agreement between the experimental data and theoretical curves for depletion of *cis* and formation of *trans* monomer was found with equal values of k_I and k_p of 2×10^{-5} sec^{-1} at 20°C. For Al/Ti > 6, less *trans* isomer was formed than would be anticipated and it was suggested that polymerization without prior isomerization may have occurred. The expression of the polymerization rate in terms of a first order rate coefficient, necessary for the derivation of a simple kinetic scheme, makes it difficult to compare with other systems. However, if the coefficient k_p is identified with k_p' [Ti], where [Ti] is the concentration of transition metal compound (23×10^{-3} mole l^{-1}), a value for k_p' of 8.7×10^{-4} l mole^{-1} sec^{-1} is obtained. This is not greatly different from the value estimated for butadiene polymerization using the same catalyst (Table 15). An interesting feature of the reaction is that chloroform inhibits the polymerization with relatively little effect on isomerization.

The intrinsic viscosity of the polymer fell with increase in Al/Ti ratio — more so with the *cis* monomer at Al/Ti ratios above 4.

In a study on the *trans* monomer with the same catalyst the activity of the catalyst was found to increase to a maximum and then steadily decline on ageing prior to the addition of monomer, producing high (76—80%) *cis* 1,4 content polymer under these conditions [311]. In the early stages of reaction or at low monomer concentrations the polymers contained much more *trans* 1,2 structure.

Monomer was polymerized by a first order reaction but it interferes with the formation of the active catalyst, and the active centre concentration falls inversely with increase in initial monomer concentration. Polymer molecular weights increase with initial monomer concentration and the monomer transfer coefficient ($k_{tr,M}/k_p$) is small (8×10^{-5}). Catalyst efficiency is low (ca. 0.05), in agreement with the observations of Dawes and Winkler [146] for the polymerization of butadiene using the same catalyst.

8.5.7 Propadiene

This monomer will polymerize to crystalline or amorphous polymers by various coordination catalysts. Kinetic studies have been reported for polymerization initiated by rhodium complexes of the compositions.

$[Rh(CO)_2 Cl]_2$, $[Rh(CO)_2 P(Ph)_3 Cl]$ and $Rh(CO)_3 Cl$ [290]

The experimental findings were as follows for polymerization conducted in ethanol solution [290].

Initiator	Reaction order	
	Monomer	Initiator
$[Rh(CO)_2Cl]_2$	2.0	0.5
$Rh(CO)_2P(Ph)_3Cl$	2.0	1.0
$Rh(CO)_3Cl$	1.0	1.0

Results were not presented in a form such that reaction rate coefficients could be calculated.

Polymerization was dependent on the ability of the solvent to produce a monomolecular catalyst species, but not to form too strong a complex with the solvent. Thus polymerization occurred in ethanol but not in strong donors such as dimethyl sulphoxide or hexamethyl phosphoramide. The results were interpreted by the following mechanism:

1. Displacement of carbon monoxide from the complexes by other ligands.

2. Initiation by hydrogen transfer from monomer to form a rhodium hydride intermediate.

3. Propagation by insertion of coordinated monomer at the $LS_2RhH(C_3H_3)$ complex (L and S are ligands, e.g. $P(Ph)_3$, solvent, etc.)

4. Termination accompanied by hydrogen abstraction from the complex.

8.5.8 Methyl methacrylate

This monomer does not polymerize with all types of Ziegler catalyst, but does so with soluble vanadium catalysts $VCl_4/AlEt_3$ [218, 219] and $VOCl_3/AlEt_3$ [220]. The former catalyst has been used in *n*-hexane and acetonitrile (Al/V = 2—3) and gives slow rates and low molecular weights. The overall rate coefficient is greater in acetonitrile than in *n*-hexane (ca. 4×10^{-4} l mole^{-1} sec^{-1} compared with ca. 5×10^{-5} l mole^{-1} sec^{-1} at $40°C$; activation energies 11 and 7.4 kcal mole^{-1}, respectively) and rates are first order in monomer and catalyst.

The mechanisms of these reactions are not completely established, but it has been suggested from the kinetic relationships, the effect of zinc and aluminium alkyls on molecular weight and polymer structure, that a coordinated anionic mechanism is probable in hydrocarbon solvents [215]. Similar conclusions have been drawn for the $Cr(acac)_3/AlEt_3$ system. For acetonitrile solution, however, the evidence is less certain and simultaneous coordinate anionic and free radical polymerization may occur.

The system with $Cu_2Cl_2/AlEt_3$ as catalyst has also been suggested to be free radical in character [221], and to explain the kinetic expression for the polymerization rate

$$R_p = K[A][T]^{1/2}[M]^2 \quad (E = 16.5 \text{ kcal mole}^{-1})$$

a complex between catalyst components and monomer was considered to be formed which on reaction with monomer decomposes to give free radicals.

The polymerization mechanisms for vinyl chloride and acrylonitrile (and also styrene) with coordination catalysts are also uncertain [222] and the copolymerization of butadiene/acrylonitrile (q.v.) also shows some features suggesting the formation of free radicals (or possibly radical-ions from charge transfer complexes). As these polar monomers can react, or form strong complexes, with the organo-metal compound it is likely that the kinetic schemes will be complex. As with styrene there is a good deal of scatter in the experimental kinetic data with these monomers which detracts from the certainty of the deductions, and much work will be required to put their polymerization by coordination catalysts on a sound mechanistic and kinetic basis.

More satisfactory data have been obtained for methyl methacrylate polymerization using soluble allylic compounds such as $Cr(\pi\text{-}C_3H_5)_3$ or $Cr(\pi\text{-}2Me\text{-}C_3H_4)_3$ [229]. With these catalysts the rate relationship in the early stages of reaction is

$$R_p = A[C]_0[M]^2/(1 + B[M])$$

or in its integrated form

$$\{1/[M] - 1/[M]_0\} + B \ln([M]_0/[M]) = A[C]_0 t$$

The constants A and B have been identified with composite terms $k_i k_p/k_t$ and k_i/k_t in a scheme in which k_i is the rate coefficient of formation of an active centre from monomer and catalyst and k_t the rate coefficient of spontaneous termination.

Values of k_p and E_p have been determined (Table 6, p. 180) and k_i/k_t was in the range 0.2—0.5 at temperatures from 0 to 5°C. The concentration of growing centres approached the initial catalyst concentration slowly. Molecular weights increased to a minor extent with monomer concentration and conversion, but were independent of catalyst concentration. Owing to the relatively fast termination reactions average DPs were low, 1—2 x 10^4 at 0°C. From an analysis of the molecular weight distribution [230] it has been proposed that, in addition to termination of growing chains, there is a spontaneous transfer reaction (k_{tr}), as well as transfer with monomer ($k_{tr,M}$). From plots of $1/\bar{P}_n$ vs. $1/[M]_0$, $k_{tr,M}/k_p$ and (with rate data) $k_{tr,M}$ have been evaluated (Table 8, p. 184).

The polymerization of methyl methacrylate by $Cr(acac)_3/AlEt_3$ is first order in chromium and monomer concentrations and was found to have an over-all activation energy of 15.9 kcal mole^{-1}. The preferred ratio of catalyst components was Al/Cr = 12/1 [256].

$VOCl_3$ with $AlEt_3$ or $AlEt_2Cl$ as co-catalyst polymerizes methyl methacrylate by reactions which are first order in monomer and catalyst.

The rate coefficients reported were 4.6 to 5.6 x 10^{-5} for $AlEt_3$ and 31.9 x 10^{-5} for $AlEt_3$, at 40°C for Al/V = 2, with $E_{over-all}$ 9.3 kcal mole^{-1} [308]. The polymers were of mixed microstructure with a preponderance of syndiotactic structure and molecular weights ranged from 5 x 10^4 to 10^5.

8.5.9 Vinyl chloride

The earliest investigations on vinyl chloride using Ziegler—Natta catalysts suggested that they were rather unsatisfactory initiators of polymerization. Yields and molecular weights were low, possibly due to participation of monomer in side reactions with catalyst components, and it was considered that polymerization resulted from free radicals produced by decomposition of unstable organo-metal compounds.

Recent investigations [259] have indicated that the polymerization is not conventional free radical in character but is likely to be coordinated anionic. In support of this view are the reactivity ratio coefficients in copolymerization of vinyl chloride with vinyl acetate and methyl methacrylate, which are different from those found with free radical initiators.

VC/MMA	(M_1 = VC)	
r_1	0.008 ± 0.004	(0.02)
r_2	7.3 ± 1.2	(15)
VC/VA	(M_1 = VC)	
r_1	0.58 ± 0.10	(1.35)
r_2	0.020 ± 0.007	(0.65)

The catalyst was $VOCl_3$/Al(i-Bu)$_3$ in THF/benzene at 30°C and the free radical coefficients given in brackets.

The reactivity ratios, although numerically different for the Ziegler and free radical catalysts, do show the same orders of monomer reactivity, vinyl acetate being less reactive and methyl methacrylate much more reactive than vinyl chloride. With the rather wide scatter in literature values for reactivity ratios — particularly with coordination systems — the data on VC/MMA cannot be regarded as completely indicative of a non-radical reaction; however, the very low value of r_2 in the VA/VC system is more definitive.

The general kinetic features of the polymerization show the rate to be first order in monomer and vanadium oxychloride concentrations, and independent of Al/V ratio above 5/1. The rate was increased in the presence of the polar solvent up to THF/V = 6 but then became independent of THF concentration. The catalyst underwent a rapid decomposition, accounting for declining polymerization rates with time, which was shown to be second order. The rate coefficient for decomposition of the catalyst, $k_d[C]_0$, was estimated to be 2.9×10^{-5} sec^{-1}, and from a plot of $\ln([M]_0/[M])$ vs. $\ln(k_d[C]_0 \, t + 1)$, k_p/k_d was estimated to be 0.28 and $k_p[C]_0 = 0.82 \times 10^{-5}$ sec^{-1}, where $[C]_0$ is the concentration of active species for the system: $VOCl_3/Al(i\text{-}Bu)_3/THF$ in benzene at $30°C$; $[M]_0$ 1.96 mole l^{-1}; $[VOCl_3] = 1.04 \times 10^{-2}$ mole l^{-1}; $Al/V = 2.5/1$; $THF/V = 6—480$. Expressed as an over-all rate coefficient for this system $k_p f = 7.2 \times 10^{-4}$ $l \, \text{mole}^{-1} \, \text{sec}^{-1}$, so that if the catalyst had an efficiency in the region of ca. 0.01 the propagation coefficient would be of the order of 10^{-1} $l \, \text{mole}^{-1} \, \text{sec}^{-1}$ at $30°C$ — low compared with most of the values given in Table 6.

The polymerization of vinyl chloride using the hydrocarbon soluble catalyst $VOCl_2,2THF/Al(i\text{-}Bu)_3/THF$ (range of composition $Al/V = 1.4—12/1$, $THF/Al = 1.4—460/1$) is first order in monomer and catalyst concentration [314]. Without THF the polymerization rate is very slow but polymer yield became independent of THF concentration at $THF/Al > 1$. A maximum in rate was observed at $Al/V = 3.8/1$.

Polymerization rates were relatively slow; typically for $Al/V = 2$, $THF/Al = 1.4$ and $[V] = 1.1 \times 10^{-2}$ mole l^{-1}, ca. 0.1% min^{-1} at $30°C$. Catalyst activity declined on ageing by a second order reaction. The active species was considered to contain V^{IV} which was transformed to V^{III} over a period of hours (ca. 20% in 4 h at $20°C$), parallel with decreasing catalyst activity.

8.5.10 Vinyl fluoride

There has been one kinetic investigation with this monomer using $VOCl_3/Al(i\text{-}Bu)_3/THF$ [260]. The polymerization was first order in $VOCl_3$, and in monomer over the concentration range 1.95—3.41 mole l^{-1}. The highest yield was obtained with an Al/V ratio of 1.6, and was independent of THF concentration for THF/V ratios above 2.3. Polymerization rates were not measured, but a conversion of ca. 10% after 48 h at $30°C$, with $[VOCl_3] = 0.175$ mole l^{-1}, $[M]_0 = 2.92$ mole l^{-1}, $Al/V = 2.27/1$ and $THF/V = 17.5$, shows them to be very low.

9. Copolymerization

Most kinetic studies on copolymerizations using coordination catalysts have been restricted to the determination of monomer reactivity ratios. There are problems both experimentally and in interpretation since the major simplification assumed to hold for most free radical initiated systems, namely that monomer incorporation is determined only by the monomer concentrations and the four rate coefficients, cannot be taken for granted. Further, catalyst activity and selectivity are influenced by the conditions of catalyst preparation including the manner and order of introduction of reagents and monomers, and there is the possibility of polymerization proceeding independently from different catalyst sites. Where this occurs it has been shown that the average reactivity ratios obtained are given by

$$\bar{r}_1 = \Sigma r_{1j} k_{12j} K_j / \Sigma k_{12j} K_j$$

$$\bar{r}_2 = \Sigma r_{2j} K_{12j} K_j / \Sigma k_{12j} K_j$$

K_j, r_{1j} and r_{2j}, and k_{12j} are the fraction, the reactivity ratios and the propagation rate coefficient corresponding to the jth type catalyst species [187]. Monomers of similar structure, such as the substituted styrenes, copolymerize normally but in most instances there are substantial differences in monomer reactivity. Some monomers, (e.g. butene-2) do not homopolymerize but form copolymers with reactive monomers such as ethylene.

9.1 ETHYLENE/PROPENE

Reactivity ratios have been obtained for these monomers with many catalysts (Table 19). The most satisfactory in giving random incorporation of monomer units are those from vanadium salts, since titanium based catalysts give mixtures of polymers with markedly variable sequence length and tacticity. A recent investigation using a variety of catalysts confirms that reactivity ratios, the published values of which are in poor agreement, do not give a true indication of polymer structure, even with the more satisfactory vanadium catalysts [187]. Those from preformed VCl_3, although they have good reactivity ratios, are not suited for the synthesis of amorphous, random copolymers because of their tendency to form isotactic propene segments. The most satisfactory are those which homopolymerize propene poorly and give atactic polymers, such as $VOCl_3/AlEt_2Cl$, $VOCl_3/Al_2Et_3Cl_3$ or $VO(OR)_3/AlEt_2Cl$. Catalysts must contain halogen groups attached either to the aluminium or the vanadium compound, and combinations of vanadium acetylacetonates or vanadyl esters with aluminium trialkyls are ineffective.

Catalysts from liquid or soluble vanadium compounds are finely divided and appear to be soluble. However, the solutions scatter light and, on standing or centrifugation, form precipitates; this indicates that the catalyst entities are aggregates, and like most heterogeneous catalysts the fraction of vanadium which is active is low (0.8—2%) [188]. These catalysts contain a proportion of unstable catalytic species. Thus, with $V(acac)_3/AlEt_2Cl$, 80—90% of the activity is lost in a few minutes at 25°C and polymer yields are about ten times higher when polymerizations are conducted at −20°C. The more stable catalyst from VCl_4/AlR_3 after ageing at 60°C will produce random polymers at steady rates for long

TABLE 19
Monomer reactivities for ethylene/propene copolymerization

Catalyst	r_1 (ethylene) M_1	r_2 (propene) M_2	$r_1 r_2$	No. of active species	Ref.
Hydrocarbon solvent					
VO(i-Bu)$_3$/AlEt$_2$Cl	16.8	0.019	0.32	2	187
VO(acac)$_3$/AlEt$_2$Cl	11.7	0.011	0.13	2	187
VO(acac)$_2$Cl/AlEt$_2$Cl	16.4	0.018	0.30	2	187
VO(acac)Cl$_2$/AlEt$_2$Cl	16.5	0.012	0.20	2	187
VO(OEt)$_3$/Al(i-Bu)$_2$Cl	15.0	0.070	1.04		189
VO(O-nBu)$_3$/Al(i-Bu)$_2$Cl	22.0	0.046	1.01		189
VOCl$_2$(OEt)/Al(i-Bu)$_2$Cl	16.8	0.055	0.93		189
VOCl(OEt)$_2$/Al(i-Bu)$_2$Cl	18.9	0.056	1.06		189
VOCl$_3$/Al(i-Bu)$_2$Cl	16.8	0.052	0.87		190
	14.8	0.037	0.55		189
VOCl$_3$/Al(i-Bu)$_2$Cl	35.3	0.027	0.95		191
VOCl$_3$/Al(i-Bu)$_2$Cl	20.3	0.022	0.45		187
VOCl$_3$/AlEt$_2$Cl	12.1	0.018	0.22	3	187
VOCl$_3$/Al$_2$Et$_3$Cl$_3$[a]	10.1	0.025	0.25	2	187
VOCl$_3$/Al(i-Bu)$_3$	28	—	—		75
VOCl$_3$/Al(C$_6$H$_{13}$)$_3$	18	0.065	1.15		193
VCl$_4$/AlEt$_3$	10.3	0.025	0.25		187
	—	—	1.00		183
VCl$_4$/Al(i-Bu)$_3$	16	—	—		75
	11.0	0.028	0.30		187
VCl$_4$/Al(C$_6$H$_{13}$)$_3$	7.1	0.088	0.68		223, 193
VCl$_4$/AlEt$_2$Cl	13.7	0.021	0.29		226
	5.9	0.029	0.14	2	187
	—	—	0.26		224, 183
VCl$_4$/Al$_2$Et$_3$Cl$_3$	9.1	0.031	0.28	1	187
VCl$_4$/AlEt$_2$Cl . anisole	—	—	0.28		183
VCl$_4$/GaEt$_3$	—	—	1.14		183
VCl$_3$/Al(C$_6$H$_{13}$)$_3$	5.6	0.145	0.81		224, 193
V(acac)$_3$/AlEt$_2$Cl	15	0.04	0.6		194
V(acac)$_3$/Al(i-Bu)$_2$Cl	16	0.04	0.64		190
AlBr$_3$/VCl$_4$/SnPh$_4$	16	~0.1	—		75
TiCl$_4$/Al(C$_6$H$_{13}$)$_3$	33.40	0.032	1.05		193
TiCl$_4$/Al(i-Bu)$_3$	37	—	—		75
TiCl$_3$/Al(C$_6$H$_{13}$)$_3$	15.7	0.110	1.7		193
TiCl$_2$/Al(C$_6$H$_{13}$)$_3$	15.7	0.110	1.7		193
ZrCl$_4$/Al(i-Bu)$_3$	61	—	—		75
HfCl$_4$/Al(i-Bu)$_3$	76	—	—		75
Chlorobenzene solvent					
VO(OEt$_3$)/AlEt$_2$Cl	26	0.039	1.02		192
VO(OR)$_3$/AlR$_2$Cl	24.4	0.041	1.0		187
VO(OBu)$_3$/AlEt$_2$Cl	19.8	0.012	0.24		187

[a] Reactivity ratios of r_1 = 2.5 and r_2 = 0.35 ($r_1 r_2$ = 0.87) have recently been reported for this catalyst [212]. These refer to monomer concentrations in the gas phase and taking a factor of 4 for the relative solubilities of ethylene and propene at 40°C, r_1 and r_2 become 10.0 and 0.087.

periods at 25°C. The catalysts produced from $VOCl_3$ or VCl_4 contain two types of active site [44, 188], both of which contain trivalent vanadium, and the presence of more than one catalytic entity is probable in many of the systems used [188] (Table 19). The differences in catalytic activity between the various vanadium systems are illustrated in Table 20. Polymerization rates are first order in total monomer and vanadium concentrations, and are linearly related to the molar composition of the monomer at constant total monomer concentration [194]★.

TABLE 20

Effect of catalyst structure on polymerization of ethylene/propene [194] n-Heptane; $(C_2H_4/C_3H_6)_{gas} = 1/4$

Catalyst	Al/V	Polymerization rate at 25°C (g polymer per mmole V per h)
$V(acac)_3/AlEt_2Cl$	3.5	8.3
$VO(acac)_3/AlEt_2Cl$	3.5	8.1
$VCl_4/Al(C_6H_{13})_3$	2.5	~200

With the $VCl_4/Al(Hex)_3$ catalyst the ratio of the rates of polymerization of ethylene and propene is ca. 1800, considerably larger than that found for more active catalysts for propene polymerization, such as $TiCl_4/AlR_3$. The large ratio for the vanadium catalysts is because most of the catalyst sites cannot initiate the polymerization of propene, although once they have added a molecule of ethylene they can subsequently add either ethylene or propene. In conformity with this view it is found that the soluble portion of the catalyst will polymerize ethylene but not propene. The overall activation energy for copolymerization with $VCl_4/Al(Hex)_3$ was found to be 6.5 kcal $mole^{-1}$ and to be the same as for the two individual monomers [194].

No details of individual rate coefficients in ethylene/propene polymerization are available, other than some approximate values for "soluble" vanadium catalysts reported by Baldwin and Ver Strate [244]. The values quoted for $k_{2,1}$ and $k_{1,2}$ are ca. 10^6 and 10^4 l $mole^{-1}$ h^{-1} compared with 10^6 and 10^3 for the homo addition reactions, $k_{1,1}$ and $k_{2,2}$. The greater ease of addition of propene to an ethylene end-group compared with a propene end-group is in agreement with the observations of Natta et al. [194] on the $VCl_4/Al(Hex)_3$ catalyst and confirms that, at least with vanadium catalysts, there is a terminal group effect in the copolymerization of ethylene and propene. The values of $k_{1,1}$ and $k_{2,1}$ for ethylene addition are substantially greater than k_p values obtained using titanium catalysts (Table 6, p. 180).

★ Georgiadis and St John Manley also report that the $VOCl_3/AlEt_2Cl$ catalyst consists of a mixture of a soluble portion and aggregates of colloidal dimensions [249].

The uncertainties and poor agreement in determinations of ethylene/ propene reactivity ratios from monomer and polymer composition [m = (M_1/M_2) monomer; $p = (M_1/M_2)$ polymer] and the copolymerization equation $p = (1 + r_1 m)/(1 + r_2/m)$ give particular interest to approaches based on the analysis of monomer unit distributions in the copolymer.

The first to attempt this were Tosi, Valvassori and Ciampelli [284] who observed a relationship between infrared methyl group absorptions in the region 900—1000 cm^{-1} and the reactivity ratio product $r_1 r_2$. These infrared bands were shown to contain characteristic propene absorptions for isolated units at 935 cm^{-1} and for sequences at 973 cm^{-1}, and a distribution index containing the rates of these absorptions was found to correlate well with the fractions of E—P(f_{12}) and E—E(f_{11}) bonds given by [285]

$$f_{12} = 1/\{r_1 m + 2 + r_2/m\}; \quad f_{11} = r_1 m/\{r_1 m + 2 + r_2/m\}$$

i.e.,

$$f_{12}/f_{11} = 1/r_1 m$$

From these and the copolymerization equation

$$r_1 r_2 = 1/p[(f_{11}/f_{12})^2 - (p - 1)(f_{11}/f_{12})]$$

For copolymers prepared with $VCl_4/AlEt_2Cl$ a value of $r_1 r_2 = 0.26 \pm 0.1$ was obtained, in excellent agreement with the direct determination of 0.27 ($r_1 = 13.6$, $r_2 = 0.02$) [226].

In this method there is some uncertainty in the origin of the infrared bands and the employment by Carman and Wilkes [286] of C_{13} NMR analysis of triad sequences of ethylene and propene represents a further step forward. The latter derived the equation

$$r_1 r_2 = 1 + p(X + 1) - (p + 1)(X + 1)^{1/2}$$

where

$$X = \frac{\% \text{ propene in sequences of two or more}}{\% \text{ isolated propene units}}$$

Data on a number of copolymers prepared with various catalysts are given in Table 19A [287]. An assumption in this work was that the small proportion of termonomer (ethylidene norbornene) did not interfere with E/P copolymerization.

From a consideration of the effects of errors the authors conclude that at 30% propene content the accuracy of propene determination must be within 2% and 3% for EPE sequences; at 60% propene content the corresponding figures are both ±1%. The $r_1 r_2$ values for samples 1, 2 and 3 are recalculations from earlier data, where estimates of 0.45, 0.40 and 0.42, respectively, are given [286].

TABLE 20A

Reactivity ratios of ethylene/propene/ethylidene norbornene terpolymers from catalysts based on V(acac)$_3$

Polymer	Mole % propene	Co-catalyst	Al/V	Polym'n temp. °C	Modifier	$r_1 r_2$
A	27.6	Ali-Bu$_2$Cl	10	—10	ZnEt$_2$	1.50
B	30.8	Al$_2$Cl$_3$Et$_3$	8	25	H$_2$	1.10
C	27.8	AlEtCl$_2$	32	0	ZnEt$_2$	1.03
D	32.2	Ali-Bu$_2$Cl	10	—10	ZnEt$_2$	0.91
1	26	AlEt$_2$Cl	—	—	—	0.92
2	34	AlEt$_2$Cl	—	—	—	0.85
3	62	AlEt$_2$Cl	—	—	—	1.16

Both the early data and the more precise values given in Table 20A differ significantly from published estimates based on monomer and polymer composition (e.g. $r_1 r_2$ = 0.60 in Table 19). As all the data relate to the, in general, more consistent "soluble" vanadium systems, this work reinforces doubts concerning the accuracy of much of the published information. A complication is that since C_2 and C_4 sequences are observed in ethylene/propene copolymers inverted head to head propene units must be present and this will reduce the accuracy of analyses of EP sequences. In copolymers prepared by VCl$_4$/AlEt$_3$ 4% of head to head propene sequences have been reported with the catalyst VCl$_4$/AlEt$_2$Cl which is syndiospecific for polypropene 8% of head to head sequences was found [295]★.

The assumption that the termonomer does not interfere in the ethylene/propene copolymerization is probably an over-simplification since it is likely to be co-absorbed on the catalyst surface to some extent with ethylene, propene, solvent and organo-metal compounds. There is, moreover, evidence to suggest that the participation of the termonomer in the polymerization is determined by the nature of the catalyst. Thus norbornene and dicyclopentadiene are polymerized by catalysts prepared from titanium tetrachloride, the former monomer giving a polymer where the bridged ring has opened (to form a *trans* double bond in the polymer chain) [312]. On the other hand vanadium compounds which dissolve in hydrocarbons (e.g. VCl$_4$ or salts of vanadium from higher fatty acids) give little or no polymer with endomethylenic (bridged ring) hydrocarbons, and it has been claimed that terpolymers with ethylene and propene do not contain sequences of the termonomer [313]. It is also worthy of note that additives to the "soluble" vanadium catalysts can markedly influence the extent of incorporation of 1,4-hexadiene into ethylene/propene copolymers [103]. Whether this results from a change in effective concentration of termonomer at the catalytic site, or more likely from an

★ For further discussion see note added in proof (A), page 247.

References pp. 249—257

increased propagation constant for its addition to an ethylene or propene terminal unit or both, has not been established. It seems reasonable to conclude that at high termonomer concentrations major effects on copolymerization are likely but at low levels there should be little or no effect.

The effect of the bridged-ring diene termonomers on polymer yield is relatively small owing to the absence of interaction of the non-polymerizing double bond with the catalyst. The monomer

contains a pair of conjugated double bonds in addition to the strained bridged-ring double bond.

Cesca et al. [317] reported a fall in ethylene/propene polymer yields with increasing concentration of this termonomer using $V(acac)_3$/$AlEt_2Cl$, $V(OBu)_3$/Et_2AlCl, and VCl_4/Et_2AlCl/anisole as catalysts. This could result from selective coordination of the conjugated diene structure at the catalyst site. Another explanation is that a small proportion of the catalyst centres react with the diene moiety forming π-allylic end-groups which are inefficient in further propagation. Incorporation levels of termonomer were not high (20—50%) and the conjugated double bonds were not significantly reactive in the polymerization, provided cationic species were not present in the catalyst (e.g. with Al/V 6—10).

Terpolymerization of dienes with ethylene and propene is of considerable complexity and few kinetic data have been reported. The topic is, however, worthy of discussion, partly in drawing attention to worthwhile areas for kinetic studies and partly because of the economic importance of the polymers.

Attempts have been made to copolymerize conjugated dienes with olefins but there are no data on polymerization rates or reactivity ratios. They are ill suited for copolymerization in that polymerization rates are markedly reduced by the presence of the conjugated diene and the copolymers are heterogeneous in composition and may be crosslinked. The reasons for this behaviour have not been established but a possible explanation is that the conjugated diene coordinates preferentially with the catalyst and so excludes the olefin, but has a slow insertion rate compared with the more reactive olefin.

Two classes of diene have been copolymerized successfully with ethylene/propene, in both of which one double bond is deactivated such that the monomer behaves as a mono-olefin. These are unconjugated diolefins, such as cis and trans 1,4-hexadiene or 3,7-dimethyl 1,6-octadiene, and compounds containing the 2-norbornene structure such as dicyclopentadiene and 5-ethylidene 2-norbornene. The most detailed study of these is by Christman and Keim [283] who expressed their reactivities (with simplifications of the instantaneous copolymerization equation) in terms of a reactivity constant r_c given by

$$([M_3]/[M_2])_{polymer} = r_c([M_3]/[M_2])_{monomer}$$

In this the rate of entry of the termonomer (3) is compared with one of the other monomers — in this instance propene.

For the vanadium catalysts $VO(Ot\text{-}Bu)_3/Al_2Et_3Cl_3$, $VOCl_3/Al_2Et_3Cl_3$, $VCl_4/Al(i\text{-}Bu)_3$ and $VCl_4/AlEt_2Cl$, r_c ranged from 5.1 to 16.0 for the norbornene derivatives, the results not being greatly affected by the type of vanadium catalyst. The most reactive monomers were, 5-ethylidene 2-norbornene, 5-isopropenyl-2-norbornene and exodicyclopentadiene. In contrast the aliphatic diolefins had values of r_c from 0.66 to 1.4 and comparable reactivities were found for tetrahydroindene and methyltetrahydroindene.

It is apparent that the bridged ring monomers in which the double bond is made more reactive by steric strain are comparable in reactivity with ethylene, whereas, as would be expected, the aliphatic dienes are more like α-olefins in reactivity. Baldwin and van Strate have critically discussed this work [244], drawing attention to the possibilities of E/P reactivity ratios being changed by the presence of the diene and of diene polymerization by cationic mechanisms. They identify r_c as

$$r_c = (k_{13}/k_{12})(1 + r_{21}/[M]_2)^{-1}$$

(the first term is the ratio of the reactivities of termonomer and propene with an ethylene terminated chain and the term in brackets becomes the mole fraction of ethylene in the terpolymer if $r_{12}r_{21} = 1$) and state that k_{13}/k_{12} is about 0.5 for the linear dienes and tetrahydroindene and ca. 20 for the bridged ring compounds — essentially the same conclusions as Christman and Keim [294].

9.2 ETHYLENE/BUTENE-1

Butene-1 is less reactive than propene but random copolymers can be prepared. Reactivity ratios for several vanadium catalysts are given in Table 21.

TABLE 21
Monomer reactivity ratios for ethylene/butene-1

Catalyst	r_1(ethylene) M_1	r_2(butene-1) M_2	Ref.
$VCl_4/Al(C_6H_{13})_3$	29.6	0.019	196
$VCl_3/Al(C_6H_{13})_3$	27.0	0.043	196
$VCl_4/AlEt_2Cl$	35.5	0.006	183
$VCl_4/Ali\text{-}Bu_2Cl$	32.5	0.019	197
$TiCl_3/MeTiCl_3$	3.6	0.160	315

9.3 PROPENE/BUTENE-1

The reactivity ratios for this monomer pair for three vanadium catalysts are given in Table 22 [184]. The rates of homopolymerization of propene and butene-1 are in the ratio of ca. 4 to 1, which is in agreement with the values of the reactivity ratios if the rates of monomer entry are determined only by the polymerizabilities of the two monomers, without any terminal unit effect, i.e., $k_{1,1} = k_{2,1}$ and $k_{2,2} = k_{1,2}$, whence $(R_p)_1/(R_p)_2 = 1/r_2 = r_1$.

TABLE 22
Monomer reactivity rates for propene/butene-1

Catalyst	r_1(propene) M_1	r_2(butene-1) M_2	Ref.
$VCl_4/Al(C_6H_{13})_3$	4.39	0.227	184
$VCl_3/Al(C_6H_{13})_3$	4.04	0.252	184
$VCl_4/AlEt_2Cl$	~0.7	~0.7	183
$TiCl_3/AlEtCl_2/HMP/H_2$	4.3	0.8	241
$TiCl_3/AlEtCl_2/HMP$	3.3	0.45	

Reactivity ratios for butene-1(M_1) and 3-methylbutene-1(M_2) have been determined for the $TiCl_3/(Ali\text{-}Bu)_3$ catalyst [316]; $r_1 = 8.5$, $r_2 = 0.013$.

9.4 STYRENE/α-d-STYRENE

Copolymers have been prepared using $(\pi\text{-}C_5H_5)_2TiCl_2/AlEt_3$. The rates of polymerization at constant concentration of catalyst and monomer increase with increase of deuterated styrene in the monomer mixture [162], but without differences in the molecular weights of the polymers. Similar findings were obtained with the $TiCl_4/AlEt_3$ (Al/Ti = 3) catalyst at 60°C [78a]. The explanation for the rate increase was considered to be an increase in the concentration of active species in the presence of the deuterated monomer.

9.5 STYRENE/β,β-d$_2$-STYRENE

These copolymerize ideally and changes in monomer composition are without effect either on rate of polymerization or polymer molecular weight [78b].

9.6 STYRENE/PROPENE

These monomers differ greatly in reactivity (Fig. 24), propene being the more reactive. Reactivity ratio values with three catalyst systems have been reported (Table 23) [199].

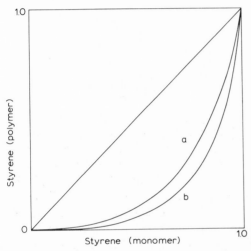

Fig. 24. Copolymerization of styrene and propene [199]. (a) VCl$_3$/AlEt$_3$; Al/V = 2 (b) TiCl$_3$/AlR$_3$; Al/Ti = 2. Temp., 40°C.

TABLE 23
Reactivity ratios for propene/styrene [199]

Catalyst	r_1(propene) M_1	r_2(styrene) M_2
TiCl$_3$/Al(i-Bu)$_3$	20.0 ± 3.5	0.2 ± 0.13
TiCl$_3$/AlEt$_3$	20.5 ± 3.0	0.3 ± 0.15
VCl$_4$/AlEt$_3$	7.2 ± 0.7	0.16 ± 0.08

9.7 STYRENE — NUCLEAR SUBSTITUTED STYRENES

These monomers copolymerize ideally. Halogen substituents reduce the rate of polymerization while methyl or ethyl groups in the para position have a slight activating effect. This is in conformity with the view that monomer entry into the polymer chain is determined only by individual reaction rates (Table 24) [200].

TABLE 24

Copolymerization of styrene (M_1) with substituted styrenes (M_2). Catalyst, Al(i-Bu)$_3$/TiCl$_4$; temp., 60°C [200].

Substituted styrene	r_2	$R_p(M_2)/R_p(M_1)$
p-Me	1.17	1.20
o-Me	0.13	0.10
m-Me	0.47	0.43
2,4-diMe	0.13	0.10
p-Et	1.05	1.10
o-F	0.18	0.20
p-F	0.72	0.74
m-F	0.47	0.50
p-Cl	0.50	0.47
m-Cl	0.42	0.40
p-Br	0.50	0.50

9.8 CONJUGATED DIENES

Most investigations on the copolymerization of conjugated diolefins have been with butadiene and isoprene, but results are mainly restricted to polymer microstructure and over-all reactivity ratios. There is little or no information on copolymerization rates in relation to those of the individual monomers, nor on molecular weight averages or molecular weight distributions. In some cases the microstructures of the copolymers are the same as in the homopolymers but in others there are significant changes in isomer ratio, e.g., a fall in the *cis* content of butadiene units in copolymers with isoprene when prepared with soluble cobalt catalysts [201]. Butadiene appears to form random copolymers with isoprene and 1,3-pentadiene, but with 2-phenyl butadiene it forms block polymers [207]. Cyclic dienes such as cyclopentadiene and 1,3 cyclohexadiene will copolymerize with open chain conjugated dienes and reactivity ratios have been reported with isoprene [203]. Reactivity ratio values for diene copolymerization are given in Tables 25 and 26.

Trans-1,3-pentadiene and isoprene copolymerize with Ti(On-Bu)$_4$/AlEt$_3$ [293]. Isoprene units in the copolymer chain are predominantly of 3,4 structure while the pentadiene units are mixed of *cis* and *trans* 1,4, 1,2 and 3,4 structure; composition was independent of Al/Ti ratio. Kinetic measurements were not reported but reactivity ratios (Table 26) show somewhat greater reactivity of the pentadiene and a tendency for alternation.

9.9 ALTERNATING COPOLYMERS

There are a number of catalysts which produce alternating copolymers, e.g., vanadium salts and aluminium alkyls for butadiene/propene and

TABLE 25

Reactivity ratios for butadiene (M_1)/isoprene (M_2) copolymerization

Catalyst	r_1	r_2	Ref.
$AlEt_3/TiCl_4$	1.6	1.1	204
$AlEt_3/TiCl_4$ (washed ppt)	1.49	1.03	204
$Al(i\text{-}Bu)_3/TiCl_4$ ($Al/Ti = 1.1/1$)	1.0 ± 0.05	1.0 ± 0.05	205
$Al(i\text{-}Bu)_2Cl/CoCl_2/EtOH$ ($Al/Co = 130/1$)	2.3	1.15	205
$AlEt_3/TiI_4$	1.88	0.55	204
$Al(i\text{-}Bu)_3/TiI_4$($Al/Ti = 8$)	2.8 ± 0.27	0.53 ± 0.05	206
$Al(i\text{-}Bu)_3/TiI_4/Bu_2S$ $Al/Ti/Bu_2S = 8/1/16$	$4.63 \pm 0.3 \times 7$	0.76 ± 0.08	206
$CoCl_2\text{—}py/AlEt_2Cl$	0.99	1.37	201
$CoCl_2/AlEt_2Cl/EtOH$	0.92	1.25	201
$CoCl_2/Al(i\text{-}Bu)_2Cl$	1.0 ± 0.1	0.9 ± 0.1	206
$CoCl_2/Al(i\text{-}Bu)_2Cl/Bu_2S$			
($Al/Bu_2S = 1/1$)	1.0 ± 0.2	1.0 ± 0.1	206
($Al/Bu_2S = 1/3$)	2.0	0.62	206
$NiCl_2$. pyridine/$AlEt_2Cl$	1.15	0.59	201
$\pi\text{-}C_4H_7NiCl/TiCl_4$ ($Ni/Ti = 1$)	6.15 ± 0.35	1.15 ± 0.08	206
$LiBu/TiCl_4$	1.6	0.8	204
$AliBu_2Cl/CoCl_2$	2.3	1.15	288
	0.92	1.25	
	0.99	1.37	
$AlEt_2Cl/Co(acac)_2$	1.0	1.0	288
$CrO_3/Al_2O_3\text{—}SiO_2$	1.0 (1.04)	0.6 (0.76)	288 (301)
$(\pi\text{-}C_3H_5)_3Cr/Al_2O_3\text{—}SiO_2$	1.06 (1.06)	0.8 (0.8)	288

EtAlCl$_2$ [209] or VOCl$_3$/ZnCl$_2$/benzoyl peroxide [210] are effective with polar monomers such as the acrylic esters. With the latter catalysts light as well as peroxide acts as an accelerator but the mechanism is not a simple free radical polymerization.

Alternating copolymerization results from coordination of one of the monomers with the metal to form a complex which interacts with the second monomer, possibly by electron transfer from the complexed monomer. The charge transfer complex when activated inserts the two monomer units into the chain. Although these systems differ from Ziegler catalysts, some formally similar combinations of organo-metal compounds and transition metal halides have high catalytic activity and have been used to prepare alternating copolymers from butadiene with acrylonitrile. Typical catalysts are VOCl$_3$/AlEtCl$_2$ and VO(OR)$_3$/AlEtCl$_2$, in which there is a high ratio of Al/V (ca. 20/1) [211], and rates varying from 60—100 g polymer per g catalyst per h have been obtained. The rates are proportional to $[Al]^{3/2}[V]^{1/2}$ with activation energies for propagation and chain transfer of 12 and 16 kcal mole^{-1}, respectively [211]. This relationship has been interpreted as resulting from interaction of the

TABLE 26
Monomer reactivity ratios for diene copolymerization

Catalyst	Monomer 1	Monomer 2	r_1	r_2	Ref.
$CoCl_2/Al(i-Bu)_2Cl$	Butadiene	2,3-dimethyl butadiene	$\begin{cases} 0.74 \pm 0.02 \\ 2.7 \end{cases}$	0.67 ± 0.05 / 1.13	206 / 288
$CoCl_2/Al(i-Bu)_2Cl/Bu_2S$ (Al/Co = 1.4, $Bu_2S/Al = 3$)	Butadiene	2,3-dimethyl butadiene	0.35 ± 0.02	0.63 ± 0.04	206
$CoCl_2/Al(i-Bu)_2Cl$	Butadiene	1,3-pentadiene	0.81 ± 0.14	0.12 ± 0.02	206
$Al(i-Bu)_3/TiI_4$ Al/Ti = 8	Butadiene	2,3-dimethyl butadiene	6.28 ± 0.21	0.15 ± 0.06	206
$Al(i-Bu)_3/TiI_4/Bu_2S$ (Al/Ti/Bu_2S = 8/1/16)	Butadiene	2,3-dimethyl butadiene	5.04 ± 0.31	0.28 ± 0.13	206
$AlHCl_2Et_2O/TiCl_4/AlI_3$ (5.5/1/1)	Butadiene	2-phenyl butadiene	1.90	2.6	207a
$TiCl_4/Al(i-Bu)_3$	Isoprene	Styrene	0.63 ± 0.39 / 11 ± 14	1.8 ± 0.9 / 0.08 ± 0.84	208 / 208

TiCl$_4$/n-BuLi (1/1)	Isoprene	1,3-cyclohexadiene	0.3 ± 0.05	0.8 ± 0.1	203
TiCl$_4$/LiAlH$_4$ (1/1)	Isoprene	1,3-cyclohexadiene	0.2 ± 0.05	1.4 ± 0.1	203
(πC$_3$H$_5$)$_4$Mo/CCl$_3$COOH	Butadiene	Styrene	30.5	0.02	288
(πC$_3$H$_5$)$_4$W/C$_2$H$_5$Cl	Butadiene	Styrene	10.8	0.12	288
(πC$_3$H$_5$)$_4$Zr	Butadiene	Styrene	34.0	0.02	288
(PhCH$_2$)$_4$Ti	Butadiene	Styrene	27.5	0.02	288
(πC$_4$H$_7$NiCl)$_2$/CCl$_3$CHO	Butadiene	Styrene	4.6	0.13	288
TiI$_4$/Al(i-Bu)$_3$	Butadiene	1,3-pentadiene (*trans*)	3.6	0.5	288
(πC$_4$H$_7$NiCl)$_2$/chloranil	Butadiene	2,3-dimethylbutadiene	2.8	0.18	288
(πC$_4$H$_7$NiCl)$_2$/chloranil	Butadiene	Methyl pentadienoate	5.8	0.05	288
(πC$_4$H$_7$NiI)$_2$	Butadiene	Cyclohexadiene	0.8	1.10	288
(πC$_4$H$_7$NiI)$_2$/(CCl$_3$COO)$_2$Ni	Butadiene	Cyclohexadiene	0.4	1.12	288
(πC$_4$H$_7$NiCl)$_2$/(CCl$_3$COO)$_2$Ni	Butadiene	1,3-cyclohexadiene	0.47 ± 0.02	0.93 ± 0.5[a]	282
(πC$_3$H$_5$NiI)$_2$	Isoprene	1,3-cyclohexadiene	0.50	1.64	288
Ti(On-Bu)$_4$/AlEt$_3$	Isoprene	1,3-pentadiene	0.3	0.5	293

[a] Similar ratios for (πC$_5$H$_9$NiCl)$_2$ and catalysts containing (CCl$_3$COO)$_2$Zn, CCl$_3$CHO or C$_6$Cl$_4$O$_2$.

active species (concentration proportional to $[Al]^{1/2}[V]^{1/2}$) with "monomer" which is regarded as a ternary complex of both monomers with the aluminium compound. The rate with respect to monomer depends on the equilibria between organo-metal compound and the two monomers. Molecular weights rise to maxima with increase in conversion, the maxima decreasing with increase in vanadium concentration.

Vanadium catalysts of specified composition produce alternating co-polymers from butadiene and propene but there are no published kinetic results [198]. In the polymerization of conjugated dienes the incorporation of monomer units in different isomeric forms is common, but the Group VIII catalysts which give alternating cis/trans 1,4 or 1,4/vinyl polymers from butadiene and isoprene belong to a special class in which the growing chain end must predispose the incoming monomer unit in a conformation which alternates at each addition step. There has so far been only one kinetic investigation on equibinary cis/trans 1,4-polybutadiene (discussed under butadiene polymerization), and this showed no outstanding differences in regard to reaction order or activation energy from the polymerization in heptane solution which gives cis 1,4-polymer.

A recent ^{13}C-NMR investigation on equibinary cis-1,4-1,2 polybutadiene prepared by $R'_3Al/MoCl_3(OR)_2$ as catalyst (Al/Mo >6) has indicated that this polymer is not alternating in structure, but is a random arrangement of cis-1,4 and 1,2 units [309].

10. Equilibrium coordination polymerization

Cycloalkenes, other than cyclohexene are polymerized by certain combinations of Lewis acids, particularly tungsten or molybdenum halides, and organo-metal compounds, by ring opening to give polyalkenamers [261]. Of particular importance is trans-polypentenamer from cyclopentene, obtained as a high molecular weight, broad distribution polymer. The configurations of the polymer double bonds in polypentenamer are determined by the catalyst type and composition. Thus with $WF_6/AlEtCl_2$ the polymer is above 80% cis with Al/W = 2, and above 90% trans with Al/W > 5 [262].

The reaction is considered to involve coordination of the cycloalkene with the transition metal complex and growth by a process of ring enlargement following opening at the double bonds [263, 264], viz.

The kinetics are undoubtedly complex and not yet studied in detail but one important feature of the polymerization is its reversible character. This is possible because of the small energy differences between polymer and monomer, and both depolymerization of preformed polymers and *cis—trans* isomerization have been demonstrated.

Kranz and Beck [268] have calculated that the enthalpies of polymerization of liquid cyclopentene to solid *cis* and *trans* polypentenamer are 4.2 and 3.2 kcal mole^{-1}, in good agreement with a calorimetric value of 4.5 kcal mole^{-1} for a 65% *trans* polymer. Other than this limited thermodynamic data the quantitative aspects of the polymerization are unknown; of particular interest would be the factors influencing molecular weight and molecular weight distribution.

Acknowledgements

With regard to Figs. 2 and 3, thanks are due to Professor J. E. Guillet and Dr L. A. M. Rodriguez.

Note added in proof

(A) Monomer feed and copolymer composition for ethylene/propene copolymerization with the catalyst $VCl_4/AlEt_2Cl$ at $-78°C$ have been shown to fit accurately empirical relationships [322]

$$p = \frac{1 + 0.0016m}{0.21 + 18.4/m} \quad \text{or} \quad p = \frac{1 + 0.010m}{1 + 8.10/m}$$

The coefficients 8.10 and 0.010 in the second equation are usually ascribed to the reactivity ratios r_1 and r_2 (Table 19). This catalyst produces polypropene consisting mainly of syndiotactic stereoblocks, together with short disordered blocks resulting from head-to-head (hh) and tail-to-tail (tt) propene enchainment and occasional isolated isotactic units, and if these features apply to copolymers prepared with vanadium catalysts, the reaction is in effect a terpolymerization. Locatelli et al. [322] derive the equation for monomer/polymer composition ratios

$$p = \frac{a/m + b + cm}{e + d/m + 1/m^2}$$

where the constants a—e are complex functions of the two monomer coordination coefficients, the nine insertion rate coefficients for ethylene and the two modes of propene entry, and the monomer concentrations.

This approximates for $m \gg 1$ to

$$p = \frac{1 + (c/b)m}{e/b + d/bm}$$

which is the same as that found experimentally. This, as indicated above, can be expressed in the form of the Lewis—Mayo copolymerization equation, but it is clear that the constants are not r_1 and r_2 ratios.

The polymerization of propene by VCl_4/$AlEt_2Cl$/anisole at $-78°C$ shows unusual kinetic features in that the polymerization rate increases with time and with monomer concentration to a maximum at $[M]/[Al]/[V] = 2 \times 10^3/30/1$ [323]. Evidently, in this system, possibly as a result of the low temperature employed, the monomer appears to occupy sites so as to preclude propagation and the rate is said to follow the relationship [322]

$$R_p = \frac{K_M k_p [M] [C^*]}{1 + K_M [M]}$$

For the rate to pass through a maximum with increase in $[M]$ it follows that $K_M[M] \gg 1$, in accord with published values of K_M for propene, since strong coordination of monomer to the sites would give rise to independence of rate on $[M]$. A fall in rate, however, implies that monomer is competing with active site formation; it is difficult to see the mechanism by which this could occur as exclusion of organometal from the catalyst surface by monomer would appear to be unlikely.

Observations on this system may not apply to other catalyst systems but the fact that abnormal addition of propene has been demonstrated in elastomeric E/P copolymers prepared under very different conditions makes it clear that, at the least, the values of r_1 and r_2 are approximations. It would be of interest to calculate (assuming, for example, that the ratio of the fractions of head-to-head and head-to-tail units equals the ratio of the corresponding rate coefficients) what differences from the true r_1 and r_2 ratios would result from the occurrence of 5—10% of abnormal addition.

(B) Cunningham and Dove [324] obtained similar results to earlier workers [35] on filtered and hydrocarbon-rinsed precipitates from $TiCl_4$/$Al(n\text{-}Pr)_3$ with or without anisole present. The work included experiments at low Al/Ti ratios (0.2/1). The main effect of anisole was to increase the amount of aluminium in the filtrate at low Al/Ti ratios and to reduce somewhat the Cl/Ti ratio in the precipitate. Perry et al. [325] show, clearly, however, that thoroughly washed precipitates lose virtually all their aluminium to the solvent, so this question may simply be one of efficiency of washing.

(C) Perry et al. [325] find that carbon disulphide retards the polymerization of isoprene by $TiCl_4$/$Al(iBu)_3$ at ratios of CS_2/Ti above 0.1/1, the effect being more marked as the Al/Ti ratio is increased. The most interesting feature, however, is the virtual elimination of low molecular weight oligomers in the presence of the carbon disulphide. It would appear that the donor selectively poisons the sites which produce oligomer but only blocks cis-1,4-polymer producing sites at much higher concentrations.

REFERENCES

1 M. Fischer, G.P. 874,215 (I. G. Farben A.G.).
2 E. Field and M. Feller, *Ind. Eng. Chem. Ind. Ed.*, *49* (1957) 1883; A. Clark, J. P. Hogan, R. L. Banks and W. C. Lanning, *Ind. Eng. Chem. Ind. Ed.*, *48* (1956) 1152.
3 K. Ziegler, E. Holzkamp, H. Breil and H. Martin, *Angew. Chem.*, *67* (1955) 426.
4 Belgian Pat. 543,292 (Dec. 1955), Goodrich Gulf Co.
5 G. Natta, Paper to 15th Ann. Tech. Conf. Soc. Plastics Eng., New York, 27 January 1959, *S.P.E. Journal*, *5* (1959) 373.
6 G. Natta, P. Pino, G. Mazzanti, U. Giannini, E. Mantica and M. Peraldo, *J. Amer. Chem. Soc.*, *79* (1957) 2975.
7 L. Porri, G. Natta and M. C. Gallazzi, *J. Polym. Sci.*, *Part C16* (1967) 2525.
8 G. Natta, *Nuovo Cimento, Ital. Fis.* (Supp. *15*. Ser X) (1960) 1.
9 J. H. DeBoer, *Trans. Faraday Soc.*, *32* (1936) 10.
10 C. F. van Heerden, *J. Polym. Sci.*, *34* (1959) 46.
11 D. F. Herman and W. K. Nelson, *J. Amer. Chem. Soc.*, *75* (1953) 3877, 3882.
12 A. Yamamoto, K. Morifuji, S. Ikeda, T. Saito, Y. Uchida and A. Misono, *J. Amer. Chem. Soc.*, *87* (1965) 4652.
13 W. Klemm and E. Krose, *Z. Anorg. Chem.*, *253* (1947) 218.
14 L. A. M. Rodriguez and H. M. van Looy, *J. Polym. Sci.*, *Part A1*, *4* (1966) 1951.
15 J. Y. Guttman and J. E. Guillet, *Macromolecules*, *3* (1970) 470.
16 R. J. L. Graff, G. Kortleve and C. G. Vonk, *J. Polym. Sci.*, *Part B8* (1970) 735.
17 V. W. Buls and T. L. Higgins, *J. Polym. Sci.*, *Part A1*, *8* (1970) 1025.
18 J. Boor, Jnr., *Ind. Eng. Chem. Prod. Res. Develop.*, *9* (1970) 437.
19 L. A. M. Rodriguez, H. M. Looy and J. A. Gabant, *J. Polym. Sci.*, *Part A1*, *4* (1966) 1905, 1917.
20 A. D. Caunt, *J. Polym. Sci.*, *Part C*, *4* (1963) 49.
21 T. P. Wilson and G. F. Hurley, *J. Polym. Sci.*, *Part C*, *1* (1963) 281.
22 C. Beerman and H. Bestian, *Angew. Chem.*, *71* (1959) 618.
23 J. Boor, Jnr. and E. A. Youngman, *J. Polym. Sci.*, *Part A1*, *4* (1966) 1861; E. J. Arlman and P. Cossee, *J. Catal.*, *3* (1964) 99.
24 P. Cossee, *Tetrahedron Lett.*, *17* (1960) 12, 17.
25 Yu. V. Kissin and N. M. Chirkov, *Eur. Polym. J.*, *6* (1970) 525.
26 H. Uelzmann, *J. Polym. Sci.*, *32* (1958) 457.
27 F. S. Dyachkovskii, A. K. Shilova and A. E. Shilov, *J. Polym. Sci.*, *Part C16* (1967) 2333.
28 W. Cooper, *Ind. Eng. Chem. Prod. Res. Develop.*, *9* (1970) 457.
29 W. Cooper, in J. C. Robb and F. W. Peaker (Eds.), *Progress in High Polymers*, Heywood, London, 1960, pp. 288—291.
30 G. Natta, G. Mazzanti, D. DeLuca, U. Giannini and F. Bandini, *Makromol. Chem.*, *76* (1964) 54.
31 W. Cooper, R. K. Smith and A. Stokes, *J. Polym. Sci.*, *Part B4* (1966) 309.
32 E. J. Badin, *J. Phys. Chem.*, *63* (1959) 1791.
33 A. Schindler and R. B. Strong, *Makromol. Chem.*, *114* (1968) 77.
34 A. Schindler, *Makromol. Chem.*, *118* (1968) 1.
35 M. L. Cooper and J. B. Rose, *J. Chem. Soc. (Lond.)*, (1959) 795.
36 R. van Helden, A. F. Bickel and E. C. Kooijman, *Tetrahedron Lett.*, *12* (1959) 18.
37 A. Malatesta, *Can. J. Chem.*, *37* (1959) 1176.
38 T. P. Wilson and E. S. Hammack, *J. Polym. Sci.*, *Part C1* (1963) 305.
39 J. Boor, Jnr., *Macromol. Rev.*, *2*. Wiley, 1967, pp. 124—7.
40 G. Natta, P. Pino and G. Mazzanti, *Gazz. Chim. Ital.*, *87* (1957) 549, 570.
41 W. M. Saltman, W. E. Gibbs and J. Lal, *J. Amer. Chem. Soc.*, *80* (1958) 5615.
42 A. A. Korotkov and Li. Tszun-Chan, *Polym. Sci., U.S.S.R.*, *3* (1963) 615.
43 Brit. Pat. 843,408, Esso Res. & Eng. Co.

44 M. H. Lehr, *Macromolecules, 1* (1968) 178.
45 J. P. Hogan, *J. Polym. Sci., Part A1, 8* (1970) 2637.
46 Yu. I. Ermakov, V. A. Zakharov and E. G. Kushnareva, *J. Polym. Sci., Part A1, 9* (1971) 771.
47 H. N. Friedlander, *J. Polym. Sci., 38* (1959) 91.
48 K. G. Miesserov, *J. Polym. Sci., Part A1, 4* (1966) 3047.
49 D. S. Breslow and N. R. Newburg, *J. Amer. Chem. Soc., 79* (1957) 5072.
50 G. Natta, G. Mazzanti, P. Corradini, U. Giannini and S. Cesca, *Atti. Accad. nazl. Lincei, Cl. Sci. Fis. Mat. Natur. Rend., 26* (1959) 150.
51 C. E. H. Bawn and R. Symcox, *J. Polym. Sci., 34* (1959) 139.
52 M. Takeda, K. Iimura, Y. Nozawa, M. Hisatome and N. Koide, *J. Polym. Sci., Part C23* (1968) 741.
53 H. Hirai, K. Hiraki, I. Noguchi and S. Makishima, *J. Polym. Sci., Part A1, 8* (1970) 147.
54 K. Hiraki, T. Inoue and H. Hirai, *J. Polym. Sci., Part A1, 8* (1970) 2543.
55 C. E. H. Bawn, A. M. North and J. S. Walker, *Polymer, 5* (1964) 419.
56 G. Satori and G. Costa, *Z. Elektrochem., 63* (1959) 108.
57 D. E. O'Reilly, C. P. Poole, Jr., R. F. Belt and H. Scott, *J. Polym. Sci., Part A2* (1964) 3257.
58 J. G. Balas, H. E. De La Mare and D. O. Schissler, *J. Polym. Sci., Part A3* (1965) 2243.
59 W. S. Anderson, *J. Polym. Sci., Part A1, 5* (1967) 429.
60 F. X. Werber, C. J. Benning, W. R. Wszolek and G. E. Ashby, *J. Polym. Sci., Part A1, 6* (1968) 743.
61 J. F. Harrod and L. R. Wallace, *Macromolecules, 2* (1969) 449.
62 E. I. Tinyakova, T. G. Golenko, B. A. Dolgoplosk, A. V. Alferov, I. A. Oreshkin, O. K. Sharaev, G. N. Chernenko and V. A. Yakovlev, *J. Polym. Sci., Part C16* (1967) 2625.
63 F. Dawans and Teyssié Ph., Organic Coatings and Plastics, Chemistry Division, ACS/CIC Meeting, 30, p. 208 (Toronto, Canada) May 1970; *Ind. Eng. Chem. Prod. Res. Develop., 10* (1971) 261.
64 U. Giannini, U. Zucchini and E. Albizzati, *J. Polym. Sci., Part B8* (1970) 405.
65 A. D. Jenkins, M. F. Lappert and R. C. Srivastava, *Eur. Polym. J., 7* (1971) 289.
66 A. Simon and G. Ghymes, *J. Polym. Sci., 53* (1961) 327.
67 E. W. Duck, D. K. Jenkins, D. P. Grieve and M. N. Thornber, *Eur. Polym. J., 7* (1971) 55.
68 T. Yoshimoto, K. Komatsu, R. Sakata, K. Yamamoto, Y. Takeuchi, A. Onishi and K. Ueda, *Makromol. Chem., 139* (1970) 61.
69 S. Otsuka, K. Mori, T. Suminoe and F. Imaizumi, *Eur. Polym. J., 3* (1967) 73.
70 G. Allegra, F. LoGiudice, G. Natta, U. Giannini, G. Faqheranzi and P. Pino, *Chem. Commun. (London),* (1967) 1263.
71 T. Miyazawa and Y. Ideguchi, *J. Polym. Sci., Part B1* (1963) 389; H. Tadokova, M. Ukita, M. Kobayashi and S. Murahashi, *J. Polym. Sci., Part B1* (1963) 405.
72 M. N. Vol'kenshtein, V. N. Nikitin and T. V. Yakovleva, *Bull. Acad. Sci., U.S.S.R. (Ser. Fiz.) 14* (1950) 471; *Chem. Abstr., 45* (1951) 3716.
73 W. Marconi, A. Mazzei, S. Cucinella and M. Cesari, *J. Polym. Sci., Part A2* (1964) 4261.
74 G. Natta, P. Pino, E. Mantica, F. Danuso, G. Mazzanti and M. Peraldo, *Chim. Ind. (Milan) 38* (1956) 124.
75 F. J. Karol and W. L. Carrick, *J. Amer. Chem. Soc., 83* (1961) 2654
76 G. Natta, L. Porri and S. Valenti, *Makromol. Chem., 67* (1963) 225.
77 G. Natta, *J. Polym. Sci., 48* (1960) 219.
78 (a) C. G. Overberger and P. A. Jarovitsky, *J. Polym. Sci., Part C4* (1963) 37. (b) A. Simon, P. A. Jarovitsky and C. G. Overberger, *J. Polym. Sci., Part A1, 4* (1966) 2513.

79 A. F. Rekasheva and L. A. Kiprianova, *Polym. Sci., U.S.S.R., 3* (1962) 987.

80 K. Ziegler, *Int. Conf. Coord. Chem. London, April 1959*, Chem. Soc., London, p. 1.

81 G. Natta, *J. Polym. Sci., 34* (1959) 21.

82 A. S. Hoffman, B. A. Fries and P. C. Condit, *J. Polym. Sci., Part C4* (1963) 109.

83 G. E. Coates, *Organo Metallic Compounds*, Methuen, Wiley, 1960, pp. 55, 133.

84 R. N. Kovalevskaya, Ye. I. Tinyakova and B. A. Dolgoplosk, *Polym. Sci., U.S.S.R., 4* (1963) 414.

85 W. L. Carrick, W. T. Reichle, F. Pennella and J. J. Smith, *J. Amer. Chem. Soc., 82* (1960) 3887.

86 J. C. W. Chien, *J. Polym. Sci., Part A1* (1963) 425.

87 C. E. H. Bawn, *Rubb. Plast. Age, 46* (1965) 510.

88 C. Longiave, R. Castelli and M. Ferraris, *Chim. Ind. (Milan), 44* (1962) 725.

89 G. Natta, L. Porri, A. Mazzei and D. Morero, *Chim. Ind. (Milan), 41* (1959) 398.

90 G. Natta, L. Porri and A. Mazzei, *Chim. Ind. (Milan), 41* (1959) 116.

91 A. Ye. Shilov, A. K. Shilova and B. N. Bobkov, *Polym. Science, U.S.S.R., 4* (1963) 526.

92 H. N. Friedlander, *J. Polym. Sci., Part A2* (1964) 3885.

93 K. Vesely, *Macromolecular Chemistry*, I.U.P.A.C. Montreal, Canada, 1961, p. 407 (Butterworths 1962).

94 K. Soga and T. Keii, *J. Polym. Sci., Part A1, 4* (1966) 2429.

95 K. Vesely, J. Ambroz and O. Hamrik, *J. Polym. Sci., Part C4* (1963) 11.

96 H. Schnecko, M. Reinmoller, K. Weirauch and W. Kern, *J. Polym. Sci., Part C4* (1963) 71.

97 D. B. Ludlum, A. W. Anderson and C. E. Ashby, *J. Amer. Chem. Soc., 80* (1958) 1380.

98 A. I. Medalia, A. Orzechowski, J. A. Trinchera and J. P. Morley, *J. Polym. Sci., 41* (1959) 241.

99 L. S. Bresler, V. A. Grechanovsky, A. Muzsay and I. Ya. Poddubnyi, *Makromol. Chem., 133* (1970) 111.

100 A. A. Shaginyan and H. C. Entskolyan, I.U.P.A.C. Symposium Budapest 1969, Preprints Vol. II, Paper 4/01 p. 167; G. V. Rakova, A. A. Shaginyan and N. S. Yeniklopyan, *Vysokomol. Soed., A9* (1967) 2578.

101 K. Vesely, J. Ambroz, R. Vilum and O. Hamrik, *J. Polym. Sci., 55* (1961) 25.

102 A. Von Gumboldt, J. Helberg and G. Schleitzer, *Makromol. Chem., 101* (1967) 229.

103 E. W. Duck and W. Cooper, *XXIII Int. Cong. Pure & Applied Chem. Supp.* Butterworths, London, *8* (1971) 215.

104 G. W. Phillips and W. L. Carrick, *J. Polym. Sci., 59* (1962) 401; Z. S. Smolyan, A. L. Graevskii, O. J. Demin, V. K. Fukin and G. N. Matveeva, *Polym. Sci., U.S.S.R., 3* (1962) 18.

105 E. K. Easterbrook, T. J. Brett, F. C. Loveless and D. N. Matthews, *XXIII Int. Cong. Pure & Applied Chem., Macromol. Preprint* Vol. II, Boston 1971, p. 712.

106 J. Boor, *J. Polym. Sci., Part C1* (1963) 257; *A3* (1965) 995; *A1, 9* (1971) 617.

107 W. Cooper, D. E. Eaves, G. D. T. Owen and G. Vaughan, *J. Polym. Sci., Part C4* (1963) 211.

108 W. Cooper, D. E. Eaves, G. Vaughan, G. Degler and R. Hank, *Elastomer Stereospecific Polymerization, Advances in Chemistry, 52* (1966) 46.

109 Yu. V. Kissin, S. M. Mezhikovsky and N. M. Chirkov, *Eur. Polymer. J., 6* (1970) 267.

110 F. P. Van de Kamp, *Makromol. Chem., 93* (1966) 202.

111 J. C. N. Chien, *J. Amer. Chem. Soc., 81* (1959) 86.

112 T. Keii, K. Soga and N. Saiki, *J. Polym. Sci., Part C16* (1967) 1507.

113 I. N. Meshkova, S. A. Kumankova, V. I. Tsvetkova and N. M. Chirkov, *Polym. Sci., U.S.S.R., 3* (1962) 1130.

114 G. Bier, A. Gumboldt and G. Schleitzer, *Makromol. Chem., 58* (1962) 43.

252

115 L. S. Bresler, L. Ya. Poddubnyi and V. N. Sokolov, *J. Polym. Sci., Part C16* (1969) 4337.
116 C. F. Feldman and E. Perry, *J. Polym. Sci., 46* (1960) 217.
117 W. Cooper, D. E. Eaves and G. Vaughan, *Makromol. Chem., 67* (1963) 229.
118 G. Natta, L. Porri, A. Carbonari and A. Greco, *Makromol. Chem., 71* (1964) 207.
119 C. W. Childers, *J. Amer. Chem. Soc., 85* (1963) 229.
120 R. E. Rinehart, *J. Polym. Sci., Part C7* (1969) 27; M. Morton, I. Piirma and B. Das, *Rubb. Plast. Age, 46* (1965) 404.
121 H. W. Coover and J. E. Guillet, *J. Polym. Sci., Part A1, 4* (1966) 2583; *C4* (1963) 1511.
122 J. F. Henderson, *J. Polym. Sci., Part C4* (1963) 233.
123 S. Tanaka and H. Morikawa, *J. Polym. Sci., Part A3* (1965) 3147.
124 H. Hirai, K. Hiraki, I. Noguchi, T. Inoue and S. Makishima, *J. Polym. Sci., Part A1, 8* (1970) 2393.
125 T. A. Zakharov and Yu. I. Ermanov, *J. Polym. Sci., Part A1, 9* (1971) 3129.
126 E. Kohn, H. J. L. Schuurmans, J. V. Cavender and R. A. Mendelson, *J. Polym. Sci., 58* (1962) 681.
127 C. W. Hoch, *J. Polym. Sci., Part A1, 4* (1966) 3055.
128 I. Pasquon, G. Natta, A. Zambelli, A. Marinangeli and A. Surico, *J. Polym. Sci., Part C16* (1967) 2501.
129 E. G. M. Tornqvist, J. T. Richardson, Z. W. Wilchinsky and R. W. Looney, *J. Catal., 8* (1967) 189.
130 A. Zambelli, G. Natta, I. Pasquon and R. Signorini, *J. Polym. Sci., Part C16* (1965) 2485.
131 E. J. Badin, *J. Amer. Chem. Soc., 80* (1958) 6545.
132 G. Bier, *Makromol Chem., 58* (1962) 1.
133 (a) M. N. Berger and B. M. Grieveson, *Makromol. Chem., 83* (1965) 80. (b) B. M. Grieveson, *Makromol. Chem., 84* (1965) 93.
134 I. N. Meshkova, G. M. Bakova, V. I. Tsvetkova and N. M. Chirkov, *Polym. Sci., U.S.S.R., 3* (1963) 1011.
135 A. Valvassori, G. Sartori, G. Mazzanti and G. Pajaro, *Makromol. Chem., 61* (1963) 46.
136 G. Bier, G. Lehman and H. J. Leugering, *Trans. Plast. Inst., 28* (6) (1960) 98.
137 G. Natta, G. Mazzanti, P. Longi and F. Bernadino, *Chim. Ind. (Milan), 41* (1959) 519.
138 A. Mujake, H. Kondo and M. Nishimo, *Angew. Chem. (Int.), 10* (1971) 802.
139 G. H. Olivé and S. Olivé, *Angew. Chem. (Int.), 10* (1971) 776.
140 J. C. McGowan and B. M. Ford, *J. Chem. Soc. (Lond.), (1958)* 1149.
141 K. Fukui, T. Shimidzu, S. Fukumoto, T. Kagiya and S. Yuasa, *J. Polym. Sci., 55* (1961) 321.
142 (a) A. Zambelli, J. di Pietro and G. Gatti, *J. Polym. Sci., Part A1* (1963) 403. (b) R. L. McConnell, A. McCall, G. O. Cash, F. B. Joyner and H. W. Coover, *J. Polym. Sci., Part A3* (1965) 2135.
143 K. Oita and T. D. Nevitt, *J. Polym. Sci., 43* (1960) 585.
144 A. V. Topchiev, B. A. Krentsel, A. I. Perelman and K. G. Miesserov, *J. Polym. Sci., 34* (1959) 129.
145 W. R. Watt, F. H. Fry and H. Pobiner, *J. Polym. Sci., Part A1, 6* (1968) 2703.
146 D. H. Dawes and C. A. Winkler, *J. Polym. Sci., Part A2* (1964) 3029.
147 H. Schnecko, W. Dost and W. Kern, *Makromol. Chem., 121* (1969) 159.
148 A. P. Firsov, N. D. Sandomirskaya, V. I. Tsvetkova and N. M. Chirkov, *Polym. Sci., U.S.S.R., 3* (1962) 943.
149 H. Feilchenfeld and M. Jeselson, *J. Phys. Chem., 63* (1959) 720.
150 (a) G. Natta, E. Giachetti, I. Pasquon and G. Pajaro, *Chim. Ind., 42* (1960) 1091. (b) J. Boer, Jr., *J. Polym. Sci. Part C1* (1963) 237.

151 V. I. Tsvetkova, O. N. Pirogov, D. M. Lisitsyn and N. M. Chirkov, *Polym. Sci.*, *U.S.S.R.*, *3* (1962) 586.

152 G. Natta, P. Pino and G. Mazzanti, *Gazz. Chim. Ital.*, *87* (1957) 570; G. Natta, I. Pasquon, A. Zambelli and G. Gatti, *J. Polym. Sci.*, *51* (1961) 387.

153 G. M. Burnett and P. J. T. Tait, *Polymer*, *1* (1960) 151.

154 P. E. M. Allen and D. Gill, *Makromol. Chem.*, *71* (1964) 33.

155 R. J. Kern, H. G. Hurst and W. R. Richard, *J. Polym. Sci.*, *45* (1960) 195.

156 F. D. Otto and G. Parravano, *J. Polym. Sci.*, *Part A2* (1964) 5131.

157 A. M. North, *Proc. Roy. Soc. (Lond.)*, *A254* (1960) 408.

158 E. V. Zabolotskaya, A. R. Gantmakher and S. S. Medvedev, *Polym. Sci.*, *U.S.S.R.*, *3* (1962) 282.

159 P. E. M. Allen and J. F. Harrod, *Makromol. Chem.*, *32* (1959) 153.

160 A. B. Deshpande, R. V. Subramanian and S. L. Kapur, *Makromol. Chem.*, *98* (1966) 90.

161 H. L. Krauss and H. Stach, *Inorg. Nucl. Chem. Lett.*, *4* (1968) 393.

162 C. G. Overberger, F. S. Diachkovsky and P. A. Jarovitsky, *J. Polym. Sci.*, *Part A2* (1964) 4113.

163 L. C. Anand, A. B. Deshpande and S. L. Kapur, *J. Polym. Sci.*, *Part A1*, *5* (1967) 2079.

164 L. C. Anand, S. S. Dixit and S. L. Kapur, *J. Polym. Sci.*, *Part A1*, *6* (1968) 909.

165 A. B. Deshpande, R. V. Subramanian and S. L. Kapur, *J. Polym. Sci. Part A1*, *4* (1966) 1799.

166 S. L. Malhotra, A. B. Deshpande and S. L. Kapur, *J. Polym. Sci.*, *Part A1*, *6* (1968) 193.

167 M. Gippin, *Ind. Eng. Chem. Prod. Res. Develop.*, *1* (1962) 32; *4* (1965) 160.

168 C. E. H. Bawn, *Rubb. Plast. Age*, *46* (1965) 510.

169 C. R. McIntosh, W. D. Stephens and C. O. Taylor, *J. Polym. Sci.*, *Part A1* (1963) 2003.

170 V. N. Zgonnik, B. A. Dolgoplosk, N. I. Nikolayev and V. A. Kropachev, *Polym. Sci. U.S.S.R.*, *2* (1966) 338.

171 H. Scott, R. E. Frost, R. F. Belt and D. E. O'Reilly, *J. Polym. Sci.*, *Part A2* (1964) 3233.

172 V. S. Bresler and S. S. Medvedev, I.U.P.A.C. Symposium Budapest 1969, Paper 4/25.

173 J. C. Marechal, F. Dawans and P. Teyssie, *J. Polym. Sci.*, *Part A1*, *8* (1970) 1993.

174 N. G. Gaylord, T. K. Kwei and H. F. Mark, *J. Polym. Sci.*, *42* (1960) 417.

175 C. D. Mason and R. J. Schaffhanser, *J. Polym. Sci.*, *Part B9* (1971) 661.

176 H. Schnecko, K. A. Jung and L. Grosse, *Macromol. Chem.*, *148* (1971) 67.

177 A. Mazzei, M. Araldo, W. Marconi and M. deMalde, *J. Polym. Sci.*, *Part A1*, *3* (1965) 753.

178 R. E. Rinehart, H. P. Smith, H. S. Witt and H. Romeyn, *J. Amer. Chem. Soc.*, *83* (1961) 4864.

179 C. E. H. Bawn, D. G. J. Cooper and A. M. North, *Polymer*, *7* (1966) 113.

180 J. Zachoval, F. Mikis and J. Krepelka, XIII Int. Cong. Pure and App. Chem., Boston 1971, *Macromol. Prep.* Vol. I, p. 164.

181 M. Morton, I. Piirma and B. Das, *Rubb. Plast. Age*, *46* (1965) 404.

182 L. A. Novokshonova, G. P. Bereseneva, V. I. Tsvetkova and N. M. Chirkov, *Vysokomol Soedin. Z.*, *9* (1967) 562; *Chem. Abstr.*, *67* (1968) 22289.

183 A. Zambelli, A. Lety, C. Tori and J. Pasquon, *Macromol Chem.*, *115* (1968) 73.

184 G. Crespi, G. Sartori and A. Valvassori, Chapter IVD in *Copolymerization*, Edited by G. E. Ham, Interscience, 1964.

185 E. J. Badin, *J. Amer. Chem. Soc.*, *80* (1958) 6549.

186 J. Boor, *XIII Int. Cong., Pure & App. Chem.*, Boston, 1971, Symp. M3, Paper 384.

254

187 C. Cozewith and G. ver Strate, *Macromolecules*, *4* (1971) 482.
188 J. S. Walker, Unpublished results, Dunlop Ltd.
189 R. I. Bushick, *J. Polym. Sci., Part A3* (1965) 2047.
190 A. P. Firsov, I. N. Meshkova, N. D. Kostrova and N. M. Chirkov, *Vysokomol. Soedin. Z.*, *8* (1966) 1860.
191 N. M. Siedov, R. M. Aliguliev and A. I. Abasov, *Vysokomol. Soedin. Z.*, *11*, (1969) 2170.
192 C. A. Lukach and H. M. Spurlin, ref. 184, p. 126.
193 G. Natta, G. Mazzanti, A. Valvassori, G. Sartori and A. Barbagallo, *J. Polym. Sci.*, *51* (1961) 429.
194 G. Natta, G. Mazzanti, A. Valvassori, G. Sartori and D. Fuimani, *J. Polym. Sci.*, *51* (1961) 411.
195 L. A. Novokshonova, V. I. Tsvetkova and N. M. Chirkov, *J. Polym. Sci., Part C16* (1965) 2659.
196 G. Natta, G. Mazzanti and A. Valvassori, *Chem. Ind.* (Milan), *41* (1959) 764.
197 N. M. Seidov, M. A. Dalin and S. M. Kyazimov, *Azerb. Khim. Zh.*, (1) 68 (1968); *Chem. Abstr.*, *69* (1968) 87824.
198 J. Furukawa, I.U.P.A.C. Symposium Boston 1971, *Macromol. Preprints*, Vol. II, p. 695.
199 N. Ashikari, T. Kanemitsu, K. Yanagisawa, K. Nakagawa, H. Okamoto, S. Kobayashi and A. Nishioka, *J. Polym. Sci., Part A2* (1964) 3009.
200 F. Danusso, *Chem. Ind. (Milan)*, *44* (1962) 611.
201 T. Saegusa, T. Naruyama, S. Kurahashi and J. Furukawa, *J. Chem. Soc. Japan, Ind. Chem. Sect.*, *65* (1962) 2082.
202 G. H. Olivé and S. Olivé, *Angew. Chem. (Int.)*, *6* (1967) 790.
203 F. Dawans, *J. Polym. Sci., Part A2* (1964) 3297.
204 J. Furukawa, K. Irako, T. Saegusa, N. Horooka and T. Naruyama, *J. Chem. Soc. Japan, Ind. Chem. Sect.*, *65* (1962) 2074.
205 L. S. Bresler, B. A. Dolgoplosk, M. F. Kolechkova and Y. N. Kropacheva, *Polym. Sci. USSR.*, *5* (1964) 1012.
206 I. N. Smirnova, B. A. Dolgoplosk and B. A. Krol, I.U.P.A.C. Meeting, Budapest, 1969, Paper 4/23.
207 (a) W. Marconi, A. Mazzei, G. Lugli and M. Bruzzone, *J. Polym. Sci. Part C16* (1965) 805. (b) J. K. Stille and E. D. Vessel, *49* (1961) 419.
208 N. Yamazaki, T. Suminoe, T. Furuhama and S. Kambara, *J. Chem. Soc., Japan, Ind. Chem. Sect.*, *64* (1961) 103.
209 M. Hirooka, H. Yabuuchi, S. Morita, S. Kawasumi, and N. Nakaguchi, *J. Polym. Sci., Part B5* (1967) 47.
210 M. Taniguchi, H. Kawasaki and J. Furukawa, *J. Polym. Sci., Part B7* (1969) 411.
211 J. Furukawa, Y. Iseda, K. Haga and N. Kataoka, *J. Polym. Sci., Part A1*, *8* (1970) 1147; J. Furukawa, E. Kobayashi and Y. Iseda, *Polym. J. (Japan) 1* (1970) 155.
212 L. Michaljlov, H. J. Cantow and P. Zugenmaier, *Polymer*, *12* (1971) 70.
213 H. Schnecko and W. Kern, *Chem. Ztg.*, *94* (1970) 229.
214 K. A. Jung and H. Schnecko, *Makromol. Chem.*, *154* (1972) 227.
215 S. L. Kappur, I.U.P.A.C. Symposium, Boston 1971. *Macromol. Preprints* Vol. 1, p. 129.
216 K. Meyer and K. H. Reichert, *Angew. Makromol. Chem.*, *12* (1970) 175.
217 H. Hocker, K. Sacki and W. Kearn, I.U.P.A.C. Symposium, Boston, 1971, *Macromol. Preprints* Vol. 1, p. 186.
218 S. S. Dixit, A. B. Deshpande and S. L. Kapur, *J. Polym. Sci. Part C31* (1970) 6.
219 S. S. Dixit, A. B. Deshpande and S. L. Kapur, *J. Polym. Sci., Part A1*, *8* (1970) 1289.

255

220 S. S. Dixit, A. B. Deshpande, L. C. Anand and S. L. Kapur, *J. Polym. Sci., Part A1, 7* (1969) 1973.

221 W. Kawai, M. Ogawa and T. Ichihashi, *J. Polym. Sci., Part A1, 9* (1971) 1599.

222 S. Inoue, T. Tsuruta and J. Furukawa, *Makromol. Chem., 49* (1961) 13.

223 G. Mazzanti, A. Valvassori, G. Sartori and G. Pajaro, *Chim. Ind. (Milan), 39* (1957) 825.

224 G. Natta, G. Mazzanti, A. Valvassori and G. Sartori, *Chim. Ind. (Milan), 40* (1958) 717.

225 Y. P. Tepenitsyna, M. I. Faberov, A. M. Kutin and G. S. Levskaia, *Polym. Sci. USSR, 1* (1960) 432.

226 H. Giachetti, P. Manaresi, R. Zacchini and S. Baccilerelli, *Chim. Ind. (Milan), 48* (1966) 1037.

227 G. Natta, I. Pasquon, J. Svab and A. Zambelli, *Chim. Ind. (Milan), 44* (1962) 621.

228 A. Schindler, *J. Polym. Sci., Part C4* (1963) 81.

229 D. G. H. Ballard and T. Medinger, *J. Chem. Soc. (Lond.) (B)* (1968) 1176.

230 D. G. H. Ballard and J. V. Dawkins, *Makromol. Chem., 148* (1971) 195.

231 D. G. H. Ballard, E. Jones, T. Medinger and A. J. P. Piolo, *Makromol. Chem., 148* (1971) 175.

232 L. L. van Reijen and P. Cossee, *Disc. Faraday. Soc., 41* (1966) 277.

233 J. P. Durand, F. Dawans and Teyssié Ph., *J. Polym. Sci., A1, 8* (1970) 979.

234 D. L. Christman, *J. Polym. Sci., Part A1, 10* (1972) 471.

235 J. A. Waters and G. A. Mortimer, *J. Polym. Sci., Part A1, 10* (1972) 895.

236 I. D. McKenzie, P. J. T. Tait and D. R. Burfield, *Polymer, 13* (1972) 307.

237 D. R. Burfield, P. J. T. Tait and I. D. McKenzie, *Polymer, 13* (1972) 321.

238 D. R. Burfield and P. J. T. Tait, *Polymer, 13* (1972) 315.

239 E. Nakamura, K. Kaguchi and T. Amemiya, *Kogyo Kagaku Zasshi, 69* (10) (1966) 1940.

240 Yu. I. Kozorezov, A. P. Rusakov and N. M. Pikalo, *Khim. Prom. (Moscow), 45*(5), pp. 343—5 (Russ).

241 H. W. Coover, R. L. McConnel, F. B. Joyner, D. F. Slonaker and R. L. Combs, *J. Polym. Sci., Part A1, 4* (1966) 2563.

242 E. M. J. Pijpers and B. C. Roest, *Eur. Polym. J., 8* (1972) 1151.

243 J. A. Waters and G. A. Mortimer, *J. Polym. Sci., Part A1, 10* (1972) 1827.

244 F. P. Baldwin and G. Ver Strate, *Rubber Chem. Technol., 45* (1972) 709.

245 I. D. McKenzie and P. J. T. Tait, *Polymer, 13* (1972) 510.

246 F. J. Karol, G. L. Karapinka, Wu Chisung, A. W. Dow, R. N. Johnson and W. L. Carrick, *J. Polym. Sci., Part A1, 10* (1972) 2621.

247 W. L. Carrick, R. J. Turbett, F. J. Karol, G. L. Karapinka, A. S. Fox and R. N. Johnson, *J. Polym. Sci., Part A1, 10* (1972) 2609.

248 P. Bourdaudurg and F. Dawans, *J. Polym. Sci., Part A1, 10* (1972) 2527.

249 T. Georgiadis and R. St John Manley, *Polymer, 13* (1972) 567.

250 J. F. Harrod and L. R. Wallace, *Macromolecules, 5* (1972) 682.

251 J. F. Harrod and L. R. Wallace, *Macromolecules, 5* (1972) 685.

252 G. Boocock and R. N. Haward, *Chemistry of Polymerization Processes*, Soc. Chem. Ind. Lond. Monograph No. 20, p. 3, 1966.

253 J. N. Hay, P. G. Hooper and J. C. Robb, *J. Organometal Chem., 28* (1971) 193.

254 F. J. Karol, G. L. Brown and J. M. Davison, *J. Polym. Sci., Part A1, 11* (1973) 413.

255 R. Warin, Ph. Teyssié, P. Bourdaudurg and F. Dawans, *J. Polym. Sci., Part B, 11* (1973) 177.

256 V. I. Klepikova, G. P. Kondratenkov, V. A. Kormer, M. I. Lobach and L. A. Churlyaeva, *J. Polym. Sci., Part B, 11* (1973) 193.

257 T. Matsumoto and J. Furukawa, *J. Polym. Sci., Part B* (1967) 935.

256

258 A. B. Deshpande, S. M. Kale and S. L. Kapur, *J. Polym. Sci., Part A1, 10* (1972) 195.
259 R. N. Haszeldine, T. G. Hyde and P. J. T. Tait, *Polymer, 14* (1973) 215, 224.
260 R. N. Haszeldine, T. G. Hyde and P. J. T. Tait, *Polymer, 14* (1973) 221.
261 G. Natta, G. Dall'Asta and G. Mazzanti, *Angew. Chem., Int. Ed. Eng., 3* (1964) 723.
262 F. Haas, K. Nützel, G. Pampus and D. Theisen, *Rubber Chem. Technol., 43* (1970) 1116.
263 A. J. Amass, *British Polym. J., 4* (1972) 327.
264 G. Dall'Asta and G. Motroni, *Eur. Polym. J., 7* (1971) 707.
265 V. W. Buls and T. L. Higgins, *J. Polym. Sci., Part A1, 11* (1973) 925.
266 M. Iguchi, *J. Polym. Sci., Part A1, 8* (1970) 1013.
267 K. Bujadoux, R. Clement, J. Jozefonvicz and J. Neel, *Eur. Polym. J, 9* (1973) 189.
268 D. Kranz and M. Beck, *Angew. Makromol. Chem., 27* (1972) 29.
269 T. Keii, *Kinetics of Ziegler-Natta Polymerization*, Kodansha, Tokyo, Chapman and Hall, London, 1973, Chapters 2, 3 and 5.
270 G. Natta and I. Pasquon, *Advan. Catal. Rel. Subj., 11* (1959) 1.
271 A. K. Ingberman, I. J. Levine and R. J. Turbett, *J. Polym. Sci., Part A1, 4* (1966) 2781.
272 K. Soga, Y. Go, S. Takano and T. Keii, *J. Polym. Sci., Part A1, 5* (1967) 2815.
273 K. Vesely, J. Ambroz, J. Mejzlik and E. Spousta, *J. Polym. Sci. Part C, 16* (1967) 417.
274 J. Ambroz, P. Osecky, J. Mejzlik and O. Hamrik, *J. Polym. Sci., Part C, 16* (1967) 423.
275 W. E. Smith and R. G. Zelmer, *J. Polym. Sci., Part A1* (1963) 2587.
276 K. H. Reichert and K. R. Meyer, *Makromol. Chem., 169* (1973) 163.
277 L. F. Borisova, E. A. Fushman, E. I. Vizen and N. M. Chirkov, *Eur. Polym. J., 9* (1973) 953.
278 D. G. H. Ballard, J. V. Dawkins, J. M. Key and P. W. van Lienden, *Makromol. Chem., 165* (1973) 173; D. G. H. Ballard, *Advances in Catalysis, 23* (1973) 263.
279 D. G. H. Ballard, E. Jones, R. J. Wyatt, R. T. Murray and P. A. Robinson, *Polymer, 15* (1974) 169.
280 D. D. Eley, C. H. Rochester and M. S. Scurrell, *Proc. Roy. Soc., A329, 361* (1972) 375.
281 D. D. Eley, C. H. Rochester and M. S. Scurrell, *J. Catal. 29* (1973) 20.
282 B. A. Dolgoplosk, S. I. Beilin, Yu. V. Korshak, G. M. Chernenko, L. M. Vardanyan and M. P. Teterina, *Eur. Polym. J., 9* (1973) 895.
283 S. S. Medvedev, *Macromol. Chem., 2* (1966) 403; I.U.P.A.C. Symposium, Prague, Aug. 30—Sept. 4, 1965, Butterworths, London.
284 C. Tosi, A. Valvassori and F. Ciampelli, *Eur. Polym. J., 4* (1968) 107.
285 G. Natta, G. Mazzanti, A. Valvassori, G. Sartori and D. Morero, *Chim. Ind., 42* (1960) 125.
286 C. J. Carman and C. E. Wilkes, *Rubber Chem. Technol., 44* (1971) 781.
287 C. E. Wilkes, C. J. Carman and R. A. Harrington, *J. Polym. Sci., Part C, 43* (1973) 237.
288 B. A. Dolgoplosk, S. I. Beilin, Yu. V. Korshak, K. L. Makovetsky and E. I. Tinyakova, *J. Polym. Sci., Part A1, 11* (1973) 2569.
289 T. Keii, Y. Doi and H. Kobayashi, *J. Polym. Sci., Part A1, 11* (1973) 1881.
290 J. P. Scholten and H. J. van der Ploeg, *J. Polym. Sci., Part A1, 11* (1973) 3205.
291 G. H. Olivé and S. Olivé, *J. Polym. Sci., Part B, 12* (1974) 39.
292 W. M. Saltman and L. J. Kuzma, *Rubber Chem. Technol., 46* (1973) 1055.
293 K. Bujadoux, M. Galin, J. Josefonvicz and A. Szubarga, *Eur. Polym. J., 10* (1974) 1.
294 D. L. Christman and G. L. Kein, *Macromolecules 1* (1968) 358.

295 A. Zambelli, G. Gatti, C. Sacchi, W. O. Crain and J. D. Roberts, *Macromolecules*, *4* (1971) 475.
296 R. N. Haward, A. N. Roper and K. L. Fletcher, *Polymer*, *14* (1973) 365.
297 Private communications from R. N. Haward.
298 D. G. Boucher, I. W. Parsons and R. N. Haward, *Macromol. Chem.*, *175* (1974) 3461.
299 I. D. McKenzie and P. J. T. Tait, *Polymer*, *13* (1972) 510.
300 D. R. Burfield and P. J. T. Tait, *Polymer*, *15* (1974) 87.
301 B. A. Dolgoplosk, E. I. Tinyakova, N. N. Stefanovskaya, I. A. Oreshkin and V. L. Shmonina, *Eur. Polym. J.*, *10*(7) (1974) 605.
302 M. N. Berger, G. Boocock and R. N. Haward, *Advan. Catal. Rel. Subj.*, *19* (1969) 211.
303 C. Mussa, *J. Appl. Polym. Sci.*, *1* (1959) 300.
304 W. D. Lansing and E. O. Kraemer, *J. Amer. Chem. Soc.*, *57* (1935) 1369.
305 H. Wesslau, *Makromol. Chem.*, *20* (1956) 111; *26* (1958) 102.
306 L. H. Tung, *J. Polym. Sci.*, *20* (1956) 495.
307 M. R. M. Swingler, Ph.D. Thesis, University of Bradford 1972.
308 V. G. Gandhi, A. B. Deshpande and S. L. Kapur, *J. Polym. Sci.*, *Part A1*, *12* (1974) 1173.
309 J. Furukawa, E. Kobayashi and T. Kawagoe, *Polym. J.* (*Japan*), *5* (1973) 231.
310 J. Y. Guttman and J. E. Guillet, *Macromolecules*, *1* (1968) 461.
311 R. Clement, K. Bujdoux, J. Josefonvicz and G. Roques, *Eur. Polym. J.*, *10* (1974) 821.
312 W. L. Truett, D. R. Johnson, J. M. Robinson and B. A. Montagne, *J. Amer. Chem. Soc.*, *82* (1960) 2337.
313 British Pat. 951,022 (Montecatini, 1964).
314 A. G. Chesworth, R. N. Haszeldine and P. J. T. Tait, *J. Polym. Sci.*, *Part A*, *12* (1974) 1703.
315 A. D. Ketley, *J. Polymer Sci.*, *Part BI* (1963) 121.
316 F. H. C. Edgecombe, *Nature*, *198* (1963) 1095.
317 A. Cesca, S. Arrighetti, A. Priola, P. V. Curanti and M. Bruzzone, *Makromol. Chem.*, *175* (1974) 2539.
318 C. C. Loo and C. C. Hsu, *Can. J. Chem. Eng.*, *52* (1974) 374, 381.
319 D. Schnell and G. Fink, *Angew. Makromol. Chem.*, *39* (1974) 131.
320 V. A. Zakharov, G. D. Bukatov, V. K. Dudchenko, A. I. Minkov and Y. I. Yernakov, *Makromol. Chem.* *175* (1974) 3035.
321 V. A. Gregor'ev, V. I. Pilipovskii, Z. V. Arkhipova, B. V. Erofeev and G. A. Fedina, *Int. Polym. Sci. Tech.*, *1*(5) (1974) T.73.
322 P. Locatelli, A. Immirzi and A. Zambelli, *Makromol. Chem.*, *176* (1975) 1121.
323 A. Zambelli, I. Pasquon, H. Signorini and G. Natta, *Makromol. Chem.*, *112* (1968) 160.
324 R. E. Cunningham and R. A. Dove, *J. Polym. Sci.*, *Part A1*, *6* (1968) 1751.
325 D. C. Perry, F. S. Farson and E. Schoenberg, *J. Polym. Sci.*, *Part A1*, *13* (1975) 1071.

Chapter 4

Polymerization of Cyclic Ethers and Sulphides

P. DREYFUSS and M. P. DREYFUSS

1. Introduction

It is just 10 years since Plesch [1] noted that the time may soon be ripe for a comprehensive treatment of oxonium ions. Indeed a book [2] about oxonium ions in chemistry has appeared since then. In addition, a wealth of work has confirmed their importance in certain cyclic ether polymerizations. Another body of work has shown that other cyclic ether polymerizations contain different active species such as carbanions or coordinate complexes. More recently the studies have been extended to cyclic sulphide compounds and an independent body of knowledge related to the polymerization of cyclic sulphur compounds has accumulated. This work has served as the source material for numerous reviews [3—12] covering hundreds of pages documented with references to a thousand or more articles. Much of the work involves kinetics carried out with varying degrees of rigour. In our opinion an attempt to include all of this work within one chapter would not only be a monumental task but would produce a report whose significance would be buried under innumerable details. Hence we have attempted to be comprehensive only in the sense that we have tried to include all types of polymerization that cyclic ethers and cyclic sulphides are known to undergo and to touch upon the many kinds of initiation that have been used. In addition, we have tried to summarize current thinking and controversies about mechanisms and kinetics of polymerization of these compounds. We have included the kinetic results obtained for the most thoroughly studied monomers and a few lesser ones also. However, we have left it to the reader to trace the historical development of these ideas through the many excellent reviews and papers to which we have only given reference.

2. Epoxides and episulphides

A truly vast literature has accumulated on the polymerization of the epoxides and episulphides. This is probably the result of their far reaching commercial importance and of their high reactivity and the multiplicity of possible reaction pathways. The investigation of their polymerization appears to continue at an almost frantic pace still today. New catalyst combinations are continually being reported; new studies, new findings, new insight into well established catalyst systems continue to appear. It is inevitable then that in this chapter, we can just survey the kinetics of epoxide and episulphide polymerizations in a selective manner.

2.1 EPOXIDES

The epoxides (alkylene oxides, 1,2-epoxyalkanes) are unique among cyclic ethers in that polymerization will occur by an anionic, a cationic, or a coordinate mechanism. All three mechanisms are not equally important, nor are they all commercially important. Still all mechanisms have been studied extensively. This discussion will be presented on the basis of type of polymerization mechanism.

2.1.1 Anionic polymerization

The base catalysed polymerization of epoxides is probably one of the oldest polymerization systems known [13]. The chemistry, an alkoxide reacting with an epoxide to form a new alkoxide one monomer unit longer, is very simple

$$RO^- + \quad \overset{\diagdown}{}C\!\!-\!\!C\overset{\diagup}{} \quad \longrightarrow \quad R\!-\!O\!-\!\overset{|}{C}\!-\!\overset{|}{C}\!-\!O^-$$

A number of studies of the kinetics and mechanism of the base catalysed reaction of epoxides with phenolic alcohols have served as background for the polymerization studies. These studies [14] showed that both the alcohol and the alkoxide participate in the rate determining step and subsequently a termolecular mechanism was proposed.

Gee et al. [15—17] studied the kinetics of the polymerization of the base initiated polymerization of ethylene oxide (EO). The initiator was sodium methoxide (NaOMe) containing a small amount of free methanol. The rate equation

$$-\mathrm{d[EO]}/\mathrm{d}t = k_2 C_0 [\mathrm{EO}] \tag{1}$$

where C_0 is the initial concentration of NaOMe, held over a wide range of initiator and monomer concentrations. The Manchester group [15] feels that there is no evidence to suggest that the reaction is termolecular. Kinetic orders less than one in catalyst concentration are explained by incomplete dissociation of the catalyst ion pair. The dependence of the rate on free alcohol concentration is thought to be due to the effect of the alcohol on the degree of dissociation of the catalyst ion pair. At 30°C the second order rate coefficient, k_2, for the polymerization of EO in 1,4-dioxane with NaOMe initiator and low alcohol concentrations was found to be about 2×10^{-5} l mole^{-1} sec^{-1} or about one tenth that with the free ion. The reaction appeared to be a "living" polymerization; no termination reaction was detected. From measurements at other temperatures an activation energy of 17.8 kcal mole^{-1} was calculated.

A similar study was carried out by the same group [17] on propene oxide (PO) polymerization. In this case unsaturation was developed in the polymer due to a transfer reaction, but for the most part a standard head to tail polymerization occurred. At 80°C using NaOMe initiator and essentially bulk monomer, the second order rate coefficient reported was approximately 2.6×10^{-4} l mole^{-1} sec^{-1}. The activation energy in this case was 17.4 kcal mole^{-1}.

However, Ishii et al. [14] feel that a termolecular mechanism is a better explanation of the data and they show that the expression (2) applied for EO polymerizations in the presence of a large excess of alcohol, viz.

$$-d[EO]/dt = k_3 C_0 [EO][ROH]_0 \tag{2}$$

where $[ROH]_0$ is the initial concentration of free alcohol. The third order rate coefficient, k_3, values varied from 1.3 to 6.5×10^{-2} kg^2 mole^{-2} sec^{-1} for reaction with EO at 40°C for a variety of alcohols. Generally, primary alcohols were more reactive than secondary ones. But all of the β-alkoxy-ethanols were about equally reactive; the k_3's were all approximately 3.3×10^{-2}. Thus, subsequent addition after initiation, as would be expected from the mechanism, seems to follow the same kinetics regardless of the initiating alcohol. The activation energy was found to be 14 kcal mole^{-1}. The corresponding k_3 values for reaction with PO were approximately 5—6.7×10^{-3} kg^2 mole^{-2} sec^{-1} and the activation energy was again 14—15 kcal mole^{-1}.

By studying several epoxides Gee et al. [15] were able to show the large steric effect of alkyl substitution on the rate of reaction. The second order rate coefficient for attack at a $-CH_2-$ (in EO or in PO), k_2, was about 10×10^{-5} l mole^{-1} sec^{-1} at 30°C while for attack at R$\overset{|}{C}$H k_2 was about 0.3—0.6×10^{-5} l mole^{-1} sec^{-1} at 30°C. Under their conditions the $-R_2 C-$ site (for example in iso-butene oxide) was completely unreactive.

A study of the base catalysed polymerization of EO was carried out by Wojtech [18]. He observed that polymer of molecular weights up to 50,000 could be obtained with this system. Optimum molecular weight required the ratio ROH/RONa to be one. Price and Carmelite [19] have also shown that molecular weights greater than 40,000 are possible using potassium t-butoxide as initiator. This is true only for EO; using PO results in molecular weights of about 1200. The rate of PO polymerization was pseudo first order. The second order rate coefficient at 30°C was 2.5×10^{-4} l mole^{-1} sec^{-1}, while for 3,3,3-trideuterated PO the rate coefficient was 1.2×10^{-4} l mole^{-1} sec^{-1}. This difference was unexpected and no explanation was offered.

References pp. 326—330

Bawn et al. [20] measured the rate of EO polymerization in dimethyl-sulphoxide (DMSO) initiated by potassium t-butoxide. At low conversions they found that the polymerization obeyed the rate expression

$$R_p = k_p [EO] [K^+O^-Bu\text{-}t] \tag{3}$$

where $k_p = 0.10 \pm 0.03$ l mole^{-1} sec^{-1}, $\Delta H_p^{\ddagger} = 6.4$ kcal mole^{-1} and $\Delta S_p^{\ddagger} = -41$ kcal mole^{-1} deg^{-1} at 25°C. They found that hydroxylic impurities had a marked effect on reaction rates. At higher conversions or with PO as the monomer the polymerizations became heterogeneous. These workers felt that the dimethylsulphoxide participated directly in the initiation reaction.

An interesting stereochemical analysis of the polymerization of t-butyl-ethylene oxide (t-BuEO) by potassium t-butoxide has been made [21]. The suggested scheme, which involves participation of the potassium ion in the propagation step makes this polymerization more coordinate anionic than anionic. At the very least it appears that polymerization occurs by way of a tight ion pair.

In a companion paper Price and Akkapeddi [22] report the kinetics of base initiated polymerization of epoxides in DMSO and hexamethyl-phosphoramide (HMPT). The initiator is potassium t-butoxide. Second order rate coefficients for (R,S)—PO were about double those for (+)—(R) or (—)—(S) monomer. They conclude that the steric factor favouring alternation of isotactic and syndiotactic placement of the t-BuEO also influences PO. Chain transfer to solvent (DMSO) was also studied. For PO polymerization in DMSO they obtain $k = 1.5 \times 10^9 \exp(-17,200/RT)$. However, due to some erratic results they are not very confident about the accuracy. In HMPT rates are about three fold faster than in DMSO; $k = 7.3 \times 10^8 \exp(-16,300/RT)$. Three other epoxides were also studied in HMPT: EO, $k = 2.75 \times 10^7 \exp(-13,300/RT)$; t-BuEO, $k = 2.0 \times 10^8 \exp(-17,100/RT)$; phenylglycidyl ether (PGE), $k = 5.4 \times 10^7 \exp(-14,700/RT)$ l mole^{-1} sec^{-1}. Thus, the approximate comparison of reaction rates is (R,S)—PO $\simeq 5$ EO $\simeq 3$ PGE $\simeq 0.1$ t-BuEO.

Price and Akkapeddi [22] note that their value for the rate of EO polymerization in HMPT is about 40-fold less than that reported for DMSO by Bawn et al. [20]. Price and Akkapeddi point out that Bawn et al. report initial rates rather than steady state rates. However, they suggest that the more important reason for the difference in rate is the large difference in catalyst concentration. The higher concentration used by Price and Akkapeddi would lead to greater association of alkoxides and to lower rate coefficients.

St. Pierre and Price [23] have also shown that anhydrous KOH is an interesting initiator of epoxide polymerization. Anhydrous RbOH and CsOH were also effective but, interestingly, NaOH and LiOH were not.

Bar-Ilan and Zilkha [24] have attempted to study the kinetics of anionic polymerization of EO initiated by anhydrous KOH. Initiation was slow and inefficient and no direct measurement of the number of active sites was made. As we shall see in later sections, such complications need to be overcome before really meaningful kinetic parameters can be obtained.

Alkali metal naphthalene complexes have also been used to initiate epoxide polymerizations. Solov'yanov and Kazanski [25] studied the polymerization of EO in tetrahydrofuran using sodium, potassium or cesium naphthalene as initiator. A "living" polymer was produced; there is no chain rupture or transfer. The rate of polymerization depends on the concentration of active centres in a complex manner. The kinetic order varies from 0.23 for Na^+ (or 0.33 for K^+ and Cs^+) up to full first order as initiator concentration decreases. The polymerization is first order in monomer, but deviations are observed at high concentrations.

2.1.2 Cationic polymerization

As we will see, cationic polymerization is the only polymerization mechanism open to the higher cyclic ethers. It is also available to epoxides, but the reactions are complex. The mechanism has not been applied commercially since the polymers which result are only of rather low molecular weight.

There have been some extensive studies of the cationic polymerization of EO [1] and of epichlorohydrin (ECH) [26]. Neither of these efforts has resulted in a clear kinetic picture. The reason seems to be that a very complex series of reactions occurs and a sensible kinetic analysis just has not emerged.

Eastham [27] studied the polymerization of EO using as initiator BF_3, $BF_3 \cdot Et_2O$ and $Et_3O^+BF_4^-$. The major product of these reactions is a depolymerization product, the 6-membered cyclic dimer, 1,4-dioxane. In addition some low molecular weight (600—700) polymer is formed. Attempts to raise the molecular weight led only to formation of more dioxane. If catalyst is added to high molecular weight polyethylene glycol (otherwise obtained), dioxane is again formed and the molecular weight is degraded to approximately the same level (600—700). Extensive kinetic studies of both the polymerization and depolymerization process have been carried out. A really satisfactory picture has not emerged. The system is simply too complex. In addition, since 1,4-dioxane does not homopolymerize, it could be argued that at equilibrium all the EO should be converted to 1,4-dioxane. The apparent equilibrium 600—700 molecular weight polymer that is formed is a mystery and has not been adequately explained.

More recently Estrin and Entelis [26] have published a series of papers on the polymerization of ECH and glycidyl nitrate initiated by BF_3 and $BF_3 \cdot Et_2O$. They have attempted to explain the low molecular weight products, invariably formed along with an apparent lack of identifiable end groups. Chain transfer obviously is important and they conclude, essentially by a process of elimination, that a large proportion of the polymer formed consists of macro-rings. They do show the coincidence in the gel chromatographs of cyclic tetramer prepared according to Kern [28] and a major portion of their products. Kern has shown that treating PO, ECH or 1-butene oxide with BF_3 or with $Et_3O^+BF_4^-$ resulted in 30—40% conversion to cyclic tetramer; larger rings and cyclic dimer were also formed.

Estrin and Entelis [26] report that the BF_3 catalysed polymerizations of ECH and glycidyl nitrate both show rapid slowing down of polymerization after 20—40% conversion. Kinetics measured at —40 to —70°C were complex. Initial rates had an order of ~1.8 with respect to monomer and were first order with respect to BF_3. Initiation was slow and chain transfer determined the molecular weight of the products. Activation energies determined were approximately 6 kcal mole^{-1} for each monomer.

A later study [26] on just ECH using $BF_3 \cdot Et_2O$ showed broadly similar results. The ether further complicated the picture. The reactions were characterized by slow initiation and an important chain transfer step. The use of $Et_3O^+BF_4^-$ was also briefly investigated. However, appreciable rates were observed only at temperatures $> 35°C$.

2.1.3 Coordinate polymerization

High molecular weight widely useful polymers were not made from epoxides until catalysts, generally referred to as coordinate catalysts, based on various metal alkoxides were discovered. The original report was by Pruitt and Baggett [29]. They reported obtaining a high molecular weight polypropene oxide using a catalyst formed from $FeCl_3$ and PO. Since then, many other catalyst systems have been reported. Most of these are based on either zinc or aluminium (derived by reaction of metal alkyl and water) but the effectiveness of many other metals has also been shown. Often these catalysts are specific for the polymerization of a particular epoxide and are completely ineffective for all other epoxides. Thus the steric interaction between the catalyst complex and the monomer becomes extremely important. Also it becomes a real challenge for the catalyst chemist to design effective systems for certain desired copolymerizations. The scope and nature of many of these various catalysts have been summarized and reviewed [4, 9, 14].

A further significant factor which comes to the fore in coordinate catalysis is stereoregular polymerization. A number of the catalyst systems are capable of producing isotactic, optically active and/or crystalline polymers. Except for Price's potassium t-butoxide system [21], this has not been observed in anionic or cationic polymerization. Thus, in addition to higher molecular weights, an important new class of epoxide polymers, the stereoregular polymers, now becomes available.

There have been a great many studies on various coordinate type polymerizations of epoxides. Almost as many suggestions have been made for the polymerization mechanism and the nature of the active sites. In some cases a monomolecular metallic propagation site was proposed and in others a bimolecular propagation process (generally involving metal—oxygen—metal sites) was assumed. In addition, it has not yet been possible to measure the number of active sites produced and so the significance of much of the kinetics is reduced. Finally, two distinct processes, coordinate cationic and coordinate anionic, are proposed depending on the particular metal and the type of coordinate catalyst being studied. Whether these are significant distinctions is still a point to be resolved.

(a) The iron catalysts

These were the first catalysts in this group discovered to give extremely high molecular weight polypropene oxide [29]. The catalysts are made by reacting ferric chloride with PO; thus they are some form of a ferric alkoxide species. Addition of small amounts of water [30, 31] increased the polymerization rate and gave a more crystalline product. The formation of FeOFe as polymerization sites was suggested in this latter case. Kinetic studies have shown first order rates with respect to both monomer and catalyst. On this basis, Gee et al. [31] concluded that the rate controlling step probably involved complexation of the monomer on the catalyst site. No definitive results seem to have emerged in spite of rather considerable effort. Most studies have involved PO as monomer, but the polymerization of other epoxides has also been reported.

(b) The zinc catalysts

This is another important group of catalysts, widely studied, especially for propene oxide polymerization [4, 9, 14]. Whereas cadmium and magnesium catalysts have also been studied and shown to be active, the zinc catalysts appear to be the most important of the Group II metal catalysts. They are formed by reaction of a zinc alkyl, most often diethyl zinc, with an active hydrogen compound such as water or alcohol or with oxygen or sulphur. Zinc alkoxy alkyls are formed. Other products proposed are zinc—oxygen—zinc bonds as well as coordination of this material to dimer, trimer and higher oligomers. A few of these species have been rather well defined, but the question unfortunately remains: are these the active polymerization sites or are they altered to polymerization sites by interaction with monomer?

Kinetic studies [32] with a diethylzinc/water catalyst showed two distinct rates. A relatively fast rate in the first 5—10% of polymerization was followed by a first order rate with respect to monomer. Addition of more monomer after all the initial charge was consumed gave further polymerization. Thus the character of a "living" polymer was indicated. Still a broad molecular weight distribution and the production of low molecular weight polymer throughout the polymerization was observed. Two catalyst species, indicated by two rates and two types of polymers, is not an uncommon observation in coordinate epoxide polymerizations.

A more recent kinetic study of PO polymerization catalysed by the related diethylzinc/alcohol system has been reported by Furukawa and Kumata [33]. They conclude that the kinetic features of the polymerization are best interpreted by assuming a dimerization equilibrium of zinc alkoxide and reversible association of the catalyst dimer with PO to form the active species for polymerization. Their conclusions are based on their observation that the dependence of the overall rate on both monomer and catalyst can vary from one to two, depending on the method by which the polymerization is carried out.

An important and significant amount of work has also been done on the asymmetric polymerization of epoxides. Much of this work was carried out with catalysts derived from diethylzinc and diethylmagnesium. The asymmetry was introduced by using optically active alcohols in the synthesis of the catalyst species. Alternately, optically active monomers were polymerized using optically inactive catalysts. A detailed discussion of these studies is beyond the scope of this chapter and the reader is referred to reviews of the subject [9, 14].

(c) The aluminium catalysts

As in the case of the zinc catalysts, active catalysts are formed by reaction of alkyl aluminium compounds with water. It is generally felt that since aluminium compounds are usually fairly strong Lewis acids, the catalysts also are somewhat more acidic in nature. Thus a coordinate cationic mechanism is generally favoured for these polymerizations. In contrast, a more anionic coordinate mechanism is usually suggested for the zinc catalysts. In fact, as will be seen in the discussion of the higher cyclic ethers, some of these catalysts are distinctly able to initiate true cationic polymerizations. However, the catalysts under discussion here as applied to epoxides are clearly considered to be coordinate.

An important step seems to be the coordination of the monomer on the metal catalyst site. Vandenberg [34] concludes that two or more aluminium atoms must be involved in the polymerization. The propagation step is then supposed to involve some charge separation and the development of a partial or full positive charge on the monomer while the counter anion is located on the catalyst complex.

In the case of the aluminium alkyl/water catalysts important differences are observed [34] if the aluminium alkyl/water reaction is carried out in a hydrocarbon solvent or in ether solvent. Further important modifications were made by the addition of a chelating agent (such as acetylacetone). Even further modification by addition of some alcohol was reported. These differences in the method of catalyst preparation are clearly seen in the catalytic activity toward certain monomers or pairs of monomers in copolymerizations and in the type and yield of polymer obtained. However, the differences in the catalyst structures, that is, just why they behave as differently as they do, has by no means been resolved.

Again, kinetic investigations using these catalysts have been rather limited to broad studies and to those required for the development of commercial processes. Good, definitive kinetic studies do not seem to have been possible as yet. For more detailed information on these catalysts and the polymerizations the reader is again referred to published reviews [4, 14, 34].

2.2 THIIRANES

These sulphur analogs of the 1,2-epoxides are variously known by the names thiiranes, thiacyclopropanes, alkylene sulphides, and episulphides. They are known to polymerize by the three catalytic mechanisms discussed for the oxygen compounds. Thus the same divisions are possible. For a discussion of some of the mechanistic aspects of these polymerizations as compared to epoxides, the reader is referred to a recent paper by Vandenberg [35].

2.2.1 Anionic polymerization

The anionic polymerization of thiiranes appears to proceed very clearly and in a well defined manner. Living systems have been found and good kinetic measurements have been possible [36]. The base catalysed polymerization of ethylene sulphide (ES) was probably one of the earliest thiirane polymerizations noted. However, the studies have been hindered because the polymer formed is highly crystalline and insoluble. Homogeneous solution studies have not been possible. Propene sulphide (PS) on the other hand gives soluble amorphous polymer and has been amenable to careful study. Indeed the formation of amorphous polypropene sulphide is the major evidence that base catalysed polymerization is ionic and proceeds by the reactions indicated by the equations

$$B^- + CH_2-CH-CH_3 \longrightarrow B-CH_2-\overset{\overset{\displaystyle CH_3}{|}}{CH}-S^-$$

$$B-CH_2-\overset{\overset{\displaystyle CH_3}{|}}{CH}-S^- + nCH_2-CH-CH_3 \longrightarrow B-CH_2-\overset{\overset{\displaystyle CH_3}{|}}{CH}-S-(CH_2-\overset{\overset{\displaystyle CH_3}{|}}{CH}-S)_n^-$$

The transfer reaction observed in the base catalysed polymerizations of PO does not occur with the sulphide analog and high molecular weight polymers are observed. In fact, under conditions of high purity [37] no termination reactions were observed and the polymerization had all the characteristics of a "living" system. The relationship

$$\overline{DP}_n = [M] / \tfrac{1}{2} [C]$$

was shown to hold fairly well. $\overline{DP}_n$ is the degree of polymerization, [M] is the initial monomer concentration, and [C] is the concentration of the initiator. Thus the presence of two active centres per growing chain was indicated. The initiator used was sodium naphthalene and the initiation mechanism proposed was

The presence of free naphthalene was demonstrated [38]. However, the scheme did not explain all the experimental data.

A kinetic study of this sodium naphthalene initiated polymerization of PS was made by Sigwalt et al. [39]. Low molecular weight living polymer was used in order to circumvent the problem of an observed induction period. If the concentration of living ends [C] was $<10^{-3}$ mole l^{-1}, the kinetics followed the equation

$$R_p = k_p [C]^{1/2} [M]$$

It was shown that an ion pair–free (active) ion equilibrium was the reason the rate expression took this form. Thus these workers demonstrated that the relationship

$$k_p' = k_{NaS}' + \frac{k_{S^-}''}{[C]^{1/2}} (K_D)^{1/2}$$

applied. Here, k_{NaS}' is the propagation rate coefficient for the ion pair, while k_{S^-}'' is the free ion coefficient; k_p' is the apparent first order rate coefficient and K_D the dissociation constant for the ion pair–free ion equilibrium. These authors did not obtain a precise measurement of k_{NaS}' because it is so small (approximately 7×10^{-3} l mole^{-1} sec^{-1}) (see Fig. 1). K_D was measured [40] and found to be 5.1×10^{-9} mole l^{-1}. Thus they

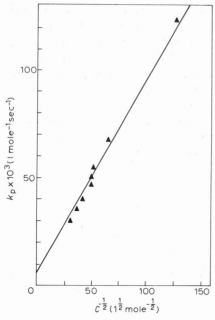

Fig. 1. Determination of the value of the propagation rate coefficients of free ions and of ion pairs in PS polymerization at $-30°C$ [39].

deduced that k''_{S^-} is about 12 l mole^{-1} sec^{-1} $(-30°C)$. It can be seen that $k''_{S^-} > 1000 k'_{NaS}$ and therefore the propagation proceeds essentially completely through free ions.

In a more recent study Sigwalt et al. [41] investigated the use of carbazyl sodium to initiate PS polymerization in tetrahydrofuran (THF). This initiator produced only one living end per chain. Excellent agreement with the results obtained on amphianionic polymer was obtained. At levels of living ends below 10^{-3} mole l^{-1} the rate coefficient found for ion pairs was 3×10^{-3} l mole^{-1} sec^{-1} and for free ions 4 l mole^{-1} sec^{-1}. The dissociation constant (determined kinetically) was 6.4×10^{-8} mole l^{-1}.

It should be noted that thiiranes polymerize under these basic conditions much more rapidly than the corresponding epoxides. Direct comparison of the rates is not possible because the same temperature range could not be used.

2.2.2 Cationic polymerization

It has long been known that acids will initiate a thiirane polymerization reaction. By analogy to the oxygen compounds it was reasonably suggested that propagation occurs via a cyclic sulphonium ion. The relatively greater ease of formation of a sulphonium ion compared to an oxonium ion, its greater stability, and, in fact, recent experimental work, are entirely consistent with the formation of cyclic sulphonium ions.

Goethals et al. [42] have studied the polymerization of PS and of thietanes (see Section 3.3.1) initiated by $Et_3O^+BF_4^-$. To date careful kinetic investigations have been largely carried out on the thietanes. The polymerization of PS is somewhat different in that a two stage conversion—time curve is observed. An initial very fast reaction, "A-stage", is non-stationary and is followed by a stationary slow "B-stage". The B-stage leads to quantitative yields of polymer. The authors propose that the A-stage is typical and proceeds, as in the case of thietane, by

$$CH_3-CH-CH_2 + Et_3O^+BF_4^- \longrightarrow Et_2O + Et-S+\underset{BF_4^-}{\overset{CH-CH_3}{\underset{CH_2}{\big|}}}$$

$$Et-S+\underset{CH_2}{\overset{CH-CH_3}{\big|}} + nPS \longrightarrow Et-(S-\overset{CH_3}{\underset{|}{CH}}-CH_2)_n-S+\underset{CH_2}{\overset{CH-CH_3}{\big|}}$$

However, in contrast to thietane, the termination reaction here seems to be monomolecular. Assuming monomolecular termination, they derive the relationship

$$\ln([M]_0/[M]_f) = (k_p/k_t)[C]_0$$

where $[M]_0$ is the initial monomer concentration, $[C]_0$ the catalyst concentration, k_p and k_t the rate coefficients of propagation and termination respectively, and $[M]_f$ is the monomer concentration at the

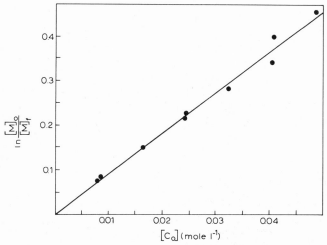

Fig. 2. Determination of k_p/k_t for the polymerization of PS [42].

end of the A-stage polymerization. The data fits the equation (see Fig. 2) and the slope of the line gives $k_p/k_t = 9.1$ l mole^{-1}. This analysis assumes that the initiation reaction is complete and immediate.

Goethals et al. [43] had earlier shown that a low molecular weight oligomer was formed in cationic PS polymerization. This oligomer was isolated and identified as the cyclic tetramer of PS. The formation of cyclic tetramer could be involved in the two stage polymerization under discussion. Goethals et al. [42] propose that in this case the termination reaction is the formation of the sulphonium ion leading to cyclic oligomer. The free tetramer is formed and polymerization is reinitiated by attack of monomer on this macrocyclic sulphonium ion. The B-stage polymerization is interpreted as this reinitiated polymerization, viz.

$$\sim CH_2-S-\overset{\overset{\displaystyle CH_3}{|}}{CH}-CH_2-S-\overset{\overset{\displaystyle CH_3}{|}}{CH}-CH_2-S-\overset{\overset{\displaystyle CH_3}{|}}{CH}-CH_2-\overset{+}{S}\big\langle{}^{CH-CH_3}_{CH_2} \xrightarrow{\text{termination}}$$

$$\overset{CH_3\diagdown}{\underset{}{}}\xrightarrow{\text{PS reinitiation}} -CH_2-\overset{+}{S}\big\langle{}^{CH-CH_3}_{CH_2} + (PS)_4$$

$$\downarrow n\text{PS}$$

polymer

2.2.3 Coordinate polymerization

The coordinate type catalysts are also effective for thiirane polymerizations. The types of systems used are also similar. Thus diethylzinc and in particular diethylzinc/water mixtures have been studied [44]. Other studies made using triethylaluminium and diethylcadmium indicated that these metal alkyls all behave similarly. The reactions seem to be rather complex, and, as also was the case with the epoxides, no well defined kinetic studies have appeared. The polymers produced are of high molecular weight and are often crystalline. Thus stereospecific polysulphides have been reported. Again the bulk of the studies involve PS. Stereoselective polymerization of racemic monomer has been accomplished [45, 46] using a catalyst prepared from diethylzinc and (+) borneol. The marked difference between PO and PS in their polymerization behaviour in response to diethylcadmium as initiator is worth noting [9]. Neither diethylcadmium/alcohol nor diethylcadmium/water induces PO polymerization, whereas these organo-cadmium systems easily polymerize PS to high molecular weight polymers. The response of PS-2-d to cadmium tartrate is particularly impressive as far as stereoselectivity is concerned. The polymer formed had at least 95% isotactic diad [47]. Optically active polymers have also been prepared by polymerizing optically active monomer [9, 48, 49].

3. Oxetanes and thietanes

Addition of one more —CH$_2$— group to the ring of the cyclic ether or sulphide changes its properties rather markedly. The oxygen atom in oxetanes becomes decidedly more basic and the ring itself becomes inert to attack by base. In fact polymerization is observed only by the cationic mechanism. A variety of substituents can be put on the ring without inhibiting polymerization. For the oxygen compounds these include bis-chloromethyl, halomethyl, dimethyl, benzenesulphonylmethyl ether, cyano and phosphorus substituents. In the sulphur series, thietanes with dimethyl and diethyl substituents, among others, have been polymerized.

3.1 OXETANES

A large number of substituted 1,3-epoxides as well as the parent oxetane (oxacyclobutane, trimethylene oxide, 1,3-epoxypropane) have been prepared and polymerized. Likewise a wide variety of substances have been used to initiate their polymerizations. Much of this work has been extensively reviewed previously [1, 3, 7] and the interested reader is referred to these earlier reviews. Here we confine ourselves to reporting representative major kinetic studies. In this section the organization is by monomer.

3.1.1 Oxetane

Rose [50] carried out one of the earliest, really thorough investigations of the kinetics of polymerization of a cyclic ether polymerization. He studied oxetane polymerizations initiated by BF$_3$. Rose was aware from the work of Farthing and Reynolds [51] that polymerization does not occur when BF$_3$ comes into contact with pure, dry monomer. However, simultaneous addition of water, ethanol or hydroxy terminated polymer is sufficient to initiate polymerization. Since Rose [50] observed polymerization in his sytem, he assumed that it was not completely dry and discussed his results assuming water is a co-catalyst. As the concentration of water increased, the rate of polymerization at first increased. The rate was first order with respect to monomer at this stage. With further increase of water, the rate passed through a maximum and finally decreased markedly. At the same time, the order of the reaction with respect to monomer changed from 1 to 2. The intrinsic viscosity of the polymer decreased steadily as the concentration of co-catalyst was increased.

The mechanism proposed was

(1) $\langle\!\!\!\square\!\!\!\rangle$ O . BF$_3$ + H$_2$O $\xrightarrow{k_i}$ $\langle\!\!\!\square\!\!\!\rangle$ +OH BF$_3$OH$^-$ $\xrightarrow{k_i}$ $\langle\!\!\!\square\!\!\!\rangle$ +O—(CH$_2$)$_3$OH
$^-$BF$_3$OH

(I)   (II)

(2) H[O(CH$_2$)$_3$]$_n$—Ȯ+$\langle\!\!\!\square\!\!\!\rangle$ + O$\langle\!\!\!\square\!\!\!\rangle$ $\xrightarrow{k_p}$ H[O(CH$_2$)$_3$]$_{n+1}$—Ȯ+$\langle\!\!\!\square\!\!\!\rangle$
$^-$BF$_3$OH $^-$BF$_3$OH

(3) H[O(CH$_2$)$_3$]$_{n-1}$—Ȯ+$\langle\!\!\!\square\!\!\!\rangle$ + HOH $\xrightarrow{k_{tr}}$ H[O(CH$_2$)$_3$]$_n$—OH + H$^+$ + BF$_3$OH$^-$
$^-$BF$_3$OH

(4) H[O(CH$_2$)$_3$]$_{n+3}$—Ȯ+$\langle\!\!\!\square\!\!\!\rangle$ $\xrightarrow{k_t}$ HO[O(CH$_2$)$_3$]$_n$—Ȯ+$\overset{\displaystyle /(CH_2)_3O(CH_2)_3\backslash}{\underset{\displaystyle \backslash(CH_2)_3O(CH_2)_3/}{}}$O
$^-$BF$_3$OH $^-$BF$_3$OH

The complex effect of water required the two step initiation process. The secondary oxonium ion (I) is much less reactive than the tertiary oxonium ion (II). The chain propagates by a nucleophilic attack of the monomer oxygen on the α-carbon of the tertiary oxonium ion according to an S$_N$2 mechanism. BF$_3$OH$^-$ is the counter-ion presumed to be formed in the initiation and is then carried along in the rest of the mechanism. The transfer step is included to explain the decrease of intrinsic viscosity as water concentration increases. The product of the termination reaction is strainless and was, therefore, presumed to be relatively unreactive until it came into contact with a terminating agent. Cyclic tetramer is then formed. Rose further points out that in addition to formation by reaction (4), cyclic tetramer can be produced without the prior formation of polymer. The reader may recall the analogous formation of cyclic oligomers in the cationic polymerization of epoxides and episulphides (Sections 2.1.2 and 2.2.2).

References pp. 326—330

Rose [50] derived kinetic equations to fit this mechanism and the experimental data. When the concentration of added water was below 0.1 of the BF_3 concentration, the kinetics were expressed by

$$-d[M]/dt = (k_i k_p/k_t)[M][BF_3]([H_2O] + C)$$

where $[BF_3]$ and $[H_2O]$ are the concentration of catalyst and water actually added to the system. The constant C was associated with the concentration of water present due to incomplete drying. In the presence of relatively large quantities of water, the rate of polymerization started to decrease and the molecular weight decreased. The rate became dependent on the square of the monomer concentration. The mechanism was then more complex.

Rose also derived heats of polymerization. The values he obtained were 20.0 kcal mole^{-1} for polymerization in methyl chloride solution at $-20°C$ and 19.3 kcal mole^{-1} for polymerization in a mixture of ethyl chloride and methyl chloride at $-9°C$. It is at once a bit unfortunate that this elegant piece of work was carried out as early as 1956 before catalysts leading to less complicated kinetics were discovered, and a tribute to Rose's careful work that he was able to sort out the many complications involved. No other detailed kinetic study of the polymerization of oxetane was made until many years later. In 1971 Saegusa et al. [52], reported a kinetic study of the polymerization of oxacyclobutane initiated by a BF_3—THF complex.

In order to fully appreciate the kinetic results obtained for oxetane polymerizations it is necessary to anticipate from a strictly historical point of view some of the work with THF. Many of the initiators used in cyclic ether polymerizations are not 100% efficient in producing active sites. In fact initiator efficiency is most often in the 20—80% range. A further complication is that the rate of initiation is frequently slow, often slower than the rate of propagation. Hence there were many instances in which the number of active sites was changing throughout the polymerization and the monomer was consumed before all the initiator was consumed. Therefore, for a long time many of the derived kinetic parameters were of questionable significance because of doubts about the number of active centres actually present at a given time during the polymerization.

There are two general methods, one direct and the other indirect, by which the concentration of growing ends can be determined [8]. In the direct method, certain physical properties of the system, such as conductivity or UV or NMR spectra, are measured during the polymerization. In the indirect method, the polymerization is quenched with a reagent that combines with the growing chain ends, and gives a product which can be conveniently measured after purification.

In 1968 Saegusa and Matsumoto [53] introduced an ingenious indirect method for determining the instantaneous concentration of propagating species at time t, [P*]. Their method consists of quantitatively converting the propagating oxonium ion into the corresponding phenyl ether by treatment with excess sodium phenoxide and then determining the

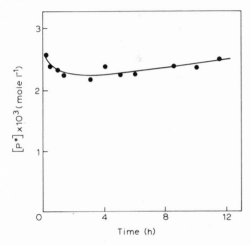

amount of the resulting phenyl ether by UV spectroscopy. Saegusa et al. are careful to point out that before [P*] can be determined by their method, these requisites must be satisfied:

(a) The molar extinction coefficient of the phenyl ether group at the

Fig. 3. Variation of [P*] with time in polymerization of oxetane by $BF_3 . THF$ [52].

end of the polymer chain must be known. This has been estimated in their work from the ϵ_{max} of phenyl alkyl ethers.

(b) The conversion of the propagating species into phenyl ether must be quantitative and instantaneous. Again these workers used model compounds to demonstrate that this requirement is reasonably well satisfied with their systems.

(c) All possible side reactions producing phenyl ether groups should be absent. A series of reference experiments were carried out to establish that side reactions are absent.

In their 1971 work Saegusa et al. [52] measured the change of [P*] in the course of oxetane polymerization at $-27.8°C$ as shown in Fig. 3. [P*] remains almost constant during the polymerization and is equal to about 80% of the initial concentration of $BF_3 . THF$. Similar [P*]—time relationships were observed at other temperatures. A propagation scheme

References pp. 326—330

similar to that proposed by Rose above led to a second order rate equation given by

$$-d[M]/dt = k_p[P^*][M]$$

Integration with respect to time, t, gives

$$\ln([M]_{t_1}/[M]_{t_2}) = k_p \int_{t_1}^{t_2} [P^*]\, dt$$

where $[M]_{t_1}$ and $[M]_{t_2}$ are the monomer concentrations at times t_1 and t_2 respectively. The cumulative value of $[P^*]$ was obtained by graphical integration of the time—$[P^*]$ curve. A typical plot of $\ln([M]_{t_1}/[M]_{t_2})$

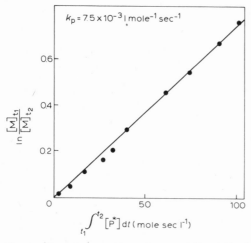

Fig. 4. Use of eqn. (8) to determine rate coefficient of polymerization of oxetane by $BF_3 \cdot THF$ [52].

versus the integrated value of $[P^*]$ from t_1 to t_2 is shown in Fig. 4. Table 1 shows the k_p values observed at four different temperatures. From these k_p values the relationship $k_p = 5.3 \times 10^{10} \exp(-14{,}000/RT)\, l\, mole^{-1}\, sec^{-1}$ was derived.

The work was later extended to include the effect of changing the solvent from the polar one, CH_2Cl_2 (dipole moment, 1.14 D), to the non-polar one, methylcyclohexane (dipole moment, zero) [54]. These data are also shown in Table 1. Now the catalyst efficiency which had been 80% in CH_2Cl_2 was reduced to 28—36% but the k_p was substantially increased, about nine-fold. Saegusa et al. [54] explain the effect on the rate in terms of relative solvation of the transition state that can be assumed to form in the S_N2 propagation reaction. They suggest that the differences in $[P^*]$ may somehow be related to low concentrations of water, comparable to the $[P^*]$ value, that were not removed in the puri-

TABLE 1

Propagation rate coefficients for oxetane polymerized by $BF_3 \cdot THF$ catalyst [52, 54]

In methylcyclohexane		In CH_2Cl_2	
Temp. (°C)	$10^2 k_p$ (1 mole^{-1} sec^{-1})	Temp. (°C)	$10^2 k_p$ (1 mole^{-1} sec^{-1})
−9.0	69	0	14.0
−20.0	18	−10.4	5.7
−30.0	9.8	−22.6	1.3
−40.0	5.4	−27.8	0.75
−50.0	0.96		

fication of the reaction system. In this connection, it should be noted that these polymerizations were carried out under a dry nitrogen atmosphere and not in high vacuum where water can be more rigorously excluded.

3.1.2 3-Methyl and 3,3-dimethyloxacyclobutanes

Rose [50] also investigated the polymerization of 2-methyl oxacyclobutane and of 3,3-dimethyloxacyclobutane enough to learn that they followed a course analogous to that of the parent compound. He also found a heat of polymerization at −9°C of 16.1 kcal mole^{-1} for 3,3-dimethyloxacyclobutane polymerized in a mixture of ethyl chloride and methyl chloride. Again Saegusa et al. [54] have extended the work to clarify the effect on the propagation rate of a methyl substituent at the 3 position of oxetane. They also examined the solvent effects of CH_2Cl_2 and methylcyclohexane for these two substituted monomers.

The kinetic analysis was carried out by the methods described above in the discussion of their oxetane study. [P*] values for these monomers were both solvent and temperature dependent. Propagation rates and some average [P*] values are given in Table 2. No data is included for polymerization of 3,3-dimethyloxetane in CH_2Cl_2 because the resulting polymer is insoluble and kinetic analysis of the heterogeneous system was found to be difficult. The k_p values are compared with those of oxetane in Table 3. Also given in Table 3 are other kinetic parameters. From the data in Table 3, Saegusa et al. conclude that:

(i) The methyl group at the 3 position of the oxetane ring enhances the polymerization rate in both solvents. Two methyl groups seem to be more effective for enhancing the rate than one.

(ii) The free energy of polymerization, ΔG_p, increases in the opposite order as k_p. If ΔG_p is taken as a measure of ring strain, then ring strain does not govern k_p.

(iii) Activation energy, $\Delta E_p^{\ddagger}$ (or $\Delta H_p^{\ddagger}$) changes in the opposite sense to the k_p value. Hence reactivity as measured by k_p cannot be explained in terms of activation energy.

TABLE 2

Propagation rate coefficients for substituted oxetanes polymerized by $BF_3 \cdot THF$ catalyst [54]

| | 3-Methyloxetane | | | | 3,3-Dimethyloxetane | |
| | In methylcyclohexane | | In CH_2Cl_2 | | In methylcyclohexane | |
Temp. (°C)	$10^2 k_p$ (1 mole^{-1} sec^{-1})	Avg. [P*] value (%)[a]	$10^2 k_p$ (1 mole^{-1} sec^{-1})	Avg. [P*] value (%)[a]	$10^2 k_p$ (1 mole^{-1} sec^{-1})	Avg. [P*] value (%)[a]
−9.3	—	—	—	—	680	—
−20.0	92	—	11	20	340	0.7
−30.0	45	5	2.8	—	100	—
−40.0	12	—	0.68	—	25	—
−50.0	4.8	20	—	—	—	—

[a] % of initial catalyst concentration.

TABLE 3
Kinetic, thermodynamic and basicity data for oxetanes [54]

Parameter	In methylcyclohexane			In CH$_2$Cl$_2$	
	(oxetane)	Me–	Me,Me–	(oxetane)	Me–
k_p at −20°C (l mole⁻¹ sec⁻¹)	0.18	0.92	3.4	0.019	0.11
E_p (kcal mole⁻¹)	11.2	11.8	12.6	14.2	16.1
$A_p \times 10^{-10}$ (l mole⁻¹ sec⁻¹)	0.13	1.5	23	5.3	830
$\Delta G_p^\ddagger$ (kcal mole⁻¹)	15.4	14.7	14.2	17.0	15.7
$\Delta H_p^\ddagger$ (kcal mole⁻¹)	10.7	11.3	12.1	13.7	15.6
$\Delta S^\ddagger$ (eu)	−18.6	−14.6	−8.5	−12.1	−1.3
ΔG_p (kcal mole⁻¹)ᵃ	−21.5	−17.7	−14.0		
$\Delta \nu_{OD}$ (cm⁻¹)ᵃ	103	106	99		
Max. [P*] (10³ mole l⁻¹)ᵃ	3.1	2.7	0.5		

ᵃ These values do not refer to methylcyclohexane solvents, but refer only to the monomer indicated in the heading.

(iv) The basicities, as measured by $\Delta\nu_{OD}$, for all three monomers are quite similar and do not change in the same order as the k_p values. Thus, in this case basicity does not govern reactivity.

(v) It appears that the frequency factor A_p (e.g. $\Delta S_p^{\ddagger}$) predominantly governs the reactivity of these cyclic monomers.

These findings represent new and significant conclusions about the relative reactivity of cyclic ethers. Prior to the work of Saegusa et al. [54], most work has suggested that a monomer's reactivity as reflected in its k_p is related to basicity and/or possibly ring strain. More will be said about these results at the end of this review after data for all the cyclic ethers and sulphides have been presented.

3.1.3 3,3-bis(Chloromethyl)oxetane (BCMO)
(a) Initiation by BF_3

Farthing [55] made the earliest report of the polymerization of BCMO. The catalysts used were BF_3 and $BF_3 \cdot Et_2O$. He did not study the kinetics and mechanism of this polymerization.

Penczek and Penczek [56] have reported a study of the kinetics and mechanism of BCMO polymerization in methylene chloride in the temperature range from $-60°$ to $+20°C$. The catalyst used was again BF_3 with water as cocatalyst. They undertook a study analogous to Rose's [50], in spite of the obvious difficulties of investigating kinetics with a system where the polymer precipitates from solution. They were able to show that even under these conditions the kinetics of polymerization are similar to those obtained by Rose for oxetane. The activation energy was 18.0 kcal mole^{-1}. The ratio of k_{tr}/k_p at 25°C for this system was 1.2×10^{-3} where k_{tr} is the rate coefficient for degradative chain transfer to polymer [57]. By degradative chain transfer Penczek and Penczek mean that the nonstrained ion formed as a result of chain transfer to polymer

$$\sim\!\!\!\oplus + O\!\!\begin{array}{c}(CH_2)_n\!-\!O\!\sim \\ (CH_2)_n\!-\!O\!\sim\end{array} \xrightarrow{k_{tr}} \sim\!\!O\!\!\begin{array}{c}\overset{+}{(CH_2)_n\!-\!O}\!\sim \\ (CH_2)_n\!-\!O\!\sim\end{array} \quad \text{for } n > 1$$

has a lower reactivity in the polymerization mixture than the strained cyclic oxonium ion,

$$O^+ \; (CH_2)_n \cdot$$

(b) Initiation by aluminium alkyls

A variety of aluminium alkyls either alone or in combination with additives has been used for the polymerization of BCMO [58]. The characteristics of most of these initiators are similar both in composition and behaviour to those described for the aluminium catalysts used for epoxides and episulphides. Only a few systems have proven to be useful for kinetic measurements.

In 1969 Penczek and Kubisa [59] reported an exhaustive study of the kinetics and mechanism of BCMO polymerization initiated by the $(i\text{-}C_4H_9)_3Al/H_2O$ system and carried out in chlorobenzene solution at 55—95°C. This system produced homogeneous conditions for the polymerization and the whole process could be described as a non-stationary reaction with slow initiation, fast propagation, and slow degradative chain transfer to polymer.

Penczek [57] has derived a series of kinetic equations that include degradative chain transfer to polymer. Both reversible and irreversible reactions are considered. He has also treated degradative transfer to transfer agents [60]. For details of the derivation and indeed for all the pertinent equations themselves, the reader is referred to the original publications [57, 59—61]. We give here only two equations. The simplest case arises if $k_i > k_p$. Then initiation is virtually completed at the very beginning of polymerization. The same result is obtained if $k_i \leqslant k_p$ and a very low starting concentration of initiator is used so that initiator is consumed before all monomer is consumed. The integrated equation is then

$$-d \ln[M]_t/dt = k_p[I]_0 - k_{tr}\{[M]_0 \ln([M]_0/[M]_t) - ([M]_0 - [M]_t)\}$$

where $[I]_0$ and $[M]_0$ are initial concentrations of initiator and monomer respectively and $[M]_t$ is the monomer concentration at time t. In the more general case, when the initiation reaction can not be neglected, but monomer consumption during initiation is insignificant in comparison with that during propagation, the equation would be

$$-d \ln[M]_t/dt = k_p[I]_0 \left\{ 1 - \exp(- k_i \int_{[M]_0}^{[M]_t} [M] \, dt) \right\}$$
$$-k_{tr}\{[M]_0 \ln([M]_0/[M]_t) - ([M]_0 - [M]_t)\}.$$

By applying these equations and carefully determining the starting concentration of active species, Penczek and Kubisa [59] obtain the following values of the absolute rate coefficients at 70°C: $k_i = 1.6 \times 10^{-3}$ l mole^{-1} sec^{-1}, $k_p = 8.5$ l mole^{-1} sec^{-1} and $k_{tr} = 1.2 \times 10^{-3}$ l mole^{-1} sec^{-1}. In the evaluation of these parameters, degradative chain transfer to polymer was treated as an irreversible process.

The catalytic activity of aluminium alkyl and of aluminium alkyl—water systems could be further enhanced by the addition of readily polymerizable oxides, usually epoxides such as ECH or PO, (promoters) [62]. These initiators are very active with BCMO but the primary kinetic studies

282

have been carried out on THF (see Section 4.1.2d) and will be discussed below.

(c) *Initiation by free radicals in the presence of maleic anhydride*

BCMO polymerization can be initiated by free radicals in the presence of maleic anhydride [63]. All the available evidence suggests that polymerization proceeds by the usual cationic mechanism. With γ-ray or UV generation of free radicals, the over-all rate of polymerization was found to be first order with respect to [M]. The activation energy was 14 kcal mole^{-1}. Peroxide generation of radicals gave similar results, E_a being 23 kcal mole^{-1}.

(d) *Initiation by γ-irradiation in the solid state*

BCMO has been much used in studies of solid state polymerizations initiated by irradiation with γ-rays or an electron beam. Discussion of these kinetics, however, is reserved for a later volume where all solid state polymerizations will be treated together.

3.1.4 Other substituted oxetanes

Campbell and Foldi [64] and Penczek and Vansheidt [65] have prepared and polymerized a number of bicyclic oxetanes. In these compounds the substitution on the 3-carbon of the oxetane ring was in the form of a hydrocarbon ring. Penczek and Vansheidt [65] made some kinetic measurements. Their rate data compared with earlier values are shown in Table 4. They point out that the strain in the second ring has no

TABLE 4

Comparison of rates of polymerization^a of oxetanes [65]

Monomer	$R_p \times 10^3$ (mole l^{-1} sec^{-1})
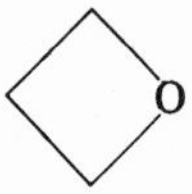	4.0
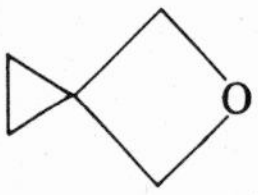	1.8
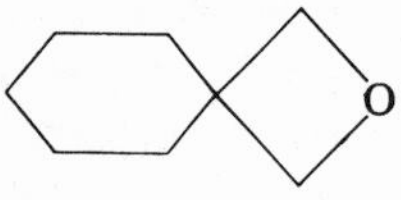	2.0
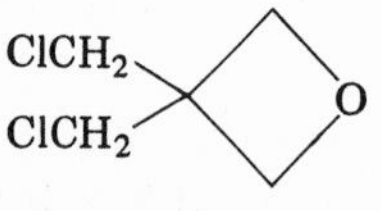 ClCH$_2$, ClCH$_2$	0.017

^{a} Polymerization at 0°C with [M] = 1.0 mole l^{-1}, [BF$_3$] = 0.1 mole l^{-1}, [H$_2$O] = 0.01 mole l^{-1}.

substantial effect on the polymerization of bicyclic oxetanes, while the electron density distribution did seem to be important. The electrophilic chloromethyl group can attract electrons from the oxygen of the oxetane ring, thereby reducing its basicity. It would be interesting to have a comparison of entropy factors for these monomers determined according to the methods of Saegusa et al. [53]. Perhaps further light would be shed on the importance of monomer basicity.

3.2 THIETANES

In comparison with oxetanes, kinetic studies of the polymerization of the corresponding sulphur compounds, the thietanes (thiacyclobutanes, trimethylene sulphides), are quite recent. Nevertheless, their polymerizability by a variety of mechanisms has been known for a long time. In a recent review, Sigwalt [36] noted that the polymerization of the parent thietane by hydrochloric acid was reported as early as 1933. Very little else was reported about these cyclic sulphides until the polymerization of 2,2-dialkyl thietanes in bulk with $BF_3 \cdot Et_2O$ at $25°C$ was described in 1964. In 1967 Stille and Empen [66] reported the use of a variety of cationic initiators, including some derived from BF_3, PF_5, and triethyl-aluminium/H_2O, for the polymerization of thietanes. Even more recently Morton and Kammereck [67] have reported the anionic polymerization of thietanes after initiation by an organo-lithium compound.

3.2.1 Cationic polymerization

The cationic polymerization of thietane is generally accepted to proceed in the same manner as that of the oxetanes. The most detailed study was carried out by Goethals et al. [42] on thietane and methyl thietanes in CH_2Cl_2 with $Et_3O^+BF_4^-$ as initiator. The initiation reaction was immediate and quantitative [68]. For these monomers propagation also occurred by nucleophilic attack of the monomer on the α-carbon of the cyclic sulphonium ion. For example, the propagation reaction for 3,3-dimethylthietane (DMT) is

Termination occurred when a sulphur atom of an existing polymer molecule reacted with the growing cyclic sulphonium ion to give a non-strained, branched sulphonium ion which was not capable of re-initiating polymerization, viz.

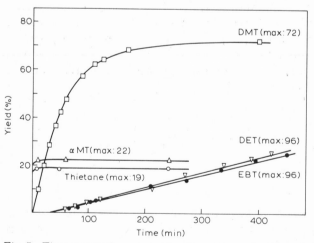

At 20°C values for k_p and k_t for DMT were found to be 6.5×10^{-3} l mole^{-1} sec^{-1} and 3.3×10^{-4} l mole^{-1} sec^{-1} respectively. The activation enthalpies, $\Delta H^{\ddagger}$, and entropies, $\Delta S^{\ddagger}$, for propagation and termination reactions of DMT determined from similar k_p and k_t values at different temperatures were $\Delta H^{\ddagger} = 12.5$ and 7.4 kcal mole^{-1} and $\Delta S^{\ddagger} = -26$ and -49 cal deg^{-1} mole^{-1} for propagation and termination, respectively, in each case. Since k_t is only an order of magnitude smaller than k_p, it is apparent that the sulphur atoms of existing polymer molecules compete very effectively with monomer. As a result the polymerization of DMT (and of other thietanes) stops before all monomer is consumed. This is illustrated in Fig. 5. Goethals et al. [42] have shown with the aid of low molecular weight compounds that the differences in rate of reaction between a sulphide and a sulphonium ion depend on the basicity and steric hindrance of the initial and final sulphides. They explain the different maximum conversions obtained with the different thietanes in terms of differences in basicity and steric hindrance between monomer and formed polymer. Table 5 shows corresponding basicity values for monomers and for model compounds of their polymers as well as k_p/k_t values for each. It appears that the sulphur atoms in the polymer chain always have higher basicity than the corresponding monomers and that the difference in basicity increases in the order DET < DMT < α-

Fig. 5. Time-conversion curve for the polymerization of thietane, α-MT, DMT, DET and 3-ethyl-3-butylthietane (EBT) in CH$_2$Cl$_2$ at 20°C [42].

TABLE 5

Correlation between relative basicity and values of k_p/k_t for different thietanes [42]

Monomer		$\Delta\nu_{OH}$ of monomer[a] (cm^{-1})	$\Delta\nu_{OH}$ of model compound for polymer[a] (cm^{-1})	$\Delta(\Delta\nu_{OH})$[b]	k_p/k_t
Structure	Abbrev.				
(thietane) S	—	241	252	11	1.1
(thietane) S, Me	α-MT	252	261	9	2.4
Me, Me (thietane) S	DMT	248	255	7	28
Et, Et (thietane) S	DET	250	255	5	450

[a] $\Delta\nu_{OH}$ is the difference of absorption maximum of the OH-function of phenol and the phenol—sulphide complex formed by hydrogen bonding.

[b] $\Delta(\Delta\nu_{OH})$ is the difference between $\Delta\nu_{OH}$ of monomer and of polymer model compounds.

MT < thietane. The values of k_p/k_t decrease in the same order. Similarly, an examination of molecular models revealed that steric hindrance is more important in the polymers than in the monomers and it increases in the order poly-thietane < poly-α-MT < poly-DMT < poly-DET. Greater steric hindrance should lead to a reduced k_t. Again the highest value of k_p/k_t would be predicted for DET and this is observed. Steric arguments are also consistent with the relative magnitudes of the $\Delta H^{\ddagger}$s and $\Delta S^{\ddagger}$s given above. In a further study in which a small amount of a low molecular weight sulphide was added to a polymerizing mixture of DMT, lower yields of polymer were obtained only if added sulphide had a higher basicity than the monomer and not if the added sulphide carried bulky groups.

In the course of their kinetic analysis of cyclic sulphide polymerizations Goethals et al. [42] were also able to determine the nature of the propagating species. They found that propagation occurs both via free ions and via ion-pairs. The amount of polymerization occurring via each route was distinctly solvent dependent. In CH_2Cl_2, polymerization of DMT via free ions was about 70 times faster than via ion pairs whereas in $C_6H_5NO_2$, the rates were nearly equal.

3.2.2 Anionic polymerization

Oxetanes are rather basic materials and are inert to attack by bases. However, as stated above, anionic polymerization of thietanes has been reported [67]. Significantly, in anionic polymerizations initiated by organo-lithium compounds the propagating end is a carbanion rather than a thiolate species; viz.

$$RLi + \quad \begin{array}{c} CH_2{-}CH_2 \\ | \quad\quad | \\ S{-}\!\!-\!\!-CH_2 \end{array} \quad \rightarrow \quad R{-}S{-}CH_2{-}CH_2{-}CH_2^- \; Li^+$$

Termination evidently does not occur as readily in anionic polymerization of thietanes as it does in cationic polymerization. Organo-lithium initiated polymerizations of thietanes lead to "living" polymers, which have been used to prepare ABA block copolymers of dienes and cyclic sulphides [69, 70]. Since the anionic polymerization of thietanes proceeds via a carbanion, thietanes can initiate vinyl polymerization and their polymerization can be initiated by vinyl monomers. Kinetic parameters of such polymerizations have not yet been reported.

3.3 AN OXETANE—THIETANE

Goethals and Du Prez [71] studied the polymerization of 2-oxo-6-thia-[3.3]spiroheptane

$$O \begin{array}{c} {}^{CH_2} \\ {}^{CH_2} \end{array} C \begin{array}{c} {}^{CH_2} \\ {}^{CH_2} \end{array} S$$

Polymerization of this compound under the influence of BF_3 gave two products

$$\left[\begin{array}{c} CH_2 \\ | \\ {-}O{-}CH_2{-}C{-}CH_2{-} \\ | \\ CH_2 \\ | \\ S \\ | \end{array} \right]_n \qquad \left[\begin{array}{c} {-}CH_2{-}C{-}CH_2{-}S{-} \\ \diagup \;\; \diagdown \\ CH_2 \;\; CH_2 \\ \diagdown \;\; \diagup \\ O \end{array} \right]_n$$

(I)   (II)

The soluble (II) was produced in about 60% yield in polymerizations carried out at $-3°C$ whereas the crosslinked polyether (I) was obtained from polymerizations at $30°C$. These authors conclude that under these experimental conditions, the thietane group polymerized more rapidly than the oxetane group but are careful to note that this does not prove that in general thietanes polymerize more rapidly than oxetanes. Instead they suggest that the sulphonium salt intermediates are more stable products than oxonium salts so that the equilibrium shown in the equation below lies almost entirely to the right. Polymerization of the oxetane group is thus inhibited by the presence of the thietane group.

$$
\begin{array}{c}
-CH_2 \\
\overset{\oplus}{O}\!-\!CH_2-\ +\ \\
-CH_2\quad X^-
\end{array}
\quad
\begin{array}{c}
-CH_2 \\
S \\
-CH_2
\end{array}
\ \rightleftharpoons\
\begin{array}{c}
-CH_2 \\
O\ +\ \\
-CH_2
\end{array}
\quad
\begin{array}{c}
-CH_2 \\
\overset{\oplus}{S}\!-\!CH_2- \\
-CH_2\quad X^-
\end{array}
$$

4. Tetrahydrofurans and tetrahydrothiophenes

4.1 TETRAHYDROFURAN

In contrast to the 1,3-epoxides where a whole variety of substituted oxetanes have been polymerized, there is only one 1,4-epoxide of major importance. This monomer is commonly called tetrahydrofuran, but it has the systematic name 1,4-epoxybutane or tetramethylene oxide. The polymer derived from it is as often called polytetramethylene oxide as polytetrahydrofuran. We have chosen to use the names tetrahydrofuran (THF) and polytetrahydrofuran (PTHF). An alternate abbreviation for THF that is found in some recent literature is "H_4 furan".

The polymerization of tetrahydrofuran has been reviewed in detail [5, 58]. There a thorough discussion of many aspects of the mechanism of polymerization and the very many initiators used was presented. Interested readers are referred to the earlier reviews for these aspects of the polymerization. Here attention is again focused primarily on the kinetics of polymerization.

4.1.1 Characteristic features of THF polymerizations

Many kinetic studies of the polymerization of THF have been carried out. From these a number of characteristic features of the polymerization have emerged.

(a) The ceiling temperature

The polymerization of THF approaches an equilibrium between monomer and polymer at every temperature. That is, for every temperature there is an equilibrium monomer concentration, $[M]_e$, that is thermodynamically determined. Also there is a temperature, T_c, above which no polymerization will occur. A value for T_c of $85 \pm 2°C$ was derived for bulk polymerization of THF using PF_6^- counter-ion [72, 73]. Dainton and Ivin [74] have related $[M]_e$ to the absolute temperature, T, and to the enthalpy change, ΔH_{ss}^0, and entropy change, ΔS^0, upon the conversion of one mole of monomer in solution to one base mole of polymer in solution, viz.

$$\ln [M]_e = \Delta H^0_{ss}/RT - \Delta S^0/R = \Delta G^0_{ss}/RT \tag{4}$$

ΔG^0_{ss} is the corresponding free energy change. By applying this theory, the value obtained [5, 58] for ΔH^0_{ss} was about -4.5 kcal mole^{-1} and for ΔS^0 was about -18 cal deg^{-1} mole^{-1}.

Some readers will be interested in the fact that Huang and Wang [75] in 1972 presented a newer theoretical treatment of the reaction kinetics of reversible polymerization in which this classic derivation of Dainton and Ivin is a special case. The thermodynamics of equilibrium polymerizations have recently been reviewed by Sawada [76].

Equation (4) holds for ideal systems but as is well-known most polymer solutions do not behave ideally. Hence the thermodynamic treatment of equilibrium polymerization has been reassessed to take this into account [77–79]. The revised theory has been applied to THF polymerizations in the 15–50°C range [80]. For the purpose of this evaluation the variation of $[M]_e$ is expressed by

$$[M]_e = A + B[P] = [M]^0_e + B[P]$$

where A and B are arbitrary parameters and $[P]$ is the equilibrium polymer concentration. A and B are determined experimentally and can be related to the thermodynamic properties of the polymerizing system by means of the next two equations, which are written in terms of ϕ_m (the monomer volume fraction), ϕ_p (the polymer volume fraction), V_m (the monomer molar volume) and V_p (the volume of one base-mole of polymer), viz.

$$\phi_m = A V_m + B(V_m/V_p)\phi_p = \phi^0_m + b\phi_p$$

$$\phi_m = \frac{-(\Delta G_{1c}/RT) + \ln \alpha + \beta}{\beta + \chi_{mp} - (1/\alpha)} + \frac{\chi_{mp} - \beta}{\beta + \chi_{mp} - (1/\alpha)} \cdot \phi_p$$

where $\beta = \chi_{ms} - \chi_{sp}(V_m/V_s)$; here χ is the free energy interaction parameter between any two components, the subscripts m, s, and p refer to monomer, solvent, and polymer, respectively, V_s is the solvent molar volume and ΔG_{1c} is the free energy change upon conversion of 1 mole of liquid monomer to 1 base mole of liquid amorphous polymer of infinite chain length. When Leonard and Maheux apply these revised formulas, they obtain $\Delta H_{1c} = -3.6 \pm 0.2$ kcal mole^{-1}, and $\Delta S_{1c} = -11.7 \pm 0.6$ eu.

Still a different approach toward determining ΔH^0_{1c} and ΔS^0_{1c} has been made by Busfield et al. [81]. They studied the equilibrium between gaseous monomer and amorphous polymer for THF between 40 and 80°C. They obtained ΔH^0_{gc} (298°K) = −9.5 ± 0.5 kcal mole^{-1} and ΔS^0_{gc} (298°K) = −26.7 ± 1.6 cal deg^{-1} mole^{-1}. From these data and thermodynamic data for the vaporization, obtained from vapour pressure measurements, they calculated ΔH^0_{lc} (298°K) = −1.8 ± 0.5 kcal mole^{-1} and ΔS^0_{lc} (298°K) = −3.9 ± 1.7 cal deg^{-1} mole^{-1}.

(b) *The general kinetic equation*

The polymerization of THF is represented formally by the scheme

The kinetics of THF polymerization are expressed in terms of the rate of reaction for an equilibrium polymerization without termination; viz.

$$-d[M]/dt = k_i[I]([M] - [M]_e) + k_p([I]_0 - [I])([M] - [M]_e) \quad (5)$$

Here, [M] is the monomer concentration and [I] is the initiator concentration; as before, $[M]_e$ is the equilibrium concentration. As long as initiation is fast and the initial rate is directly proportional to $[I]_0$, [I] will be very small. The rate is then closely approximated by

$$-d[M]/dt = k_p[I]_0([M] - [M]_e) \quad (6)$$

Of course, when the number of active centres is measured directly, $[I]_0$ becomes equal to the concentration of active centres and the simplified form (6) applies at once. After integration, eqn. (6) becomes

$$k_p t = \frac{1}{[I]_0} \ln \frac{[M]_0 - [M]_e}{[M] - [M]_e} \quad (7)$$

This is the form to which the experimental data is most frequently fitted. In the work of Saegusa et al. where the change of active centres, [P*], with time is taken into account, eqn. (6) becomes on integration

$$\ln \frac{[M]_{t_1} - [M]_e}{[M]_{t_2} - [M]_e} = k_p \int_{t_1}^{t_2} [P^*] \, dt \quad (8)$$

290

where $[M]_{t_1}$ and $[M]_{t_2}$ are the monomer concentrations at times t_1 and t_2 respectively.

(c) Initiator effects

The species which initiates THF polymerization comprises a cation and an anion portion. The nature of both portions greatly affects subsequent kinetic measurements.

The nature of the counter-ion X^- and the temperature at which the polymerization is carried out are important. For example, in a study of THF polymerizations at 30°C initiated by equivalent amounts of triethyl-oxonium salts with different counter-ions, Dreyfuss and Dreyfuss [82] observed differences in both conversion and rates of viscosity change with time of polymerization. In the cases of BF_4^- and $SbCl_6^-$, the final viscosities and conversions that were attained were lower than when SbF_6^- or PF_6^- counter-ions were used. The viscosity of BF_4^- polymers remained constant after constant conversion was reached, so termination can be inferred. The viscosity of $SbCl_6^-$ polymers continued to decrease even after constant conversion was attained. With the $SbCl_6^-$ counter-ion, both termination and transfer occurred. In a comparison of rates of THF polymerization at 0°C initiated by $Et_3O^+BF_4^-$ and $Et_3O^+AlCl_4^-$, Saegusa and Matsumoto [83] confirmed the termination inferred by Dreyfuss and Dreyfuss [82] and even earlier by Vofsi and Tobolsky [86] in a study with only the BF_4^- counter-ion at 0°C. Saegusa and Matsumoto applied the [P*] method [53] for determining active centres described in Section 3.1.1. The termination reaction was less obvious at 0°C than at 30°C; nevertheless, it was clearly evident. Further they found that termination

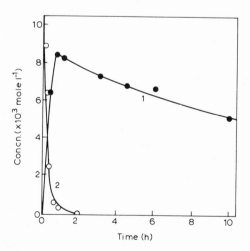

Fig. 6. Variation of [P*] (Curve 1) and of $[Et_3O^+]$ (Curve 2) in polymerization of THF by $Et_3O^+BF_4^-$ in CH_2Cl_2 at 0°C [83].

with the $AlCl_4^-$ counter-ion is even more serious. The comparison of the $[P*]$ and $[Et_3O^+]$ curves is shown in Figs. 6 and 7.

The importance of both the cationic and anionic portions of the initiator was revealed by a study of various initiators for THF polymerization having different cations and different counter-ions. In this study, Yamashita et al. [84, 85] concluded that the k_p value with CH_2Cl_2 solvent at $0°C$ is almost independent of counter-ion when triethyloxonium ions are used as initiators. They found much slower apparent rates when the different cation initiators, acetyl hexafluoroantimonate and 2-methyl-1,3-dioxolenium perchlorate, were used. They explained the slower rates by decreased rates of initiation. The apparent k_p's can be increased to the

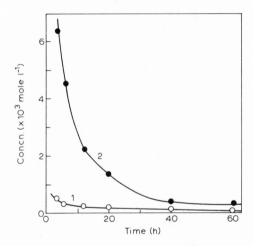

Fig. 7. Variation of $[P*]$ (Curve 1) and of $[Et_3O^+]$ (Curve 2) in polymerization of THF by $Et_3O^+AlCl_4^-$ in CH_2Cl_2 at $0°C$ [83].

usual value by the addition of ECH or trimethylene oxide but only at the expense of some transfer early in the polymerization. The rate data are given in Table 6.

4.1.2 Kinetics with specific initiators

(a) Triethyloxonium tetrafluoroborate

Vofsi and Tobolsky [86] reported the first careful kinetic study of THF polymerization. They used $Et_3O^+BF_4^-$ as initiator and carried out the polymerizations at $0°C$ in dichloroethane. By using ^{14}C labelled catalyst, they were able to show that initiation was fast but not immediate. The rate of initiator disappearance was measured. The initial propagation rate was directly proportional to initial monomer concen-

TABLE 6
Kinetic parameters for THF polymerization

Initiator	Temp. (°C)	Solvent	E_p (kcal mole⁻¹)	$10^{-8} A_p$ (l mole⁻¹ sec⁻¹)	$10^2 k_i$ (l mole⁻¹ sec⁻¹)	$10^3 k_p$ (l mole⁻¹ sec⁻¹)	$10^2 k_d$ (sec⁻¹)	$10^5 k_t$ (sec⁻¹)	Ref.
Et₃O⁺BF₄⁻	0	(—CH₂Cl)₂			1.3—1.7[a]	4.83			86
Et₃O⁺BF₄⁻	0	CH₂Cl₂			0.02[b]	3.7		1[b]	83
Et₃O⁺BF₄⁻	0	CH₂Cl₂				4.83			84
Et₃O⁺SbCl₆⁻	20	Bulk	15						89
Et₃O⁺PF₆⁻	0	CH₂Cl₂				4.19			84
Et₃O⁺SbF₆⁻	0	CH₂Cl₂				5.89			84
Et₃O⁺AlCl₄⁻	0	CH₂Cl₂			0.04			1000	83
Et₃O⁺ClO₄⁻	0	CH₂Cl₂				4.10			84
CH₃CO⁺SbF₆⁻	0	CH₃NO₂				1.47[c]			84
CH₃C⁺◯⌐ ClO₄⁻	0	CH₃NO₂				1.65[c]			84
BF₃ . Et₂O—ECH	0—40	Bulk	13.3	1.64					87
BF₃ . Et₂O—ECH	20	Et₂O				16.6	4.67		87
BF₃ . THF—ECH	0	Bulk				4.5			91
BF₃ . THF—ECH	0	CH₂Cl₂	12	0.11		4.1			91, 118
SnCl₄ . THF—ECH	0	Bulk				6.2[d]			91
AlEtCl₂ . THF—ECH	0	Bulk				7.8			91
AlEt₃—H₂O—ECH[e]	0	Bulk				6.6			94
AlEt₃—H₂O—PO[e]	0	Bulk				6.0			94
AlEt₃—H₂O—BPL[e]	0	Bulk				6.7			94

[a] Initial catalyst in range 1.53—6.1 × 10⁻² mole l⁻¹; [M]₀ = 6 l mole l⁻¹.
[b] Initial catalyst 9 × 10⁻³ mole l⁻¹.
[c] Calculated assuming rapid initiation which was shown to be invalid.
[d] Average value, see Table 7.
[e] See Table 8.

tration. The fit to eqn. (7) was good. The values obtained for k_i and k_p are given in Table 6. They did not determine a termination rate coefficient but realized that a complete description of the system would require one. As stated above, Saegusa and Matsumoto [83] were able to get an estimate of the importance of the termination term by using the [P*] method. The rate coefficients they found are given in Table 6.

Rozenberg et al. [87] generated their catalyst in situ from ECH and $BF_3 . Et_2O$. They carried out their polymerizations in bulk and in Et_2O solution. The experimental data are reported to fit eqn. (6). They define $[I]_0$ in terms of the catalyst components charged, assuming instantaneous and quantitative conversion to catalyst, but do not make any direct measurements of active centres. Their kinetic parameters are given in Table 6.

Before leaving $Et_3O^+BF_4^-$ initiated polymerizations of THF one other study needs to be mentioned. In polymerizations of thietanes above we saw that free sulphonium ions did not propagate very much more rapidly than ion pairs (Section 3.2.1). Sangster and Worsfold [88] have found that in THF polymerizations with BF_4^- counter-ions in CH_2Cl_2 the free ion rate is only a factor of 7 greater than that of the ion pair rate.

(b) Triethyloxonium hexachloroantimonate

Lyudvig et al. [89] studied the kinetics of bulk THF polymerizations initiated by $Et_3O^+SbCl_6^-$ at 20°C. In another study using this initiator, Stejny [90] examined the effect of simultaneous polymerization and crystallization of THF. He found that at his reaction temperatures, −25° and −45°C, at the onset of crystallization the polymerization rate was somewhat accelerated.

(c) Lewis acid . THF . ECH

In this kinetic study of bulk THF polymerizations at 0°C Saegusa and Matsumoto [91] examined the effect of changing the Lewis acid in the system Lewis acid . THF . ECH. [P*] was again determined by the phenoxyl end-capping procedure. Kinetics were fitted to eqn. (8) and k_p's were determined in the usual way. The pattern of [P*] change during polymerization varied considerably with the Lewis acid component. [P*] increased slowly in the beginning with the $BF_3 . ECH$ system and then remained almost constant. In polymerizations initiated by $SnCl_4 . ECH$, initiation was very rapid and the [P*] decreased very rapidly due to a termination reaction. Conversion to polymer was low. The course of the polymerization with initiation by the $AlEtCl_2 . ECH$ system was similar to that initiated by the $SnCl_4 . ECH$ system, except that initiation was somewhat slower. In spite of these differences in initiation and over-all conversion, there was not a wide difference in the k_p's obtained with the

TABLE 7

Effect of Lewis acid on propagation rate coefficient of bulk
THF polymerization at 0°C [91]

Lewis acid		$[\text{ECH}]_0$ (mole 1^{-1})	$10^3 k_p$ (1 mole^{-1} sec^{-1})
Type	Conc. (mole 1^{-1})		
BF_3	0.011	0.010	4.5
BF_3	0.021	0.020	4.6
$BF_3{}^a$	0.025	0.020	4.1
$SnCl_4$	0.028	0.029	5.8
$SnCl_4$	0.055	0.053	6.7
$AlEtCl_2$	0.054	0.051	7.8

[a] This polymerization was carried out in CH_2Cl_2 solution at $[M]_0 = 6.3$ mole 1^{-1}.

different initiators. The data are shown in Table 7. Further these data are
compared with those using other initiators in Table 6.

(d) *Triethylaluminium—water—promoter system*

Saegusa et al. [92—4] carried out studies of the polymerization of THF
initiated by the system $AlEt_3$—$H_2O(2:1)$—promoter. ECH, PO, and
β-propiolactone (BPL) were employed as promoters. The concentration of
active sites was determined by the phenoxyl end-capping method in
studies after 1968 or was calculated from yield and molecular weight
[92]. The former gave more reliable results. Kinetics could be fitted to
eqn. (8). The rate coefficients observed are shown in Table 8. The k_p
values for the three promoters are almost the same. They are compared
with k_p values using other initiators in Table 6. The data show that the

TABLE 8

Effect of promoter on propagation rate coefficients of bulk
THF polymerizations[a] at 0°C [94]

Promoter		$[\text{AlEt}_3]_0{}^b$ (mole 1^{-1})	$10^3 k_p$ (1 mole^{-1} sec^{-1})
Type	Conc. (mole 1^{-1})		
ECH	0.02	0.18	6.6
PO	0.03	0.15	6.0
BPL^c	0.005	0.30	6.7

[a] Initiated by $AlEt_3$—$H_2O(2:1)$ promoter.
[b] Amount of $AlEt_3$ used in preparation of catalyst.
[c] β-propiolactone.

nature of the promoter does not affect the propagation reaction, although it does affect the initiation reaction and the overall rate. In the case of PO, a termination reaction was thought to be responsible for a slower rate of conversion.

(e) Triisobutyl aluminium—water—allyl glycidyl ether system

Zak et al. [95] derived a mathematical description of the kinetics of polymerization of THF with 2% allyl glycidyl ether initiated by triisobutylaluminium and water. The approach is quite different from that reported for these polymerizations by others. In their calculations they regard their system as a homopolymerization of THF and introduce the corresponding simplifying assumptions into their calculations of the kinetic parameters. They consider the usual propagation reaction and conclude that their polymerization can be adequately represented by the equation

$$ x = \left(1 - \frac{m_e}{m_0}\right)\left\{1 - e^{(k_g/k_d)} e^{-k} d^{[t - t_0]} - 1\right\} $$

where x is the degree of conversion m_e is the equilibrium and m_0 the initial monomer concentration, k_g is $k_g' k C_0$ (k_g' is the polymer chain growth rate coefficient, k is a proportionality factor, C_0 is the initial triisobutyl aluminium concentration), k_d is the coefficient for deactivation of active centres, t is the polymerization time and t_0 is the initial time. Values of kinetic coefficients observed and calculated according to this scheme were in good agreement.

(f) Strong proton acids

Pruckmayr and Wu [96] reported a direct method for determining the concentration of growing chain ends in an oxonium ion polymerization (cf. Section 3.1.1 for other methods). High resolution proton NMR spectroscopy was used to investigate the solution polymerization of cyclic ethers. The method is most clearly illustrated by their report on THF polymerization initiated by strong proton acids such as trifluoromethyl sulphonic acid, fluorosulphonic acid, and fuming sulphuric acid. No kinetic parameters are reported in this elegant study but in the present authors' opinions the method gives so much qualitative information about the changes occurring during polymerization that it deserves to be included in some detail in this chapter nonetheless. Equivalent information is difficult to determine unequivocally by other methods.

The usual scheme of polymerization needs to be modified only slightly in these polymerizations, viz.

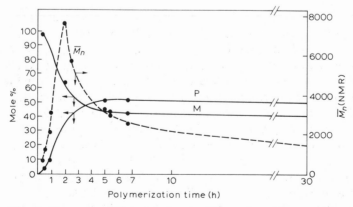

All the reactants and products shown in the equation can be identified in the NMR spectra. The changes in concentration of all the different species can be followed easily by comparing the intensities of the different NMR signals. An example is shown in Fig. 8 for a THF polymerization at 0°C. The time to reach equilibrium and indeed the conversion that will be obtained depends as usual on the type and concentration of initiator, polymerization temperature, monomer concentration, and solvent polarity. Once more the lack of termination and the "living" character of the system is demonstrated. Note that as the authors point out a maximum is attained in the molecular weight, M_n. They suggest that the maximum is due to an increase in the number of growing chains and does not represent an actual depolymerization. Their evidence is that addition of monomer to an equilibrated system does not lead to another maximum but only produces an increase in molecular weight with additional time of polymerization.

Extension of these studies to strong Lewis acid initiators such as $SbCl_5$ gave similar results except that a high concentration of THF . Lewis acid complex was formed very early and remained high throughout the polymerization. Lyudvig et al. [89] reached a similar conclusion in an

Fig. 8. Monomer (M) and polymer (P) concentrations and molecular weights ($\overline{M}_n$) during polymerization of THF by strong proton acid in benzene at 0°C [96].

earlier kinetic study. Unfortunately, corresponding studies with pre-formed trialkyl oxonium ion salts are not reported and NMR confirmation of the seemingly uniquely rapid initiation believed to occur in these systems, and demonstrated in the [P*] and other studies, is lacking.

Once again the importance of carefully controlling the whole polymer-ization must be emphasized. All the work above clearly indicates that THF will polymerize well in the presence of appropriate strong acids and

the NMR evidence shows that HO^{+}⟨⟩ initiates polymerization even though

rather slowly. Lyudvig et al. [97] state that THF does not polymerize in the presence of acids. They observed that their $Et_3O^+SbCl_6^-$ initiated polymerizations were terminated by added water and explained the

termination on the basis of the formation of inert HO^{+}⟨⟩. This explana-

tion cannot be correct in view of what has just been discussed. The counter-ion in their study was $SbCl_6^-$. A more likely explanation is that hydrolysis of the counter-ion leads to chloride ion and that this ion (known to be too nucleophilic to be an acceptable counter-ion for polymerization) terminates the polymerization. In Pruckmayr and Wu's systems, the counter-ions are sulphonates. For trifluoromethyl sulphonic acid, the counter-ion is $CF_3SO_2O^-$ and for fluorosulphonic acid, FSO_2O^-. It will be noted that these are not anions that we have discussed elsewhere. They are very satisfactory counter-ions for THF polymer-izations. This has been nicely demonstrated in the work of Smith and Hubin [98]. The chemistry is very interesting, but there is, however, no kinetic data yet published on these counter-ions★.

(g) Electrochemical initiation

Since about 1966, repeated reports of polymerization of THF electro-chemically in the presence of salts such as $(C_4H_9)_4NClO_4$ [99, 100], $LiClO_4$, or $(C_2H_5)_4NClO_4$ [101] have appeared. The PTHF formed was found in the anode compartment of a divided cell. In 1971 Nakahama et al. [102] reported a kinetic study carried out in order to elucidate the relationship between the amount of current and the number of propa-gating ends formed in the electrochemical process. The polymerization was carried out in bulk at 30°C using $(C_4H_9)_4NClO_4$ as an electrolyte. Their experimental results were fitted to the usual THF kinetic eqn. (7) in the form

$$\ln \frac{[M]_0 - [M]_e}{[M] - [M]_e} = k_p [P^*] t$$

★ See note added in proof (p. 326).

where the terms have the same meaning as before. Assuming one active species is formed by one electron oxidation, these workers then write [P*] in the form

$$[P^*] = f \cdot \frac{Q}{V}$$

where f is the current efficiency, Q is the amount of current, and V is the volume of the system. The reasonable fit to the derived equation is shown in Fig. 9.

The yield of polymer increased with reaction time and reached about 70%, the yield that would be expected from previous data of conversion

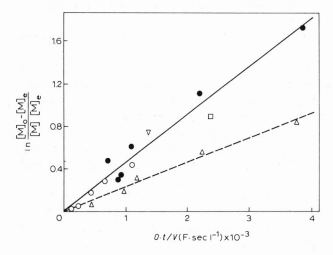

Fig. 9. Effect of amount of current and current density on the rate of electrolytic polymerization of THF [102]. Amount of current and current density: $\triangledown$, 0.16 mF, 1 mA cm^{-2}; $\square$, 0.30 mF, 1 mA cm^{-2}; $\bullet$, 1.2 mF, 1 mA cm^{-2}; $\circ$, 3.7 mF, 1 mA cm^{-2}; $\triangle$, 0.75 mF, 0.5 mA cm^{-2}.

versus temperature [72]. The molecular weight increased proportionately to the yield of polymer, as would be expected for a "living" system. The polymer was found to have about two moles of hydroxyl end groups per mole of polymer. This result indicates initiation by a proton. From the results obtained the initiation mechanism, involving the perchlorate radical and a proton, proposed was

(1) $ClO_4^- -e \longrightarrow ClO_4 \cdot$

(2) $ClO_4 \cdot + O\langle\hspace{1em}\rangle \longrightarrow HClO_4 + \overset{\cdot}{O}\langle\overset{H}{\hspace{1em}}\rangle$

(3) $HClO_4$ + ⬠O ⟶ H—O+⬠ ClO_4^-

(4) HO+⬠ + ⬠O ⟶ $HO(CH_2)_4$—O+⬠

The presence of a radical species was indicated by the electrolytic copolymerization of methylmethacrylate and acrylonitrile, which were not oxidized to form radicals in the anodic reaction.

(h) Other initiators

Lyudvig et al. [89] also reported a study of the kinetics of polymerization of THF in the presence of the reaction product of CH_3COCl and $SbCl_5$, presumably $CH_3CO^+SbCl_6^-$. The rate curves were S-shaped but there was a point of maximum rate and a steady state was reached. Their results should be compared with the more recent ones of Yamashita et al. [84], who showed that initiation with the corresponding SbF_6^- salt is slow.

PF_5 is an important and well known initiator for cyclic ether polymerizations. However, its mode of initiation seems complex and is not well documented. Thus attempts to measure kinetics have given only complex results [103]. Sims [103] has studied initiation by PF_5 promoted with ECH. The kinetic results were also complex.

Bawn et al. [104] have compared the rates of THF polymerization initiated by $Ph_3C^+SbCl_6^-$ with those initiated by $Ph_3C^+PF_6^-$. They conclude that there is not much difference in the rate of polymerization as a result of replacing $SbCl_6^-$ by PF_6^-. There now seems to be general agreement that k_p is not much affected by the counter-ion. Differences in overall polymerization rate arise rather from differences in k_i and k_t.

Very recently the spontaneous, simultaneous homopolymerization of vinylidene cyanide and cyclic ethers including THF has been reported [105]. Polymerizations were initiated via a donor—acceptor complex but further investigation of the mechanism of initiation is continuing. It is still too early for kinetics to have been reported but they should be interesting when they do come.

4.2 SUBSTITUTED TETRAHYDROFURANS

Polymerization of substituted 1,4-epoxides to high molecular weight polymers is not expected on a thermodynamic basis. Indeed high molecular weight polymers have not yet been prepared. However, 3-MeTHF [106] and 2-MeTHF [96] have both been polymerized to give oligomers.

1,4-Epoxycyclohexanes, which can be considered to be 2,5-disubstituted THF's, polymerize quite readily [90, 107—9]. For example, from studies of the mechanism of polymerization of *exo-* and *endo*-2-methyl-7-oxabicyclo [2.2.1] heptanes, Kops and Spanggaard [109, 110] found that the ring opening appears to proceed exclusively through nucleophilic attack on the activated ring structure at C-4 for both isomers. A similar conclusion was reached about the same time by Saegusa et al. [111]. The latter group of authors also showed by NMR techniques that both the *exo* and *endo* derivatives lead to polymers in which succeeding ether groups are *trans* to each other, viz.

This means that the C-4 carbon atom selectively undergoes inversion of configuration.

Kinetic analyses of the polymerizations of these *exo* and *endo* isomers have been reported by Guisti et al. [112] and by Saegusa et al. [113, 114]. The former made their kinetic analysis on the basis of the initial velocity observed in the conversion—time curve. As usual, the latter group used their phenoxyl end-capping method. The initiator used was the BF_3—ECH system. Polymerizations of both isomers were characterized by slow initiation and fairly rapid termination. In contrast to THF polymerizations, the depolymerization reaction was insignificant and the kinetics were analysed on the basis of the kinetic equation

$$\ln \frac{[M]_{t_1}}{[M]_{t_2}} = k_p \int_{t_1}^{t_2} [P^*] \, dt$$

in which the $[M]_t$'s are instantaneous monomer concentrations and $\int_{t_1}^{t_2} [P^*] \, dt$ was obtained by graphical integration. The kinetic parameters

TABLE 9

Kinetic parameters for *exo-* and *endo*-2-methyl-7-oxabicyclo[2.2.1] heptane [113, 114]

Parameter	Temp. (°C)	*exo*-MOBH	*endo*-MOBH
$10^3 k_p$ (l mole^{-1} sec^{-1})	−40	0.18	0.85
	−30	0.87	4.0
	−20	2.6	8.4
	−10	9.4	27
E(kcal mole^{-1})		15	14
$10^{-9} A$ (l mole^{-1} sec^{-1})		41	7

are shown in Table 9. In this case the difference in kinetic activity can be ascribed to the activation energy rather than to the pre-exponential factor which was important in the oxetane series. The implication is that the *endo* isomer is more strained than the *exo* isomer and thus polymerizes more readily.

4.3 TETRAHYDROTHIOPHENES

In a study of the ring-opening polymerization of cyclic sulphides containing three-, four-, and five-membered rings, Stille and Empen [66] observed that tetrahydrothiophene and 7-thiabicyclo [2.2.1] heptane would not polymerize by any of the methods employed in their study. Those methods included using initiators known to be among the most effective for cyclic oxide and other ring size cyclic sulphide polymerizations. The difference in reactivity did not seem to be related to difference in basicity. Rather it was explained by a comparison of the strain energies of the cyclic compounds. The strain energy [66] for a 5-membered cyclic oxide is 5.63 kcal mole^{-1} and that for the corresponding sulphide only 1.97 kcal mole^{-1}. This lower strain energy was arrived at in part from the large difference in length of the C—S bond, 1.81 Å, compared to the length of the C—O bond, 1.43 Å.

5. Cyclic ethers and cyclic sulphides with more than five members in the ring

The parent member in the 1,5-epoxide series is tetrahydropyran. Calculations [66, 75, 115] show that the ring opening polymerization of tetrahydropyran would have a positive free energy of about 1.3 kcal mole^{-1} and a strain energy of 1.16 kcal mole^{-1}. In agreement with this, the ring opening polymerization of tetrahydropyran has not been observed even by the sensitive NMR technique of Pruckmayr and Wu [96]. The strain energy reported [66] for the six-membered ring sulphur compound is -0.27 kcal mole^{-1} and one can expect it to be a very stable compound with respect to polymerization. Apparently, introducing further strain into the ring by means of a bicyclic bridge as in 2-oxabicyclo [2.2.2] octane is sufficient to make polymerization possible. Polymerization of the latter bicyclo compound to "sticky solids" has been reported by Saegusa et al. [116].

The polymerization of 2,3-dihydropyran has been reported. However, this compound is a vinyl ether. The polymerization occurs mainly through the double bond and not by ring opening [3]. Kamio et al. [117] found that the polymer obtained by polymerizing 2,3-dihydropyran with

BF_3 . Et_2O had mainly the structure, $(-\overline{CH-CH_2-CH_2-CH_2-OC}H-)_n$.

Calculations [75, 115] suggest that higher epoxides such as oxepanes (1,6-epoxides) and oxecanes (1,7-epoxides) should be capable of polymerization by a ring opening mechanism. To date, kinetic data has been reported for only oxepane [118, 119]. The polymerization was carried out in CH_2Cl_2 using the initiator BF_3 . THF—ECH and again [P*] was determined by the phenoxyl end-capping procedure. The rate coefficient was calculated from eqn. (8) but the value of $[M]_e$ was so small, 8×10^{-2} mole l^{-1} at 3° and 6×10^{-2} mole l^{-1} at 10°C that this quantity could be left out in practical calculations. At some temperatures the termination rate coefficient was significant but it was less important than with THF. The kinetic parameters are given in Table 10.

Polydecamethylene oxide has been prepared by polycondensation of the corresponding glycols [120, 121] but not by ring-opening polymerization.

TABLE 10

Kinetic parameters for ring-opening polymerization of oxepane[a] [118]

Temp. (°C)	$10^5 k_p$ (l mole^{-1} sec^{-1})	$10^6 k_t$ (sec^{-1})	E_p (kcal mole^{-1})	$10^{-8} A$ (l mole^{-1} sec^{-1})
−10	0.33	—	18	19
0	1.5	2.9		
10	3.1	11		
20	9.1	24		
30	43	72		

[a] Solution polymerization in CH_2Cl_2 with BF_3 . THF—ECH as initiator (each 0.057 mole l^{-1}); $[M]_0 = 2.9$ mole l^{-1}.

6. Cyclic formals, thials, and oxathiolanes

These compounds can be considered to be cyclic co-dimers of formaldehyde or thial with a cyclic ether or a cyclic sulphide. The thials are stable only as cyclic compounds whereas the monomeric aldehydes are well known. Polymerization of all three types of monomers in this group is known to occur [36, 122]. Cyclic formals except the six-membered ones, are polymerizable [122, 123]. Again in the 5-membered ring group, substitution seems to reduce or eliminate polymerization and 2-methyl and 2,4-dimethyl-1,3-dioxolanes do not polymerize [124, 125]. However, 4-phenyl-1,3-dioxolane has been polymerized to a high molecular weight polymer [126]. Introducing strain into the dioxolane ring by means of a 2,4-disubstituted bridge as in 6,8-dioxabicyclo [3.2.1]octane again enhances polymerizability and under the right conditions high polymer can be formed [125]. Again, as with the cyclic ethers and cyclic sulphides, numerous studies of the "kinetics" of polymerization of the formals have been made but there are few if any studies of the corresponding thio and oxo-thio compounds. We shall discuss kinetics for only three monomers in this group: 1,3-dioxolane, 1,3-dioxepane, and trioxane.

6.1 1,3-DIOXOLANE

Presently the member of the group that is being most intensively studied seems to be 1,3-dioxolane. However, controversy about the mechanisms involved still abounds. The variety of structures proposed for the active chain end is surprising for such a simple molecule. Each group that studies the polymerization seems to have a favourite scheme and presents data to show why the suggestion of another group cannot be correct. Rather than try to decide the issue on the basis of currently available evidence, we choose to present several schemes and some of the criticisms of each of these schemes and hope that in time at least one of them will be generally accepted. We restrict ourselves to recent publications on the assumption that these represent the most encompassing thought in the current controversy.

6.1.1 Triethyloxonium tetrafluoroborate initiation

In 1968, Yamashita et al. [127] reported a study of the kinetics of 1,3-dioxolane polymerization initiated by $Et_3O^+BF_4^-$ and carried out in CH_2Cl_2 solution. They concluded that the mechanism, deduced from the structure of the initiation product and the molecular weight of the polymer, is simply

The mechanism of initiation is based on an examination of reaction products in an early stage of the polymerization. After short-stopping the polymerization by addition of sodium methoxide/methanol solution, the main product subsequently identified by gas chromatography was $C_2H_5OCH_2CH_2OCH_2OCH_3$. The rate of polymerization is presumed to follow the standard eqn. (6) for an equilibrium polymerization without termination and with rapid initiation. It is acknowledged that there is an induction period (presumably due to reaction of catalyst with adventitious water, since rigorous drying reduced the induction time to only

TABLE 11
Kinetic parameters for dioxolane polymerizations

Initiator	Temp. (°C)	Solvent	k_p (l mole⁻¹ sec⁻¹)	E_p (kcal mole⁻¹)	$10^5 A_p$ (l mole⁻¹ sec⁻¹)	ΔH_p (kcal mole⁻¹)	ΔS_p (cal deg⁻¹ mole⁻¹)	$10^2 k_d$ (sec⁻¹)	E_d (kcal mole⁻¹)	$10^{-10} A_d$ (sec⁻¹)	Ref.
$Et_3O^+BF_4^-$	30	CH_2Cl_2	0.043[a]	11.7		-3.6	-14	6.5			127, 128
$Et_3O^+BF_4^-$	25	CH_2Cl_2	0.167[a,b]								129
$Et_3O^+SbCl_6^-$	20	EtCl	0.067[c]	6.7	2.7	-6.5	-22.4		13.2	1.4	97
$Et_3O^+SbF_6^-$	25	CH_2Cl_2	~20[a,b]								129
$HClO_4$	25	CH_2Cl_2	21[a,b]								129
$HClO_4$	20	CH_2Cl_2	8.3								135
$Ph_3C^+SbF_6^-$	25	CH_2Cl_2	50[c]								137
$Ph_3C^+AsF_6^-$	25	CH_2Cl_2	50[c]								137

[a] There was an induction period before this rate was achieved.
[b] A dependence of first order rate coefficient k_1 on concentration c of initiator was found so the rate quoted here is really k_1/c.
[c] There was no induction period.

minutes) before conversion—time curves become linear. Although high molecular weight polymer is prepared, the molecular weight is 60—70% of what would be expected from a truly "living" system with 100% efficient initiation, and no termination or transfer reactions [128]. This lower than expected molecular weight suggests that some transfer or termination reaction may be occurring. Furthermore, Yamashita et al. note that, in copolymerization studies, 1,3-dioxolane will initiate polymerization of both monomers that propagate only via carbenium ions (for example, styrene) and monomers that propagate only via oxonium ions (for example, THF). It will be recalled that normally an oxonium ion such as that derived from THF will not initiate the polymerization of vinyl monomers. They further suggest that the propagating species in the cationic polymerization of cyclic formals is something between a carbenium ion and an oxonium ion. The kinetic parameters derived are shown in Table 11.

A somewhat later study of the polymerization of 1,3-dioxolane by $Et_3O^+BF_4^-$ was reported by Jones and Plesch [129] in 1969. Their studies were carried out with CH_2Cl_2 solvent, but they used a reactor fitted with Pt electrodes and followed conductivity and polymerization simultaneously. They agreed that the formation of active centres is slow and found that the rate obtained from the linear portion of the conversion—time curve varied with initiator concentration. Their measurements showed a sharp rise in conductivity initially, then a steady fall to a very low constant value. They conclude from further studies that the concentration of propagating ions is much lower than that of the ions present initially. They suggest that the following reactions occur when $Et_3O^+BF_4^-$ is added to dioxolane, viz.

$$Et_3O^+BF_4^- + \underset{O\diagdown\diagup O}{\bigcirc} \xrightarrow{\text{fast}} Et_2O + EtH + \underset{\underset{H\quad BF_4^-}{\overset{|}{C}}}{\underset{O\diagup{}^+\diagdown O}{\bigcirc}}$$

$$Et_3O^+BF_4^- \longrightarrow EtF + Et_2O + BF_3$$

However, Jones and Plesch do not believe that either of these reactions leads to the initiating species. They allow that the dioxlenium salt may be able to initiate polymerization, though very slowly and inefficiently. They suggest that the true initiation reaction may indeed be the ethylation suggested by Yamashita et al. [127], but point out that the concentration of growing centres cannot be $[I]_0$ as the latter group assumed. The disagreement here again underscores the desirability of making direct measurements of the number of growing centres in polymerizations involving cyclic oxygen compounds.

6.1.2 Triethyloxonium hexafluoroantimonate initiation

It is most interesting to note that, when Jones and Plesch [129] carried out an exploratory study with $Et_3O^+SbF_6^-$, this salt seemed to react rapidly and cleanly to give a high yield of growing centres which seemed to be ethylated dioxolane. It's a pity that the more detailed studies were reported for the BF_4^- counter-ion, which time and again is being shown to be inferior for polymerizations of cyclic oxygen compounds.

6.1.3 Triethyloxonium hexachloroantimonate initiation

In 1971 Lyudvig et al. [97] reported an investigation of the polymerization of 1,3-dioxolane using "precision purification of reactants" and $Et_3O^+SbCl_6^-$ as initiator. Under these conditions polymerization with different initial monomer concentrations are reported to take place without any induction period. Lyudvig et al. report that the polymerization is first order with respect to both monomer and initiator concentration. Polymerization of 1,3-dioxolane is a reversible process. The final kinetic equation takes a form similar to eqn. (6). A UV spectroscopic method was used to investigate the nature of the active centre in the polymerization. Very briefly, what these workers found was that the maximum they observed for the polymerizing mixture was different from that which could be attributed to a simple cyclic oxonium ion. Hence they propose that the active centre has the "polymeric" tertiary oxonium ion structure

$$
\begin{array}{c}
\vdots \\
CH_2 \\
| \\
CH_2 \\
| \\
\text{---O---CH}_2\text{---O+ } SbCl_6^- \\
| \\
CH_2 \\
| \\
O \\
\vdots
\end{array}
$$

They suggest two possible pathways for the polymerization reaction, viz.

$$(2) \quad \sim O-CH_2-\overset{+}{\underset{\displaystyle \overset{|}{\underset{\displaystyle \overset{CH_2}{|}}{O}}}{\underset{\displaystyle \overset{|}{\underset{\displaystyle \sim}{O}}}{O}}}{\overset{\displaystyle \overset{CH_2}{|}}{\overset{\displaystyle \overset{|}{CH_2}}{}}} \; + \; \overset{O \diagdown \diagup O}{\bigcirc} \; \longrightarrow$$

$$\sim OCH_2\overset{+}{O} \quad \overset{O \cdot O}{\underset{\displaystyle \overset{|}{\underset{\displaystyle \overset{CH_2}{|}}{O}}}{\underset{\displaystyle \overset{|}{\underset{\displaystyle \sim}{O}}}{}}}{\overset{\displaystyle \overset{CH_2}{|}}{\overset{\displaystyle \overset{|}{CH_2}}{}}} \longrightarrow$$

$$\sim OCH_2-OCH_2CH_2OCH_2-\overset{+}{\underset{\displaystyle \overset{|}{\underset{\displaystyle \overset{CH_2}{|}}{O}}}{\underset{\displaystyle \overset{|}{\underset{\displaystyle \sim}{O}}}{O}}}{\overset{\displaystyle \overset{CH_2}{|}}{\overset{\displaystyle \overset{|}{CH_2}}{}}}$$

Lyudvig et al. [97] believe that in the presence of water the mechanism of the reactions are different. An induction period is then observed, i.e., polymerization is inhibited.

6.1.4 Borontrifluoride etherate initiation

Yamashita [128] reports that the kinetics of polymerization of 1,3-dioxolane initiated by $BF_3 \cdot Et_2O$ are quite complex and that initiation is not a simple reaction. In 1973 Rozenberg et al. [130] reported a kinetic study of such a polymerization in CH_2Cl_2 solution. They again found an induction period, but this time stated that the acceleration is the result of the autocatalytic action of the macromolecules formed. Their conclusions are based (i) on the relationship of the rate coefficients of $BF_3 \cdot Et_2O$ reactions with cyclic monomer or with polymer chain and (ii) on the decrease of the induction period on addition of polymer or methylal to a polymerizing mixture. The mechanism of initiation suggested is

308

$$\text{Et}_2\text{O} \cdot \text{BF}_3 + \underset{O\diagdown\diagup O}{\bigcirc} \rightleftharpoons \underset{CH_2-CH_2}{\overset{O-CH_2}{\bigg|}} \diagdown O : BF_3 + \text{Et}_2\text{O}$$

$$\underset{CH_2-CH_2}{\overset{O-CH_2}{\bigg|}} \diagdown O : BF_3 + \underset{O\diagdown\diagup O}{\bigcirc} \longrightarrow CH_3OCH_2CH_2OBF_3^- \quad \underset{CH_2-CH_2}{\overset{CH-O}{O\diagup}} \diagup$$

$$\underset{CH_2-CH_2}{\overset{O-CH}{\bigg|}} \diagdown \overset{+}{O} + \underset{O\diagdown\diagup O}{\bigcirc} \longrightarrow \underset{O}{\overset{H}{\diagdown}} C-O-CH_2CH_2\overset{+}{O} \underset{CH_2-CH_2}{\overset{CH_2-O}{\diagup}} \bigg|$$

$$\sim\!\!\text{OCH}_2\!-\!\underset{\overset{..}{BF_3}}{O}\!-\!\text{CH}_2\text{CH}_2\!\sim \longrightarrow \sim\!\!\overset{+}{O}=\text{CH}_2\ \bar{B}F_3OCH_2CH_2\!\sim$$

6.1.5 Perchloric acid initiation

In 1968 Plesch and Westermann [131] published a report of 1,3-dioxolane polymerization in their adiabatic calorimeter under vacuum in CH_2Cl_2 with anhydrous perchloric acid as initiator. They confirmed that the polymerization can be an equilibrium involving only monomer and polymer without side reactions. For a one molar solution the ceiling temperature was found to be $1°C$, $\Delta S_{ss}^0 = -18.6 \pm 1.2$ cal deg^{-1} mole^{-1} and $\Delta H_{ss}^0 = -5.2 \pm 0.2$ kcal mole^{-1}. The polymer consisted, as was previously found, of a regularly alternating sequence of formaldehyde and ethylene oxide units; that is, it was a homopolymer of 1,3-dioxolane. The yield and degree of polymerization were independent of initiator concentration and depended only on the initial monomer concentration and the temperature. It was suggested that transfer to monomer occurs. The polymer prepared in the presence of water had terminal OH groups, but as far as Plesch and Westermann could determine the polymer prepared under strictly anhydrous conditions had no detectable end groups, and was presumably cyclic. The mechanisms of initiation and propagation proposed are:

Initiation

$$\underset{O\diagdown\diagup O}{\bigcirc} + HClO_4 \rightleftharpoons \underset{O\diagdown\diagup \overset{+}{O}H}{\bigcirc} \quad ClO_4^-$$

Propagation

Note that the propagating species here is a secondary oxonium ion. Transfer presumably involves the transfer of a proton from one cyclic species to another. Conductivity studies [129] of a $HClO_4$ initiated polymerization in a dilatometer indicate a polymerization with no induction period, but with an acceleration stage, and a first order stage.

Shortly after Plesch and Westermann's work was published, Jaacks et al. [132] also reported a study of the polymerization of 1,3-dioxolane by $HClO_4$ in CH_2Cl_2 at 20°C. Again the need for the most rigorous absence of water is emphasized [133]. They found a gradual but quantitative initiation reaction with no kinetic termination reaction. Eventually from further studies [134] they concluded that the slow initiation involves two reactions

(1) H^+ + [dioxolane ring $O{\smallfrown}O$] $\longrightarrow$ $HO\overset{+}{\smile}O$ $\xrightarrow{\text{dioxolane}}$ $HOCH_2CH_2OCH_2{-}\overset{+}{O}{\smile}O$

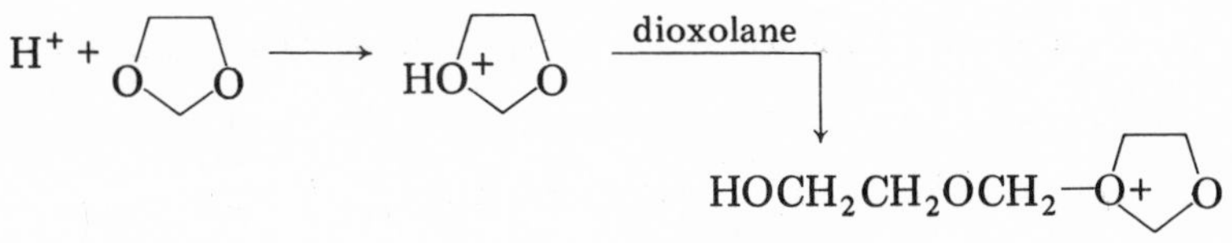

and

(2)

$$\text{---}OCH_2CH_2{-}\overset{H}{\underset{+}{O}}{-}CH_2OCH_2CH_2OCH_2\text{---}$$
$$\uparrow$$
$$\text{---}CH_2OCH_2CH_2{-}\underline{O}{-}CH_2OCH_2CH_2O\text{---}$$

$\longrightarrow$

$\text{---}OCH_2CH_2{-}OH$

+

$\text{---}CH_2OCH_2CH_2\overset{+}{\diagup}\overset{CH_2OCH_2CH_2OCH_2\text{---}}{\underset{O{-}CH_2OCH_2CH_2O\text{---}}{}}$

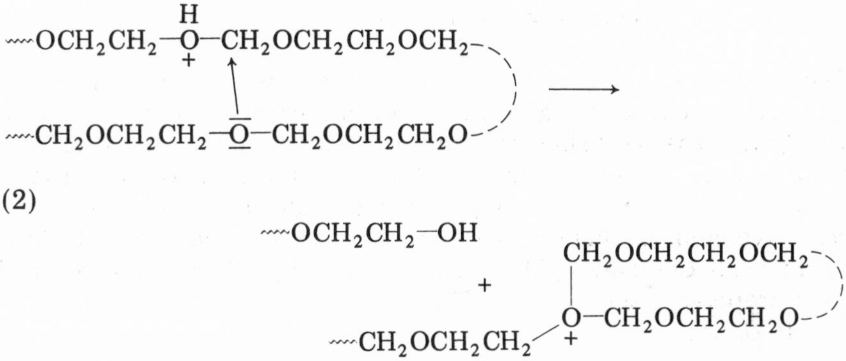

One involves protonation of the monomer and the other protonation of the polymer. They conclude that the second reaction must have a higher rate coefficient. Further, when initiation is finally complete, propagation occurs exclusively through tertiary oxonium ions. If monomer is added to such a system, they propose that the conversion—time curve is no longer S-shaped but linear, represented by the equation

$$-\frac{d[M]}{dt} = k_w [P^+][M]$$

where k_w is the propagation rate coefficient, $[P^+]$ is the oxonium ion concentration, and $[M]$ is the monomer concentration. Their rate coefficient is given in Table 11.

Jaacks et al. terminated the polymerizations by adding sodium alkoxide

310

and then determined the alkoxy endgroups in order to estimate the concentration of growing centres. They considered three possible propagating species in their first paper. Later [135], in one of the most detailed analyses of the mechanism to date, they included as a fourth possible species the "polymeric" tertiary oxonium ion proposed by Lyudvig et al. [97] (Section 6.1.3).

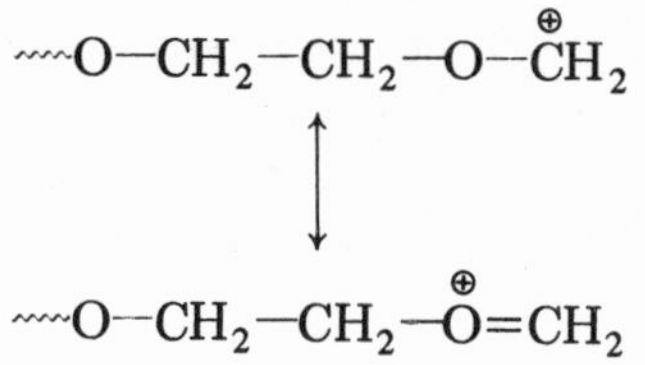

$$\text{~~O—CH}_2\text{—CH}_2\text{—O—}\overset{\oplus}{\text{C}}\text{H}_2$$

$$\updownarrow$$

$$\text{~~O—CH}_2\text{—CH}_2\text{—}\overset{\oplus}{\text{O}}\text{=CH}_2$$

carboxonium ions

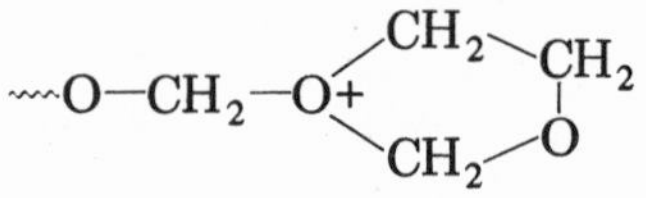

tertiary oxonium ions

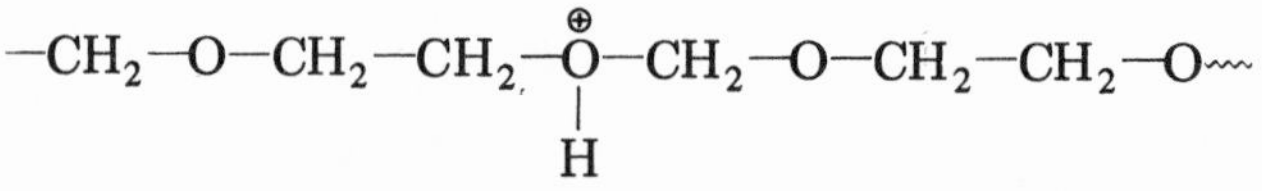

$$\text{—CH}_2\text{—O—CH}_2\text{—CH}_2\text{—}\overset{\oplus}{\underset{\text{H}}{\text{O}}}\text{—CH}_2\text{—O—CH}_2\text{—CH}_2\text{—O~~}$$

secondary oxonium ions

Since they found about 0.95 ethoxy end groups in the polymer per mole of $HClO_4$ used, they argue in favour of tertiary oxonium ions as the propagating species. (They point out that carboxonium ions seem unlikely on the basis of other studies. Also copolymerization parameters in copolymerization with dioxacycloheptane can be explained by this tertiary oxonium mechanism and not by others.) They suggest that the macrocycles observed by Plesch are formed by a backbiting mechanism. Their proposed reactions [132] are:

Initiation:

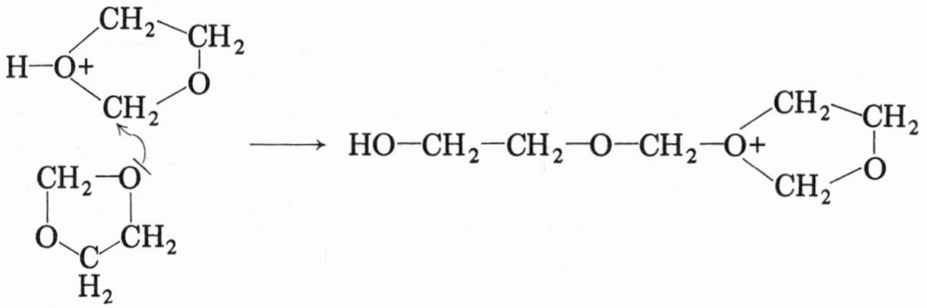

$$\longrightarrow \text{HO—CH}_2\text{—CH}_2\text{—O—CH}_2\text{—}\overset{+}{\text{O}}$$

Transfer:

$$\text{~~OCH}_2\text{—O—CH}_2\text{CH}_2\text{—(OCH}_2\text{OCH}_2\text{CH}_2)_n\text{—OCH}_2\text{—}\overset{+}{\text{O}}$$

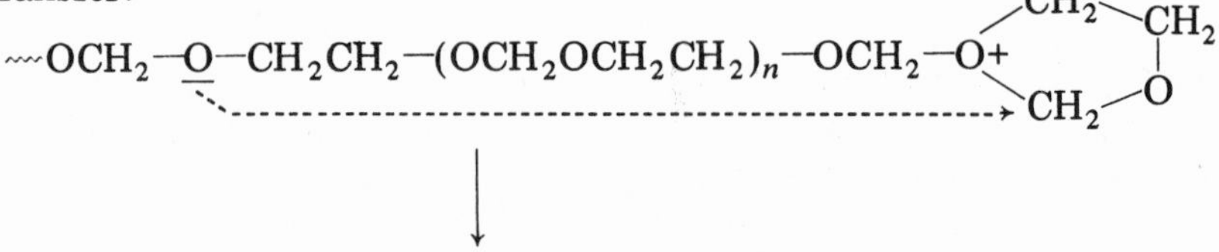

$$\sim\!\!OCH_2\!-\!\overset{+}{O}\!-\!CH_2CH_2\!-\!(OCH_2OCH_2CH_2)_n\!-\!OCH_2\!-\!OCH_2CH_2$$

(structure: backbiting oxonium ion with pendant CH_2—O ring)

$\downarrow$

$$\sim\!\!OCH_2\!-\!\overset{+}{O}\underset{CH_2}{\overset{CH_2\diagdown CH_2}{\diagup}}\!\!O \;+\; O\!-\!CH_2CH_2\!-\!(OCH_2OCH_2CH_2)_n\!-\!OCH_2OCH_2CH_2$$

In fact, Kelen et al. [134] have presented data and a theoretical approach indicating that the backbiting mechanism seems to be the most probable interpretation of ring-formation in the cationic polymerization of 1,3-dioxolane.

Jaacks et al. [135] allow that most of the experimental results are in conformity with the "polymeric" tertiary oxonium ion proposed by Lyudvig et al. [97] (Section 6.1.3) but they point out that this mechanism is inconsistent with the copolymerization results.

Unfortunately, due to the untimely passing of V. Jaacks in a climbing accident, we must accept these as the last words [136], except for posthumous publications of his associates, that he can contribute to the controversy.

6.1.6 Triphenylmethyl hexachloroantimonate initiation

In 1973 Penczek and Kubisa [137] published a detailed study of the interaction of triphenylmethyl hexachloroantimonate with 1,3-dioxolane. Polymerization occurs as a result of the interaction but the triphenylmethyl salt itself does not initiate the polymerization. Instead the sequence of reactions

(1) $Ph_3C^+SbCl_6^- + $ (1,3-dioxolane) $\longrightarrow Ph_3CH + $ (dioxolan-2-ylium ion) $SbCl_6^-$

(2) (dioxolane···Cl···SbCl$_5$ complex) $\longrightarrow ClCH_2CH_2O\overset{O}{\overset{\|}{C}}\!-\!H + SbCl_5$

(3) 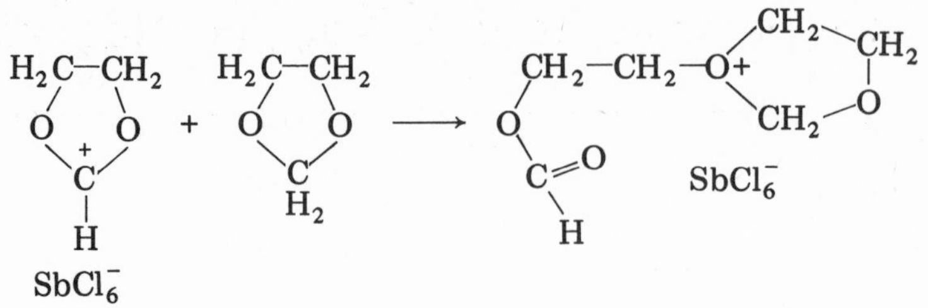 [structural reaction scheme: two cyclic oxocarbenium/dioxolane species with $SbCl_6^-$ counterion react to give open-chain and cyclic oxonium product with $SbCl_6^-$]

(4) 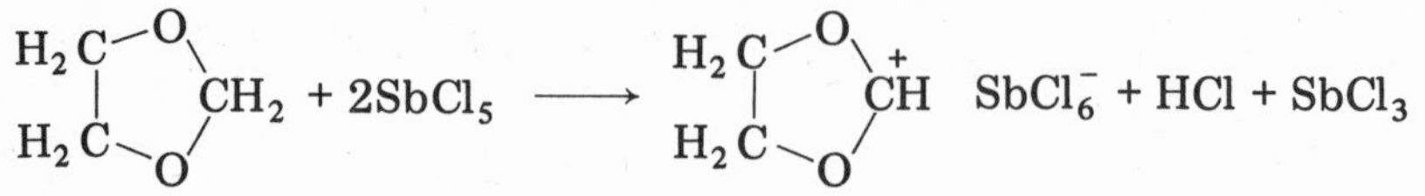 [1,3-dioxolane structure] $+ 2SbCl_5 \longrightarrow$ [dioxolenium cation] $SbCl_6^- + HCl + SbCl_3$

occurs. Many of the by-products are noninitiating. Only the dioxolenium salt initiates polymerization and, as shown above, it tends to decompose into β-chloroethyl formate and $SbCl_5$, two moles of which will initiate polymerization by forming more dioxolenium salt and unreactive by-products. By studying triphenylmethyl salts with other counter-ions, Kubisa and Penczek were able to show that the degree of decomposition of the dioxolenium salt depends on the nature of the counter-ion and is dominant with $SbCl_6^-$. Still it was found that polymerization rates are proportional to initial initiator concentration, $[I]_0$, for toluene solution and proportional to $[I]_0^2$ for dichloromethane solution. Polymerization was first order in monomer and proceeded without an induction period. The k_p value found is given in Table 11. Note that the k_p was determined with a triphenylmethyl salt initiator having a more stable counter-ion such as SbF_6^- or AsF_6^- and assuming only one kind of growing species, namely tertiary oxonium salts.

One other interesting observation reported by Penczek and Kubisa is shown in Table 12, where a comparison is made of the amount of

TABLE 12

Amount[a] of β-chloroethyl formate formed in the polymerization[b] of 1,3-dioxolane using several initiators [137]

Initiator	$[I]_0$ (10^3 mole l^{-1})	$[M]$ (mole l^{-1})	β-Chloroethyl formate found	
			(10^3 mole l^{-1})	moles per mole of initiator
$Ph_3C^+SbCl_6^-$	5.36	2.43	10.7	2.0
$Ph_3C^+SbCl_6^-$	7.5	2.9	18.7	2.5
$SbCl_5$	14.5	2.9	14.5	1.0
$Et_3O^+SbCl_6^-$	12.5	2.9	25.0	2.0

[a] Determined by gas chromatography.
[b] In CH_2Cl_2 solution at 25°C.

β-chloroethyl formate formed by interaction of 1,3-dioxolane and different polymerization initiators. According to the table approximately two moles of the formate are formed for every molecule of initiator except that only one is formed in the case of SbCl$_5$, as would be expected from the sequence of reactions above. This would suggest that the use of SbCl$_6^-$ counter-ion leads to inefficient initiation. In fact, UV measurements suggest that 95% of the initiator is converted to inactive product. The data in the table show that this is true even of the triethyloxonium hexachloroantimonate initiator.

6.1.7 Summary

In conclusion, in the kinetics of dioxolane polymerizations with many catalysts, the initiation mechanism is complex and inefficient. The degree of efficiency seems to be related both to the cation and to the anion. Again as in the case of cyclic ethers and cyclic sulphides, an independent measurement of the number of active sites seems essential for precise kinetics. The most probable k_p for the polymerization seems to be of the order of $10-50 \, l \, mole^{-1} \, sec^{-1}$. With careful choice of polymerization conditions a kinetically reversible polymerization occurs, but the molecular weight of the polymer produced is not related to the initiator concentration, probably as a result of a transfer reaction.

6.2 1,3-DIOXEPANE

Plesch and Westermann [123] have studied the polymerization of 1,3-dioxepane, $\overline{CH_2-O-CH_2-O-CH_2-CH_2-CH_2}$. They propose a ring expansion mechanism similar to the one they suggested for 1,3-dioxolane. Again they find an equilibrium polymerization, this time with a ceiling temperature of $+27°C$ for a 1 molar solution. $\Delta H_{ss}^0 = -3.6 \pm 0.3 \, kcal \, mole^{-1}$ and $\Delta S_{ss}^0 = 11.5 \pm 1.5 \, cal \, deg \, mole^{-1}$.

6.3 TRIOXANE AND TRITHIETANE

Trioxane is the cyclic trimer of formaldehyde and it can be polymerized to yield polyoxymethylene having the same structure as polyformaldehyde. Polymerization has been carried out with or without catalyst in the liquid, solid, and sublimed states. All polymerizations appear to proceed by a cationic mechanism and the usual type of cationic initiators are effective [122, 138—141]. The structure of the cationic chain ends is not clear and two types of propagating centres have been proposed [142], namely, tertiary oxonium ions and carbenium ions. Their propagation reactions are

(1) by tertiary oxonium ions (S$_N$2 type propagation)

$$\sim\text{OCH}_2\text{-O}^+\underset{\text{CH}_2\text{-O}}{\overset{\text{CH}_2\text{-O}}{\diagdown}}\text{CH}_2 \longrightarrow \sim(\text{OCH}_2)_4\text{-O}^+\underset{\text{CH}_2\text{-O}}{\overset{\text{CH}_2\text{-O}}{\diagdown}}\text{CH}_2$$

$$\text{H}_2\text{C}\overset{\text{O}}{\underset{\text{O}}{\diagup}}\underset{\text{CH}_2}{\overset{\text{CH}_2}{\diagdown}}\text{O}$$

(2) by linear carbenium ions formed by spontaneous ring opening of cyclic tertiary oxonium ions (S_N1 type reaction)

$$\sim\text{OCH}_2\text{-O}^+\underset{\text{CH}_2\text{-O}}{\overset{\text{CH}_2\text{-O}}{\diagdown}}\text{CH}_2 \longrightarrow \sim(\text{OCH}_2)_3\text{-O-}\overset{+}{\text{C}}\text{H}_2$$

$$\updownarrow$$

$$\sim(\text{OCH}_2)_3\text{-}\overset{+}{\text{O}}\text{=CH}_2$$

$$\sim(\text{OCH}_2)_4\text{-O}^+\underset{\text{CH}_2\text{-O}}{\overset{\text{CH}_2\text{-O}}{\diagdown}}\text{CH}_2 \xleftarrow{\text{trioxane}}$$

The evidence for the presence of at least some carbenium ends is that hydride transfer does occur forming CH_3 end groups [143] and formaldehyde is eliminated during polymerization. Jaacks et al. [132] have used their sodium alkoxide method, which was referred to above in Section 6.1.5, to study the number of active centres in trioxane polymerizations. Large discrepancies were observed between concentrations of initiators ($SnCl_4$, $BF_3 . Et_2O$, $HClO_4$, $Et_3O^+BF_4^-$) and the concentration of polyoxymethylene cations which reacted with alkoxide ion. Furthermore, the polymer precipitates during polymerization. According to Jaacks et al. [132] quantitative results in this heterogeneous system are not reliable. It can be anticipated that trioxane polymerizations will have similar kinetic characteristics to those of formaldehyde polymerizations, which are reviewed later in this volume and to which the reader is referred. (See Chapter 5.)

Recently, some kinetic coefficients have been determined in electrochemically initiated polymerizations [144]. Trioxane, dissolved in three different solvents (acetonitrile, benzonitrile, and nitrobenzene) with tetrabutylammonium perchlorate as background electrolyte, was polymerized in the anodic compartment. Definite solvent dependence was shown. When the current was turned off, polymerization in acetonitrile ceased; in benzonitrile, polymerizations continued but the rate gradually decreased; in nitrobenzene, polymerizations seemed to continue at the initial rate. The reaction mechanism was not defined. The results were fitted to a kinetic equation of the form

$$-\frac{d[M]}{dt} = \frac{k_p}{k_t} fQ[M]$$

where k_p is the specific rate coefficient of propagation, f is current efficiency, Q is current intensity, k_t is the specific rate coefficient of termination, and [M] is the monomer concentration. As more is learned about reaction mechanisms in the various solvents, the coefficients given will probably be revised, so we will not quote them here. Nevertheless, it is significant that trioxane is another example of an electrochemically initiated polymerization of a compound containing oxygen atoms.

In some recent studies [145] of the kinetics of the polymerization of trioxane initiated by $BF_3 . Et_2O$ and by $FeCl_3 . Ph_3CCl$, the course of the polymerization appeared to be quite complex. Rate and equilibrium parameters were evaluated and their significance was discussed. The number of active centres was not determined independently, so again these results must be considered preliminary and will not be quoted here.

In contrast to the polymerization of trioxane which has been studied a great deal, few studies of the polymerization of trithietane have been reported [36]. It has been polymerized using many of the same initiators as trioxane and the mechanism is probably just as complex. As far as we know kinetic studies have not been published.

7. Copolymers

A large number of copolymers of cyclic ethers, cyclic sulphides and cyclic formals have been prepared. Many cyclic compounds that will not homopolymerize do copolymerize readily [7, 146, 147]. Some cyclic compounds will copolymerize with lactones, cyclic anhydrides, or vinyl monomers. Very many commercially important materials have resulted from these copolymerizations.

However, the preparation of useful polymers does not mean that the chemistry and kinetics of these copolymerizations are well understood. It is known that the copolymerizations proceed by the usual mechanisms that are available to the individual monomers. Sometimes more than one mechanism operates simultaneously; that is, more than one type of catalyst site is present. All the problems inherent in the interpretation of homopolymerization are still present and new problems are generated as a result of the simultaneous presence of two polymerizable monomers and their interaction with each other. The familiar problems from homo-polymerizations include doubts about the nature and number of growing centres in a polymerization, solvent effects [148], catalyst effects, termination reactions that produce low molecular weight polymers or cyclic oligomers, relative reactivity of ions formed from monomer and from polymer, and transfer reactions of all kinds mentioned before. In addition, some catalysts were shown to give polymers with block co-polymer character or even mixtures of homopolymer and copolymer. Backbiting [149] and effects of the penultimate unit have been found [122].

In studies of the kinetics of copolymerization of cyclic compounds the Mayo—Lewis equations [150] for kinetics of copolymerization have been applied, often with deserved caution. Many monomer reactivity ratios have been derived in this way. A large number of them have been summarized previously [7, 151] and we will not repeat them here nor attempt to update the lists. Instead we shall concentrate on some of the factors that seem to be important in regulating the copolymerizations and on some of the newer approaches that have been suggested for dealing with the complicated kinetics and give only a few examples of individual rate studies.

7.1 INFLUENCE OF BASICITY AND RING STRAIN

The factors most commonly mentioned as important in studies of copolymerization of these cyclic compounds are relative basicity and ring strain. It seems fairly well established that relative basicity plays an important role in controlling the relative reactivity of cyclic ethers. This lends further support to the belief that propagation proceeds by nucleophilic attack of monomer oxygen in the cationic copolymerizations. Basicity seems to be less important than other factors such as ring strain in determining relative reactivity of cyclic sulphides.

Relative basicities of cyclic ethers have been determined by several methods [146, 152, 153]. By one method Wirth and Slick [152] obtained relative basicities by infrared from a study of the equilibrium constant at $27^{\circ}C$ for the distribution of BF_3 between two ethers in benzene solution. In another case, Yamashita et al. [146], following a method similar to that of Gordy and Stanford [154], measured the infrared absorption spectra of cyclic ethers in the presence of 0.1 mole l^{-1} methanol-d. They then calculated the basicity of the monomers from the shift value of O—D stretching band. Nuclear magnetic resonance measurements of the equilibrium constant between $BF_3 . Et_2 O$ and the cyclic ether were also used [146].

Similarly, base strengths of cyclic sulphides have been determined by several methods. Stille and Empen [66] used the position of the wavelength of maximum absorption in the ultraviolet as one measure of electron density of the sulphur atom. In this study it was assumed that the sulphur atom with the highest electron density would have the longest wavelength absorption. They also examined the NMR of α- and β-protons for the same purpose by assuming that deshielding at the α-proton must be compensated for by a high electron density on the sulphur atom. However, they preferred complex formation with iodine, as measured by UV or visible charge transfer spectra, or with phenol, as measured in the IR region by the proton shift upon complexation.

Yamashita et al. [146] point out some features of the basicity of cyclic ethers. The basicity is affected by chemical structure, ring size, and substituents. The ring size affects basicity in the order $4 > 5 > 6 > 3$. In 5-membered rings the basicity order is ether $>$ lactone $>$ formal. Methyl substitution increases the basicity and chloromethyl substitution decreases basicity. The relative basicities of the monomers are broadly in agreement with their reactivity in copolymerization experiments [58, 122].

The order of basicity found for the cyclic sulphides on the basis of ring size is $5 > 6 > 4 > 3$. This order is different from that in cyclic ethers. Stille and Empen [66] state that this difference has been ascribed to the differences in heteroatom size, differences in ring size (ring strain), and also to differences in polarizability between oxygen and sulphur atoms. Basicity did not correlate well with reactivity in the sulphide series. Ring strain seemed to be more important. However, it should be noted that the reactivity measured in the sulphide case was in homopolymerizations. Very few copolymerization studies have been carried out so far.

Tanaka [155] has attempted to make a quantitative estimate of the contribution of ring strain and basicity to reactivity of cyclic ethers in cationic copolymerization. Free energy of polymerization was used as a measure of ring strain. The relationship he derived related the logarithm of relative reactivity, $1/r_n$, of m-membered ring ethers with i substituents to n-membered ring compounds with j substituents to a linear combination of the differences in basicity, $\Delta(pK_b)_{m,i-n,j}$ and in free energy, $\Delta(\Delta G)_{m,i-n,j}$, viz.

$$\log(1/r_n) = \alpha\Delta(\Delta G)_{m,i-n,j} + \beta\Delta(pK_b)_{m,i-n,j} + \gamma$$

where α, β, and γ are constants. He compared calculated values of $1/r_n$ with observed values taken from the literature and found good agreement.

7.2 INDIVIDUAL RATE STUDIES

As we said, we do not propose to include an updated list of monomer reactivity ratios. We feel the most recent progress in this area is best illustrated by three specific examples.

7.2.1 Copolymerization of trioxane with 1,3-dioxolane

One of the most complex copolymerization systems in the field of cyclic ethers is that of the cationic copolymerization of trioxane with dioxolane. As was pointed out by Jaacks [136, 142, 156], who reported the study, in this case the conventional Mayo—Lewis method for determining reactivity ratios fails, and even if the reactivity ratios were known, copolymer compositions could not be calculated. The reasons are threefold:

(i) There is a polymerization—depolymerization equilibrium of formaldehyde with the active centres. The formaldehyde can add to either trioxane or dioxolane end groups. The actual monomer trioxane is not involved in these reactions. Once equilibrium is established, the polymerization is really a terpolymerization and it happens that, since formaldehyde is the most reactive monomer, it exerts a strong influence on the course of the polymerization.

(ii) Cleavage of cyclic formals from the cationic chain ends occurs. Oxacyclic compounds such as tetroxane are formed.

(iii) Chain transfer by polymer occurs.

Furthermore, as we have seen, there may be two different kinds of active centres in polymerization of cyclic formals and thus two different kinds of propagation reaction. One polymerization occurs via tertiary oxonium ions and the other via linear carbenium ions. Jaacks has proposed a novel method for determining reactivity ratios for this and similar systems. He suggests obtaining each reactivity ratio individually from one experiment or a series of similar experiments. Copolymerizations are run with a large enough excess of monomer (M_1) so that almost pure poly(M_1) is obtained containing only a few and single M_2 units. Then $\sim m_1^+$ comprises a large majority of the active centres. Under these conditions, the reactivity ratio for trioxane, r_1, can be determined from

$$\log \frac{[M_1]_t}{[M_1]_0} = r_1 \log \frac{[M_2]_t}{[M_2]_0}$$

Here $[M_1]_0$ and $[M_2]_0$ are initial monomer concentrations and $[M_1]_t$ and $[M_2]_t$ are concentrations of unreacted monomers after termination of the copolymerization. By analogy it should be possible to determine r_2, the reactivity ratio for dioxolane, by using a large enough excess of dioxolane. The method may be applicable to some systems. However, in practice there were severe limitations with trioxane—dioxolane copolymerizations. The usual trioxane polymerization is heterogeneous and polymerization occurs mainly in the crystalline phase. Polydioxolane is soluble. Thus the most Jaacks could hope to obtain for the system was r_1. In addition, it turned out that even in a copolymerization containing 98.2% trioxane, dioxolane was consumed much faster than trioxane at first and produced soluble, high dioxolane polymer until about 50% of the dioxolane was consumed. Then consumption of trioxane began and an insoluble polymer containing 98 mole % trioxane was produced. Jaacks concluded that, in this heterogeneous copolymerization, two different r_1's would be needed to characterize the system — one for dissolved and the other for crystalline copolymer cations.

7.2.2 Copolymerization of BCMO with THF below T_c for THF

Yamashita et al. [157] have derived a copolymer composition equation that includes the depropagation reaction such as might be expected in the cationic copolymerization of BCMO and THF. They consider two models. For the first one it is assumed that monomer M_2 adds reversibly to both active chain ends m_1^* and m_2^* and that depropagation by detachment of an M_1 unit is neglected. The elementary reactions are then

(1) $\sim m_1^* + M_1 \xrightarrow{k_{11}} \sim m_1 m_1^*$

(2) $\sim m_1^* + M_2 \underset{k'_{12}}{\overset{k_{12}}{\rightleftharpoons}} \sim m_1 m_2^*$

(3) $\sim m_2^* + M_1 \xrightarrow{k_{21}} \sim m_2 m_1^*$

(4) $\sim m_2^* + M_2 \underset{k'_{22}}{\overset{k_{22}}{\rightleftharpoons}} \sim m_2 m_2^*$

They define the parameters r_1, r_2, ϕ, ρ, δ, and α as follows:

$$r_1 = k_{11}/k_{12}, \quad r_2 = k_{22}/k_{21}, \quad \phi = k_{22}/k'_{22}, \quad \rho = k_{12}/k'_{12}$$

$$\delta = k_{11}/k_{22}, \quad \text{and} \quad \alpha = [(m_2)^*_{n+1}]/[(m_2)^*_n] \, (n \geqslant 1).$$

The final copolymer composition is given by

$$\left(\frac{\{M_1\}}{\{M_1\} + \{M_2\}} \right)_{\text{copoly}} = \frac{(1-\alpha) + r_1(1-\alpha)\dfrac{[M_1]}{[M_2]} + (1-\alpha)^2 \dfrac{\delta r_2}{\rho[M_2]}}{(2-\alpha) + r_1(1-\alpha)\dfrac{[M_1]}{[M_2]} + (1-\alpha)^2 \dfrac{\delta r_2}{\rho[M_2]}}$$

The second model was similar, except that both M_1 and M_2 monomer add to the active chain end (m_2^* reversibly) and depropagation of the active end m_1^* is neglected. That is, elementary reactions (1) and (2) are considered to be irreversible and elementary reactions (3) and (4) are considered to be reversible. Two additional parameters are defined

$$\sigma = k'_{21}/k_{21}, \quad \beta = \frac{[(m_2)_n(m_1)^*_1]}{[(m_2)^*_n]} \quad (n \geqslant 1)$$

References pp. 326–330

where $[(m_2)_n (m_1)_1^*]$ denotes the concentration of chain ends which have a terminal M_1 unit with n units of M_2 preceding it, i.e.

$$\sim m_1 \underbrace{m_2 \cdots m_2}_{n} m_1^*$$

The final equation is then

$$\left(\frac{\{M_1\}}{\{M_1\} + \{M_2\}}\right)_{copoly} = \frac{(1 - \alpha)(1 + r_1 X)}{(1 - \alpha)(1 + r_1 X) + 1}$$

where $X = [M_1]/[M_2]$.

By defining M_1 as BCMO and M_2 as THF, Yamashita et al. obtain a good fit to the second model, for polymerizations conducted at 30—50°C with $Et_3 O^+ BF_4^-$ as catalyst and in $CH_2 Cl_2$ solution. They conclude that their study demonstrates the effect of the penultimate THF monomer unit on the depropagation reaction. It should be noted that Yamashita et al. do not attempt to determine the concentration of growing chains or the degree of termination occurring in their copolymerizations. With the BF_4^- counter-ion, substantial termination would be expected for polymerizations at 50°C (see Section 4.1.1c). They do report a definite effect of initiator and counter-ion on the course of the copolymerization and point out that different initiators lead to varying degrees of alternating copolymerization.

7.2.3 Copolymerization of substituted oxetanes with THF above T_c for THF

A more favourable case for studying the kinetics of copolymerization of cyclic ethers was reported by Kubisa and Penczek [158]. They studied the copolymerization of oxetanes with a large excess of THF above the ceiling temperature of THF. The catalyst used was triisobutylaluminium and only monomer was used as solvent. Highly ordered copolymers were formed. If the oxetane was 3,3-dimethyloxetane (DMO) or 3,3-bis(fluoromethyl)oxetane (BFMO), the limiting composition [thf]/[ox] of the polymer tended towards one mole of THF for every mole of oxetane as the excess of THF monomer grew larger. If the oxetane was 3,3-bis(chloromethyl)oxetane (BCMO), the limiting composition was close to two moles of THF for every mole of oxetane. In order to explain their results Kubisa and Penczek recall the reversibility of ionic copolymerizations, especially above the ceiling temperature of one of the monomers. They suggest that if a depropagation reaction involving a THF molecule is to occur a transition state such as

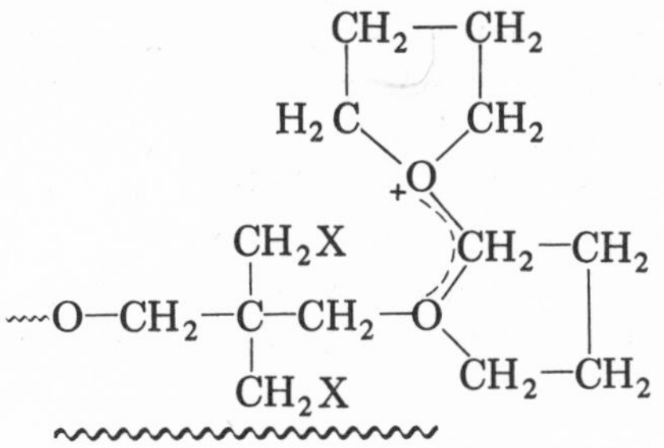

$$CH_2\text{—}CH_2$$

(structure)

(X = H, F or Cl)

needs to form. Thus, before a monomer molecule splits off the chain end, a partial bond within a penultimate unit is formed in the transition state. If the antepenultimate unit in the depropagation step, underlined by a wavy line in the formula, does not disturb the properties of the active centres which have THF molecules at their end, there is no reason to expect any interference with the propagation step. This seems to be the case when X = H or F. However, in the copolymerization of the BCMO—THF pair, the influence of the penultimate unit structure (or antepenultimate unit in the depropagation step) cannot be ignored. Analysis of Stuart molecular models showed that the chloromethyl groups of BCMO in these units are frozen and cannot rotate freely. Thus the depropagation reaction proceeds much more slowly in this case. Thermodynamically, the depropagation reaction proceeds with a much more negative entropy of activation than the same reaction with the other monomers. The kinetics of the copolymerization are also affected. The scheme proposed for these polymerizations was

(1) $\sim m_1^* + M_1 \xrightarrow{k_{11}} \sim m_1^*$

(2) $\sim m_1^* + M_2 \xrightarrow{k_{12}} \sim m_1 m_2^*$

(3) $\sim m_1 m_2^* + M_2 \underset{k_{-22}}{\overset{k_{22}}{\rightleftharpoons}} \sim m_2 m_2^*$

(4) $\sim m_1 m_2 m_2^* + M_2 \underset{k_{-222}}{\overset{k_{222}}{\rightleftharpoons}} \sim m_2 m_2 m_2^*$

(5) $\sim m_1 (m_2)_n m_2^* + M_1 \rightarrow \sim m_1 (m_2)_{n+1} m_1^*, \quad n \geqslant 0.$

The symbols m_2 and M_2 always refer to the THF, polymer and monomer, respectively. Positive subscript k's refer to the forward reactions and negative subscript k's to reverse reactions as usual. The scheme is based on several facts and assumptions:

(i) The chain propagation of oxetanes is considered to be practically irreversible. This is a reasonable assumption based on experience with oxetanes in homopolymerizations.

(ii) Addition of a THF molecule to an active centre bearing an oxetane at its end is irreversible since depropagation would require cyclization to a 4-membered ring.

(iii) Addition of THF to its own active centre is reversible.

(iv) Addition of oxetane to a THF active centre is irreversible.

Some elements of the scheme were previously presented by Yamashita et al. [157]. The general kinetic solution of the scheme is given by the equation

$$\frac{m_1}{m_2} = \frac{\left(r_1 \dfrac{[M_1]}{[M_2]} + 1\right)(1 + \alpha + \beta)}{1 + 2\alpha + 3\beta}$$

where

$$\alpha = \frac{[\sim m_1 m_2 m_2^*]}{[m_1 m_2^*]} = \frac{[M_2]}{\dfrac{1}{K_1} + \dfrac{1}{r_2}[M_1] + \dfrac{K_2^2 r_2 [M_2]}{r_2 + K_2 [M_1]}}$$

$$\beta = \frac{[\sim m_1 m_2 m_2 m_2^*]}{[\sim m_1 m_2 m_2^*]} = \frac{[M_2]}{\dfrac{1}{K_2} + \dfrac{1}{r_2}[M_1]}$$

$r_1 = k_{11}/k_{12}$ and $r_2 = k_{22}/k_{21}$. K_1 is the equilibrium constant of the propagation—depropagation equilibrium of a sequence such as (-bcmo)—

—(thf)—O+ ⟨ ⟩ , and K_2 is the equilibrium constant of the

propagation—depropagation equilibrium of a sequence such as -(thf)——

(thf)—O+ ⟨ ⟩ .

For the copolymerization involving DMO and BFMO, $\alpha = \beta = O$ and $K_1 = K_2 = O$. Hence the general solution reduces to

$$\frac{m_1}{m_2} = 1 + r_1 \frac{[M_1]}{[M_2]}$$

Thus r_1 can be determined from a plot of m_1/m_2 against $[M_1]/[M_2]$. For the copolymerization involving BCMO, $\beta = O$ and $K_2 = O$ but $\alpha \neq O$. In this case r_1 was measured as the slope of the tangent to the experimental plot relating m_1/m_2 and $[M_1]/[M_2]$ when $[M_2]/[M_1] \rightarrow O$. To determine K_1 and r_2 a knowledge of the dependence of α on $[M_1]/[M_2]$ is required. If α is written as a function of $[M_1]/[M_2]$ and m_1/m_2

$$\alpha = \frac{(m_1/m_2) - 1 - r_1[M_1]/[M_2]}{2(m_1/m_2) - 1 - r_1[M_1]/[M_2]}$$

the values of α can be determined from the same plot that gave r_1. Furthermore,

$$\frac{[M_2]}{\alpha} = \frac{1}{K_1} + \frac{[M_1]}{r_2}$$

Thus a plot of $[M_2]/\alpha$ against $[M_1]$ allows determination of the value of K_1 from the intercept and r_2 from the slope. The resulting kinetic parameters are summarized in Table 13. The comparison with the basicity values, here given in terms of pK_b, again shows that in a copolymerization of cyclic ethers, reactivity correlates well with basicity.

TABLE 13

Kinetic parameters in copolymerization of oxetanes and THF,[a] basicities of monomer [158]

Oxetane	r_1	r_2	K_1 (mole l^{-1})	K_2 (mole l^{-1})	pK_b
DMO	4.5	—	0	0	3.13[b]
BFMO	0.25	—	0	0	—
BCMO	0.35	9.0	0.8	0	5.65

[a] For copolymerizations carried out above T_c for THF.
[b] Taken to be approximately equal to the value for oxetane.
  Note that the pK_b for THF is 5.00.

7.3 CONCLUDING COMMENTS

All the above discussion has centred on cationic polymerizations. It should be realized that all the other types of homopolymerization mentioned for the monomers can occur in copolymerizations as well [14, 159]. Even cyclopolymerizations [160] and charge transfer reactions [161, 162] are known. But sorting out the exact reactions that are occurring and the efficiency with which they occur has a long way to go.

In summary, although copolymerizations of many cyclic ethers and cyclic sulphides have been reported, really meaningful determination of the kinetic parameters is still in its infancy. These studies are still at the stage where it is most important to recall the precept reiterated recently by Plesch [8]: "First the Chemistry, then the Kinetics."

8. Comparisons

"Industry, Perseverance, and Frugality, make Fortune yield." (B. Franklin)

After years of effort on the part of many scientists some reasonably well-substantiated kinetic parameters to describe the polymerization of some cyclic ethers and of some cyclic sulphides have evolved. It has been demonstrated that the choice of solvent (or lack of it) and of initiator are critical. The need for an independent means of determining the number of active sites has been proven again and again. The necessity of understanding the chemistry before embarking on a detailed kinetic analysis has been shown.

At the present time the available data seems to allow a few over-all conclusions to be drawn about the influence of the chemical structure of the monomer on its polymerizability and its rate of polymerization.

8.1 EFFECT OF RING SIZE

The effect of number of chain atoms in the ring is somewhat different for the cyclic oxides and sulphides. This is not surprising since compared to the oxygen atom the sulphur atom is much larger. The large atom naturally leads to a longer C—S bond, which would tend to allow greater C—S—C bond angle distortion in the cyclic sulphide molecule and relieve some of the strain imposed by the small ring system [66]. A comparison of strain energies is shown in Table 14.

TABLE 14
Strain energy and polymerizability of cyclic ethers and sulfides[a]

Ring size	Strain energy (kcal mole^{-1})		Polymerizability	
	Ether	Sulphide	Ether	Sulphide
3	27.2	19.8	+	+
4	25.5	19.6	+	+
5				
Unsubstd.	5.6	2.0	+	—
Substd.	n.a.	n.a.	±	—
Bicyclic	n.a.	n.a.	+	—
6	1.2	−0.3	—	—

[a] Most of this data is taken from Stille and Empen [66]; n.a. = not available.

Saegusa et al. [118, 163] have compared kinetic parameters determined for cyclic oxides of different ring sizes. Their comparisons are shown in Table 15. All of these parameters were determined under similar conditions and using the same techniques so that the comparison is significant. The order of reactivity is oxetane > THF > oxepane. Saegusa et al. attribute the difference in reactivity of oxetane and THF to a difference in pre-exponential factor. Both rings are essentially planar and strained but the oxetane ring is more strained. In contrast, the difference in the reactivity of THF and oxepane seems to come from the larger activation

TABLE 15

Propagation rate coefficients and activation parameters [118, 163] of cyclic ether polymerizations[a]

Parameter	Temp. (°C)	Ring size		
		4	5	7
$k_p \times 10^3$ (l mole^{-1} sec^{-1})	−28	7.5	—	—
	−23	13	—	—
	−10	57	1.7	0.0033
	0	140	4.1	0.015
	10	—	8.4	0.031
	20	—	—	0.091
	30	—	—	0.43
$\Delta G^{\ddagger}$(kcal mole^{-1} at 0°C)		17	22	27
E_p(kcal mole^{-1})		14	12	18
$10^{-7}A$(l mole^{-1} sec^{-1})		5300	1.1	190

[a] Solution polymerization in CH_2Cl_2. For oxetane: $[M]_0 = 3.1$ mole l^{-1}, $[BF_3 . THF]_0 = 0.003$ mole l^{-1}. For THF: $[M]_0 = 6.3$ mole l^{-1}, $[BF_3 . THF]_0 = [ECH]_0 = 0.01$ mole l^{-1}. For oxepane: $[M]_0 = 2.9$ mole l^{-1}, $[BF_3 . THF]_0 = [ECH]_0 = 0.057$ mole l^{-1}.

energy for oxepane. The reason for the larger activation energy may again be ring strain. The 7-membered ring takes a puckered form. Thus the strain in the 7-membered ring oxonium ion can be relieved by small deformations of the angles to other bonds. It is interesting to note that the free energy increases in the order $4 < 5 < 7$. From these data it would seem that activation energy and pre-exponential factor are better measures of reactivity than free energy.

8.2 BASICITY

In much of the early work basicity emerged as the most important factor relating relative reactivity of the cyclic ethers. It is still considered an important factor in many polymerizations. We have already seen that during polymerizations of thietanes, the relative basicity of the sulphonium ion of both the monomers and the polymers plays an important role. In this case steric hindrance was also important.

In copolymerizations of cyclic ethers, too, basicity of the monomers seems to correlate well with the reactivity ratios determined from copolymer composition. Recently, Saegusa et al. [54] have pointed out that the apparent values of the monomer reactivity ratios in cyclic ether copolymerizations may be more influenced by the exchange

$$\sim\!\!\!O^+ \quad M_1 + O \quad M_2 \; \rightleftharpoons \; \sim\!\!\!O^+ \quad M_2 + O \quad M_1$$

References pp. 326—330

than by kinetic reactivity of the ring opening process of a propagation reaction. We have also seen from the data of Kubisa and Penczek [158] on the copolymerization of THF and BCMO that the depropagation reaction may play an important role in copolymerization kinetics and that even penultimate and antepenultimate units may be important.

It seems fair to say that at the time of writing the factors influencing reactivity of cyclic ethers and sulphides are still not completely known. We can look for greater insight as more detailed and carefully controlled kinetic studies are made.

Note added in proof

Since this review was written a wealth of kinetic data on THF polymerizations initiated by esters of trifluoromethyl sulphonic acid, fluorosulphonic acid and other "super acids" has appeared [164—168]. It has correctly been pointed out [167, 169] that secondary oxonium ions should not be observable separately under the experimental conditions of Pruckmayr and Wu [96] and that some of their NMR assignments are wrong. Instead, it has been verified that the observed spectrum corresponds to a very solvent sensitive equilibrium between a macro-ester and a macro-ion as proposed earlier by Smith and Hubin [98], viz.

$$\sim O^+\!\!\bigcirc \quad SO_3CF_3^- \;\rightleftharpoons\; \sim OCH_2\,CH_2\,CH_2\,CH_2\,OSO_2\,CF_3$$

Thorough kinetic analyses have been made for THF and several other cyclic ethers by using 60 MHz ^{1}H NMR [169], ^{19}F NMR [165, 169] and 300 MHz ^{1}H NMR [170] spectroscopy. Unfortunately, incorporation of this important body of new results in the present chapter at this stage is not practical and the reader is referred to the original papers, which we have cited.

REFERENCES

1 P. H. Plesch, *The Chemistry of Cationic Polymerization*, Macmillan, New York, 1963.
2 H. Perst, *Oxonium Ions in Organic Chemistry*, Verlag Chemie Academic Press, Mouton, The Hague, Netherlands, 1971.
3 N. G. Gaylord (Ed.), High Polymers, Vol. XIII, *Polyethers Part I. Polyalkylene Oxides and Other Polyethers*, Wiley—Interscience, New York, 1963.
4 J. Furukawa and T. Saegusa, *Polymerization of Aldehydes and Oxides*, Wiley—Interscience, New York, 1963.
5 P. Dreyfuss and M. P. Dreyfuss, *Advan. Polym. Sci.*, 4 (1967) 528.
6 A. Ledwith and C. Fitzsimmonds, in J. P. Kennedy and E. G. M. Törnqvist (Eds.), *Polymer Chemistry of Synthetic Elastomers, Part I*, Interscience, New York, 1968, p. 377.

7 K. C. Frisch and S. L. Reegen (Eds.), *Ring-Opening Polymerization*, Marcel Dekker, New York, 1969.
8 P. H. Plesch, *Advan. Polym. Sci.*, 8 (1971) 137.
9 T. Tsuruta, *Macromol. Rev. J. Polym. Sci. D.*, 6 (1972) 179.
10 L. A. Korotneva and G. P. Belonovskaya, *Russ. Chem. Rev.*, 41 (1972) 83.
11 O. Vogl and J. Furukawa, *Polymerization of Heterocyclics*, Marcel Dekker, New York, 1973.
12 P. Dreyfuss, *Chem. Tech.*, 3 (1973) 356.
13 E. Roithner, *Monatsh. Chem.*, 15 (1894) 679.
14 Y. Ishii and S. Sakai, Chapter 1 in ref. 7.
15 G. Gee, W. C. E. Higginson, P. Levesley and K. J. Taylor, *J. Chem. Soc., London,* (1959) 1338.
16 G. Gee, W. C. E. Higginson and G. T. Merrall, *J. Chem. Soc., London,* (1959) 1345.
17 G. Gee, W. C. E. Higginson, K. Taylor and M. W. Trenholme, *J. Chem. Soc., London,* (1961) 4298.
18 B. Wojtech, *Makromol. Chem.*, 66 (1966) 180.
19 C. C. Price and D. D. Carmelite, *J. Amer. Chem. Soc.*, 88 (1966) 4039.
20 C. E. H. Bawn, A. Ledwith and N. McFarlane, *Polymer*, 10 (1969) 653.
21 C. C. Price, M. K. Akkapeddi, B. T. DeBona and B. C. Furie, *J. Amer. Chem. Soc.*, 94 (1972) 3964.
22 C. C. Price and M. K. Akkapeddi, *J. Amer. Chem. Soc.*, 94 (1972) 3972.
23 L. E. St. Pierre and C. C. Price, *J. Amer. Chem. Soc.*, 78 (1956) 3432.
24 A. Bar-Ilan and A. Zilkha, *J. Macromol. Sci., Chem.*, 4 (1970) 1727.
25 A. A. Solov'yanov and K. S. Kazanskii, *Polym. Sci. USSR*, 12 (1970) 2396.
26 Ya. I. Estrin and S. G. Entelis, *Polym. Sci. USSR*, 10 (1968) 3006; 11 (1969) 1286; 13 (1971) 1862.
27 A. M. Eastham, Chapter 10 in ref. 1.
28 R. J. Kern, *J. Org. Chem.*, 33 (1968) 388.
29 M. E. Pruitt and J. M. Baggett (to Dow Chemical Co.), U.S. Patent 2,706,181 (1955).
30 R. O. Colclough, G. Gee, W. C. E. Higginson, J. B. Jackson and M. Litt, *J. Polym. Sci.*, 34 (1959) 171.
31 G. Gee, W. C. Higginson and J. B. Jackson, *Polymer*, 3 (1962) 231.
32 C. Booth, W. C. Higginson and E. Powell, *Polymer*, 5 (1964) 479.
33 J. Furukawa and Y. Kumata, *Makromol. Chem.*, 136 (1970) 147.
34 E. J. Vandenberg, *J. Polym. Sci., Part A-1*, 7 (1969) 525.
35 E. J. Vandenberg, *J. Polym. Sci., Part A-1*, 10 (1972) 329.
36 P. Sigwalt, Chapter 4 in ref. 7.
37 S. Boileau, G. Champetier and P. Sigwalt, *Makromol. Chem.*, 69 (1963) 180.
38 S. Boileau, G. Champetier and P. Sigwalt, *J. Polym. Sci., Part C16* (1967) 3021.
39 J. C. Favier, S. Boileau and P. Sigwalt, *Eur. Polym. J.*, 4 (1968) 3.
40 S. Boileau and P. Sigwalt, *Eur. Polym. J.*, 3 (1967) 57.
41 P. Guerin, P. Hemery, S. Boileau and P. Sigwalt, *Eur. Polym. J.*, 7 (1971) 953.
42 E. J. Goethals, W. Drijvers, D. Van Ooteghem and A. M. Boyle, *J. Macromol. Sci., Chem.*, 7 (1973) 1375.
43 L. Lambert, D. Van Ooteghem and E. J. Goethals, *J. Polym. Sci., Part A1*, 9 (1971) 3055.
44 J. P. Machon and P. Sigwalt, *C. R. Acad. Sci.*, 260 (1965) 549.
45 J. Furukawa, N. Kawabata and A. Kato, *J. Polym. Sci., Part B*, 5 (1967) 1073.
46 N. Spassky and P. Sigwalt, *C. R. Acad. Sci.*, 265 (1967) 624.
47 K. J. Ivin, E. D. Lillie, P. Sigwalt and N. Spassky, *Macromolecules*, 4 (1971) 345.
48 N. Spassky and P. Sigwalt, *Tetrahedron Lett.*, 32 (1968) 3541.
49 N. Spassky and P. Sigwalt, *Bull. Soc. Chim. Fr.*, (1967) 4617.

328

50 J. B. Rose, *J. Chem. Soc., London*, (1956) 542, 546.
51 A. C. Farthing and R. J. W. Reynolds, *J. Polym. Sci., 12* (1954) 503.
52 T. Saegusa, J. Hashimoto and S. Matsumoto, *Macromolecules, 4* (1971) 1.
53 T. Saegusa and S. Matsumoto, *J. Polym. Sci., Part A1, 6* (1968) 1559.
54 T. Saegusa, H. Fujii, S. Kobajashi, H. Ando and R. Kawase, *Macromolecules, 6* (1973) 26.
55 A. C. Farthing, *J. Chem. Soc., London*, (1955) 3648.
56 I. Penczek and S. Penczek, *Makromol. Chem., 67* (1963) 203.
57 S. Penczek, *Bull. Acad. Pol. Sci., Ser. Sci. Chim., 18,*(1970) 53.
58 P. Dreyfuss and M. P. Dreyfuss, Chapter 2 in ref. 7.
59 S. Penczek and P. Kubisa, *Makromol. Chem., 130* (1969) 186.
60 S. Penczek, *Bull. Acad. Pol. Sci., Ser. Sci. Chim., 18* (1970) 47.
61 S. Penczek, *Makromol. Chem., 134* (1970) 299.
62 T. Saegusa, H. Imai and J. Furukawa, *Makromol. Chem., 65* (1963) 60.
63 K. Takakura, K. Hayashi and S. Okamura, *J. Polym. Sci., Part A1, 4* (1966) 1731, 1747.
64 T. W. Campbell and V. S. Foldi, *J. Org. Chem., 26* (1961) 4654.
65 S. Penczek and A. A. Vansheidt, *Polym. Sci. USSR, 4* (1963) 927.
66 J. K. Stille and J. A. Empen, *J. Polym. Sci., Part A1, 5* (1967) 273.
67 M. Morton and R. F. Kammereck, *J. Amer. Chem. Soc., 92* (1970) 3217.
68 E. J. Goethals and W. Drijvers, *Makromol. Chem., 136* (1970) 73.
69 M. Morton, R. F. Kammereck and L. J. Fetters, *Macromolecules, 4* (1971) 11; *Brit. Polym. J., 3* (1971) 120.
70 M. Morton and S. L. Mikesell, *J. Macromol. Sci., Chem., 7* (1973) 1391.
71 E. J. Goethals and E. Du Prez, *J. Polym. Sci., Part A1, 4* (1966) 2893.
72 M. P. Dreyfuss and P. Dreyfuss, *J. Polym. Sci., Part A1, 4* (1966) 2179.
73 D. Sims, *J. Chem. Soc., London*, (1964) 864.
74 F. S. Dainton and K. J. Ivin, *Quart. Rev. Chem. Soc., 12* (1958) 61.
75 C. Huang and H. Wang, *J. Polym. Sci., Part A1, 10* (1972) 791.
76 H. Sawada, *J. Macromol. Sci.-Revs. Macromol. Chem., C8* (1972) 235.
77 S. Bywater, *Makromol. Chem., 52* (1962) 120.
78 J. Leonard, *Macromolecules, 2* (1969) 661.
79 K. J. Ivin and J. Leonard, *Eur. Polym. J., 6* (1970) 331.
80 J. Leonard and D. Maheux, *J. Macromol. Sci., Chem., 7* (1973) 1421.
81 W. K. Busfield, R. M. Lee and D. Merigold, *Makromol. Chem., 156* (1972) 183.
82 P. Dreyfuss and M. P. Dreyfuss, *Advan. Chem. Ser., 91* (1969) 335.
83 T. Saegusa and S. Matsumoto, *J. Macromol. Sci., Chem., 4* (1970) 873.
84 Y. Yamashita, S. Kozawa, M. Hirota, K. Chiba, H. Matsui, A. Hirao, M. Kadama and K. Ito, *Makromol. Chem., 142* (1971) 171.
85 Y. Yamashita, H. Matsui, G. Hattori, S. Kozawa and M. Hirota, *Makromol. Chem., 142* (1971) 183.
86 D. Vofsi and A. V. Tobolsky, *J. Polym. Sci., Part A, 3* (1965) 3261.
87 B. A. Rozenberg, O. M. Chekhuta, E. B. Lyudvig, A. R. Gantmakher and S. S. Medvedev, *Polym. Sci. USSR, 6* (1964) 2246.
88 J. M. Sangster and D. J. Worsfold, *Macromolecules, 5* (1972) 229.
89 E. B. Lyudvig, B. A. Rozenberg, T. M. Zuereva, A. R. Gantmakher and S. S. Medvedev, *Polym. Sci. USSR, 7* (1965) 296.
90 J. Stejny, *J. Macromol. Sci., Chem., 7* (1973) 1435.
91 T. Saegusa and S. Matsumoto, *Macromolecules, 1* (1968) 442.
92 H. Imai, T. Saegusa, S. Matsumoto, T. Tadasa and J. Furukawa, *Makromol. Chem., 102* (1967) 222.
93 T. Saegusa, H. Imai and S. Matsumoto, *J. Polym. Sci., Part A1, 6* (1968) 459.
94 T. Saegusa, S. Matsumoto and Y. Hoshimoto, *Polym. J., 1* (1970) 31.
95 A. V. Zak, E. Zh. Menligaziev, V. M. Breitman and Yu. A. Gorin, *J. Appl. Chem. USSR, 44* (1971) 786.

96 G. Pruckmayr and T. K. Wu, *Macromolecules*, *6* (1973) 33.
97 E. B. Lyudvig, E. L. Berman, Z. N. Nysenko, V. A. Ponomarenko and S. S. Medvedev, *Polym. Sci. USSR*, *13* (1971) 1546.
98 S. Smith and A. J. Hubin, *J. Macromol. Sci., Chem.*, *7* (1973) 1399.
99 S. N. Bhadani, Ph.D. Thesis, University of Manitoba, 1966, p. 120.
100 N. Yamazaki, *Advan. Polym. Sci.*, *6* (1969) 377.
101 C. F. Heins, *J. Polym. Sci., Part B*, *7* (1969) 625.
102 S. Nakahama, S. Hino and N. Yamazaki, *Polym. J.*, *2* (1971) 56.
103 D. Sims, *Makromol. Chem.*, *98* (1966) 235, 245.
104 C. E. H. Bawn, R. M. Bell, C. Fitzsimmons and A. Ledwith, *Polymer*, *6* (1965) 661.
105 N. Oguni, M. Kamachi and J. K. Stille, *Macromolecules*, *6* (1973) 146.
106 R. Chiang and J. H. Rhodes, *J. Polym. Sci., Part B*, *7* (1969) 643.
107 E. L. Wittbecker, H. K. Hall, Jr. and T. W. Campbell, *J. Amer. Chem. Soc.*, *82* (1960) 1218.
108 P. Guisti, U. Fiorentino, G. Turchi, F. Andruzzi and P. L. Magagnini, *Makromol. Chem.*, *128* (1969) 1.
109 J. Kops and H. Spanggaard, *Makromol. Chem.*, *151* (1972) 21.
110 J. Kops and H. Spanggaard, *J. Macromol. Sci., Chem.*, *7* (1973) 1455.
111 T. Saegusa, M. Motoi, S. Matsumoto and H. Fujii, *Macromolecules*, *5* (1972) 233.
112 M. Baccaredda, P. Guisti, F. Andruzzi, P. Cerrai and M. Dimaina, *J. Polym. Sci., Part C31* (1970) 157.
113 T. Saegusa, S. Matsumoto, M. Motoi and H. Fujii, *Macromolecules*, *5* (1972) 236.
114 T. Saegusa, S. Matsumoto, M. Motoi and H. Fujii, *Macromolecules*, *5* (1972) 815.
115 P. A. Small, *Trans. Faraday Soc.*, *51* (1955) 1717.
116 T. Saegusa, T. Hodaka and H. Fujii, *Polym. J.*, *2* (1971) 670.
117 K. Kamio, K. Meyersen, R. C. Schulz and W. Kern, *Makromol. Chem.*, *90* (1966) 187.
118 T. Saegusa, T. Shiota, S. Matsumoto and H. Fujii, *Macromolecules*, *5* (1972) 34.
119 T. Saegusa, T. Shiota, S. Matsumoto and H. Fujii, *Polym. J.*, *3* (1972) 40.
120 L. Lal and G. S. Trick, *J. Polym. Sci.*, *50* (1961) 13.
121 T. P. Hobin and R. T. Lowson, *Polymer*, *7* (1966) 217, 223.
122 J. Furukawa and K. Tada, Chapter 3 in ref. 7.
123 P. H. Plesch and P. H. Westermann, *Polymer*, *10* (1969) 105.
124 M. Okada, Y. Yamashita and Y. Ishii, *Makromol. Chem.*, *80* (1964) 196.
125 H. Sumitomo, M. Okada and Y. Hibino, *J. Polym. Sci., Part B*, *10* (1972) 871.
126 B. Krummenacher and H. G. Elias, *Makromol. Chem.*, *150* (1971) 271.
127 Y. Yamashita, M. Okada and H. Kasahara, *Makromol. Chem.*, *117* (1968) 256.
128 Y. Yamashita, *Advan. Chem. Ser.*, *91* (1969) 351.
129 F. R. Jones and P. H. Plesch, *Chem. Commun.*, (1969) 1230.
130 B. A. Rozenberg, B. A. Komarov, T. I. Ponomareva and N. S. Enikolopyan, *J. Polym. Sci., Polym. Chem. Ed.*, *11* (1973) 1.
131 P. H. Plesch and P. H. Westermann, *J. Polym. Sci., Part C16* (1968) 3837.
132 V. Jaacks, K. Boehlke and E. Eberius, *Makromol. Chem.*, *118* (1968) 354.
133 K. Boehlke and V. Jaacks, *Makromol. Chem.*, *145* (1971) 219.
134 T. Kelen, D. Schlotterbeck and V. Jaacks, *XXIII IUPAC Macromolecular Preprint*, Boston, *2* (1971) 649.
135 K. Boehlke, P. Weyland and V. Jaacks, *XXIII IUPAC Macromolecular Preprint*, Boston, *2* (1971) 641.
136 V. Jaacks, *Makromol. Chem.*, *161* (1972) 161.
137 S. Penczek and P. Kubisa, *Makromol. Chem.*, *165* (1973) 121.
138 V. Jaacks, K. Boehlke and W. Kern, *Makromol. Chem.*, *165* (1973) 51.
139 W. Kern, H. Cherdron and V. Jaacks, *Angew. Chem.*, *73* (1961) 177.
140 W. Kern, *Chem.-Ztg., Chem. App.*, *88* (1964) 623.

330

141 K. Weissermel, E. Fischer, K. Gutweiler, H. D. Hermann and H. Cherdron, *Angew. Chem.*, *79* (1967) 512.
142 V. Jaacks, *Advan. Chem. Ser.*, *91* (1969) 371.
143 V. Jaacks, H. Frank, E. Grünberger and W. Kern, *Makromol. Chem.*, *118* (1968) 290.
144 G. Mengoli and G. Vidotto, *Makromol. Chem.*, *165* (1973) 137, 145.
145 P. S. Raman and V. Mahadevan, *Makromol. Chem.*, *165* (1973) 153.
146 Y. Yamashita, T. Tsuda, M. Okada and S. Iwatsuki, *J. Polym. Sci., Part A1, 4* (1966) 2121.
147 T. Tsuda, T. Nomura and Y. Yamashita, *Makromol. Chem.*, *86* (1965) 301.
148 E. J. Alvarez, V. Hornof and L. P. Blanchard, *J. Polym. Sci., Part A1, 10* (1972) 2237.
149 Y. Yamashita, F. Inoue, G. Hattori and K. Ito, *Makromol. Chem.*, *151* (1972) 91.
150 F. R. Mayo and F. F. Lewis, *J. Amer. Chem. Soc.*, *66* (1944) 1594.
151 P. Dreyfuss and M. P. Dreyfuss, *Encycl. Polym. Sci. Technol.*, *13* (1970) 670.
152 H. E. Wirth and P. L. Slick, *J. Phys. Chem.*, *66* (1962) 2277.
153 S. Iwatsuki, N. Takigawa, M. Okada, Y. Yamashita and Y. Ishii, *J. Polym. Sci., Part B, 2* (1964) 549.
154 W. Gordy and S. C. Stanford, *J. Chem. Phys.*, *7* (1939) 93; *8* (1940) 170; *9* (1941) 204, 215.
155 Y. Tanaka, *J. Macromol. Sci. Chem.*, *1* (1967) 1059.
156 V. Jaacks, *Makromol. Chem.*, *105* (1967) 289.
157 Y. Yamashita, H. Kasahara, K. Suyama and M. Okada, *Makromol. Chem.*, *117* (1968) 242.
158 P. Kubisa and S. Penczek, *J. Macromol. Sci., Chem.*, *7* (1973) 1509.
159 C. C. Price, Y. Atarashi and R. Yamamoto, *J. Polym. Sci., Part A1, 7* (1969) 569.
160 C. L. McCormick and G. B. Butler, *J. Macromol. Sci. Revs. Macromol. Chem.*, *C8* (1972) 201.
161 G. Butler, J. T. Badgett and M. Sharabash, *J. Macromol. Sci. Chem.*, *4* (1970) 51.
162 A. Cardon and E. J. Goethals, *J. Macromol. Sci. Chem.*, *5* (1971) 1021.
163 T. Saegusa, *J. Macromol. Sci. Chem.*, *6* (1972) 997.
164 K. Matyjaszewski, P. Kubisa and S. Penczek, *J. Polym. Sci., Part A-1, 12* (1974) 1333.
165 S. Kobayashi, H. Danda and T. Saegusa, *Macromolecules, 7* (1974) 415.
166 T. Saegusa and S. Kobayashi, *ACS Symposium Series, No. 6, Polyethers*, 1975, p. 150.
167 S. Kobayashi, H. Danda and T. Saegusa, *Bull. Chem. Soc. Jap.*, *46* (1973) 3214.
168 K. Matyjaszewski and S. Penczek, *Macromolecules, 7* (1974) 173.
169 T. K. Wu and G. Pruckmayr, *Macromolecules, 8* (1975) 75.
170 K. Matyjaszewski and S. Penczek, *J. Polym. Sci., Part A-1, 12* (1974) 1905.

Chapter 5

Kinetics of Aldehyde Polymerization

OTTO VOGL

1. Introduction

Aldehyde polymers are probably the oldest synthetic polymers [1—8]. Polyoxymethylene the polymer of formaldehyde, was first described by Butlerov in 1859 and the polymer of chloral was first prepared in 1832 by Liebig. This was about one century before the concept of linear macromolecules was developed. Polyformaldehyde and polychloral were isolated because they were stable at room temperature and above.

The true understanding of the polymerization of aliphatic higher aldehydes began with the clear understanding of the importance of the ceiling temperature for these polymerizations. The polymers of aliphatic aldehydes have ceiling temperatures substantially below room temperature (under atmospheric pressure). Ceiling temperature increases with pressure and at pressures of several thousand atm the ceiling temperature of aliphatic aldehyde polymerization is above room temperature [9].

In this review the polymerization of formaldehyde, higher aliphatic aldehydes and haloaldehydes will be discussed with particular emphasis on the kinetics of the polymerization. As will be apparent the kinetics of aldehyde polymerization have not been studied as extensively as the kinetics of more conventional polymerizations, for example, the free radical bond opening polymerizations of styrene, vinyl chloride or methylmethacrylate or the ring opening polymerizations of tetrahydrofuran or ethylene oxide. One reason is that polyoxymethylene is the only polyaldehyde produced commercially and much of our knowledge on formaldehyde polymerization is proprietary information. Another is that the polymerization systems are very complex and the polymers precipitate during polymerization.

Polyformaldehyde can also be prepared by polymerization of trioxane, the cyclic trimer of formaldehyde. Trioxane polymerizes by ring opening polymerization and cationic initiators are the only effective initiators. Formaldehyde is always present when trioxane is polymerized because the growing polyoxymethylene chains by depropagation may lose one monomer unit, which is formaldehyde not trioxane. In spite of the fact that formaldehyde plays an (as yet incompletely understood) role in trioxane polymerization, which is a cyclic ether polymerization like dioxolane or tetrahydrofurane [5], trioxane will not be discussed in this review.

The polymerization of aldehydes is initiated by ionic initiators and the polymerization proceeds by ionic propagation. No radical polymerization of aldehydes has been documented yet. In the case of anionic polymerizations the growing ion is an alkoxide ion. The cationic polymerization has as the propagating species an oxonium ion. Most recent experimental results have shown that haloaldehydes, such as chloral polymerize exclusively by an anionic mechanism.

Aldehyde polymerizations are carried out in aprotic anhydrous media. Even small amounts of protonic impurities cause efficient chain transfer reactions and low molecular weight polymer is formed.

In some cases, with stannous acylates as initiators, formaldehyde polymerization [10] can be carried out to very high molecular weight polyethymethylene even in the presence of 2 mole % of water. Unlike most polyoxymethylenes made under scrupulously anhydrous conditions with common cationic or anionic initiators, which have a most probable molecular weight distribution, polyoxymethylenes made with stannous acrylates as the initiators have a very broad molecular weight distribution.

Formaldehyde can also polymerize in hydroxylic media [11] to high molecular weight. In water or methanol, formaldehyde forms first methylene glycol or hemi-formal and then low polymers. This tendency of formaldehyde has been utilized to develop conditions where high molecular weight polymers were obtained at a reasonable rate and of rather narrow molecular weight distribution. This process is based on the understanding of crystallization during the polymer formation and the morphology of the polymer formed. In fact the major driving force for this polymerization is the crystallization of the polymer.

This brings us to an important point in aldehyde polymerization, the problem of precipitation or crystallization of the polymer during polymerization. In all cases of aldehyde polymerization where crystalline polymers are formed, in formaldehyde polymerization and higher aldehyde polymerization to isotactic polymers, precipitation occurs during the polymerization.

Rapid polymerization of acetaldehyde to amorphous polymer has been observed with highly active initiators at very low temperatures in solvents, where the polymer is not soluble. Since the polymer is below the T_g at reaction temperature it precipitates as a glass. The rate of polymer formation is almost explosive. With weakly cationic initiators acetaldehyde can be polymerized to low molecular weight polymer where the polymer remains in solution.

Most aldehyde polymerizations are carried out in solvents of low dielectric constant. The solubility of aldehydes in low dielectric constant solvent is limited. At room temperature, formaldehyde is soluble in pentane only to the extent of 0.5% and in toluene of 2%. Nevertheless the polymerization often proceeds as fast as the monomer can be supplied. Acetaldehyde is miscible with pentane in all proportions above $-30°C$ and *n*-butyraldehyde above $-90°C$. The limited solubility of aldehydes, especially the lower members, suggests a strong tendency of aldehydes to associate in these solvents. It has actually been found that in hydrocarbons at low temperatures acetaldehyde forms dipolar associates averaging 3—5 molecules [5, 12].

In solvents of high dielectric constant, (e.g. dimethylformamide), formaldehyde polymerized sluggishly and polymers were formed in low yield. In similar solvents, aliphatic aldehydes could not be polymerized. Precipitated aldehyde polymers have the tendency to absorb monomers. The monomer concentration near the propagating site may be much higher than that in the surrounding solution. This occurs with *n*-butyraldehyde polymerizations in pentane [5]; the polymer precipitated during the polymerization is highly swollen by the monomer.

Such observations demonstrate that the physical state and surface area of the crystallized polymer have a pronounced influence on the progress of polymerization. Availability of growing polymer ends and diffusion of the monomer to the active site play an all important role in polymerizations where the polymer crystallized during polymerization. This has been demonstrated in the polymerization of formaldehyde in hydroxylic media. When precipitation of polyformaldehyde in the methanolic polymerization of formaldehyde is allowed to occur indiscriminately about 80% of potentially active sites are excluded within the polymer crystal. Acetic anhydride can only acetylate about 20% of the free hydroxyl end groups. If, however, polymerization is carried out under optimum conditions with formation of the proper polymer crystals, almost 100% of the ends are "available". In this case, the molecular weight distribution is nearly similar to that of a living polymer and the polymer is almost 100% crystalline. The growth of the polymer crystals can be influenced by solvent, concentration of monomer, temperature and proper nucleation. Other factors which might influence the crystal habit of the polymers is the type of initiator, the structure of the growing polymer chain, transfer reactions and possible impurities.

Induction periods of various lengths have been reported for anionic and cationic polymerizations of formaldehyde. It is apparent that the compound that is added "as initiator" is rarely the actual initiator. Tetraalkyl ammonium acetate used for the polymerization of formaldehyde is a slow initiator but capable of initiating formaldehyde polymerization. Methanol, other alcohols or water, always present in the polymerization mixture, are responsible for the high rates of polymerization. They act as efficient chain transfer agents and alkoxide or hydroxide ions are the actual initiators; they initiate by a factor of several powers of ten more efficiently than acetate.

Nothing is known about the actual initiators of aldehyde polymerization using Lewis acids. It is, however, believed that a stable counter-ion is always essential for the formation of high molecular weight polymers. An example of initial formation of the actual initiator is chloral polymerization with triphenyl phosphine. Triphenyl phosphine reacts instantaneously with one mole of chloral to give triphenyl dichlorovinyloxy phosphonium chloride which initiates chloral polymerization by chloride initiation.

Solubility of the initiator in the reaction medium can have a pronounced effect on rate of polymerization, molecular weight of the polymer and bulk density of the polymer formed. Extensive initiation in a true solution of the catalyst has given polymer suspensions which are hard to stir and form polymers of low bulk density. Associated initiators in colloidal solutions often give polymers of high bulk density. Relatively small changes in the structure of the initiator or the counter-ion (an added CH_2 group or branching in cation or anion) may change drastically the behaviour of the initiator.

In cationic polymerizations of aldehydes the growing cation can always be solvated by the acetalic oxygens in the polyoxymethylene chain, viz.

$$\text{---}CH_2\overset{(+)}{-}O{=}CH_2 + \underset{R'}{\overset{R'}{O}} \leftrightarrows \text{---}CH_2{-}O{-}CH_2\overset{R'}{\underset{R}{\overset{|(+)}{O}}} \qquad R' = \text{polyoxymethylene chain}$$

Very little is known about the effect of this interaction and how important this equilibrium is for the cationic polymerization, especially in solid/liquid interface reactions. Triethyloxonium fluoroborate, an excellent initiator for formaldehyde polymerization, can be visualized as an ethylcarbonium ion solvated by one mole of diethylether.

It is known that the transacetalization of polyoxymethylene occurs readily with acid catalyst and dioxalane or polydioxolane may be incorporated randomly into the polyoxymethylene chain. It is also known that electrophilic compounds, Lewis acids, formic acid, etc., are readily absorbed on the polyoxymethylene chains.

We have presented all these points in order to make the reader realize why relatively few papers in the literature are concerned with the kinetics of aldehyde polymerizations. It is almost impossible to take into consideration all the facts that have been discussed in this introduction in each experiment. Consequently, most authors report simply the time versus conversion curve of the polymerization without a detailed scrutiny of the individual factors. In addition, aldehyde polymerizations are fast, in some cases almost explosive with poor temperature control, and many aldehyde polymerizations are carried out in a semibatch process with continuous addition of monomers, although we know commercial processes are carried out in continuous reaction.

2. Formaldehyde

Formaldehyde polymerization has been studied in the liquid state, in "solution" of protic or aprotic solvents and in the gaseous state where gaseous formaldehyde forms directly crystalline polymer. It has been studied with anionic and cationic initiators and by high energy radiations. Although there are more than 100 MM lbs. of polyformaldehyde produced per year, very few papers have been published that are actually concerned with the kinetics of formaldehyde polymerizations. The reason for this lack of detail is understandable when one realizes how difficult it is to obtain pure formaldehyde (with impurities of less than 100 p.p.m). Even pure formaldehyde undergoes side reactions and self condensation which cause new introduction of impurities.

Early kinetic studies with gaseous formaldehyde [13] showed that the polymerization may be initiated by relatively small amounts of formic acid and can become almost explosive at higher formic acid concentrations. The rate of polymerization was found to be greater at lower temperatures and at $100°C$, even with high HCOOH concentrations, no polymerization was observed, indicating an early observance of the now well established ceiling temperature phenomena.

In the 1950's and 1960's a great number of patents [14] were issued suggesting hundreds of initiators for the polymerization of formaldehyde. No actual kinetic data were published, however, and even now the kinetic understanding of formaldehyde polymerization is limited.

2.1 ANIONIC POLYMERIZATIONS

Anionic polymerization of anhydrous formaldehyde has been disclosed in numerous patents, including the use of amines, phosphines, onium compounds, etc., as initiators. Mechanistically the propagation of the anionic polymerization of formaldehyde proceeds via the polymeric alkoxide ion,

$$\sim(CH_2-O)_n-CH_2-O^{(-)}$$

The termination of the polymer chain may be by chain transfer agents or by occlusion of the active chain ends. An example for the anionic chain transfer with water is

$$\sim(CH_2-O)_n-CH_2-O^{(+)} + H_2O \rightarrow$$

$$\sim(CH_2-O)_n-CH_2-OH + HO^-$$

$$HO^- + CH_2O \rightarrow HOCH_2-O^{(-)}$$

The initiation with acylates, for example tetraalkyl ammonium acetates, is fairly straightforward: acylate anion initiates formaldehyde by adding to the carbonyl carbon of the aldehyde, viz.

$$RCOO^- \overset{+}{N}R_4' + CH_2O \rightleftharpoons RCOOCH_2-O^{(-)}\overset{+}{N}R_4'$$

It depends primarily on the nucleophilicity of the anion of the onium salt how efficiently the polymerization of formaldehyde is initiated. It is believed that initiation of the polymerization of formaldehyde with acylate is slow and the over-all polymerization rate is determined by chain transfer to trace impurities of alcohols and water in the polymerizing system.

A more difficult question is the initiation of formaldehyde with amines, notably tertiary amines. Kern et al. [15, 16] recently discussed the initiation of formaldehyde with Lewis bases. They favour initiation of formaldehyde polymerization by direct addition to the nucleophilic end of the amine

$$R_3N + CH_2O \rightarrow R_3\overset{+}{N}-CH_2-O^{(-)}$$

The suggested formulation of a polymeric zwitter ion is always difficult to visualize because of the change separation in the zwitter ions in subsequent propagation steps.

The authors studied the polymerization of formaldehyde with amines including tertiary amines at $-78°C$ in various solvents (Table 1), and determined the conversion after 15 min reaction time. Tertiary amines are highly reactive initiators for formaldehyde polymerizations even at the level of 10^{-6} mole l^{-1} per mole l^{-1} of formaldehyde. The reactivity of the amine is related to its pK_a value but also to the branching of the aliphatic side chains of the substituents on the nitrogen atom. Branched amines, especially when the branching is on the α-carbon atom as in the case of a tertiary butyl group, are less effective initiators than tertiary amines with n-alkyl chains. The pK_a of the amine is not the essential feature for an efficient tertiary amine initiator, because pyridine was almost as effective as tri-n-butylamine but quinoline, with a similar pK_a as pyridine, is almost inactive (Table 1).

Primary and secondary amines have been found to be much less active as initiators for formaldehyde polymerization, presumably because these compounds react with formaldehyde to form methylol compounds of the amines which are much less basic.

It is interesting to note that triphenylphosphine is more effective as initiator than tertiary amines. In general it can be stated that the initial rate of formaldehyde polymerization with tri-n-butylamine is proportional to the amine concentration and the initial rate is also proportional to the monomer concentration (Figs. 1 and 2). The temperature dependence of

TABLE 1

Polymerization of monomeric formaldehyde at −78°C in solution (diethyl ether, acetone, toluene) with various amines as initiators
(Conversion after a reaction time of 15 min)

Conditions

Formaldehyde concentration, mole l⁻¹	2.76	4.85	3.2	5.3	4.4
Initiator concentration, mole l⁻¹	9.5×10^{-4}	9.5×10^{-4}	9.5×10^{-4}	2×10^{-4}	9.1×10^{-6}
Solvent	Acetone	Toluene	Ether	Ether	Ether
Dielectric constant of solvent, −78°C	34.5	2.6	7.5	7.5	7.5

Initiator	pK_a	Conversion %	Conversion %	Conversion %	Conversion %	Conversion %
Primary aliphatic amines:						
n-Butyl amine	10.4	0.6	11.1	0.3	19.2	0.2
tert-Butyl amine	10.4		0	0		0
Cyclohexyl amine	10.6		0	0		0
Secondary aliphatic amines:						
Di-n-butylamine	11.2	45.1	12.3	27.1	56.8	10.0ᵃ
Pyrrolidine	11.3	35.4	12.3	29.6		
Tertiary aliphatic amines:						
Tri-n-butyl amine	10.9	49.2	14.4	41.4	67.6	13.2ᵃ
Dimethyl-tert-butyl amine	10.5					4.8ᵃ
Dimethylcyclohexyl amine		25.4	12.3	26.8	48.4	8.4ᵃ
Aromatic N-Heterocycles:						
Pyridine	5.2	25.3		26.6	53.2	7.7ᵃ
Quinoline	4.9			2.3	5.0	
Arylamines:						
Aniline	4.6	0	0	0	0	
Monomethylaniline	4.8	0	0	0	0.2	
Dimethylaniline	5.1	2.6	9.0	0	1.3	
Diphenylamine	0.85	0	0	0	0	
Triphenylamine		0	0	0	0	

ᵃ Molecular weight of polymers between 225,000 and 330,000.

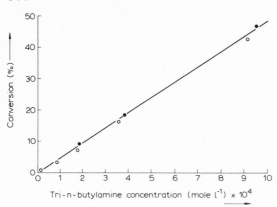

Fig. 1. Conversion (in %) of anionic polymerization of formaldehyde in diethyl ether at $-78°C$ as a function of the tri-n-butylamine concentration. Polymerization time: 15 min ($\bullet$) 3.52 mole l^{-1} CH_2O; ($\circ$) 3.64 mole l^{-1} CH_2O.

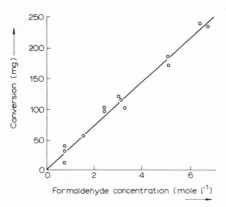

Fig. 2. Anionic polymerization of formaldehyde in diethyl ether at $-78°C$ with tri-n-butylamine.

the initial rate gives a low over-all activation energy of 3.9 kcal mole^{-1} (Kern et al. [16]) or more generally 3—4 kcal mole^{-1} (Enikolopyan [17]).

The suggestion of direct nucleophilic attack of the nitrogen on the carbonyl carbon to form the initiating zwitter ion is in contrast to the earlier and more reasonable suggestion that protic impurities act as co-catalyst in the following manner

$$R_3N + H_2O \rightleftharpoons R_3\overset{+}{N}H \, \overset{-}{O}H$$

$$R_3\overset{+}{N}H\overset{-}{O}H + CH_2O \rightarrow HO{-}CH_2\overset{(-)}{O} \, \underset{\underset{H}{|}}{\overset{+}{N}R_3}$$

Machacek et al. [18—22] and Vesely and Mejzlik [23, 24] studied the kinetics of formaldehyde polymerization in ether dilatometrically at −58°C. They took into account the effect of impurities, chiefly water and formic acid. The rate of polymerization was chosen by varying the concentration of initiator and monomer so that the initial over-all rate did not exceed 5% conversion per min. Monomer concentrations in diethylether ranged from 2.7×10^{-6} to 25×10^{-6} mole l^{-1}.

As seen in Fig. 3, the conversion curves especially at higher conversions have the character of reactions of higher order which seems to indicate that the polymerization becomes diffusion controlled. In actual fact, precipitation of polymer occurred during polymerization. The reaction rate decreases sharply in the initial stages of the polymerization. With

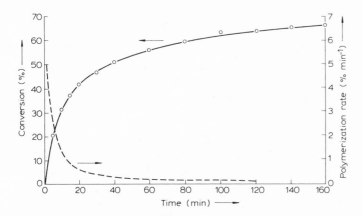

Fig. 3. Plot of (———) dependence of degree of conversion on time; (- - -) derivative of this function. Abscissa: time, min.; ordinates (left): polymerization, %, (right): polymerization rate, % min^{-1}.

increasing n-butylamine concentration (and constant water concentration), the over-all polymerization rate increased.

It was also observed that at constant n-butylamine concentration the over-all rate of polymerization increased with increasing water concentration. In the experiments illustrated in Fig. 3, the water concentration was 2.5×10^{-3} mole l^{-1}, about 1000 times higher than the initiator concentration. From extrapolation of the rate data to zero, it was noticed that some impurities were present in the polymerization mixture which used up part of the initiator (Fig. 4). The two sets of data were thought by the authors to be due to the fact that they had to prepare a new batch of monomer for each series of experiments. Formic acid or CO_2 were inhibitors for anionic formaldehyde polymerization.

In another set of experiments the authors showed that the initial rate of formaldehyde polymerization was not affected when the water concentration was varied from 2.5×10^{-3} to 104×10^{-3} mole l^{-1} at constant

340

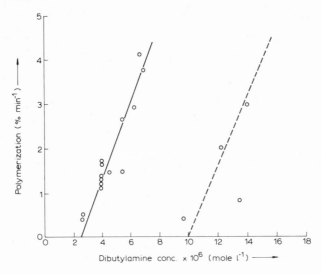

Fig. 4. Dependence of rate of polymerization on concentration of dibutylamine. Water concentration range 2.5×10^{-3} -3×10^{-3} mole 1^{-1}; formaldehyde concentration 4.7 mole 1^{-1}; solvent, diethyl ether; temperature $-58 \pm 0.1^{\circ}$C.

dibutylamine concentration (4×10^{-6} mole 1^{-1}). However, it must be pointed out that this concentration of water is more than 1000 times that of the initiator. It would be interesting to know how the rate would be influenced when the water concentration was of the same order of magnitude as the initiator concentration, an experimentally almost impossible achievement. As expected, the molecular weight of polyoxymethylene at increased water levels decreased, due to chain transfer reactions.

In further work, formic acid was added up to 8×10^{-3} mole 1^{-1}, while the monomer concentration (4.7 mole 1^{-1}), initiator concentration (dibutylamine, 8×10^{-6} mole 1^{-1}) and water (2.5×10^{-3} mole 1^{-1}) were held constant. With increasing formic acid concentration the initial rate of formaldehyde polymerization decreased and the molecular weight of the polymer also decreased.

The kinetics of formaldehyde polymerization in toluene solutions (80%) in the presence of tetrabutylammonium laurate, triethyl amine and calcium stearate were also studied. The initiator activity of these compounds decreased in the order tetrabutylammonium laurate $>$ triethyl amine $>$ calcium stearate. It was found that with triethyl amine no "spontaneous polymerization" was observed. Spontaneous polymerization was apparently an anionic polymerization and was inhibited by CO_2 or formic acid. In our opinion this is an indication that tertiary amines need a co-initiator for formaldehyde polymerization. In the case of water as the co-initiator HO^- was the initiating anion which was inhibited by CO_2.

Enikolopyan et al. [17, 25—27] showed that the formaldehyde polymerization at —30° in toluene with calcium stearate and tetrabutyl ammonium laurate was first order and no co-initiator was needed (Fig. 5). The polymerization with triethyl amine was higher than second order and a co-initiator was required. The dependence of the initial first order rate coefficient on initiator concentration showed that active centres were not consumed during the reaction, and indicated that the reaction was first order in monomer as well.

Chain termination in this formaldehyde polymerization occurred by transfer reactions. The average degree of polymerization was one order of

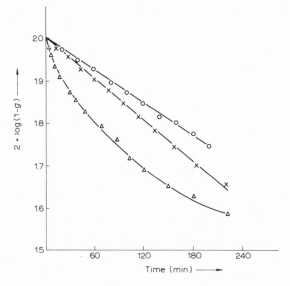

Fig. 5. Semilogarithmic plots for polymerization of formaldehyde at —30°C: (1) catalyst calcium stearate, concentration 4×10^{-4} mole l^{-1}; (2) catalyst tetrabutyl-ammonium laurate, concentration 2×10^{-4} mole l^{-1}; (3) catalyst triethylamine, concentration 3×10^{-4} mole l^{-1}. g is the degree of conversion.

magnitude lower than one would have expected from monomer/initiator ratio but the DP increased with conversion. Separate experiments showed that the molecular weight increased when the water in the mixture had been consumed. As was demonstrated by Machacek et al., water does not influence the reaction rate but decreases the DP.

Studies of potential chain transfer agents which were added to the anionic formaldehyde polymerization demonstrated that water, methanol and methylformate were good transfer agents [28]. The first two compounds terminated the chain growth by donating a proton to the polymeric alkoxide to give hydroxyl terminated polymer. HO^-, or CH_3O^- reinitiated the polymerization. In the case of methyl formate, the formyl group was the terminating group and methoxide the reinitiating group.

References pp. 376—377

With formic and carbonic acid as potential chain transfer agents, termination would involve the proton of the undissociated acid, but $HCOO^-$ and $HOCOO^-$ were very poor nucleophiles and did not reinitiate the polymerization. This amounted to an overall inhibition or retardation of the formaldehyde polymerization. 2 mole % of CO_2 inhibited formaldehyde polymerization completely [29], as did other inorganic or organic acids which gave poorly nucleophilic anions. Higher organic acid concentration may cause cationic polymerization.

2.2 CATIONIC POLYMERIZATION

Formaldehyde polymerization can also be carried out in anhydrous media with cationic initiators. This polymerization is much more complicated than anionic polymerization and may be characterized by the following basic reactions. The electrophilic initiator adds to the carbonyl oxygen with the formation of an oxonium ion

$$R^+ + O{=}CH_2 \rightarrow R{-}\overset{+}{O}{=}CH_2 \rightleftharpoons R{-}O{-}CH_2^+$$

This oxonium ion may in some cases react in its resonance form, the carbonium ion. The propagation step could be by electrophilic attack of the electrophilic carbonyl atom of the methylene group on the oxygen atom of the carbonyl group of the highly polar formaldehyde. This mechanism is similar to that advanced for the polymerization of trioxane or tetrahydrofuran. A further refinement of the mechanism takes into account the likely possibility that this oxonium ion

$$\text{\textasciitilde}O^+{=}CH_2 \leftrightarrow \text{\textasciitilde}O{-}CH_2^+$$

can be further solvated, for example with an already formed polyoxymethylene chain to give the species

$$\text{\textasciitilde}O{-}CH_2{-}\overset{+}{O}\underset{\displaystyle CH_2{-}O\text{\textasciitilde}}{\overset{\displaystyle CH_2{-}O\text{\textasciitilde}}{\Big\langle}}$$

The idea suggests some intriguing variations for further chain growth. While there is no direct evidence at this time for this kind of propagation, indirect evidence is available from acid catalysed random incorporation of oxyethylene units into polyoxymethylene (transacetalization).

Much work on cationic polymerization has been done with Lewis acids as initiators. We must realize that Lewis acids by themselves are not the actual initiators but a reaction product of unknown structure is formed either with the monomer or with impurities in the system. Some of the Lewis acids are extremely active initiators and are active at very low concentrations. Other Lewis acids apparently act differently, as judged by the formaldehyde homo- and co-polymers obtained. $SnCl_4$ and $SnBr_4$ were found to be good formaldehyde initiators and gave poly-formaldehydes of similar molecular weight distributions and type of endgroups to the polymer obtained by initiation with anionic ion-pair initiation. On the other hand $AlBr_3$, $FeBr_3$ and aluminium iso-propoxide gave polyoxymethylenes with a high content of formate and methoxy endgroups and broad molecular weight distributions. The simultaneous production of formate and methoxy endgroups appears to be typical for this group of initiators and a coordinative cationic insertion mechanism has been proposed for these polymerizations [11].

Unlike anionic initiators or anionically growing alkoxide chains which can only grow (or terminate), cationic initiators (Lewis, Brönsted acids or preformed initiators) or the cationically growing chain may cause acetal-interchange reactions. These reactions are also called transacetalization and cause rearrangement in the molecular weight distribution in homo-polymers. The rates of transacetalization are relatively slow compared to that of polymerization except at high temperatures. In the presence of cyclic ethers or cyclic formals, for example, dioxolane, polyformaldehyde can incorporate randomly the co-monomer polyoxyethylene units into the polymer under transacetalization conditions.

Formaldehyde was polymerized in ether, toluene or methylene chloride with cationic initiators [30]. The activity of the initiators was quite independent of the solvent. The individual initiators showed a remarkable difference [15]. In ether at $-78°C$ BF_3 initiated formaldehyde polymer-ization reached very quickly a final low conversion of 30% while $SnCl_4$ initiation caused a very high conversion to polymer (Fig. 6). Qualitatively the effectiveness of initiators in cationic formaldehyde polymerization at

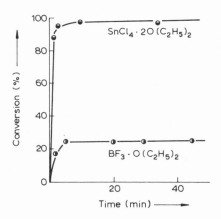

Fig. 6. Conversion in cationic polymerization of formaldehyde (1.45 mole l^{-1}) in diethyl ether at $-78°C$ as a function of time. Initiator concentration, 2.0×10^{-3} mole l^{-1}.

344

−78° in toluene, diethyl ether or methylene chloride can be arranged as follows with protonic acids the most active initiators: Protonic acids: H_2SO_4, H_3PO_4. Lewis acids: $SnCl_4$, $FeCl_3$, $BF_3 \cdot OEt_2$. Stable cations: $CH_3^+COClO_4^-$. The activity of acetyl perchlorate was less than that of $SnCl_4$ but more than that of $FeCl_3$.

A more extensive study of the cationic polymerization and particularly the kinetics of cationic polymerization was more recently carried out by Jaacks et al. [30—32]. These workers studied the influence of monomer concentration, type and concentration of initiator, type of solvent, and (to a limited extent) temperature, concentration and availability of growing ends, with full recognition of the fact that polyoxymethylene precipitated during the polymerization.

The cationic polymerization of formaldehyde can be visualized in the following manner, where R^+ may be H^+, a Lewis acid or their electrophilic reaction products capable of addition to formaldehyde or it may be a cationically growing polyoxymethylene chain; viz.

$$R^+ + O=CH_2 \rightarrow \quad \begin{array}{c} R-\overset{+}{O}=CH_2 \\ \updownarrow \\ R-O-CH_2^+ \end{array}$$

In the case of a Lewis base type solvent like diethyl ether or when polymer has already been formed, additional structures must be taken into account, such as

$$R-O-CH_2-\overset{+}{O}\!\!\diagup^{\displaystyle C_2H_5}_{\displaystyle \diagdown C_2H_5} \quad \text{or} \quad R-O-CH_2-\overset{+}{O}\!\!\diagup^{\displaystyle CH_2-polymer}_{\displaystyle \diagdown CH_2-polymer}$$

The effect of initiators on the cationic formaldehyde polymerization was first investigated with toluene as the solvent [30]. A typical polymerization conversion—time curve with $SnCl_4$ at −78°C is given in Fig. 7. After

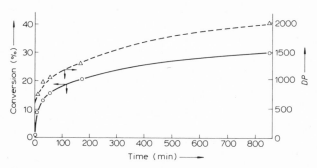

Fig. 7. Cationic polymerization of formaldehyde in toluene at −78°C with $SnCl_4$ as initiator. Dependence of conversion (—○—) and *DP* (- -△- -) of polyoxymethylenes on reaction time. $[CH_2O]_0 = 6.5$ mole l^{-1}; $[SnCl_4] = 1.65 \times 10^{-3}$ mole l^{-1}.

20—30% conversion the polymerization slowed down substantially and continued at slower rates; the degree of polymerization at the same time continued to increase, relatively faster than the conversion to polymer. The decrease in polymerization rate was not caused by a decrease of available monomer but by the fact that it became increasingly difficult for the monomer to diffuse to the reaction sites. The authors recognized that it was important to find a method to determine the reaction polymerization sites of the growing polymer. The method which was developed and the important results of these studies will be discussed later. It should be pointed out that the polymer precipitated as a voluminous gel from toluene, but as a granular sandy powder from diethyl ether, another solvent in which the formaldehyde polymerization was studied. Conversion/time curves at higher monomer concentrations and higher temperatures showed a smaller decrease in rate of polymerization; the DP's of the polyoxymethylene increase during the polymerization as the conversion increases. At $-78°$ with $SnCl_4$ in toluene, it was found that the kinetic chain lengths (L_{kin}) were about half of the viscosity molecular weights of the polyoxymethylene formed, which indicated that chain transfer had occurred. Chain transfer increased with temperature and at the lower temperature of $-30°$, the kinetic chain length was about twice the DP.

A linear relation was found between initiator concentration and conversion at lower conversions up to 1 h and in the temperature range between $-78°$ and $-30°C$ (Fig. 8). The plots did not go through the origin which indicated deactivation of initiator possibly by reaction with impurities. The DP increased with increasing initiator concentration at relatively low conversions. Similar relationships were found at $-30°C$ but

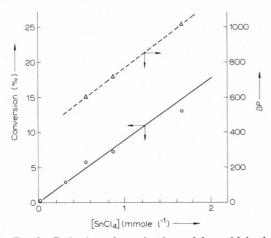

Fig. 8. Cationic polymerization of formaldehyde in toluene at $-78°C$ with $SnCl_4$ as initiator. Dependence of conversion (—○—), and DP (- -△- -) of polyoxymethylene, on initiator concentration. $[CH_2O]_0 = 6.5$ mole l^{-1}; $t = 15$ min. $DP \sim L_{kin}$.

the increase in *DP* with increasing initiator concentration was much smaller (*DP* = 5000 with 0.04 mmole $SnCl_4$ l^{-1} at 7% conversion and 7000 with 0.5 mmole $SnCl_4$ l^{-1} at 83% conversion).

The influence of the monomer concentration on the rate of formaldehyde polymerization and *DP* was investigated. At constant initiator concentration and polymerization time (but below 15% conversion) at −78°C and −30°C the polymer yield increased linearly with formaldehyde concentration (5−30 mole % formaldehyde in toluene). Thus, the polymerization was first order with respect to monomer. Again the *DP* did not increase as readily at −30° as it did at −78°C. Kern et al. [30] also reported slightly different results for formaldehyde polymerization with

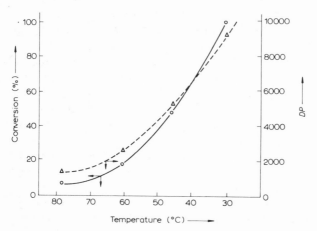

Fig. 9. Cationic polymerization of formaldehyde in toluene with $SnCl_4$ as initiator. Dependence of conversion (—○—) and *DP* (- -△- -) of polyoxymethylene on polymerization temperature. $[CH_2O]_0$ = 4.8 mole l^{-1}; $SnCl_4$ = 0.4 mmole l^{-1}; *t* = 60 min.

TABLE 2

Comparison of effectiveness of various initiators (I) for the polymerization of formaldehyde in toluene
Polymerization conditions: (a) −78°C, $[CH_2O]_0$ = 9 mole l^{-1}, [I] = 1.75 × 10^{-4} mole l^{-1}; (b) −30°C, $[CH_2O]_0$ = 2.4 mole l^{-1}, [I] = 4.8 × 10^{-4} mole l^{-1}

	Conversion (%) after 15 min			Conversion (%) after 15 min	
	a	b		a	b
$SnCl_4$	18	70	$TiCl_4$	1.1	
CH_3COClO_4	15	50	$FeCl_3$	0.9	<1
$HClO_4$	8	20	$SbCl_5$	0.5	3
(70% in H_2O)			H_2SO_4	0.4	1.5
$AlBr_3$	1.4	60	Cl_3CCOOH	<0.2	<0.2
$BF_3 \cdot O(C_2H_5)_2$	1.2	0.4	I_2	<0.1	<0.1

BF$_3$ as initiator in ether, but otherwise under conditions similar to the work with SnCl$_4$ in toluene at $-78°$C. A higher order (1.3) with respect to monomer was found. The conversion increased rapidly after 1 h; (previous work involved reaction times of 15 min). A *DP* increase with increasing reaction temperature was also found; (Fig. 9 shows data with SnCl$_4$ as initiator). Typical conversions after a standard time of 15 min are given for a number of initiators at $-78°$C and $-30°$C in Table 2. No polymers could be obtained either at $-78°$ or $-30°$ in ether or toluene with AlCl$_3$, ZnCl$_2$, or AgClO$_4$.

Kern et al. [30] calculated activation energies for the temperature range $-78°$ to $-30°$ for various initiators from data like that in Fig. 9. They varied from 1 to about 10 kcal mole^{-1} (Table 3).

TABLE 3

Activation energies of formaldehyde polymerization in toluene for various initiators

Initiator	E_a(kcal mole^{-1})
AlBr$_3$	9.9
SbCl$_5$	7.1
SnCl$_4$	5.1
CH$_3$COClO$_4$	5.1
H$_2$SO$_4$	4.9
HClO$_4$	3.9
BF$_3$. O(C$_2$H$_5$)$_2$	1.2

Solvents influenced the polymerization of formaldehyde to varying degrees. In general solvents of higher dielectric constant gave higher rates of polymerization. The rate was, however, not directly related to the polarity nor to the basicity of the solvents. It must be pointed out that the monomer formaldehyde itself is highly polar, probably highly associ-ated in nonpolar solvents and influences at least at low conversion the rate of polymerization. In addition, solvation of the growing chain ends and the already formed formaldehyde polymer must be taken into account. This would mean that the effective local monomer concentration is much higher than the over-all monomer concentration in the solution. In solvents of very high dielectric constant the rate of polymerization may be lower because the monomer is displaced from the growing site. In nitroethane the formaldehyde polymerization rate with SnCl$_4$ was found to be slower than in CH$_2$Cl$_2$ or acetone. Table 4 shows kinetic data for the SnCl$_4$ initiated formaldehyde polymerization at $-78°$ and $-30°$C in various solvents.

All time—conversion curves are approximately linear up to 15 min. Fig. 10 shows the linear time—conversion curve for polymerization of formal-dehyde in methylene chloride, with SnCl$_4$ as initiator at significantly lower concentration than for the system with toluene as solvent. The *DP*

TABLE 4

Cationic polymerization of CH_2O with $SnCl_4$ in various solvents[a]

Solvent	Conversion (%) after 15 min		Minimum values k_p(l mole^{-1} min^{-1})		E_a (kcal mol^{-1})	Shape of time-conversion curve	
	−78°C	−30°C	−78°C	−30°C		−78°C	−30°C
Toluene	2	16	10	140	5.1	Strong bending	Slight bending
Diethyl ether	5	7	11	150	3.3	Bending after 15 min	Slight bending
THF	13		50			Bending from beginning	
CH_2Cl_2	60	39	130	320	1.8	Linear	Linear to 15 min
Acetone	80	>68	~300	~300	~0	Linear	Linear
Nitropentane	31	28	80	~150	2.6	Linear to 15 min	

[a] $[CH_2O]_0 = 4.5$ mole l^{-1}; $SnCl_4$ at −78°C = 0.36 mole l^{-1}; at −30°C = 0.09 mole l^{-1}.

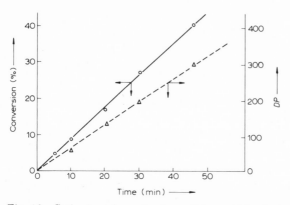

Fig. 10. Cationic polymerization of formaldehyde in methylene chloride with $SnCl_4$ as initiator at −78°C. Dependence of conversion (—○—) and DP (- -△- -) of polyoxymethylene on polymerization time. $[CH_2O]_0 = 4.0$ mole l^{-1}; $SnCl_4 = 0.09$ mmole l^{-1}; $DP \sim 4\,L_{kin}$.

did not increase in the same way. It was considered possible that trace amounts of H_2O in CH_2Cl_2 caused significant chain transfer reactions. Polymerization of formaldehyde in ether with $SnCl_4$ (0.1 mmole l^{-1}) at −78°C showed a linear time/conversion curve to beyond 60% conversion. Chain transfer was, however, very pronounced in diethyl ether. In contrast to $SnCl_4$ as initiator with which formaldehyde polymerization proceeded to high conversion, with BF_3 in ether the polymerization stopped after 10—20 min. Chain transfer in cationic polymerization of formaldehyde was most pronounced with solvents in which the polymerization rate was high (acetone, methylene chloride, tetrahydrofuran). Acetone was particularly effective in chain transfer and each kinetic chain transferred more than 1000 times. Acetone fragments have not been found in the polymers, although there are potentially 6 active hydrogen atoms in the acetone molecules which could act as transfer agents. In toluene and pentane (in the latter formaldehyde has low solubility) chain transfer is minimal.

In order to determine the possible influence of additives, water, formic acid, methanol, acetone and acetic anhydride were added to the cationic polymerization of formaldehyde at $-78°C$ with $SnCl_4$ in toluene. These reagents were added in small amounts to the polymerization mixture prior to addition of the initiator. Conversion and DP of the polyoxymethylenes were determined after 60 min reaction time. Up to a molar ratio of 1 : 1 of additive to $SnCl_4$, conversion and DP increased and went through a maximum; the rate also increased. At higher levels of additives, conversion and DP decreased.

In the case of methanol addition, it was attempted to determine the amount of methoxyl end-groups in the polymer. It was found that one mole of methanol per mole of initiator was incorporated into the polymer and no more. It could not be ascertained from the data whether the methoxyl groups were ends of the polymer chain, or whether a reaction product of $SnCl_4$ with one mole of methanol was quantitatively absorbed on the polymer chain. Acetate end-groups could not be introduced by using acetic anhydride as a potential chain transfer agent. Formic acid in small amounts, unlike water and methanol, also did not cause any decrease in DP in larger than 1 : 1 amounts (with respect to initiator).

In order to obtain information about initiation and termination in these polymerizations, attempts were made to determine the rates of these processes. For the quantitative determination of the concentration of polyoxymethylene cations the polymers were allowed to react with a large excess of amyl alcohol, viz.

$$-O-CH_2-\overset{+}{O}=CH_2 + C_5H_{11}OH \rightarrow$$

$$--O-CH_2-O-CH_2-O-C_6H_{11} + H^+ \rightarrow \text{no initiation}$$

This method determined the cations "available" for propagation. Under the conditions under which the amyl alcohol treatment was carried out, no initiation or chain transfer was possible and excessive amounts of pentoxyl groups were not introduced into the polyoxymethylene chains. The polymers were purified, degraded and the resulting amyl alcohol determined by gas chromatography. In addition, the initiator content of the polymer was analysed to determine the fate of the initiator in the early part of the polymerization. Naturally, the amyl alcohol method for the determination of active cations can only determine the available and not the occluded cations; these occluded sites are also unavailable for further polymerization by addition of formaldehyde. This method gives consequently, only minimal values of active ends.

The amyl alcohol method showed that at $-78°C$ only diethyl ether/$SnCl_4$, diethyl ether/$HClO_4$ and THF/$SnCl_4$ gave fast and complete initiation. After a polymerization time of 2—5 min and a conversion of less than 10%, 75% of the active cations could be determined. Under these conditions initiation is much faster than propagation. The amount of available cations in tetrahydrofuran did not change in hours and even with diethyl ether only a very small decrease was detectable. In contrast, with BF_3. etherate in diethyl ether at $-78°C$ only 10—20% active polymeric cations could be detected, and with toluene/$SnCl_4$ only 2—4%, after 5 min.

In order to follow the fate of $SnCl_4$ during the initiation of formaldehyde polymerization at $-78°C$ in toluene, the polymer suspension was filtered at low conversion. After 2 min reaction time, no tin could be detected in the filtrate, all the tin being adsorbed on the polymer. The same result was obtained in BF_3. etherate initiated formaldehyde polymerization in toluene; all the boron was adsorbed by the polymer. This result could mean that initiation is almost instantaneous and all the initiator was used for initiation. Alternatively, the initiator could have been partially used for initiation but the rest of the initiator was quantitatively adsorbed by the polymer formed. Adsorption experiments showed that $SnCl_4$ at $-78°$ in toluene is only adsorbed to an extent of 25% after 120 min on high surface area polyformaldehyde. These experiments still do not exclude the possibility that a substantial portion of all $SnCl_4$ in toluene is coordinated to the initially formed polymer and only a small portion is transformed into the actual initiator.

In a few cases of cationic formaldehyde polymerization (THF/$SnCl_4$) it was found by terminating rapidly with amyl alcohol that a high concentration of available polymer cations are formed rapidly and that these cations are stable and do not terminate although they react with amyl alcohol. Nevertheless, the conversion/time curves bend after relatively small monomer conversions. During the progress of polymerization the rate decreased by more than one order of magnitude. In a normal second order growth reaction the rate is given by $-d[CH_2O]/dt = k_p$ $[CH_2O][POM^+]$. Since $[POM^+]$ does not change and k_p can be assumed to be constant, the decrease of the rate must depend on the decrease of $[CH_2O]$ near the polymerization site; this means the rate of monomer diffusion to the active centre must decrease and the polymerization rate becomes dependent on the diffusion of monomer to the heterogeneous polymeric cation. The rate of diffusion must be influenced by the solvation or swelling of the polymer chain, because in methylene chloride, THF and acetone high conversions were obtained. Higher temperature and the addition of polar additives showed also a beneficial effect on the achievement of higher conversion. Lastly the type of the gegenion must also affect the rate of diffusion of the monomer to the active site, since substantial differences were noted with different gegenions.

A more detailed study of the cationic formaldehyde polymerization carried out in diethyl ether at $-78°C$ with $SnCl_4$ and $HClO_4$ was reported [31]. The conclusions reached earlier needed no modifications. The accurate estimation of the polymer cation concentration allowed an accurate determination of apparent k_p, although it was evident that the assumption of a homogeneous reaction was not fulfilled.

The time—conversion curve (Fig. 11) shows that there was a rapid initial rate of polymerization which decreased with time [32]. The number of active centres, as determined by the amyl alcohol method was initially high but decreased rapidly in accordance with the general decrease in rate of polymerization, indicating inclusion of active centres. Data accumulated and tabulated in Table 5 allowed the calculation of the

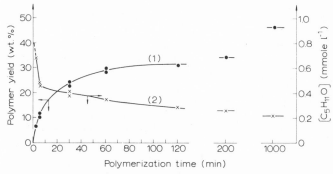

Fig. 11. Cationic polymerization of formaldehyde in diethyl ether with $SnCl_4$ at $-78°C$. Dependence of polymer yield (curve 1) and concentration of pentoxyl end groups (curve 2) on polymerization time. $[CH_2O]_0 = 7.94$ mole l^{-1}; $[SnCl_4] = 0.86$ mmole l^{-1}.

TABLE 5

Cationic polymerization of formaldehyde at $-78°C$

(Calculation of the propagation rate coefficient, k_p, from the overall polymerization rate, V. $[CH_2O]_0 = 7.94$ mole l^{-1} in diethyl ether with $[SnCl_4] = 0.86$ mmole l^{-1}.)

Polymerization time t (min)	Polymer yield (mole l^{-1})	V (mole l^{-1} min^{-1})	$[CH_2O]_t{}^a$ (mole l^{-1})	$[C_5H_{11}O^+]$ (mmole l^{-1})	k_p (l mole^{-1} min^{-1})
2	0.52	0.1410	7.42	0.68	28.0
5	0.84	0.0810	7.10	0.45	25.5
30	1.83	0.0250	6.11	0.39	10.0
60	2.37	0.0072	5.57	0.35	3.7
120	2.48	0.0035	5.46	0.29	2.2
250	2.73	0.0015	5.21	0.26	1.1
1000	3.68	~0.0005	4.26	0.22	~0.5

^{a} Formaldehyde concentration at time t.

References pp. 376—377

apparent k_p. Its value of 30 l mole^{-1} min^{-1} measured after 2 min seemed to be close to the true value of the k_p for cationic formaldehyde polymerization.

Perchloric acid was considered the most active initiator for formaldehyde polymerization. Consequently, fast initiation and long life time of the active centres could be expected. Results in Table 6 show that at short reaction times the number of active polyoxymethylene cations determined by the amyl alcohol method agreed well with the number of initiator molecules, indicating fast and complete initiation and stable active centres [33, 34]. The apparent rate decreased very rapidly. The k_p had a somewhat higher value of 40 l mole^{-1} min^{-1}. These values of k_p for CH_2O were substantially lower than those for cationic polymerizations of other monomers measured in homogenous medium (see ref. 35, p. 113). The DP of $HClO_4$ initiated polymerization was lower by a factor of 10 than the kinetic chain lengths calculated from the monomer/initiator ratio. In addition the apparent activation energy of formaldehyde polymerization with $HClO_4$ in ether between $-78°$ and $-30°C$ was determined as 3.5 kcal mole^{-1}.

Recently, the cationic polymerization of formaldehyde in CO_2 (60% formaldehyde) has been studied [36, 37]. The monomer/CO_2 mixture was prepared by decomposing α-polyoxymethylene at 150—180°C in a CO_2 atmosphere. About 0.8 wt. % methanol, and 1.2 wt. % water were present as impurities. The experimental results and theoretical conclusion must be measured with the knowledge of this background. The "uncatalysed" reaction was studied (under pressure) from 20° to 50°C. Polymer yields after 10 min increased from 3 wt. % to 25 wt. % and the DP from 300 to 720. The authors interpreted these results as indicating an increase in the initiating species and the promotion of the rate of propagation.

Acidic compounds increased the rate of formaldehyde polymerization and also the DP as seen in Table 7. It may be seen from the pK_a values that stronger acids gave better results in the polymerization.

Methyl formate had no effect as a chain transfer agent but methanol and water reduced the molecular weight of the resulting polymer. From the rate data for the "uncatalysed" polymerization of formaldehyde, Fukui et al. [36] concluded that the rate of initiation was considerably faster than the rate of propagation, that the amount of initiating species increased with temperature and that the rate of chain transfer was slower than that of propagation. It was also concluded that this "uncatalysed" polymerization of formaldehyde in CO_2 was, in fact, acid catalysed. Addition of acetic acid or dichloracetic acid as initiator did not change the general picture developed for the "uncatalysed" reaction substantially. Naturally, acetic acid initiator which was added at 0.1 mole % caused higher rates of polymerization and higher molecular weights. Dichloroacetic acid, being the stronger acid polymerized formaldehyde in CO_2 at a greater rate.

TABLE 6

Cationic polymerization of a 10 ml formaldehyde solution in diethyl ether with $HClO_4$ (70%) at $-78°C$
$[CH_2O]_0 = 6.0$ mole l^{-1}, $[HClO_4] = 0.5$ mmole l^{-1}. Termination by addition of 10 ml n-amyl alcohol at $-78°C$ (reaction time; 5 min at $-78°C$).)

Polymerization time, t (min)	Polymer yield		Pentoxyl content of polymer (wt. %)	$[C_5H_{11}O^+]$ (mmole l^{-1})	$\dfrac{[-CH_2O-]}{[C_5H_{11}O^+]}$	$[CH_2O]_t^a$ (mole l^{-1})	V^b (mole l^{-1} min^{-1})	k_p (l mole^{-1} min^{-1})
	wt. %	mole %						
5	9.1	0.55	0.23	0.41	1270	5.45	0.086	38.5
15	14.2	0.85	0.14	0.38	2140	5.15	0.067	34.5
30	30.0	1.79	0.06	0.34	5050	4.21	0.026	18.2
60	37.4	2.24	0.04	0.29	7410	3.76	0.012	10.9
200	49.1	2.93	0.02	0.20	14000	3.07	0.0014	2.6
1000	57.2	3.42	0.01₅	0.17	19900	2.58	<0.001	<2.0

^{a} Formaldehyde concentration at time, t.
^{b} Overall polymerization rate.

354

TABLE 7
Influence of addition of various acids on formaldehyde polymerization[a]

Acid	pK_a	Polymer yield, (wt. %)	DP
CF$_3$COOH	0.23	64.3	1500
CCl$_3$COOH	0.66	59.6	1500
CHCl$_2$COOH	1.25	35.6	1100
CH$_2$ClCOOH	2.87	29.2	1050
HCOOH	3.75	21.6	900
CH$_3$COOH	4.75	19.0	850
None		5.8	380

[a] Monomer solution, 10 g; monomer concentration, 60 wt. %; temperature, 30°C; time, 10 min; reaction vessel, 30 ml.

Incidentally, as a matter of speculation, if one assumes that 0.025 to 0.04 mole % of formic acid were present in the original formaldehyde solution one could expect polymerization results of the "uncatalysed" polymerization in terms of conversion, molecular weight and rate similar to the ones Fukui et al. [36] reported.

2.3 RADIATION INDUCED POLYMERIZATION

Radiation induced polymerization of formaldehyde has been studied by a number of investigators [38—43].

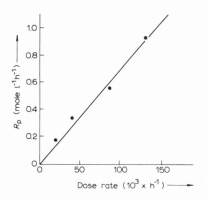

Fig. 12. Effect of dose rate on the rate of γ-induced polymerization of CH$_2$O. Monomer conc., in diethyl ether, 14.4 mole l^{-1}.

Formaldehyde polymerization in diethyl ether initiated by γ-rays from a ^{60}Co source [40], was first order with respect to the dose rate (Fig. 12). There was an induction period which was inversely proportional to the dose rate. The polymerization was second order with respect to the monomer concentration in methylene chloride (Fig. 13); and proceeded

355

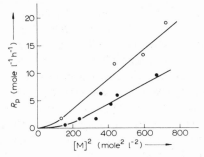

Fig. 13. Effect of monomer concentration on the rate of polymerization in methylene chloride. Dose rate: ($\bullet$), 1.96×10^3 rad h^{-1}; ($\circ$), 4.39×10^3 rad h^{-1}.

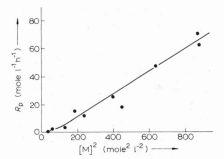

Fig. 14. Effect of monomer concentration on the rate of polymerization in toluene. Dose rate, 1.78×10^3 rad h^{-1}.

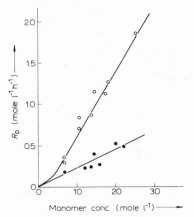

Fig. 15. Effect of monomer concentration on the rate of polymerization in diethyl ether. Dose rate: ($\bullet$) 3.88×10^3 rad h^{-1}; ($\circ$) 1.33×10^3 rad h^{-1}.

without induction period. The polymerization was also second order in monomer concentration in toluene (Fig. 14). In diethyl ether solution the radiation induced formaldehyde polymerization was first order in monomer concentration (Fig. 15).

References pp. 376—377

From these results Okamura et al. [40] concluded that in toluene and methylene chloride the formaldehyde polymerization occurred by a cationic mechanism, but in diethyl ether by an anionic mechanism. This is in contrast to the earlier but less unlikely proposal by Chachaty et al. [42, 43] that the radiation induced polymerization of formaldehyde proceeded by a radical process. It was also shown by Okamura that the molecular weights of the formaldehyde polymers were independent of the monomer concentration, but that they were consistently higher when the radiation polymerization was carried out in ether as compared to those polymers obtained in toluene or methylene chloride.

In the presence of water (3.8 to 243 x 10^{-3} mole l^{-1}), at constant dose rate and monomer concentration the rate of polymerization was found to increase slightly and the molecular weight to decrease (as expected) with increasing water concentration. It was suggested that water probably took part in the initiation step.

Fukui et al. [37] had found that CO_2 above 2 mole % inhibited the spontaneous polymerization of formaldehyde. A kinetic study of the γ-ray polymerization of formaldehyde was carried out in the presence of CO_2 (no copolymerization between formaldehyde and CO_2 could be detected) at temperatures from +13 to $-17°C$. Experimentally the spontaneous contribution could be separated from the γ-ray initiated polymerization, presumably a cationic polymerization.

The rate of γ-ray polymerization was proportional to the square root of the CO_2 concentration; the molecular weight was unaffected. It was also believed that CO_2 did not decrease the termination rate or increase the propagation rate but caused an increase in the initiation rate in the γ-ray initiated formaldehyde polymerization. The rate of polymerization was proportional to the square root of monomer concentration, but not of higher order as earlier suggested. Nakashio et al. [41] had apparently observed substantial spontaneous polymerization in their work. The temperature dependence of the rate found by Fukui et al. allowed an activation energy of 10.3 kcal $mole^{-1}$ to be calculated for this polymerization.

2.4 POLYMERIZATION OF GASEOUS FORMALDEHYDE

Very little work has been done on the polymerization of liquid, neat formaldehyde, although it had long been known that it is almost impossible to keep liquid formaldehyde from polymerizing. It was early suspected that bases especially amines and relatively weak acids were active initiators for formaldehyde polymerization. Water, which is the most important impurity in any formaldehyde system, was an active initiator, later shown to be an anionic initiator which caused initiation by the HO^- ion.

A simpler system than polymerization of formaldehyde in the liquid state is the polymerization of formaldehyde from the gaseous state. This was investigated in three distinct periods: the early period [44—46] was followed by the elegant work of Norrish et al. [13, 47]. Recently some additional work on the kinetics of gaseous formaldehyde was published which adds but little extra knowledge to that from Norrish's work.

The early work on polymerization from the gas phase was done on what was then thought to be pure formaldehyde. Solid polymer formed on the cold surfaces but the data were very irreproducible. Unknown kinds and unknown amounts of impurities in the monomer and on the glass surfaces made polymerization results erratic. The polymerization of formaldehyde from the gas phase has one advantage over polymerization in solution for kinetic studies. The rate of monomer disappearance can be followed readily manometrically; in addition, additives can be added simply and very accurately to gaseous formaldehyde.

When a few percent of formic acid was added to gaseous formaldehyde at about 500 mm pressure a rapid polymerization was observed, the velocity was some hundredfold greater than with pure formaldehyde. It appeared that formic acid was a powerful initiator of formaldehyde polymerization under these conditions. The polymerization was confined to the surface of the vessel and the kinetics were those of a heterogeneous system. Because of the much faster formaldehyde polymerization promoted by formic acid the purity of the formaldehyde became less important. The erratic results of earlier investigators were best explained by varying degrees of purity of earlier preparations of formaldehyde monomer.

Carruthers and Norrish [13] found that no polymerization occurred if the whole system was held above 100°C, even at 30 torr formic acid pressure. If one portion of the apparatus was cold, polymerization occurred in this portion. If the cold portion was cooled throughout the polymerization, this went to completion; if the cooling was insufficient and the surface was exposed to the bombardment of hot gases depolymerization occurred and an equilibrium between polymer and monomer was established.

An apparatus was designed which allowed the experimenters to prepare mixtures of formaldehyde and various gaseous initiators without any undesired polymerization. The whole apparatus was heated to 100°C or slightly higher. The actual polymerization portion of the apparatus was kept at 20°C, the commonly used temperature at which the polymerization of gaseous formaldehyde was studied.

It can be seen (Fig. 16) that the rate of formaldehyde polymerization increased rapidly with increasing formic acid partial pressure up to 40 torr with 500 torr of formaldehyde. In the kinetic scheme developed the initiation rate was proportional to the partial pressure of formic acid and

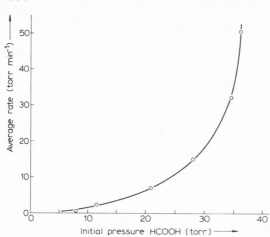

Fig. 16. Effect of formic acid upon the polymerization of 500 torr gaseous formaldehyde (plotted from results of Carruthers and Norrish [13]).

formaldehyde; the rate of polymer growth was proportional to the partial pressure of formaldehyde and number of active centres and the termination rate was also proportional to the partial pressure of formaldehyde and number of active centres. The rate of kinetic branching, a purely kinetic concept, was proportional to the number of active centres and the partial pressure of formaldehyde. The number of active centres in the steady state was considered to be constant.

The mechanism was developed on the basis of knowledge, available in 1935, of addition of formic acid to formaldehyde and of formaldehyde to $\sim$OCH—CH$_2$OH. Today one would interpret the data on the basis of a cationic mechanism with a growing oxonium ion as the propagating species. The termination could be by recombination of the formate anion with the growing polyoxymethylene cation or by the occlusion of the growing end. The physical meaning of the kinetic observation of branching has not yet been explained satisfactorily. One explanation was that the initiator may cleave already formed chains to cause the formation of 2 growing chains.

In addition to the initiation of gaseous formaldehyde with formic acid, HCl, boron trifluoride and stannic chloride were studied and found to be more active than formic acid (Fig. 17). The rate of gaseous formaldehyde polymerization with the initiator was measured under the same conditions as the formic acid initiated polymerization by determining the decrease in pressure of formaldehyde. HCl as initiator (at about 3—4% in the mixture) caused kinetic branching, i.e. a rapid increase in the rate of polymerization.

The actual mechanism of these gaseous formaldehyde polymerizations with HCl and formic acid as initiators is not very well understood because the work was analysed only from the kinetic point of view. It is not

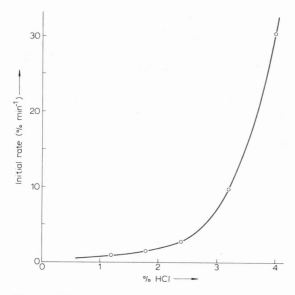

Fig. 17. Effect of hydrogen chloride upon the polymerization of gaseous formaldehyde.

known whether the "used up" initiator is incorporated into the polymer or only adsorbed on the polymer. However, when the polymerization was carried out with 5% HCl as initiator until all formaldehyde was polymerized some HCl was still left. It is unfortunate that the early investigators could not determine the molecular weight of the polyoxymethylene formed. The authors commented on the kinetic chain branching of the gaseous formaldehyde polymerization. It was, however, not considered probable that impurities in the monomer had a profound influence on the acceleration of the formaldehyde polymerization.

Initiators other than HCl and formic acid, for example, BF_3 or $SnCl_4$ were substantially more active. To give the same rate of polymerization only 1/50 molar amount of BF_3 as compared to HCl was required. These initiators unlike formic acid and HCl were either incorporated into the polymer or at least converted to non-volatile products.

Formaldehyde polymerization data have also been obtained by Sauterey [48] and in the early work of Toby and Rutz [49]. More recently, Boyles and Toby [50, 51] reinvestigated the kinetics of "noncatalysed" gaseous formaldehyde polymerization on the basis of careful attention to the purity of the monomer formaldehyde, a study of the surface to volume ratio of the vessel and the purity of the surface on which the polymer was deposited.

Pure formaldehyde was prepared by decomposing α-polyoxymethylene (99.7%—99.9%) at 50—100°C under reduced pressure and condensing it at −196°C. The monomer was then distilled successively from −86° to a

360

—196°C trap, sizeable foreshots being discarded each time, and analysed by gas chromatography. The analyses are given in Table 8.

Toby et al. studied the polymerization of gaseous formaldehyde, using formaldehyde containing a known level of impurities and in apparatus made from virgin quartz which had been flamed out and cooled in vacuum. No grease on connectors other than glass seals were used in this apparatus. When the apparatus was rinsed with detergent, the results became erratic and unreliable. Emphasis in Toby's work was on the surface to volume ratio (S/V) of the reaction vessel, which was 1.0, 2.6 and 5.0 cm^{-1}. Basically two types of surface of the reaction vessel were used: one the flamed out surface, the other the surface of already deposited polymer. In other words the polymer deposit was not removed between runs.

Generally speaking, the rate of formaldehyde polymerization increased with increasing monomer pressure for both types of surface at a given temperature. The polymerization rate increases with increased S/V ratio at a given initial pressure and temperature (Fig. 18). When the polymer

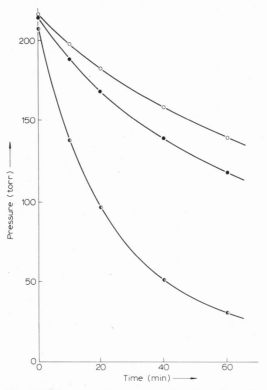

Fig. 18. Pressure—time curves showing the effect of S/V for bare-wall runs at 30°C and constant initial monomer pressure. (○) $S/V = 1.0$ cm^{-1}; (●) $S/V = 2.6$ cm^{-1}; (◐) $S/V = 5.0$ cm^{-1}.

deposit was allowed to increase, the rate of polymerization decreased progressively from run to run, but did not seem to reach a limiting value.

The initially fast polymerization rate decreased as the equilibrium pressure was approached, and during the polymerization the pressure decreased until the equilibrium pressure was reached. The initial polymerization rate increased with increasing temperature. However, the authors did not study the polymerization close to the ceiling temperature of the polymerization, only up to 60°C.

The authors recognized that the polymerization should be considered a three step sequence; (a) The propagation of active centres, (b) the depropagation of active centres and (c) disappearance of active centres. This model required the assumption that the growing centres were present and available at the beginning of the polymerization. They identified three stages in the reaction mechanism. In the initial stage of bare-wall polymerizations active centres were in high concentration and mobile. In the intermediate stage the termination rate coefficient began to decrease because the lower mobility of active centres was not favourable for termination. In the third stage all active centres were effectively trapped; the termination rate was now at a low and possibly constant value and the concentration of active centres was essentially constant.

Toby et al. also determined activation energies for gaseous formaldehyde polymerization from the Arrhenius plots (Fig. 19). The slopes in

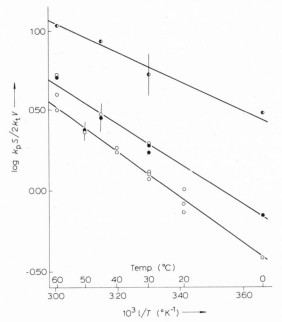

Fig. 19. Arrhenius plot of log $(k_p S/2k_t V)$ vs. $1/T$. (○) S/V = 1.0 cm^{-1}, (●) S/V = 2.6 cm^{-1}, (◑) S/V = 5.0 cm^{-1}. Error limits shown are typical.

Fig. 19 yield $E_p - E_t$ and the intercepts give A_pS/ZA_tV, where E_p and E_t are the activation energy for polymerization and termination reaction and A_p and A_t the Arrhenius A factors for the same reactions. The data are listed in Tables 9 and 10, together with values from other work.

TABLE 8
Impurities in formaldehyde

Distillation	Methyl formate (%)	Methanol (%)	Water
Direct from polymer	0.35	1.39	2.68
1, Distillation (at $-86°C$)	0.04	0.04	0.06
2, Distillation (at $-86°C$)	0.03	0.04	—
3, Distillation (at $-86°C$)	0.01	0.05	0.04
After 3 months at $-196°C$	0.01	0.02	0.02

TABLE 9
Activation-energy differences and A factor ratios derived from Arrhenius plots [51]

S/V (cm^{-1})	$E_p - E_t$ (kcal mole^{-1})	$(A_pS/2A_tV) \times 10^{-4}$ (dimensionless)
1.0	6.4 ± 0.7	5.2 ± 0.7
2.6	5.7 ± 0.8	7.4 ± 1.4
5.0	4.4 ± 1.9	20 ± 9

The nature of the active sites is open to discussion. Toby et al. chose to follow a possible earlier suggestion and used the addition of formaldehyde to a neutral polymer chain as the propagation mechanism. Because of a slight inhibition of the polymerization by oxygen, a radical mechanism was not completely ruled out. Looking at the formaldehyde polymerization as a whole and accepting some of the observations of Toby et al., it must be concluded that their formaldehyde polymerization was a cationic polymerization. Active centres, or active sites were actually oxonium

TABLE 10
Activation energies for gaseous formaldehyde polymerizations [51]

E	kcal mole^{-1}
$E_p - E_t$	6.4 ± 0.7
$E_p - E_d$	-12.35 ± 0.05
$E_d - E_t$	18.8 ± 0.7
E_d	27.1
E_d	26
E_p	14.2 ± 0.6
E_t	7.8 ± 0.9

cations $-\overset{+}{O}=CH_2$. Anionic polymerization of formaldehyde would not be affected by oxygen and would be inhibited by CO_2. A cationic formaldehyde polymerization would not be affected by CO_2 and may be slightly affected by oxygen.

It was recognized that this polymerization occurred by deposition of a solid polymer. Consequently, the active centres became less mobile as the reaction progressed and in addition more and more active centres were occluded and consequently became unavailable for further polymerization. In addition, chain transfer was possible, generating new sites or better more available sites. Without further discussion here it is suggested that the reader compare results of Toby's work with explanations described in the section on the polymerization of formaldehyde in hydroxylic media. It appears that the physical and mechanistic explanation in Toby's work should be reinterpreted in the light of our present knowledge on polymer crystal growth and more sophisticated and real mechanistic knowledge on aldehyde polymerization.

2.5 POLYMERIZATION OF FORMALDEHYDE IN PROTIC MEDIA

The polymerization of formaldehyde in water or methanol, but especially in water [1], is of great historic value; much of Staudinger's work [1] on formaldehyde polymerization, the preparation of α-, β- and γ-polyoxymethylene, are based on this investigation. It was remarkable what insight into polymerization reaction these earlier investigators had, although many principles about polymerization and structures of polymers, which are very familiar to us today, had not been developed or were in the process of being developed. From the practical point of view polyoxymethylenes of the 20's and early 30's were of relatively low molecular weight and of limited usefulness. The full understanding of monomer/polymer equilibrium, side reactions in solutions, end groups of polymers and the morphology of the polymers has developed very recently.

Polyoxymethylenes prepared in hydroxylic media are produced by a fundamentally different mechanism from that in anhydrous media. Formaldehyde reacts in water in the form of methylene glycol and with methanol to give the hemiacetal. Even in dilute solutions there is a small amount of free formaldehyde present, but, for the purpose of this discussion, the polymerization of formaldehyde in aqueous medium is discussed in the traditional manner as the polymerization of methylene glycol. This means that this polymerization is mechanistically a chain growth polymerization but the polymer formed is a condensation polymer.

When considering the polymerization of formaldehyde in aqueous media, it must be realized that methylene glycol is not stable and, even in relatively dilute solution, establishes complicated equilibria between a very small amount of free formaldehyde, and low molecular weight oligomers of polyoxymethylene glycols. In dilute solution the main components are polyoxymethylene glycols where the oxymethylene units are 1–3. Increasing the formaldehyde concentration, defined as bound formaldehyde in the mixture, increases the number of the higher oligomer units with respect to the lower units and CH_2O (Fig. 20). Increasing temperature causes dissociation of these oligomeric polyoxymethylene glycols to lower oxymethylene units and free formaldehyde is also increased. Iliceto and Brezzi [52, 53] have calculated the distribution of the oligomeric species. The equilibria between the oligomeric species are reversible but catalysed by acids and bases and the neutral point is at around pH 4. At room temperature and at a formaldehyde concentration of about 35% the mixture is still homogeneous. Increasing the formaldehyde concentration or decreasing the temperature causes precipitation

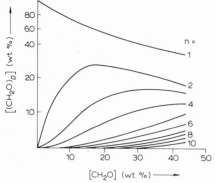

Fig. 20. Composition of aqueous solutions of formaldehyde as a function of the concentration of CH_2O.

of solid crystalline polyoxymethylene glycols. The homogenous equilibria are [11]

$$HOH \underset{-CH_2O}{\overset{+CH_2O}{\rightleftharpoons}} HO-CH_2-OH \underset{-CH_2O}{\overset{+CH_2O}{\rightleftharpoons}}$$

$$HO(CH_2O)_2H \underset{-n\,CH_2O}{\overset{+n\,CH_2O}{\rightleftharpoons}} HO(CH_2O)_{n+2}H$$

When a DP of approximately 10 is reached, the oligomeric polyoxymethylene glycols are no more soluble, precipitation occurs and further polymerization proceeds in the heterogeneous, crystalline phase. The heterogeneous equilibria are

Nucleation

$$[HO(CH_2O)_nH]_{soluble} \rightleftharpoons [HO(CH_2O)_nH]_{cryst.}$$

Growth

$$[HO(CH_2O)_nH]_{cryst.} + HOCH_2OH \rightleftharpoons [HO(CH_2O)_{n+1}H] + HOH$$

TABLE 11

Thermodynamic data for polymerization in aqueous solution

	ΔH(kcal mole^{-1})	ΔG(kcal mole^{-1})
Crystallization of polymer	−1.7	− 0.32
Liquid phase reactions	−0.8	−0.04
Total	−2.5	−0.36

Once crystallization occurs further polymerization proceeds exclusively in the crystalline phase and the driving force of the polymerization is the crystallization of the polymer (see Table 11).

A schematic representation of the solid—liquid relationship is given in Figure 21. The position of the formaldehyde equilibrium concentration

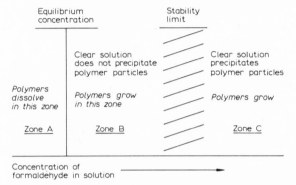

Fig. 21. A schematic representation of the solid—liquid relationship.

depends on the temperature of the system and the width of the individual zones also depends on the temperature. As indicated before, at higher temperatures higher concentrations of formaldehyde are required for these reactions. In principle no catalyst is required for the reactions, but, for practical purposes in order to work with reasonable rates, acid or base catalysis is used.

In Zone A, at monomer concentrations below the equilibrium concentration solid polymer dissolves. This is equivalent to the dissolution of a crystalline low molecular wt. compound in a good solvent. In Zone C the solution is super-saturated, it is above the stability limit, spontaneous nucleation occurs and polymer precipitates from the clear solution. There is a Zone B, about 4—8% above the equilibrium concentration of formaldehyde, where super-saturation is insufficient to cause spontaneous precipitation of polyoxymethylene but where seeds or nuclei of polyoxymethylene can grow when added to the clear solution and can increase in weight and molecular weight. This is the desirable range for the preparation of high molecular weight polyoxymethylene in hydroxylic media.

As can be seen in Fig. 22 successfully polymerizing formaldehyde at a reasonable rate in aqueous media depends on the full understanding of the factors involved in the polymerization; the growth rate of the polymer crystals, the nucleation rate and the limiting molecular weight. The growth rate is nearly linear with % CH_2O and goes through the origin at the equilibrium concentration. The nucleation rate is zero until the stability limit is reached and increases rapidly until it equals the growth rate at which point only low molecular weight polyoxymethylene is formed. As nucleation occurs, which means spontaneous precipitation of low molecular weight polymer, the limiting molecular weight decreases and the molecular weight distribution broadens. The best formaldehyde concentration for polymerization is that with which the growth rate is already substantial but the nucleation rate is still zero.

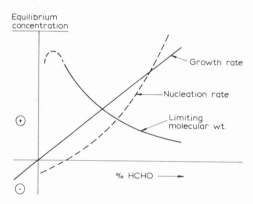

Fig. 22. Polymerization of formaldehyde.

Another problem must be overcome to produce high molecular weight polyoxymethylene. The seed crystal must have the proper surface in order to grow properly. The ideal polymer crystal is born by spontaneous nucleation and can then grow exclusively by addition to the chain ends which are aligned and available at the face. The polydispersities of polymers prepared under these conditions are near unity.

If polymerization is carried out under conditions of high nucleation, polymer chains are deposited rapidly and indiscriminately and the active sites necessary for growth are occluded. If growth is carried out near the equilibrium concentration, depolymerization becomes important and again the active sites become occluded.

High molecular weight polyoxymethylenes prepared under the right conditions have more than 80% of "available" end groups as judged by acetylation at 60°C with acetic anhydride, whereas β-polyoxymethylene has almost no detectable "available" end groups. It has been indicated that acids and bases increase the polymerization reaction rate; acids, however, cause side reactions, predominantly transacetalization. In methanolic media acetal formation is the most undesirable side reaction. The basic catalysts are good catalysts for the Cannizarro reaction. Nevertheless weak bases such as sodium formate are still respectable catalysts for formaldehyde polymerization in hydroxylic media [54]. One of the very important discoveries for the perfection of the formaldehyde polymerization in hydroxylic media was that the polymerization can be carried out at elevated temperatures, e.g. 95°C [55]. These conditions favour growth over nucleation and side reactions. Another important development was the use of the quick-chill seed preparation. Nucleation could be accomplished by dropping the temperature of the mixture from 95° to 60°C, the seeds which had been formed were reintroduced under growth condition. These crystals were desirable for successful growth to high molecular weight polyoxymethylene.

The common crystal form of polyoxymethylene is the hexagonal form. Mortillaro et al. [56—58] found that another crystalline form of polyoxymethylene was produced when the polymerization of aqueous formaldehyde was carried out at 20° to 35°C at pH > 10 in high salt concentrations (>20%). It is shown in Table 12 that high molecular weight polymer is only achieved below 35°C. The rate of polymerization is slow; it takes about 10 days to obtain maximum molecular weight. The formation of the proper seed crystal is important; almost any crystalline polyoxymethylene regardless of the crystal structure can be used as seed. The type of salt used as catalyst is critical to obtain orthorhombic polyoxymethylene of reasonable molecular weight (Table 13).

TABLE 12
Polymerization of aqueous formaldehyde, HCOONa as catalyst (34—38%); pH 10.4

Temp. (°C)	Steady state	
	η_{red}	Orthorhombicity (%)
20	0.8	95
35	1.0	90
50	<0.1	<5
60	<0.1	<1

TABLE 13
Polymerization of aqueous formaldehyde, salt and pH effect (35°C, pH 10.4, 30 to 40% salt, unless noted)

Salt		η_{red}	Orthorhombicity (%)
KF	pH 9.4	0.1	10
KF	—	1.0	80—85
HCOOK	—	0.8	80—85
HCOOK (ex KOH)	5% max.	0.4	80—85
CH_3 ⟨⟩ SO_3Na	—	<0.1	<0.1

Hexagonal polyoxymethylene of polydispersity 1 and high molecular weight can also be prepared in methanol instead of water. The equilibria are analogous to those in the aqueous system. Basic catalysts are also preferred in this system. The polymerization is best carried out at 110°C and a highly concentrated solution of 85% formaldehyde can be used. Under these conditions the polymerization must be carried out under pressure, and the methanol is distilled overhead while new formaldehyde is added. Small amounts of water in the system induced side reactions, mainly the formation of methyl formate. In one reported run 22% of polyoxymethylene of a molecular weight $\bar{M}_n$ of 46,000 was obtained after 90 h. The loss of formaldehyde by side reactions was less than 1%.

3. Higher aliphatic aldehydes

Higher aldehydes, for example acetaldehyde or n-butyraldehyde, have much less tendency to polymerize compared to formaldehyde [5, 6]. Reasons have been given in thermodynamic terms by referring to the lower enthalpy of polymerization (about -7 kcal mole^{-1}) as compared to formaldehydes (-12 kcal mole^{-1}), which results in ceiling temperatures of $-40°C$. In terms of reactivity, aliphatic aldehydes undergo hydration and hemiacetal formation to an extent of about 50%.

All polymerizations of higher aldehydes must be carried out under anhydrous conditions and at low temperature. Strong nucleophiles, like alkoxides, hydrides if appropriately soluble in the medium, but also electrophiles, like Lewis acids and protic acids with a pK_a of <2 are good initiators for higher aldehyde polymerization. These initiators may be used as such, but some effective cationic initiators have also been deposited on carriers such as aluminium oxides, silica gel or even polymers, such as polyoxymethylene, polypropene.

The anionic polymerization is mechanistically analogous to the anhydrous formaldehyde polymerization, viz.

$$R^- + \underset{H}{\overset{R'}{C}}=O \longrightarrow R-\underset{H}{\overset{R'}{C}}-O^- \xrightarrow{+nR'CHO} R-\left[\underset{H}{\overset{R'}{C}}-O-\right]_n \underset{H}{\overset{R'}{C}}-O^-$$

where R may be alkoxide, alkyl etc., and R$'$ an aliphatic chain, for example ethyl. The cationic polymerization may be exemplified with a protic acid HA as initiator; viz.

$$H^+A^- + O=\underset{H}{\overset{R'}{C}} \rightarrow H-\overset{+}{O}=\underset{A^-\ H}{\overset{R'}{C}} \xrightarrow{nR'CHO} H\left[-O-\underset{H}{\overset{R'}{C}}-\right]_n -\overset{+}{O}=\underset{A^-\ H}{\overset{R'}{C}}$$

A⁻ is the counter-ion of the growing oxonium ion.

It should be mentioned that under acidic conditions higher aldehydes, especially acetaldehyde with H_2SO_4, undergo very readily cyclo-trimerization to paraldehyde; at temperatures below 0°C a small amount of the cyclic tetramer, metaldehyde, is also formed. The reaction is very fast and is accompanied by a rapid rise of temperature. As a consequence of this, the cationic acetaldehyde polymerization to linear polymers must always be carried out under conditions to avoid the trimerization using solvents of low dielectric constant, very low temperatures and initiators of appropriate activity. High dielectric constant solvents, high temperature, and very active initiators cause exclusively trimer formation.

Aldehydes like α-olefins are capable of undergoing stereoregular polymerization. Polyaldehydes of high isotacticity have been prepared under appropriate conditions. Most of the work on higher aldehyde polymerization was concerned with studies to prepare polymers of high stereoregularity. Detailed work on the kinetics of higher aldehyde polymerization has not been done. The reasons why these systems have not been studied are: (a) difficulty to prepare monomers in high purity, (b) low ceiling temperature of polymerization and (c) precipitation of polymer with subsequent occlusion of growing polymer ends and formation of polymers of varying stereoregularity.

We will discuss the scarce data that are available on the rates of polymerization in an attempt to critically review their usefulness at the present time. The subject is divided into four types of polymerization that lead to higher aldehyde polymers: (a) the so-called "crystallization" polymerization (b) cationic polymerization (c) anionic polymerization and (d) polymerization with aluminium alkyls and related compounds.

Crystallization polymerization was originally used to prepare elasto-meric polyacetaldehyde of high molecular weight. It is now established that this polymerization was acid catalysed and crystallization of the monomer was a fortuitous way to assist the polymerization. Impurities in glass surfaces and monomer provided the initiators for the cationic crystallization polymerization of acetaldehyde. Some early kinetic studies have been made by Bevington and Norrish [47]. They found that active centres for acetaldehyde polymerization were formed as acetaldehyde was condensed on the cold surface and that the polymer grew by addition of monomer molecules from the gas phase. The number of active centres increased from zero to a stationary value when the rates of formation and destruction were equal. The authors observed straight line weight of polymer versus distillation time plots which did not go through the origin. At slower distillation times the rate of polymerization was lower. Molecular weight of the initial polymer was also lower. This seemed to indicate that an inhibitor (which might very well be water) codistilled with the monomer, preferably in the earlier part of the distillation, and prevented the formation of active centres for the polymerization. Much of the early work on acetaldehyde polymerization, particularly the crystalliz-ation polymerization, must now be reinterpreted as cationic polymeriz-ation. For practical purposes crystallization polymerization is still a useful way to make small amounts of pure and stable elastomeric polyacetal-dehyde of very high molecular weight.

The kinetics of cationic polymerization of acetaldehyde or higher aldehydes in solution are even less understood. No rate data on acetal-dehyde polymerization with BF_3 in ether are available [6]. It is only known that after an induction period of 5—15 min a vigorous polymeriz-ation occurs which is completed in a few minutes. No attempts were made to control the temperature during this uncontrolled reaction and polymer precipitated. Other cationic polymerizations of acetaldehyde with "less

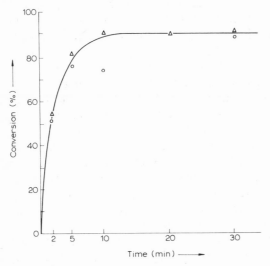

Fig. 23. Polymerization of n-butyraldehyde at $-78°C$ with $KOPh_3$ in pentane.

active" initiators were reportedly slower, but the literature reports of reaction times of 2 to 16 h appear to be "standard" times or times of convenience rather than actual reaction times (16 h often means "over-night"). No rates can be estimated from the data.

The only useful conversion—time curve for an anionic aldehyde polymerization has been obtained for n-butyraldehyde (25% solution in pentane) at $-78°C$ with potassium triphenyl methoxide (Fig. 23) [59]. The conversion was determined from polymer weight and monomer remaining in solution. It was shown that the polymerization is very rapid and an 80% conversion is reached in 5 min.

The literature of aldehyde polymerization, particularly acetaldehyde polymerization, with modified aluminium alkyls as initiators is volumi-nous. The earliest studies were done by Ishida [60]. He found that there were remarkable differences in polymerization rate behaviour when the

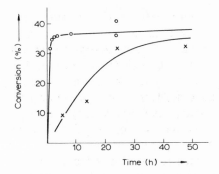

Fig. 24. Course of polymerization of acetaldehyde (0.227 mole in 30 cm³ n-hexane) with 0.00439 mole AlEt₃. (○), catalyst addition at 20°C; (x), catalyst addition at −78°C.

initiator was added at room temperature or at low temperature (Fig. 24). In both cases a limited conversion value of about 40% was reached. When the initiator was added at room temperature, this value was reached in a short time. At the lower temperature the initial polymerization rate was much slower. Ishida found also that relatively high initiator concentration (2 wt. %) gave maximum conversion. In another set of experiments it was found that a lower polymerization temperature gave a faster polymerization rate and a larger fraction of crystalline polymer (Fig. 25). An apparent activation energy of −8.5 kcal mole⁻¹ was calculated for this system.

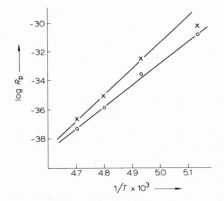

Fig. 25. Variation of the rate of polymer formation (R_p) with temperature. (x), rate of formation of total polymer; (○), rate of formation of crystalline polymer. Charge: n-hexane 30 cm³, acetaldehyde 0.227 mole, AlEt₃ 0.00439 mole. Catalyst addition at 20°C.

4. Haloaldehydes

Polymerizations of halo aldehydes are often discussed separately from those of the aliphatic higher polyaldehydes [61]. Chloral and fluoral have reaction behaviours which sometimes resemble those of formaldehyde. They form stable hydrates and hemiacetals but do not form polymers in hydrophylic media. We have seen in the case of formaldehyde that the dehydration is intermolecular and low molecular weight polyoxymethylene hydrates are formed. Only under exceptional circumstances, with strong dehydrating agents and anhydrous conditions can polychloral be formed from chloral hydrate. Usually chloral hydrate is dehydrated to chloral but is stable in water. The state of our knowledge on haloaldehyde polymerization has been summarized by Rosen [61] in 1966.

The best studied haloaldehyde polymerization is that of chloral. No accurate kinetic study of chloral polymerization has been undertaken because chloral polymer is insoluble. When polymerization was initiated, the initiator solution reacted instantaneously with the monomer to form insoluble polymer which prevented thorough mixing of the initiator with the monomer solutions prior to polymerization.

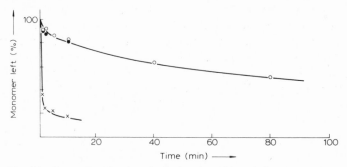

Fig. 26. Rate of chloral polymerization (NMR). Initiator concentration, 0.2 mole %: (○) Ph_3P; (×), $LiOC(CH_3)_3$; (●), collidine. No solvent. Mixing temp., 60°C; cooling bath temp., 0°C.

We found recently [62] that chloral can be polymerized readily and reproducibly in a two step process. Monomer and initiator are first mixed above the polymerization temperature to form a homogenous solution. The initiated monomer polymerized rapidly and in the case of neat monomer the reaction was complete in a few minutes. In dilute solution the polymerization temperature was lower.

Alkali metal salts with nucleophilic anions are notably good initiators for chloral anionic polymerization (Fig. 26). The most studied initiator is lithium *tert*-butoxide. When 0.2 mole % of lithium *tert*-butoxide (based on chloral) was added to neat chloral monomer at 60°C the alkoxide $(CH_3)_3COCH(CCl_3)O^-$ Li^+ was formed instantaneously, but no further addition of chloral occurred. This reaction was observed by an NMR study of the system and confirmed by the chemical reactions of the product alkoxide, which acted as the initiator. Tertiary amines such as pyridine and NR_3 where R is an alkyl group have been found to be good initiators for chloral polymerization. They are slower initiators than lithium *tert*-butoxide and NMR studies with tertiary amine initiators did not show identifiable initiating species in detectable concentrations.

Chloride initiation of chloral polymerization could be readily achieved with tetraalkyl ammonium chlorides, such as tetrabutyl ammonium chloride, or with trialkyl sulphonium chlorides as initiators. Chloral polymerization initiated with R_4NCl behaved very similarly to that with tertiary amine initiation. It is likely that the actual initiator of chloral polymerization with tertiary amines was chloride ion, which was presumably formed by chloride abstraction from chloral by the amine. The ease of chloride exchange in chloral reactions was demonstrated by initiation studies with $^{36}Cl^-$ as initiator.

Extensive studies of chloral polymerization were carried out with tertiary phosphines and with phosphonium chlorides as initiators. Tertiary phosphines, for example triphenyl phosphine, reacted instantaneously and quantitatively with one mole of chloral to form $PH_3P^+-OCH=CCl_2Cl^-$. The phosphonium chloride had the same ability to initiate chloral polymerization as triphenylphosphine.

The rate of chloral polymerization was studied by estimating the monomer disappearance by NMR. Figure 26 shows a comparison of the rate of polymerization of chloral initiated by lithium *tert*-butoxide, a quaternary ammonium chloride and triphenyl phosphine. The rate of chloral polymerization initiated with lithium *tert*-butoxide was much faster than the chloride initiated polymerization. This may be explained by the much more efficient and essentially quantitative initiation with the butoxide and consequently the greater number of growing polychloral chains.

Chloral polymerization to give uniform samples of polychloral cannot be carried out by an isothermal polymerization because the polymer is not soluble in the monomer. All previous work on rates and conversions of chloral polymerization is very questionable. The present author's data [62] are good for conversions; the rates are semiquantitative but internally self-consistent. The cooling bath temperature was accurately controlled, but the rate of heat generation caused by the polymerization, the heat dissipation and consequently the internal temperature of the polymerization mixture varied as the polymerization progressed, but was between 48° and 50°C during most of the polymerization. The polymerization was carried out under quiescent conditions. After about 2% conversion a self-supporting gel was formed, but the polymerization was then allowed to go to completion.

Kambe et al. [63] studied quite recently equilibria in chloral polymerization unfortunately with only one initiator, sodium naphthalene. Unlike the chloral polymerization without solvent where the polymerization temperature is 58°C [62], in the case of one molar monomer concentration it was 12°C [63], the ceiling temperature of chloral polymerization. In Kambe's work the initiator was always added below the polymerization

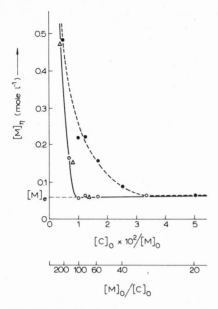

Fig. 27. Residual monomer concentration in the polymerization of chloral versus catalyst—monomer ratio. (- - ● - -), direct polymerization (—○—), repolymerization; (△), block copolymerization.

temperature and caused incomplete initiation. The portion of the polymerization mixture that is initiated at low temperatures has most characteristics of a "living" polymer (it can be acetylated with acetic anhydride, etc.). The polymer precipitated at low temperatures; on warming, the polymer depolymerized, cooling caused repolymerization. This cycle could be repeated several times without loss of polymerizability (Fig. 27). Under the reported experimental conditions an equilibrium amount of monomer concentration in the polymer is always reached in each cycle. The curve (---), representing what is called "direct polymerization", differs from the other curve because the authors initiated the polymerization below the ceiling temperature which means the polymerization was incompletely initiated. In Fig. 28 polymer yield (conversion) of chloral polymerization at various temperatures is given as a function of initiator/monomer concentration. At the highest temperature the conversion levelled off at about 40% while at lower temperatures a conversion of 80% could be reached. It should be pointed out that in this work Kambe et al. [63] were interested in determining the equilibria in chloral polymerization and no attempts have been made to study the rates.

In addition to chloral the polymerization of other chlorinated aldehydes was studied by Kambe et al., for example that of α-chloro iso-butyraldehyde. It was also found in this case that the initiation was almost instantaneous and the growing polymer ends could be acetylated when acetylation was carried out immediately at the low polymerization

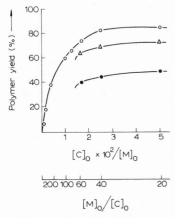

Fig. 28. Polymer yield versus catalyst—monomer concentration ratio (—○—), −78°C; (—△—), −48°C; (—●—), −30°C.

temperature without warming the mixture to room temperature. When the polymerizations were carried out at higher temperatures with neat monomers no acetyl end groups could be introduced, indicating that the growing polymer ends were occluded and unavailable for a chemical reaction. Alternatively at higher temperatures stable ends were formed by an as yet unexplained termination reaction. The course of these reactions probably depends on the initiator.

Polymers with different end groups usually also have differences in thermal stability. It has been found that tertiary amines or quaternary ammonium salts initiated chloral polymers have very poor thermal stability. Triphenyl phosphine initiated polychloral samples have good thermal stability and the lithium tertiary butoxide initiated polychloral has a thermal stability close to that of the phosphine initiated polymers.

The polymerization of fluoral [61] has been studied in detail by Busfield [64]. It was found that fluoral polymerized very slowly in the presence of 2 mole % of formic acid but extremely quickly in the presence of as little as 0.2 mole % of a tertiary amine such as pyridine or trimethyl amine. Gasphase polymerization of fluoral was studied with trimethyl amine and it was found that the ceiling temperature was 73°C. The half life (50% conversion to polymer) of fluoral polymerization with formic acid was 12 h but for pyridine or trimethyl amine it was about 30 sec.

Busfield [64] reported that the time/conversion curve under his experimental conditions did not follow any simple rate laws. In the range of 0.1—3% of the total pressure, the initiator pressure had little effect on the rate of polymerization at lower concentrations but an induction period of up to $\frac{1}{2}$ min was sometimes observed. After the induction period a fast polymerization ensued and the pressure approached a value independent of initiator and initial monomer concentration but dependent on the temperature. The pressure continued to fall very slowly until

376

an equilibrium pressure was reached after 5—15 h. Unfortunately no analysis of this monomer was reported; consequently, small amounts of impurities could have influenced the results.

Polyfluoral made by pyridine or trimethylamine initiation was thermally not very stable. It depolymerized above 70°C in three minutes to monomer and initiator. Evidently the polymer—monomer equilibrium is very mobile, suggesting that the growing ends retained their activity.

REFERENCES

1 H. Staudinger, *Die Hochmolekularen Organischen Verbindungen*, Springer, Berlin, 1932; New Edition, Springer, Berlin, 1960.
2 J. F. Walker, *Formaldehyde*, 3rd edn., Reinhold, London, 1964.
3 J. Furukawa and T. Saegusa, *Polymerization of Aldehydes and Oxides*, Wiley—Interscience, New York, 1963.
4 J. C. Bevington, in N. G. Gaylord (Ed.), *Polyethers I High Polymers, Vol. XIII*, Wiley—Interscience, New York, 1963.
5 O. Vogl, *Polyaldehydes*, M. Dekker, New York, 1967.
6 O. Vogl, in J. P. Kennedy and E. Tornquist (Eds.), *Polymer Chemistry of Synthetic Elastomers*, Wiley—Interscience, New York, 1968.
7 S. J. Barker and M. B. Price, *Polyacetals*, Elsevier, New York, 1971.
8 G. F. Pregaglia and M. Binaghi, in A. D. Ketley (Ed.), *The Stereochemistry of Macromolecules*, M. Dekker, New York, 1967.
9 B. W. Bridgeman and J. B. Conant, *Proc. Acad. Sci. U.S.*, 15 (1929) 680.
10 French Pat. 1322460 (1963), 1,371,065 (1964) and ff-Farbenfabriken Bayer.
11 N. Brown, *J. Macromol. Sci., Chem.*, 1 (1967) 209.
12 M. Goodman and J. Brandrup, *J. Polym. Sci., Part A*, 3 (1965) 327.
13 T. E. Carruthers and R. G. W. Norrish, *Trans. Faraday Soc.*, 32 (1936) 195.
14 R. N. MacDonald, French Pat. 1082519 (1954), British Pat. 748836 (1953).
15 W. Kern, *Chem. Ztg. Chem. App.*, 88 (1964) 623.
16 Kilnzel, A. Gieffer and W. Kern, *Makromol. Chem.* 96 (1966) 17.
17 N. S. Enikolopyan, *J. Polym. Sci.*, 58 (1961) 1301.
18 Z. Machacek, J. Mejzlik and J. Pac, *J. Polym. Sci.*, 52 (1961) 309.
19 Z. Machacek, J. Mejzlik and J. Pac, *Vysokomol. Soedin*, 3 (9) (1961) 1421.
20 J. Mejzlik, J. Mencikova and Z. Machacek, *Vysokomol. Soedin*, 4 (1962) 769, 776.
21 Z. Machacek and J. Mencikova, *Chem. Prum.*, 13 (38) (1963) 433.
22 Z. Machacek and E. Foniokova, *Chem. Prum.*, 14 (39) (1964) 75, 252.
23 K. Vesely and J. Mejzlik, *Vysokomol. Soedin. Zh.*, 5 (9) (1963) 1425.
24 K. Vesely, *Pure Appl. Chem.*, 4 (1962) 407.
25 M. F. Proshlyakova, F. F. Sanaya and N. S. Enikolopyan, *Vysokomol. Soedin. Zh.*, 5 (1963) 1632.
26 W. J. Irzack, L. M. Romanov and N. S. Enikolopyan, *Vysokomol. Soedin. Zh.*, 5 (1963) 1638.
27 L. P. Bobkova, V. S. Korsakov, L. M. Romanov and N. S. Enikolopyan, *Vysokomol. Soedin. Zh.*, 5 (1963) 1653.
28 British Patent 796,862 (1958).
29 H. Yokota, M. Kondo, T. Kagiya and K. Fukui, *J. Polym. Sci., Part A1*, 5 (1967) 3129.
30 W. Kern, E. Eberius and V. Jaacks, *Makromol. Chem.*, 141 (1971) 63.
31 V. Jaacks, K. Boehlke and E. Eberius, *Makromol. Chem.*, 118 (1968) 354.
32 K. Boehlke and V. Jaacks, *Makromol. Chem.*, 142 (1971) 189.
33 Y. Yamaguchi, T. Kawasake and H. Matsuda, *Makromol. Chem.*, 112 (1968) 40.

34 W. Fukada and H. Kakiuchi, *J. Chem. Soc. Jap., Ind. Chem. Sect., 65* (1962) 2054.
35 P. H. Plesch, *The Chemistry of Cationic Polymerization*, Pergamon, Oxford, 1963.
36 H. Yokoda, M. Kondo, T. Kagiya and K. Fukui, *J. Polym. Sci., Part A1, 6* (1968) 425, 435.
37 T. Kagiya, M. Izu and K. Fukui, *Bull. Chem. Soc. Jap., 40* (1967) 1045, 1049.
38 S. Okamura, S. Nakashio and K. Hayashi, *Doitai To Hoshasen, 3* (1960) 242.
39 Y. Tsuda, *J. Polym. Sci., 49* (1961) 369.
40 H. Yamaoka, K. Hayashi and S. Okamura, *Makromol. Chem., 76* (1964) 196.
41 S. Nakashio, M. Ueda and K. Takahashi, *Makromol. Chem., 83* (1965) 23.
42 C. Chachaty, M. Magat and L. T. Minassian, *J. Polym. Sci., 48* (1960) 139.
43 C. Chachaty, *C. R. Acad. Sci., 251* (1960) 385.
44 M. Trautz and E. Ufer, *J. Prakt. Chem., 113* (1926) 105.
45 T. R. Bates and R. Spence, *J. Amer. Chem. Soc., 53* (1931) 1689.
46 R. Spence, *J. Chem. Soc.,* (1933) 1193.
47 J. C. Bevington and R. G. W. Norrish, *Proc. Roy. Soc., London, A205* (1951) 516.
48 R. Sauterey, *Ann. Chim. (Paris),* (12), 7 (1952) 5.
49 S. Toby and E. R. Rutz, *J. Polym. Sci., 60* (1962) 541.
50 J. G. Boyles and S. Toby, *J. Polym. Sci., Part B, 4* (1966) 411.
51 J. G. Boyles and S. Toby, *J. Polym. Sci., Part A1, 5* (1967) 1705.
52 A. Iliceto, *Gazz. Chim. Ital., 81* (1951) 786, 915.
53 A. Iliceto and S. Brezzi, *Chim. Ind. (Milan), 33* (1951) 212.
54 L. Mortillaro, G. Galiazzo and S. Brezzi, *Gazz. Chim. Ital., 94* (1964) 109.
55 C. H. Manwiller and J. B. Thompson, U.S. Pat. 3,193,533 (1965), French Patent 1,367,281 (1964).
56 L. Mortillaro, G. Galiazzo and S. Brezzi, *Chim. Ind. (Milan), 46* (1964) 39, 144.
57 L. Mortillaro, G. Galiazzo and A. Bandel, *Chim. Ind. (Milan), 46* (1964) 1143.
58 L. Mortillaro and G. Galiazzo, *Chim. Ind. (Milan), 46* (1964) 1148.
59 O. Vogl, *J. Polym. Sci., Part A2* (1964) 4607; *J. Macromol. Sci., Part A1* (1967) 243.
60 S. Ishida, *J. Polym. Sci., 62* (1962) 1; *Kobunshi Kagaku, 18* (1961) 187.
61 I. Rosen, *J. Macromol. Sci., Part A1* (1967) 267.
62 O. Vogl, U.S. Patent 3,454,527 (1969). See also French Patent 1,528,327 (1968) and 1,567,895 (1968).
63 I. Mita, I. Imai and H. Kambe, *Makromol. Chem., 97* (1970) 143.
64 W. K. Busfield, *Polymer, 7* (1966) 541.

Chapter 6

Lactams

J. ŠEBENDA

1. Introduction

The formation of polyamides by ring-opening polymerization of lactams can proceed through several types of reversible transacylation reactions in which the lactam amide bond is cleaved while a polymer amide bond is formed, viz.

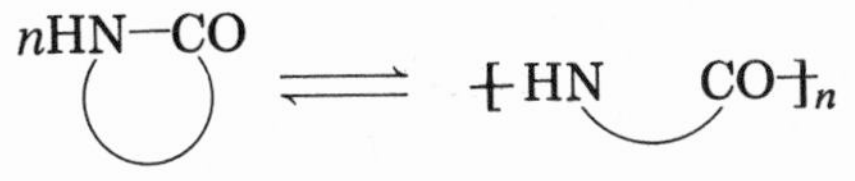

$$n\text{HN—CO} \rightleftharpoons +\text{HN} \quad \text{CO}\}_n$$

In most cases, these reactions involve activated lactam molecules and/or activated end groups (growth centres) acting as nucleophiles or acylating agents. These activated species display an increased reactivity (acylating ability or nucleophilicity) as compared with unactivated end groups or monomer (Table 1). The kind of activated species involved in the initiation and growth reactions is determined by the nature of the initiator and by reaction conditions. According to the prevailing kind of activated species and also for historical reasons, lactam polymerizations are classified as anionic, cationic and hydrolytic processes (see Sections 4—6). In the absence of catalysts a direct transamidation between two neutral monomer or polymer amide groups is extremely slow and usually does not contribute to polymer formation [1].

Usually, lactams are referred to by their trivial names such as pyrrolidone, piperidone, caprolactam, enantholactam, capryllactam and laurinlactam for the 5-, 6-, 7-, 8-, 9-, and 13-membered lactam.

Both industrial demand and academic interest stimulated detailed research of the synthesis of lactam polymers and elucidation of the polymerization mechanisms. Until recently, nylon 6 has been the only industrially produced lactam polymer and, therefore, most of the extensive investigations are dealing with the polymerization of caprolactam [2]. The mechanisms of the main elementary reactions derived for caprolactam are applicable to other lactams as well, except to the *N*-substituted ones.

TABLE 1
Species involved in initiation and propagation reactions

Acylating species		Nucleophile	
	Unsubstituted lactams		
HN—CO (ring)		$^{\ominus}$N—CO (ring)	Activated lactam
H$_2$$\overset{\oplus}{\text{N}}$—CO (ring)	Activated lactam	HN—CO (ring)	
—CON—CO (ring)	Activated end group	—NH$_2$	
—CO—$\overset{H}{\underset{}{\overset{\oplus}{N}}}$—CO (ring)	Activated end group	—COO$^{\ominus}$	
(—CO)$_2$O	Activated end group		
—COOH			
	N-Substituted lactams		
H$\overset{R}{\overset{\oplus}{N}}$—CO (ring)	Activated lactam	RN—CO (ring)	
RN—CO (ring)		RNH—	
—COX	Activated end group	—COO$^{\ominus}$	
	(X = —Cl, —Br, —OCO)		

2. Equilibria in lactam polymerization

Most of the transacylation reactions proceeding during lactam polymerizations are reversible and cyclization is competitive with linear polymerization. Complicated equilibria are established between all components, provided that side reactions do not interfere and the reaction medium is homogenous.

The following symbols are used in this chapter:

L   cyclic lactam
—L—   linear monomer unit
L_n cyclic n-mer
AB   initiator yielding two residues which may be incorporated as end groups or terminating units (or which can act as growth centres)
$A—L_n—B$ linear n-mer terminated by initiator residues A and B (degree of polymerization $P_n = n$)
[L]   lactam concentration
[—NHCO—]   concentration of linear amide groups linking monomer units
[A] or [B]   concentration of A (or B) end groups (e.g. $—NH_2$, $—COOH$, $RCO—$, $RNH—$)
[S]   concentration of linear polymer molecules (linear chains)
$[S_1]$ up to $[S_n]$ concentration of molecules containing 1 to n monomer units terminated with initiator residues ($A—L—B$ up to $A—L_n—B$)

Subscripts 0 and e indicate initial and equilibrium values.

Lactam polymerizations are usually carried out at high temperatures in the absence of solvents and the problem of volume contraction is circumvented by expressing the concentrations in moles or equivalents per kg.

2.1 TYPES OF EQUILIBRIA

The polymerization of any lactam starts with the ring opening of the monomer (or activated monomer) by the initiator

$$AB + L \rightleftharpoons A—L—B \tag{1}$$

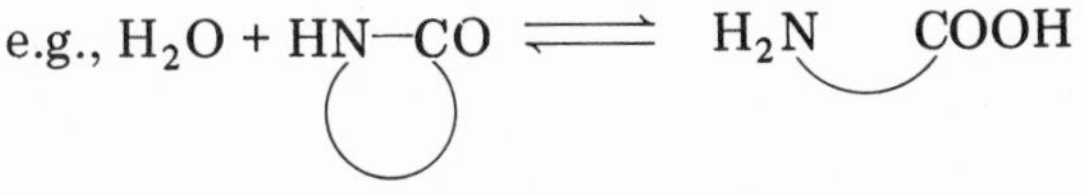

e.g., $H_2O + HN—CO \rightleftharpoons H_2N \quad COOH$

with

$$K_1 = \frac{[S_1]_e}{[AB]_e[L]_e}$$

The growth of polymer chains then proceeds either through additions between terminal groups of two linear chains

$$A-L_m-B + A-L_n-B \rightleftharpoons A-L_{m+n+1}-B \tag{2a}$$

e.g., NH_2⁓⁓CON—CO + NH_2⁓⁓CON—CO $\rightleftharpoons$

NH_2⁓⁓CONH   CONH⁓⁓CON—CO

or condensation between two linear molecules

$$A-L_n-B + A-L_m-B \rightleftharpoons AB + A-L_{m+n}-B \tag{2b}$$

e.g., NH_2⁓⁓COOH + NH_2⁓⁓COOH $\rightleftharpoons$ H_2O + NH_2⁓⁓CONH⁓⁓COOH

or addition of monomer (or activated monomer) to the active end of the polymer molecule

$$A-L_{n-1}-B + L \rightleftharpoons A-L_n-B \tag{3}$$

e.g., A⁓⁓CON—CO + $^\ominus$N—CO $\rightleftharpoons$ A⁓⁓CON$^\ominus$ CON—CO

Similar reactions of chains composed of more than one monomer unit lead to the following equilibria involving cyclic oligomers:

$$A-L_n-B \rightleftharpoons AB + L_n \tag{4}$$

e.g., NH_2 / COOH (ring) $\rightleftharpoons$ H_2O + NH / CO (ring)

$$A-L_{m+n}-B \rightleftharpoons A-L_m-B + L_n \tag{5}$$

e.g., CONH⁓⁓ / NH_2 (ring) $\rightleftharpoons$ CO / NH (ring) + NH_2⁓⁓

Transacylations between polymer molecules are classified as exchange reactions which do not alter the number of molecules and affect only the molecular weight distribution, e.g.,

$$A\text{\large$\sim\sim$}NHCO\text{\large$\sim\sim$}B \qquad A\text{\large$\sim\sim$}NH_2$$

$$+ \qquad\qquad \rightleftharpoons \qquad + \qquad\qquad (6)$$

$$C\text{\large$\sim\sim$}NH_2 \qquad\qquad C\text{\large$\sim\sim$}NHCO\text{\large$\sim\sim$}B$$

In homogenous media, most of the transacylation reactions are reversible and as soon as the first polymer amide groups are formed, the same kind of reactions can occur both at the monomer and at the polymer amide groups. Unless the active species are steadily formed or consumed by some side reaction, a set of thermodynamically controlled equilibria is established between monomer, cyclic as well as linear oligomers and polydisperse linear polymer. The existence of these equilibria is a characteristic feature of lactam polymerizations and has to be taken into account in any kinetic treatment of the polymerization and analysis of polymerization products. The equilibrium fraction of each component depends on the size of the lactam ring, substitution and dilution, as well as on temperature and catalyst concentration.

2.2 RING—CHAIN EQUILIBRIA

The most important equilibrium is the ring—chain equilibrium (3) which expresses the relative thermodynamic stabilities of the lactam and the open-chain monomer unit inside the polymer chain. We have

$$K_3 = \frac{[A-L_n-B]}{[L]_e[A-L_{n-1}-B]} = \frac{[S_n]}{[L]_e[S_{n-1}]} \qquad (7)$$

For statistical molecular weight distributions

$$[A-L_n-B]/[A-L_{n+1}-B] = 1/(1-1/P_n) \qquad (8)$$

so that the following general relation [3—9] should be valid between the equilibrium monomer concentration and the average number of monomer units per macromolecule (P_n), viz.

$$K_3 = \frac{1}{[L]_e} \cdot \frac{P_n - 1}{P_n} \qquad (9)$$

For sufficiently high molecular weights, $P_n \simeq P_n - 1$ so that

$$K_3 = \frac{1}{[L]_e} \qquad (10)$$

For very low degrees of polymerization, the contribution of the incorporated initiator has to be taken into account for the calculation of P_n from the actual molecular weight (M_n), viz.

$$P_n = \frac{M_n - M_{AB}}{M_1} = \frac{[L]_0 - [L]_e}{[AB]_0 - [AB]_e} \tag{11}$$

(where M_1 and M_{AB} are the molecular weights of the monomer and initiator) and for $M_{AB} \simeq M_1$, $P_n = M_n/M_1 - 1$. However, Elias and Fritz [10, 11] found that the equation

$$K_3 = \frac{1}{[L]_e} \cdot \frac{P_n}{P_n + 1} \tag{12}$$

yields more constant values for K_3 than eqn. (9) (Table 2). This could be due to the fact that at very high initiator concentrations and very low molecular weights, such important quantities as the dielectric constant and activity coefficients are different from those with the high molecular weight system. The influence of the composition and dielectric constant on the activity coefficients and equilibria in lactam polymerizations suggested by Wiloth [12] was confirmed by Kruissink [13] as well as Giori and Hayes [14]. Therefore, the most reliable value of K_3, which is independent of the nature of the catalyst and reaction mechanism, can be obtained by extrapolation of K_3 to zero initiator concentration or by measurement of the monomer—polymer equilibrium at very low initiator concentration.

TABLE 2

Polymerization equilibria with laurinlactam (L) and lauric acid (AB) [10] (Concentrations in mole kg^{-1})

Temp. ($^\circ$C)	$[AB]_0$	$[AB]_e$	$[L]_0$	$[L]_e$	K_3 (mole kg^{-1})		
					eqn. (12)	eqn. (9)	eqn. (10)
280	0.499	0.050	4.561	0.100	9.06	8.97	9.97
280	1.248	0.275	3.801	0.082	9.70	9.03	12.24
280	2.496	0.998	2.534	0.062	10.00	6.32	16.05
300	0.499	0.049	4.561	0.122	7.46	7.38	8.22
300	1.248	0.257	3.801	0.110	7.15	6.64	9.07
300	2.496	1.011	2.534	0.084	7.45	4.71	11.96

When the structure of the initiator is very similar to that of the surroundings of the active end group, then the value of K_1 is very close to the value of K_3. Such a case occurs in the polymerization of laurinolactam initiated with lauric acid [10, 11], the acidity of which is very close to that of the carboxyl group of a polymer molecule:

T (°C)	K_1 (kg mole^{-1})	K_3 (kg mole^{-1})
280	9.05	9.51
300	7.00	7.40

Since K_3 is related to the free energy of polymerization, the monomer—polymer equilibrium will be treated together with other thermodynamic aspects in Section 3.

Similar equilibrium constants can also be derived for cyclic oligomers [15]. In most cases, the amount of cyclic oligomers cannot be neglected and has to be taken into account for calculations of equilibrium constants and degrees of polymerization. The fraction of cyclic oligomers depends on the ring size of the monomer and for unstable monomers, e.g. medium and large lactams, the fraction of the cyclic dimer can be even higher than that of the monomer (Table 3). Until now, cyclic oligomers have been determined in the extractable fraction only [16—20]. With respect to their very low solubility, the fraction of cyclic oligomers remaining in the polymer increases with the molecular weight of the cyclic compound. Precise equilibrium values for cyclic oligomers require accurate determination of the latter in the equilibrium polymerization product. Such data could be soon made available, at least in some favourable cases, by GPC measurements of the whole polymerization product.

TABLE 3
Content of extractable cyclic oligomers and monomer in equilibrium polymers [16]

Monomer	Content of cyclics (wt %)					
	L_1	L_2	L_3	L_4	L_5	ΣL_n
Caprolactam	8.5	0.71	0.43	0.22	0.13	10.09
Laurinlactam	0.33	0.94	0.25			1.52

The aforementioned equilibria may be attained only in the liquid state, i.e., in the melt or solution. When the temperature is low enough to allow crystallization of the polymer, then the fraction of crystallized polymer does not take part in the equilibrium. Only the amorphous fraction of the polymer is involved in the monomer—polymer equilibrium and as a result, the monomer content decreases with increasing crystallinity [21]. In this way, the monomer content may be lowered substantially. The effect of crystallization or phase separation is very important in the polymerization of five- and six-membered lactams which can thus be forced to polymerize to higher yields. Dilution and lowering of the melting temperature, on the other hand, increases the equilibrium monomer content.

2.3 MOLECULAR WEIGHTS

As long as all transacylation equilibria can be attained, the molecular weight distribution approaches the most probable one. The existence of a statistical distribution for equilibrium caprolactam polymers has been well established [3, 22—24], and this is an indirect proof of the independence of the reactivity of functional groups on the chain length.

The equilibria (1) and (2) between the initiator and the monomer and polymer amide groups determine the equilibrium concentration of linear polymer molecules [S] and the molecular weight of the polymer (M_n), respectively. The position of these equilibria and, consequently, the molecular weight of the resulting polymer may be affected by the concentration and nature of the initiator.

The molecular weight can be controlled by adding low molecular weight substances forming a pair of end groups, e.g. alkyl amides

$$RCONHR' + \text{---}CONH\text{---} \quad \rightleftharpoons \quad RCONH\text{---} + \text{---}CONHR' \qquad (13a)$$

or other substances entering into the condensation equilibrium, e.g. primary amines or carboxylic acids

$$RCOOH + \text{---}CONH\text{---} \quad \rightleftharpoons \quad RCONH\text{---} + \text{---}COOH \qquad (13b)$$

Assuming that the reactivities of end groups or growth centres are equal and independent of the chain length, then

$$K_2 = \frac{[AB]_e[S_{m+n}]}{[S_m][S_n]} = \frac{[AB]_e[-NHCO-]_e}{[A]_e[B]_e} \qquad (14)$$

and for

$$[A]_e = [B]_e = [S]_e = \frac{1000}{M_n}$$

$$K_2 = \frac{[AB]_e[-NHCO-]_e}{[S]_e^2} = \frac{([AB]_0 - [S]_e)[-NHCO-]}{[S]_e^2} \qquad (15)$$

In order to determine $[-NHCO-]$, the equilibrium concentrations of the monomer $[L]_e$ and cyclic oligomers $[L_n]_e$ have to be estimated and then

$$K_2 = \frac{([AB]_0 - [S]_e)([L]_0 - [L]_e - n[L_n]_e)}{[S]_e^2} \qquad (16)$$

As will be shown later, the concentration of the initiator and its incorporated fragments does not always remain constant and/or the corresponding equilibria need not necessarily be established.

For non-equilibrium polymerizations, the average degree of polymerization of linear molecules is related to the instantaneous monomer concentration [L] through equation

$$P_n = \frac{[L]_0 - [L] - n[L_n]}{[AB]_0 - [AB]} \tag{17}$$

where $[AB]_0 - [AB] = S$.

The values of K_1 and K_2 depend on the nature of the initiator and, therefore, the corresponding equilibria will be considered together with the corresponding reaction mechanisms.

3. Polymerizability of lactams

Lactam polymerization comprises the conversion of a cyclic lactam unit into a linear one without the formation of any new chemical bonds. The term polymerizability involves both the thermodynamic feasibility and a suitable reaction path to convert the cyclic monomer into a linear polymer. Sometimes, a slight confusion arises when the term polymerizability is used as a synonym for both the rate of polymerization and the thermodynamic instability of the lactam. Due to the reversible nature of the polymerization of most lactams, eqns. (1)—(3), their polymerizabilities cannot be expressed in terms of the rate of polymerization only, but the rate of both polymerization and monomer reformation must be compared.

Frequently, the order of rates of polymerization of various lactams is different from that of the thermodynamic parameters for polymerization. For example, the initial rates of hydrolytic polymerization were practically the same for capro-, enantho- and capryllactam [25—27], whereas the corresponding heats of polymerization differed significantly for these monomers [27] ($-\Delta H_p$ = 3.3, 5.3 and 7.8 kcal mole^{-1}, respectively). Similarly, for substituted caprolactams the sequence of free energies of polymerization is just opposite to the order of rates of anionic polymerization (Fig. 1).

Moreover, the sequence of rates of polymerization may vary with the reaction conditions, i.e. temperature and catalyst. Hence the reactivity of the lactam amide group under the given reaction conditions should be always distinguished from the thermodynamic feasibility of polymerization. Generally, the reactivity of the reacting lactam ring depends on its conformation which may be different under different reaction conditions. For example, the conformation of the lactam anion can differ from that of the protonated lactam so that the order of reactivities of different lactams can be influenced by the nature of catalyst. As long as the catalyst does not change the nature of the major fraction of monomer and polymer, the thermodynamic values should be independent of the re-

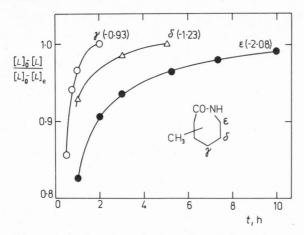

Fig. 1. Anionic polymerization of methylcaprolactam isomers. Catalyst: 0.5 mole % sodium caprolactam and 0.5% of imide [28—30]. Temperature: 172°C (γ- and ϵ-isomer) and 175°C (δ-isomer); the corresponding ΔG values are given in parentheses.

action mechanism. Therefore, only ΔH and ΔS values should be used for the comparison of the polymerizabilities of various lactams.

3.1 THERMODYNAMICS OF LACTAM POLYMERIZATION

The free energy change accompanying the incorporation of one mole of lactam into a high molecular weight polymer is given by the difference of the molar free energies of the monomer segment of an amorphous high molecular weight polymer and of the liquid monomer, viz.

$$\Delta G_p = G_{p,a} - G_{m,1} = \Delta H_p - T\Delta S_p \tag{18}$$

A negative value of ΔG_p for a lactam only indicates that polymerization is possible. Sometimes, such lactams do not polymerize because of the lack of a suitable reaction path, or they undergo other kinetically or thermodynamically more favoured reactions (see Section 8).

Irrespective of the reaction mechanism, the polymerization of lactams leads to an equilibrium between monomer, cyclic oligomers and polymer. Tobolsky and Eisenberg [9] showed that the thermodynamic parameters are independent of the reaction mechanism, so that the polymerizability may be rationalized in terms of the ease of formation of the cyclic monomer, or, its opening into a linear chain unit. The simple relation between the equilibrium monomer concentration $[L]_e$, temperature, and standard heat and entropy of polymerization,

$$RT \ln[L]_e = \Delta H_p^0 - T\Delta S_p^0 = -RT \ln K \tag{19}$$

is valid only if the monomer and polymer behave ideally. Taking into account the interaction between monomer and polymer, Ivin and Leonard [31] were able to show that the following expression is valid for the free energy of polymerization, viz.

$$\Delta G_p = RT[\ln \phi_1 - (\ln \phi_2)/n + 1 - 1/n + \chi(\phi_2 - \phi_1)] \tag{20}$$

where ϕ_1 and ϕ_2 are the volume fractions of monomer and polymer and χ is the polymer—monomer interaction parameter. At sufficiently high degrees of polymerization (n), the simplified equation

$$\Delta G_p = RT[\ln \phi_1 + 1 + \chi(\phi_2 - \phi_1)] \tag{21}$$

can be used.

3.1.1 Heat of polymerization

At normal pressure, the $p\Delta V$ term in $\Delta H = \Delta E + p\Delta V$ is negligible and the enthalpy change is almost equivalent to the change in internal energy of the monomer. Hence, the heat of polymerization can be considered as a measure of the strain energy in the cyclic compound [32]. However, the enthalpy difference between the cyclic compound and its polymer does not always reflect the actual ring strain, since the polymerization of a strained lactam into an open chain polyamide does not necessarily release all strain existing in the cyclic monomer. Namely, in polymers of highly substituted cyclic monomers there can be some strain imposed on the monomer unit inside the polymer chain as compared to the isolated open chain monomer unit. Such interactions have to be taken into account when calculating ΔH_p values from the heats of combustion of the monomer and low molecular weight open chain amides [33].

More reliable ΔH_p values can be obtained from direct measurements of the heat evolved during polymerization [34—39], or from the difference between the heats of combustion of the amorphous polymer and the liquid monomer. With respect to the very low ΔH_p values, as compared with the heats of combustion, the latter method is very inaccurate. The evaluation of calorimetric data is more difficult for partially crystalline polymers, for which the crystallinity as well as heat of crystallization must be known [38]. Equations (19)—(21) can be applied to such monomer—polymer equilibria for which the equilibrium monomer concentrations at different temperatures are available with sufficient precision [28—30, 40]. The latter method is also limited to completely amorphous polymers because the crystalline ordered areas do not take part in the monomer—polymer equilibrium [21].

The ring strain in lactams arises from:

(a) bond angle distortion (angle strain);

(b) bond stretching or compression;

(c) repulsion between eclipsed hydrogen atoms or substituents on neighbouring ring atoms (conformational strain, bond torsion, bond opposition);

(d) nonbonded interaction between atoms or substituents attached to different parts of the ring (transannular strain, compression of van der Waals radii);

(e) inhibition or reduction of amide group resonance.

The contribution of each type of strain to the total molecular strain depends on the ring size, substitution and nature of ring atoms.

Severe distortion of bond angles is the major source of the high strain in three- and four-membered lactams. Bond opposition forces arising from eclipsed conformations are responsible for the strain in five-membered lactams. In medium sized rings, strain arises primarily from nonbonded interactions and bond opposition as well as prevention of resonance. In very large rings, transannular interactions can be avoided completely by arranging the ring atoms into two almost parallel chains [41], e.g.,

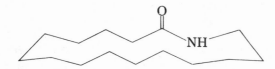

This arrangement resembles open chain compounds and the strain of such lactams approaches zero ($\Delta H_p \rightarrow O$).

With increasing ring size, the heat of polymerization passes through a minimum for five- and six-membered lactams and after attaining a

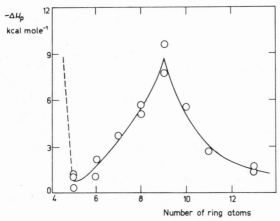

Fig. 2. Effect of ring size on the heat of polymerization of unsubstituted lactams $(CH_2)_n CONH$ [10, 27, 33, 35, 37—39, 42, 43].

maximum for the nine-membered lactam approaches zero for very large lactams (Fig. 2). Analogous to the thermodynamic parameters for cyclo-paraffins, the $-\Delta H$ values pass through a minimum for the six-membered lactam. However, lactams should be compared rather to cycloalkene rings because two adjacent trigonal atoms are present and the amide bond has a double bond character due to the strong nitrogen carbonyl interaction [44].

3.1.2 Entropy of polymerization

The accumulation of a great number of small particles into one polymer chain is an aggregation process resulting in a decrease in the translational entropy of the system. In the polymerization of cyclic monomers, the decrease of translational entropy is partially counterbalanced by the increase in rotational and vibrational entropy resulting from the conversion of a more or less rigid cyclic monomer into a flexible monomer unit inside a polymer chain. Thus the net entropy of polymerization of lactams is more positive (e.g. -3 eu for seven-membered lactams) than the entropy of polymerization of vinyl monomers (-25 to -30 eu).

The highest rigidity of the cyclic monomer can be expected for medium rings with 8—11 ring atoms. Conversion of such rigid lactams to a polymer results in a large increase in the rotational and vibrational entropy because of the enhanced flexibility of the open chain monomer unit. Therefore, the contribution of the entropy term to the free energy of polymerization is expected to increase very steeply for medium rings. Calculations of ΔS_p^0 from specific heats indicated that the entropy of polymerization increases linearly from the five-membered to the eight-membered lactam [39]. When extrapolating linearly to large rings, Bonetskaya and Skuratov [43] arrived at $\Delta S_p^0 = 30$ eu for laurinlactam (Fig. 3). This value appears too high with respect to $\Delta S_p^0 = -3.75$ eu calculated from monomer—polymer equilibria [10]. It has to be borne in mind, that the mobility of ring atoms is very restricted only in lactams up to the nine-membered ring since there is almost no freedom of rotation for any of the ring atoms. In larger lactams, the mobility of ring atoms increases with increasing ring size so that the polymerization entropy should not increase as steeply as for smaller rings. The author is inclined to assume that for large rings the value of ΔS_p should follow a line similar to that derived (Fig. 3) from the content of cyclic oligomers (assuming $\Delta H_p = 0$).

Hence, entropy changes become increasingly important with increasing ring size. Whereas ΔH_p makes the main contribution to ΔG_p for strained lactams up to the eight-membered ring, the polymerization entropy makes the main contribution to ΔG_p for more than twelve-membered lactams for which ΔH_p approaches zero. For the 9-, 10-, and 11-membered ring the values of ΔH_p and ΔS_p contribute equally to the free energy of polymerization. The entropy changes become more favourable in co-

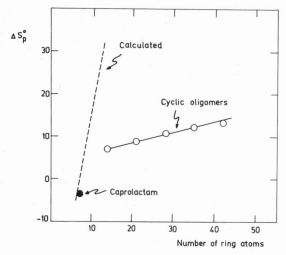

Fig. 3. Ring size and entropy of polymerization. Calculated values for lactams [43] compared with values obtained from the equilibrium content of caprolactam and its oligomers [16].

polymerization (of two different lactams) and, therefore, pyrrolidone and piperidone may be copolymerized even at higher temperatures at which homopolymerization does not proceed.

Skuratov et al. [39] tried to estimate the polymerization entropy from specific heat measurements of the monomer and polymer. However, the calculated value of ΔS^0_{298}, e.g. for caprolactam (+1.1 eu), is too positive because the authors disregarded the fact that the polymer was not completely crystalline. Making allowance for the partial crystallinity, the value of ΔS^0_p becomes more negative and approaches the value -3.2 eu calculated from the monomer—polymer equilibria [45].

3.2 AMIDE GROUP CONFORMATION

The *trans* conformation predominates in straight chain amides [46] and in more than nine-membered lactams [47]. Up to the eight-membered lactam the amide group is constrained to adopt the *cis* conformation [47] which is 1.4 kcal mole^{-1} less stable than the *trans* form [48]. This amount of energy is released in the conversion of the *cis*-amide group (in lactams up to enantholactam) into the *trans* form of the polymer and significantly contributes to the free energy of polymerization. From this point of view, the planar cis-amide group imposes some strain on the five- and six-membered lactams. This enthalpic factor is the main driving force in the polymerization of the almost unstrained six-membered lactam and the experimental value [39] of the heat of polymerization (ΔH_p = -1.1 kcal mole^{-1}) is very close to the enthalpy difference between the cis- and trans-amide group.

The highest resonance stabilization due to delocalization of the carbonyl π-orbital and the nitrogen p-orbital may be achieved only when the amide group is planar. Any deviation from planarity of the lactam amide group lowers its resonance stabilization and increases ring strain. This results, in most cases, in a higher polymerizability. The maximum resonance energy should be released in the polymerization of bicyclic lactams with the amide nitrogen at the bridgehead in which the amide group cannot be planar [49]. The preference for coplanarity renders such lactams very unstable and some of them polymerize even at room temperature [50].

In a more detailed investigation, Hallam and Jones [51] were able to show that there exist slight differences of the amide group conformation among the lactams of each group. Pyrrolidone cannot adopt a fully planar conformation without introducing some strain into the ring due to the repulsion of eclipsed hydrogens. In addition, the tendency of retaining an exocyclic double bond in the five-membered ring [52] results in a lower resonance stabilization of the amide group in pyrrolidone. This effect increases the ring strain caused by bond opposition so that pyrrolidone should polymerize easier than piperidone. The amide group of the latter six-membered lactam may easily attain a planar conformation and also the tendency of formation of an endocyclic double bond leads to an increased resonance stabilization, viz.

The planar cis-amide conformation in the seven-membered lactam already can impose some strain on the ring and thus increase its polymerizability.

The nine-membered lactam representing an intermediate between cis and trans lactams was assumed to be sufficiently flexible to allow the more stable trans amide group to co-exist with the strainless cis form [47]. However, a strainless planar trans form is not possible and the actual conformation is rather a skew structure which drastically reduces the p—π overlap and decreases resonance stabilization of the amide group. Therefore, an increased amount of delocalization energy is released in the conversion of the skew (or *cis*) conformation of the lactam into the fully resonance stabilized *trans* form of the polymer amide group, so that the large ring strain in the nine-membered lactam arises both from strong transannular interaction and inhibition of resonance stabilization.

In the ten-membered and higher lactams, the planar *trans* conformation can be strainless so that the contribution of conformational changes of the amide group to the ΔG_p value will be minimum. Only the 10-membered lactam contains about 5% of *cis* form and a corresponding amount of energy will be released during polymerization.

Irrespective of the ring size, the resonance stabilization of a planar amide group in *N*-substituted lactams is the same as in the open chain *N*-substituted lactam unit inside a polymer chain. Therefore, the heat of polymerization of *N*-substituted lactams (up to the eight-membered ring) should be by about 1.4 kcal mole^{-1} lower than the heat of polymerization of unsubstituted *cis* lactams. As a matter of fact, the calorimetric method which was extensively used by Muromova et al. [53] revealed that *N*-methylation decreases the exothermicity of polymerization of enantholactam by about 1.4 kcal mole^{-1}. Similar differences between the heats of polymerization of *N*-substituted and unsubstituted lactams were estimated for the six- and seven-membered lactams [53] (1.7 and 1.5 kcal mole^{-1}). *N*-Substitution of large lactam rings, on the other hand, should not alter the value of ΔH_p as significantly as in the case of small rings, except that the substituent at the nitrogen increases transannular strain.

3.3 SUBSTITUTION

As expected, the number, size and location of substituents has a pronounced effect on the rate and equilibrium of polymerization. Generally, substituents may affect the kinetic and thermodynamic factors differently. The rate determining step is very sensitive to steric effects in the vicinity of the amide group [54], whereas the free energy of polymerization is affected more by substitution at tetragonal atoms favouring ring closure. For example, in the series of methylcaprolactams, the equilibrium is attained earlier with the γ-isomer than with the ϵ-isomer (Fig. 1), whereas the equilibrium polymer content is the lowest for the γ-isomer and the highest for the α- and ϵ-isomers (Fig. 4). The effect of the position of the substituent is even more pronounced for bis-lactams (Table 4). Whereas the α,ϵ'- and ϵ,ϵ'-bis(caprolactam) polymerized to high yield, the γ,γ'-bis(caprolactam) did not polymerize at all [56, 57]. The effect of the γ-bonded lactam ring on the ring stability obviously exceeds that of an isopropyl and *tert*-butyl group.

Substituents affect both the heat and entropy of polymerization, mainly through conformational effects [58—61]. Steric repulsions between substituents as well as between atoms of the monomer unit and substituents usually do not change the enthalpy of the cyclic lactam. Only

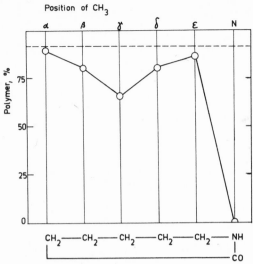

Fig. 4. Equilibrium polymer content for methylcaprolactam isomers at 250°C [28–30, 55]. (Dashed line: unsubstituted caprolactam.)

TABLE 4

Polymerization of bis-caprolactams [56] with 2% of aminocaproic acid at 260°C

R	Position	Polymer at equilibrium (%)
—	ϵ,ϵ'	99.9
$-CH_2-$	ϵ,ϵ'	99.5[a]
$-CH_2-$	α,ϵ'	99.7
$-CH-$ $\quad$ CH$_3$	ϵ,ϵ'	99.6
$-(CH_2)_4-$	ϵ,ϵ'	99.2
—	γ,γ'	0
CH$_3$ $-C-$ CH$_3$	γ,γ'	0[b]

[a] From the bislactam isomer, m.p. 232–233°C.
[b] Also at 250°C with anionic, cationic and hydrolytic catalysts.

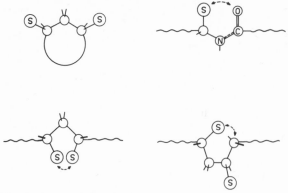

Fig. 5. Interactions of substituents in cyclic and linear monomer units.

in some medium and large rings, substituents may increase the ring strain through transannular interactions. New interactions arising from substituents increase the enthalpy of the linear monomer unit relative to the cyclic one [41] and thus decrease the value of $-\Delta H_p$ (Fig. 5).

Similarly, substitution only changes the entropy of the lactam very slightly, but decreases considerably the entropy of the linear monomer unit inside a polymer chain by restricting its rotation. In this way, substitution shifts the enthalpy and entropy of polymerization in favour of the cyclic structure and decreased polymerizability.

The effect of repulsion of substituents may be assumed even for monosubstituted lactams with a substituent in the vicinity of the nitrogen atom (Fig. 5). The preferred *trans* conformation of the amide group imposes some rigidity on the amide bond and, because the CH_2-N bond has the highest mobility, a substituent next to the nitrogen reduces the entropy of the open chain unit more than a substituent in the β-position. In this way the more negative value of ΔS_p^0 found for ϵ-methylcaprolactam [29] can be rationalized.

Cubbon [61] has attempted to correlate the effect of substitution on the polymerizability of caprolactam in terms of the change of the number of gauche interactions and the conformation of the amide bond. This simplified approach consists of the estimation of the increase in the number of gauche interactions per monomer unit resulting from ring closure. Each additional gauche interaction was assumed to increase ring strain by 0.8 kcal mole^{-1}, the changes in gauche interactions being related to the planar zig-zag form of the linear monomer unit. However, for the substituted backbone of the polymer chain this conformation needs not to be the conformation with the lowest energy. Measurements of the fibre identity period of polyamides from optically active methylenantholactam revealed that the substituted monomer unit is shorter than the unsubstituted one [62]. This indicates that not even in the solid state is the substituted unit present in a planar zig-zag conformation. The conformation of the polymer chain will be certainly influenced by the position of the substituent and this fact will be reflected both in the enthalpy and entropy of the polymer chain. As a matter of fact, the actual effect of methylation on the heat of polymerization of the seven-membered lactam, $(\Delta H_p)_u - (\Delta H_p)_s$, suggests that the contribution of one additional gauche conformation is rather $0.3-0.5$ kcal mole^{-1} than the estimated value 0.8 kcal mole^{-1} and depends on the position of the methyl group. Similar low values of ΔH_p and ΔS_p were obtained from calculations based on the number of rotational isomers of the linear monomer unit [58].

With increasing length of a linear hydrocarbon substituent next to the nitrogen, the equilibrium yield of polymer passes a minimum at C_3-C_4 (Fig. 6) and a maximum around C_6. This maximum could be interpreted

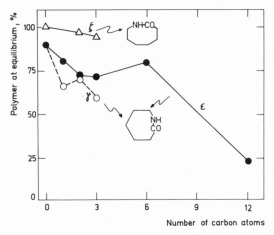

Fig. 6. Effect of chain length of substituent on the monomer—polymer equilibrium of seven- and eight-membered lactams. Linear alkyl groups in position $\epsilon(\bullet)$ and $\xi(\triangle)$ at 260°C and in the γ-position ($\circ$) at 254°C [28, 29, 63—68].

by increased transannular interaction. The fairly high heats of polymerization found for ω-substituted lactams (Table 5) suggest that substituents could increase the enthalpy of the monomer through transannular interaction. For very long substituents the yield of polymer approaches zero (Fig. 6). If the effect of substituent were dependent only on its length, then polymeric substituents should prevent polymerization of a seven-membered lactam completely. Polymerization of ϵ,ϵ'-bis-(caprolactam), however, gives an insoluble polymer in almost 100% yield. It has to be borne in mind that bis-lactams are tetrafunctional so that already the polymerization of either lactam ring yields 100% of polymer. The polymerization of the second ring does not change the yield and only increases branching and crosslinking. The formation of insoluble polymer from the bis-lactam indicates that the very long polyamide chain in the

TABLE 5
Heat of polymerization of substituted lactams

Caprolactam	HN—CO ϵ α δ β γ			Enantholactam	HN—CO ξ α ϵ β		
Substituent	Position	$-\Delta H_p$, (kcal mole^{-1})	Ref.	Substituent	Position	$-\Delta H_p$, (kcal mole^{-1})	Ref.
—	—	3.6	37	—	—	5.3	65
CH_3	α	3.0	55	CH_3	N	3.9	53
C_2H_5	α	3.8	63	C_2H_5	β	5.2	64
C_3H_7	α	3.9	63	C_2H_5	ϵ	5.1	64
CH_3	β	2.9	30	C_2H_5	ξ	6.8	66
C_2H_5	β	3.5	64	C_3H_7	ξ	4.9	66
CH_3	ϵ	4.0	30				
C_2H_5	ϵ	3.5	63				
C_3H_7	ϵ	3.5	63				

ϵ-position of the second lactam ring still allowed its polymerization. This suggests that the effect of substitution on the polymerizability depends not only on the size and bulkiness of the substituent but also the nature of the substituent, e.g. presence of polar groups and heteroatoms.

The effect of substitution on the heat of polymerization depends not only on the size and position of the substituent. It follows from Table 5 that the size of the lactam ring also plays an important role, so that substituents affect the polymerizability in a complex manner. Hence, in a general treatment, the effect of the substituent on the enthalpy and entropy of both the corresponding linear and cyclic unit must be taken into account.

Even more complicated relations exist in the series of bridged bi-, tri-, and tetracyclic lactams. The extensive pioneering work of Hall [49, 69, 70] on the influence of structure on the polymerizability of lactams containing six-membered rings revealed that lactams in which the cyclohexane ring occurs in the boat form (I) polymerize [71]. On the other hand, most lactams with the amide group in a six-membered chair (II) failed to polymerize [70] and the same result can be expected for lactams in which two stable chair forms are fused together (III), viz.

(I)   (II)   (III)

The only exception are lactams with nitrogen at the bridgehead (IV)

(IV)

in which resonance stabilization is largely suppressed. Although the lactam is free of H—H crowding, the deviation of the amide group from coplanarity causes a sufficient ring strain to render the lactam unstable [70].

3.4 HETEROATOMS

Replacement of a methylene group by a heteroatom changes the thermodynamic parameters of the cyclic as well as linear monomer unit. Carbon—heteroatom bonds differ from C—C bonds with respect to bond length and bond angle as well as to ionic character of the bond [72]. The fundamental studies of Ogata et al. [73—75] revealed that introduction of —O— or —S— groups in seven-membered lactams increases their polymerizability. For example, substituted seven-membered lactam ethers and thioethers polymerized to higher yields than the corresponding caprolactam derivatives (Table 6). Contrary to β,δ-dimethylcaprolactam which

TABLE 6
Polymerization of seven-membered lactam ethers [73] and lactam thioethers [74, 75]

Monomer:	HN—CO, O	HN—CO, CH_3—O—CH_3	HN—CO, CH_3—S—CH_3
Polymer (%)	80	95	45
Temp. (°C)	207	207	184
Time (h)	24	44.5	5
Catalyst (%)	H_3PO_4 (3)	H_3PO_4 (8)	$AlEtCl_2$ (1)

Monomer:	HN—CO, S	HN—CO, CH_3—S	HN—CO, S—CH_3
Polymer (%)	99	91.6	87
Temp. (°C)	184	200	200
Time (h)	5	20	20
Catalyst (%)	$AlEtCl_2$ (1)	H_3PO_4 (1)	H_3PO_4 (1)

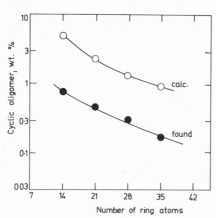

Fig. 7. Calculated [81] and estimated [16] content of cyclic oligomers in caprolactam polymers.

failed to polymerize [76], the corresponding dimethyllactam ether and thioether could be polymerized [73, 75]. Increased polymerizability of the latter lactams most probably arises from increased ring strain.

A similar effect arises from the introduction of an additional amide group into the lactam ring. Recently, Haberthür and Elias [77] found that 2,5-dioxopiperazine polymerized even at high temperatures at which piperidone could not be polymerized. Also in copolymerization, 2,5-dioxopiperazine was more active [78] than piperidone [79]. Similarly, the heat of polymerization of cyclo-di-β-alanyl (IV) was higher [80] (ΔH_p = -7.4 kcal mole^{-1}) than that of enantholactam (ΔH_p = -5.3 kcal mole^{-1}). Hence, it can be assumed that the additional amide group increases the ring strain because of the larger bond angles at two ring atoms and the stiffness of the amide group.

When calculating the equilibrium concentration of macrocyclics in caprolactam polymers, the heat of polymerization was assumed to be zero [81]. For the aforementioned reasons, the actual polymerizability is certainly higher and, therefore, the estimated content of the individual cyclic oligomers is lower than the calculated one (Fig. 7).

3.5 EFFECT OF PRESSURE

Homogenous lactam polymerizations usually proceed with a decrease of volume. At atmospheric pressure the value of the $p\Delta V$ term is negligible and the heat of polymerization can be considered as a measure of the difference in internal energy between the linear and cyclic monomer unit. At very high pressures, however, the effect of volume contraction during polymerization on the monomer—polymer equilibrium cannot be neglected. For a precise comparison of the ring-strain, the values of $\Delta H - p\Delta V$ should be used.

With increasing pressure, the $p\Delta V$ term shifts the free energy of polymerization to more negative values and polymerizations which are thermodynamically impossible at normal pressures can proceed at very high pressures. The effect of very high pressure on the monomer—polymer equilibrium has been demonstrated for piperidone [82]. The fairly stable six-membered lactam has a very unfavourable monomer—polymer equilibrium under normal pressure and can be polymerized only at low temperatures. At a pressure of 20,000 atm., piperidone could be polymerized to a high yield (82%) even at 160°C. In other words, extremely high pressures increase the ceiling temperature.

In this connection it has to be stressed that the ceiling temperature concept should be applied to the polymerization of lactams very carefully. At elevated temperatures, the reversible transacylation reactions are accompanied by decomposition of the polymer and/or monomer and by other side reactions. With respect to the low ΔS_p values (as compared to vinyl monomers), very high ceiling temperatures can be expected for most lactams. Therefore, the calculated values of T_c lie high above the decomposition temperature of the polymer and monomer except for five-, six- and substituted seven-membered lactams. In addition, for very large rings the heat of polymerization approaches zero and the polymerization entropy becomes positive, so that the value of $T_c = \Delta H_p / \Delta S_p$ becomes meaningless.

4. Anionic polymerization

4.1 MECHANISM OF INITIATION AND PROPAGATION

Under substantially anhydrous conditions, the anionic polymerization of lactams possessing an unsubstituted amide group may be initiated by any strong base capable of forming the free lactam anion

$$B^\ominus + HN{-}CO \; \rightleftharpoons \; BH + \left[N{=}C \overset{O}{\diagup} \right]^\ominus \qquad (22)$$

The acceleration at the start of the polymerization (Fig. 8, curve 1) represents an induction period in which certain active species are formed gradually. It has been established that the active chain growth centre is represented by an N-acylated lactam [44, 83, 85—94] which is formed in the slow disproportionation reaction

402

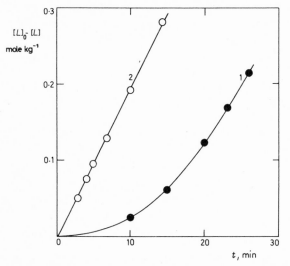

$$HN-CO + {}^{\ominus}N-CO \rightleftharpoons HN-\overset{\overset{\displaystyle O^{\ominus}}{|}}{C}-N-CO \rightleftharpoons HN^{\ominus} \quad CO-N-CO$$

$$HN-CO + HN^{\ominus} \quad CO-N-CO \rightleftharpoons {}^{\ominus}N-CO + H_2N \quad CO-N-CO \quad (23)$$

When an acyllactam is added at the beginning of polymerization, the induction period is absent (Fig. 8, curve 2). Hence, the anionic polymer-

Fig. 8. Anionic polymerization of caprolactam with sodium caprolactam (0.083 mole kg^{-1}) at 160°C (curve 1) or with equimolar amounts of sodium caprolactam and acetylcaprolactam (0.0177 mole kg^{-1}) at 100°C (curve 2); data from refs. [83] and [84].

ization starts with a nucleophilic attack of the lactam anion on the cyclic carbonyl group of an N-acylated lactam

$$-CO-N-\overset{\overset{\displaystyle O}{||}}{C} + {}^{\ominus}N-CO \rightleftharpoons$$

$$-CO-N-\overset{\overset{\displaystyle O^{\ominus}}{|}}{C}-N-CO \rightleftharpoons -CO-N^{\ominus} \quad CO-N-CO \quad (24)$$

The polymer amide anion is in equilibrium with monomeric lactam and lactam anions are regenerated

$$-CON^{\ominus} \quad CON-CO + HN-CO \rightleftharpoons -CONH \quad CON-CO + {}^{\ominus}N-CO$$

$$(25)$$

The net result of reactions (24) and (25) is the incorporation of one linear monomer unit into the polymer chain with regeneration of both the active end group and the lactam anion.

The high speed of anionic polymerization is due to the fact that both reacting species are chemically activated and hence highly reactive: the lactam anion represents an activated monomer with increased nucleophilicity and the terminal N-acylated lactam unit represents an activated end group (growth centre) with increased acylating ability.

As soon as polymer amide groups are formed, they can take part in both types of transacylation reactions (23) and (24). The disproportionation reactions involving polymer amide groups and/or anions produce acyllactam and diacylamine structures entering into the polymerization process

$$\begin{array}{c} \sim CO \\ | \\ \sim NH \end{array} + {}^{\ominus}N-CO \rightleftharpoons \sim CO-N-CO + \sim\sim NH^{\ominus} \qquad (26)$$

$$\begin{array}{c} \sim CO \\ | \\ \sim NH \end{array} + \begin{array}{c} {}^{\ominus}N \sim \\ | \\ CO \\ \sim \end{array} \rightleftharpoons \begin{array}{c} \sim CO-N \sim \\ | \\ CO \\ \sim \end{array} + \sim\sim NH^{\ominus} \qquad (27)$$

$$\begin{array}{c} \sim N^{\ominus} \\ | \\ \sim CO \end{array} + CO-NH \rightleftharpoons \begin{array}{c} \sim N-CO \quad NH^{\ominus} \\ | \\ \sim CO \end{array} \qquad (28)$$

Transacylations between polymer amide anions and acyllactam represent either depolymerization (29) or incorporation of one monomer unit (30)

$$\sim CO-N-CO + \begin{array}{c} {}^{\ominus}N \sim \\ | \\ CO \\ \sim \end{array} \rightleftharpoons \begin{array}{c} \sim CO-N \sim \\ | \\ CO \\ \sim \end{array} + {}^{\ominus}N-CO \qquad (29)$$

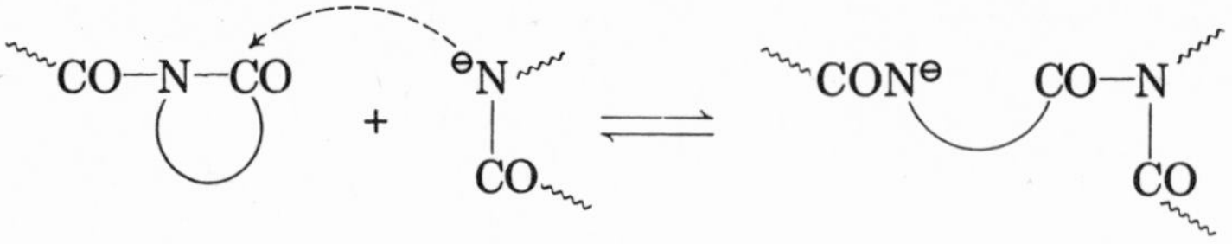

$$\text{CO-N-CO} \quad + \quad {}^{\ominus}\!\text{N} \quad \rightleftharpoons \quad \text{CON}^{\ominus} \quad \text{CO-N} \tag{30}$$

Acylation of a lactam anion with diacylamine, i.e. the reverse of reaction (29),

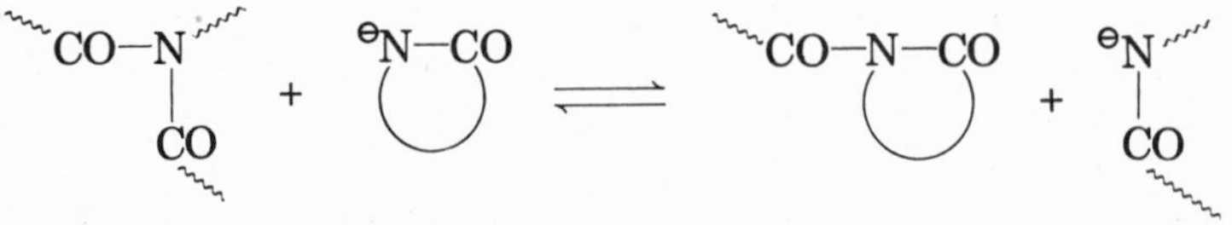

$$\text{CO-N} \quad + \quad {}^{\ominus}\!\text{N-CO} \quad \rightleftharpoons \quad \text{CO-N-CO} \quad + \quad {}^{\ominus}\!\text{N} \tag{31}$$

results in the incorporation of one lactam unit and can be regarded as the second path of propagation [95], provided that reaction (30) precedes reaction (31). The growth reactions discussed so far can be summarized in scheme 32, according to which polymerization can proceed both through reaction (24) or through the sequence of reactions (30) and (31).

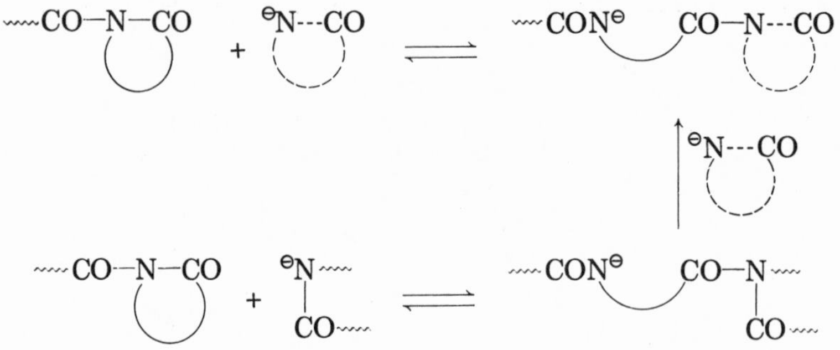

$$\tag{32}$$

Transacylation reactions between diacylamine groups and polymer amide anions represent only an exchange reaction

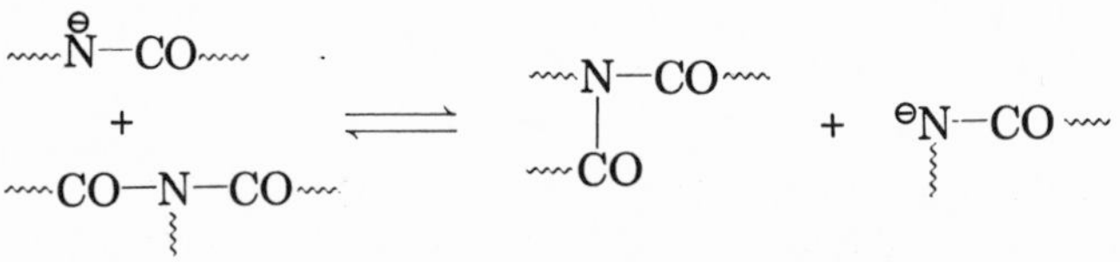

$$\tag{33}$$

The sequence of disproportionation reactions (23) or (26) and the bimolecular aminolysis at the cyclic carbonyl group

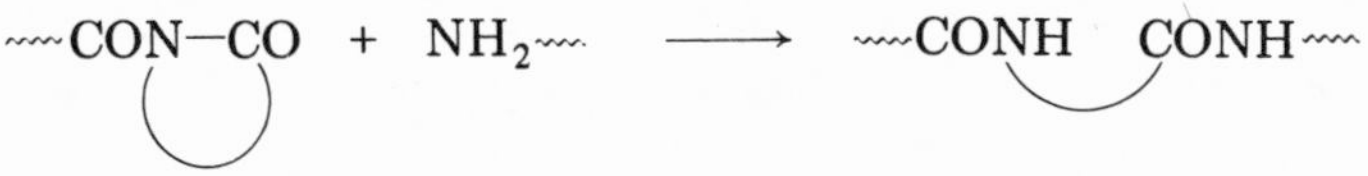

$$\text{CON-CO} + \text{NH}_2 \longrightarrow \text{CONH} \quad \text{CONH} \tag{34}$$

does not contribute significantly to the lactam consumption because the disproportionation is very slow as compared with the over-all rate of polymerization.

The most important features of the anionic lactam polymerization are that:

(a) the growing species is a neutral group with increased acylating ability (activated end group);

(b) only activated monomer molecules (lactam anions) with increased nucleophilicity take part in the main propagation reaction;

(c) the lactam anion incorporated into the polymer through acylation reactions (24), (26) or (28) is not yet incorporated into the polymer as a linear monomer unit, but

(d) it is always the terminal N-acylated lactam unit which is converted in reactions (24) and (30) into a linear chain unit [94].

It has to be stressed that the number of polymer molecules is increased only in reactions (23) and (26). Reactions (27) and (28) increase the number of polymer chains only when followed by a reaction with a lactam anion, reaction (31).

4.2 CATALYSTS

It follows from the reaction mechanism of propagation that two active species are involved in the chain growth reaction, namely, the lactam anion and the N-acylated lactam at the growing end, reaction (24). In the literature, several terms are used for the two active species: the source of lactam anions is designated as initiator, activator or catalyst, whereas the growth centres (or their sources) are called co-catalyst, promoter, initiator or activator. Although the lactam anions are regenerated to a large extent during the polymerization process, the concentration of these anions decreases due to the consumption of lactam and, in most cases, also due to side reactions (see Section 4.3) so that the term "catalyst" would be incorrect. Since the lactam anion itself may produce N-acylated lactam, reaction (23), and thus initiate an anionic polymerization even without the addition of growth centres, the author is inclined to reserve the term "initiator" for the metal lactamate or its source (e.g., NaH).

Acylation of the monomer activates the latter towards the nucleophilic attack by a lactam anion. Hence, it is rational to use the term "activator" for lactam derivatives capable of growing (e.g., N-acyllactams) or such compounds which produce the latter much faster than the initiator alone (e.g., acid chloride). In most cases, the activator or its fragments become part of the polymer molecule. Anionic polymerizations in which an activator has been added are designated as activated polymerizations, whereas in non-activated polymerizations the growth centres are produced by the initiator, reaction (23).

A great number of substances have been suggested as initiators, e.g. alkali metals [96—99], metal hydrides [44, 100], organometallic [101] and Grignard [102] compounds, alkali metal alkoxides [100, 103] or hydroxides [104], and easily decomposing salts of carboxylic acids [105]. Lactam anions may be generated also by electrolysis of lactam solutions of neutral salts [106]. In the reaction of alkali metals with caprolactam, significant amounts of amine, aminoalcohol and other by-products are formed [99]. Therefore, purified lactam salts (e.g. crystallized sodium caprolactam [83, 94]) are preferred for kinetic studies since traces of impurities can interfere seriously with the non-activated polymerization by consuming the slowly forming growth centres.

A great variety of substances is capable of acting as activators. The first group comprises N-substituted lactams with polar substituents at the nitrogen. These N-substituted lactams must be able to acylate lactam anions with opening of the lactam ring of the activator at a rate comparable to the rate of polymerization. The main representatives of this group are acyllactams [84, 89, 107—110] (V), N-substituted [111] (VI) or N,N-disubstituted [112] carbamoyllactams (VII), N-carboxylic acid esters of lactams [112] (VIII) and salts of lactam-N-carboxylic acids [107] (IX), viz.

$$RCO-N-CO \qquad RNHCO-N-CO \qquad {R \atop R}{>}NCO-N-CO$$

$$(V) \qquad\qquad (VI) \qquad\qquad (VII)$$

$$ROCO-N-CO \qquad\qquad Na\left[OCO-N-CO\right]$$

$$(VIII) \qquad\qquad\qquad (IX)$$

The second group of activators are substances which react with the lactam or its anion to yield an N-substituted lactam [107, 113—116] in a sufficiently fast reaction e.g., isocyanates, anhydrides, esters, ketenimines. The activating effect of N-substituted lactams was found to increase with increasing electronegativity of substituents [84, 111, 117]. Another group of activating substances are precursors of compounds of the second group. For example, N-2,2-trisubstituted β-oxoamides [118—120] decompose at elevated temperatures very easily into ketone and isocyanate, which is a very effective activator.

Some compounds can act both as initiator and activator simultaneously. For example, sodium salts of N-2-disubstituted β-oxoamides decompose at elevated temperatures in the presence of lactam to yield isocyanate (activator) and sodium lactam [119]

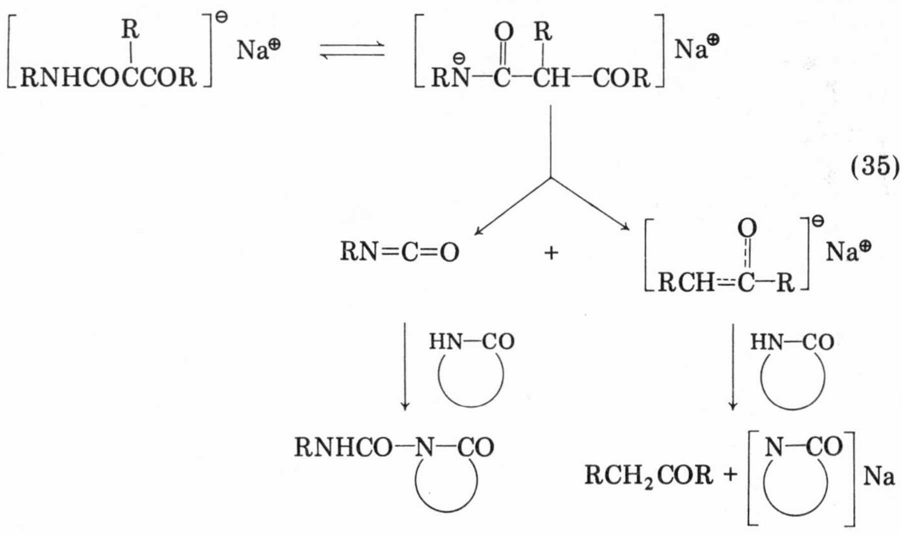

$$\left[\begin{array}{c} R \\ | \\ \text{RNHCOCCOR} \end{array} \right]^{\ominus} Na^{\oplus} \rightleftharpoons \left[\begin{array}{c} O \quad R \\ \| \quad | \\ R\overset{\ominus}{N}-C-CH-COR \end{array} \right] Na^{\oplus}$$

(35)

$$RN{=}C{=}O \quad + \quad \left[\begin{array}{c} O \\ \| \\ RCH{=}C-R \end{array} \right]^{\ominus} Na^{\oplus}$$

$$\downarrow \; HN-CO \qquad\qquad \downarrow \; HN-CO$$

$$RNHCO-N-CO \qquad RCH_2COR + \left[N-CO \right] Na$$

4.3 SIDE REACTIONS

If only the indicated disproportionation and transacylation reactions proceeded, then the catalytic activity of amide anions and growth centres should be maintained and the system should behave as "living". However, the activated monomer (lactam anion) and growth centres of increased reactivity (N-acyllactam and diacylamine) give rise to a variety of side reactions in which both growth centres and lactam anions are consumed [89, 94, 95, 103, 119, 121—137]. It was found that the content of acyllactam and imide groups in the polymerization product is always lower than in the initial reaction mixture [94, 123, 125, 127] (Table 7). Impressive evidence of the high speed of the decay of active species is represented in Fig. 9. Although the initial rate of polymerization increases with increasing concentration of activator, the monomer—polymer equilibrium is attained much sooner with a low activator:

TABLE 7

Decay of imide groups (acyllactam and diacylamine structures) during anionic caprolactam polymerization [125]

Sodium caprolactam (mmole kg^{-1})	Acetylcaprolactam (mmole kg^{-1})	Temp. (°C)	Time (min)	Decrease of imide groups	
				(mmole kg^{-1})	(%)
152	762	80	2	122	16
77	770	100	2	150	19.5
152	762	80	20	242	32
38.6	773	100	2	93	13.6

References pp. 465—471

408

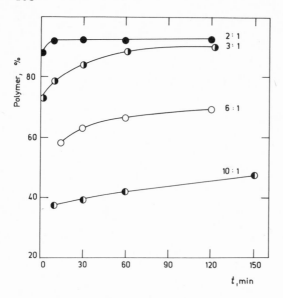

Fig. 9. Anionic polymerization of caprolactam at 210°C and constant concentration of sodium caprolactam (0.0437 mole kg^{-1}). Concentration of butyrylcaprolactam 0.0874 (●), 0.1310 (◐), 0.262 (○) and 0.4365 mole kg^{-1} (◑); the figures indicate the activator : initiator ratio [135].

initiator ratio. This indicates that the monomer anions disappear in some very fast reactions involving the activated growth centres.

In the presence of a strong base, e.g., lactam anions, the imide groups containing at least one α-hydrogen undergo a Claisen type condensation analogous to the condensation of esters or ketones [121, 125], viz.

$$-\text{CONCOCH}_2- \ + \ ^\ominus\text{N}-\text{CO} \ \rightleftharpoons \ -\text{CONCO}\overset{\ominus}{\text{CH}}- \ + \ \text{HN}-\text{CO}$$

$$-\text{CONCO}\overset{\ominus}{\text{CH}} + \overset{\text{O}}{\overset{\|}{\text{C}}}-\text{N}-\overset{\text{O}}{\overset{\|}{\text{C}}}- \ \rightleftharpoons \ \left[-\text{CONCOCH}-\overset{\overset{\ominus}{\text{O}}}{\overset{|}{\text{C}}}-\text{N}-\overset{\text{O}}{\overset{\|}{\text{C}}}- \right] \ \rightleftharpoons$$

$$-\text{CONCOCHCO}- \ + \ ^\ominus\text{N}-\text{CO}- \qquad (36)$$

(X)

The open chain structure (X) is formed in the condensation of diacylamine groups. The *N*-acyllactam growth centres can undergo condensation at the cyclic as well as exocyclic methylene and carbonyl groups

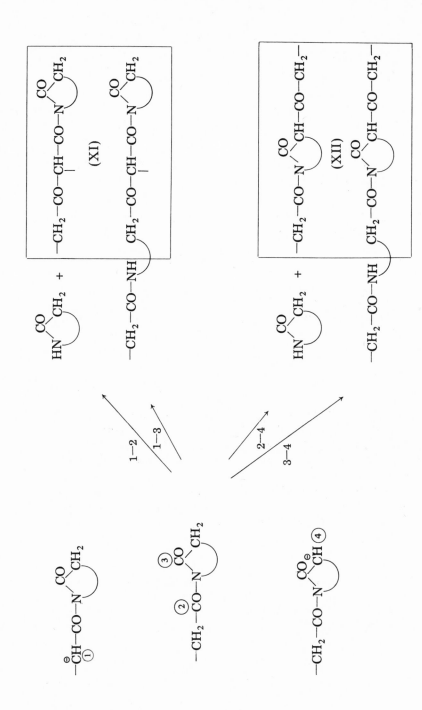

Scheme 37. Condensation of growth centres.

yielding two types of N-acylated β-keto amides, (XI) and (XII) in scheme (37). The neutral N-acyllactam structures (XI) react as growth centres, viz.

$$\text{\textasciitilde COCHCON-CO} \ + \ ^{\ominus}\text{N-CO} \ \rightleftharpoons \ \text{\textasciitilde COCHCON}^{\ominus} \ \text{CON-CO}$$

$$\text{(XI)} \qquad\qquad \text{(XIII)} \qquad\qquad \text{(38)}$$

and also acylation of lactam anions with structures (X) and (XII) regenerates acyllactam growth centres

$$\text{-COCHCONCO-} \ + \ ^{\ominus}\text{N-CO} \ \rightleftharpoons \ \text{-COCHCON}^{\ominus} \ + \ \text{-CON-CO}$$

$$\text{(X)} \qquad\qquad \text{(XIII)} \qquad\qquad \text{(39)}$$

$$\text{-CON}\overset{\text{CO}}{\diagup}\text{CHCO-} \ + \ ^{\ominus}\text{N-CO} \ \rightleftharpoons \ ^{\ominus}\text{N}\overset{\text{CO}}{\diagup}\text{CHCO-} \ + \ \text{-CON-CO}$$

$$\text{(XII)} \qquad\qquad\qquad \text{(XIV)} \qquad\qquad \text{(40)}$$

Similar reactions involving polymer amide anions yield the corresponding diacylamine structures along with substituted β-oxoamides, keto amides (XIII) and (XIV). In this way, the sequence of condensation and transacylation results in the formation of five types of β-keto acid derivatives (X)—(XIV).

Substituted keto amides (X)—(XIV) having one α-hydrogen are much more acidic than polymer or lactam amide groups and they lower the concentration of lactam anions through the equilibria

$$\text{-CONCOCHCO-} \ + \ ^{\ominus}\text{N-CO} \ \rightleftharpoons \ \text{-CON-}\overset{\text{O}}{\overset{\|}{\text{C}}}\text{-}\overset{\ominus}{\text{C}}\text{-}\overset{\text{O}}{\overset{\|}{\text{C}}}\text{-} \ + \ \text{HN-CO}$$

$$\text{(X)-(XII)} \qquad\qquad\qquad\qquad\qquad\qquad (41)$$

$$\text{-CON-}\overset{\text{O}^{\ominus}}{\text{C}}\text{=C-}\overset{\text{O}}{\overset{\|}{\text{C}}}\text{-} \ \longleftrightarrow \ \text{-CON-}\overset{\text{O}}{\overset{\|}{\text{C}}}\text{-C=}\overset{^{\ominus}\text{O}}{\text{C}}\text{-}$$

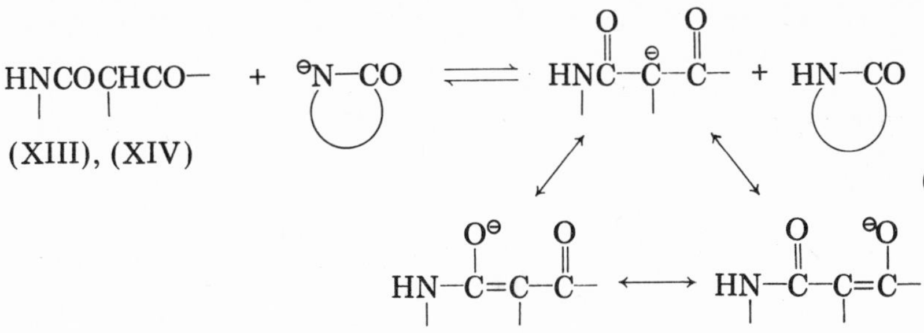

$$\text{HNCOCHCO--} \ + \ ^{\ominus}\text{N--CO} \ \rightleftharpoons \ \text{HNC--C--C--} \ + \ \text{HN--CO}$$

(XIII), (XIV)

(42)

$$\text{HN--C=C--C--} \ \longleftrightarrow \ \text{HN--C--C=C--}$$

in which resonance stabilized enolate anions of substituted keto amides are formed. Hence, the reactions (36)—(42) decrease the concentration of both lactam anions and growth centres.

The aforementioned reactions can occur with lactams having a methylene group next to the carbonyl and yield α-monosubstituted keto amides (X)—(XIV),

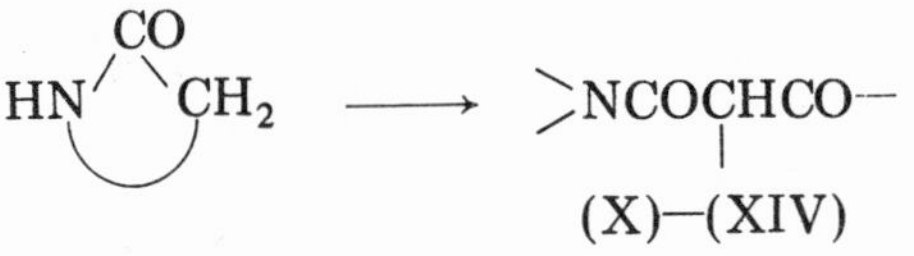

$$\text{HN}\overset{\text{CO}}{\diagup}\overset{}{\diagdown}\text{CH}_2 \ \longrightarrow \ \text{>NCOCHCO--}$$

(X)—(XIV)

(43)

whereas α-monosubstituted lactams yield α,α-disubstituted keto amide derivatives (XV)

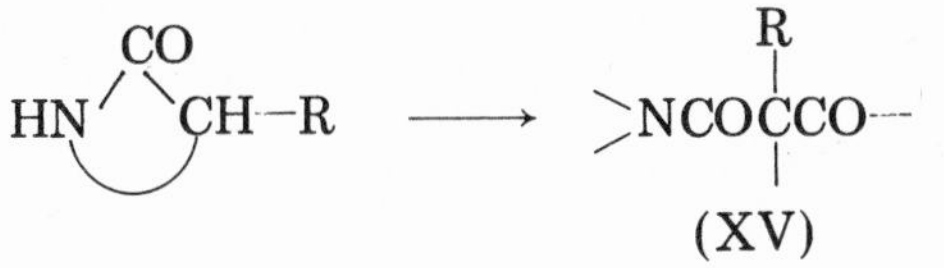

$$\text{HN}\overset{\text{CO}}{\diagup}\overset{}{\diagdown}\text{CH--R} \ \longrightarrow \ \text{>NCOCCO--}$$

(XV)

(44)

Only α,α-disubstituted lactams are expected to polymerize without side reactions unless deactivation occurs with impurities or activator residues.

Like other β-dicarbonyl compounds, the substituted keto amides (X)—(XV) are very reactive and represent the key intermediates in a complex set of side reactions [95]. A considerable amount of research has been devoted to the elucidation of the reactions of these compounds under polymerization conditions [118—120, 124, 125, 127, 128, 130, 131, 134—143]. With respect to the high reactivity of carbonyl groups and the adjacent $-CH_2-$ or $>CH-$ groups, the structures like (X)—(XV) are unstable at elevated temperatures and even less stable in the presence of a strong base. Bukač and Šebenda [137] showed that keto amides with one hydrogen atom at the nitrogen decompose very easily in the presence of base into isocyanate which takes part in many subsequent reactions, scheme (45). The most important consequence of side reactions is the fact that water is present even if the anionic polymerization is started under

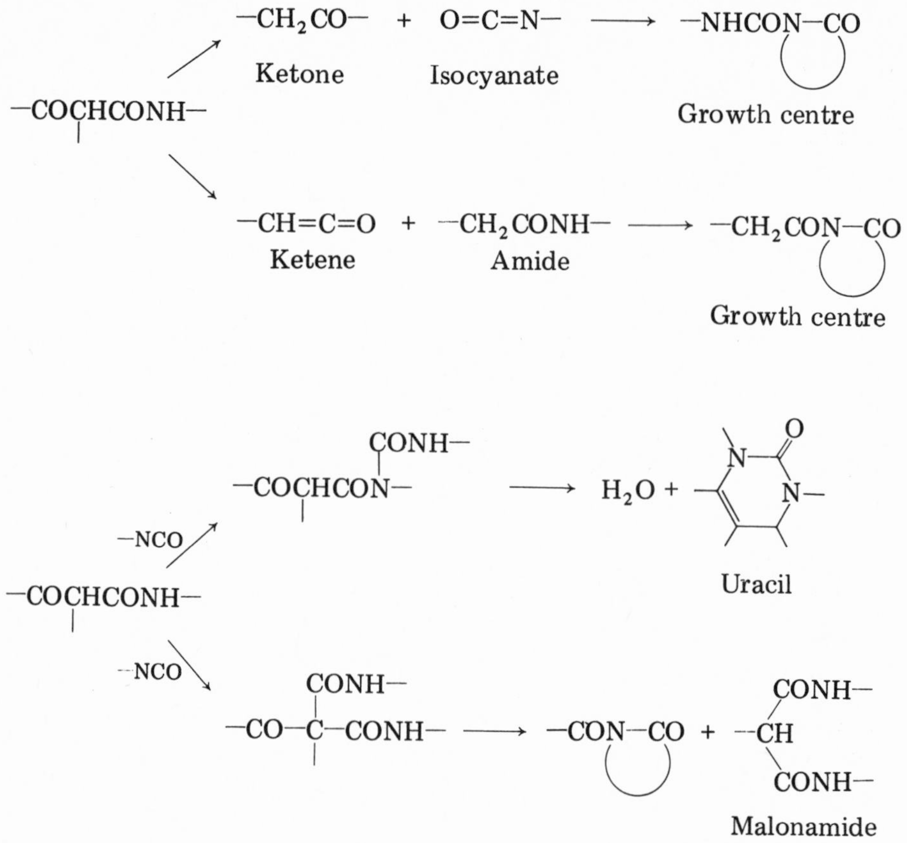

$$-COCHCONH- \nearrow \begin{array}{c} -CH_2CO- \ + \ O=C=N- \ \longrightarrow \ -NHCON-CO \\ \text{Ketone} \qquad \text{Isocyanate} \qquad\qquad \text{Growth centre} \end{array}$$

$$\searrow \begin{array}{c} -CH=C=O \ + \ -CH_2CONH- \ \longrightarrow \ -CH_2CON-CO \\ \text{Ketene} \qquad \text{Amide} \qquad\qquad \text{Growth centre} \end{array}$$

$$-COCHCONH- \begin{array}{c} \nearrow -NCO \\ \\ \searrow -NCO \end{array}$$

$$\begin{array}{c} \overset{CONH-}{-COCHCON-} \ \longrightarrow \ H_2O \ + \ \text{Uracil} \end{array}$$

$$\begin{array}{c} \overset{CONH-}{-CO-C-CONH-} \ \longrightarrow \ -CON-CO \ + \ \overset{CONH-}{-CH} \\ \hspace{8em} CONH- \\ \text{Malonamide} \end{array}$$

Scheme 45. Products from keto amides (XIII).

strictly anhydrous conditions. The very fast base catalysed hydrolytic reactions consume both active species, i.e. lactam anions and growth centres,

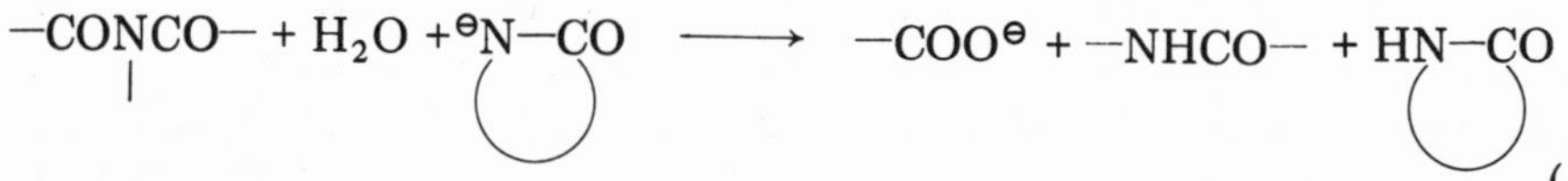

$$-CONCO- + H_2O + {}^{\ominus}N-CO \ \longrightarrow \ -COO^{\ominus} + -NHCO- + HN-CO \tag{46}$$

$$-CONCOCHCO- + H_2O + {}^{\ominus}N-CO \ \longrightarrow$$
$$-COO^{\ominus} + -NHCOCHCO- + HN-CO \tag{47}$$

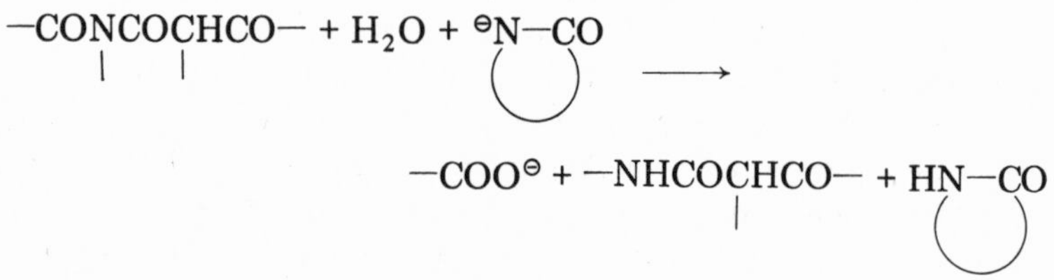

$$-NHCOCHCO- + H_2O + {}^{\ominus}N-CO \longrightarrow$$

(with a ring attached below the CHCO group and a cyclopentane-like ring on the N—CO)

$$-NHCOCH_2- + -COO^{\ominus} + HN-CO \qquad (48)$$

$$-NHCOCHCO- + 2H_2O + 2{}^{\ominus}N-CO \longrightarrow$$

$$-NH_2 + CO_3^{2\ominus} + -CH_2CO- + 2HN-CO \qquad (49)$$

The structures, groups and salts indicated in reactions (36)—(49) and in scheme (45) have been identified in model experiments and most of them also in polymerization products [118—120, 124—128, 130, 131, 136—144].

However, it follows from the processes discussed so far that in some reactions growth centres are also regenerated. For example, decomposition of the anion of a substituted keto amide yields lactam anions as well as isocyanate which is an effective activator, reaction (35). Acyllactam growth centres are also generated from keto amide and isocyanate, scheme (45). At high temperatures, lactam anions are generated also from carbonate [130, 145, 146]

$$CO_3^{2-} + 2HN-CO \longrightarrow CO_2 + {}^{\ominus}N-CO + NH_2 \quad COO^{\ominus} \qquad (50)$$

so that the active species do not disappear completely and a low catalytic activity is maintained for long periods.

The complicated set of reversible and irreversible side reactions summarized in scheme (51) indicates that some irregular structures resulting from side reactions can regenerate both active species. In this way, both lactam anions and growth centres are steadily consumed and regenerated in several reaction paths. It has to be pointed out, however, that from the very beginning of polymerization the concentration of active species is always much lower than the initial values but, on the other hand, catalytic activity (although at a very low level) is maintained for very long periods.

According to this simplified scheme of reactions occurring during anionic polymerization, the initiation and polymerization reactions constitute only a minor part of the complex set of reactions and one can get the impression that the polymerization is a sort of side reaction, as well. As a matter of fact, the anionic polymerization of some lactams proceeds

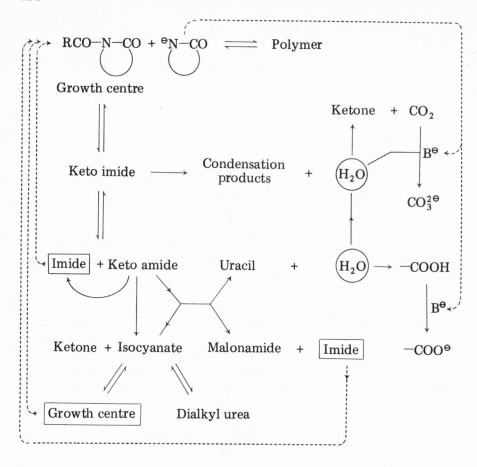

Scheme 51. Decay and regeneration of active species in the anionic polymerization of lactams.

with a high speed, till the equilibrium, only in a limited range of concentrations of active species (Fig. 9).

The pattern of side reactions in the polymerization of α-substituted lactams is slightly different. The keto amide structures (XV) formed in the Claisen type condensation have no acidic α-hydrogen so that the concentration of lactam anions cannot decrease so dramatically as with the acidic keto amides (X)—(XIV). Consequently, the position of the condensation equilibrium is shifted in favour of the initial components. Also the main products formed from α,α-disubstituted keto amides [136], scheme (52), are different from those of the monosubstituted keto amides, scheme (45).

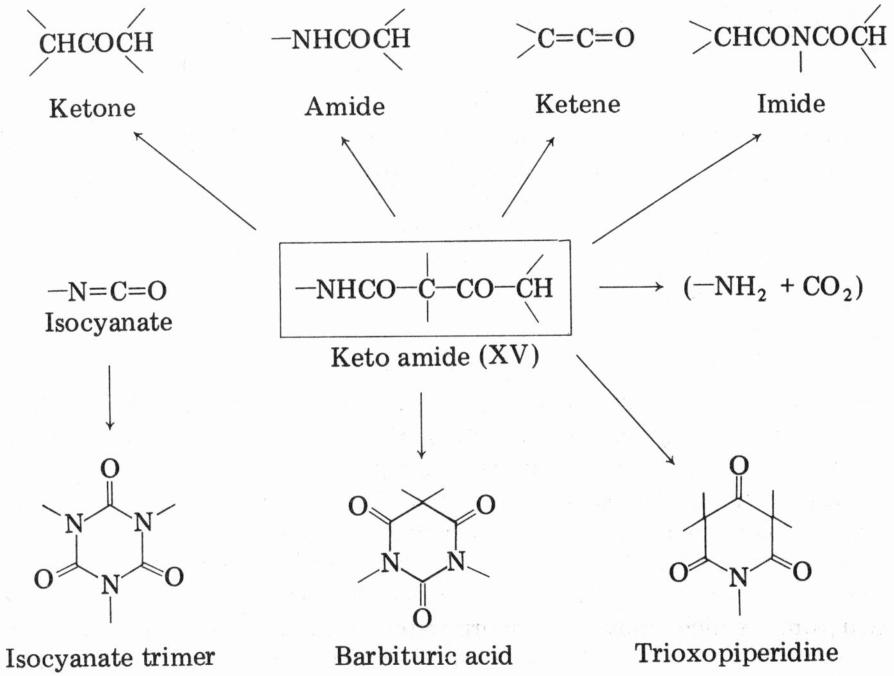

$$\overset{\diagdown}{\underset{\diagup}{C}}HCO\overset{\diagup}{C}H \qquad -NHCO\overset{\diagup}{C}H \qquad \overset{\diagdown}{C}{=}C{=}O \qquad \overset{\diagdown}{C}HCONCO\overset{\diagup}{C}H$$

Ketone Amide Ketene Imide

$$-N{=}C{=}O \qquad \boxed{-NHCO-\overset{|}{\underset{|}{C}}-CO-\overset{\diagup}{C}H} \longrightarrow (-NH_2 + CO_2)$$

Isocyanate Keto amide (XV)

Isocyanate trimer Barbituric acid Trioxopiperidine

Scheme 52. Products from keto amides (XV).

4.4 STRUCTURE OF POLYMER MOLECULES

In the activated polymerization, the activating group is incorporated into the growing polymer molecule in the first propagation step [94] and activator residues have been identified in the polymer for many types of activators [147, 148]. Sekiguchi [149] has shown that in the polymerization of pyrrolidone with benzoylpyrrolidone, all activating groups (R) were incorporated into the polymer even at low conversions. Unless additional growth centres are formed by disproportionation reactions or side reactions start to interfere, all molecules should be linear with an acyllactam structure at the second end, viz.

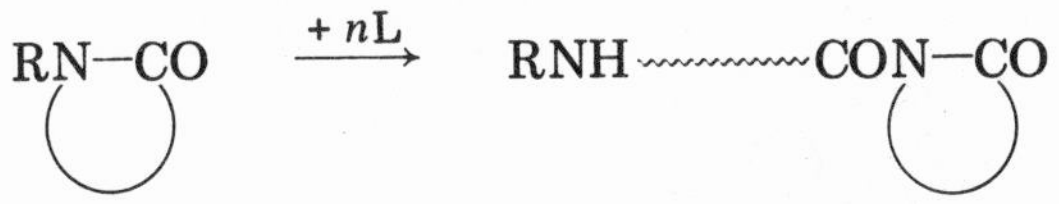

$$RN{-}CO \xrightarrow{\ +n\mathrm{L}\ } RNH \text{------} CON{-}CO$$

(XVI)

Oligomers of this structure were found in polymers prepared at low temperature [92, 150]. Bifunctional activators, e.g. terephthaloyl-bis-caprolactam, yield molecules with two growing end groups

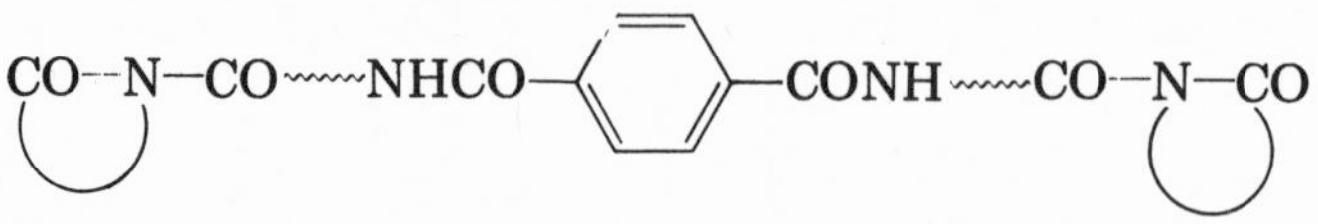

$$CO-N-CO \sim NHCO-\langle\bigcirc\rangle-CONH \sim CO-N-CO$$

Acylation of polymer amide anions with acyllactam groups, reactions (29) and (30), yields branched polymer molecules

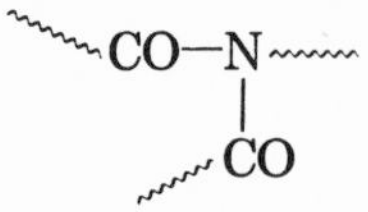

$$\sim CO-N \sim$$
$$|$$
$$\sim CO$$

(XVII)

Since lactam anions are acylated by diacylamine groups much faster than with cyclic acyl groups of acyllactams [94], the extent of such branching should be unimportant as long as a few percent of monomer are present. It has been confirmed that the majority of imide groups in anionic polycaprolactam are present as acyllactam [123, 150].

As a matter of fact, structures (XVI) and (XVII) represent only a minor fraction of polymer molecules. Due to the great number of irregular structures which may be incorporated into the polymer molecules (Section 4.3), a great variety of types of macromolecules can be present in anionic polymers [95]. The nature of irregular structures formed during anionic lactam polymerization is primarily determined by the type and concentration of catalytic species and temperature, as well as by ring size and substitution of the lactam. Only polymers of α,α-disubstituted lactams are free of irregular structures, and should be composed only of macromolecules of type (XVI) and (XVII).

Polymers derived from other lactams contain irregular units such as indicated in the schemes (45) and (52) so that both linear and branched macromolecules are present [134, 151, 152]. Structures (XVIII)—(XXII) containing substituted keto amide units will prevail at lower temperatures, viz.

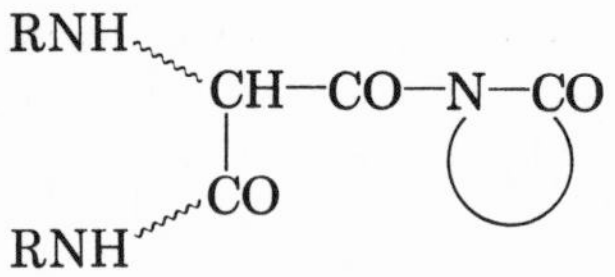

$$RNH \diagdown$$
$$\qquad CH-CO-N-CO$$
$$\qquad | \qquad \bigcirc$$
$$\qquad CO$$
$$RNH \diagup$$

(XVIII)

$$RNH \diagdown$$
$$\qquad CH-CO-NH \sim CON-CO$$
$$\qquad | \qquad\qquad \bigcirc$$
$$\qquad CO$$
$$RNH \diagup$$

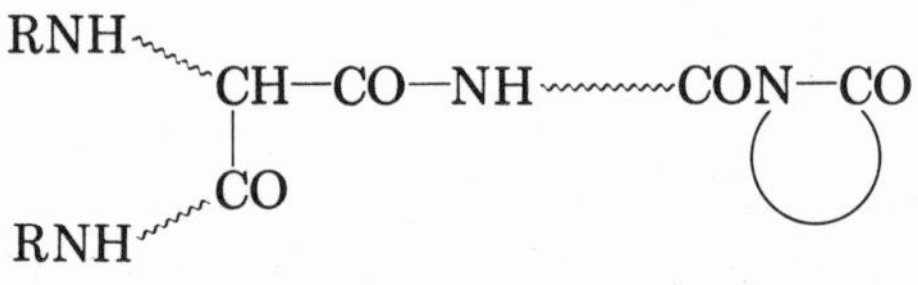

(XIX)

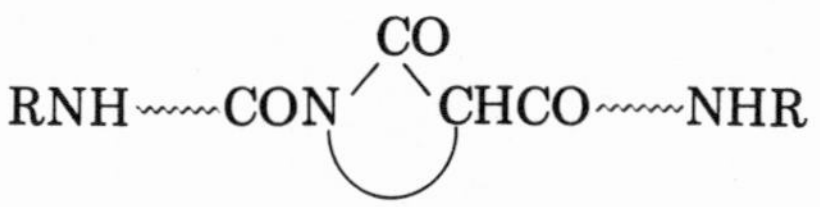

$$\text{RNH}\sim\text{CON}\underset{\smile}{\overset{\text{CO}}{\diagup}}\overset{\diagdown}{\text{CHCO}}\sim\text{NHR}$$

(XX)

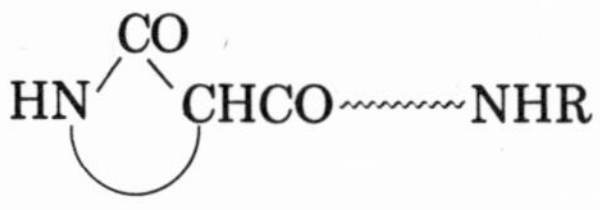

$$\text{HN}\underset{\smile}{\overset{\text{CO}}{\diagup}}\overset{\diagdown}{\text{CHCO}}\sim\text{NHR}$$

(XXI)

$$\begin{array}{c}\text{RNH}\sim\\ \quad\quad\text{CH-CO-N}\sim\text{CON-CO}\\ \text{RNH}\sim^{\text{CO}}\quad\quad\text{CO}\sim_{\text{NHR}}\end{array}$$

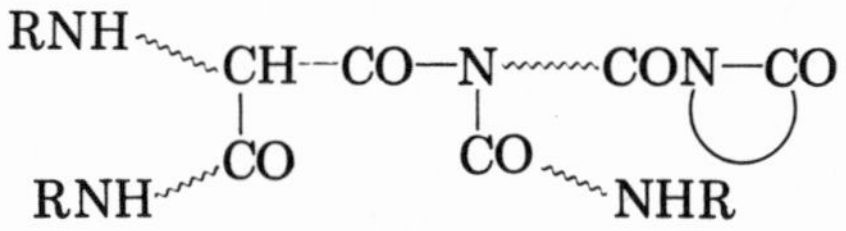

(XXII)

At elevated temperatures, the base catalysed dissociation of structure (XIX) at the keto amide unit (XIII) yields two linear molecules

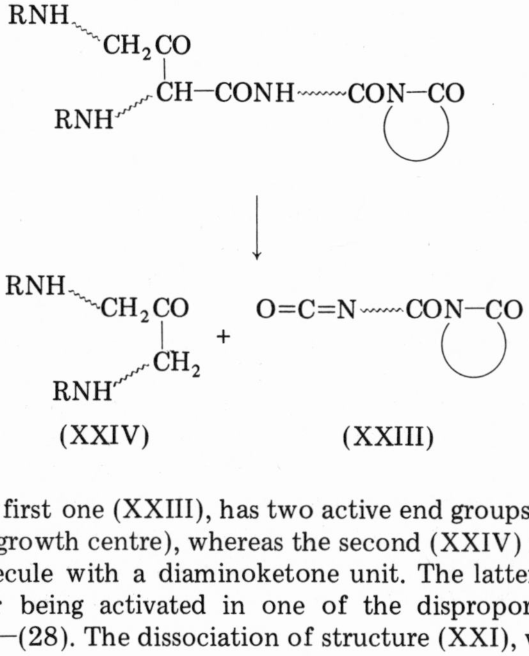

$$\begin{array}{c}\text{RNH}\sim\\ \quad\text{CH}_2\text{CO}\\ \quad\quad\text{CH-CONH}\sim\text{CON-CO}\\ \text{RNH}\sim\end{array}$$

↓   (53)

$$\begin{array}{c}\text{RNH}\sim\\ \quad\text{CH}_2\text{CO}\quad\quad\text{O=C=N}\sim\text{CON-CO}\\ \quad\quad\quad+\\ \quad\text{CH}_2\\ \text{RNH}\sim\end{array}$$

(XXIV)   (XXIII)

The first one (XXIII), has two active end groups (isocyanate is a precursor of a growth centre), whereas the second (XXIV) represents a dead polymer molecule with a diaminoketone unit. The latter may start growing again after being activated in one of the disproportionation reactions (23), (26)—(28). The dissociation of structure (XXI), viz.

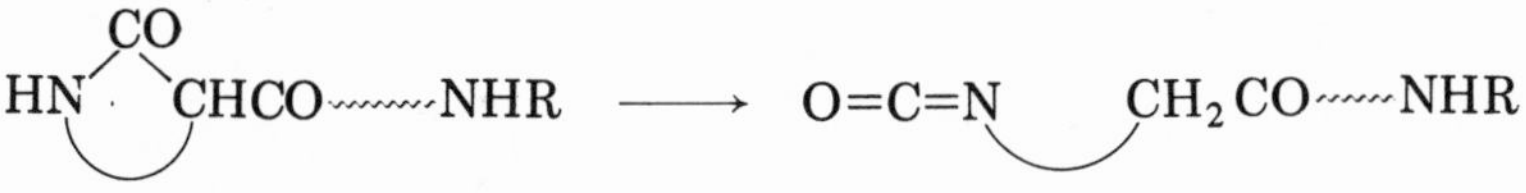

$$\text{HN}\underset{\smile}{\overset{\text{CO}}{\diagup}}\overset{\diagdown}{\text{CHCO}}\sim\text{NHR}\longrightarrow\text{O=C=N}\underset{\smile}{\quad}\text{CH}_2\text{CO}\sim\text{NHR}\quad(54)$$

References pp. 465—471

yields one linear molecule with one active end group. Additional condensation reactions involving keto amides and their decomposition products, schemes (45) and (52), yield tri- and tetrafunctional branching units, e.g., substituted malonamide, uracil, trioxotriazine, oxoglutarimide and barbituric acid [128, 136, 137], which are incorporated in the polymer [124, 127, 134, 138, 141].

In polymer molecules free of irregular structures inside the chain, all monomer units are arranged in the same direction, viz.

$$\overset{\longrightarrow}{\sim\sim\sim NH} \quad \overset{\longrightarrow}{CONH} \quad \overset{\longrightarrow}{CONH} \quad \overset{\longrightarrow}{CONH} \quad \overset{\longrightarrow}{CONH} \quad CO\sim\sim\sim$$

(XXV)

As soon as any of the irregular structures is incorporated inside the polymer, the direction of monomer units is altered at the foreign unit, viz.

$$\overset{\longrightarrow}{\sim\sim NH} \; \overset{\longrightarrow}{CONH} \quad \overset{\longrightarrow}{CONH} \quad CON \overset{CO}{\diagup} CHCO \quad \overset{\longleftarrow}{NHCO} \quad \overset{\longleftarrow}{NHCO} \quad \overset{\longleftarrow}{NH\sim\sim}$$

(XXVI)

It has been established, that on the average one to two branches per molecule are present in anionic caprolactam polymers [134, 152].

In non-activated polymerization in which the growth centres are formed in the slow disproportionation reactions, (23) and (26)—(28), similar linear as well as branched polymer molecules are formed which are terminated with primary amine instead of with RNH— groups.

However, even activated polymers contain basic (amine) groups resulting both from disproportionation and from side reactions [135, 151], e.g., reaction (49). The end groups in anionic polymers are represented by primary amine, activator residue, acyllactam (both N- and α-) and carboxyl groups [134]. In addition, some basic groups other than primary amine are present [126]. A certain fraction of basic groups are ketimine structures resulting from intramolecular cyclization of a terminal amino ketone unit [124]

$$\sim\sim\sim CONH \quad \overset{\overset{O}{\parallel}}{C} \quad NH_2 \quad \xrightarrow{-H_2O} \quad \sim\sim\sim CONH \quad \overset{\overset{N}{\parallel}}{C} \qquad (55)$$

4.5 MOLECULAR WEIGHT AND DISTRIBUTION

In the non-activated caprolactam polymerization, the number of growth centres (and consequently of polymer molecules) steadily increases as long as lactam anions are present (or, until the disproportionation equilibrium is attained) and the growth centres enter into the polymerization process successively. During the major part of the polymerization, the redistribution reaction between polymer molecules is not much faster than the polymerization and the redistribution through depolymerization is even slower. Therefore, the growth centres formed at the very beginning of polymerization attain a much higher molecular weight than those formed towards the end of polymerization. As a result, the molecular weight distribution is very broad [110, 154, 155] (Fig. 10)

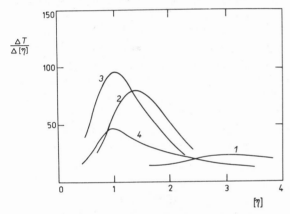

Fig. 10. Molecular weight distribution in anionic polymerization of caprolactam [155]. Temperature, 246°C; concentration of sodium caprolactam, 0.0292 mole kg^{-1}: time; 1 (1), 22 (2), 120 (3) and 500 h (4).

and the molecular weight increases during the major part of the polymerization. When the polymerization approaches the monomer—polymer equilibrium, redistribution due to depolymerization, (25) and (29), and polymer—polymer transacylation reactions, (29), (30) and (33), is becoming more important. In addition, the number of polymer molecules steadily increases through disproportionation reactions involving or, followed by, transacylation with lactam anions, (23), (26)—(28) and (31), so that the molecular weight decreases and the molecular weight distribution is approaching the statistical one [154, 155] (Fig. 10). After passing the viscosity maximum [104, 153—156], the molecular weight slowly approaches a distinct final value (M_f) which is related to the initial concentration of metal lactamate (initiator) [155, 157, 158] by

$$1/M_f = K[I]_0^{0.5} \qquad (56)$$

420

The value of K depends on the nature of initiator, of lactam and on the temperature. The dependence of the number of polymer molecules on the square root of the initiator concentration (56) indicates that the value of M_f is related to the disproportionation equilibrium (23) and the dissociation of the lactamate [157]. However, at temperatures above 200°C, sodium lactamates are completely dissociated (Section 4.6.1) and the final molecular weight is rather a consequence of the equilibrium between the formation and decay of polymer molecules both through disproportionation and side reactions [134] (Section 4.3). The broadening of the distribution observed after extended heating at high temperature [155] might be due to the formation of tri- and tetra-functional branching units leading to slightly crosslinked polymers.

In the activated polymerization, the number of polymer molecules is primarily determined by the concentration of added activator molecules but additional growth centres are formed in the slow disproportionation reactions (23) and (26)—(28) as long as lactam anions are present. Simultaneously, side reactions of the growth centres start to interfere (Section 4.3), so that the number of polymer molecules is not equal to the number of added (or formed) growth centres. It has been established [134] that in the anionic polymerization of caprolactam, both activator (A) and initiator (I) are consumed in a ratio $\Delta[A]/\Delta[I] = 2.5$ already within the first minutes (at temperatures above 190°C). At lower ratios of initial concentrations, the surviving strong base increases the number of

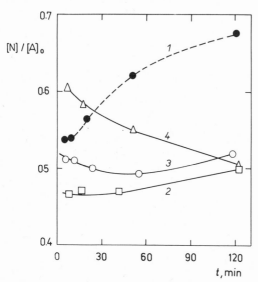

Fig. 11. Number of polymer molecules formed per acyllactam molecule ($[N]/[A]_0$) in the anionic polymerization of caprolactam [134]. Temperature, 210°C; concentration of butyrylcaprolactam, 0.0360 mole kg^{-1}: concentration of sodium caprolactam; 0.036 (1), 0.0182 (2), 0.014 (3) and 0.012 mole kg^{-1} (4).

polymer molecules through disproportionation (Fig. 11, curve 1), whereas at $[A]_0/[I]_0 > 2.5$, only traces of lactam anions are present and condensation reactions lowering the number of molecules predominate (Fig. 11, curve 4).

The extent of the fast side reactions depends on the temperature and on the initial concentration of catalytic species as well as on the nature of the lactam. Hence, the molecular weight at the end of polymerization is a complicated function of reaction conditions [110, 134, 159, 160]. So far, only empirical relations can be derived between the reaction conditions and the molecular weight of the polymer [110, 134, 159—161], which

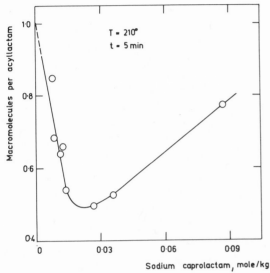

Fig. 12. Number of polymer molecules formed per one activator molecule in the anionic polymerization of caprolactam [134]. Temperature, 210°C; concentration of acyllactam, 0.0088—0.0360 mole kg^{-1}; time, 5 min.

can be controlled by the choice of catalytic species and their concentration as well as by regulators, e.g. amines [162]. Since the rate of most side reactions is governed primarily by the concentration of polymer and lactam anions, the number of molecules formed per acyllactam growth centre is determined rather by the initial concentration of metal lactamate (Fig. 12).

As expected, the molecular weight distribution also depends strongly on the concentration of amide anions. In the polymerization of capro-lactam around 200°C with an activator/initiator ratio $[A]_0/[I]_0 < 2.5$, the concentration of lactam anions is lowered from the very beginning of polymerization by the value $\Delta[I] = [A]_0/2.5$. The remaining initiator $([I] = [I]_0 - [A]_0/2.5)$ produces additional growth centres entering successively into the polymerization. Similarly, as in the non-activated polymerization, the number of polymer molecules increases during extended heating and shifts the molecular weight distribution to lower molecular weights (Fig. 13, curves 1, 2). After very long periods, branching side reactions increase the high molecular weight fractions [163]. At $[A]_0/[I]_0 \geqslant 2.5$, the concentration of active base is very low since the beginning of polymerization and the formation of new growth centres during the polymerization is negligible. Hence, all acyllactam

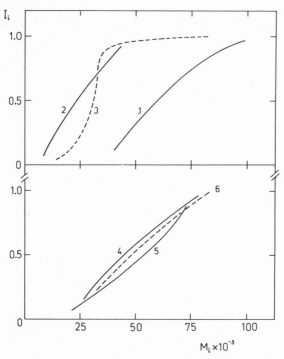

Fig. 13. Molecular weight distribution in anionic caprolactam polymerization at 220°C. Concentration of butyrylcaprolactam, 0.0265 (curves 1—3) and 0.0248 mole kg^{-1} (curves 4—6): concentration of sodium caprolactam; 0.0265 (curves 1—3) and 0.0088 mole kg^{-1} (curves 4—6) [163]: time; 5 min (4), 30 min (1, 5), 10 h (2, 6) and 113 h (3).

growth centres start growing simultaneously and due to the low extent of redistribution reactions a fairly narrow molecular weight distribution results both towards the end of polymerization and after extended heating [163] with $M_w/M_n = 1.2$ (Fig. 13, curves 4—6).

A slightly different situation is met with in polymerizations at temperatures at which the polymer crystallizes either during or near the end of polymerization. In this case, the polymer—polymer exchange reactions and also the depolymerization reactions are largely suppressed. Also the disproportionation is too slow to contribute significantly to the number of growth centres in the activated polymerization. Therefore, even at equimolar concentrations of initiator and activator (when a large fraction of strong base survives for a long period) the polymerization consists essentially in a step-addition of lactam anions to a constant number of growth centres resulting in a narrow molecular weight distribution. Such a narrow molecular weight distribution with an M_w / M_n ratio close to unity has been found for anionic caprolactam polymers prepared below the melting point of the polymer [151].

4.6 KINETICS OF POLYMERIZATION

Due to the relatively fast side reactions consuming both initiator and growth centres, the evaluation of the kinetics of anionic polymerization becomes very difficult. We are dealing with a system of varying concentration of both active species which, according to schemes (45), (51) and (52), can be not only consumed but also regenerated in the complicated set of side reactions. Hence, the key problem of the anionic lactam polymerization consists in the determination of the instantaneous concentrations of lactam anions and growth centres.

4.6.1 Dissociation of lactamates

At the beginning of polymerization, the concentration of lactam anions is determined by the dissociation of the lactam salt added as initiator, viz.

$$
\left[\begin{array}{c} N\text{--}CO \\ \bigcirc \end{array} \right] M \ \rightleftharpoons \ \left[\begin{array}{c} N\text{=-}C\text{=-}O \\ \bigcirc \end{array} \right]^{\ominus} + \ M^{\oplus} \tag{57}
$$

TABLE 8
Dissociation of sodium and potassium caprolactams
in caprolactam [164]

Temp. (°C)	Dissociation constant, K (mole kg^{-1})	
	Sodium salt	Potassium salt
110.4		0.018
120.5	0.00049	0.036
130.8	0.0021	0.13
130.8[a]	0.011	
130.8[b]	0.12	
141.0	0.029	0.47
151.2	0.10	

[a] Addition of 0.025 mole N-ethylbutyramide per mole caprolactam.
[b] Addition of 0.10 mole N-butylcaproamide per mole caprolactam.

The degree of dissociation of metal lactamates depends on the nature of both metal cation and lactam. As expected, the dissociation of lactamates increases with increasing electropositivity of the metal cation [164] (Table 8). In the series of alkali metal salts of caprolactam, the degree of dissociation increases in the sequence Li < Na < K. From published values of dissociation constants [164] (between 120 and 150°C) we arrive at

$$K_{\text{sodium caprolactam}} = 1.2 \times 10^{30} \exp(-60,300/RT) \text{ mole kg}^{-1}$$

$$K_{\text{potassium caprolactam}} = 1.1 \times 10^{21} \exp(-40,600/RT) \text{ mole kg}^{-1}$$

Hence, almost complete dissociation (at catalytical concentrations) of sodium caprolactam occurs only above 190°C.

Lactamates of highly electropositive metals, e.g., aluminium or chromium dissociate only very slightly and, therefore, such salts reveal a very low initiating activity [165]. As a matter of fact, chromium lactamate is unable to initiate the polymerization of caprolactam up to 240°C.

The acidity of *cis* amide groups (of small and medium lactams) is much lower than that of the trans amide groups in the corresponding linear polymer. As soon as polymer amide groups are formed, the concentration of lactam anions is lowered due to the equilibrium

$$\tag{58}$$

However, the total concentration of anions is increased appreciably (Table 8) and an increasing fraction of anions arises from *trans* amides whereas the concentration of lactam anions decreases steeply. It has been estimated [164], that the formation of 2.5% of polymer lowers the concentration of lactam anions by 30% whereas the total concentration of amide anions doubles.

4.6.2 Initiation

The initiation reaction involves formation of acyllactam growth centres of at least the same reactivity as the growing end of the polymer molecule. The growth centres are formed either in the slow base catalysed disproportionation reactions (23) and (26) or in the reaction of the activator with the lactam or its anion, e.g.

$$\tag{59}$$

The acylation of the lactam anion with the cyclic carbonyl of an N-substituted lactam, viz.

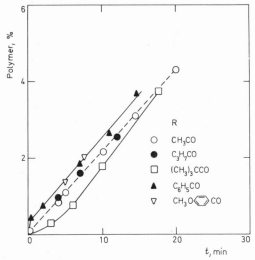

$$\text{RNHCON--CO} + {}^{\ominus}\text{N--CO} \rightleftharpoons \text{RNHCON}^{\ominus} \quad \text{CON--CO} \tag{60}$$

should also be considered as an initiation reaction, unless the rate of this reaction is equal to the rate of propagation. In most cases the rate of the initiation reaction with the activator is comparable with the rate of polymerization. Only for slowly reacting activators or acyllactams, has the

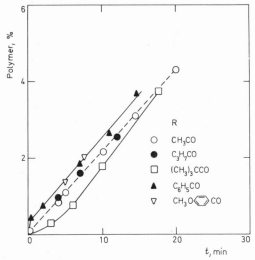

Fig. 14. Anionic polymerization of caprolactam activated with various acyllactams [84]. Temperature, 100°C; concentration of sodium caprolactam = concentration of acyllactam = 0.0177 mole kg^{-1}.

rate of the initiation reaction to be included in the kinetic equation. This is the case, for example, with the highly sterically hindered reaction of pivaloylcaprolactam with the caprolactam anion (Fig. 14) or with such activators as esters or amides. Highly polar acyllactams, on the other hand, react with the lactam anion much faster than the growing chain; for example, benzoyllactams reacted much faster than acetyllactams [84, 117] (Fig. 14).

In the non-activated polymerization, the acyllactam growth centres are formed in the very slow disproportionation reactions (23) and (26). The rate of the disproportionation reaction has been estimated from the rate of formation of amine groups in the model reaction of an open chain N-alkylamide (N-butylcaproamide [83] or N-ethylacetamide [140]) with its sodium salt [83], viz.

$$
\begin{array}{ccc}
\text{R'CO} & \overset{\ominus}{\text{NCOR'}} & \overset{k_{11}}{\underset{k_{-11}}{\rightleftarrows}} & \text{R'CONCOR'} + \text{R}\overset{\ominus}{\text{N}}\text{H} \\
| \quad + \quad | & & | \\
\text{RNH} & \text{R} & & \text{R}
\end{array} \qquad (61)
$$

$$\text{(D)}$$

$$
\text{R}\overset{\ominus}{\text{N}}\text{H} + \text{R'CONHR} \quad \underset{k_{-12}}{\overset{k_{12}}{\rightleftarrows}} \quad \text{RNH}_2 + \text{R'CO}\overset{\ominus}{\text{N}}\text{R} \qquad (62)
$$

The neutralization reaction (62) is extremely fast and $k_{12} \gg k_{-12}$ so that the over-all rate of disproportionation is

$$
v_1 = d[\text{NH}_2]/dt = d[\text{D}]/dt = k_{11}[\text{—CONH—}][\text{—CO}\overset{\ominus}{\text{N}}\text{—}] - k_{-11}[\text{NH}_2][\text{D}]
$$

$$(63)$$

where —CONH—, $\text{—CO}\overset{\ominus}{\text{N}}\text{—}$, NH_2 and D are amide, amide anion, amine and diacylamine, respectively. Extrapolation of rates (measured at low conversions) to $[\text{—NH}_2] = 0$, eliminates the backward aminolysis as well as side reactions. From the initial rate thus obtained, the rate coefficient k_{11} can be calculated, assuming complete dissociation of the sodium salt of N-ethylacetamide [140], to be

$$
(k_{11})_{\text{linear amide}} = 3.8 \times 10^8 \exp(-29,000/RT) \text{ kg mole}^{-1} \text{ sec}^{-1}
$$

When the metal lactamate (I) is not dissociated completely, then the concentration of lactam anions calculated from

$$
[\text{L}^-] = (0.25 \, K^2 + K[\text{I}]_0)^{0.5} - 0.5 \, K \approx (K[\text{I}]_0)^{0.5} \qquad (64)
$$

has to be inserted into eqn. (63). Measurements of the rate of amine group formation [83] in the polymerizing solution of sodium caprolactam in caprolactam yielded

$$
k_{11} = 1.6 \times 10^7 \exp(-26,700/RT) \text{ kg mole}^{-1} \text{ sec}^{-1}
$$

At 190°C, identical values of k_{11} were found for the disproportionation of both caprolactam and N-butylcaproamide with the corresponding sodium salt [83].

If the dissociation constant K is not known, only the product $k_{11}K^{0.5}$ (or $k_{11}K$) can be calculated from the initial rates of disproportionation or polymerization (see Section 4.6.3).

4.6.3 Propagation and depolymerization

In the anionic lactam polymerization it is always the lactam anion which is incorporated into the polymer molecules either in reaction (24)

or in the sequence of reactions (30) and (31). Accordingly, depolymerization proceeds as an acylation with the exocyclic carbonyl of an
acyllactam both as a monomolecular elimination at the polymer amide
anion next to the acyllactam unit, reverse reaction (24), and as a
bimolecular reaction with any polymer amide anion, reaction (29). These
fast acylation reactions of polymer or lactam anions with acyllactam or
diacylamine groups representing propagation, depolymerization and exchange reactions are summarized in Scheme 65, viz.

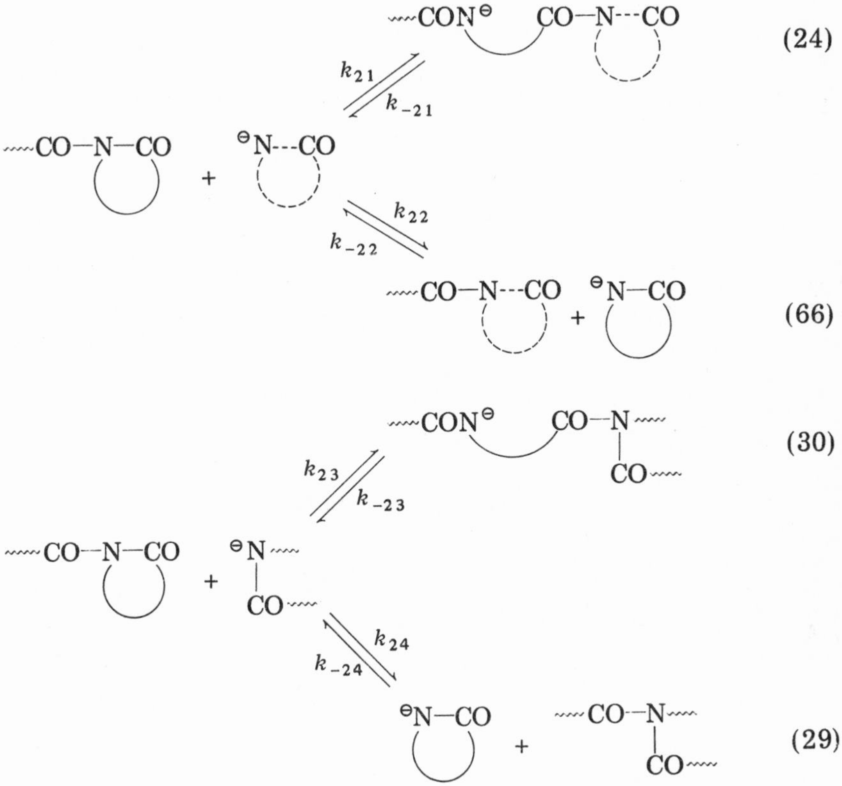

Scheme 65. Polymerization, depolymerization and transacylation reactions.

Even without taking into account side reactions changing the concentrations of the active species, a great number of rate and equilibrium
constants are required in order to describe the whole course of polymerization. So far, no data are available on equilibrium (25) from which the
concentrations of lactam and polymer amide anions could be estimated.
Therefore, the individual rate coefficients can be obtained only from
measurements of the initial rates of the isolated reactions. In this case, the
participation of the reaction products in subsequent reactions can be
neglected.

References pp. 465—471

At the very beginning of polymerization, the linear monomer units are incorporated into the polymer in reaction (24) which is followed by the fast proton exchange (25) regenerating lactam anions. The proton exchange between pyrrolidone and its anion

$$\text{HN—CO} + {}^{\ominus}\text{N---CO} \underset{k_0}{\overset{k_0}{\rightleftharpoons}} {}^{\ominus}\text{N—CO} + \text{HN---CO} \qquad (67)$$

was found to proceed 10^4 times faster (at $30°C$, $k_0 = 10^5$ l mole^{-1} sec^{-1}) than the propagation [166]. Hence, the rate determining step of polymerization is the acylation of lactam anions (24). At the beginning of polymerization and at low temperatures, the depolymerization can be neglected for most lactams (except the five- and six-membered) and the rate of polymerization is given by [94]

$$-\mathrm{d}[L]/\mathrm{d}t = k_{21}[L^-][A] \approx k_{21}K^{0.5}[I]^{0.5}[A] \qquad (68)$$

where A represents acyllactam. It can be assumed that the reactivity of the acyllactam structure at the growing end is independent of the length of the polymer. The value of k_{21} for the growing polymer chain should be very close to the value of k_{21} for the addition of the first lactam anion to an acyllactam with a substituent similar to the monomer unit (e.g. butyryl- or caproyllactam). It has to be emphasized, that the nature of activator can influence the rate of polymerization by changing the concentration of lactam anions. For example, with carbamoyllactams of type (VI), the rate of polymerization decreases with increasing electronegativity of the substituent R. Due to the increased acidity of the terminal urea groups, the concentration of lactam anions is decreased after the addition of the first lactam unit [111], viz.

$$\text{RNHCONH}\text{\textasciitilde} + {}^{\ominus}\text{N—CO} \rightleftharpoons \overset{\ominus}{\text{RN}}\text{CONH}\text{\textasciitilde} + \text{HN—CO} \qquad (69)$$

Similarly, any compound consuming the acyllactam growth centres (e.g. primary and secondary amine [162, 170]) decreases the rate of polymerization. So far, the rate coefficients k_{21} of the addition of the first lactam anion to the growth centre have been estimated only for a few monomers. In most cases, however, the published rate coefficients were calculated from the second order rate equation

$$-\mathrm{d}[L]/\mathrm{d}t = k_p[A][I] \qquad (70)$$

without taking into account the incomplete dissociation of the lactam salt (Table 9). It has been shown, that even at $50°C$, the rate of side reactions

TABLE 9
Rate coefficients of reaction (24), k_p, and (66), k_e, in tetrahydrofuran at 25°C
(The figure in the lactam anion indicates the number of ring atoms.)

Acyllactam	Lactamate	k_p	k_e	Ref.
		($l\ mole^{-1}\ sec^{-1}$)		
Acetylpyrrolidone	[N—CO (5)] MgBr	0.028		117
Acetylpiperidone	[N--CO (7)] MgBr	0.02	0.20	167
Acetylcaprolactam	[N—CO (7)] MgBr	0.0075		167
Acetylcaprolactam	[N—CO (5)] MgBr	0.05	0.60	167
Benzoylpyrrolidone	[N—CO (5)] MgBr	0.095		117
Acylpyrrolidone[a]	[N—CO (5)] K	9^b		168

[a] Acyl = $C_6H_5CO[NH(CH_2)_3CO]_n$.
[b] In pyrrolidone at 30°C.

amounts to 20% of the rate of polymerization [94]. Therefore, the only reliable data are those obtained after very short reaction periods. At high temperatures, the active species disappear so rapidly, that the rate coefficients calculated from the initial concentrations of initiator and activator are too low (Table 10, compare k_{21} for 100 and 150°C).

As the concentration of polymer amide groups increases during the polymerization, the equilibrium (25) decreases the concentration of lactam anions and increases the concentration of polymer amide anions (P^-). Hence, an increasing fraction of polymer is formed in the sequence of reactions (30) and (31) whereas the contribution of reaction (24) to the chain growth decreases. Similarly, the contribution of the bimolecular depolymerization reaction (29) will be increasing as compared with the monomolecular depolymerization in reaction (24). The rate of the monomolecular depolymerization is proportional to the concentration of amide anions in the vicinity of the acyllactam, viz.

$$(v_d)_1 = d[L]/dt = k_{-21}[P^-]/n \tag{71}$$

TABLE 10

Rate coefficients of reactions (24), k_{21}, and (66), k_{22}. Source of lactam anions; sodium caprolactam. The value of k_{21} calculated using $K = 1.2 \times 10^{30}$ exp(−60,300/RT) mole kg^{-1}

Acyllactam	Solvent	Temp. (°C)	$k_{21}K^{0.5}$ $k_{22}K^{0.5}$ $(kg^{0.5}\ mole^{-0.5}\ sec^{-1})$		k_{21} $(kg\ mole^{-1}\ sec^{-1})$	Ref.
Butyrylenantholactam	Dimethyl-formamide	50	0.005	0.027		94
Acylcaprolactam[a]	Capro-lactam	100	0.14[d]		40	84
Acylcaprolactam[a]	Capro-lactam	150	1.71[d]		5	169
Acylpyrrolidone[b]	Pyrro-lidone	30	1.4[c]			168

[a] Acyl = $CH_3CO[NH(CH_2)_5CO]_n$.
[b] Acyl = $C_6H_5CO[NH(CH_2)_3CO]_n$.
[c] Reaction with potassium pyrrolidone.
[d] See preceding page.

and the rate of bimolecular depolymerization may be expressed by

$$(v_d)_2 = d[L]/dt = k_{24}[A][P^-] \tag{72}$$

With increasing ring strain and degree of polymerization (n), the bimolecular depolymerization will prevail, while the monomolecular reaction is important only at the very beginning of polymerization. For cis-lactams reaction (31) can be assumed to proceed faster than reaction (30), so that the rate determining reactions can be summarized in the equation

$$-d[L]/dt = [A](k_{21}[L^-] + k_{23}[P^-] - k_{24}[P^-]) \tag{73}$$

The basicity of anions derived from small and medium lactams is higher than that of polymer amide anions, so that $k_{21} > k_{23}$ and $k_{22} > k_{24}$. When the lactamate is not dissociated completely, then the formation of the more acidic polymer amide groups (trans amide) increases the total concentration of anions taking part in the propagation and depolymerization reactions. The decreased reactivity of polymer amide anions as compared with lactam anions (k_{23}/k_{21}) is being compensated for by the increasing concentration of polymer anions. In addition, the decreasing basicity may shift the complicated equilibria of side reactions in favour of the regeneration of growth centres. The observed constant rate of polymerization up to conversions of 15% (Fig. 15) indicates that the decreasing concentration of lactam anions as well as the lower reactivity of polymer amide anions are counterbalanced by the increasing concen-

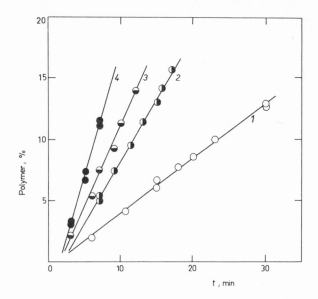

Fig. 15. Anionic polymerization of caprolactam [169]. Initiator: sodium caprolactam (0.0088 mole kg^{-1}). Activator: tetraacetylhexamethylenediamine (0.0044 mole kg^{-1}). Temperature: 140 (1), 150 (2), 160 (3) and 170°C (4).

tration of growth centres and polymer amide anions. This is the reason, why the course of most polymerizations may be described satisfactorily by the equation [171]

$$\log ([L]_0/[L] = k_p[A]_0[I]_0^{0.5}t \tag{74}$$

It has to be stressed, however, that only a fraction of the initial concentration of acyllactam and lactamate is present (in the polymerization of lactams having at least one α-hydrogen) so that the value of k_p in eqn. (74) is not identical with the rate coefficients k_{21} or k_{23} (or $k_{21}K^{0.5}$ or $k_{23}K^{0.5}$, respectively). The decay of amide anions proceeds very rapidly [132, 135] (Fig. 16) and is very sensitive to the ratio $[A]_0/[I]_0$. Therefore, all rate equations derived without taking into account the decay of active species or empirical equations like

$$-d[L]/dt = k[A]_0[I]_0^{0.5} \tag{75}$$

are valid in a rather narrow range of the activator/initiator ratio (Fig. 17).

In the non-activated polymerization, the acyllactam growth centres (A) are formed in the slow disproportionation reactions. Assuming a constant rate of initiation and neglecting the side reactions to the first approximation we have

References pp. 465—471

$$d[A]/dt = k_{11}[L^-][L] = k_{11}K^{0.5}[I]^{0.5}[L] \tag{76}$$

and

$$[A] = k_{11}K^{0.5}[I]^{0.5}[L]t \tag{77}$$

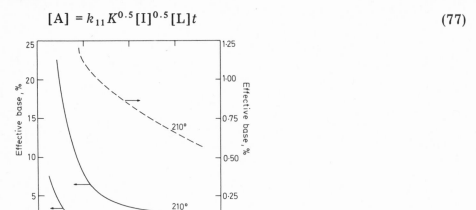

Fig. 16. Decay of catalytically active base in the anionic polymerization of capro-lactam [135]. The concentration of effective base is expressed in percent of the initial concentration of sodium caprolactam. Concentration of benzoylcaprolactam, 0.0360 mole kg^{-1}; initial concentration of sodium caprolactam, 0.0360 (———) or 0.0090 mole kg^{-1} (- - - -).

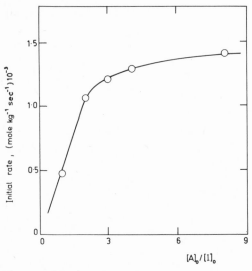

Fig. 17. Effect of the initial ratio of activator/initiator ($[A]_0/[I]_0$) on the initial rate of anionic polymerization of caprolactam [169]. Concentration of sodium capro-lactam $[I]_0 = 0.0044$ mole kg^{-1}; $[A]_0$, the initial concentration of activator (tetraacetylhexamethylenediamine), expressed in moles of diacylamine groups per kg; temperature, $150°C$.

The initial rate of polymerization through reaction (24), neglecting depolymerization, is given by

$$-d[L]/dt = k_{21}[A][L^-] = k_{21}k_{11}K[I][L]t \qquad (78)$$

and

$$\ln([L]_0/[L]) = k_{21}k_{11}K[I]t^2 \qquad (79)$$

Published data [83] for caprolactam follow equation (79) up to conversions of 15% (Fig. 18) and to initial concentrations of sodium caprolactam of 0.16 mole kg^{-1} with

$$k_{21}k_{11}K = 10^{19}\exp(-51,300/RT)\ \text{kg mole}^{-1}\ \text{sec}^{-2}$$

Due to the fast decay of acyllactam growth centres at elevated temperatures and high concentrations of base, the value of k_{11} describes the actual concentration of growth centres and is different from k_{11} estimated from the rate of formation of amine groups (Section 4.6.2). Only at low temperatures, when side reactions do not interfere to a larger extent, can the actual rate of disproportionation be calculated from the initial rate of polymerization. Using published data [118] for k_{21} and conversions for polymerizations of pyrrolidone with MgBr-pyrrolidonate in THF at 25°C, we obtain from eqn. (79)

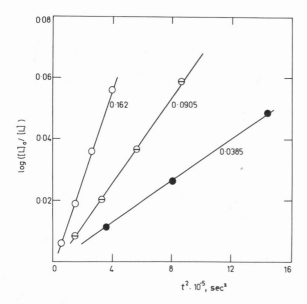

Fig. 18. Non-activated anionic polymerization of caprolactam [83]. Concentration of sodium caprolactam; 0.0385 (●), 0.0905 (⊖) and 0.162 mole kg^{-1} (○): temperature, 180°C.

$$k_{11}K = 5 \times 10^{-11} \ \text{sec}^{-1}.$$

Hence, the initiation is slower by many orders of magnitude than the propagation reaction.

In the bulk polymerization of most unsubstituted lactams at low temperatures, the crystallization of the polymer starts long before the polymerization is finished. For example, in the polymerization of pyrrolidone, the polymer starts separating at very low conversion [149, 168, 172]. Almost all acyllactam growth centres are incorporated into the separated polymer and the rate of polymerization decreases considerably. During this "heterogenous period", the rate of polymerization is slowed down both by the decreased rate of diffusion of lactam anions to the active ends in the swollen precipitated polymer and by "physical termination" of the active ends by the mutual impingement of the growing crystallites. The latter process was evaluated by assuming an Avrami-type occlusion of the growing chains, and the derived rate equation [172]

$$-\mathrm{d}[\mathrm{L}]/\mathrm{d}t = kN([\mathrm{L}]_0 - [\mathrm{L}]) \tag{80}$$

(N = number of active chains) was in a very good agreement with the experimental results.

4.6.4 Anionic copolymerization

The composition of the copolymer formed at the beginning of polymerization is generally different from the composition of the equilibrium polymerization product [167, 173, 174]. Whereas kinetic factors are decisive for the former, the heat and entropy of polymerization determine the composition of the equilibrium copolymer.

In the copolymerization of lactams of different ring size, the relative rate of incorporation of the two lactams is not necessarily determined by the reaction in which the lactam ring is cleaved. Vofsi et al. [167] showed that in the anionic copolymerization of caprolactam and pyrrolidone (Table 9), the acylation of lactam anions with the exocyclic carbonyl of the growing acyllactam structure (i.e. exchange of monomer units) occurs faster than acylation with the cyclic carbonyl (propagation), viz.

Therefore, the copolymer composition is determined by the relative acidities of both lactams and by the nucleophilicities of the corresponding anions in the transacylation reactions. The distribution of lactam anions is given by the equilibrium constant

$$K_a = \frac{[L_2][L_1^-]}{[L_1][L_2^-]} \tag{81}$$

and the exchange (transacylation) equilibrium which determines the relative rates of incorporation is then given by

$$d[L_1]/d[L_2] = K_e[L_1^-]/[L_2^-] = K_aK_e[L_1]/[L_2] \tag{82}$$

where L_1 and L_2 (L_1^- and L_2^-) represent the two lactams (and their anions). For the given pair of monomers, pyrrolidone was found more acidic ($K_a = 0.4$) and more reactive in the transacylation ($K_e = 0.3$) than caprolactam [167].

4.6.5 Adiabatic polymerization

The high speed of anionic polymerization as well as the low heat of polymerization of most lactams allows bulk polymerizations to be carried out under adiabatic conditions. Vice versa, the adiabatic polymerization is convenient for the estimation of the heat of polymerization [37, 38].

The kinetic treatment of the adiabatic polymerization is very complicated with respect to the variation with increasing temperature of rate coefficients and equilibrium constants. Equations derived with a series of simplifying assumptions (e.g. heat capacity of the monomer ($c_{p, m}$) = heat capacity of the amorphous polymer ($c_{p, p}$) = constant = c_p) lead to [2]

$$-d[L]/dt = \frac{dT}{dt} \cdot \frac{c_p}{\Delta H_p} \tag{83}$$

provided that no crystallization occurs during polymerization. For non-adiabatic conditions, a term including the heat transfer constant of the equipment used (α) as well as the wall temperature (T_w) are added [2], viz.

$$-d[L]/dt = \frac{dT/dt + \alpha(T - T_w)}{\Delta H_p/c_p} \tag{84}$$

When the monomer—polymer equilibrium is attained at T_e, then the instantaneous monomer concentration is related to the initial (T_0) and instantaneous temperature (T) by the equation [175]

$$\frac{[L]_0 - [L]}{[L]_0} = \frac{T - T_0}{T_e - T_0} \tag{85}$$

The value of

$$[L] = [L]_0 \frac{T_e - T}{T_e - T_0} \tag{86}$$

can then be inserted in any of the rate equations. In this way, the course of polymerization has been described in terms of changes of temperature [175] up to conversions of 60% by the expression

$$\ln\left(\frac{dT}{dt} \cdot \frac{1}{T_e - T}\right) = A + \ln[A]_0[I]_0 - E/RT \tag{87}$$

with $A = 18.4$ and $E = 36.6$ kcal mole^{-1}. The activation energy estimated from half-times of polymerization, i.e. $t_{0.5}$ at which $\Delta T = 0.5 (T_e - T_0)$, is much lower (16.8 kcal mole^{-1}) and agrees with that calculated from isothermal measurements (17.5 kcal mole^{-1}) [37, 169]. In a limited range of $[A]_0/[I]_0$, the empirical equation

$$\log t_{0.5} = K_1 - \log[A]_0 + K_2/(T_0 + 14) \tag{88}$$

is valid [37]. For a more precise evaluation, the variation of c_p with temperature has to be taken into account [175].

At low temperatures, at which crystallization occurs during polymerization many simplifications have had to be used [2, 37], to arrive at equations which could be tested experimentally.

5. Cationic polymerization

This type of lactam polymerization is initiated under anhydrous conditions with acids or acid salts which do not split off water at the polymerization temperature (e.g., lactam or amine hydrochloride) as well as with some Lewis acids [176, 177]. The activated species is the monomer cation which takes part both in the initiation and propagation reactions.

5.1 INITIATION AND PROPAGATION

The acidic initiator increases the reactivity of the lactam carbonyl towards nucleophiles. Although protonation of amides occurs preferentially at the oxygen atom [47] a small amount of N-protonated lactam is assumed to be present in the tautomeric equilibrium [178—181]

$$
\begin{array}{ccc}
\text{(XXVII)} & & \text{(XXVIII)}
\end{array}
\tag{89}
$$

(XXVII) ⟷ (XXVIII), below (XXIX)

Because of the lack of resonance stabilization, the amidium ion (XXIX) has an increased acylating ability and may acylate the nucleophiles present. At the beginning of the polymerization initiated with lactam salts of strong acids (e.g. lactam hydrochloride), the neutral lactam is a stronger nucleophile (pK_b = 13.6) than the chloride anion (pK_b = 21). Acylation of the lactam with the amidium cation (XXIX) results in the formation of an aminoacyllactam [178—181]

$$
\text{CO–NH} + \text{CO–}\overset{\oplus}{\text{NH}}_2 \underset{k_{-31}}{\overset{k_{31}}{\rightleftharpoons}} \text{CO–N–CO} \quad \overset{\oplus}{\text{NH}}_3
\tag{90}
$$

This reaction is similar to the disproportionation (23) occurring in the anionic polymerization except that the nucleophiles are activated in the latter case. The protonated primary amine group is in equilibrium with the lactam, viz.

$$
\overset{\oplus}{\text{NH}}_3 + \text{CO–NH} \rightleftharpoons \text{NH}_2 + \text{CO–}\overset{\oplus}{\text{NH}}_2
\tag{91}
$$

and a small fraction of protonated lactam is regenerated. The neutral amine group represents the strongest nucleophile and acts as the growth centre in two kinds of propagation reactions.

Acylation of the amine group with lactam cations (XXIX), viz.

$$
\text{NH}_2 + \text{CO–}\overset{\oplus}{\text{NH}}_2 \rightleftharpoons \text{NHCO} \quad \overset{\oplus}{\text{NH}}_3
\tag{92}
$$

438

leads to the incorporation of one monomer unit and regeneration of the growth centre. The second propagation reaction consists in the bimolecular aminolysis of aminoacyllactam molecules [182—186]. This reaction involving neutral as well as protonated amine groups occurs both at the cyclic as well as exocyclic carbonyl of a terminal acyllactam, viz.

$$HCl \cdot NH_2 \quad CONH \quad CON{-}CO + HCl \cdot HN{-}CO$$

$$k_{32} \Big/ k_{-32} \tag{93a}$$

$$HCl \cdot NH_2 \quad CON{-}CO + HCl \cdot NH_2 \quad CON{-}CO$$

$$k_{33} \Big\backslash k_{-33}$$

$$HCl$$

$$HCl \cdot NH_2 \quad CONH \quad CONH \quad CON{-}CO \tag{93b}$$

$$HN{-}CO$$

$$HCl \cdot NH_2 \quad CONH \quad CONH \quad CON{-}CO + HCl \cdot HN{-}CO$$

The consumption of one amine group in reaction (93) increases the acidity of the medium. In order to establish the equilibria (90) and (91), new amine groups and acyllactam structures are formed. As soon as at least one lactam molecule or lactam cation is involved in the disproportionation reaction (90), the sequence of disproportionation (90) and bimolecular aminolysis (93) results in the incorporation of one or two monomer units into the polymer molecule. The participation of this type of chain growth in cationic lactam polymerization, suggested by Doubravszky and Geleji [182—184], has been confirmed both for polymerization [185—188] as well as for model reactions [189, 190]. The heating of an equimolar mixture of acetylcaprolactam with cyclohexylamine hydrochloride (175°C for 15 min) resulted both in ring opening and deacylation of acetylcaprolactam [189], viz.

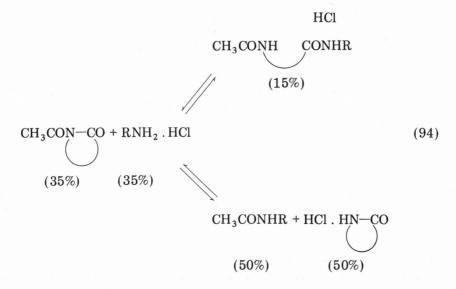

$$CH_3CON-CO + RNH_2 . HCl \qquad (94)$$

(35%)   (35%)

The composition indicates that the cyclic and exocyclic acyl groups reacted at a ratio of 1 : 3, so that polymerization via reactions (90) and (93) is feasible.

Under polymerization conditions, both the disproportionation and bimolecular aminolysis were found to proceed at a comparable rate (Figs.

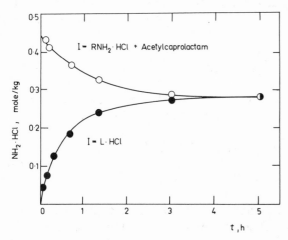

Fig. 19. Concentration of amine hydrochloride during cationic polymerization of caprolactam [186]. Initiator: caprolactam hydrochloride (0.44 mole kg⁻¹), or a mixture of benzylamine hydrochloride (0.47 mole kg⁻¹) + acetylcaprolactam (0.47 mole kg⁻¹). Temperature, 160°C.

References pp. 465—471

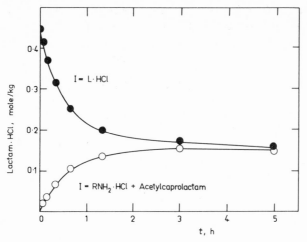

Fig. 20. Concentration of caprolactam hydrochloride during cationic caprolactam polymerization [186]. For conditions see Fig. 19.

19 and 20) [186]. At sufficiently low temperatures at which side reactions do not interfere (Section 5.2), the same equilibrium concentrations are approached from the left as well as right-hand side of (90). Hence, the initiation reaction (90), as well as the polymerization reaction (93), should follow the rate equation for reversible bimolecular reactions

$$-d[L \cdot HCl]/dt = d[A]/dt = k_{31}[L \cdot HCl][L] - k_{-31}[A][NH_2 \cdot HCl] \quad (95)$$

(where A represents acyllactam). Since the concentration of free amino groups is negligible and when no other basic groups are formed in side reactions, then the sum

$$[NH_2 \cdot HCl] + [L \cdot HCl] = [L \cdot HCl]_0$$

and the concentration of amine groups is equal to the concentration of acyllactam units ($[NH_2 \cdot HCl] = [A]$). Equation (95) then becomes

$$-d[L \cdot HCl]/dt = d[A]/dt = k_{31}[L \cdot HCl][L]$$
$$-k_{-31}([L \cdot HCl]_0 - [L \cdot HCl])^2 \quad (96)$$

The equilibrium (90) can be attained only at low temperatures, when side reactions do not interfere significantly (Section 5.2). From published data on the polymerization of caprolactam with caprolactam hydrochloride [186] at 160°C (Fig. 19) we obtain $k_{31} = 8 \times 10^{-5}$ mole kg^{-1} sec^{-1} and $k_{-31} = 1.5 \times 10^{-3}$ mole kg^{-1} sec^{-1}, and the corresponding equilibrium constant

$$K_{31} = \frac{k_{31}}{k_{-31}} = \frac{[NH_2 \cdot HCl][A]}{[L \cdot HCl][L]} = 0.054$$

Due to its high speed, the cationic initiation reaction contributes to the over-all lactam consumption, in contrast to the anionic polymerization in which the disproportionation is usually very slow as compared with the fast propagation reactions involving acyllactam growth centres. Neutral and protonated polymer amide groups can also take part in the disproportionation, viz.

$$(97)$$

$$(98)$$

$$(99)$$

These reactions are similar to those describing the initiation and polymerization reactions in the anionic polymerization. It follows from this scheme, that only reactions involving the protonized lactam molecule, (90) and (99), can contribute to chain growth. Reaction (98) results in the incorporation of one monomer unit, only when followed by aminolysis (93b) at the cyclic carbonyl of the acyllactam.

Due to the high rate of both disproportionation and aminolytic reactions at elevated temperatures (above 200°C for caprolactam), the equilibrium (90) is attained during the first minutes of polymerization [186, 187]. Hence, the initiation with lactam hydrochloride is practically equivalent to the initiation with an equimolar mixture of an acyllactam with a primary amine hydrochloride (provided that the amine is similar to the growing end group with respect to steric effects and basicity).

Kinetic treatment of the whole course of the cationic polymerization is not yet possible because of the fast side reactions which change the concentration of all active species. So far, only the initial (maximum) rates of polymerization with various kinds of initiators (I) have been described by the following empirical equation [176, 182, 183, 191—193]:

$$-d[L]/dt = k[L][I]_0^a \tag{100}$$

The apparent order of reaction with respect to the initiator was found to vary between 0.3 and 1.5, dependent on the type of initiator [182, 191—193] (Fig. 21). In addition, a great variation of the value of a was observed during the polymerization [176, 192].

Polymerizations initiated with salts of strong acids with primary and secondary amines start with a high concentration of growth centres and addition of protonated lactam to the amine group (92) is the prevailing growth reaction at the beginning of polymerization. The rate of addition of the first lactam unit depends on the structure and basicity of the amine. Since the concentration of amidium cations required for the disproportionation increases with decreasing basicity of the amine, equilibrium (91), the relative participation of both propagation reactions as

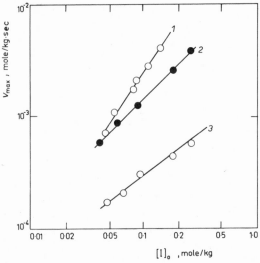

Fig. 21. Maximum rate of cationic polymerization of caprolactam with different initiators [193]. Caprolactam hydrochloride (1), benzylamine hydrochloride (2) and triethylamine hydrochloride (3) at 256°C.

well as the fraction of incorporated amine will depend on the nature of the amine. Hence, the structure and basicity of the amine can influence the whole course of polymerization.

In polymerizations initiated with salts of tertiary amines, the growth centres (i.e. primary amine groups and acyllactam structures) must be formed first, before propagation can start. Due to the low concentration of protonated lactam, the rate of polymerization is much lower than in polymerizations initiated with salts of lactams or primary and secondary amines (Fig. 21).

Polymerizations initiated under anhydrous conditions with carboxylic acids [194—196] should follow a similar mechanism as polymerizations with strong acids. Lactam cations (XXIX) formed in the equilibrium

$$-COOH + \underset{\underset{\displaystyle \smile}{}}{CO-NH} \;\rightleftharpoons\; -COO^{\ominus} + \underset{\underset{\displaystyle \smile}{}}{CO-\overset{\oplus}{N}H_2} \tag{101}$$

can acylate not only neutral lactam molecules, reaction (90), but also carboxylate anions, viz.

$$-COO^{\ominus} + \underset{\underset{\displaystyle \smile}{}}{CO-\overset{\oplus}{N}H_2} \;\underset{k_{-41}}{\overset{k_{41}}{\rightleftharpoons}}\; -CO-O-CO \underset{\underset{\displaystyle \smile}{}}{\;\;NH_2} \tag{102}$$

The subsequent fast intramolecular rearrangement of the amino anhydride [197, 198]

$$-CO-O-CO \underset{\underset{\displaystyle \smile}{}}{\;\;NH_2} \;\overset{k_{42}}{\longrightarrow}\; -CONH \underset{\underset{\displaystyle \smile}{}}{\;\;COOH} \tag{103}$$

or its bimolecular aminolysis

$$-CO-O-CO \underset{\underset{\displaystyle \smile}{}}{\;\;NHCO-} \;+\; NH_2 \underset{\underset{\displaystyle \smile}{}}{\;\;COOH} \text{ or its bimolecular aminolysis}$$

$$k_{43}$$

$$-CO-O-CO \underset{\underset{\displaystyle \smile}{}}{\;\;NH_2} + NH_2 \underset{\underset{\displaystyle \smile}{}}{\;\;CO-O-CO-} \tag{104}$$

$$k_{43}$$

$$-CO-O-CO \underset{\underset{\displaystyle \smile}{}}{\;\;NHCO} \underset{\underset{\displaystyle \smile}{}}{\;\;NH_2} + HOCO-$$

result in the incorporation of one or two monomer units.

A low concentration of anhydride and amine groups can also be established through the elimination of water from two carboxyl groups which is known to proceed at elevated temperatures [199]. With respect to the very low concentration of amine groups, the rate of the bimolecular reaction (104) will be much lower than that of the intramolecular rearrangement (103) i.e., $k_{42} \gg k_{43}$ [NH$_2$]. The initial rate of polymerization via reactions (102) and (103) is then [197]

$$-d[L]/dt = \frac{k_{41} \cdot k_{42}}{k_{42} + k_{-41}} \cdot K_4[L][COOH] \tag{105}$$

where

$$K_4 = \frac{[COO^{\ominus}][CO\overset{\oplus}{N}H_2]}{[COOH][CONH]} \tag{106}$$

The polymerization of lactams initiated with carboxylic acids has been found to be first order with respect to the lactam [10, 194, 200]. However, the order of reaction with respect to the initiating acid, eqn. (100), varied from 0.5 for caprolactam [194, 196] to 0.8 for capryllactam [200]. The deviation of the apparent order of reaction from unity can be due to variation in the activity coefficients of the reacting species which are affected by the dielectric constant of the medium [197]. The different polarity and basicity of caprolactam and capryllactam is one of the reasons for the different apparent order of reaction with respect to the initiator concentration.

5.2 DEACTIVATION OF CATALYTIC SPECIES

Surprisingly, the initially fast cationic polymerization of caprolactam slows down before reaching the monomer—polymer equilibrium [176, 182—184, 188, 192, 193] and after a certain period the rate of polymerization again increases (Fig. 22). This peculiar course of polymerization indicates that the catalytic species are deactivated in some side reaction and that other active species are formed in a slow reaction.

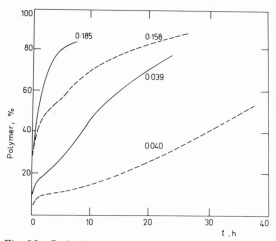

Fig. 22. Cationic polymerization of caprolactam at 256°C. Initiator, caprolactam hydrochloride [183] (- - - -) and benzylamine hydrochloride [192] (———); concentrations in mole kg^{-1}.

Amine hydrochlorides are known to react at elevated temperature with diacylamines and amides with the formation of amidines [201]

$$-\text{CONHCO}- + \text{RNH}_2.\text{HCl} \xrightarrow{140°} [\text{RN}=\underset{|}{\text{C}}-\text{NH}-]\text{HCl} \qquad (107)$$

Schlack et al. [202—207] were able to identify amidine groups in cationic caprolactam polymers prepared with caprolactam or amine hydrochloride. In polymers prepared at elevated temperatures, primary amine groups were substantially absent and the majority of basic groups was represented by the strongly basic amidine groups decreasing the acidity of the medium. Amidine hydrochlorides cannot initiate a fast polymerization [206, 208, 209], so that the reduced concentration of lactam cations decreases the concentration of primary amine as well as acyllactam groups through the equilibria (90) and (91). In this way the formation of amidine groups decreases the rate of polymerization.

In the polymerization with caprolactam hydrochloride, the maximum rate of polymerization coincides with the period of maximum concentration of amine and acyllactam groups (Fig. 23). The drastic reduction in the rate occurs during the period in which the concentrations of primary amine and imide groups decrease sharply due to amidine formation, and the rate of polymerization decreases roughly in parallel with the concentration of caprolactam hydrochloride (Fig. 23). Small amounts of strongly acidic amide hydrochloride are present in cationic polymers even after long periods [188]. This residual acidity is effective in proton-catalysed nucleophilic carbonyl substitution reactions resulting in a slow polymeriz-

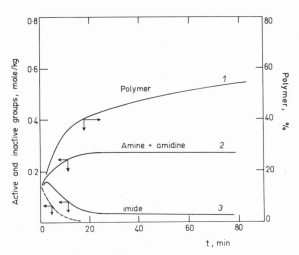

Fig. 23. Concentration of active species and polymer in the cationic polymerization of caprolactam at 256°C. Initial concentration of caprolactam hydrochloride, 0.233 mole kg^{-1} (curve 1) [183] and 0.270 mole kg^{-1} (curves 2, 3 and dashed curve) [186]; the dashed curve indicates strong acidity (L . HCl, amide . HCl).

References pp. 465—471

ation with an almost steady concentration of active species (amide hydrochloride, amine and imide), and the rate of polymerization is governed predominantly by the lactam concentration.

There is an increasing amount of evidence that the amidine groups are present in the semicyclic form resulting either from intramolecular cyclization of the terminal amine hydrochloride [202, 204, 206, 210], viz.

$$\sim\!\!NHCO\underbrace{}NH_2 \cdot HCl \xrightarrow{\ k_{51}\ } H_2O + \sim\!\!NH\!-\!\underset{\underbrace{}}{C}\!\!=\!\!N \qquad (108)$$

or from bimolecular reactions of amine hydrochloride with lactam [211] or acyllactam, viz.

$$\sim\!\!NH_2 \cdot HCl + \underset{\underbrace{}}{CO\!-\!NH} \xrightarrow{\ k_{52}\ } H_2O + \left[\sim\!\!NH\!-\!\underset{\underbrace{}}{C}\!\!=\!\!N\right] HCl \qquad (109)$$

$$\sim\!\!CO\!-\!\underset{\underbrace{}}{N\!-\!CO} + HCl \cdot NH_2\!\sim\!\! \xrightarrow{\ k_{53}\ } \left[\sim\!\!CO\!-\!\underset{\underbrace{}}{N\!-\!C}\!\!=\!\!N\!\sim\!\!\right] HCl + H_2O$$

$$\qquad (110)$$

$$\downarrow$$

$$\sim\!\!COOH + \left[\underset{\underbrace{}}{N\!\!=\!\!C\!-\!NH}\!\sim\!\!\right] HCl$$

So far, no kinetic data are available for the estimation of k_{51}, k_{52} and k_{53}. The decay of the strongly acidic amide hydrochloride follows the equation [184]

$$-d[L \cdot HCl]/dt = k_5[L \cdot HCl]^2 \qquad (111)$$

with $k_5 = 3.6 \times 10^{-3}$ kg mole^{-1} sec^{-1} at 218°C.

The preferential formation of semicyclic amidine groups was supported by the fact that amidine groups were not found in cationic polymers of capryl- and laurinlactam [203, 206], where formation of the corresponding unstable cycles is very improbable. Detailed studies revealed that, under certain conditions, cationic capryllactam polymers contain semicyclic and non-cyclic amidine groups [252].

Irrespective of the fact that the polymerization has been started under anhydrous conditions, water is formed very soon in reactions

(108)—(110). Hydrolysis occurs preferentially at the acyllactam or acylamidine groups but hydrolysis of amide groups cannot be omitted from consideration either. The resulting carboxyl and amine groups then take part in the polymerization process so that more than one polymerization mechanism is operative in the second stage of the cationic lactam polymerization. In addition, the reaction of amidine and carboxyl groups [92] results in the regeneration of the strongly acidic amide hydrochloride, viz.

$$\text{~~~COOH} + \text{HCl} \left[\begin{array}{c} \text{N=C—NH~~~} \\ \bigcirc \end{array} \right] \longrightarrow \text{~~~CONH} \overset{\text{HCl}}{\underset{\smile}{\text{CONH~~~}}} \qquad (112)$$

which again enters into the sequence of initiation, polymerization and side reactions. Reaction (112) could be responsible for the low concentration of amide hydrochloride surviving for long periods.

5.3 END GROUPS AND MOLECULAR WEIGHT

The individual oligomers, arising from disproportionation (90) and chain growth (92—94), viz.

$$\text{H[}-\text{NH(CH}_2)_5\text{CO]}_n\text{N—CO} \qquad (113)$$

were identified in cationic caprolactam polymers [181, 190] (up to $n = 6$). Analogous oligomers were identified also in polymers of capryllactam prepared with a series of initiators [212]. Due to the relatively fast side reactions (108)—(110), semicyclic amidine groups may be present, instead of amine groups at one end and carboxyl groups instead of acyllactam groups at the other, viz.

$$\text{N=C—NH~~~~~COOH} \qquad (114)$$

Accordingly, the molecular weight of cationic polymers calculated from viscosity measurements corresponds to one basic group per macromolecule [184]. The number of polymer molecules (as well as of basic groups) is very close to the initial initiator concentration and remains constant over a long period, except the very beginning of polymerization [176, 177, 180, 188].

Whereas structure (113) prevails in cationic caprolactam polymers prepared at low temperature and short reaction periods, the fraction of amidine groups increases with temperature and time of polymerization. However, only small fractions of carboxyl groups were found in cationic polymers [180]. Only fractionated polymers contained equivalent amounts of basic and carboxylic groups which could also arise from hydrolysis of acyllactam groups [213].

The molecular weight distribution of cationic polymers is broader than the statistical one because new growth centres are formed during the polymerization. The presence of two maxima in the later stages of polymerization indicates the superimposition of two reaction mechanisms [213].

6. Hydrolytic polymerization

The water initiated polymerization of lactams represents the classical, industrially widely used process and, therefore, a large effort has been devoted to the investigation of the kinetics and mechanism of the individual reactions. In order to establish the accepted reaction mechanism, the concentration of all components involved in the complex set of reactions had to be followed during polymerization and determined at equilibrium. Thorough studies by Hermans, Heikens, Kruissink, Reimschuessel, Staverman and by van der Want as well as by Wiloth elucidated the complex scheme of equilibrium reactions involving water, cyclic and open chain amide groups, amine and carboxyl groups [1, 2, 4, 5, 12, 13, 15, 22, 23, 177, 214–235].

6.1 REACTION MECHANISM

The water initiated polymerization is characterized by the following three principal equilibrium reactions: hydrolytic ring-opening,

$$HN{-}CO + H_2O \underset{k_{-1}}{\overset{k_1}{\rightleftharpoons}} NH_2 \quad COOH \qquad (115)$$

$$(L) \qquad\qquad (S_1)$$

formation of open chain amide groups by condensation of amine and carboxyl groups,

$$\sim\!\!\sim COOH + NH_2\!\!\sim\!\!\sim \underset{k_{-2}}{\overset{k_2}{\rightleftharpoons}} H_2O + \sim\!\!\sim CONH\!\!\sim\!\!\sim \qquad (116)$$

$$(S_m) \qquad (S_n) \qquad\qquad\qquad (S_{m+n})$$

and stepwise addition of lactam molecules either at the amine

$$\text{---NH}_2 + \underset{\bigcirc}{\text{CO—NH}} \underset{k_{-3}}{\overset{k_3}{\rightleftharpoons}} \text{---NHCO} \diagdown \diagup \text{NH}_2 \qquad (117)$$

or carboxyl group, reactions (102) and (103).

The hydrolysis of the lactam represents the initiation reaction providing end groups at which lactam molecules can be added, reaction (117), as well as the amino acid which can enter into the polymer through bimolecular condensation (116).

The polymerization can proceed both as a stepwise addition (117) and condensation (116). The contribution of the individual processes to the over-all lactam consumption depends on the nature of the monomer and on the reaction conditions. In the polymerization of caprolactam, the prevailing fraction of lactam enters into the polymer through the stepwise addition [1, 215, 231, 233], and only a few percent of monomer units are incorporated into the polymer through hydrolysis (115) and bimolecular condensation (116) [1, 236].

The specific role of either amine or carboxyl groups in the polyaddition reaction has been established by Heikens et al. [216] by following the incorporation of added amine or carboxylic acid during the hydrolytic caprolactam polymerization. These experiments revealed that the lactam is added at the amine end group and that the reaction is catalysed by carboxyl groups.

Several detailed reaction paths have been proposed for the addition process (117). The most powerful acylating agent present is the protonated lactam (XXIX), p. 437, and the strongest nucleophile is the neutral amine group so that the fastest addition should be reaction (92). Since the equilibrium (91) results in a low concentration of lactam cations and neutral amine groups, most amine groups are present as ammonium ions or undissociated salt ($-\overset{\ominus}{\text{C}}\text{OO}\overset{\oplus}{\text{N}}\text{H}_3-$). Therefore, a two-step or concerted mechanism could also be operative in which the N-protonation of the lactam by an ammonium ion is accompanied or followed by acylation of the amine and intramolecular proton transfer [220], viz.

$$(118)$$

$$\text{---NH—CO} \diagdown \underset{\overset{\oplus}{\text{NH}}_3}{}$$

Kruissink [13, 220] demonstrated that the kinetics of the polyaddition is described satisfactorily by assuming two reactions, i.e. an uncatalysed addition of the lactam at the ammonium ion, reaction (118), and the carboxyl group catalysed addition of lactam at the neutral amine group. The catalysed reaction can be visualized as (92), in which the reactive species can be represented also by the undissociated lactam salt (or complex) [220] of increased acylating ability, viz.

$$\text{$\sim\!\!\sim$COOH.HN$-$CO} + NH_2\text{$\sim\!\!\sim$} \;\rightleftharpoons\; \text{$\sim\!\!\sim$COOH.NH}_2 \quad \text{CONH$\sim\!\!\sim$} \tag{119}$$

The same mechanism applies to the reverse reaction, i.e., formation of lactam or cyclic oligomers by intramolecular cyclization from the amine end of the polymer molecule, reaction (5).

Incorporation of lactam units into the polymer chain through the uncatalysed transamidation between cyclic and linear amide groups, viz.

$$\begin{array}{ccc} \text{$\sim\!\!\sim$CO$-$NH$\sim\!\!\sim$} & & \text{$\sim\!\!\sim$CO} \quad \text{NH$\sim\!\!\sim$} \\ + & \rightleftharpoons & \text{NH} \quad \text{CO} \\ \text{HN$-$CO} & & \end{array} \tag{120}$$

can be only of minor importance, at least in the polymerization of caprolactam [1, 231].

6.2 KINETICS

The polymerization starts with the slow uncatalysed hydrolysis (115). At the beginning of polymerization

$$d[S_1]/dt = k_1[L][H_2O] - k_{-1}[S_1] = d[COOH]/dt = d[NH_2]/dt \tag{121}$$

As soon as carboxyl groups are formed, the acid catalysed reaction described by

$$d[S_1]/dt = {_c}k_1[L][H_2O][COOH] - {_c}k_{-1}[S_1][COOH] \tag{122}$$

becomes increasingly important so that the initiation reaction is an autocatalytic process (Fig. 24). The rate coefficients of the catalysed and uncatalysed hydrolysis can be obtained from the initial rate of formation of end groups (Fig. 25) [1, 237]. For caprolactam, the rate coefficients of the uncatalysed (k_1) and catalysed lactam hydrolysis (${_c}k_1$) are [233]

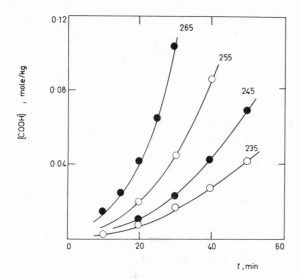

Fig. 24. Formation of carboxyl groups in the hydrolytic polymerization of caprolactam [237]. Initial concentration of water, 1.1 mole/kg. The figures on the curves indicate the temperature.

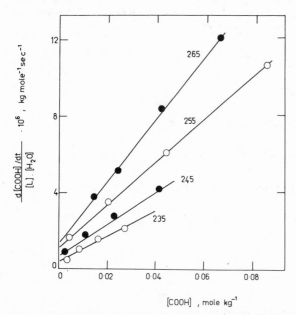

Fig. 25. Evaluation of the rate coefficients k_1 and $_c k_1$ for the hydrolytic polymerization of caprolactam [237]. For conditions see Fig. 24.

$$k_1 = 4.7 \times 10^2 \exp(-21{,}400/RT) \text{ kg mole}^{-1} \text{ sec}^{-1}$$

$$_c k_1 = 1.14 \times 10^4 \exp(-18{,}750/RT) \text{ kg}^2 \text{ mole}^{-2} \text{ sec}^{-1}$$

The catalysed and uncatalysed addition of lactam at the amine group (as well as depolymerization from all chains except S_1) is represented by

$$-d[L]/dt = (k_3 + _ck_3[S])[S][L] - (k_{-3} + _ck_{-3}[S])([S] - [S_1])$$

(123)

The over-all rate of lactam consumption in the initiation (115) and polyaddition (117) reactions is then given by

$$-d[L]/dt = (k_1 + _ck_1[S])[L]([H_2O]_0 - [S]) - (k_{-1} + _ck_{-1}[S])[S_1]$$

$$+ (k_3 + _ck_3[S])[S][L] - (k_{-3} + _ck_{-3}[S])([S] - [S_1])$$

(124)

The rate coefficients for caprolactam are [233]

$$k_3 = 7.3 \times 10^5 \exp(-21,270/RT) \text{ kg mole}^{-1} \text{ sec}^{-1}$$

$$_ck_3 = 6.6 \times 10^6 \exp(-20,400/RT) \text{ kg}^2 \text{ mole}^{-2} \text{ sec}^{-1}$$

The main depolymerization reaction consists of a carboxyl group catalysed elimination of lactam from the amine end group, and the rate of lactam formation follows the equation [217]

$$d[L]/dt = _ck_{-3}[NH_2][COOH]$$

(125)

For polycaprolactam at 250°C

$$_ck_{-3} = 1.2 \times 10^{-2} \text{ kg mole}^{-1} \text{ sec}^{-1}$$

The rate of the condensation reaction (116) is expressed by

$$-d[S]/dt = (k_2 + _ck_2[S])[S]^2 - (k_{-2} + _ck_{-2}[S])([H_2O]_0 -$$

$$[S])([L]_0 - [L] - [S]) - (k_1 + _ck_1[S])[L]([H_2O]_0 - [S]) +$$

$$(k_{-1} + _ck_{-1}[S])[S_1]$$

(126)

and for caprolactam [223]

$$k_2 = 2.4 \times 10^5 \exp(-22,550/RT) \text{ kg mole}^{-1} \text{ sec}^{-1}$$

$$_ck_2 = 6.5 \times 10^6 \exp(-20,670/RT) \text{ kg}^2 \text{ mole}^{-2} \text{ sec}^{-1}$$

The rate equation for the formation of the amino acid (S_1) may then be written as

$$d[S_1]/dt = (k_1 + {}_ck_1[S])[L]([H_2O]_0 - [S]) - (k_{-1} + {}_ck_{-1}[S])[S_1]$$

$$- 2(k_2 + {}_ck_2[S])[S][S_1] + 2(k_{-2} + {}_ck_{-2}[S])([H_2O]_0 - [S]) \times$$

$$([S] - [S_1]) - (k_3 + {}_ck_3[S])[S_1][L] + (k_{-3} + {}_ck_{-3}[S])[S_2] \quad (127)$$

(to a first approximation $[S_2] \simeq [S_1]$).

The equations (121)—(127) have been derived from the complex course of polymerization as well as from the concentrations of chains and amino acid [1, 2, 217, 219, 223, 231—233]. A typical set of experimental data

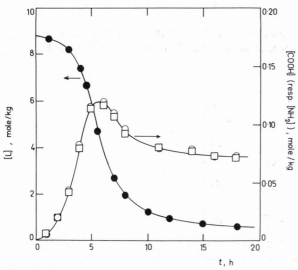

Fig. 26. Hydrolytic polymerization of caprolactam [1]. Initial concentration of water, 0.59 mole kg^{-1}. Temperature 221.5°C. Concentration of caprolactam (●), and of carboxyl (○) and amine (□) end groups.

required for the calculation of the rate coefficients is represented in Figs. 26 and 27. For lactams forming appreciable amounts of cyclic oligomers, more complicated equations are required to describe the variation of the concentration of all components.

The observed increase of the rate coefficients of polyaddition with increasing conversion indicates that either some other reaction is involved in the polymerization, or that the activities or relative proportions of the active species are changing. Kruissink [13, 220] provided a refined analysis of the polyaddition reaction by taking into account changes of the medium during polymerization. In the hydrolytic polymerization of caprolactam, the dielectric constant decreases from 15 to about 5. As a consequence, the degree of ionization (α) of end groups drastically decreases during polymerization from $\alpha = 0.8$ to $\alpha < 0.001$. Inspection of

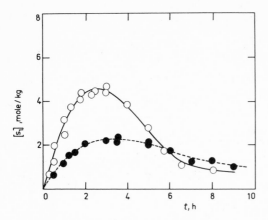

Fig. 27. Concentration of aminocaproic acid ($[S_1]$) during hydrolytic polymerization of caprolactam [1]. Initial concentration of water; 0.59 (●) resp. 0.87 mole kg^{-1} (○); temperature: 221.5°C.

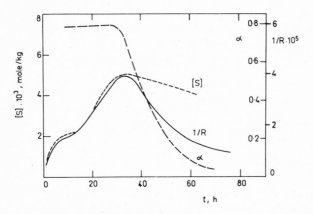

Fig. 28. Changes of the degree of ionization during hydrolytic polymerization of caprolactam at 220°C with 1 mole % of water. Conductivity of the polymerizing mixture (———) and content of end groups (- - - -) [228]. The values of α were calculated from the conversions given in ref. [231] and data in ref. [13].

Fig. 28 shows that the number of chains estimated from the conductivity of the polymerizing mixture starts to deviate from the actual content of end groups from the moment at which the value of α starts to decrease steeply. Kruissink suggested that the lactam reacts with ammonium groups ($[\overset{\oplus}{N}H_3] = \alpha[S]$) in an uncatalysed reaction, the rate being given by

$$-d[L]/dt = k_3[\overset{\oplus}{N}H_3][L] = k_3\alpha[S][L] \tag{128}$$

and with the uncharged amine group ($[NH_2] = (1-\alpha)[S]$) in a carboxyl group catalysed addition, the rate of which is

$$-d[L]/dt = {}_ck_3[NH_2][COOH][L] = {}_ck_3(1-\alpha)^2[S]^2[L] \qquad (129)$$

As the degree of ionization decreases during the polymerization, the relative participation of the uncatalysed reaction decreases, whereas the contribution of the fast catalysed reaction increases. Hence, the rate of the polyaddition is related to the degree of ionization by the expression

$$-d[L]/dt = \{k_3\alpha + {}_ck_3(1-\alpha)^2[S]\}[L][S] \qquad (130)$$

With suitable values of k_3 and ${}_ck_3$, eqn. (130) describes the whole course of polyaddition very satisfactorily.

Addition of amine decreases the concentration of the unionized carboxyl groups so that the uncatalysed reaction becomes increasingly important [13]. In the polymerization initiated with amines in the absence of water and carboxylic acid, the polymerization proceeds very slowly [194]. Both an uncatalysed addition of lactam at the uncharged amine group as well as at the ammonium group can be assumed to participate in the polymerization process (with the lactam as the protonating agent). In the polymerization of cis-lactams, the acidity of the medium increases with increasing conversion so that the fraction of protonated amine groups increases. This could account for the induction period observed in the polymerization of caprolactam initiated with amines [194].

Wiloth [231, 232] arrived at an excellent agreement between experimental and calculated data by introducing an additional reaction between the amino acid and the lactam. Integration of a complicated set of rate equations revealed that the reaction is very fast and it was suggested that the zwitterion of the amino acid plays an important role in the addition reaction [231, 232].

Heikens and Hermans [215] further improved the agreement between the experimental and calculated course of caprolactam polymerization by taking into consideration the volume contraction (ca. 9%) occurring during the polymerization.

6.3 EQUILIBRIA

The principal equilibrium reactions involved in the hydrolytic polymerization are: hydrolysis of lactam, cyclic oligomers and linear polymer, as well as addition of lactam or oligomers at the end groups, i.e., the reversible reactions (4), (5), (115)—(117). For a sufficiently high molecular weight of the polymer, the reactivities of all end groups are equal. Using concentrations instead of activities, the corresponding equilibrium constants are defined as

$$K_1 = \frac{[S_1]_e}{[L]_e[H_2O]_e} = \frac{k_1 + {}_ck_1[S]_e}{k_{-1} + {}_ck_{-1}[S]_e} \tag{131}$$

$$K_2 = \frac{[S_{m+n}][H_2O]_e}{[S_m][S_n]} = \frac{[-CONH-]_e[H_2O]_e}{[COOH]_e[NH_2]_e} = \frac{k_2 + {}_ck_2[S]_e}{k_{-2} + {}_ck_{-2}[S]_e} \tag{132}$$

$$K_3 = \frac{[S_{n+1}]_e}{[S_n]_e[L]_e} = \frac{k_3 + {}_ck_3[S]_e}{k_{-3} + {}_ck_{-3}[S]_e} \tag{133}$$

Assuming that the principle of equal reactivity applies even for the amino acid, then

$$K_2 = \frac{[S_{n+1}]_e[H_2O]_e}{[S_n]_e[S_1]_e} \tag{134}$$

and $K_3 = K_1K_2$. Hence, all three equilibria are described by two independent equilibrium constants.

Evaluation of the equilibrium concentrations of water, aminocaproic acid, chains and lactam in the hydrolytic polymerization of caprolactam [2, 3, 5, 8, 12, 221, 235, 238] resulted in the following values of ΔH_i and ΔS_i in the relation $-RT \ln K_i = \Delta H_i - T\Delta S_i$:

$$\Delta H_1 = 2.11 \text{ kcal mole}^{-1}; \quad \Delta S_1 = -7.9 \text{ eu}$$

$$\Delta H_2 = -6.14 \text{ kcal mole}^{-1}; \quad \Delta S_2 = 0.9 \text{ eu}$$

$$\Delta H_3 = -4.03 \text{ kcal mole}^{-1}; \quad \Delta S_3 = -7.0 \text{ eu}$$

It has to be pointed out, however, that the values of K_1, K_2 and K_3 depend on the initial concentration of water [2, 3, 5, 12, 221, 225, 232, 234, 235] (Figs. 29 and 30) even when the formation of cyclic oligomers is taken into account [15]. Giori and Hayes [14] demonstrated that the variation of the equilibrium constants with changing initial composition is due to the variation of the activity coefficient of water. Substitution of the water activity for the molar concentration in eqn. (132) yields much lower but fairly constant values of K_2 (Fig. 30). Besides this, the variation of the dielectric constant with changing water content will affect the degree of ionization of end groups and, consequently, the proportion of the catalysed and uncatalysed reactions (115)—(117).

Generally it has been assumed that, in the water initiated polymerization, the concentrations of carboxyl and amine groups are equal. This assumption is not valid at high temperatures (above 265°C for caprolactam polymers) when degradation reactions change the concentration of

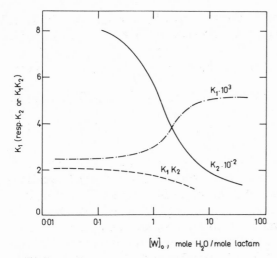

Fig. 29. Equilibrium constants in the hydrolytic polymerization of caprolactam [15]. Temperature, 220°C; initial concentration of water, $[W]_0$, in mole H_2O per mole caprolactam.

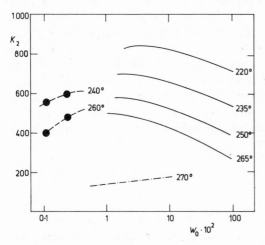

Fig. 30. Equilibrium constant K_2 in the hydrolytic polymerization of caprolactam. Apparent equilibrium constants (220—265°C) and equilibrium constant corrected for the water activity coefficient (270°C). Data from refs. [2], [14] and [233].

end groups [222]. However, there are indications that even at lower temperatures, the actual concentration of carboxyl groups can be lower than that of amine groups, probably, as a consequence of the formation of diacylamine groups [239].

The relations between the initial composition of the reaction mixture and the final molecular weight are very complicated, when the polymerization is started in the presence of molecular weight regulators. For example, in the polymerization initiated with salts of primary amines and carboxylic acids, the following equilibria have to be taken into account:

$$RCOOH \cdot NH_2R' \rightleftharpoons RCONHR' + H_2O \qquad (135)$$

$$RCONH\text{---} + \text{---}COOH \rightleftharpoons RCOOH + \text{---}CONH\text{---} \qquad (136)$$

$$R'NHCO\text{---} + \text{---}NH_2 \rightleftharpoons R'NH_2 + \text{---}NHCO\text{---} \qquad (137)$$

Assuming equal equilibrium constants for various types of regulators, Reimschuessel and Dege [224] as well as Majury [240] derived equations for the calculation of the equilibrium molecular weight of hydrolytic polymers involving the presence of molecular weight regulators, e.g. mono-, di- and trifunctional organic acids, amines and their salts.

Data for the individual rate coefficients and equilibrium constants were applied by Reimschuessel and Nagasubramanian [223] to the optimization of the hydrolytic caprolactam polymerization. With respect to the exothermic nature of the condensation and addition reactions, the equilibrium conversion and molecular weight decrease with increasing temperature. The temperature range that is of practical interest is rather narrow (220—265°) and, therefore, the influence of temperature on the yield and molecular weight is less important than the influence of temperature and of the initial concentration of water on the kinetics of the polymerization. The optimal processes with respect to the desired final molecular weight and conversion are essentially two stage processes, characterized by a high initial water concentration, polymerization to a conversion close to equilibrium (0.85—0.90) and subsequent, almost complete, rapid removal of water.

7. N-Substituted lactams

Substitution at the nitrogen atom decreases the polymerizability of small and medium lactams much more than substitution at any other ring atom (see Section 3) and, therefore, only the highly strained four-, eight- and nine-membered N-substituted lactams have been polymerized so far [53, 241—243] (Table 11). Because of the lack of a dissociable hydrogen at the amide group, N-substituted lactams cannot form the chemically activated species required for the anionic polymerization, i.e. lactam anions and N-acylated lactam; an anionic propagation could proceed only via dialkylamine anions ---ṆR. The polymerization of N-substituted lactams initiated with water or mixtures of carboxylic acids with primary or secondary amines [241] should not differ essentially from the water initiated polymerization of unsubstituted lactams but, at the present time, no kinetic measurements are available.

TABLE 11

Equilibrium monomer content for N-methyllactams at 260°C

Lactam	$[L]_e$ (%)	Ref.
N-Methylcaprolactam	100	53
N-Methylenantholactam	19	53
N-Methylcapryllactam	~1	243

In the polymerization initiated with strong acids under anhydrous conditions the major part of monomer is incorporated into the polymer via highly reactive intermediates [243]. In the presence of strong acids, protonation of N-substituted lactams can occur both at the oxygen and nitrogen (see Section 3). Similarly as with the unsubstituted lactams (Section 5.1), the N-protonated form (XXXI) resulting from the equilibrium

$$RN{-}CO + HCl \rightleftharpoons \left[RN{-}CO\right]HCl \rightleftharpoons \overset{\overset{H}{|}{\oplus}}{RN}{-}CO + Cl^\ominus \qquad (138)$$

$$(XXX) \qquad\qquad (XXXI)$$

represents the activated species of increased acylating ability. However, in contrast to the unsubstituted compound, the N-substituted lactam cannot be acylated at the nitrogen and the only nucleophile present is the anion of the initiating acid. In the hydrogen chloride initiated polymerization, the acylation reaction then yields the aminoacyl chloride (XXXII), viz.

$$\overset{\overset{H}{|}{\oplus}}{RN}{-}CO + Cl^- \underset{k_{-71}}{\overset{k_{71}}{\rightleftharpoons}} RNH \quad COCl \qquad (139)$$

$$XXXII$$

$$\left[RN{-}CO\right]HCl \underset{k_{-72}}{\overset{k_{72}}{\rightleftharpoons}} RNH \quad COCl \qquad (140)$$

which is immediately neutralized

$$RNH \quad COCl + HCl \longrightarrow RNH \overset{HCl}{\overset{\cdot}{\quad}} COCl \qquad (141)$$

460

The acyl chloride represents an activated growth centre which reacts with the strongly nucleophilic amine group either in the monomolecular process (139), (140) or in the bimolecular condensation reaction

$$
\begin{array}{ccc}
\text{HCl} & & \text{HCl} \quad \text{HCl} \\
\bullet & k_{73} & \bullet \quad \bullet \\
2 \ \text{RNH} \quad \text{COCl} & \underset{k_{-73}}{\rightleftharpoons} & \text{RNH} \quad \text{CONR} \quad \text{COCl} + \text{HCl} \qquad (142)
\end{array}
$$

The latter reaction represents the main chain growth process with regeneration of the strong acid which again enters into the initiation reactions (139) or (140). Hence, the incorporation of lactam units into the polymer occurs in the sequence of initiation (139), (140) and condensation (142) reactions. This propagation mechanism has been confirmed by the following facts [243]:

(a) acyl chloride is formed from dialkylamide hydrochloride at elevated temperature,

(b) acyl chlorides or dialkylamine hydrochlorides alone initiate a slow polymerization only, and

(c) the rate of polymerization initiated with mixtures of acyl chloride and dialkylamine hydrochloride increases with increasing amine/acylchloride ratio (Q) very steeply up to $Q = 1$ (Fig. 31) and an excess of amine hydrochloride has a much smaller effect.

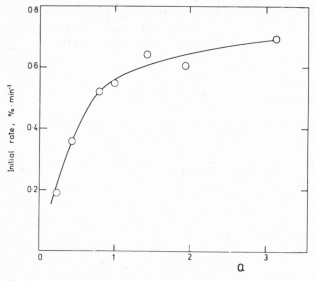

Fig. 31. Polymerization of N-methylcapryllactam [243]. Temperature, 185°C; concentration of butyryl chloride, 0.03 mole kg^{-1}; ratio of butylamine hydrochloride to butyryl chloride = Q.

Ogata [244] found that the basicity of N-substituted lactams is close to the basicity of the linear *trans* amide group. Therefore, it can be assumed that the basicity of the dialkylamide group and the ionization constant of its salt is also the same for both the N-alkyllactam and the corresponding open chain unit. The concentration of the lactam salt (XXX) or protonated lactam (XXXI) is then proportional to the lactam concentration, viz.

$$[L \cdot HCl] = K_7[L][HCl]_0 \tag{143}$$

Assuming that the condensation reaction (142) is much faster than the ring opening (139) and (140), $k_{73} \gg k_{71}$ and k_{72}, then the rate determining step is the formation of the aminoacyl chloride (XXXII). For large rings, the depolymerization can be neglected ($k_{71} \gg k_{-71}$ and $k_{72} \gg k_{-72}$) and the rate of polymerization should obey the expression

$$-d[L]/dt = k_{71}[LH^{\oplus}][Cl^{\ominus}] + k_{72}[L \cdot HCl] = k_7 K_7[L][HCl]_0 \tag{144}$$

The polymerization of N-methylcapryllactam initiated with hydrogen chloride [243] has been found to follow eqn. (144) with

$$k_7 K_7 = 1.1 \times 10^{13} \exp(-32{,}700/RT) \text{ kg mole}^{-1} \text{ sec}^{-1}$$

8. Polymerization with isomerization

The polymerization of lactams does not necessarily lead to polyamides composed of the corresponding open chain units. Reimschuessel [245] introduced a new category of lactam derivatives the polymerization of which does not yield polyamides. Certain carboxyl substituted lactams prefer to polymerize in an isomerization—polycondensation process resulting in the formation of linear polyimides [245—251], viz.

$$(145)$$

$$(146)$$

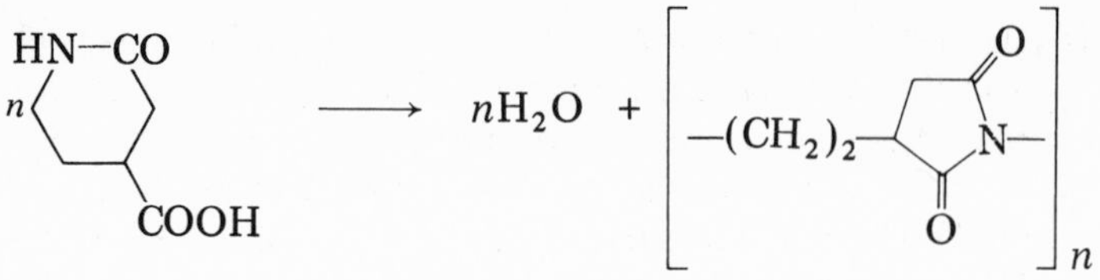

$$
n \; \substack{HN-CO \\ \text{(ring with COOH)}} \longrightarrow nH_2O + \left[-(CH_2)_2 \underset{O}{\overset{O}{\text{(imide)}}} N- \right]_n \tag{147}
$$

$$
n \; \substack{HN-CO \\ \text{(ring with COOH)}} \longrightarrow nH_2O + \left[-CH_2 \underset{O}{\overset{O}{\text{(imide)}}} N- \right]_n \tag{148}
$$

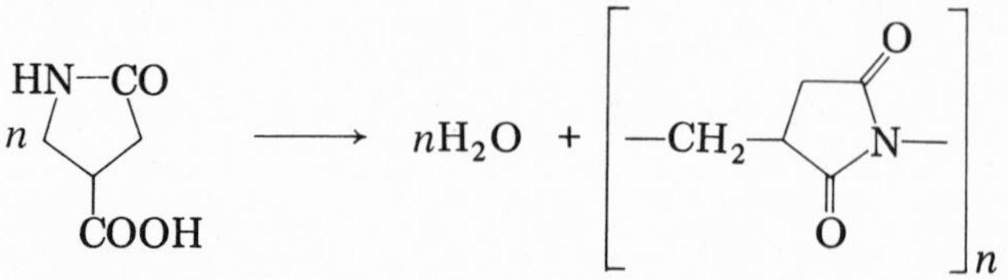

The repeating units (i.e. cyclic imides) could result either from dehydration of the polyamide formed in a ring opening polymerization

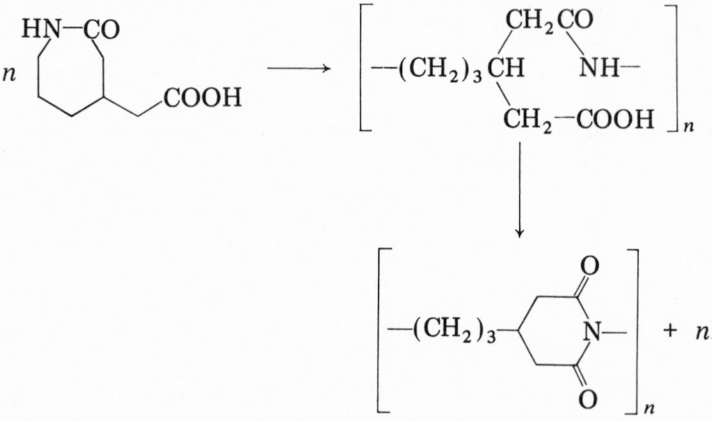

$$
n \; \substack{HN-CO \\ \text{(ring with COOH)}} \longrightarrow \left[-(CH_2)_3 CH \overset{CH_2CO}{\underset{CH_2-COOH}{\big\langle}} NH- \right]_n \tag{149}
$$

$$
\downarrow
$$

$$
\left[-(CH_2)_3- \underset{O}{\overset{O}{\bigcirc}} N- \right]_n + nH_2O
$$

or from transcyclization of the polycondensation product, viz.

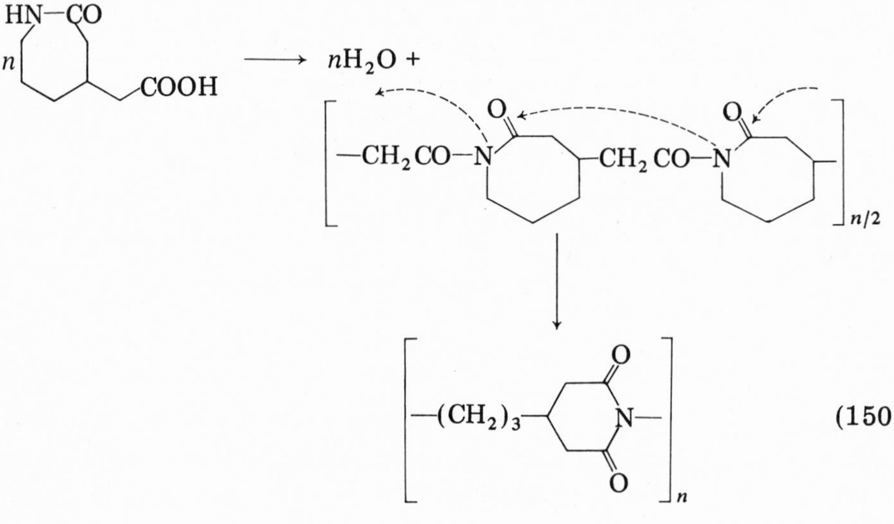

$$
n \; \substack{HN-CO \\ \text{(ring with COOH)}} \longrightarrow nH_2O +
$$

$$
\left[-CH_2CO-N \overset{O}{\big\langle} -CH_2CO-N \overset{O}{\big\langle} \right]_{n/2}
$$

$$
\downarrow
$$

$$
\left[-(CH_2)_3- \underset{O}{\overset{O}{\bigcirc}} N- \right]_n \tag{150}
$$

or from the polycondensation of the isomerized lactam, viz.

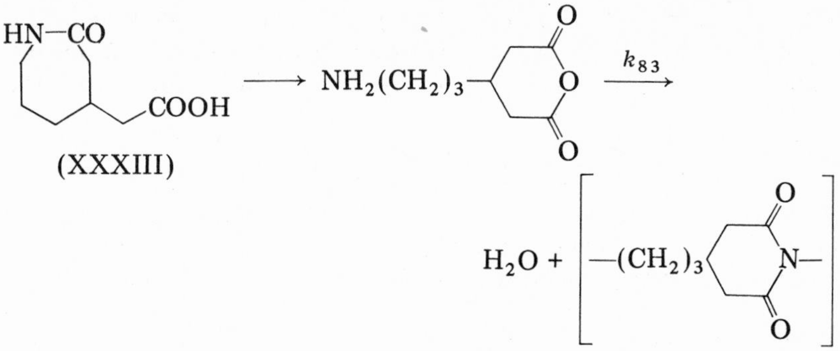

$$\text{(XXXIII)} \quad \longrightarrow \quad NH_2(CH_2)_3- \quad \xrightarrow{k_{83}} \quad H_2O + \left[-(CH_2)_3- \right] \quad (151)$$

The polymerization reaction followed second-order kinetics which was rationalized by assuming that the bimolecular rearrangement involving two monomer molecules was the rate determining step. It has been assumed that the isomerization results in the formation of the amino anhydride (XXXIV) or its dimer (XXXV), viz.

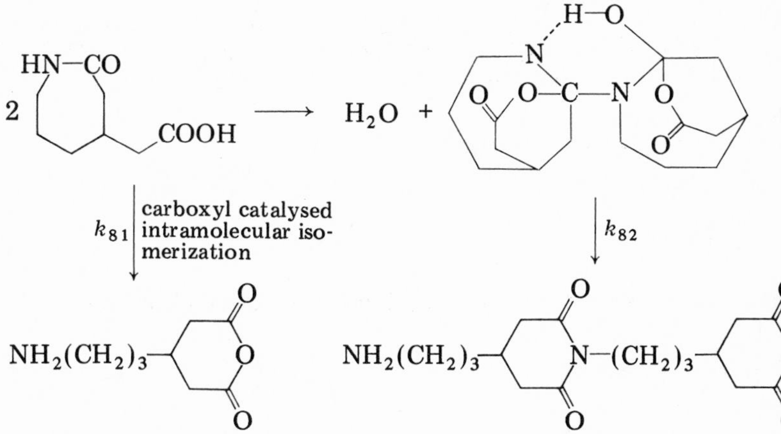

$$2 \quad \longrightarrow \quad H_2O + \quad \quad (152)$$

k_{81} carboxyl catalysed intramolecular isomerization

k_{82}

$$NH_2(CH_2)_3- \qquad NH_2(CH_2)_3-\!\!-N-(CH_2)_3-$$

(XXXIV)   (XXXV)

which yields a polyimide in the subsequent polycondensation involving the amino group and the anhydride moiety [247], viz.

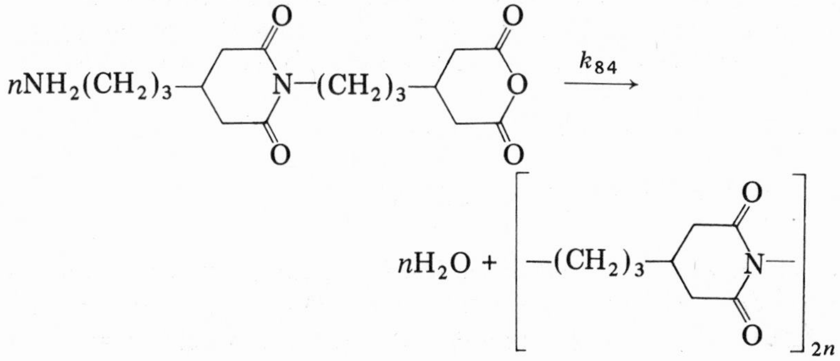

$$nNH_2(CH_2)_3-\!\!-N-(CH_2)_3- \quad \xrightarrow{k_{84}} \quad nH_2O + \left[-(CH_2)_3-\!\!-N- \right]_{2n} \quad (153)$$

References pp. 465—471

The polymerization of the carboxy-lactam (XXXIII) proceeded with a distinct induction period [247] (t_i) which was taken into account in the evaluation of the conversion data. The monomer is consumed in the isomerization reaction (152) with the formation of the amino anhydride (S) the rate being

$$-\text{d}[L]/\text{d}t = \text{d}[S]/\text{d}t = k_{81}[L]^2 \text{ (or } k_{82}[L]^2) \tag{154}$$

The polycondensation reactions (151) and (153) resulting in a decrease of the concentration of the linear species (S) follow the rate equation

$$-\text{d}[S]/\text{d}t = k_{83}[S]^2 \text{ (or } k_{84}[S]^2) \tag{155}$$

The extent of reaction (i.e. conversion of lactam into imide linkages) is expressed by

$$[I] = [-\overset{|}{\text{CONCO}}-] = 1 - [L] - [S]$$

so that the rate of formation of isomer in the sequence of isomerization and polymerization is given by the equations

$$\text{d}[I]/\text{d}t = -k_{81}[L]^2 - k_{83}[S]^2 \tag{156}$$

and

$$\text{d}[I]/\text{d}t = -k_{82}[L]^2 - k_{84}[S]^2 \tag{157}$$

Integration between t_i and t (and $2t_i$ and t, respectively) leads to [247]

$$1/[L] = k_{81}(t - t_i) \tag{158}$$

and

$$1/[S] - 1/[S]_{max} = k_{83}(t - 2t_i) \tag{159}$$

where $[S]_{max}$ is the maximum concentration of chains found at $2t_i$.

In all equations, the concentrations are expressed in moles per mole of monomer. From plots of $1/[L]$ and $1/[S]$ versus time, the following rate coefficients were estimated for 4-carboxymethyl caprolactam [247]:

$$k_{81}(\text{or } k_{82}) = 7.5 \times 10^8 \exp(-23,800/RT)$$

$$k_{83}(\text{or } k_{84}) = 2.4 \times 10^9 \exp(-23,800/RT)$$

A similar isomerization preceding or following the polymerization has been observed with methylene-bis-caprolactam. Due to the preferential formation of stable five- and six-membered rings, the polymerization of

α,ϵ'-methylene-bis-caprolactam results in bifunctional structural units [56], viz.

$$\tag{160}$$

rather than in branched tetrafunctional ones.

In all cases described the driving force of the rearrangement is the formation of stable five- or six-membered rings.

REFERENCES

1 P. H. Hermans, D. Heikens and P. F. van Velden, *J. Polym. Sci.*, *30* (1958) 81.
2 H. K. Reimschuessel, in K. C. Frisch and S. L. Reegen (Eds.), *Ring-opening Polymerization*, Marcel Dekker, New York, 1969, pp. 303—326.
3 A. B. Meggy, *J. Chem. Soc.*, (1953) 796.
4 F. Wiloth, *Makromol. Chem.*, *14* (1954) 156.
5 P. H. Hermans, *J. Appl. Chem.*, *5* (1955) 493.
6 A. V. Tobolsky, *J. Polym. Sci.*, *25* (1957) 220.
7 A. V. Tobolsky, *J. Polym. Sci.*, *31* (1958) 126.
8 A. V. Tobolsky and A. Eisenberg, *J. Amer. Chem. Soc.*, *81* (1959) 2302.
9 A. V. Tobolsky and A. Eisenberg, *J. Amer. Chem. Soc.*, *82* (1960) 289.
10 H.-G. Elias and A. Fritz, *Makromol. Chem.*, *114* (1968) 31.
11 A. Fritz, Dissertation No. 4046, Eidgenössische Technische Hochschule, Zürich, (1967), pp. 78—86.
12 F. Wiloth, *Makromol. Chem.*, *15* (1955) 98.
13 Ch. A. Kruissink, Symposium on Macromolecules, Wiesbaden 1959, Short Communications, Sec. III/C 2.
14 C. Giori and B. T. Hayes, *J. Polym. Sci.*, *Part A1*, *8* (1970) 351.
15 F. Wiloth, *Z. Phys. Chem. (Frankfurt am Main)*, *5* (1955) 66.
16 S. Mori, M. Furusawa and T. Takeuchi, *Anal. Chem.*, *42* (1970) 661.
17 H. Zahn and H. Spoor, *Z. Anal. Chem.*, *168* (1959) 190.
18 M. Rothe, *J. Polym. Sci.*, *30* (1958) 227.
19 G. Reinisch, K.-D. Schwenke and G. Rafler, *Faserforsch. Textiltech.*, *16* (1965) 425.
20 S. Mori and T. Takeuchi, *J. Chromatogr.*, *49* (1970) 230.
21 T. M. Cawthon and E. C. Smith, *Polym. Prepr.*, *Amer. Chem. Soc., Div. Polym. Chem.*, *1* (1960) 98.
22 P. H. Hermans, D. Heikens and P. F. van Velden, *J. Polym. Sci.*, *16* (1955) 451.
23 D. Heikens and P. H. Hermans, *Makromol. Chem.*, *28* (1958) 246.
24 W. Griehl and H. Lückert, *J. Polym. Sci.*, *30* (1958) 399.
25 R. C. P. Cubbon, *Polymer*, *4* (1963) 545.
26 N. Ogata, *J. Polym. Sci.*, *Part A1* (1963) 3151.
27 A. K. Bonetskaya, S. M. Skuratov, N. A. Lukina, A. A. Strel'tsova, K. E. Kuznetsova and M. P. Lazareva, *Vysokomol. Soedin. Ser. B*, *10* (1968) 75.
28 P. Čefelín, D. Doskočilová, A. Frydrychová and J. Šebenda, *Collect. Czech. Chem. Commun.*, *29* (1964) 485.

466

29 P. Čefelín, A. Frydrychová, P. Schmidt and J. Šebenda, Collect. Czech. Chem. Commun., 32 (1967) 1006.
30 P. Čefelín, J. Labský and J. Šebenda, Collect. Czech. Chem. Commun., 33 (1968) 1111.
31 K. J. Ivin and J. Leonard, Polymer, 6 (1965) 621.
32 F. S. Dainton, K. J. Ivin and D. A. G. Walmsley, Trans. Faraday Soc., 56 (1960) 1784.
33 A. A. Strepikheev, S. M. Skuratov, O. N. Katchinskaya, R. S. Muromova, E. P. Brykina and S. M. Shtekher, Dokl. Akad. Nauk SSSR, 102 (1955) 105.
34 S. M. Skuratov, A. A. Strepikheev and E. M. Kanarskaya, Kolloid. Zh., 14 (1952) 185.
35 S. M. Skuratov, V. V. Voevodskii, A. A. Strepikheev, E. N. Kanarskaya, R. S. Muromova and N. V. Fok, Dokl. Akad. Nauk SSSR, 95 (1954) 591.
36 J. Králíček and J. Šebenda, Chem. Prum., 13 (1963) 545.
37 O. Wichterle, J. Tomka and J. Šebenda, Collect. Czech. Chem. Commun., 29 (1964) 610.
38 O. Riedel and P. Wittmer, Makromol. Chem., 97 (1966) 1.
39 V. P. Kolesov, I. E. Paukov and S. M. Skuratov, Zh. Fiz. Khim., 36 (1962) 770.
40 H. Yumoto, K. Ida and N. Ogata, Bull. Chem. Soc. Jap., 31 (1958) 249.
41 H. Sawada, J. Macromol. Sci. Rev. Macromol. Chem., 5 (1) (1970) 151.
42 K. Dachs and E. Schwartz, Angew. Chem., 74 (1962) 541.
43 A. K. Bonetskaya and S. M. Skuratov, Vysokomol. Soedin., Ser. A, 11 (1969) 532.
44 H. K. Hall, Jr., J. Amer. Chem. Soc., 80 (1958) 6404.
45 Z. Bukač, P. Čefelín, D. Doskočilová and J. Šebenda, Collect. Czech. Chem. Commun., 29 (1964) 2615.
46 R. A. Russel and H. W. Thompson, Spectrochim. Acta, 8 (1956) 138.
47 R. Huisgen, H. Brade, H. Waltz and I. Glogger, Chem. Ber., 90 (1957) 1437.
48 H. S. Gutowsky and C. H. Holm, J. Chem. Phys., 25 (1956) 1228.
49 H. K. Hall, Jr., J. Amer. Chem. Soc., 82 (1960) 1209.
50 H. Pracejus, Chem. Ber., 92 (1959) 988.
51 H. E. Hallam and C. M. Jones, J. Mol. Struct., 1 (1967—68) 413.
52 H. C. Brown, J. Chem. Soc., (1956) 1248.
53 R. S. Muromova, A. A. Strepikheev and Z. A. Rogovin, Vysokomol. Soedin., 5 (1963) 1096.
54 C. G. Overberger and G. M. Parker, J. Polym. Sci., Part C, 22 (1968) 387.
55 P. Čefelín, A. Frydrychová, J. Labský, P. Schmidt and J. Šebenda, Collect. Czech. Chem. Commun., 32 (1967) 2787.
56 J. Králíček, J. Kondelíková and V. Kubánek, Internat. Conference on Chemical Transformations of Polymers IUPAC, Bratislava 1971, preprint P-70.
57 H. Hopff, H. Bracher and H.-G. Elias, Makromol. Chem., 91 (1966) 121.
58 H. Yumoto, J. Chem. Phys., 29 (1958) 1234.
59 R. F. Brown and N. M. van Gulick, J. Org. Chem., 21 (1956) 1064.
60 Ya. L. Gol'dfarb and L. I. Belen'kii, Usp. Khim., 29 (1960) 470.
61 R. C. P. Cubbon, Makromol. Chem., 80 (1964) 44.
62 C. G. Overberger and T. Takekoshi, Macromolecules, 1 (1968) 7.
63 O. B. Salamatina, A. K. Bonetskaya, S. M. Skuratov, B. P. Fabrichnyi, I. F. Shalavina and Ya. L. Gol'dfarb, Vysokomol. Soedin., Ser. B, 10 (1968) 10.
64 A. K. Bonetskaya, T. V. Sopova, O. B. Salamatina, S. M. Skuratov, B. P. Fabrichnyi, I. F. Shalavina and Ya. L. Gol'dfarb, Vysokomol. Soedin., Ser. B, 11 (1969) 894.
65 M. P. Kozina and S. M. Skuratov, Dokl. Akad. Nauk SSSR, (1959) 127.
66 O. B. Salamatina, A. K. Bonetskaya, S. M. Skuratov, B. P. Fabrichnyi, I. F. Shalavina and Ya. L. Gol'dfarb, Vysokomol. Soedin., 7 (1965) 485.

67 J. Králíček, J. Kondelíková and V. Kubánek, *Collect. Czech. Chem. Commun.*, *37* (1972) 1130.
68 A. V. Volokhina, B. P. Fabrichnyi, I. F. Shalavina and Ya. L. Gol'dfarb, *Vysokomol. Soedin.*, *4* (1962) 1829.
69 H. K. Hall, Jr., *J. Amer. Chem. Soc.*, *80* (1958) 6412.
70 H. K. Hall, Jr., *Polym. Prepr. 6* (2) (1965) 535.
71 A. V. Volokhina, G. I. Kudryavtsev and S. S. Tuzhikova, *Vysokomol. Soedin.*, *9* (1967) 683.
72 H. R. Allcock, *Heteroatom Ring Systems*, Academic Press, New York, 1967, p. 34.
73 N. Ogata, T. Asahara and S. Tohyama, *J. Polym. Sci., Part A1, 4* (1966) 1359.
74 N. Ogata, K. Tanaka and M. Inagaki, *Makromol. Chem.*, *113* (1968) 95.
75 N. Ogata, K. Tanaka and K. Takayama, *Makromol. Chem.*, *119* (1968) 161.
76 W. Ziegenbein, A. Schäffler and R. Kaufhold, *Chem. Ber.*, *88* (1955) 1906.
77 U. Haberthür and H.-G. Elias, *Makromol. Chem.*, *144* (1971) 183.
78 U. Haberthür and H.-G. Elias, *Makromol. Chem.*, *144* (1971) 193.
79 A. V. Volokhina, G. I. Kudryavtsev and M. V. Shablygin, *Chem. Prum.*, *17* (1967) 594.
80 U. Haberthür and H.-G. Elias, *Makromol. Chem.*, *144* (1971) 213.
81 J. A. Semlyen and G. R. Walker, *Polymer, 10* (1969) 597.
82 F. Korte and W. Glet, *J. Polym. Sci., Part B4* (1966) 685.
83 E. Šittler and J. Šebenda, *Collect. Czech. Chem. Commun.*, *33* (1968) 3182.
84 J. Stehlíček, J. Labský and J. Šebenda, *Collect. Czech. Chem. Commun.*, *32* (1967) 545.
85 S. Murahashi, H. Yuki and H. Sekiguchi, *Sen'i-Kagaku Kenkyûsho Nempô* (*Osalia*), *10* (1956) 88, 93.
86 W. O. Ney and M. Crowther, U.S. Pat. 2,739,959 (1956).
87 R. E. Noble, *Diss. Abstr., 17* (1957) 2823.
88 W. R. Nummy, C. E. Barnes and W. O. Ney, Abstracts, 133rd Meeting Amer. Chem. Soc., San Francisco, 1958, p. 22R.
89 J. Šebenda and J. Králíček, *Collect. Czech. Chem. Commun.*, *23* (1958) 766.
90 J. Šebenda and J. Králíček, IUPAC Symp. on Macromolecules, 1959, Wiesbaden, Short Commun. Section III/C2.
91 E. H. Mottus, R. M. Hedrick and J. M. Buttler, U.S. Pat. 3,017,391 (1962).
92 K. Gehrke, *Faserforsch. Textiltech.*, *13* (1962) 95.
93 G. Reinisch, *Faserforsch. Textiltech.*, *15* (1964) 472.
94 J. Šebenda, *J. Polym. Sci., Part C23* (1968) 169.
95 J. Šebenda, in O. Vogl and J. Furukawa (Eds.), *Polymerization of Heterocyclics*, Marcel Dekker, New York, 1973, p. 153.
96 R. M. Joyce and D. M. Ritter, U.S. Pat., 2,251,519 (1941).
97 H. Yumoto and N. Ogata, *Bull. Chem. Soc. Jap., 31* (1958) 907.
98 N. Yoda and A. Miyake, *J. Polym. Sci., 43* (1960) 117.
99 A. Ciaperoni, L. Mariani and G. B. Gechele, *Chim. Ind.* (*Milan*), *50* (1968) 772.
100 P. Čefelín and J. Šebenda, *Collect. Czech. Chem. Commun.*, *26* (1961) 3028.
101 H. Tani and T. Konomi, *J. Polym. Sci., Part A1, 4* (1966) 301.
102 S. Schaaf, *Faserforsch. Textiltech.*, *10* (1959) 257.
103 O. Wichterle and V. Gregor, *J. Polym. Sci., 34* (1959) 309.
104 J. Saunders, *J. Polym. Sci., 30* (1958) 479.
105 O. Wichterle, J. Králíček and J. Šebenda, *Collect. Czech. Chem. Commun.*, *24* (1959) 755.
106 H. Gilch and D. Michael, *Makromol. Chem.*, *99* (1966) 103.
107 J. Stehlíček, J. Šebenda and O. Wichterle, *Collect. Czech. Chem. Commun.*, *29* (1964) 1236.
108 R. P. Scelia, S. E. Schonfeld and L. G. Donaruma, *J. Appl. Polym. Sci., 8* (1964) 1363.

468

109 R. P. Scelia, S. E. Schonfeld and L. G. Donaruma, *J. Appl. Polym. Sci.*, *11* (1967) 1299.
110 W. Griehl and S. Schaaf, *Makromol. Chem.*, *32* (1959) 170.
111 J. Stehlíček, K. Gehrke and J. Šebenda, *Collect. Czech. Chem. Commun.*, *32* (1967) 370.
112 G. Falkenstein and H. Dörfel, *Makromol. Chem.*, *127* (1969) 34.
113 T. Yasumoto, *J. Polym. Sci., Part A3* (1965) 3301.
114 T. Yasumoto, *J. Polym. Sci., Part A3* (1965) 3877.
115 I. Tanaka, A. Suzuki and T. Yoshida, *Int. Symp. Macromol. Chem. Tokyo-Kyoto, 1966*, Prepr. of Scientific Papers I, p. 255.
116 A. Mattiussi and G. B. Gechele, *Eur. Polym. J.*, *4* (1968) 695.
117 S. Barzakay, M. Levy and D. Vofsi, *J. Polym. Sci., Part A1*, *4* (1966) 2211.
118 E. Bäder and H. Amann, *Makromol. Chem.*, *124* (1969) 10.
119 Z. Bukač, J. Tomka and J. Šebenda, *Collect. Czech. Chem. Commun.*, *34* (1969) 2057.
120 Z. Bukač and J. Šebenda, *Collect. Czech. Chem. Commun.*, *36* (1971) 1995.
121 O. Wichterle, *Makromol. Chem.*, *35* (1960) 174.
122 O. Wichterle, E. Šittler and P. Čefelín, *J. Polym. Sci.*, *53* (1961) 249.
123 J. Šebenda and B. Mikuľová, *Collect. Czech. Chem. Commun.*, *29* (1964) 738.
124 G. Nawrath, *Int. Symp. on Macromol. Chem.*, Prague 1965, Prepr. of Scientific Communications, p. 335.
125 J. Šebenda, *Collect. Czech. Chem. Commun.*, *31* (1966) 1501.
126 J. Šebenda and B. Masař, *Collect. Czech. Chem. Commun.*, *31* (1966) 3331.
127 J. Šebenda, B. Masař and Z. Bukač, *J. Polym. Sci., Part C16* (1967) 339.
128 Z. Bukač and J. Šebenda, *Collect. Czech. Chem. Commun.*, *32* (1967) 3537.
129 G. B. Gechele, A. Ciaperoni, A. Mattiussi and G. Stea, *Chimica Ind. (Milan)*, *8* (1968) 904.
130 B. Masař and J. Šebenda, *Collect. Czech. Chem. Commun.*, *34* (1969) 2598.
131 J. Stehlíček and J. Šebenda, *Collect. Czech. Chem. Commun.*, *35* (1970) 1188.
132 T. Konomi and H. Tani, *J. Polym. Sci., Part A1*, *8* (1970) 1261.
133 S. Schaaf, *Faserforsch. Textiltech.*, *10* (1959) 224.
134 J. Šebenda and J. Kouřil, *Eur. Polym. J.*, *7* (1971) 1637.
135 J. Šebenda and J. Kouřil, *Eur. Polym. J.*, *8* (1972) 437.
136 Z. Bukač and J. Šebenda, *Collect. Czech. Chem. Commun.*, *38* (1973) 3610.
137 Z. Bukač and J. Šebenda, *J. Polym. Sci. Part C,42* (1973) 345.
138 J. Stehlíček, P. Čefelín and J. Šebenda, *Collect. Czech. Chem. Commun.*, *37* (1972) 1926.
139 P. Čefelín, J. Stehlíček and J. Šebenda, *J. Polym. Sci., Part C*, *42* (1973) 79.
140 J. Šebenda and J. Králíček, *Collect. Czech. Chem. Commun.*, *29* (1964) 1017.
141 J. Stehlíček, P. Čefelín and J. Šebenda, *J. Polym. Sci., Part C*, *42* (1973) 89.
142 F. Wiloth and E. Schindler, *Chem. Ber.*, *100* (1967) 2373.
143 F. Wiloth and E. Schindler, *Chem. Ber.*, *103* (1970) 757.
144 D. Heikens, *Makromol. Chem.*, *18/19* (1956) 62.
145 O. Wichterle and J. Šebenda, *Chem. Listy*, *49* (1955) 1293.
146 H. Yumoto and N. Ogata, *Bull. Chem. Soc. Jap.*, *31* (1958) 913.
147 S. Chrzczonowicz, B. Ostaszewski and W. Reimschüssel, *Bull. Acad. Pol. Sci., Ser. Sci., Chim.*, *12* (1964) 691.
148 P. Čefelín, J. Stehlíček and J. Šebenda, *Collect. Czech. Chem. Commun.*, *37* (1972) 3861.
149 H. Sekuguchi, *Bull. Soc. Chim. Fr.*, (1960) 1831.
150 J. Šebenda and J. Stehlíček, *Collect. Czech. Chem. Commun.*, *28* (1963) 2731.
151 G. Stea and G. B. Gechele, *Eur. Polym. J.*, *1* (1965) 213.
152 C. V. Goebel, P. Čefelín, J. Stehlíček and J. Šebenda, *J. Polym. Sci., Part A1*, *10* (1972) 1411.
153 S. Schaaf, *Faserforsch. Textiltech.*, *10* (1959) 328.
154 W. Griehl, *Faserforsch. Textiltech.*, *7* (1956) 207.

155 J. Králíček and J. Šebenda, *J. Polym. Sci.*, *30* (1958) 493.
156 W. E. Hanford and R. M. Joyce, *J. Polym. Sci.*, *3* (1948) 167.
157 G. Champetier and H. Sekiguchi, *J. Polym. Sci.*, *48* (1960) 309.
158 J. Králíček and P. Starý, unpublished results.
159 G. B. Gechele and G. Stea, *Eur. Polym. J.*, *1* (1965) 91.
160 G. B. Gechele and G. F. Martinis, *J. Appl. Polym. Sci.*, *9* (1965) 2939.
161 J. Králíček, L. Červený and L. Zimmer, *Scientific Papers of the Inst. of Chem. Technology, Pardubice*, *C9* (1966) 5.
162 E. H. Mottus, R. M. Hedrick and J. M. Butler, *Polym. Prepr., Amer. Chem. Soc., Div. Polym. Chem.*, *9* (1) (1968) 390.
163 M. Kapuscinska, E. Šittler and J. Šebenda, *Collect. Czech. Chem. Commun.*, *38* (1973) 2281.
164 E. Šittler and J. Šebenda, *J. Polym. Sci., Part C16* (1967) 67.
165 R. Puffr and J. Šebenda, *Eur. Polym. J.*, *8* (1972) 1037.
166 S. Barzakay, M. Levy and D. Vofsi, *J. Polym. Sci., Part B3* (1965) 601.
167 S. Barzakay, M. Levy and D. Vofsi, *J. Polym. Sci., Part A1*, *5* (1967) 965.
168 G. Champetier and H. Sekiguchi, *C. R. H. Acad. Sci.*, *249* (1959) 108.
169 E. Šittler and J. Šebenda, *Collect. Czech. Chem. Commun.*, *33* (1968) 270.
170 G. Stea, G. B. Gechele and C. Arena, *J. Appl. Polym. Sci.*, *12* (1968) 2691.
171 R. Z. Greenley, J. C. Stauffer and J. E. Kurz, *Macromolecules*, *2* (1969) 561.
172 T. Komoto, M. Iguchi, H. Kanetsuna and T. Kawai, *Makromol. Chem.*, *135* (1970) 145.
173 F. Kobayashi and K. Matsuya, *J. Polym. Sci., Part A1* (1963) 111.
174 Yu. L. Pankratov and G. I. Kudryavtsev, *Vysokomolek. Soedin.*, *6* (1964) 1862.
175 P. Wittmer and H. Gerrens, *Makromol. Chem.*, *89* (1965) 27.
176 G. M. van der Want and Ch. A. Kruissink, *J. Polym. Sci.*, *35* (1959) 119.
177 F. Wiloth, *Makromol. Chem.*, *27* (1958) 37.
178 M. Rothe, G. Reinisch and W. Jäger, *Faserforsch. Textiltech.*, *12* (1961) 448.
179 G. Reinisch and W. Jäger, *Faserforsch. Textiltech.*, *13* (1962) 79.
180 M. Rothe, G. Reinisch, W. Jaeger and I. Schopov, *Makromol. Chem.*, *54* (1962) 183.
181 M. Rothe, H. Boenisch and W. Kern, *Makromol. Chem.*, *67* (1963) 90.
182 S. Doubravszky and F. Geleji, *Int. Symp. on Macromol. Chem., Prague 1965*, Prepr. p. 364.
183 S. Doubravszky and F. Geleji, *Makromol. Chem.*, *105* (1967) 261.
184 S. Doubravszky and F. Geleji, *Makromol. Chem.*, *110* (1967) 246.
185 S. Doubravszky and F. Geleji, *Makromol. Chem.*, *113* (1968) 270.
186 S. Doubravszky and F. Geleji, *Makromol. Chem.*, *143* (1971) 259.
187 M. Rothe and J. Mazánek, *Makromol. Chem.*, *145* (1971) 197.
188 G. Stea, G. B. Gechele, L. Mariani and F. Manescalchi, *J. Appl. Polym. Sci.*, *12* (1968) 2697.
189 G. Reinisch and W. Jaeger, *Faserforsch. Textiltech.*, *16* (1965) 583.
190 M. Rothe, H. Boenisch and D. Essig, *Makromol. Chem.*, *91* (1966) 24.
191 F. Geleji and S. Doubravszky, *Faserforsch. Textiltech.* *14* (1963) 298.
192 S. Doubravszky and F. Geleji, *Makromol. Chem.*, *110* (1967) 257.
193 S. Doubravszky and F. Geleji, *Makromol. Chem.*, *111* (1968) 259.
194 T. G. Majury, *J. Polym. Sci.*, *31* (1958) 383.
195 J. N. Hay, *J. Polym. Sci., Part B5* (1965) 577.
196 G. M. Burnett, A. J. MacArthur and J. N. Hay, *Eur. Polym. J.*, *3* (1967) 321.
197 K. G. Wyness, *Makromol. Chem.*, *38* (1960) 189.
198 Wang Pao-Jen, Pi Hsien-Tung and Wang Yu-Huai, *Acta Chimica Sinica*, *22* (1956) 225.
199 D. Davidson and P. Newman, *J. Amer. Chem. Soc.*, *74* (1952) 1515.
200 R. Puffr and J. Šebenda, *J. Polym. Sci., Part C*, *42* (1973) 21.
201 K. Brunner, W. Seeger and St. Dittrich, *Monatsh. Chem.*, *45* (1925) 69.

470

202 P. Schlack, *Z. Gesamte. Textilind.*, *65* (1963) 1052.
203 P. Schlack, *Melliand Textilber.*, *12* (1964) 1406.
204 P. Schlack, *Abh. Deut. Akad. Wiss. Berlin, Kl. Chem.*, *Geol. Biol. 3* (1965) 9.
205 P. Schlack, *Chemiefasern, 15* (1965) 64.
206 G. Falkenstein, Thesis, Technische Hochschule, Stuttgart, 1965.
207 P. Schlack and J. Rieker, *Angew. Makromol. Chem.*, *15* (1970) 203.
208 Z. Csürös, I. Rusznák, G. Bertalan, L. Trézl and J. Körösi, *Makromol. Chem.*, *137* (1970) 9.
209 Z. Csürös, I. Rusznák, G. Bertalan and J. Körösi, *Makromol. Chem.*, *137* (1970) 17.
210 M. Rothe, *Angew. Chem.*, *80* (1968) 245.
211 Z. Csürös, G. Bertalan, J. Nagy, L. Trézl and J. Körösi, *J. Polymer Sci.*, *Part C16* (1968) 3175.
212 G. G. Gabra, M. H. Nosseir and N. N. Messiha, *J. Chem. U.A.R.*, *12* (1970) 569.
213 G. B. Gechele, G. Stea and F. Manescalchi, *Eur. Polym. J.*, *4* (1968) 505.
214 D. Heikens, P. H. Hermans and H. A. Veldhofen, *Makromol. Chem.*, *30* (1959) 154.
215 D. Heikens and P. H. Hermans, *J. Polym. Sci.*, *44* (1960) 429.
216 D. Heikens, P. H. Hermans and G. M. van der Want, *J. Polym. Sci.*, *44* (1960) 437.
217 P. H. Hermans, D. Heikens and S. Smith, *J. Polym. Sci.*, *38* (1959) 265.
218 P. H. Hermans, *Chem. Weekbl.*, *56* (1960) 122.
219 Ch. A. Kruissink, G. M. van der Want and A. J. Staverman, *J. Polym. Sci.*, *30* (1958) 67.
220 Ch. A. Kruissink, *Chem. Weekbl.*, *56* (1960) 141.
221 H. K. Reimschuessel, *J. Polym. Sci.*, *41* (1959) 457.
222 H. K. Reimschuessel and G. J. Dege, *J. Polym. Sci.*, *A1*, *8* (1970) 3265.
223 H. K. Reimschuessel and K. Nagasubramanian, *Chem. Eng. Sci.*, *27* (1972) 1119.
224 H. K. Reimschuessel and G. J. Dege, *J. Polym. Sci.*, *Part A1*, *9* (1971) 2343.
225 F. Wiloth, *Makromol. Chem.*, *15* (1955) 106.
226 F. Wiloth, *Kolloid Z.*, *144* (1955) 58.
227 F. Wiloth, *Kolloid Z.*, *143* (1955) 129.
228 F. Wiloth, *Kolloid Z.*, *143* (1955) 138.
229 F. Wiloth, *Makromol. Chem.*, *21* (1956) 50.
230 F. Wiloth, *Kolloid Z.*, *151* (1957) 176.
231 F. Wiloth, *Z. Phys. Chem.*, *Neue Folge, 11* (1957) 78.
232 F. Wiloth, *Kolloid Z.*, *160* (1958) 48.
233 F. Wiloth, *Makromol. Chem.*, *30* (1959) 189.
234 F. Wiloth, *Makromol. Chem.*, *144* (1971) 329.
235 P. F. van Velden, G. M. van der Want, D. Heikens, Ch. A. Kruissink, P. H. Hermans and A. J. Staverman, *Rec. Trav. Chim. Pays-Bas*, *74* (1955) 1376.
236 V. V. Korshak, R. V. Kudryavtsev, V. A. Sergeev and L. B. Icikson, *Izv. Akad. Nauk SSSR, Otd. Khim. Nauk*, (1962) 1468.
237 C. Giori and B. T. Hayes, *J. Polym. Sci.*, *Part A1*, *8* (1970) 335.
238 O. Fukumoto, *J. Polym. Sci.*, *22* (1956) 263.
239 G. Reinisch and U. Gohlke, *Faserforsch. Textiltech.*, *24* (1973) 282.
240 T. G. Majury, *J. Polym. Sci.*, *24* (1957) 488.
241 T. Kagiya, H. Kishimoto, S. Narisawa and K. Fukui, *J. Polym. Sci.*, *Part A3* (1965) 145.
242 T. Haug, F. Lohse, K. Metzger and H. Batzer, *Helv. Chim. Acta*, *51* (1968) 2069.
243 J. Šebenda and B. Masař, *Collect. Czech. Chem. Commun.*, *40* (1975) 93.
244 N. Ogata, *Bull. Chem. Soc. Jap.*, *34* (1961) 248.
245 H. K. Reimschuessel, *J. Polym. Sci.*, *Part B4* (1966) 953.

246 H. K. Reimschuessel, L. G. Roldan and J. P. Sibilia, *J. Polym. Sci., Part A2, 6* (1968) 559.
247 H. K. Reimschuessel, *Advan. Chem. Ser., 91* (1969) 717.
248 H. K. Reimschuessel, K. P. Klein and G. J. Schmitt, *Macromolecules, 2* (1969) 567.
249 K. P. Klein and H. K. Reimschuessel, *J. Polym. Sci., Part A1, 9* (1971) 2717.
250 H. K. Reimschuessel and K. P. Klein, *J. Polym. Sci., Part A1, 9* (1971) 3071.
251 H. K. Reimschuessel, *Trans. N.Y. Acad. Sci., Ser. II, 33* (2) (1971) 219.
252 R. Puffr and J. Šebenda, *J. Polym. Sci., Part A1, 12* (1974) 21.

Chapter 7

The Kinetics of Polycondensation Reactions

J. H. SAUNDERS and F. DOBINSON

1. Introduction

With the continual discovery of new reactions leading to the formation of polymers, and with the increased understanding of the mechanisms of those reactions, it is natural that certain definitions of descriptive terms will change with time. The term "polycondensation" is used here to denote those polymerization reactions which proceed by a propagation mechanism in which an active polymerization site disappears every time one monomer equivalent reacts. This may be illustrated for one propagation step in polyesterification by

$$\sim\text{COOH} + \text{HO}\sim \quad \rightarrow \quad \sim\overset{\displaystyle O}{\overset{\|}{\text{C}}}\text{O}\sim + \text{H}_2\text{O}$$

While this particular reaction fulfils one of the early definition requirements of the term "condensation polymerization", i.e., that another molecule was eliminated as a by-product, polyurethane formation

$$\sim\text{NCO} + \text{HO}\sim \quad \rightarrow \quad \sim\text{NH}\overset{\displaystyle O}{\overset{\|}{\text{C}}}\text{O}\sim$$

fits our current definition, but not the earlier.

These reactions are in contrast to the propagation step of an "addition" polymerization reaction, when the reactive site is regenerated by the addition of one more monomer unit

$$\sim\underset{\displaystyle \text{C}_6\text{H}_5}{\text{CH}_2\text{CH}\cdot} + \text{CH}_2{=}\text{CHC}_6\text{H}_5 \quad \rightarrow \quad \sim\underset{\displaystyle \text{C}_6\text{H}_5}{\text{CH}_2\text{CH}}\underset{\displaystyle \text{C}_6\text{H}_5}{\text{CH}_2\text{CH}\cdot}$$

It is generally characteristic of the polycondensation reactions that the number of growing chains decreases steadily with time, that the molecular weight distribution initially includes only a range of low molecular weight

References pp. 575—581

products which steadily increase in molecular weight, and that free monomer is largely consumed early in the reaction sequence. The term polycondensation is thus used to describe those polymerization reactions which have also been designated "step-growth polymerization" by Lenz [1].

Some of the most familiar reactions falling into the polycondensation class are those leading to polyamides derived from dicarboxylic acids and diamines, polyesters from glycols and dicarboxylic acids, polyurethanes from polyols and polyisocyanates, and polyureas from diamines and diisocyanates. Similar polymer formations utilizing bifunctional acid chlorides with polyols or polyamines also fall into this class. The condensations of aldehydes or ketones with a variety of "active hydrogen" compounds such as phenols and diamines are in this group. Some of the less familar polycondensation reactions include the formation of polyethers from bifunctional halogen compounds and the sodium salts of bis-phenols, and the addition of bis-thiols to diolefins under certain conditions.

The practical significance of an understanding of the polycondensation reactions is immediately clear. The nylon (polyamide) fibre and moulding resin market is in the multi-billion pound range, world-wide, as is the polyester fibre market. The polyurethane market (foams, elastomers, coatings, adhesives, fibres and others) also is reaching the billion (10^9) pound range. Phenolic resins are among the oldest and best known of the large-scale synthetic polymers. In addition to the large volume products, many newer, more specialized products are obtained by polycondensation reactions: polycarbonates, polysulphones, polyimides and aromatic poly-amides.

Just as the products of polycondensation are greatly varied, so are the reaction conditions used in their production. Some are produced in the melt (many polyamides and polyesters), some initially in the melt but with extensive polymerization continuing in the solid state (polyurethane foams and elastomers), in solution (some polyurethane fibres) or in non-homogeneous liquid systems (some polycarbonates, very high melting polyamides).

This chapter will attempt to survey critically the major areas of experimentally determined kinetic data which are available on polycondensation reactions and their mechanisms, and to emphasize the mechanistic similarities of many of the reactions. The statistics of polycondensation reactions will be touched on only to the extent that it helps understand the reactions and their kinetics. The general approach to the subject of kinetics is designed to be of primary use and interest to those polymer chemists who are concerned with the synthesis of products having desired properties, and with the understanding of their synthetic processes.

2. Statistical treatment of polycondensation reactions

The rates of polycondensations in the melt, in solution and in emulsion are controlled primarily by the kinetics of the reactions between the functional groups involved. Interfacial and solid-state polycondensation are largely dependent on the rate of diffusion of reacting species to the site of the reaction. The following discussion of the statistical development of equations related to molecular weight, molecular weight distribution and gelation applies to those polycondensations which are limited by the kinetics of the chemical reactions, rather than by the rate of diffusion in the system. The equations are independent of the actual kinetics of the particular reaction, however. The development of these equations is due largely to Flory [2, 3], who has given an extensive treatment of the subject.

A basic assumption in this approach is that the rate of reaction of a functional group is independent of the size of the molecule to which it is attached. This may not seem reasonable at first glance, since the frequency of collisions between molecules at a given temperature is normally inversely proportional to the square root of the mass. In the case of polymerization, however, it is the frequency of collision of reactive end groups that is important. While increasing molecular weight may reduce the ease with which reactive ends come together, the large molecules also serve as a "cage", reducing the ease with which reactive ends diffuse away from each other. Except at very high molecular weights in the melt polycondensations, the two effects seem essentially to cancel, so that the assumption of equal reactivity agrees reasonably well with experimental results. Rate equations must be expressed on the basis of the concentration of reactive groups, of course, thus compensating for the diluting effect of the large polymer chains.

In the statistical analysis of polycondensation the extent of reaction, p, is defined as the probability that a functional group has reacted at time t. (In kinetic treatments p is often used as the equivalent expression, the fraction of functional groups which has reacted.) It follows that $(1-p)$ is the probability of finding a functional group unreacted.

A molecule containing x repeating units may be designated an x-mer. Such a molecule contains $(x-1)$ reacted functional groups of a particular type, e.g. carboxyl groups in the esterification of a dibasic acid with a glycol. In addition, it will have one of the functional groups in question (carboxyl) on an end. The probability of finding a reacted carboxyl group in the molecule is p, and that of finding $(x-1)$ of them in the same molecule is p^{x-1}. The probability of finding the complete molecule, then, is $p^{x-1}(1-p)$, where $(1-p)$ is the probability of finding the unreacted end group. The fraction of all the molecules which are x-mers is the same as the probability of finding the complete molecule.

If the total number of units (in this case, carboxyls) is N_0, then the number of polymer molecules (N) is

$$N = N_0(1-p)$$

The number of x-mers (N_x) is

$$N_x = Np^{x-1}(1-p)$$
$$= N_0(1-p)^2 p^{x-1}$$

This last equation is the statistical number-distribution function for a linear polycondensation reaction at the extent of reaction p.

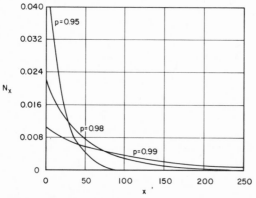

Fig. 1. Number or mole-fraction distribution in a linear condensation polymer for several extents of reaction p (ref. 2).

In a similar way one can develop the expression for the weight fraction (W_x) of x-mers, viz.

$$W_x = xN_x/N_0$$
$$= x(1-p)^2 p^{x-1}$$

This is the statistical weight-distribution function for a linear poly-condensation reaction at the extent of reaction p. The number-distribution and weight-distribution functions are illustrated in Figs. 1 and 2 for values of p.

The equations for the weight- and number-average distribution functions can be used to calculate the different average degrees of polymerization. The number-average degree of polymerization ($\bar{x}_n$) is

$$\bar{x}_n = N_0/N = 1/(1-p)$$

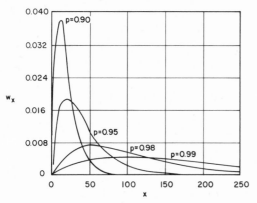

Fig. 2. Weight-fraction distribution in a linear condensation polymer for several extents of reaction p (ref. 2).

while the weight-average degree of polymerization ($\bar{x}_w$) is

$$\bar{x}_w = (1 + p)/(1 - p)$$

Because of the random interaction of all polymer molecules that generally occurs in many polycondensation reactions, most polymer samples contain a broad distribution of molecular weight species. The ratio of weight-average degree of polymerization to number-average degree of polymerization, $\bar{x}_w/\bar{x}_n$, is a measure of the heterogeneity of the polydisperse system. The distribution of molecular weights in poly-condensations, where all functional groups are assumed to have equal probability of reaction, is termed the most probable distribution. In this case, $\bar{x}_w/\bar{x}_n = 2$ at high degrees of polymerization.

High molecular weights are obtained when the reactants are present in exactly equivalent amounts, assuming no side reactions. The effect of monomer ratio on the molecular weight may be calculated as follows. We assume monomers A and B are present with B in excess, and N_A and N_B represent the number of moles of A and B, respectively. If $N_A/N_B = r$, the total number of repeating units AB in the system would be

$$N_{AB} = \tfrac{1}{2}(N_A + N_B) = \tfrac{1}{2}N_A(1 + 1/r)$$

At extent of reaction p (for A groups, rp for B groups), the total number of end groups in the system is

$$N_A(1 - p) + N_B(1 - rp) = N_A[1 - p + (1 - rp)/r]$$

Since each polymer molecule has two end groups, the number of molecules will be one-half this value.

References pp. 575—581

The degree of polymerization ($\bar{x}_n$) is the ratio of the number of reacted molecules to the number of polymer molecules formed

$$\bar{x}_n = \frac{\frac{1}{2}N_A(1 + 1/r)}{\frac{1}{2}N_A[1 - p + (1 - rp)/r]} = \frac{1 + r}{1 + r - 2rp}$$

As p approaches one, this simplifies to

$$\bar{x}_n = (1 + r)/(1 - r)$$

One mole % excess of B will limit the value for $\bar{x}_n$ thus:

$$\bar{x}_n = [1 + (100/101)]/[1 - (100/101)] = 201$$

One means of controlling the molecular weight obviously is to have one monomer in slight excess ($r \neq 1$). A preferred method is usually the addition of a desired amount of a monofunctional reactant. The effect of a monofunctional reactant B' on the degree of polymerization is indicated by the equation

$$\bar{x}_n = \frac{2N_B}{N_{B'}} + 1$$

or if $r' = N_{B'}/N_B$,

$$\bar{x}_n = \frac{2}{r'}$$

The molecular weight is thus inversely proportional to the concentration of monofunctional reactant.

When polyfunctional reactants (e.g., glycerol) are included, the formation of a crosslinked polymer, causing gelation, may be expected. An important feature of such a system is that at the onset of gelation the number-average molecular weight is relatively low, while its weight-average molecular weight becomes infinite. The statistical treatments of such systems are somewhat in error in assuming equal reactivity of all functional groups (they are not equal in glycerol) and that all reactions occur between functional groups on different molecules (intramolecular reactions can occur in some cases, and cyclization reactions). Nevertheless, the theory serves well, and experimental results are usually moderately close to theoretical predictions.

In calculating the gel point, or the degree of reaction at which gelation occurs, one utilizes a term "branching coefficient", designated α. This is the probability that a given functional group on a branch unit (a reactant molecule with functionality greater than 2) is connected to another branch unit. One may deduce the value of α at which gelation occurs in the following way. Assume bifunctional A—A and B—B units are present, along with polyfunctional A_f units with functionality f. The resulting polymer may be expected to have segments

$$A_{f-1} - A[B-BA-A]_i B-BA-A_{f-1}$$

Other segments may also be present with one or no branch units on their ends. Gel formation will occur when at least one of the $(f-1)$ segments radiating from the end of the segment of the type shown is in turn connected to another branch unit. The probability that this will occur is $1/(f-1)$. Thus the critical value of α for gelation is

$$\alpha_c = 1/(f-1)$$

Here f is the functionality of the branch units, or the average functionality of all branch units if more than one type is present.

One may relate α to the extent of reaction. If the extents of reaction for A and B groups are p_A and p_B, and the ratio of A groups on branch units to all A groups is y, the probability that a B group has reacted with a branch unit is $p_B y$; with a bifunctional A, $p_B(1-y)$. The probability that a segment of the type shown is formed is

$$p_A[p_B(1-y)p_A]^i p_B y$$

The summation for all values of i gives

$$\alpha = \frac{p_A p_B y}{1 - p_A p_B (1-y)}$$

By utilizing the previous term $r = N_A/N_B$, one can eliminate either p_A or p_B since $p_B = rp_A$. One can then obtain

$$\alpha = \frac{rp_A^2 y}{1 - rp_A^2(1-y)} = \frac{p_B^2 y}{r - p_B^2(1-y)}$$

In the special case of A = B, $r = 1$ and $p_A = p_B = p$

$$\alpha = \frac{p^2 y}{1 - p^2(1-y)}$$

When there are no A—A units, $y = 1$ and

References pp. 575—581

$$\alpha = rp_A^2 = \frac{p_B^2}{r}$$

When both special conditions apply, $\alpha = p^2$. When only branch units are present, $\alpha = p$.

The agreement of these equations with experimental results is generally fairly good. Thus, in a study of the reaction of glycerol with dibasic acids, gelation occurred at $p = 0.765$ [4]. Here one would calculate $f = 3$, $\alpha_c = p^2 = 0.50$. The experimental value for $\alpha_c = p^2$ was 0.58. Experimental values for α_c are frequently somewhat higher than the calculated values.

The calculation of molecular weights and molecular weight distributions for three-dimensional polymers is considerably more difficult and complicated than that for linear polymers. Flory's calculation of the

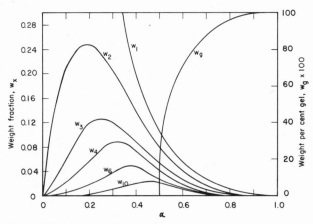

Fig. 3. Weight fractions of various molecular species in a trifunctional polycondensation as a function of $\alpha = p$ (ref. 2).

weight fractions for various species is shown in Fig. 3 for the special case of $\alpha = p$ (only branch units present). It is noteworthy here that the weight fraction of the monomer (w_1) is always greater than that of any other species which has not gelled. The extent of reaction for which the weight fraction of any species reaches its maximum never occurs beyond the gel point ($\alpha = 0.5$). Details of these and other calculations for three-dimensional systems may be found in Flory's publications [2, 3].

3. Physical classification of reaction systems

Polycondensation reactions are conducted under a variety of reaction conditions. The initial features of kinetic control may vary considerably according to the system used. For example, some are limited primarily by the kinetics of the chemical reactions, whereas others have as the limiting rate processes the diffusion of the reactants. The former are those in which the chemical rate coefficients are considerably smaller than the diffusion rate coefficients, of course. The inverse is true for the latter reactions.

In most practical situations the reaction system is not chosen because of its influence on the reaction kinetics. Economic feasibility is most often the dominating factor in this choice. Where adequate process control can be obtained, melt or a combination of melt and solid-state polymerization may be preferred for obvious reasons. When solvents must be used, they are often selected primarily on a combination of polymer solubility and cost.

Each of the main types of polycondensation systems will be reviewed briefly, with emphasis on the primary kinetic features. Additional details may be found in earlier publications [5, 6].

Polymerization in the melt is widely used commercially for the production of polyesters, polyamides, polycarbonates and other products. The reactions are controlled by the chemical kinetics, rather than by diffusion. Molecular weights and molecular weight distributions follow closely the statistical calculations indicated in the preceding section, at least for the three types of polymers mentioned above. There has been much speculation as to the effect of increasing viscosity on the rates of the reactions, without completely satisfactory explanations or experimental demonstrations yet available. Flory [7] showed that the rate of reaction between certain dicarboxylic acids and glycols was independent of viscosity for those materials, in the range studied. The viscosity range had a maximum of 0.3 poise, however, far below the hundreds of thousands of poises encountered in some polycondensations.

Solution polycondensation, i.e., where the monomers and polymer are soluble in the solvent(s), is also kinetically controlled. It is closely related mechanistically to reactions in the melt, but the choice of the solvent may influence both the rate and equilibrium, and hence the molecular weight. This process is utilized commercially in the production of a variety of organic coatings and certain solution-spun fibres, e.g., certain spandex types and also very high-melting aromatic polymers.

The effect of the solvent on the kinetics of organic reactions has been studied in detail for the reactions of a host of model compounds, and for some polymerizations. Several generalizations are readily apparent. The most powerful solvents, i.e., those giving the fastest reaction rate, are often those which may be considered to function as mild catalysts themselves. The dipoles in highly polar solvents such as dimethyl formamide or dimethyl sulphoxide may actually function as weak acids or bases, catalysing the reactions. In contrast, solvents of intermediate polarity may complex the active hydrogen compound to such an extent (at least at low to moderate temperatures) that they shield it from reaction, thus retarding the reaction. Solvents of very low polarity may permit faster reactions in these cases. For example, the reaction between phenyl isocyanate and methanol at 20°C is about seventy times faster in toluene than in acetonitrile [8].

Other general effects of solvents on reaction kinetics have been summarized by a number of authors, such as Gould [9]. The solvent effects are usually greatest when ionic charges are either increased or decreased considerably as a result of the reaction. Reactions in which charges are created usually go fastest in polar solvents, since those solvents can aid in dispersing the charges. Conversely, reactions in which charges are reduced will be favoured by solvents of low polarity.

Because of the solvent influences on kinetics, the composition of copolymers may be dependent on the solvent choice. In the case of the reaction of diethyl dichlorosilane and phenyltrichlorosilane with water, the solvent choice affected both the copolymer composition and molecular weight distribution [10].

In addition to kinetic effects, the choice of solvent may influence the equilibrium in many cases. As an illustration, if water is evolved as a by-product in a reversible polycondensation, a solvent system which has a low solubility for water will be favourable for high molecular weight. When hydrogen chloride is evolved, an increase in the dielectric constant of the medium often favours higher molecular weight.

Emulsion systems, while widely used in the polymerization of unsaturated monomers, are used rarely for polycondensation. The emulsion system is one in which two (or more) liquid phases are present, and in which polymerization occurs entirely in the bulk of one of the phases and is almost exclusively kinetically controlled. It thus represents a transition from solution polymerization to interfacial polymerizations. In the case of polycondensation reactions, emulsion polymerization has not been studied in detail. Results thus far indicate that molecular weight and molecular weight distribution are subject to the same statistical considerations as apply to solution and melt polymerizations.

Interfacial polycondensation has been studied in considerable detail in recent years, since this technique is quite useful for preparing high-melting polymers for fibre and other applications. This polymerization takes place in a two-phase system, with the propagation reaction occurring at or very near the interface. The mechanism is essentially diffusion controlled.

A typical interfacial system employs an aqueous solution of one reactant, such as a diamine, and an organic solution of the other reactant, such as a bifunctional acid chloride. In this case the aqueous phase would also include an acid acceptor. In most such polymerizations the polymer is formed on the organic side of the interface, of course. The interface may function primarily in controlling the diffusion of the diamine into the immediate polymerization zone. The aqueous phase is a carrier for the diamine and a stronger base to neutralize the evolved hydrogen chloride, and aids in removing hydrogen chloride from the reaction zone. The acid chloride will normally have little solubility in the aqueous phase, while the diamine will tend to distribute itself between the aqueous and organic phases according to its partition coefficient. The proper matching of solvent with diamine may be necessary to provide reaction control. The rate coefficient for reactions of this type may be in the range of $10^2 - 10^6$ l mole^{-1} sec^{-1}, so that it is easy to achieve diamine diffusion as the limiting process.

Since this process is controlled by the diffusion of one monomer across the interface, the polymerization does not respond in the same way to process variables as do those reactions which are kinetically controlled. For example, the monomer ratio in the reactor has much less effect on molecular weight than in a melt or solution polymerization. The molecular weight distribution may be different, too, often with much higher molecular weight fractions being present in the interfacial polymer. Any factor influencing the balance between the chemical kinetics and the limiting diffusion of the monomer across the interface may affect the molecular weight, the yield, and other characteristics of the process. Sokolov [11] has listed the following as the major factors.

Factors influencing the rate of chemical reaction are: surface tension; polarity of the organic solvent; acid-base properties of the aqueous phase; relative rate of hydrolysis and other side reactions (salt formation, etc.); rate of separation of polymer out of solution; rate of removal of side products of the reaction.

Factors influencing the rate of monomer diffusion are: surface tension; distribution coefficient of monomers between the two phases; rate of monomer transfer from phase to phase; solvent power of organic phase for the polymer; permeability of polymer film to monomers; adsorption of monomers by the polymer film; viscosity of the system.

Interfacial systems have been described in detail, especially by Morgan [6], Morgan and Kwolek [12], and Sokolov [5].

Other diffusion-controlled processes which have been used, but studied in relatively little detail, include solid-state polymerizations and gas— liquid interfacial polymerizations. Both processes have some commercial importance.

4. Chemical classification of reaction mechanisms

Most polycondensation reactions can be grouped into two categories: carbonyl addition reactions and displacement reactions. Each behaves much like the corresponding reactions of its monofunctional analogs. The best data showing the relationships between reactant structure and its reactivity, the role of catalysts and reaction medium are available from studies of these monofunctional compounds. This section will outline the basic features of the major reactions involved in polycondensation, emphasizing similarities within a group of related reactions. Later sections will give more specific details on the same reactions.

References pp. 575—581

4.1 CARBONYL ADDITION REACTIONS

A wide range of carbonyl addition reactions is available to the polymer synthesis chemist. To the extent that reactants having functionalities of two or higher are available, these reactions may be suitable for poly-condensation. A simplified scheme is

$$
\underset{\substack{\parallel \\ R-C-X}}{O} + H-Y\colon \;\rightleftarrows\; \left[\underset{\substack{\uparrow \\ \cdot\cdot \\ Y-H}}{\overset{\substack{\cdot\cdot \\ :O: \\ \mid}}{R-C-X}} \right] \;\rightleftarrows\; \underset{\substack{\parallel}}{O}{RC-Y} + HX
$$

Typical carbonyl compounds are carboxylic acids, acid halides, esters, amides, ureas and urethanes. The "active hydrogen" compounds may be alcohols, amines, thiols, carboxylic acids, phenols and compounds with the active hydrogen on carbon if appropriate substituents increase the lability of that hydrogen sufficiently. In general the component Y carries at least one unshared pair of electrons, though in special cases that pair may be generated by removal of the active hydrogen as a proton. The Lewis base, Y, attacks the electron-poor carbonyl carbon reversibly. The reaction product is formed by the elimination of X, taking with it the pair of electrons from the original C—X bond. When the base Y has carried the active hydrogen with it, as shown, HX is eliminated.

The formation of an amide by the reaction of an ester with an amine will illustrate this sequence more specifically, viz.

$$
\underset{\substack{\parallel \\ RCOR'}}{O} + R''NH_2 \;\rightleftharpoons\; \left[\underset{\substack{\mid \\ R''NH \\ H}}{\overset{\substack{O \\ \mid}}{R-C-OR'}} \right] \;\rightleftharpoons\; \underset{\substack{\parallel}}{O}{RCNHR''} + R'OH
$$

A reaction in which the attacking agent Y carries no hydrogen on the Y can be illustrated by the reaction of a phenoxide ion with an acid chloride

$$
\underset{\substack{\parallel \\ RCCl}}{O} + ArO^- \;\rightleftharpoons\; \left[\underset{\substack{\mid \\ OAr}}{\overset{\substack{O \\ \mid}}{RC-Cl}} \right]^- \;\rightleftharpoons\; \underset{\substack{\parallel}}{O}{RCOAr} + Cl^-
$$

One can outline a similar sequence for the reaction of "internal an-hydrides", such as isocyanates or ketenes

$$RN=C=O + R'OH \rightleftharpoons \left[\begin{array}{c} R-N=C-O \\ | \\ HOR' \end{array} \right] \rightarrow RN-\overset{\overset{\displaystyle O}{\|}}{C}-OR' \\ H$$

In this case, no X has been eliminated, taking the pair of electrons which had been in the C—X bond orbitals. Somewhat analogously, however, one pair of electrons from the carbon—nitrogen double bond has been eliminated from the orbitals shared by carbon, being displaced to the nitrogen. A similar sequence can be written for ketenes.

The similarities outlined, if valid, should mean that the rates of the many reactions indicated should be governed by some balance of the same list of structural influences relating monomer structure to reactivity, and the effect of the nature of the reaction medium. The following guiding principles are outlined briefly. More detailed discussions of factors affecting the rates of carbonyl additions, as well as mechanistic details, are given in several references [13—15] and in later sections of this chapter.

In the general mechanism, one expects that a decrease in electron density at the carbonyl carbon should increase the ease of approach by the attacking electron donor. This decrease may be facilitated by the presence of electron-withdrawing groups on R, and by an increase in the electron withdrawing character of X.

An increase in the base strength of Y should promote the approach of Y to the carbonyl carbon. Also, an increase in the basicity of Y normally makes the elimination of X, with its pair of electrons from the C—X bond, easier.

A complicating feature is that a decrease in the electron density of the carbonyl carbon, while aiding the approach of Y, reduces the ease of elimination of X with its pair of electrons. Thus one could find a sequence of reactions in which electron-withdrawing substituents on R would promote the reaction, and another in which they would retard the reaction.

A high electron affinity by X (low basicity) usually favours the elimination of X. (See the next section on displacement reactions for more details of the effect of the structure of X on its leaving ability.)

Hammett [14] has developed quantitative expressions relating the effect of substituents in either R or X on the rate of the reaction, when steric factors are negligible. He defined a reaction constant ρ by the equation

$$\rho = \frac{\log k - \log k^0}{\sigma}$$

where σ is dependent on the substituent and independent of the reaction series, k^0 is the rate coefficient for the reaction of an unsubstituted reference compound, such as benzoic acid, and k is the rate coefficient for the same reaction with various substituents in the *meta* and *para* positions of the benzoic acid molecule. A positive value for ρ means that the reaction rate is increased by electron-withdrawing substituents such as nitro groups, and a negative value for ρ means that the reaction is promoted by electron-donating substituents such as alkyl or alkoxyl. The magnitude of the value for ρ indicates the sensitivity of the rate change to changes in electron density at the site of the reaction. Hammett summarized ρ values for a number of reactions, and Jaffe [16] tabulated many more.

In a later table Hammett ρ values will be summarized for a variety of carbonyl addition type reactions, along with other features of these reactions.

In addition to the electronic effects of substituents, steric factors are also important in controlling the rates of reactions. In the case of aromatic reactants, bulky substituents in the 2- and 6-positions, relative to the reaction site, may retard a reaction so drastically that it is no longer practical. In the case of aliphatic reactants substituents on the *beta* carbon may be even more effective in retarding a reaction than those on the *alpha* carbon. This effect has been described by Newman [17] as the "rule of six", which states that those atoms which are most effective in providing steric hindrance are separated from the attacking atom in the transition state by a chain of four atoms. Thus, in esterification, if either the attacking atom or the carbonyl oxygen is designated "1", the blocking atom will be in the "6" position, viz.

Steric factors may also influence the effectiveness of a catalyst. Since a catalyst must approach the site of reaction as closely as a reactant itself, the ease of approach and extent to which the approach is possible will be influenced by steric relations between the catalyst and one or more of the reactants.

Catalysis of the polycondensation reaction is often crucial for the practical success of the preparative method. Reasonable prediction and experience show that catalysts may be suitable acids, bases or multivalent metals. The acids may serve by promoting electron withdrawal from the carbonyl carbon

$$R-\overset{\overset{\displaystyle O}{\|}}{C}-OR' + H^+ \; \rightleftharpoons \; \left[R-\overset{\overset{\displaystyle OH}{|}}{C}-OR' \right]^+$$

One might expect that such an acid-catalysed reaction could be more likely to show a negative Hammett ρ effect than uncatalysed or base-catalysed reactions. Such seems to be the case in the acid-catalysed esterification shown in Table 1.

TABLE 1

Similarities in carbonyl addition reactions utilized in polycondensations

Reaction type	Transition state	Demonstrated catalysts	Hammett ρ value	
			R	R'
A Esterification				
1 Direct	$R-\overset{\overset{\displaystyle O}{\|}}{\underset{\underset{\displaystyle R'-O-H}{\|}}{C}}-OH$	Acids, metals	-0.5^a $+0.5^b$	
2 Transesterification	$R-\overset{\overset{\displaystyle O}{\|}}{\underset{\underset{\displaystyle R'-O-H}{\|}}{C}}-OR''$	Acids, bases, metals		
3 From acid chloride	$R-\overset{\overset{\displaystyle O}{\|}}{\underset{\underset{\displaystyle R'-O-H}{\|}}{C}}-Cl$		$+1.5^c$	
B Amidation				
1 Direct	$R-\overset{\overset{\displaystyle O}{\|}}{\underset{\underset{\displaystyle R'-NH_2}{\|}}{C}}-OH$	Acids		-1.2^d
2 Ester plus amine	$R-\overset{\overset{\displaystyle O}{\|}}{\underset{\underset{\displaystyle R'-NH_2}{\|}}{C}}-OR''$		$+0.5^e$	

TABLE 1—*continued*

Reaction type	Transition state	Demonstrated catalysts	Hammett ρ value R	R'
3 Transamidation	$$R-\overset{\overset{\displaystyle O}{\|}}{\underset{\underset{\displaystyle R'-NH_2}{\|}}{C}}-NHR''$$			
4 From acid chloride	$$R-\overset{\overset{\displaystyle O}{\|}}{\underset{\underset{\displaystyle R-NH_2}{\|}}{C}}-Cl$$		$+1.2^f$	-2.7^g
C Urethane formation 1 From isocyanate	$$R-N=C-O$$ $$\underset{\displaystyle H-O-R'}{\|}$$	Acids, bases, metals	$+1.69^h$	
2 From chloroformate	$$R-\overset{\overset{\displaystyle H}{\|}}{\underset{\underset{\displaystyle H}{\|}}{N}}-\overset{\overset{\displaystyle Cl}{\|}}{\underset{\underset{\displaystyle OR'}{\|}}{C}}-O$$			
D Carbonate formation 1 Ester interchange	$$R-O-\overset{\overset{\displaystyle H\ \ OR''}{\|\ \ \ \|}}{\underset{\underset{\displaystyle OR''}{\|}}{C}}-O$$	Acids, bases, metals		
2 From chloroformate	$$R-O-\overset{\overset{\displaystyle O}{\|}}{\underset{\underset{\displaystyle R'OH}{\|}}{C}}-Cl$$	Tertiary amines		
E Urea formation 1 From isocyanate	$$R-N=C-O$$ $$\underset{\underset{\displaystyle H}{\|}}{H-N-R'}$$	Acids, bases, metals	+	
2 From carbamoyl chloride	$$R-\overset{\overset{\displaystyle H\ \ O}{\|\ \ \|}}{\underset{\underset{\underset{\displaystyle H}{\|}}{\displaystyle R'-N-H}}{N}}-C-Cl$$			

Basic catalysts are often tertiary amines or other bases which are not destroyed as the reaction proceeds. For polyurethane formation, Baker et al. [18—20] proposed initial coordination with the carbonyl carbon, with subsequent displacement by alcohol

$$
RNCO + B\colon \rightleftarrows \left[\begin{array}{c} R-N=C-O \\ \uparrow \\ \colon\colon \\ B \end{array} \right] \xrightarrow{R'OH} R-NH\overset{\overset{\displaystyle O}{\|}}{C}OR' + B\colon
$$

In the case of tertiary amine catalysis of the reactions of acid chlorides, phosgene and chloroformates, an initial complex formation has been widely thought to be likely, viz.

$$
R\overset{\overset{\displaystyle O}{\|}}{C}Cl + NR'_3 \rightleftarrows \left[R-\overset{\overset{\displaystyle O}{\|}}{C}NR'_3 \right]^+ Cl^-
$$

This may also be looked upon as being similar to the isocyanate example

$$
\left[\begin{array}{c} \overset{\displaystyle O}{|}{}^{(\delta-)} \\ R-C-Cl \\ \uparrow \\ \colon\colon \\ NR'_3 \\ {}_{(\delta+)} \end{array} \right]
$$

In either case, the tertiary amine is apparently displaced by the active-hydrogen compound (alcohol, amine, etc.).

In some cases, the base may serve to maintain a low but steady concentration of a highly reactive intermediate, as in the case of isocyanate/thiol reactions

---

[a] Esterifications of benzoic acids with ethanol, N HCl; ref. 14, p. 189.
[b] Reaction of benzoic acids with cyclohexanol, acid catalyst; ref. 16.
[c] Reaction of benzoyl chlorides with ethanol; ref. 14, p. 189.
[d] Reaction of anilines with formic acid; ref. 14, p. 190.
[e] Reaction of methyl benzoates with aniline; ref. 16.
[f] Reaction of benzoyl chlorides with aniline; ref. 14, p. 189.
[g] Reaction of anilines with benzoyl chloride; ref. 14, p. 189.
[h] Reaction of phenyl isocyanates with an alcohol; ref. 22.

$$R'SH + :B \rightleftharpoons R'\overset{-}{S:} + B\overset{+}{:}H$$

$$RNCO + R'\overset{-}{S:} \rightarrow \left[\begin{array}{c} O \\ \parallel \\ RN\!-\!CSR \end{array} \right]^{-} \overset{B\overset{+}{:}H}{\longrightarrow} \begin{array}{c} O \\ \parallel \\ RNHCSR \end{array} + B:$$

Multivalent metals may serve as Lewis acids in some cases, and metal oxides may function as Lewis bases. At least in some reactions, such as the isocyanate/alcohol reactions, the metal atom may act as an acid, and also hold both reactants in essentially ideal proximity for reaction, viz.

$$RNCO + R'OH + MX_2 \rightarrow \left[\begin{array}{cc} X_2M & \leftarrow O \\ \uparrow & | \\ R'O & C \\ H & \parallel \\ & NR \end{array} \right] \rightarrow RNHCOOR' + MX_2$$

A similar transition state can be drawn for metal catalysis of esterification, as shown later, and for many other such reactions.

The very powerful catalytic effect of certain metal compounds suggests a similarity to enzymatic reactions, which often lead to polycondensations or degradation of polycondensation-type polymers. Koshland [21] has made an interesting new proposal concerning the unusual effectiveness of enzymatic catalysts; his suggestions may have some bearing on the role of metal catalysts in polycondensations, as well. He proposed that, in addition to bringing the reactants into close proximity in coordination with the catalyst, the enzyme lines up the bonding orbitals of the reactants in the very precise manner which is most favourable for immediate reaction to occur.

Table 1 summarizes similarities for many of the familiar carbonyl addition reactions utilized in polycondensations. Indications of probable transition states for some of the most common reactions (omitting participating solvent molecules), types of catalysis actually demonstrated, and Hammett ρ values, or at least the sign of the value, are included in many cases. These values are given for reactions in which either R or R' has been altered by substituents.

In addition to these reactions, aldehyde-phenol, aldehyde-urea, aldehyde-melamine and many other aldehyde-based polymers are prepared by polycondensation processes. The first step in a typical reaction of this type is

$$\begin{matrix} & \overset{\displaystyle O}{\underset{\displaystyle \parallel}{}} \\ H_2NCNH_2 \end{matrix} + CH_2O \;\rightleftarrows\; \left[\begin{matrix} \overset{\displaystyle O}{\underset{\displaystyle \parallel}{}} & \overset{\displaystyle H}{\underset{\displaystyle |}{}} & \overset{\displaystyle H}{\underset{\displaystyle |}{}} \\ H_2NC\!-\!N & \rightarrow & C\!-\!O \\ \underset{\displaystyle H}{\underset{\displaystyle |}{}} & & \underset{\displaystyle H}{\underset{\displaystyle |}{}} \end{matrix} \right] \;\rightarrow\; \begin{matrix} \overset{\displaystyle O}{\underset{\displaystyle \parallel}{}} \\ H_2NCNHCH_2OH \end{matrix}$$

4.2 DISPLACEMENT REACTIONS

With the increasing availability of suitable monomers and the desire for polymers having improved heat resistance combined with higher levels of mechanical properties, "displacement" (or "substitution") reactions have become of increased interest for polycondensation. A suitable example with monofunctional reactants would be the Williamson synthesis of ethers

$$ROH + R'X \xrightarrow{\text{Na}} ROR' + NaX + \tfrac{1}{2}H_2$$

It is possible for displacement or substitution reactions (here X has been displaced by RO, or RO has been substituted for X) to proceed by one or both of two general mechanisms, often designated S_N1 ("substitution, nucleophilic, first order") or S_N2 ("substitution, nucleophilic, second order"). In the S_N1 sequence it is considered that reaction proceeds as

$$R'X \xrightarrow{\text{slow}} [R']^+ + X^- \xrightarrow[\text{fast}]{RO^-} R'OR$$

The slow step is the ionization of $R'X$, and the rate is dependent on the concentration of $R'X$. Combination of the positive ion $[R']^+$ with an electron-donating ion or molecule is fast, and is not rate-determining. The S_N2 reaction is visualized as occurring by the mechanism

$$RO^- + R'X \;\rightarrow\; \left[RO\!-\!\overset{\diagdown \diagup}{\underset{|}{C}}\!-\!X \right] \;\rightarrow\; ROR' + X^-$$

where the carbon atom shown in the brackets is the carbon in R' which is attached to X. In this case the rate is dependent on the concentration of both RO^- and $R'X$, so that the kinetics are second order. In some cases both mechanisms apparently are operative simultaneously, so that the kinetics may be more complicated.

Both the S_N1 and the S_N2 reactions have the common feature that X leaves the RX molecule, taking the bonding pair of electrons with it. The term "leaving group" is often used to describe X. For ease of leaving, the bond between R' and X should be relatively weak, with X having a high affinity for the electrons in the R'X bond. In the case of C—X bonds it generally follows that the less basic the substituent X⁻, the more easily it is pulled off by solvent ($S_N 1$) or pushed off by an attacking nucleophile ($S_N 2$). For displacement from aliphatic carbon atoms the following order of increasing ease of displacement is usually observed [23]

$$-NR_2 < -OH \sim -OR < -NR_3^+ < -OAc \sim -F < -Cl <$$

$$Br < I < -OSO_2-\left\langle\bigcirc\right\rangle-CH_3 < OSO_2-\left\langle\bigcirc\right\rangle-Br$$

In some cases one can circumvent this sequence. For example, if —OH or —OR must be displaced, one can usually promote it with an acid catalyst. When protons are available, the group displaced will be $-OH_2^+$ or $-OHR^+$, which are much better leaving roups than —OH or —OR.

With $S_N 2$ reactions, the nature of the attacking group (RO⁻ in our example) is also of major importance. This attacking group is serving to donate a pair of electrons as a part of the process of forming the new bond. The ability to do this frequently parallels the basicity of the attacking group. However, "basicity" is usually measured against a proton-bearing acid, whereas in the displacement reaction the electron acceptor is usually carbon. Thus it is not surprising that the order of reactivity of a series of attacking reagents is not always the same as their order of basicity. The term "nucleophilicity" is frequently used to describe the reactivity of an electron donating agent which attacks the carbon atom in a displacement reaction. One often finds the following series of decreasing nucleophilicity [23]:

$$R_3C\overset{..}{:}{}^- > R_2N\overset{..}{:}{}^- > RO^- > F^-$$

and for a series in which oxygen is always the attacking atom:

$$OH^- > OPh^- > CO_3^= > OAc^- > H_2O > ClO_4^-$$

Steric hindrance obviously will be a very important consideration in the displacement reactions. Hindrance at R' will make attack by RO⁻ difficult, and a bulky character in R will reduce its own attacking efficiency. Steric hindrance between R' and X may facilitate elimination of X.

We see from this brief outline of essentially self-evident steric factors much of the effect which the structure of R' has on the rate of reaction. It also happens that substitution of the *alpha* carbon in R' by electron donating groups, such as alkyl, tends to stabilize the carbonium ion formed in an $S_N 1$ dissociation

$$(CH_3)_3CBr \rightleftarrows [(CH_3)_3C]^+ \quad Br^-$$

Other substituents which would permit resonance stabilization of the carbonium ion would also favour this mechanism. In general, one finds that the effect of the structure of R$'$ on the reaction is such that if X is attached to a primary carbon atom, the displacement reaction is most likely to proceed by an S_N2 mechanism. When X is attached to a tertiary carbon, S_N1 reaction is most likely. When X is attached to a secondary carbon the situation is often borderline, with both mechanisms occurring to a significant extent.

Excellent, detailed treatments of the kinetics of displacement reactions can be found in a number of references [23—25].

Hammett ρ values have been established for the displacement reaction leading to ether formation. In the case of ether formation from ROH and R$'$CH$_2$X, the ρ value was negative for substituent changes in R, and positive for changes in R$'$ [26].

Some displacement reactions which are of importance in polycondensation are illustrated below.

Polysulphone formation, e.g.

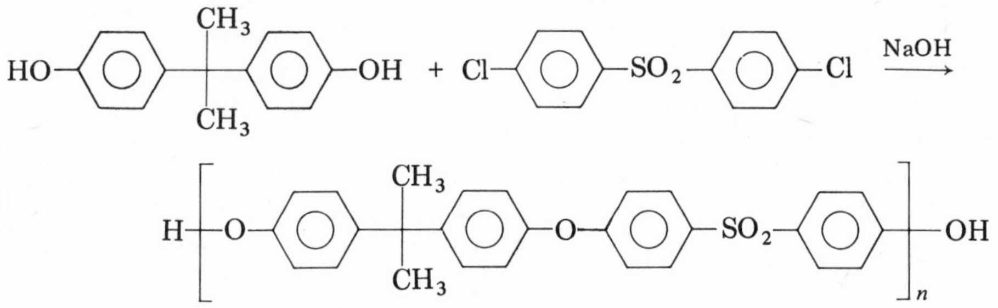

$$HO-\langle C_6H_4 \rangle-C(CH_3)_2-\langle C_6H_4 \rangle-OH + Cl-\langle C_6H_4 \rangle-SO_2-\langle C_6H_4 \rangle-Cl \xrightarrow{NaOH}$$

$$H\left[O-\langle C_6H_4 \rangle-C(CH_3)_2-\langle C_6H_4 \rangle-O-\langle C_6H_4 \rangle-SO_2-\langle C_6H_4 \rangle\right]_n OH$$

Polyphenylene sulphide formation, e.g.

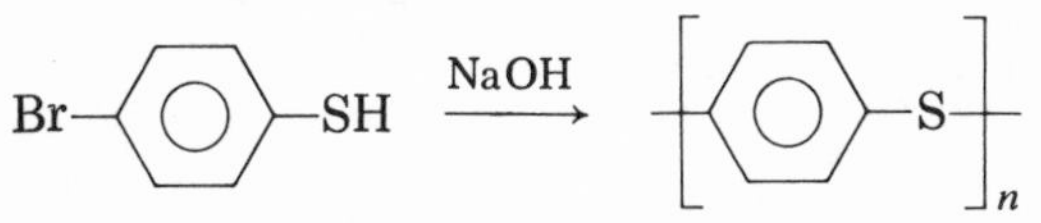

$$Br-\langle C_6H_4 \rangle-SH \xrightarrow{NaOH} \left[\langle C_6H_4 \rangle-S\right]_n$$

Polyaryl ether formation, e.g.

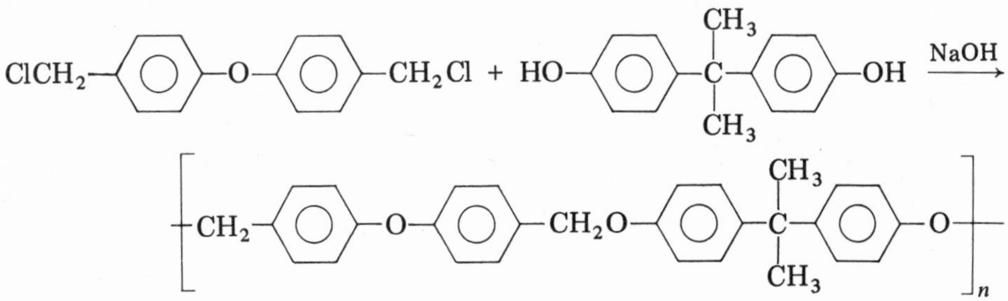

$$ClCH_2-\langle C_6H_4 \rangle-O-\langle C_6H_4 \rangle-CH_2Cl + HO-\langle C_6H_4 \rangle-C(CH_3)_2-\langle C_6H_4 \rangle-OH \xrightarrow{NaOH}$$

$$\left[CH_2-\langle C_6H_4 \rangle-O-\langle C_6H_4 \rangle-CH_2O-\langle C_6H_4 \rangle-C(CH_3)_2-\langle C_6H_4 \rangle-O\right]_n$$

One may question whether reactions of acid halides are displacement reactions

$$RCOCl + R'OH \longrightarrow \left[\begin{array}{c} H \\ R' \end{array} \!\! >\!\! O \rightarrow \!\! \begin{array}{c} O \\ \| \\ C-Cl \\ R \end{array} \right] \longrightarrow$$

$$\left[\begin{array}{c} H \\ R' \end{array} \!\! >\!\! O-C \!\! \begin{array}{c} O \\ \diagup\diagdown \\ R \end{array} \right] Cl^- \longrightarrow \quad R'O\overset{O}{\overset{\|}{C}}R + HCl$$

as has been proposed by some [13], rather than carbonyl addition reactions [25]

$$RCOCl + R'OH \; \rightleftarrows \; \left[\begin{array}{c} O \\ | \\ R-C-Cl \\ \uparrow \\ R'-O-H \end{array} \right] \rightarrow R\overset{O}{\overset{\|}{C}}OR' + HCl$$

Indeed, the same question can be raised, though perhaps with slightly less justification, for most reactions leading to esters and amides, as well. For those reactions of acids, esters, amides, and acid halides which are not truly first order (such as the alcoholysis of mesitoyl chloride), however, the authors prefer to consider these as "carbonyl addition" reactions, rather than as "displacement" reactions. The fundamental differences appear to be which of the carbon bond orbitals is attacked first by the nucleophile, and the number of steps in the reaction. In the displacement reaction the nucleophile is considered to approach the back side of the orbitals of the single bond between carbon and the leaving group. In a single step the new bond is made and the leaving group is expelled. In the carbonyl addition mechanism, however, the nucleophile attacks the back side of an orbital in the carbon—oxygen double bond. This reaction involves two distinct steps, viz. addition to the double bond, followed by an elimination. Reliable experimental distinction between the two mechanisms is difficult. The effects of the structures of the reactants on the rate of the reaction is essentially the same in either case.

4.3 MISCELLANEOUS REACTIONS

Several additional reactions have been used to prepare condensation polymers, although relatively little is known about their kinetics. Some are indicated here, to illustrate more fully the scope of polycondensations.

Friedel—Crafts reactions have been used to prepare low-molecular-weight and also crosslinked polymers. A typical reaction is that of benzyl chloride

Free radicals may also be intermediates in polycondensation reactions, in special cases, such as

A variety of unusual polymerizations falling into this class has been reviewed by Sokolov [5].

5. Polyesterification

In 1847, Berzelius [27] reacted glycerol with tartaric acid, thus probably producing the first synthetic polyester, which was most likely crosslinked. A linear polymer from ethylene glycol and succinic acid was prepared sixteen years later [28], though at that time such products were believed to be cyclic or macrocyclic esters. Carothers [29] prepared the first linear polyesters of high molecular weight, recognizing their linear structure. Today several types of polyester are of commercial significance. The polyterephthalates (from dimethyl terephthalate or terephthalic acid) of ethylene glycol or 1,4-cyclohexane-dimethanol, the adipic acid polyesters of ethylene, 1,2-propylene, and 1,4-butylene glycols, and the polycarbonate of 2,2-bis(4-hydroxyphenyl)propane (Bisphenol A) are the most important linear polyesters. Alkyd resins, the condensation products of dicarboxylic acids (or anhydrides) and polyhydric alcohols, find extensive commercial applications as plasticizers and in paints and other coatings.

5.1 SYNTHETIC ROUTES

Polyesters can be prepared by the direct esterification of carboxyl groups with alcoholic hydroxyl groups in the same or different compounds. Similarly, polyesters can be prepared from the reaction of carboxylic esters and alcoholic hydroxyl groups in the same or different compounds, by transesterification (also known as ester exchange, ester interchange, or ester alcoholysis). Related reactions are those of acidolysis (between, for example, an acid and a glycol ester) and esterolysis (interchange between a carboxylic ester and a glycol ester).

Two different synthetic routes to polyesters are through the reaction of glycols or other polyfunctional hydroxy-compounds and anhydrides such as phthalic anhydride, and by the reaction of acid chlorides with glycols or bisphenols. These two reactions may be regarded as irreversible. In the case of polyester-forming reactions that involve the use of acid chlorides (Section 5.7), the reaction proceeds better when the hydrochloric acid produced in the polycondensation is effectively removed, by neutralization, for example.

Reversible reactions involve both formation of polyester and the reverse reaction of the polymer with the liberated low-molecular-weight product. The so-called polycondensation equilibrium occurs, of course, when the rates of the forward and reverse reactions are the same. This equilibrium determines the extent of conversion to polymer, and the position of the equilibrium determines the molecular weight of the product. In the case of monofunctional reactants, the equilibrium governs only the yield of the product.

The molecular weight of any product that is formed by these reversible polycondensations is profoundly affected by removal of the volatile side products from the reaction zone. In practice, such removal may be carried out in various ways, including sweeping away volatile materials by means of a stream of inert gas or by applying a vacuum.

Under certain conditions, the equilibrium constant for the formation of poly(ethylene terephthalate) (PET) was shown [30] to be 9.6; that is, under the conditions employed the polycondensation reaction coefficient leading to the formation of PET was 9.6 times greater than the rate coefficient for glycolysis of PET.

The equilibrium constant of polyesterification is typically equal to that of the analogous model reaction between monofunctional compounds. This can be explained by the proposition [3] that the reactivity of a functional end-group in any growing polymer chain is independent of the degree of polymerization or chain-length. There have been several experimental verifications of this theory. Thus the equilibrium constant for the reaction between dimethyl terephthalate (DMT) and ethylene glycol at 280°C was found [31] to be 4.9 under certain conditions. Under identical conditions, the equilibrium constant for the model reaction was 5.0.

Another demonstration of the independence of the reactivity of functional groups with chain length was given by determining the degree of polycondensation for polymers derived from a homologous series of acids, each acid reacting with 1,6-hexanediol under identical conditions [32]

	Degree of polycondensation
Succinic acid/1,6-hexanediol	48
Adipic acid/1,6-hexanediol	44
Sebacic acid/1,6-hexanediol	49

In several instances, discussed in more detail later in this section, experimental evidence strongly indicates that the principle of equal reactivity of functional groups is not obeyed for all chain lengths, and that the equilibrium constant may itself be a function of the degree of polycondensation.

The stepwise growth mechanism of polyesterification results in polymer molecules of widely varying molecular weights. Flory's hypothesis [3] that the reactive end-groups of a growing polymer chain show equal reactivity predicts that the molecular weights vary in a geometric fashion, referred to as the "most probable" distribution. For essentially all polycondensations examined in the melt, this distribution has been shown to be at least approximately correct. In the case of PET there are reported indications (see Section 5.3.2a) that this distribution is not followed perfectly for low-molecular-weight products. Polyesterification in solution also appears to give a "most probable" distribution of molecular weights. In reactions where the polymer precipitates rapidly, as in certain interfacial polycondensations, the molecular weight distribution does not usually follow the statistical form. Sometimes, as in the case of certain polycarbonates made interfacially by a two-step process, the non-statistical distribution is a result of the technique used to prepare the polymer (see Section 5.7). In general, interfacial polymerization in a non-solvent gives an unusually narrow distribution of molecular weights; a possible explanation may be that continual fractional precipitation occurs during the reaction with higher-molecular-weight fractions being precipitated, allowing further polymerization of the soluble oligomers until they too become insoluble [6]. Rates of mixing and concentrations of monomers can also influence molecular weight distributions in interfacial polymerizations. Dilute solutions tend to result in the formation of greater amounts of cyclic products [6].

In all the synthetic approaches reviewed here, and indeed in all polycondensation reactions, various side reactions can influence the formation of polymer to a greater or lesser extent. Among these side reactions are loss of functionality as in decarboxylation or in the oxidation of an aliphatic hydroxyl group, the formation of volatile intermediates, the formation of unreactive intermediates, crosslinking, and other reactions. These reactions are examined in more detail in the relevant sub-sections.

Where possible in each sub-section, the mechanism of the reaction is described, together with the effect of catalysts, factors governing selection of catalyst, and the mechanism of catalysis. Structural considerations that may affect the progress of the reaction will also be discussed where relevant.

In later sections an important distinction is made between the reactions of dicarboxylic acids and glycols, whether in equimolar quantities or not, where the glycol is essentially non-volatile under the conditions of the polycondensation, and the reactions of dicarboxylic acids with an excess of a glycol that is volatile under the conditions required for polycondensation. In the former case (Section 5.2 below) esterification and polycondensation proceed by essentially identical paths with the elimination of a molecule of water for each ester linkage formed. In the second case the reaction initially follows the same steps, including the formation of low-molecular-weight polymer, but the product is glycol-terminated due to the excess used. Once esterification is thus substantially complete, reaction conditions are made more strenuous and polycondensation then becomes a transesterification reaction with the now volatile glycol being eliminated. This type of reaction is described in Section 5.3.

5.2 DIRECT ESTERIFICATION

Linear polyesters prepared by direct esterification of carboxyl groups with hydroxyl groups can be derived from one compound, viz.

$$n \text{ HO—A—COOH} \rightleftharpoons \text{—[A—C—O]}_n + n \text{ H}_2\text{O}$$
$$\overset{\|}{\text{O}}$$

or from two compounds (or more in the case of copolymers)

$$n \text{ HOOC—A—COOH} + n \text{ HO—G—OH} \rightleftharpoons \text{—[C—A—C—OGO]}_n + 2n \text{ H}_2\text{O.}$$
$$\overset{\|}{\text{O}} \quad \overset{\|}{\text{O}}$$

Throughout this section the symbol —A— is used for the non-functional monovalent or bivalent portion of a mono- or dicarboxylic acid; —G— represents the bifunctional organic residue attached to the hydroxyl groups of a glycol; and P— or —P— represent polymeric residues not taking part in the reaction under consideration. A and G also symbolize the acid and glycol residues within a polymer chain.

In commercial practice today reactions of the second type frequently utilize an excess of glycol, increasing the rate of esterification. Later stages of such polycondensations, as in the preparation of polyethylene terephthalate, take place by a transesterification mechanism with liberation of glycol. The direct esterification reaction may be catalysed by a second molecule of the carboxylic acid (self-catalysis) or by an independent acidic catalyst (catalysed esterification).

Following Ingold's classification [33] of the reversible reactions of hydrolysis of esters, eight possible mechanisms can be put forward for the formation of an ester linkage from a carboxylic acid and an alcohol. The mechanisms depend on whether catalysis is by acid or base, whether the reaction is unimolecular or bimolecular, and whether the reaction takes place through acyl cleavage or alkyl cleavage. All eight mechanisms are $S_N 1$, $S_N 2$, or tetrahedral. The four acid-catalysed hydrolyses are symmetrical and totally reversible and are therefore mechanisms for acid-catalysed esterifications.

The $A_{AL} 1$ reaction (acid-catalysed, alkyl cleavage, unimolecular) is of the $S_N 1$ type

$$R'OH \; \underset{}{\overset{H^+}{\rightleftharpoons}} \; R'OH_2 + \; \underset{H_2O}{\overset{\text{slow, RCOOH}}{\rightleftharpoons}} \; R\!-\!\underset{O}{\overset{\|}{C}}\!-\!OH + R'^+ \; \underset{\text{slow}}{\rightleftharpoons}$$

$$R\!-\!\underset{O \;\; H}{\overset{+}{\underset{\|}{C}}}\!-\!\overset{+}{O}R' \; \rightleftharpoons \; R\!-\!\underset{OH}{\overset{+}{\underset{|}{C}}}\!-\!OR' \; \underset{H^+}{\overset{}{\rightleftharpoons}} \; R\!-\!\underset{O}{\overset{\|}{C}}\!-\!OR'$$

and is perhaps the most common mechanism for cases where a stable carbonium ion R'^+ can be formed. The corresponding $S_N 2$ type of reaction ($A_{AL} 2$) has not been observed.

The acid-catalysed reactions that do not proceed through formation of a stable carbonium ion derived from R'OH are almost entirely of the $A_{AC} 2$ type (acid-catalysed, acyl cleavage, bimolecular), which is tetrahedral

$$R\!-\!\underset{O}{\overset{\|}{C}}\!-\!OH \; \overset{H^+}{\rightleftharpoons} \; R\!-\!\underset{OH}{\overset{+}{\underset{|}{C}}}\!-\!OH \; \underset{}{\overset{R'OH}{\underset{\text{slow}}{\rightleftharpoons}}} \; R\!-\!\underset{\underset{a}{OH}}{\overset{\overset{OH}{|}}{\underset{|}{C}}}\!-\!\overset{+}{O}R' \; \rightleftharpoons \; R\!-\!\underset{\underset{b}{OH}}{\overset{\overset{+}{OH_2}}{\underset{|}{C}}}\!-\!OR' \; \underset{H_2O}{\overset{}{\underset{\text{slow}}{\rightleftharpoons}}}$$

$$R\!-\!\underset{OH}{\overset{+}{\underset{|}{C}}}\!-\!OR' \; \underset{H+}{\overset{}{\rightleftharpoons}} \; R\!-\!\underset{O}{\overset{\|}{C}}\!-\!OR'$$

The internal proton transfer is (a—b in the mechanism above) may take place directly or through the reaction medium.

The remaining acid-catalysed reaction, $A_{AC} 1$, is quite rare and is found only where R (acyl function) is bulky. It is an $S_N 1$ type of reaction. Again, the proton transfer (a—b in the mechanism below) may occur directly or through the solvent.

References pp. 575—581

$$R\overset{\|}{\underset{O}{C}}OH \overset{H+}{\rightleftharpoons} R\overset{+}{\underset{OH}{C}}OH \rightleftharpoons R\overset{+}{\underset{\|}{\underset{O}{C}}}OH_2 \underset{H_2O}{\overset{slow}{\rightleftharpoons}} R\overset{+}{\underset{\|}{\underset{O}{C}}} \overset{R'OH}{\underset{slow}{\rightleftharpoons}}$$

$$R\overset{+}{\underset{\|\ |}{\underset{O\ H}{C}}}OR \rightleftharpoons R\overset{+}{\underset{OH}{C}}OR \overset{}{\underset{H+}{\rightleftharpoons}} R\overset{\|}{\underset{O}{C}}OR'$$

a   b

During polycondensation, reaction conditions are generally far more complex than in the case of a simple esterification reaction between monofunctional starting materials. However, Reinisch et al. [34] proposed that the catalysed polycondensation of aromatic and aliphatic bisglycol esters followed an $A_{AC}2$ mechanism.

Solomon et al. [35] concluded that the observed high-order kinetics for esterification and polyesterification reactions ruled out unimolecular $A_{AC}1$ and $A_{AL}1$ mechanisms, and that the $A_{AC}2$ mechanism was the most likely.

Side reactions that might occur during direct polyesterification include loss of functionality by decarboxylation of an acidic end group, thermal degradation of ester linkages, the formation of ether linkages, and cyclization reactions.

Decarboxylation becomes an important degradative reaction at elevated temperatures. Table 2 lists the decarboxylation temperatures [32] for the first ten dicarboxylic acids of the homologous series that begins with oxalic acid. A dicarboxylic acid containing an odd number of carbon atoms decarboxylates at a lower temperature than either of the two neighbouring acids having an even number of carbon atoms in the chain.

Pyrolytic degradation of esters in the vapour phase [36] occurs by a first-order unimolecular reaction

TABLE 2

Decarboxylation temperatures of dicarboxylic acids[a]

Acid	Decarboxylation temperature (°C)
Oxalic	160—180
Malonic	140—160
Succinic	290—310
Glutaric	280—290
Adipic	300—320
Pimelic	290—310
Suberic	340—360
Azelaic	320—340
Sebacic	350—370

[a] Taken from ref. 32.

$$\underset{\substack{| \\ O-C-CH_3 \\ \| \\ O}}{Ar-CH-CH_3} \;\rightleftharpoons\; \underset{\substack{| \\ C---O \\ | \\ CH_3}}{\overset{Ar}{\underset{O \qquad H}{CH{=}CH_2}}} \;\rightleftharpoons\; \begin{array}{l} Ar-CH{=}CH_2 \\ + CH_3COOH \end{array}$$

The pyrolysis mechanism has been elucidated by Taylor et al. [37].

Goodings [38] used ethylene dibenzoate as a model compound for pyrolysis studies on PET. His results indicated that an important pyrolytic reaction in molten PET is cleavage of an internal ester linkage to form a carboxylic acid end group and a vinyl ester end group

$$\underset{\substack{\| \\ O}}{P-C}-O-CH_2CH_2-O-\underset{\substack{\| \\ O}}{C}-P \;\rightarrow\; \underset{\substack{\| \\ O}}{P-C}-O-CH{=}CH_2 + HO-\underset{\substack{\| \\ O}}{C}-P$$

During polymerization, ester interchange between the vinyl ester and a glycol-terminated molecule eliminates acetaldehyde, the stable keto-form of the vinyl ester function. Activation energies for the thermal degradation of PET range [39] between 32 and 62 kcal mole^{-1}.

In the melt preparation of polycarbonates, the Kolbe reaction can occur [40], leading to chain-branching and crosslinking

Melt-spun PET always contains small concentrations of copolymerized diethylene glycol (3-oxa-1,5-pentanediol). The major mode of formation of these ether linkages does not apparently take place through a simple dehydration reaction between two glycol molecules, catalysed by traces of acidic impurities or by the free carboxylic acid of terephthalic acid (TA) or its proton. The reaction appears to involve the ester linkage [41]. In addition to diethylene glycol moieties, some dioxan (about one-tenth the proportion of diethylene glycol) was also produced. The rate of formation of these ether compounds was proportional to the concentration of hydroxyl and ester, with a rate coefficient k of about 33

$\times 10^{-8}$ 1 mole^{-1} sec^{-1}. The proposed reaction is analogous to the ester-interchange reaction, viz.

$$P\text{—}\overset{\overset{O}{\|}}{C}\text{—}O\text{—}G\text{—}OH \;+\; HO\text{—}G\text{—}OH \;\rightarrow\; P\text{—}\underset{OH}{\overset{\overset{O}{\|}}{C}} \;+\; \underset{G\text{—}OH}{O\text{—}G\text{—}OH}$$

The acid-catalysed etherification of glycol, between glycol and a hydroxy-ethyl-terminated molecule, or between two hydroxy-terminated polyester molecules cannot be totally discounted, however, for the rate of formation of diethylene glycol during the preparation of polyester by direct esterification of TA is greater than when DMT is used, and the former rate can be markedly lowered using certain buffers [42].

Dioxan is believed to be formed from a terminal ester group containing a diethylene glycol residue [41], viz.

$$P\text{—}\underset{\overset{\|}{O}}{C}\text{—}O\text{—}CH_2\text{—}CH_2 \cdots \longrightarrow\; P\text{—}\overset{\overset{\|}{O}}{C}\text{—}OH \;+\; \underset{CH_2\text{—}CH_2}{\overset{CH_2\text{—}CH_2}{O\qquad O}}$$

Various cyclic intermediates can be formed during the direct esterification of dicarboxylic acids with glycols. During the preparation of PET, thermal or acid-catalysed dehydration of ethylene glycol can occur, leading to the formation of acetaldehyde, most of which is lost by immediate volatilization. Further amounts of acetaldehyde are formed through cleavage of vinyl esters. Some of the acetaldehyde reacts with glycol to form the cyclic acetal, which can be detected in the off-gases from the reaction, viz.

$$HOCH_2CH_2OH + CH_3CHO \longrightarrow CH_3\text{—}\underset{O\text{—}CH_2}{\overset{O\text{—}CH_2}{CH}}\Big|$$

A major reaction occurring in the polyester equilibrium is the formation of cyclic oligomers. This reaction is of particular importance during the preparation of polyethylene terephthalate. Goodman and Nesbitt [43] isolated the cyclic trimer, tetramer and pentamer of ethylene terephthalate from polyester chips, along with the cyclic oligomer from one mole of ethylene glycol, one mole of diethylene glycol, and two moles of terephthalic acid. They found 1.3—1.7 wt. % of these cyclic oligomers in polyester chips.

Three distinct mechanisms could account for cyclooligomerization: (1) the cyclization of linear oligomers; (2) cyclodepolymerization proceeding from the ends of a polymer chain; (3) exchange-elimination reactions occurring within a chain or between nearby chains. The two exchange-elimination reactions are

$$
\begin{array}{c}
\text{--G--A--G--A--G} \\
| \\
\text{--A--G--A--G--A}
\end{array}
\rightleftharpoons
\begin{array}{c}
\text{--G--A--G} \\
| \\
\text{--A--G--A}
\end{array}
+
\begin{array}{c}
\text{A--G} \\
| \quad | \\
\text{G--A}
\end{array}
$$

$$
\begin{array}{c}
\text{--G--A--G--A--G--A--G--} \\
+ \\
\text{--G--A--G--A--G--A--}
\end{array}
\rightleftharpoons
\begin{array}{c}
\text{G} \\
\diagdown \text{A} \diagdown \text{G} \\
| \quad | \\
\text{G} \diagup \text{A}
\end{array}
+
\begin{array}{c}
\text{A--G} \\
| \quad | \\
\text{G--A}
\end{array}
+
\begin{array}{c}
\text{A} \diagup \text{G} \\
| \\
\text{G} \diagdown \text{A}
\end{array}
$$

Solvent-extracted polyester yarn, free from linear or cyclic oligomers, formed cyclic oligomers on remelting, making the first mechanism unlikely in the opinion of Goodman and Nesbitt [43], particularly as linear oligomers of ethylene terephthalate show no tendency to cyclize on heating [44]. The second mechanism was dismissed [43] on the grounds that there were insufficient hydroxyl groups present, thus indicating that exchange-elimination was the primary mode. Peebles et al. [45] argued that the second mechanism ("back-biting") was a more likely one. They isolated 2.0—3.8 wt. % cyclic oligomers from PET and additionally identified a further component as the cyclic pentamer containing one diethylene glycol residue.

In other polyesters, extensive depolymerization by formation of cyclic oligomers will occur when five- or six-membered rings can be formed; seven- to nine-membered rings will be formed only in small amounts, and larger rings will not be formed in significant quantities. For these reasons it is thus essentially impossible to prepare high-molecular-weight polyesters from lactic acid, glycolic acid, 4-hydroxybutyric acid, 5-hydroxypentanoic acid, or 6-hydroxycaproic acid by high-temperature polycondensation.

The elimination of a cyclic intermediate can serve to drive polycondensation forward in special cases. Thus heating the polycarbonate copolymer prepared from ethylene glycol bis(chlorocarbonic acid) and bisphenol A at 280°C in a vacuum leads to elimination of cyclic ethylene carbonate and the formation of the polycarbonate of bisphenol A [46].

For a consideration of more strictly kinetic aspects of polyesterification by direct esterification, it is convenient to discuss separately the cases of self-catalysis and catalysed esterification.

5.2.1 Self-catalysis

Hinshelwood et al. [47] first showed that esterification of monobasic acids with alcohols was second order with respect to acid in the absence of

504

catalysts, though these simple kinetics were not followed in all monomeric esterifications studied.

Flory [3, 48] found that the esterification reactions between model compounds on the one hand and polyfunctional reactants on the other are substantially identical. Thus the reaction of two monofunctional compounds (lauric acid, lauryl alcohol), of a bifunctional compound with a monofunctional one (adipic acid, lauryl alcohol) and two bifunctional compounds (adipic acid, decamethylene glycol) followed essentially third-order kinetics. In the absence of added strong-acid catalyst a second molecule of the carboxylic acid functions as catalyst. Thus when the concentrations (C) of the reacting groups are identical, the rate is given by:

$$-\mathrm{d}C/\mathrm{d}t = kC^3$$

where k is the rate coefficient. Integrating

$$2kt = 1/C^2 - \text{Constant}$$

If the reactions are third order from the outset the constant of integration is $1/C_0^2$ where C_0 is the starting concentration. Ignoring the small volume change caused by loss of water during esterification, the concentration at time t may be replaced by $C_0(1-p)$ where p is the extent of reaction, and so

$$2C_0^2 kt = 1/(1-p)^2 - \text{Constant}$$

Flory's data are plotted in Fig. 4. The model reactions (monofunctional reactants) are somewhat slower, but the differences are due primarily to

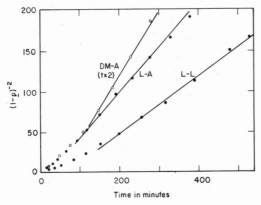

Fig. 4. Decamethylene glycol—adipic acid (DM—A), lauryl alcohol—adipic acid (L—A) and lauryl alcohol—lauric acid (L—L) reactions at 202°C (from ref. 48). Time values for the DM—A system have been multiplied by 2.

differing equivalent weights of functional groups in the three systems. In the cases shown and typically in all polyesterification reactions, the early stages do not follow third-order kinetics. Flory attributed this primarily to the rather pronounced changes in the dielectric properties of the reaction medium during the initial phase of the reaction.

Solomon et al. [35] carried out simple esterification reactions and polyesterification reactions in a medium of ester or polyester product, essentially removing from consideration changes in properties of the reaction medium. These workers extended Flory's polyesterification of adipic acid with decamethylene glycol up to 96.9% esterification of initial acidic groups; Flory's study stopped at 92.7% esterification. They confirmed Flory's earlier findings [48], and their data supported the hypothesis of equal reactivity of functional groups in polycondensation systems. Third-order rate coefficients were unchanged for various ratios of acid to alcohol. At very high extents of reaction, these uncatalysed esterifications became extremely slow. They considered the most likely mechanism to be $A_{AC}2$.

In Ingold's $A_{AC}2$ mechanism the slow step is the bimolecular reaction

$$RCOOH_2^+ + R'OH \rightleftharpoons RCOOHR'^+ + H_2O$$

where $RCOOH_2^+$ could be either a free ion or a component of an ion pair. In the self-catalysed reaction, the oxonium ions arise from autoprotolysis

$$2\ RCOOH \rightleftharpoons RCOO^- + RCOOH_2^+$$

which takes place simultaneously with protonation of the alcohol groups

$$RCOOH + R'OH \rightleftharpoons RCOO^- + R'OH_2^+$$

In reaction media of low dielectric constant, such as esters and polyesters, the concentrations of RCOOH and R'OH are essentially equal to the total concentrations of acid and alcohol, and the ions are probably largely associated as ion pairs. Under these conditions, the concentration of the ion pairs would be simply proportional to the square of the concentration of the acid RCOOH, and the kinetics would then be

$$\text{Rate} \propto [RCOOH]^2 [R'OH]$$

which agrees with experimental findings [35].

Other orders of reaction have been deduced from the results reported [48] by Flory, including second order [49], second order followed by third order [49], and a 2.5-order scheme [50]. These conflicting interpretations may be a consequence of ignoring or inadequately taking into consideration the complexities of kinetics under typical conditions of polycondensation. For example, reactions are often carried out in the absence of added solvent, so that the reaction medium changes from an equimolar mixture of starting reactants to the ester product, with a corresponding change in the polarity of the medium. The reaction solutions studied are almost certainly thermodynamically non-ideal owing to the high concentrations involved. Most meaningful polyesterification data probably come from studies at rather high conversions, where the acid and alcohol may be regarded as forming relatively ideal solutions in the polyester or ester product.

Tang and Yao [50] proposed a 2.5-order reaction scheme for polyesterification in the absence of catalysts. However, they were selective in the data points plotted, and the full plot departs considerably from linearity. The justification for the 2.5-order plot came from assuming that the reaction was proton-catalysed and that the proton arose from ionization of the acid

$$A-COOH \rightleftharpoons A-COO^- + H^+$$

For neutrality the concentrations of protons and anions are identical, so that if K_e is the ionization constant, then

$$[H^+][A-COO^-] = K_e[A-COOH] = [H^+]^2$$

Therefore,

$$[H^+] = K_e^{1/2}[A-COOH]^{1/2}$$

The kinetic equation then becomes

$$-d[A-COOH]/dt = k[COOH]^{1.5}[OH]$$

Huang et al. [51] believed that the ionization proposed by Tang was unlikely under conditions found in polyesterifications, though they supported a 2.5 kinetic order. Their mechanism involved a complex series of interactions between acid dimers.

Kemkes [51a] studied the esterification of terephthalic acid in a medium of oligomeric ethylene terephthalate. The rate coefficient for this esterification was greater than when the reaction was carried out in a medium composed primarily of ethylene glycol, where superatmospheric pressures were necessary. The former reaction is performed at atmospheric pressure. For these reasons, rate and pressure, commercial PET is most advantageously made from terephthalic acid by reacting the acid and glycol in a medium of oligomeric reaction product. Many patents describing this type of process are currently appearing in the patent literature. Kemkes [51a] assumed second order kinetics, arriving at values of the rate coefficient for the esterification of TA of 1.63×10^{-4}, 2.33×10^{-4}, 3.09×10^{-4}, and 4.31×10^{-4} kg equiv^{-1} sec^{-1} at 230, 240, 250 and 260°C, respectively. Catalysis by TA or its proton was not discussed.

Mares et al. [52] recently concluded that hydrogen-ion catalysis during this esterification was not likely and that catalysis by undissociated acid occurred together with a reaction in which, kinetically at least, no form of catalysis was involved. On the other hand, Vansco-Szmercsanyi et al. [53] concluded that polyesterification of maleic, fumaric and succinic acids with ethylene glycol or 1,2-propylene glycol was catalysed by protons. Much work clearly remains to be done on defining the detailed mechanism of catalysis by the acidic species during esterifications.

TABLE 3

Rate coefficients for uncatalysed esterification and polyesterification[a]

Reactants	Molar ratio COOH:OH	Reaction temp. (°C)	k_3[b]	Activation energy (kcal mole^{-1})
Lauryl alcohol, lauric acid[c]	1:1	195	31×10^{-5}	12.9
Lauryl alcohol, lauric acid[c]	1:1	163	11×10^{-5}	12.9
Lauryl alcohol, lauric acid[c]	1:2	195	33×10^{-5}	—
Lauryl alcohol, lauric acid[c]	2:1	195	28×10^{-5}	—
1,10-decanediol, adipic acid	1:1	190	22×10^{-5}	14.2
1,10-decanediol, adipic acid	1:1	161	7.5×10^{-5}	14.2
Diethylene glycol, adipic acid	1:1	190	3.5×10^{-5}	7.6
Diethylene glycol, adipic acid	1:1	161	2×10^{-5}	7.6

[a] Taken from ref. 35.
[b] k_3 in equiv^2 kg^{-2} sec^{-1}.
[c] In lauryl laurate reaction medium.

Table 3 lists determined third-order rate coefficients for self-catalysed esterifications and polyesterifications.

5.2.2 Catalysed esterification

In the presence of an added catalyst such as p-toluenesulphonic acid, simple esterification reactions and polyesterification reactions are second order [48]. Thus the kinetics of the catalysed reaction of lauric acid and lauryl alcohol in a medium of lauryl laurate closely parallels those of the polymer-forming reaction between adipic acid and 1,10-dodecanediol in a medium of polyester product. Second-order rate coefficients for the two reactions were [35], respectively, 45×10^{-4} equiv kg^{-1} sec^{-1} and 16×10^{-4} equiv kg^{-1} sec^{-1}.

The catalysed reaction can be analysed in a general fashion, allowing for simultaneous occurrence of catalysed and uncatalysed reactions. Thus, if

k_2 and k_3 are the rate coefficients for the second- and third-order reactions, and k_2 is a function of catalyst concentration but not of time, then

$$-dC/dt = k_3(C_0-C)^3 + k_2(C_0-C)^2$$

Integrating

$$k_2 t = \frac{k_3}{k_2} \ln \left[\frac{C_0-C}{k_3(C_0-C)+k_2} \cdot \frac{k_3 C_0 + k_2}{C_0} \right] - \frac{1}{C_0} + \frac{1}{C_0-C}$$

In cases where this equation was applied [35], an approximate value was first found for k_2 by treating the reaction as if it were pure second order. This value was then combined with values for k_3 found independently, at the same temperature and in the absence of catalyst. The result was that the first term in the integrated equation was insignificant in comparison with the last two, so that the reaction could justifiably be treated as a simple second-order process.

The polycondensation of adipic acid and diethylene glycol catalysed by p-toluenesulphonic acid was thus shown to be second order over a wide range of molecular weights [48, 54]. During the initial portions of the reaction the kinetics were more complex, perhaps because of changes in the character of the reaction medium.

Similarly, the self-condensation of 12-hydroxystearic acid in the molten state, catalysed by p-toluenesulphonic acid, obeys simple second-order kinetics for about 85% of the reaction [55]. The tailing-off noted in the latter stages of polycondensation, observed also in other systems, was ascribed [48] to depletion of the catalyst; at high degrees of conversion the concentrations of end groups from carboxylic acid and glycol and of sulphonic acid groups become comparable, and catalyst can therefore function as a chain-terminating additive. Bawn and Huglin [55] confirmed this by demonstrating continued linearity in the rate plot beyond 85% reaction when further catalyst was added.

5.3 TRANSESTERIFICATION

Transesterification (also known as ester exchange, ester interchange, or ester alcoholysis) is the most important of the esterification reactions. Polyesters of relatively low molecular weight (around 2000) prepared from aliphatic dicarboxylic acids and glycols are most commonly made by direct esterification (Section 5.2) with or without addition of an external acidic catalyst. Phosphoric acid, p-toluenesulphonic acid, and antimony pentafluoride have been used as catalysts. The preparation of PET by the direct esterification of TA with ethylene glycol was not practical until the last few years, when acid having the extremely high purity required for commercial production of acceptable polyester became available. The use of added acidic catalysts in this case is undesirable as they promote side reactions leading to the incorporation of ether linkages in the polymer chain (Section 5.2), and may in addition cause discolouration or poor stability of the polymer or of dyes on the polymer in fibre form. Accordingly, most PET is manufactured starting with dimethyl terephthalate, a relatively low-melting ester that can be purified with ease.

The transesterification of a diester and a glycol is customarily carried out using an excess of glycol to drive the equilibrium

$$n \text{ ROOC—A—COOR} + (n + 1)\text{HOGOH} \rightleftharpoons$$

$$\text{HOGO—[—}\underset{\substack{\| \\ O}}{C}\text{—A—}\underset{\substack{\| \\ O}}{C}\text{—OGO—]}_n\text{H} + 2n \text{ ROH}$$

to the right. Catalysts are usually employed for this step, where substantially all of the starting diester is converted to hydroxy-terminated monomer and oligomer ($n = 1$ to ~5).

After the more volatile alcohol ROH has been eliminated, further reaction proceeds by elimination of glycol, usually at higher temperatures and subsequently under vacuum, viz.

$$2 \text{ HOGO—}\underset{\substack{\| \\ O}}{C}\text{—P—}\underset{\substack{\| \\ O}}{C}\text{—OGOH} \rightleftharpoons$$

$$\text{HOGO—}\underset{\substack{\| \\ O}}{C}\text{—P—}\underset{\substack{\| \\ O}}{C}\text{—OGO—}\underset{\substack{\| \\ O}}{C}\text{—P—}\underset{\substack{\| \\ O}}{C}\text{—OGOH} + \text{HOGOH}.$$

It is convenient to consider the preparation of polyesters by transesterification as taking place in two stages (Sections 5.3.1 and 5.3.2) as shown by the last two reactions. The last, of course, also applies to the process of direct esterification when a two-fold or greater excess of volatile glycol is used. The glycolysis of DMT (Section 5.3.1) and the polycondensation stage (Section 5.3.2) are fundamentally the same, especially in regard to catalysis [56].

Most PET is prepared by catalysed ester-interchange between DMT and glycol, followed by catalysed polycondensation. In the case of PET prepared directly from TA, catalysts are customarily used only in the polycondensation stage. The choice of the best catalyst or catalysts is obviously of extreme importance in the commercial preparation of PET. Rates of reaction must be as fast as can practically be achieved, and colour-forming side reactions, degradations, or reactions that lead to the copolymerization of excessive amounts of diethylene glycol must be minimized.

References pp. 575—581

Zimmermann [57] roughly divided polyester catalysts into four groups of metal-containing compounds (Table 4). Groups I and II were the most catalytically active. Some side reactions were more seriously promoted by Group II catalysts, as measured by the carboxylic end group concentration, which can be taken as a measure of degradation, through chain splitting. In addition, zinc salts catalyse the formation of diethylene glycol more than antimony or manganese compounds, see for example ref. 42. Zimmermann's Group III catalysts promoted few chain-splitting degradations, but their catalytic activity is inadequate for the requirements of commercial production. Nickel salts (Group IV) were relatively poor catalysts and promoted extensive degradation.

TABLE 4
Effect of catalysts on the characteristics of PET[a]

Group	Catalyst	η_{Rel}	COOH End-groups (milliequiv g^{-1})	Polymer colour
I	Ti(OCH$_3$)$_4$	1.420	33	Yellow
	Fe(III) acetylacetonate	1.408	36	Brown
	Pb(OAc)$_2$	1.378	34	Yellow
	Sb(OCH$_3$)$_3$	1.372	27	Light grey
	Mn(OAc)$_2$	1.365	31	Yellow to brown
II	Zn(OAc)$_2$	1.375	75	Colourless to yellow
	Co(OAc)$_2$	1.368	43	Violet
	Cd(OAc)$_2$	1.328	35	Yellow to brown
III	Cr(III) acetylacetonate	1.216	24	Yellow
	GeO$_2$	1.168	15	Colourless to yellow
	LiOAc	1.262	14	Colourless to yellow
	NaOAc	1.162	16	Colourless to yellow
	Mg(OAc)$_2$	1.286	36	Colourless to yellow
	Ca(OAc)$_2$	1.126	12	Colourless to yellow
	H$_3$BO$_3$	1.138	23	Colourless to yellow
IV	Ni(OAc)$_2$	1.255	112	Green to brown

[a] From ref. 57. Products analysed after polycondensation at 280°C for three hours, using catalyst concentration of 1.2×10^{-3} mole per mole of ester.

Yoda et al. [58] assumed that the metal compounds induce a positive charge on the carbonyl carbon atom of the ester group of DMT, facilitating attack of the glycol. In model reactions, catalysts (M) more strongly activated 2-hydroxyethyl esters, perhaps through formation of a chelate [57]

Yoda [59] showed that in a series of metal acetates, those salts of metals having an electronegativity [60] lying between 1.0 and 1.7 (that is, calcium, manganese, zinc, cadmium, lead and cobalt) were most catalytically active.

Fontana [61] suggested that the mechanism of esterification or polyesterification involved a six-membered transition state

$$
\underset{\text{Reactants}}{
\begin{array}{c}
\text{R} \\
| \\
\text{O} \quad \text{O} \\
\diagdown \quad \diagup \\
\text{R-C} \quad\quad \text{H} \\
| \\
\text{R-O} \quad\quad \text{O} \\
\diagdown_{\text{M}} \quad {}^{\diagup}\text{R}
\end{array}}
\quad \rightleftharpoons \quad
\underset{\text{Transition State}}{
\begin{array}{c}
\text{R} \\
| \\
\text{O} \quad \text{O} \\
\diagdown \quad {\cdots} \\
\text{R-C}\cdots\text{H} \\
| \\
\text{R-O}\cdots \quad \text{O} \\
{}^{\cdots}\text{M}^{\cdots}\diagdown\text{R}
\end{array}}
\quad \rightleftharpoons \quad
\underset{\text{Products}}{
\begin{array}{c}
\text{R} \\
| \\
\text{O} \quad \text{O} \\
\diagdown \quad \diagdown \\
\text{R-C} \quad\quad \text{H} \\
\\
\text{R-O} \quad \diagup\text{O-R} \\
\quad \text{M}
\end{array}}
$$

R represents any long or short polymer chain or simple organic radical, M represents an atom of the catalytic metal. Fontana [62] has proposed similar "aromatic transition states" for the polymerization of olefins.

Though the transesterification and polyesterification reactions are not fundamentally different [56], the catalysts are. Thus Hovenkamp [56] distinguishes between highly active transesterification catalysts like compounds of manganese, lead or zinc, and antimony compounds which catalyse glycol elimination only. The transesterification catalysts are equally active in media that contain low or high concentrations of hydroxyl groups, but their catalytic activity is poisoned by small concentrations of carboxyl groups. Antimony compounds become catalytically most active at the later stages of polycondensation, where the hydroxyl concentration is relatively small, and their activity is insensitive to the presence of carboxyl groups.

In commercial practice, all PET is made using an antimony compound for the final polycondensation stage. The transesterification reaction between DMT and the glycol is catalysed by salts of manganese, zinc, calcium, cobalt, or other metals. At the end of the ester-interchange stage, when essentially all of the methanol has been evolved, the transesterification catalyst is converted to a catalytically inactive and substantially colourless form by reaction with a phosphorus compound such as triphenyl phosphate or phosphite. Polyesters of 1,4-cyclohexanedimethanol and DMT or TA are made using complex titanium catalysts.

Transesterifications of aliphatic carbonate esters with glycols are catalysed by alkali metal alkoxides. No catalyst is needed for the transesterification of diaryl carbonates with aliphatic diols. Alkyl carbonate esters and p-xylylene glycol undergo transesterification reactions when certain titanium compounds are used as catalysts. The preparation of aromatic polycarbonates by transesterification is best carried out with at least a slight excess of diphenyl carbonate using strongly basic catalysts such as lithium hydride [40]. The side reactions that can occur during transesterification reactions in the melt include loss of functionality, thermal degradation, formation of volatile and/or co-polymerizable ether-containing compounds, and the formation of cyclic materials such as cyclic oligomers. These reactions were discussed in Section 5.2.

The more precisely kinetic aspects of transesterification are divided into two sections in the following pages: ester-interchange between esters and glycols, and the polycondensation reaction. The latter is subdivided into sections on uncatalysed and catalysed polycondensation.

5.3.1 Ester-interchange between esters and glycols

Griehl and Schnock [63] reported that the precondensation reaction of DMT with ethylene glycol was first order in ester group alone, though the reaction was also proportional to catalyst concentration. Peebles and Wagner [64] found data for the same reaction consistent with two competitive, second-order reactions but not sufficiently inconsistent with single first-order kinetics to permit a decision. This interpretation of the reaction as two consecutive second-order reactions leads to the unlikely conclusion, however, that the methyl ester function on methyl 2-hydroxy-ethyl terephthalate is roughly three times more reactive than the ester group of DMT. Challa [65] later showed these reactivities to be the same; he found the equilibrium constants for the reactions

$$HOC_2H_4OH + 1,4\text{-}CH_3OCOC_6H_4COOCH_3 \rightleftharpoons$$

$$1,4\text{-}HOC_2H_4OCOC_6H_4COOCH_3 + CH_3OH$$

$$HOC_2H_4OH + 1,4\text{-}HOC_2H_4OCOC_6H_4COOCH_3 \rightleftharpoons$$

$$1,4\text{-}HOC_2H_4OCOC_6H_4COOC_2H_4OH + CH_3OH$$

to be 0.33 and 0.30, identical within experimental error.

Volume changes during polycondensation are frequently small and customarily ignored. However, during the reaction of DMT with an excess of glycol relatively large volume changes occur. To allow for this, Fontana [61] took into account the change in catalyst concentration (usually assumed constant). Treating the catalyst as a reagent in this manner led to the assignment of third-order kinetics to the reaction.

Fontana also reinterpreted some earlier data [63], and suggested the existence of a reversible dissociation reaction

$$\underset{\underset{O}{\|}}{P-C}-O-C_2H_4-O-\underset{\underset{O}{\|}}{C-P} \rightleftharpoons \underset{\underset{O}{\|}}{P-C}-OCH=CH_2 + HOOC-P$$

This interpretation was recently shown to be erroneous [66]. No evidence exists for the reversible dissociation, though there is little doubt that at elevated temperatures ester linkages can cleave to form carboxylic acid end-groups and vinyl esters (Section 5.2).

Neglecting catalyst concentration, then, the transesterification reaction between esters and glycols appears to be of the second order, with the rate proportional to ester and glycol concentrations.

In the case of polycarbonates, Turska and Wrobel [67] showed that the transesterification reaction between bisphenol A and diphenyl carbonate was third order, including a term for catalyst, when the reaction was carried out above 230°C. Below 230°C there were anomalous results.

5.3.2 Polycondensation by transesterification

(a) Uncatalysed reaction

Challa [68, 69] studied the uncatalysed polymerization of bis(2-hydroxyethyl)terephthalate and its oligomers to PET. The equilibrium constant varied with degree of conversion as shown in Fig. 5, but only slightly with temperature (Fig. 6).

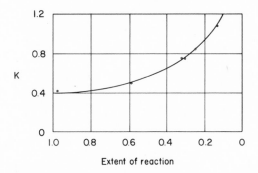

Fig. 5. Equilibrium constant for the polycondensation of bis(2-hydroxyethyl tereph-thalate) as a function of the extent of reaction (from ref. 68).

It is difficult to suppress the reverse (glycolysis) reaction that occurs during the preparation of polyesters, though Griehl and Schnock [63] virtually eliminated it. Their results and later data of Challa [69], who took the reverse reaction into account, agree quite well. Griehl and Schnock proposed first-order kinetics; however, if account is taken of the gradual increase in over-all rate coefficient, their data fit second-order kinetics better [69].

The second-order rate coefficients k and k', for the polycondensation and glycolysis reaction of PET, showed roughly the same dependence on

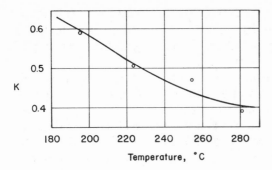

Fig. 6. Equilibrium constant for the polycondensation of bis(2-hydroxyethyl tereph-thalate) as a function of temperature (from ref. 68).

temperature [69], corresponding to an activation energy of about 22 kcal mole^{-1}. The rate coefficient for the forward reaction (Table 5) increased with degree of polymerization but the glycolysis rate was unchanged. Thus the ester linkages within a polyester chain are equally reactive in the reverse reaction, but all end groups are not equally reactive in polycondensation.

The principle of equal reactivity of polymer end groups has earlier been proposed by Flory [3, 48]. The intrinsic chemical characteristics of a functional group can often be altered by substitution but only when the substituent is introduced near the functional group. If the point of substitution is removed from the functional group by more than six atoms and assuming no conjugation intervenes, the effect is negligible. Thus the reactivity of the carboxyl group of lactic acid is probably altered somewhat by esterification of its hydroxyl group with a second molecule of lactic acid, but the effect of third and subsequent esterifications on the reactivity of this end group would be negligible. It is clear, however, that in the earliest stages of polycondensation or polymerization, the intrinsic reactivity of the functional group may change. Flory's discussions [48] include this possible exception to generalized equal reactivity, but the point has often been overlooked in later publications.

Challa [69] found that the monomer content of the polyesters was greater than that predicted by the Flory—Schulz distribution function, the so-called "most probable distribution" of molecular species. Challa proposed that the monomer molecule — in this case bis(2-hydroxy-ethyl)terephthalate, though the conclusion could be general for all polycondensations — lost more entropy on entering the transition state than did the longer molecules.

Support for this proposal was given in another paper by Challa [70], where the redistribution reaction between end groups and chain bonds of polyesters was examined, viz.

TABLE 5
Rate coefficients and activation energies for uncatalysed and catalysed polycondensations by transesterification

Product	Reaction temp. (°C)	Catalyst	Rate coefficient	Activation energy (kcal mole^{-1})	Ref.
PES[a]	250	Mn(OAc)$_2$; 5 x 10^{-4} mole equiv.	1.68 x 10^{-3}[g]	7.7 ± 1.6	34
PES/PET[a,b]	250	Mn(OAc)$_2$; 5 x 10^{-4} mole equiv.	1.52 x 10^{-3}[g]	—	34
PET[a]	250	Mn(OAc)$_2$; 5 x 10^{-4} mole equiv.	1.10 x 10^{-3}[g]	9.7 ± 1.4	34
PET	221	Sb$_2$O$_3$; 0.022 wt %	1.00 x 10^{-5}[h]	29	71
PET[c]	231	Sb$_2$O$_3$; 0.025 wt %	4.03 x 10^{-5}[h]	29	71
PET[c]	241	Sb$_2$O$_3$; 0.025 wt %	3.97 x 10^{-5}[h]	29	71
PET[c]	251	Sb$_2$O$_3$; 0.025 wt %	7.83 x 10^{-5}[h]	29	71
PET[d]	275	Sb$_2$O$_3$; 0.025 wt %	10 x 10^{-3}[h]	14	72
PET[d]	275	None	0.5 x 10^{-3}[h]	45	72
PET[e]	223	None	0.39 x 10^{-5}[h]	22	69
PET[e]	254	None	1.31 x 10^{-5}[h]	22	69
PET[f]	223	None	0.58 x 10^{-5}[h]	22	69
PET[f]	254	None	2.28 x 10^{-5}[h]	22	69

[a] Films, 3—4 mm thick, 0.70 < p < 0.97. PES is poly(ethylene sebacate).
[b] Equimolar amounts of sebacate and terephthalate.
[c] Mean equilibrium constant 0.36, p = 0.70.
[d] Films (1 mil) on rhodium.
[e] Mean equilibrium constant 0.49, p = 0.43.
[f] Mean equilibrium constant 0.73, p = 0.67.
[g] In g mmole^{-1} sec^{-1}.
[h] In l mole^{-1} sec^{-1}.

$$
\begin{array}{c}
\underset{\displaystyle\overset{\displaystyle O}{\|}}{P-C}-OGO-\underset{\displaystyle\overset{\displaystyle O}{\|}}{C}-P \\[4pt]
+ \\[2pt]
P-\underset{\displaystyle\underset{\displaystyle O}{\|}}{C}-OGOH
\end{array}
\quad \rightleftharpoons \quad
\begin{array}{c}
\underset{\displaystyle\overset{\displaystyle O}{\|}}{P-C} \\[6pt]
P-\underset{\displaystyle\underset{\displaystyle O}{\|}}{C}-O-G-O
\end{array}
\quad + \quad
\begin{array}{c}
\underset{\displaystyle\underset{\displaystyle H}{|}}{O}-G-O-\underset{\displaystyle\overset{\displaystyle O}{\|}}{C}-P
\end{array}
$$

The redistribution rate coefficient was calculated from the rate of formation of monomer distilled from molten monomer-free precondensate, making allowance for monomer formation through the glycolysis (reverse) reaction, which is not included in the redistribution reaction. In this way an energy of activation of about 31 kcal mole^{-1} was calculated for the redistribution reaction. The entropy of activation for the redistribution reaction was $\Delta S^{\ddagger} = -23.4$ cal deg^{-1} mole^{-1}, whereas for the glycolysis reaction [69], $\Delta S^{\ddagger} = -37.8$ cal deg^{-1} mole^{-1}, indicating that the very mobile molecule of glycol loses far more entropy in reacting with an ester linkage than the polymer molecules do in the redistribution interchange. The increase in the rate of reaction with decreasing concentration of hydroxyl groups was recently confirmed [56].

(b) Catalysed reactions

Most of the reported work on the kinetics of catalysed polycondensation by transesterification is quite recent and in the main deals with the preparation of PET, though related work on other esters from ethylene glycol has been reported. Reinisch et al. [34] found that the melt polycondensation leading to poly(ethylene sebacate) in 3—4 mm films under vacuum, with manganous acetate as catalyst, was second order from $p = 0.70$ to $p = 0.97$. Rate coefficients are listed in Table 5 for this reaction, for the formation of poly(ethylene sebacate-co-terephthalate) and for PET.

Stevenson and Nettleton [71] studied the polycondensation of low-molecular-weight PET over the range 221—251°C in the presence of antimony trioxide catalyst (Table 5), essentially paralleling Challa's earlier research [68, 69, 70] on the uncatalysed reactions. The catalysed reactions were faster, but the rate coefficient for polycondensation was relatively low even though the reverse reaction had been taken into account. They considered that a vacuum-volatile constituent of the reaction mixture was responsible for deactivation of the catalyst, thus slowing the reaction. Antimony trioxide does not vigorously catalyse the polycondensation of the pure monomer, bis(2-hydroxyethyl)-terephthalate, and Stevenson and Nettleton concluded that the latter substance was responsible for the proposed deactivation of the catalyst.

Later work by Stevenson [72] supported this hypothesis. The preparation of PET catalysed by antimony trioxide was studied in thin films on metal surfaces that were carefully selected to avoid catalysis by surface effects or by dissolved metal; as mentioned earlier, a large number of metals and their oxides, salts or other derivatives catalyse the polyesterification reaction. On inactive surfaces like silver or rhodium the catalysed polycondensation rate increased with decrease in film thickness. In the absence of added catalyst there was no tendency for the rate to increase with decreasing film thickness. Stevenson proposed that in thin films the catalyst-deactivating component was more readily lost, thereby increasing the reaction rate.

Cefelin and Malek [73] polymerized bis(2-hydroxyethyl)terephthalate in α-methyl naphthalene at 260°C, showing that the reaction was second order in the molecular-weight region of 3000—30,000 using antimony trioxide as catalyst. The reaction curve was non-linear below a molecular weight of 3000 even when 0.25, 1.0 or 5.0 mole % of 2-hydroxyethyl benzoate was present in the original reaction mixture. Copolymerization of 2.5, 5.0 or 10.0 mole % of bis(2-hydroxyethyl)isophthalate did not influence the progress of the reaction, though the melting point of the product was lowered 2.5°C per mole % copolymer unit. Glycol was removed as an azeotrope with the solvent. They concluded that side reactions were a major factor governing the shape of the kinetic curve in the early stages of polycondensation.

Table 5 shows data for catalysed and uncatalysed transesterification-polymerization reactions. In summary, there is general agreement that this reaction is second order, but there are quite wide variations in reported data. These variations may result from one or more causes. Reverse reactions such as glycolysis and hydrolysis [74], which are particularly difficult to suppress fully in those reactions studied at atmospheric pressure, are often neglected. The effects of stirring rate (or the absence of stirring) and vessel geometry on the rate of loss of water, glycol or other volatile components such as low-molecular-weight glycol esters are not thoroughly understood though they clearly affect data measured. Changes in volume are sometimes disregarded. In spite of recent careful research, there is clearly still room for further studies on polycondensation by transesterification.

A chemically interesting variant of the transesterification reaction is the use of enol esters, where formation of the stable keto-form serves to drive the reaction forward. For example

$$n \; \underset{CH_2}{\overset{CH_3}{\diagdown}} C-OCO(CH_2)_8 COOC \underset{CH_2}{\overset{CH_3}{\diagup}} + n \; HOCH_2 C_6 H_{10} CH_2 OH \; \rightleftharpoons$$

$$-[-OCO(CH_2)_8 COOCH_2 C_6 H_{10} CH_2-]_{\overline{n}} + n \; CH_3 COCH_3.$$

5.4 ACIDOLYSIS

Esters can be prepared by acidolysis, as in the reaction of a carboxylic acid with an acetate ester leading to the elimination of acetic acid. Acidolysis is also, of course, one of the many reactions occurring during conventional polyesterification in the melt, viz.

$$
\begin{array}{l}
\text{P--C--OH} + \text{HO--C--A--C--OG--O--P} \;\rightleftharpoons \\
\quad \Vert \qquad\qquad \Vert \quad\;\; \Vert \\
\quad \text{O} \qquad\qquad \text{O} \;\;\;\; \text{O} \\[4pt]
\text{P--C--O--G--O--P} + \text{HOOC--A--COOH} \\
\quad \Vert \\
\quad \text{O}
\end{array}
$$

This acid exchange parallels the alcoholysis reaction (Section 5.3) but in practice the latter is more important owing to the volatility of the glycol. In general the acidolysis reaction shown above does not contribute to the build-up of molecular weight.

Bisphenol esters react with dibasic acids to form polyesters. In a perhaps related fashion, bisphenols and dibasic acids can be polymerized in an excess of acetic anhydride [40].

No major study has been made of this mode of polyesterification. Reports have been published [32] of model compound studies on the acidolysis of ethyl stearate in acetic acid and of ethyl stearate by acetic acid in trioxane solution.

5.5 ESTEROLYSIS

Glycol esters and carboxylic esters can also react to form polyester

$$
n\,\text{R'OCO--A--COOR'} + n\,\text{RCOOGOCOR} \;\rightleftharpoons
$$

$$
-[-\text{CO--A--COO--G--O--}]_{\overline{n}} + 2n\,\text{RCOOR'}
$$

Thus PET has been prepared by reaction of DMT and diacetoxyethane. Similarly, polycarbonates have been made from the bis(alkyl carbonates) or bis(aryl carbonates) of bisphenols. No important kinetic studies have been made of the esterolysis (or double ester-exchange) reaction.

5.6 POLYESTERS FROM ANHYDRIDES

The reaction of glycols with either maleic or phthalic anhydride has been commercially applied to the preparation of polyesters. The reactions involve two distinct steps with the second, the esterification of the free carboxyl group, being slower, viz.

$$O=C \overset{\overset{\displaystyle O}{}}{\diagdown} C=O$$

(benzene ring) + HOGOH $\longrightarrow$

$$\overset{HOGO}{\diagdown} C \overset{\overset{\displaystyle O}{\diagup}}{\diagup} \quad COOH$$

(benzene ring) + H_2O

$$n \; HOGO \diagdown C \overset{\overset{\displaystyle O}{\diagup}}{} COOH$$

(benzene ring) $\rightleftharpoons$

$$\left[\overset{\overset{\displaystyle O}{\parallel}}{C} \overset{\overset{\displaystyle O}{\parallel}}{C} - O - G - O \right]_n + n H_2O$$

(benzene ring)

More usually, phthalic anhydride is reacted with a polyfunctional material such as glycerol to form a network polyester (a glyptal). This class of polymer, modified with monofunctional additives and known as alkyd resins, is very important in the paint and plastics industry. Consequently most of the theoretical studies have been concerned with the formation of infinite networks ("microgel") and the distinction between this and the so-called gel-point [75].

Pentaerythritol reacts with phthalic anhydride at a rate roughly five times greater than with tetrachlorophthalic anhydride; the respective activation energies [32] are 24.0 and 27.3 kcal mole^{-1}. Other polyesters based on phthalic anhydride have been surveyed [32].

5.7 POLYESTERS FROM ACID CHLORIDES

Diacyl chlorides react with glycols or other diols such as bisphenols without catalyst, forming polyesters in the presence of an acid acceptor; removal of the hydrochloric acid formed is important to the equilibrium

$$n \; Cl-\underset{\underset{O}{\parallel}}{C}-A-\underset{\underset{O}{\parallel}}{C}-Cl + n \; HOG-OH \; \rightleftharpoons$$

$$-[-\underset{\underset{O}{\parallel}}{C}-A-\underset{\underset{O}{\parallel}}{C}-O-G-O-]_n + 2n \; HCl$$

These reactions can be carried out in the melt [76], in solution using inert solvents, or by interfacial polymerization.

At the higher temperatures used in melt polycondensations it is necessary to remove the liberated hydrogen chloride by bubbling inert gas through the reaction medium. Side reactions are common, and this polymerization method is used only rarely. No major kinetic studies have been made.

References pp. 575—581

During polyesterifications in solution, starting with acid chlorides, the hydrogen chloride that is evolved is generally removed by a stream of inert gas; the progress of the reaction can be followed by titration of an alkaline solution of the effluent gas. Side reactions also limit this technique. Entelis et al. [77] studied the reaction of terephthaloyl chloride and a very large excess of ethylene glycol in dioxane or dioxane/acetonitrile mixtures; the glycol concentration was comparable with that of the inert solvent. The reaction was second order. The rate coefficient increased with increasing dielectric constant of the medium. The second-order kinetics indicated that there was no ionization of the acid chloride as a preliminary, rate-determining step as Ingold [33] had suggested for similar reactions. Entelis et al. proposed that the reaction took place in two steps, with the second being the rate-determining stage

$$Cl-\underset{\underset{O}{\|}}{C}-\langle C_6H_4 \rangle-\underset{\underset{O}{\|}}{C}-Cl + HO-G-OH \longrightarrow$$

$$ClC(=O)-\langle C_6H_4 \rangle-C(=O)-OGOH + HCl$$

$$HOGOC(=O)-\langle C_6H_4 \rangle-C(=O)-Cl + HO-G-OH \longrightarrow$$

$$HO-G-OC(=O)-\langle C_6H_4 \rangle-C(=O)-\overset{+}{O}(H)-G-OH \cdot Cl^-$$

The size of the "active portion" of the activated complex in the rate-determining step was estimated [77] to be 2 ± 0.1 Å, supporting the proposal that the reaction product first exists as a protonated molecule.

Interfacial polycondensation is performed at low or moderate temperatures. The reaction of aliphatic glycols with acid chlorides is too slow relative to the competing hydrolysis by hydroxyl ions, or even water, for the formation of high-molecular-weight polymers. Thus this technique is limited to the preparation of polyesters from aromatic diols such as the bisphenols. Aliphatic diacid chlorides can be used, but aromatic diacid chlorides give products of higher molecular weight [78]. Reaction takes place at or near the interface between the aqueous alkaline solution of the bisphenol (thus present as the phenoxide ion) and the organic phase containing the dissolved acid chloride. Excess of base in the aqueous phase serves as acceptor for the hydrochloric acid produced. The water--immiscible organic solvent should preferably be a solvent or swelling agent for the polymeric product. When the dibasic acid chloride is phosgene, the products are polycarbonates. Most polycarbonates are made by interfacial polycondensation in this manner.

When the polycarbonate of bisphenol A was made interfacially in a non-solvent for the polymer the distribution of molecular weights showed two clear maxima and was unusually broad [40]. This was probably a consequence of the two-step process often employed in the preparation of polycarbonates; here, the first step yields low-molecular-weight polymer from phosgene and the aqueous alkaline solution of the aromatic diol, then an accelerator salt and more alkali and phosgene are added in the second step. Polycarbonates prepared in the melt and in solution show the expected essentially statistical distribution of molecular weights [40]. Polycarbonates have also been prepared from bisphenols and bisphenol bischloroformates; studies of this reaction in nitrobenzene solution have shown it to be second order [79].

Many factors affect the progress of the interfacial polymerization reaction, including choice of solvent, the use of detergents as emulsifiers, temperature, rate of addition of one monomer, and rate of stirring [6]. However, no major study has yet been made of reaction rates, the extent of formation of unreactive cyclic oligomers, and the nature and extent of various possible side reactions in this class of polycondensation.

5.8 MISCELLANEOUS REACTIONS

Polyesters have been synthesized by the reaction of glycols with the diammonium or bis(triethylammonium) salts or N-methylpyrrolidone complexes of dibasic acids. Dicarboxamides or dinitriles also react with glycols to form polyesters. Alkylene halides and salts of terephthalic acid react to give polyesters. In general, however, all these reactions give low-molecular-weight products, and no kinetic studies have been made.

6. Polyamidation and related polycondensations

Nylon-6,6, poly(hexamethylene adipamide), prepared by reacting substantially equimolar quantities of adipic acid and hexamethylene diamine, has for many years been the world's major wholly synthetic fibre. The rapid rate of growth of polyester production indicates that nylon will soon be supplanted as the major fibre by poly(ethylene terephthalate).

Nylon-6,6, discovered by Carothers in 1929, is by far the most important polyamide prepared by condensation polymerization. Nylon 6, prepared by the ring-opening (non-condensation) polymerization of ε-caprolactam, will not be discussed in this chapter. Other commercially important polyamides prepared by condensation techniques include nylon-6,10, from hexamethylene diamine and sebacic acid, poly(m-phenylene isophthalamide) from *m*-phenylene diamine and isophthaloyl chloride, and the new polyamide Qiana, which is probably made from 4,4'-methylenedicyclohexylamine and dodecanedioic acid.

Polyureas, which are strictly speaking polyamides of carbamic acid (and which can indeed be prepared by reacting diamines with phosgene) differ in many ways from the other polyamides. This class of polymer is discussed briefly in Section 8.

6.1 SYNTHETIC ROUTES

The most important method used in the preparation of polyamides is direct amidation, usually through the intermediate formation of a salt of the diamine and dicarboxylic acid, but without it in the case of aminoacids or for pairs of monomers that do not readily form a salt. Esters can react with diamines to form polyamides with liberation of alcohol or phenol. Diamines can be reacted with diamides yielding polyamides and freeing ammonia. Polyamides have been prepared by acidolysis of acyl derivatives of diamines (compare Section 5.4 for acidolysis in polyester preparation). Bis-anhydrides react with diamines to form polyamides and, if reacted further, polyimides. The low-temperature reaction of acid chlorides with diamines has been used, interfacially or as a solution technique, to prepare certain polyamides (compare Section 5.7 for related reactions in polyester synthesis).

Flory [3] proposed that the distribution of molecular weights resulting from the random interactions of growing polymer molecules would follow the "most probable distribution". This distribution is a consequence of Flory's hypothesis [3] of equal reactivity of polymer end groups. In general, polyamides prepared by melt processes contain proportions of various molecular weights that follow this predicted geometric distribution. Solution polycondensation in the presence of a solvent for the polymeric product also generally produces this statistical distribution of molecular weights. When polymer is precipitated during interfacial polymerizations using a non-solvent for the polymer, the distribution of molecular weights tends to be broader. Little work has been reported on this subject for the case of polyamides. The number-average molecular weights of nylon-6,10 samples made interfacially in stirred and unstirred systems were lower than the number-average molecular weights of melt-prepared nylon-6,10 samples having the same intrinsic viscosity [6]. Many additional factors govern the progress of an interfacial polyamidation, including stirring rate and monomer concentration. It is possible that under conditions of extremely high-speed stirring, interfacial polyamidation could become kinetically controlled, and the distribution of molecular weights would be statistical. Dilute solutions would be expected to produce relatively larger amounts of cyclic monomer, though in some cases better yields of polymer and higher molecular weights were obtained at greater dilutions [6].

6.2 DIRECT AMIDATION

Nylon-6,6 is made commercially from adipic acid and hexamethylene diamine with the intermediate formation of a salt (see below). Isolation of this salt provides a means of further purification to the stringent levels required for polymerization to high molecular weight.

$$n \text{ HOOC(CH}_2)_4\text{COOH} + n \text{ H}_2\text{N(CH}_2)_6\text{NH}_2 =$$

$$n \text{ } (\bar{\text{O}}\text{OC(CH}_2)_4\text{COO}^-)(\text{H}_3\text{N}^+(\text{CH}_2)_6\text{N}^+\text{H}_3)$$

$$n \text{ } (\bar{\text{O}}\text{OC(CH}_2)_4\text{COO}^-)(\text{H}_3\text{N}^+(\text{CH}_2)_6{}^+\text{NH}_3) \rightleftharpoons$$

$$\text{+CO(CH}_2)_4\text{CONH(CH}_2)_6\text{NH+}_n + 2n \text{ H}_2\text{O}$$

Polyamides from long-chain aminoacids can be made directly. For example, nylon-11 is synthesized from 11-aminoundecanoic acid

$$n \text{ H}_2\text{N(CH}_2)_{10}\text{COOH} \rightleftharpoons \text{+NH(CH}_2)_{10}\text{CO+}_n + n \text{ H}_2\text{O}$$

Shorter-chain aminoacids form more or less major quantities of cyclic product (lactam) under polymerization conditions.

In certain other cases, where the diamine and dicarboxylic acid do not readily form a salt, as in the case of 4,4'-methylenedianiline and sebacic acid, direct amidation is used to prepare the polymer [80].

The polycondensation reaction may be written

$$\text{P}-\text{NH}_2 + \text{P}-\text{COOH} \rightleftharpoons \text{P}-\text{CONH}-\text{P} + \text{H}_2\text{O}$$

If ΔH is the enthalpy change and B is the temperature-independent part of the equilibrium constant K, then

$$K = \frac{[-\text{CONH}-][\text{H}_2\text{O}]}{[-\text{COOH}][-\text{NH}_2]} = B \exp(-\Delta H/RT)$$

There is experimental evidence [81] that K decreases with increasing concentration of water, but for ranges of molecular weight of particular interest, K is essentially constant. For at least a wide range of polyamides, K has a value of roughly 250 to 350 after extrapolation to 280°C [81]. The equilibrium water concentration in a particular polyamide held under fixed conditions of steam pressure is characteristic of that polymer. At atmospheric pressure and 280°C, this equilibrium concentration is about 0.16% for nylon-6,6. If K is taken as 250, and the concentration of amide groups is about 8800 equiv per 10^6 g, the equilibrium value for $[-COOH][-NH_2]$ is 3140 (equiv per 10^6 g)^2.

For a fixed molecular weight the total number of end groups is fixed. In the case of A—B polyamides, derived from amino-acids, the end group concentrations of $-NH_2$ and $-COOH$ are clearly identical. In the case of AA—BB classes of polymers, such as nylon-6,6, the actual concentrations of amine and carboxylic end groups will vary depending on several factors. Most important is the deliberate or accidental use of non-stoichiometric amounts of diamine and dicarboxylic acids; accidental imbalance could result from the presence of unrecognized impurities. End-group imbalance can arise during the course of the reaction by side reactions of various kinds, as briefly discussed later in this section.

The mechanism of the direct amidation reaction between diamines and dicarboxylic acids is not clear, though it certainly proceeds through a tetrahedral intermediate. It is not known whether the amide-formation step involves the substituted ammonium salt or whether small equilibrium amounts of free acid and free amine that are always present interact. The latter would seem more likely.

Attempts to prepare polyamides by direct amidation can fail for various reasons. Thus, lactams are readily formed from γ- or δ-amino acids. For example

When α-amino acids are heated, diketopiperazines are formed

Direct amidation at room temperature is sometimes possible. For instance, peptides have been synthesized directly from amines and acids at room temperature using dehydrating agents such as dicyclohexyl-carbodiimide [82]. N,N′,-carbonyldiimidazole, viz.

has also been employed in peptide synthesis [83]. In both cases, the acid anhydride is formed as an intermediate.

As in other polycondensations at high temperatures, there are several side reactions that occur during the melt polycondensation of diamines and dicarboxylic acids. One has already been mentioned in Section 5.2: decarboxylation. As shown in the list of decarboxylation temperatures for a homologous series of dicarboxylic acids (Table 2), adipic acid begins to decarboxylate at around $300°C$. In commercial practice, the maximum temperature that nylon-6,6 experiences during its preparation is about $290°C$, so that decarboxylation is one of the most important degradation reactions of nylon-6,6. Nylon-6 and nylon-6,10 are more thermally stable than nylon-6,6.

The reaction of two carboxylic end groups with the liberation of water and carbon dioxide has been proposed as a side reaction that could occur during polyamidation, viz.

$$2\ P\!-\!(CH_2)_n COOH\ \rightarrow\ P\!-\!(CH_2)_n CO(CH_2)_n\!-\!P + H_2O + CO_2$$

No evidence has been found for this reaction, but the parallel reaction between amine ends is known to occur. Bis(6-aminohexyl)amine has been found among the hydrolysis products of nylon-6,6, viz.

$$2\ P\!-\!\underset{\underset{O}{\|}}{C}\!-\!NH(CH_2)_6\!-\!NH_2\ \rightarrow$$

$$P\!-\!\underset{\underset{O}{\|}}{C}\!-\!NH(CH_2)_6 NH(CH_2)_6 NH\underset{\underset{O}{\|}}{C}\!-\!P\ +\ NH_3$$

hydrolysis

$$2\ P\!-\!\underset{\underset{O}{\|}}{C}\!-\!OH + H_2N(CH_2)_6 NH(CH_2)_6 NH_2$$

The secondary amine formed in this way during polymerization is a site for branching

$$P\!-\!NH\!-\!P + P\!-\!COOH\ \rightarrow\ P\!-\!\underset{\underset{P}{\overset{|}{\underset{|}{C=O}}}}{N}\!-\!P\quad +\ H_2O$$

This thermal degradation of amine-ended polymer and the later chain branching and cross-linking may be a major cause of the "gels" of insoluble, highly crosslinked material that form to greater or lesser extents during commercial preparation of nylon-6,6.

The weakest bond in the aliphatic polyamide chain is said to be that between the carbon atom of the carbonyl and the nitrogen atom of the —NH— function [84]. Cleavage of such bonds anywhere within the molecular chain of nylon-6,6 would produce a fragment that could cyclize to form carbon monoxide and cyclopentanone

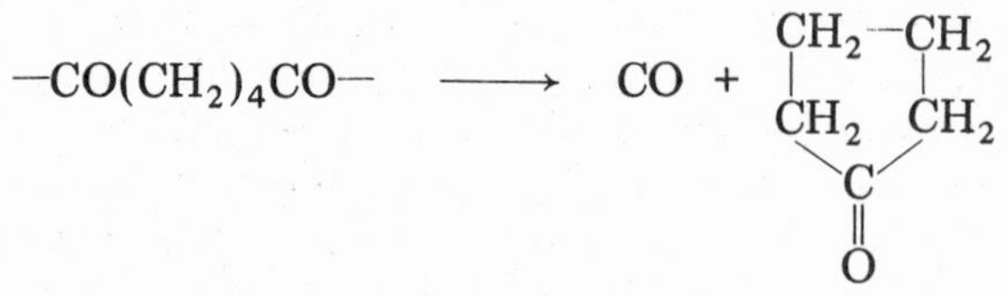

$$-CO(CH_2)_4CO- \longrightarrow CO + \begin{matrix} CH_2-CH_2 \\ | \quad\quad | \\ CH_2 \quad CH_2 \\ \diagdown C \diagup \\ \| \\ O \end{matrix}$$

Scission near the end of a polymer chain would give the fragment —$CO(CH_2)_4COOH$, which could also cyclize to form cyclopentanone, together with carbon dioxide.

More carbon dioxide is formed during polyamidation at high temperatures than would be expected from the simple cleavages just described. Most likely this additional carbon dioxide is formed by decarboxylation of acid-terminated polymer chains.

Straus and Wall [85] believed that the bonds (I, II) β to the carbonyl function in a polyamide were the weakest, with the carbon—nitrogen bond (II) being the weaker, viz.

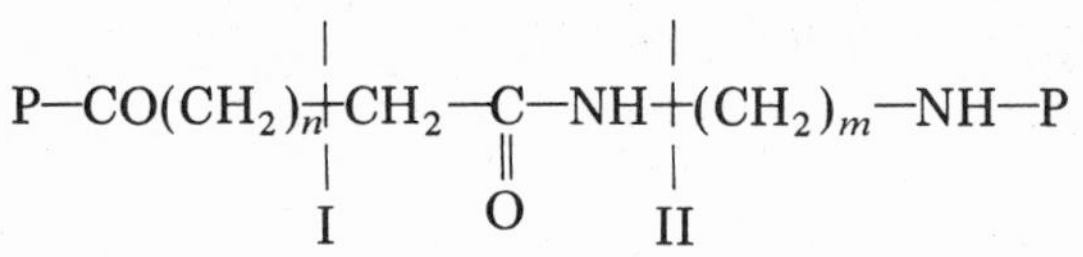

$$P-CO(CH_2)_n\overset{|}{+}CH_2-\underset{\underset{O}{\|}}{C}-NH\overset{|}{+}(CH_2)_m-NH-P$$
$$\quad\quad\quad\quad I \quad\quad O \quad II$$

Cleavage at II could produce radicals that could disproportionate as shown below for nylon-6,6

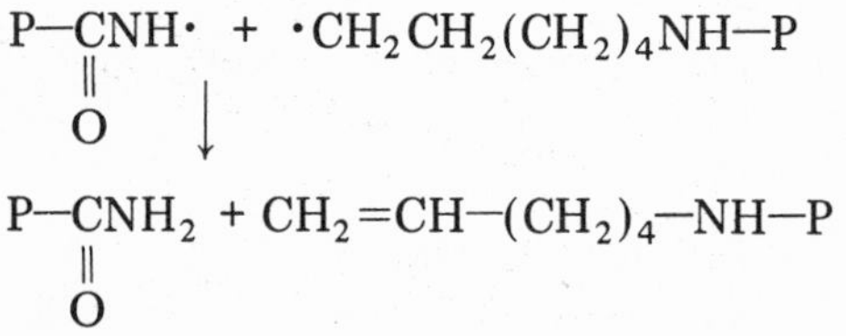

$$P-\underset{\underset{O}{\|}}{C}NH\cdot + \cdot CH_2CH_2(CH_2)_4NH-P$$
$$\quad\quad\quad\quad\quad \downarrow$$
$$P-\underset{\underset{O}{\|}}{C}NH_2 + CH_2=CH-(CH_2)_4-NH-P$$

Unsaturated end groups do indeed occur in nylon-6,6. The amide could dehydrate to form a nitrile end group. Such end groups have been detected in nylon-6,6.

Another type of disproportionation would lead to an end group containing a saturated hydrocarbon function and an intermediate that would rearrange to form an isocyanate

$$P-\underset{\underset{O}{\|}}{C}NH\cdot \ + \ \cdot CH_2CH_2(CH_2)_4NH-P$$

$$\downarrow$$

$$P-\underset{\underset{O}{\|}}{C}N\diagdown \ + \ CH_3(CH_2)_5NH-P$$

$$\downarrow$$

$$P-NCO$$

Again, these end groups have been detected in nylon-6,6.

A side reaction often occurring in polycondensation reactions that take place in the molten state is the formation of cyclic oligomers. As already mentioned (Section 5.2), this is an important group of reactions in the preparation of poly(ethylene terephthalate). The reactions that lead to cyclooligomerization are, of course, not degradative; the cyclic material is in equilibrium with linear polymer molecules.

In the case of nylon-6,6 roughly 1—2 wt. % of cyclic oligomers are formed. The ring size of the smallest oligomer is large, viz.

$$\underset{HN\diagdown \underset{(CH_2)_6}{}\diagup NH}{\overset{O\diagdown \overset{(CH_2)_4}{C}\diagup O}{}}$$

For nylon-6,10 the smallest cyclooligomer contains 18 atoms in a ring. However, for nylons prepared from amino-acids, much smaller rings are possible. Nylon-6 at 270°C contains about 8 wt. % of caprolactam, the cyclic monomer [81]. Nylon-4 degrades thermally to form the very stable cyclic monomer, pyrrolidone, and the attempted preparation of nylon-4 at conventional high temperatures produces very little polymer [86].

The polyamidation reaction when carried out in the melt appears to follow second-order kinetics [87—90]. Thus, for equal concentrations (C) of $-NH_2$ and $-COOH$ groups, the second-order equation applies, viz.

$$-dC/dt = kC^2$$

The integrated form of this equation is

$$C = \frac{C_0}{1 + C_0 kt}$$

When end group concentrations are relatively low, as at conversions greater than roughly 90%, a third-order reaction becomes increasingly important. Under these conditions the formation of the amide group seems to be catalysed by carboxyl, though there is still a strong

References pp. 575—581

second-order contribution to the rate of reaction in some cases [91]. For the third-order reaction

$$\frac{-d[-COOH]}{dt} = k_1[-COOH]^2[-NH_2]$$

For equimolar quantities of diamine and dicarboxylic acid, this equation simplifies to

$$\frac{-d[-COOH]}{dt} = k_1[-COOH]^3$$

Integrating, we have

$$\frac{1}{[-COOH]^2} - \frac{1}{[-COOH]_0^2} = 2\,k_1 t,$$

where $[-COOH]_0$ is the concentration of carboxyl group at time $t = 0$.

The carboxyl-catalysed reverse reaction (hydrolysis) must be taken into account. If k_2 is the rate coefficient for hydrolysis, then

$$-d[-COOH]/dt = k_1[-COOH]^2[-NH_2] -$$

$$k_2[-COOH][-CONH-][H_2O]$$

For kinetic studies at high conversion, when the concentration of amide groups may be regarded as constant, the substitution

$$k_2[-CONH-][H_2O] = k_1[-COOH]_{equ}[-NH_2]_{equ}$$

can be introduced. Using this substitution, the third-order equation for the carboxyl-catalysed reaction is

$$-d[-COOH]/dt = k_1[-COOH]$$

$$([-COOH][NH_2] - [-COOH]_{equ}[-NH_2]_{equ})$$

where the subscript equ denotes equilibrium concentrations.

For balanced systems, where equimolar quantities of diamine and diacid were used, substitution and integration gives

$$\ln\frac{[-COOH]^2}{[-COOH]^{-2} - [COOH]_{equ}^2} = 2\,k_1[-COOH]_{equ}^2 t + \text{Constant}$$

Reported values for the activation energy of the amidation reaction vary. For the melt polycondensation of aminoundecanoic acid the activation energy was found to be [87] 11 kcal mole^{-1}; for the poly-condensation of the same acid as a 30% solution in m-cresol, the reported [92] activation energy is 31 kcal mole^{-1}. The activation energy of the second-order melt polycondensation of p-aminophenylalkanoic acids was reported [93] to be between 17 and 19 kcal mole^{-1}.

Typical values of rate coefficients reported [92] for polyamidation in m-cresol include 7.7×10^{-4} g mmole^{-1} sec^{-1} for the preparation of poly-(m-xylylene adipamide) and 4.2×10^{-4} g mmole^{-1} sec^{-1} for poly-(p-xylylene sebacamide) (Table 6).

TABLE 6

Rate coefficients and activation energies for polyamide formation in m-cresol[a]

Salt or amino acid	Temp. °C	Rate coefficient[b]	Activation energy[c]
m-Xylylene diammonium adipate (I)	175	7.7×10^{-4}	25.3
Aminoundecanoic acid	176	1.8×10^{-4}	31.3
I and aminoenanthic acid[d]	170	2.5×10^{-4}	22.0
I and aminoundecanoic acid[d]	170	3.2×10^{-4}	25.0
Aminoenanthic acid	165	0.8×10^{-4}	29.2
p-Xylylene diammonium sebacate (II)	174	4.2×10^{-4}	19.4
II and aminoundecanoic acid[d]	172	3.8×10^{-4}	18.3

[a] From ref. 92; 30% solutions in m-cresol.
[b] In g mmole^{-1} sec^{-1}.
[c] In kcal mole^{-1}.
[d] Equimolar quantities.

Ogata [94a] showed that the value of the rate coefficient of polyamida-tion depended on the concentration of water when the reaction was carried out in a sealed tube in the presence of fixed amounts of water. He regarded the concentration of water as being constant under the condi-tions used, and the reaction then appeared to follow second-order kinetics, though the rate coefficient clearly must contain a term for the water concentration.

Exchange reactions occur in the molten polyamide system as they do in molten polyesters. In the case of a homopolyamide, of course, such exchange reactions do not increase the number of molecules present and thus have no effect on the molecular weight. Exchange reactions may take place between an amide linkage and an amine-terminated polymer mole-cule

$$\text{P—CONH—P}' + \text{H}_2\text{N—P}'' \rightleftharpoons \text{P—CONH—P}'' + \text{H}_2\text{N—P}'$$

between an amide linkage and the end group of an acid-terminated polyamide molecule

$$P-CONH-P' + HOOC-P'' \rightleftharpoons P''-CONH-P' + HOOC-P$$

and between amide linkages in different polymer molecules or within the same molecule

$$P-CONH-P' + P''-CONH-P''' \rightleftharpoons P''-CONH-P' + P-CONH-P'''$$

Included in these reactions are those transamidations that result in the formation of cyclic oligomers.

The exchange reactions involving amine-terminated or acid-terminated polymer molecules proceed at a faster rate than exchange reactions between amide linkages. Korshak et al. [94b] showed that formation of block copolymers from mixed homopolyamides having chemically blocked end groups was much slower than when the homopolyamides had free carboxyl and free amine end groups.

Additional support [95] was provided by a study of the amide-amide interchange between the cyclic monomer of nylon-6,6 and N,N'-diacetyl hexamethylene diamine. Assuming the process to be second order, the rate coefficient was found to be 4.5×10^{-7} l mole^{-1} sec^{-1} at 260°C, and 20×10^{-7} l mole^{-1} sec^{-1} at 290°C. Thus the activation energy was 6.9 kcal mole^{-1}. For the reaction between the same cyclic monomer and the monoacetylated monomer $CH_3 CONH-(CH_2)_6 NHCO(CH_2)_4 COOH$, the rate coefficients were very much greater: k(amide-acid) = 2.2×10^{-4} l mole^{-1} sec^{-1} at 260°C, 7.0×10^{-4} l mole^{-1} sec^{-1} at 290°C; activation energy, 5.6 kcal mole^{-1}.

Considerable scatter exists in the reported data for rate coefficients and activation energies for the direct amidation process. Values of the activation energy for the polymerization of several amino acids varied between 11 and 19 kcal mole^{-1}. Rate coefficients and activation energies are given in Table 6 for a series of polymers and copolymers, as measured in m-cresol [92]. Again in m-cresol, the activation energies for the reactions between hexamethylene diamine and adipic acid and azelaic acid were found [96] to be 16.8 and 20.9 kcal mole^{-1}, respectively. In the melt, the latter polyamidation was shown [96] to have an activation energy of 24 kcal mole^{-1}. These wide discrepancies may be the consequence of great experimental difficulties certainly involved in studying polyamidation kinetics. They may also arise from unspecified experimental variations, some of which were mentioned earlier for the case of polyesterification (page 506).

Simple factors such as stirring or vessel shape may influence experimental results. Certainly whether or not the system was open or closed could have a profound effect. In addition, in the case of polyamides, there is the additional possibility of catalysis.

No systematic study of catalysis of the polyamidation reaction has been made, unlike the case of polyester. The polycondensation reaction that leads to the formation of poly(ethylene terephthalate) (Section 5) will not proceed at a satisfactory rate without catalysts, whereas polyamidation will. However, there are some reports of catalysis of the polyamidation reaction. Flory [88] stated that the reaction between aromatic diamines and aliphatic dicarboxylic acids was third order and that added non-carboxylic catalysts such as phosphoric acid were needed to obtain satisfactory reaction rates. Later work [80] has shown that this is not necessarily true, and that certain aromatic-aliphatic polyamides can be made by melt processes without added catalysts. Other materials that have shown some catalytic effect include magnesium oxide [97], boric acid [97], phosphoric acid [98] and zinc chloride [98].

It may be that trace quantities of materials inadvertently present as impurities in the monomers or solvents used, or contaminating the walls of experimental vessels and the like, can have a major catalytic influence on the polyamidation reaction in the melt. In the case of poly(ethylene terephthalate), Stevenson [72] studied catalysis on plates of various materials in order to find a surface with substantially no catalytic activity. The use of an inert plate allowed a study of catalysis by antimony trioxide to be made in thin polyester films (see Section 5.3.2b). A systematic study of the catalysis of polyamidation in the melt would seem to be a worthwhile study, both for commercial reasons, where catalysts could increase productivity, and in the area of kinetics, where such a study might permit more precise and meaningful measurements to be made.

6.3 POLYAMIDES FROM ESTERS

The amidation of a diester with a diamine

$$n\ ROOC-A-COOR + n\ H_2N-R-NH_2 \rightleftharpoons$$
$$-\!\!\left[CO-A-CONH-R-NH\right]_n + 2n\ ROH$$

is sometimes employed when direct amidation is not possible. Phenyl esters are frequently used, as phenol often acts as a plasticizer in these systems, and the phenyl group is a good leaving group. Oxamides have been made in this manner [99]; oxalic acid itself decomposes too readily (see Table 2, page 500). The use of alkyl esters often results in partially N-alkylated polymers having less desirable properties such as lower melting-point. Ester-interchange catalysts have been reported to have no influence on this reaction [100]. Repeating units of polyamides made in this way must be of such a size that cyclic imides are not formed.

No major kinetic studies have been made on this polyamidation reaction. The mechanism, at least as regards monofunctional model compounds [101], is essentially $B_{AC}2$ (basic or neutral medium, acyl-oxygen cleavage, bimolecular [33])

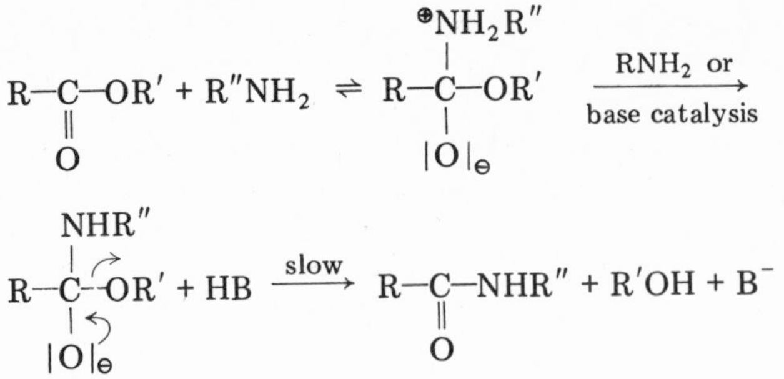

$$R-\underset{\underset{O}{\|}}{C}-OR' + R''NH_2 \rightleftharpoons R-\underset{\underset{|O|_\ominus}{|}}{\overset{\overset{\oplus NH_2R''}{|}}{C}}-OR' \xrightarrow[\text{base catalysis}]{RNH_2 \text{ or}}$$

$$R-\underset{\underset{|O|_\ominus}{|}}{\overset{\overset{NHR''}{|}}{C}}-OR' + HB \xrightarrow{\text{slow}} R-\underset{\underset{O}{\|}}{C}-NHR'' + R'OH + B^-$$

6.4 POLYAMIDES BY ACIDOLYSIS

As in the case of polyesterification (Section 5.4), acidolysis is one of the reactions taking place during conventional polyamidation in the molten state, viz.

P—COOH + HOOC—A—CONH—P ⇌

P—CONH—P + HOOC—A—COOH

Again as in the case of polyesters, this reaction does not contribute to the build-up of molecular weight.

Nylon-6,6 has been made [102] by the acidolysis of N,N'-diacetyl hexamethylene diamine with adipic acid.

No major kinetic studies of this reaction have been reported.

6.5 POLYAMIDES AND POLYIMIDES FROM ANHYDRIDES

Polyimides, first reported in 1908 [103], have been prepared from dianhydrides and diamines or by self-condensation of an AB monomer such as 4-aminophthalic anhydride.

Aliphatic diamines react with a diester-diacid derived from a dianhydride to form an ammonium salt which can be polymerized, usually in the melt, to a polyamic acid. Further heat converts the polyamic acid to the polyimide

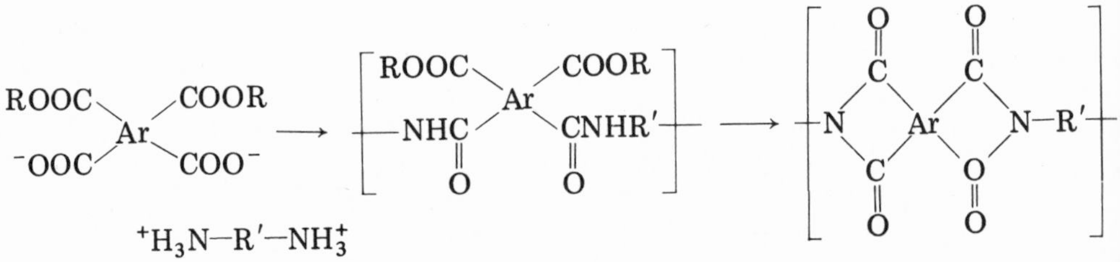

Aliphatic polyimides are in general fusible and soluble in many solvents, and they can often be made in one step. Aromatic polyimides, however, are usually made in two steps.

The reaction between an aromatic diamine and an aromatic dianhydride such as pyromellitic dianhydride is fast and exothermic. It is usually carried out in a solvent for the polyamic acid, which is, of course, a polyamide. Both *meta*- and *para*-diamide units are formed

In the case of pyromellitic dianhydride, the rate of polyamidation is limited only by the rate of solution of the dianhydride [104].

Cyclization of the polyamic acid to the polyimide occurs rapidly at temperatures above 150°C in the solid state, with elimination of water. The rate coefficient of imidization decreases in the solid phase with increasing degree of imidization [105]. The imidization reaction proceeds by attack of the amide nitrogen on the adjacent carboxylic acid group

6.6 POLYAMIDES FROM ACID CHLORIDES

The Schotten–Baumann reaction of an acid chloride and a diamine

$$n\ H_2N-R-NH_2 + n\ ClCO-A-COCl =$$
$$-[NH-R-NHCO-A-CO]_n + 2nHCl$$

is rapid enough to be employed as a synthetic method in interfacial and solution polycondensation.

There are two basic methods by which this reaction may be applied, referred to as interfacial polycondensation and solution polycondensation. Morgan [6] describes these processes in detail.

6.6.1 Interfacial polyamidation

The general requirement for the achievement of high-molecular-weight polycondensation polymers, implicit throughout this chapter, is an essential equivalence of reacting monomers. Molecular weights in commercial practice are often "stabilized" by employing a small excess of one of the monomers. In certain cases, such as in the melt preparation of poly-(ethylene terephthalate), a large excess of volatile monomer is employed. In this case the purpose is to create a new monomer or group of monomers: bis(2-hydroxyethyl)terephthalate and its oligomers. The reaction then proceeds by transesterification by elimination of glycol. However, in the case of the low-temperature technique of interfacial polycondensation, a strict equivalence of reactive functions is not a prime prerequisite.

Interfacial reactions have been employed using the reactions of dicarboxylic acid chlorides with diamines to form polyamides, with hydrazine and hydrazides to form polyhydrazides; disulphonyl halides have been used to prepare polysulphonamides; diisocyanates and glycols have been used to prepare polyurethanes interfacially (Section 7) and diisocyanates and diamines have been reacted interfacially to form polyureas [6] (Section 8). Polyesters prepared by this technique were discussed in Section 5.7.

In interfacial polyamidation the reaction takes place close to the interface between a solution or suspension (usually aqueous) of the diamine and a solution of the diacyl or disulphonyl chloride in an immiscible organic solvent. The aqueous layer often contains an acid acceptor such as triethylamine, pyridine, sodium hydroxide or sodium carbonate, to neutralize the hydrochloric acid produced in the reaction. Often an emulsifier is used to increase interfacial area. The reaction is believed to occur just within the organic solvent layer. The rate of reaction is so fast that the process becomes diffusion-controlled.

The nature of the organic solvent used to dissolve the acid chloride can affect the course of the reaction. Factors that may be involved include the rate of the principal reaction, rates of side reactions, the rate of polymer precipitation, the permeability of the polymer film to monomers, etc. All these factors doubtless influence the molecular weight of the polymer produced. In one example [106], the viscosity of poly(p-phenylene terephthalamide) made interfacially varied with the organic solvent as follows

Solvent	$[\eta](dl\ g^{-1}, H_2SO_4)$
Ethyl ether	0.06
Dichloromethane	0.37
Chloroform	0.41
p-Xylene	0.48
Benzene	0.54
Chlorobenzene	0.57
Dibutyl ether	0.61
Carbon tetrachloride	0.67
Octane	0.70

The solvent power of the organic phase can be varied by including in it an organic liquid that is a solvent for the polyamide. Sokolov and Turetskii [107] showed that for mixtures of tricresol (polymer solvent) and dibutyl ether (non-solvent) there was an optimum ratio of the two solvents for maximum polymer viscosity and therefore for maximum molecular weight (Fig. 7). The content of tricresol in those experiments also influenced the yield of polymer [107] (Fig. 8).

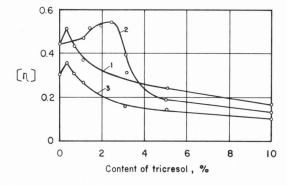

Content of tricresol , %

Fig. 7. Viscosity of polyamides as a function of solvent power (from ref. 107). 1, poly(hexamethylene sebacamide), 2, poly(ethylene terephthalamide), 3, poly(p-phenylene terephthalamide).

As in all polycondensation reactions, the molecular weight of the product of interfacial polymerization may be limited by the incorporation of monofunctional substances. The effects of the latter depend to a large extent on their distribution between the organic and aqueous phases [6]. Thus chain-terminating additives that are more soluble in the organic phase will cause a greater lowering of molecular weight than additives more soluble in the aqueous phase. This provides further evidence that the polymer-forming reaction takes place primarily on the organic side of the interface. Figure 9 shows a schematic representation of concentrations at the interface [6].

536

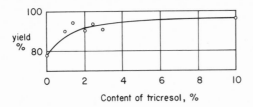

Fig. 8. Yield of poly(p-phenylene terephthalamide) as a function of solvent power (from ref. 107).

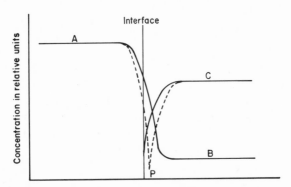

Fig. 9. Schematic diagram of concentrations in an interfacial polyamidation. Amine groups (A) are in the water phase. Acid chloride (C) and amine groups at partition equilibrium (B) are in the organic phase. P and the dotted lines show the relation of the reactive groups when the first incremental layer of polymer forms (from ref. 6).

The interfacial process in general produces higher molecular weights than are found in other polycondensation reactions, primarily as a consequence of the rate-controlling process, diffusion. Under conditions of extremely high stirring it is possible that the rates become kinetically controlled. Little work has been reported on reaction rates of interfacial polyamidation, though rate coefficients have been estimated to be in the range 10^2-10^6 l mole^{-1} sec^{-1}. Nor has adequate study been made of the influences of solvent polarity, the effects of emulsifiers and other additives, and the extents of various side reactions. The formation of branched structures has been reported to occur during the interfacial preparation of nylon-6,10 through reaction of the acid chloride at the amide group to form an imide structure [6].

6.6.2 Solution polyamidation

Bifunctional acid chlorides and diamines can react in a single organic phase to form polyamides. The term "solution polycondensation" is used to describe two major processes that differ in their product, though the reactions probably follow similar mechanisms. In one group of polycondensations, polymer is precipitated as it forms or as it achieves a range of molecular weights at which it becomes insoluble. Polymer must then be isolated and redissolved if it is to be further fabricated into fibres or films, for the polymers formed by these processes are customarily those that decompose below their melting point or melt at too high a temperature for practical purposes.

In the other form of solution polycondensation the polymer remains in solution as it is formed. The reaction solution can then be used for preparation of fibres or films of the product. For practical reasons, this method is clearly to be preferred. Solutions of polyamides have been made in this way and have been spun to strong, thermally stable fibres. In these polymers the carbonamide groups were linked by phenylene rings [108], fused or multiple rings [109], or heterocyclic structures [110].

As with interfacial polycondensation an acid-acceptor is necessary to neutralize the hydrochloric acid formed in the reaction. These low-temperature polycondensation reactions are irreversible, and the acid-acceptor is necessary only to keep the reacting diamine free for reaction with the acid chloride. N,N-Dimethylacetamide and related solvents are often employed. N,N-Dimethylformamide cannot be used as it reacts with the acid chloride, and only low-molecular-weight polymer results. These amide solvents form loose complexes with the hydrochloric acid produced during the polymerization, and no additional acid-acceptor is needed. However, the final solutions are usually neutralized to minimize corrosion of metallic equipment during later steps such as spinning, and to provide small amounts of water often found necessary for the long-term stability of the polymer solutions [111].

The rate coefficient for the reaction of aniline and benzoyl chloride depended to some extent on the solvent used [112], as shown in Table 7.

Other relationships have been examined in model compound studies. For example, the frequencies of symmetric and asymmetric vibrations of

TABLE 7

Kinetic parameters of the reaction between aniline and benzoyl chloride in various solvents at 25°C[a]

Solvent	Dielectric constant e	Rate coefficient $(l\ mole^{-1}\ sec^{-1})10^2$	Activation energy E $(kcal\ mole^{-1})$
Carbon tetrachloride	2.23	0.80	7.10
m-Xylene	2.37	4.77	8.20
p-Xylene	2.27	4.84	7.35
Toluene	2.29	4.87	8.10
Benzene	2.23	6.77	7.46
Anisole	4.33	29.4	4.30
Nitrobenzene	36.40	106	6.30
Acetone	21.45	433	5.75

[a] From ref. 112.

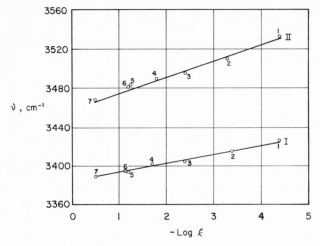

Fig. 10. Rate coefficients of the reactions between substituted anilines and benzoyl chloride as a function of vibration frequencies of NH-group: symmetric (I) and asymmetric (II) (from ref. 112). 1, p-nitroaniline; 2, m-nitroaniline; 3, m-chloroaniline; 4, p-chloroaniline; 5, 4-aminodiphenyl; 6, aniline; 7, p-toluidine.

NH-bonds in substituted anilines correlated well with the rate coefficient of the reaction [112], as illustrated in Fig. 10.

As in the case of interfacial polycondensation, there is very little reported work on the kinetic and related aspects of low-temperature solution polycondensation.

6.7 MISCELLANEOUS REACTIONS

There are many other ways by means of which polyamides have been synthesized, not all of them polycondensation reactions. Thus the ring-opening polymerization of ϵ-caprolactam produces nylon-6 by a chain-growth [1] process

$$n \, (CH_2)_5 \Big\langle \begin{matrix} C {\scriptstyle \diagup\!\!\diagdown} O \\ | \\ NH \end{matrix} \quad \longrightarrow \quad {+}(CH_2)_5{-}CONH{+}_n$$

Other chain-growth reactions that lead to polyamides are the addition of amines to active double bonds

$$n \, CH_2{=}CR{-}CONH_2 \;\rightarrow\; {+}CH_2CHRCONH{+}_n$$

and the polymerization of isocyanates

$$n\ R-N=C=O\ \rightarrow\ -[N-C]_n$$
$$\hspace{5cm}|\quad||$$
$$\hspace{5cm}R\quad O$$

Finally, there are several other polycondensation reactions that have been used to prepare polyamides. No major kinetic studies have been made, but they are briefly summarized here for the sake of completeness.

Diketenes react with diamines to form polyamides. Thus carbon suboxide reacted interfacially with p-xylylene diamine in toluene/water [113]

$$O=C=C=C=O + H_2NCH_2-\langle\bigcirc\rangle-CH_2NH_2 \longrightarrow$$

$$\left[\begin{array}{c} -C-CH_2-C-NH-CH_2-\langle\bigcirc\rangle-CH_2NH- \\ \quad||\qquad\quad|| \\ \quad O\qquad\quad O \end{array}\right]$$

Formaldehyde forms polyamides when reacted with dinitriles [114]

$$n\ HCHO + n\ NCRCN\ \rightarrow\ -[NHCORCONHCH_2]_n$$

The reaction of a dicarboxylic acid and a diisocyanate (or diisothiocyanate) first produces a mixed carboxylic—carbamic anhydride which decomposes on heating, losing carbon dioxide [115]

$$n\ OCN-R-NCO + n\ HOOC-A-COOH\ \rightarrow$$
$$-[NH-R-NHCO-A-CO]_n + 2n\ CO_2$$

Polyamides have also been prepared by the condensation reaction between a diamine and a diamide with liberation of ammonia

$$n\ H_2NCO-A-CONH_2 + n\ H_2N-R-NH_2\ \rightleftharpoons$$
$$-[NHCO-A-CONH-R]_n + 2n\ NH_3$$

7. Polyurethane formation

Polyurethanes vary widely in their chemical composition, the starting materials which are used in their synthesis, the final polymer structure, and their uses. The one chemical feature which all types have in common, and hence their name, is an abundance of urethane groups, $-NHCOO-$. A polyurethane may be formed from a simple glycol and a diisocyanate

$$R(NCO)_2 + G(OH)_2\ \rightarrow\ -[OCNH-R-NHCO-G]_n$$
$$\hspace{4cm}||\qquad\qquad\quad||$$
$$\hspace{4cm}O\qquad\qquad\quad O$$

The glycol may be replaced by many other raw materials which terminate in hydroxyl groups, for example, polyethers, polyesters, and castor oil. The preparation may also utilize a reaction between a bis(carbamoyl chloride) and a glycol

$$R(R'NCOCl)_2 + G(OH)_2 \xrightarrow{-HCl} -\!\!\left[OCNR'\!\!-\!\!R\!\!-\!\!NR'CO\!\!-\!\!G\right]_n$$
$$\overset{\shortparallel}{O} \qquad\qquad\qquad \overset{\shortparallel}{O}$$

a bis(chloroformate) and a diamine

$$R(NH_2)_2 + G(OCOCl)_2 \xrightarrow{-HCl} -\!\!\left[OCNH\!\!-\!\!R\!\!-\!\!NHCO\!\!-\!\!G\right]_n$$
$$\overset{\shortparallel}{O} \qquad\qquad\qquad \overset{\shortparallel}{O}$$

a bis-urethane and a glycol

$$R(NHCOOR'')_2 + G(OH)_2 \xrightarrow{-R''OH} -\!\!\left[OCNH\!\!-\!\!R\!\!-\!\!NHCO\!\!-\!\!G\right]_n$$
$$\overset{\shortparallel}{O} \qquad\qquad\qquad \overset{\shortparallel}{O}$$

and other similar reactions. The procedure most frequently used is the reaction between a diisocyanate and a polyhydroxy compound.

7.1 FROM ISOCYANATES AND HYDROXYL COMPOUNDS

An extensive review of the reactions of isocyanates with hydroxy compounds, including a thorough survey of the catalysis and kinetics of the reactions, was published in 1962 [116]. Other relatively recent surveys of the subject of special interest include those of Farkas and Mills [117] and Lyman [118]. Other reviews which are of special supplemental value with respect to control of the reactions in polyurethane formation include those of Johnson [119] and Saunders and Frisch [120].

This survey will repeat only that information in earlier reviews which is judged to be necessary for an over-all understanding of the subject. The major emphasis will be on data on polyurethane formation (rather than reaction of monofunctional model compounds) which have been published within the period of 1962 to 1969, with partial coverage of 1970. For more details on earlier publications the reader is directed to the references given above.

A satisfactory understanding of polyurethane formation requires an understanding that in addition to the main reaction of urethane formation, other reactions of isocyanates are possible, especially in the presence of certain catalysts. With the proper selection of temperature and catalyst these side reactions can usually be avoided, but the experimentalist should be aware of them. One common side reaction is with traces of water in the system, viz.

$$2RNCO + H_2O \rightarrow RNHCONHR + CO_2$$

This reaction is discussed in more detail in the later section on polyurea formation. Its importance includes a consideration of the subsequent reaction to form a biuret

$$RNCO + RNHCONHR \rightarrow \underset{\underset{\displaystyle CONHR}{|}}{RNCONHR} \quad \text{(biuret)}$$

a reaction which may cause crosslinking. The urethane group may react similarly, leading to an allophanate branch point

$$RNCO + RNHCOOR \rightarrow \underset{\underset{\displaystyle CONHR}{|}}{RNCOOR} \quad \text{(allophanate)}$$

Carboxyl groups, which may be present in certain polyesters, give primarily amides

$$RNCO + R'COOH \rightarrow RNHCOR' + CO_2$$

These in turn can lead to acyl amides as branch points

$$RNCO + RNHCOR' \rightarrow \underset{\underset{\displaystyle CONHR}{|}}{RNCOR'} \quad \text{(acyl amide)}$$

Finally, the isocyanate may react with itself to give a dimer, a trimer, or a carbodiimide

$$2\,RNCO \rightleftharpoons \quad \text{(dimer)}$$

$$3\,RNCO \longrightarrow \quad \text{(trimer)}$$

$$2\,RNCO \longrightarrow RN=C=NR + CO_2 \quad \text{(carbodiimide)}$$

Details of these reactions have been reviewed, and indications of catalysts favouring one or another reaction have been given [116]. In the kinetic studies reported the experimental conditions have generally been chosen so as to give essentially complete urethane formation, free from large amounts of side reactions, except as noted.

The isocyanate group is of such a structure that it may be expected to reflect the resonance forms

$$[R-\overset{..}{\underset{..}{N}}-C=\overset{..}{\underset{..}{O}} \leftrightarrow R-\overset{..}{N}=C=\overset{..}{\underset{..}{O}} \leftrightarrow R-\overset{..}{N}=C-\overset{..}{\underset{..}{O}}:]$$

These resonance structures suggest the probability of ionic reactions, with electron donors attacking the carbonyl carbon, and electron acceptors attacking the oxygen or nitrogen. Catalysis by Lewis acids and bases should be common.

One may visualize the reaction of an isocyanate with an alcohol as involving the following steps, when a basic catalyst is present,

$$RNCO + :B \rightleftharpoons \begin{bmatrix} RN=C-O \\ \uparrow \\ B \end{bmatrix} \xrightarrow{ROH}$$

$$\begin{bmatrix} \overset{H}{\diagdown}\,\overset{R}{\diagup} \\ O \\ \downarrow \\ RN-C-O \\ \uparrow \\ B \end{bmatrix} \longrightarrow \overset{H}{\underset{}{RN-COR}} + :B$$

An acid catalyst could function in the following way

$$RNCO + HA \rightleftharpoons [R-N=C-\overset{..}{\underset{..}{O}}: \rightarrow HA] \xrightarrow{R'OH} RNHCOOR' + HA$$

Such a reaction appears straightforward at first glance, probably being first order in alcohol and also in isocyanate, and with the rate also proportional to the catalyst concentration. Indeed, this is often the case. However, if one wishes to look more closely, it is apparent that fine ramifications are also possible, and many have been found experimentally.

The basic catalyst, for example, may be any suitable electron donor, including tertiary amines frequently utilized, also the alcohol itself and the product of the reaction, the urethane. While the preferred role of the catalyst may be as shown, it can also form hydrogen bonds with the alcohol, thus affecting its reactivity. Other, more complex catalysts may be used which complex with both the isocyanate and the hydroxyl compound. The solvent, too, may associate with each reactant, with the catalyst, and with the products of reaction. Reversal of the reaction is possible, especially with phenolic reactants. For this reason and because of side reactions, deviations from purely second-order kinetics should be expected, and are often found. Details of molecular weight distributions of soluble polyurethanes are not available; these may not follow Flory's calculations, again because of the usual occurrence of some side reactions.

Baker et al. [18—20] published the first detailed kinetic treatment of the isocyanate/alcohol reaction. They proposed the existence of both a catalysed reaction

$$[R-N=C=O \leftrightarrow RN=\overset{+}{C}-\overset{-}{O}] + :B \underset{k_2}{\overset{k_1}{\rightleftharpoons}} \left[RN=\underset{\underset{+}{\overset{\uparrow}{B}}}{C}-\overset{-}{O} \right] \overset{R'OH}{\underset{k_3}{\longrightarrow}}$$

$$\overset{H}{RNCOOR' + :B}$$

and an uncatalysed one

$$RNCO + R'OH \overset{k_0}{\longrightarrow} RNHCOOR'$$

Assuming the stationary state condition, the concentration of the complex, RNCO:B, was given by the equation

$$[complex] = \frac{k_1 [RNCO][B]}{k_2 + k_3 [R'OH]}$$

and the rate of disappearance of isocyanate and rate of formation of the product was

$$\frac{-d[RNCO]}{dt} = k_0 [R'OH][RNCO] + \frac{k_1 k_3 [RNCO][R'OH][B]}{k_2 + k_3 [R'OH]}$$

From this, the experimentally observed rate coefficient for any one set of conditions would be

$$k_{exp.} = k_0 + \frac{k_1 k_3 [B]}{k_2 + k_3 [R'OH]}$$

If the value of k_2 is much greater than $k_3 [R'OH]$, or if $[R'OH]$ is constant in a set of experiments, this may be further simplified to

$$k_{exp.} = k_0 + k_c [B]$$

where k_c, the "catalytic rate coefficient" is given by

$$k_c = \frac{k_1 k_3}{k_2 + k_3 [R'OH]}$$

The reaction of phenyl isocyanate with methanol in dibutyl ether at 20°C was found to be second order, but the value of the rate coefficient increased slowly with time, indicating catalysis by the product of the reaction. The addition of methyl carbanilate to the initial reaction mixture increased the rate of reaction, confirming catalysis by the urethane, a weak base. The rate coefficient for the initial uncatalysed reaction was 0.28×10^{-4} l mole^{-1} sec^{-1}.

Furthermore, the alcohol itself was found to function as a weak catalyst. Thus the value of the second order rate coefficient was found to increase as the alcohol/isocyanate ratio was increased.

Tertiary amines such as triethylamine were catalysts, with the observed rate of reaction directly proportional to the catalyst concentration. The base strength of the amine was a major factor in its catalytic strength, with stronger bases generally being stronger catalysts. Dialkylanilines were not catalytic, however, presumably due to steric hindrance. The fact that the dialkylanilines were not catalytic was cited as evidence for the association of isocyanate and catalyst as being the important role of the catalyst, although infrared evidence for alcohol-amine association was also found.

Other important features of the reaction were also established. Primary alcohols reacted faster than secondary, which were much more reactive than tertiary alcohols. Electron-withdrawing substituents on the benzene ring of phenyl isocyanate enhanced the reactivity, while electron-releasing groups retarded it. The reaction was faster in non-polar than in polar solvents. Alcohol-solvent association was found when dibutyl ether was the solvent, based on infrared data. This association was considered to reduce the rate primarily by reducing the magnitude of k_3. The activation energy for the uncatalysed reaction of phenyl isocyanate with methanol in dibutyl ether was 10 kcal mole^{-1}, and that for the triethylamine-catalysed reaction was 5 kcal mole^{-1}.

Many other investigators have studied the isocyanate/alcohol reaction and have extended the findings of Baker et al. Some have found systems with somewhat different balances of the kinetic parameters, so that their results differed somewhat from those cited above. Farkas and Strohm [121] have reviewed briefly most of the variations which have been found in the amine catalysed reaction. In general, however, the early studies of Baker et al. have provided a sound basis for understanding the kinetics of these reactions.

7.1.1 *Reactions of monoisocyanates with alcohols, as model compounds*

The kinetics of urethane formation from monoisocyanates and alcohols have been studied in extensive detail, and serve as a useful guide for understanding the kinetics of polyurethane formation. Certain obvious differences must be remembered, especially when one is primarily interested in polymerization in the melt or solid state, however.

The structure of the isocyanate has been found to have a great effect on the rate of urethane formation. Bailey et al. [122] and later Kaplan [123] studied the reactions of numerous aromatic isocyanates with a large excess of 2-ethylhexanol in benzene at 28°C, following the reaction with infrared. Additional comparative data were published by Burkus and Eckert [124], who observed the reaction of several isocyanates with n-butyl alcohol in toluene, with triethylamine catalyst. These results show remarkably good agreement [116], with electron-withdrawing substituents increasing the reaction rate, and electron-donating substituents retarding it.

Brock [22] showed that the reaction of aryl isocyanates with alcohols could be correlated by the Hammett [14] linear free energy relationship, $\log k/k_0 = \rho \Sigma \sigma$. Using data from published sources, with reaction temperatures in the range of 20—39.69°C and using several solvents, the value for ρ was calculated to be 1.69. Substituent constants (σ) were given by Brock and also by Kaplan [123].

Aliphatic isocyanates are generally much less reactive than phenyl isocyanate. The work of O'Brien and Pagano [125] showed an activation energy of 16 kcal mole^{-1} for the reaction between tert-octyl isocyanate and ethanol at 25—40°C, and 22 kcal mole^{-1} for reaction with 2-butanol. These relatively high activation energies are apparently typical for aliphatic isocyanates.

The effect of a variety of solvents on the rate of the isocyanate/alcohol reaction was confirmed by Ephraim et al. [8]. The change from acetonitrile to benzene as a solvent was reported to change the rate of the phenyl isocyanate/methanol reaction at 20°C by a factor of 71 in this series of experiments. In general, the rate of reaction decreased approximately in order of the increasing dielectric constant of the solvent and the ability of the solvent to hydrogen bond with the alcohol. This behaviour is in agreement with Baker's suggestion that association of the solvent with the alcohol slows the reaction of alcohol with isocyanate. Others have observed similar effects from changing the reaction medium.

Although reactions of aromatic isocyanates with alcohols generally follow second-order kinetics or modified second-order kinetics as outlined by Baker and Holdsworth [19], Sato [126] reported that aliphatic and alkenyl isocyanates were much more subject to alcohol and urethane catalysis. The reactions of these compounds were found to agree with the expression

$$dx/dt = k_1(a-x)(b-x)^2 + k_2(x)(a-x)(b-x) + k_3(\text{cat.})(a-x)(b-x)$$

This equation is an expanded form of Baker's equation for the catalysed reaction, but does not include a term for an uncatalysed reaction. The first term of Sato's equation corresponds to alcohol catalysis, the second to urethane catalysis, and the third to the effect of an added catalyst, e.g., tertiary amine.

7.1.2 The reactions of diisocyanates with alcohols

The reactions of diisocyanates are usually more complicated kinetically than are those of monoisocyanates. The reactivity of a diisocyanate initially is similar to that of a monoisocyanate substituted by an activating group, in this case the second isocyanate group. As soon as one isocyanate group has reacted with an alcohol, the remaining isocyanate group has a reactivity similar to that of a monoisocyanate substituted by a urethane group. A urethane group in the *meta* or *para* position has only a very mild activating effect, much less than an isocyanate group in the *meta* or *para* position [22, 123]. The reactivity of a diisocyanate having both iso-cyanate groups on one aromatic ring should decrease significantly as the reaction passes approximately 50% completion. This decrease in reactivity may be even greater if another substituent is present *ortho* to one isocyanate group. An example illustrating these effects is 2,4-tolylene diisocyanate (2,4-TDI)

The most reactive group should be the 4-position isocyanate group, which is activated by the 2-position isocyanate group. The 2-position group has similar activation initially by the 4-position isocyanate group, but compensating deactivation by the 1-position methyl group. After the 4-position isocyanate group has reacted with an alcohol the 2-position isocyanate group is even less reactive than initially because of strong deactivation by the 1-methyl, far overshadowing the very slight activating tendency of the 4-position urethane group. Brock [127] and DiGiacomo [128] have reported data of this type for reactions of 2,4-TDI with alcohols.

In diisocyanates where the two isocyanate groups are on different aromatic rings or where they are separated by aliphatic chains the effect of one isocyanate or urethane group on a second isocyanate group is less pronounced. The effect becomes still less as the aromatic rings are separated further and further from each other, for example, by progressively longer aliphatic chains.

The reactions of several diisocyanates with a large excess of 2-ethylhexanol have been reported by Bailey et al. [122]. Reactions were run in benzene with the extent of reaction being followed by loss of infrared absorption at 4.4 μm, characteristic for the isocyanate group. A sharp decrease in the rate of reaction of 2,4-tolylene diisocyanate and the 80:20 isomer ratio of tolylene diisocyanate at approximately 50% reaction was found. In contrast, 2,6-tolylene diisocyanate, 4,4'-diphenylmethane diisocyanate, p-phenylene diisocyanate and m-phenylene diisocyanate showed only a slight decrease in rate as the reaction proceeded.

A similar study of several diisocyanates, and for comparison, three monoisocyanates, was reported by Burkus and Eckert [124]. Triethylamine was included as a catalyst in all experiments. Both energy and entropy of activation were calculated for the first group to react (rate coefficient, k_1) and for the second group (rate coefficient, k_2). It is noteworthy that in calculating values for k_1 and k_2 for 2,4-tolylene diisocyanate the initial reactivity of the 4-position isocyanate group was assumed equal to that of the 2-position group, an assumption which does not appear justified. Data are shown in Table 8. The activation energies are shown, for example for m-phenylene diisocyanate, as 2.4/3.4, the 2.4 value applying to the reaction of the first isocyanate group to react, the 3.4 value to the second. Entropy of activation values are shown similarly.

While nearly all previous rate data were obtained by following the rate of disappearance of isocyanate, Kogon [129] followed the rate of formation of urethane, using infrared absorption at 6750 cm^{-1}. The second-order plot of the uncatalysed reaction of 2,4-tolylene diisocyanate was linear to about 60% reaction, and that of 2,6-tolylene diisocyanate to 30% reaction.

To throw additional light on the relative reactivity of diisocyanates at higher temperatures Cunningham and Mastin [130] measured rates of reactions with alcohols at 115°C. To provide both primary and secondary hydroxyl groups, 1- and 2-octanol were used (Table 9). At 115°C the

TABLE 8

The reaction of isocyanates with n-butanol in toluene, triethylamine catalysed (0.05M isocyanate, 0.10M butanol, 0.03M triethylamine)

Isocyanate	$k \times 10^4$ (l mole^{-1} sec^{-1} at 24.41°C)		$k \times 10^4$ (l mole^{-1} sec^{-1} at 39.69°C)		E_a (kcal mole^{-1})	$-\Delta S^{\ddagger}$ (eu)
	k_1	k_2	k_1	k_2		
m-Phenylene diisocyanate	597	64.8	723	86.2	2.4/3.4	58.1/59.1
p-Phenylene diisocyanate	392	39.2	525	57.2	3.5/4.6	55.2/56.3
2,6-Tolylene diisocyanate	122	18	147	23.8	2.3/3.4	61.8/61.8
Dürene diisocyanate			3.65	1.38		
Diphenylmethane diisocyanate			160	55		
3,3'-Dimethyl-4,4'-diphenylmethane diisocyanate			27.5	11.7		
2,4-Tolylene diisocyanate	255	20.5	330	27.7	3.1/3.6	57.4/60.7
Hexamethylene diisocyanate			0.83	0.42		
Phenyl isocyanate	52		67.7		3.1	60.7
p-Tolyl isocyanate	26		35		3.4	60.9
o-Tolyl isocyanate	7.6		19		4.3	61.4

secondary alcohol was still less reactive than the primary, usually being about one-half to one-third as reactive. The relative effect of the catalyst was not consistent, but it often promoted the slower reactions more than the faster ones.

The differences in reactivities of isocyanate groups in diisocyanates wherein one group was influenced by strong steric hindrance were shown by Case [131]. The structures studied included 1-t-butyl-2,4-phenylene diisocyanate, 1-phenoxy-2,4-phenylene diisocyanate, and others. As expected, a large group *ortho* to an isocyanate group greatly reduced its reactivity. Later publications showed similar effects in other unsymmetrical diisocyanates [132, 133]. The greatest difference was obtained when one isocyanate group was in an unhindered position on the aromatic ring and the other was on an aliphatic carbon with bulky β-substituents.

Rate data comparing the reactivity of aryl mono- and diisocyanates with benzyl-type mono- and diisocyanates were reported by Ferstandig

TABLE 9

Second order reaction rate coefficients for diisocyanates and alcohols at $115°C \times 10^4$ (l $mole^{-1}$ sec^{-1})
(0.1 N reactants, chlorobenzene)

Diisocyanate	With 1-octanol		With 2-octanol	
	Catalysed[a]	Uncatalysed	Catalysed[a]	Uncatalysed
m-Phenylene, k_1	125	23	35	14
k_2	77	20	24	7.0
2,4-Tolylene, k_1	51	9.0	17	5.5
k_2	26	4.5	7.7	1.2
4,4-Diphenylmethane[b]	49	6.7	14	5.2
3,3'-Dimethyl-biphenylene	17	1.8	5.8	1.2

[a] 0.01 N tri-n-butylamine; [b] 97% pure.

and Scherrer [134]. The benzyl-type isocyanates were somewhat less reactive than corresponding aryl isocyanates. In diisocyanates of the xylylene type the second isocyanate group to react did so at a rate about one-half to one-third that of the first group, similar to the behaviour of diphenylmethane diisocyanate. Similar studies have also been reported by Mandasescu et al. [135].

While activation energies have been published for many isocyanate/ alcohol reactions, relatively few reports have been made of the heat of reaction. Bayer [136] reported a heat of reaction of 52 kcal $mole^{-1}$, or 26 kcal $equiv^{-1}$, for the hexamethylene diisocyanate/1,4-butanediol reaction. Lovering and Laidler [137] measured heats of reactions for the butyl alcohol isomers with several aromatic isocyanates. Values ranged from 18.5 to 25 kcal $equiv^{-1}$.

7.1.3 Catalysis of the isocyanate/hydroxyl reaction

The preceding sections have shown the effect of changes in structure of isocyanate and hydroxyl compound, solvent and temperature on the isocyanate/hydroxyl reaction, and have indicated the role of basic catalysts. The largest commercial polyurethane processes utilize catalysed reactions, especially the preparation of foams. For this reason the catalysis of the isocyanate/hydroxyl reaction has been the object of extensive research.

The initial work of Baker and Holdsworth [19], discussed in Section 7.1, outlined the role of a basic catalyst and demonstrated that an increase in base strength was accompanied by increased catalytic strength except when steric hindrance interfered. Evidence for association of the tertiary amine base with the isocyanate was given. Later workers have confirmed this [138, 139]. In a special case, the addition product of a sulphonyl isocyanate with a tertiary amine has been isolated [140]. Oberth and Bruenner [141], like Baker et al., reported infrared evidence for hydrogen bond formation between alcohol and tertiary amine.

While most tertiary amine catalysts are effective approximately in proportion to their base strength, an exception is triethylene diamine (1,4-diaza[2.2.2]bicyclo-octane). As shown by Farkas et al. [142, 143] this catalyst is much more powerful than would be predicted from its base strength, being five times stronger as a catalyst than N,N'-dimethyl-piperazine, which has slightly greater basicity. It was first suggested that the explanation may be the complete lack of steric hindrance in the structure.

Later work by Farkas and Strohm [121], however, related the catalytic strength of an amine to its $E_{1/2}$ value, the potential at the half-neutralization point in a potentiometric titration in ethyl acetate. In this analysis a high potential indicates low basicity. The $E_{1/2}$ value should be a better measure of the basic strength of the amine as a catalyst for the isocyanate/hydroxyl reaction than pK measures of base strength, since the pK value is obtained in aqueous media. The $E_{1/2}$ values do, indeed, correlate well with catalytic effect, and explain the unusual catalytic strength of triethylene diamine, as shown in Table 10. The rate data in this table are for the reaction between phenyl isocyanate and 2-ethyl-hexanol.

The mild effect of acidic catalysts, hydrogen chloride and boron trifluoride etherate, was demonstrated by Tazuma and Latourette [144]. When a large excess of alcohol was present the acids had little or no catalytic effect.

While tertiary amines are adequate catalysts for many reactions, a variety of metal catalysts has been found to be far more powerful. Organotin compounds are now used commercially in many polyurethane processes. The discovery of these catalysts was made independently by Hostettler and Cox [145] and by Britain and Gemeinhardt [146], all of whom were searching for more effective catalysts for foam formation.

TABLE 10

Catalyst activity, pK, and basicity in ethyl acetate ($E_{1/2}$) for various amines

Amine	k_c	Relative activity	pK_a	$E_{1/2}$
Triethylenediamine	5600	1.00	8.60	124
2-Methyl-1,4-diaza-(2.2.2)-bicyclooctane	5130	0.92	8.86	168
Triethylamine	500	0.19	10.64	186
N-Ethylmorpholine	315	0.05	7.15	291

The tremendously powerful effect [145] of certain tin compounds compared to the more familar catalysts is shown in Table 11. It can readily be seen that tin compounds such as dibutyltin dilaurate cannot function as simple acids or bases. The strongly basic quaternary ammonium hydroxide and the more acidic stannic chloride were much weaker catalysts.

TABLE 11
Catalysis of the phenyl isocyanate/methanol reaction in dibutyl ether at 30°C

Catalyst	Mole % catalyst used	Relative activity at 1.0 mole % catalyst
None	—	1.0
Triethylamine	1.00	11
Cobalt naphthenate	0.93	23
Benzyl trimethyl ammonium hydroxide	0.47	60
Tetra-n-butyltin	1.01	82
Stannic chloride	0.30	99
Dimethyltin dichloride	0.020	2100
Di-n-butyltin dilaurate	0.0094	37000

A synergistic effect was also found when using a combination of amine and tin catalysts [145]. A much faster reaction occurred between phenyl isocyanate and butanol with mixed tin and tertiary amine catalysts than would be expected from the individual catalyst concentrations. The synergistic effect of tertiary amine on tin catalyst has been confirmed by Wolfe [147] and by Willeboordse et al. [148].

An ingeniously simple screening method was used by Britain and Gemeinhardt [146] to evaluate catalysts for the isocyanate/hydroxyl reaction. To approximate as closely as possible actual polymerization conditions, the 80:20 ratio of 2,4- and 2,6-tolylene diisocyanate (80:20 TDI) isomers and a polyether triol of 3000 molecular weight were mixed at NCO:OH ratio of 1.0. A 10% solution of catalyst in dry dioxane was added, the final catalyst concentration being 1% of the weight of polyether. The time for the mixture to gel at 70°C was noted as an indication of catalytic strength. This technique used the same reactants employed in one-shot flexible polyether-based foam systems, almost completely eliminated solvent, and was used to screen quickly hundreds of possible catalysts.

Many metallic compounds were found to be catalysts for the isocyanate/hydroxyl reaction. A list of the type compounds in a roughly descending order of catalytic activity is: Bi, Pb, Sn, triethylene diamine, strong bases, Ti, Fe, Sb, U, Cd, Co, Th, Al, Hg, Zn, Ni, trialkyl amines, Ce,

Mo, Va, Cu, Mn, Zr and trialkyl phosphines. Arsenic, boron, calcium and barium compounds did not show any catalytic activity within the limits of the screening test used.

Similar gelation tests were also used with m-xylylene diisocyanate and with hexamethylene diisocyanate [146]. In these tests the order of strength of the catalysts was found to be different from when tolylene diisocyanate was used. Table 12 lists the results of some catalyst screening tests with the three different diisocyanates.

The catalysts tested were classified into three general groups. The first group of tertiary amine catalysts did not greatly affect the relative reaction rates of the different isocyanates. The second group included stannous, lead, bismuth and organo-tin compounds which activated the aliphatic diisocyanates more than tolylene diisocyanate so that the relative rates of the diisocyanate reactions were approximately equal. The third group had a much larger effect on the aliphatic diisocyanates so that these diisocyanates were faster to react with hydroxyl groups than tolylene diisocyanate. This last group included zinc, cobalt, iron, stannic, antimony and titanium compounds.

A reaction mechanism for metal catalysts proposed is

This coordination effect, which proposes that the hydroxyl group enters on the metal side of the complex and attaches in close proximity to the isocyanate nitrogen, can explain the remarkable catalytic actions of the metals. Obviously the sequence of reactions leading to the complex could be the reverse. It was suggested that those catalysts which caused the aliphatic isocyanates studied to react faster than tolylene diisocyanate could perform in this manner because the aliphatic diisocyanates used were not sterically hindered; whereas, the 2- and 6-positions on the

TABLE 12
Catalyst tests with three different isocyanates

Compound tested	Gelation time (min) at 70°C with		
	Tolylene diisocyanate	m-Xylylene diisocyanate	Hexamethylene diisocyanate
Blank	> 240	> 240	> 240
Triethylamine	120	> 240	> 240
Triethylene diamine	4	80	> 240
Stannous octoate	4	3	4
Dibutyltin di(2-ethylhexoate)	6	3	3
Lead 2-ethylhexoate (24% Pb)	2	1	2
Sodium o-phenylphenate	4	6	3
Bismuth nitrate	1	1/2	1/2
Tetra(2-ethylhexyl)titanate	5	2	2
Zinc naphthenate (14.5% Zn)	60	6	10

tolylene diisocyanate are hindered by the methyl group in the 1-position. It was also suggested that if the metal does not have the property to complex with the incoming hydroxyl compound and the isocyanate group in such a way that the hydroxyl group and the isocyanate nitrogen are brought close together, then the only catalytic activity observed will be due to the acid or base reactions of the compound tested.

Subsequent investigations have examined various details of the role of the metal catalysts, and most have been in general agreement with the proposal of coordination with both reactants. The subject has been reviewed briefly by Reegen and Frisch [149], who also showed cryoscopic evidence for complexing with both reactants.

Looking further into the role of tin catalysts, Bloodworth and Davies [150] found that tin(IV) alkoxides react with isocyanates to give N-stannyl carbamates

$$RNCO + Sn(OR')_4 \rightarrow R-\underset{\underset{Sn(OR')_3}{|}}{N}-COOR'$$

These are readily alcoholized to give the carbamate and the regenerated tin alkoxide

$$R\underset{\underset{Sn(OR')_3}{|}}{N}COOR' + R'OH \rightarrow RNHCOOR' + Sn(OR')_4$$

554

Research by O'Brien and Pagano [125] showed that with some metal catalysts, at least, the choice of catalyst may affect the two groups in a diisocyanate differently. With 1,8-menthane diisocyanate, the isocyanate group in the 1-position was 1.5 times more reactive toward 1-butanol than was the 8-position group when copper naphthenate was used as a catalyst. With lead naphthenate the ratio was 5.8. It might be expected that the isocyanate group with less steric hindrance could coordinate with certain catalysts more easily than one with more steric hindrance, thus accounting for the difference.

Several types of additives have been found which reduce or eliminate the effects of various catalysts in polyurethane formation. When the catalyst is a simple base, an acid or acid precursor such as an acid chloride will reduce the rate of reaction due to neutralization of the basic catalyst. This is often most desirable, since many of the bases stronger than tertiary amines promote side reactions such as trimerization, carbodiimide formation and allophanate formation. The beneficial effect of acid retarders in a variety of systems was illustrated by Heiss et al. [151], as well as by others.

Metal catalysts present a somewhat more difficult problem of deactivation. In general, chelating agents such as citric acid may be beneficial. A compound of the structure $(HOCH_2)_2P(O)OH$ has been reported to be especially useful [152].

7.1.4 Polyurethane formation from diisocyanates

One of the first kinetic investigations of the reaction of diisocyanates with difunctional hydroxyl compounds was that of Bailey et al. [153]. Several diisocyanates were studied, using hydroxyl-terminated diethylene glycol adipate of 1800 molecular weight. Data obtained using 2,4-tolylene

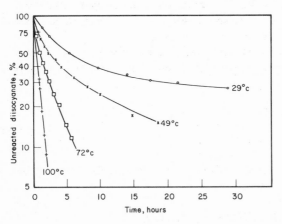

Fig. 11. Reaction of 2,4-tolylene diisocyanate with diethylene glycol adipate polyester, in chlorobenzene (ref. 153).

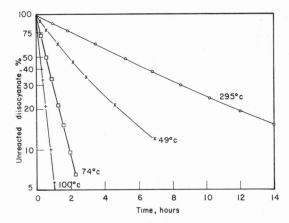

Fig. 12. Reaction of 4,4'-diphenylmethane diisocyanate with diethylene glycol adipate polyester in chlorobenzene (ref. 153).

diisocyanate are shown in Fig. 11, and those using 4,4'-diphenylmethane diisocyanate in Fig. 12. Other data were reported with *m*-phenylene and 3,3'-dimethyl-4,4'-biphenylene diisocyanate. Calculated rate coefficients and activation energies are shown in Table 13.

The decrease in rate after 50% reaction is quite apparent in the reaction of tolylene diisocyanate with the polyester at 29°C (Fig. 11). This change in rate illustrates the reduced reactivity of the 2-position isocyanate group, having steric hindrance from the *ortho* methyl substituent, as well as the lesser activating influence of a *meta* urethane substituent compared to a *meta* isocyanate substituent. An increase in reaction temperature favours the slow reaction more than the fast, as would be expected if differences in activation energy accounted for at least part of the difference in rates. Thus at 100°C there was little decrease in rate of reaction with TDI. In the case of 4,4'-diphenylmethane diisocyanate (Fig. 12), there was little change in rate after 50% reaction at any of the temperatures studied.

The effect of several catalysts on the reaction between 80:20-TDI and a ten-molar excess of diethylene glycol adipate was also reported by Bailey et al. [153]. *o*-Chlorobenzoyl chloride was a slight retarder; tertiary amines and cobalt naphthenate were catalytic.

The reaction between TDI and polyols has been simulated in a computer study [154]. The objective was to prepare NCO-terminated prepolymers containing a minimum of monomeric TDI. Twelve simultaneous and/or consecutive reactions were treated mathematically. The calculated predictions were in good agreement with experimental results.

Some data on the effect of structure of the hydroxyl compound on reactivity with *p*-phenylene diisocyanate were given by Cooper et al. [155]. The system was not specified except that the temperature was

TABLE 13

The reactivity of diisocyanates with diethylene glycol adipate
(0.2M ester, 0.02M isocyanate in monochlorobenzene)

Diisocyanate		Rate coefficient $k \times 10^4$ (l mole^{-1} sec^{-1}) at						Activation energy (kcal mole^{-1})
		29–31.5°C	40°C	49–50°C	60°C	72–74°C	100–102°C	
m-Phenylene	k for 1-NCO	1.4	1.9	2.8	4.0			7.5
	k for 3-NCO	0.7	1.0	1.5	2.3			8.4
2,4-Tolylene	k for 2-NCO	0.057		0.18		0.72	3.2	12.6
	k for 4-NCO	0.45		1.2		3.4	8.5	9.3
4,4'-Diphenylmethane	k (total)	0.34		0.94		3.6	9.1	10.5
4,4'-Tolidine	k (total)	0.048				0.74	3.2	13.1

TABLE 14

The reactivity of diols with p-phenylene diisocyanate at $100°C$

Diol type	$k \times 10^4$ (l mole^{-1} sec^{-1})
Polyethylene adipate	36
Polytetrahydrofuran	10—32
1,4-Butanediol	9.0
1,4-cis-Butenediol	4.0
1,5-Bis(β-hydroxyethoxy)naphthalene	2.5
1,4-Butynediol	0.6

$100°C$. The data are very interesting for a comparison of the reactivity of several hydroxyl compounds of commercial importance (Table 14).

The rates of reactions of several familiar diisocyanates with poly-(ethylene adipate) and other active hydrogen compounds were also reported [155]. The solvent and reactant concentration were not described; it is assumed that no catalyst was used. The method of calculating rate coefficients was not indicated, but all were shown as second order rate coefficients. The data which are given in Table 15 are of particular

TABLE 15

Reactions of isocyanates with active hydrogen compounds at $100°C$

Isocyanate	Reactive grouping				
	Hydroxyl[a]	Water	Urea[b]	Amine[c]	Urethane[d]
p-Phenylene diisocyanate					
k(l mole^{-1} sec^{-1}) $\times 10^4$	36.0	7.8	13.0	17.0	1.8 at $130°C$
E_a(kcal mole^{-1})	11.0	17.0	15.0	7.9	
2-Chloro-1,4-phenylene diisocyanate					
k(l mole^{-1} sec^{-1}) $\times 10^4$	38.0	3.6	13.0	23.0	
E_a(kcal mole^{-1})	7.5	6.4	15.0	3.4	
2,4-Tolylene diisocyanate					
k(l mole^{-1} sec^{-1}) $\times 10^4$	21.0	5.8	2.2	36.0	0.7 at $130°C$
E_a(kcal mole^{-1})	7.9	10.0	17.0	9.5	
2,6-Tolylene diisocyanate					
k(l mole^{-1} sec^{-1}) $\times 10^4$	7.4	4.2	6.3	6.9	
E_a(kcal mole^{-1})	10.0	24.0	11.8	9.0	
1,5-Naphthalene diisocyanate					
k(l mole^{-1} sec^{-1}) $\times 10^4$	4.0	0.7	8.7	7.1	0.6
E_a(kcal mole^{-1})	12.0	7.7	13.0	12.0	
Hexamethylene diisocyanate					
k(l mole^{-1} sec^{-1}) $\times 10^4$	8.3	0.5	1.1	2.4	2×10^{-5} at $130°C$
E_a(kcal mole^{-1})	11.0	9.2	17.0	17.0	

[a] Polyethylene adipate; [b] Diphenylurea; [c] 3,3′-Dichlorobenzidine; [d] p-Phenylene dibutylurethane.

value as comparative indications of relative reactivities at a temperature frequently used in commercial polymerizations.

This is an interesting comparison with regard to control of polymer crosslinking by biuret- or allophanate-forming side reactions. For example, 1,5-naphthalene diisocyanate showed the greatest tendency toward allophanate formation, and hexamethylene diisocyanate the least.

The rate coefficients for the reactions of hexamethylene diisocyanate at least with hydroxyl appear large in relation to those for the aromatic isocyanate. Others have found the aliphatic isocyanates consistently less reactive than unhindered aromatic isocyanates. The wide variation in the reported rates of the other reactions, not always consistent with structure of the reactants, and variations in activation energies in any given series are unexpected. Several of these data are not in agreement with what

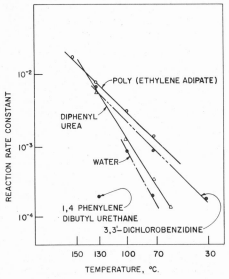

Fig. 13. Effect of temperature on reactivity of p-phenylene diisocyanate with model compounds (ref. 155).

would be expected from other published results. More information concerning the experimental data would be necessary in order to interpret these results satisfactorily.

The effect of temperature on the reactions of p-phenylene diisocyanate with the model active hydrogen compounds of Table 15 is shown in Fig. 13 [155].

It has been pointed out earlier that one should follow the rate of appearance of urethane groups, as well as the rate of disappearance of isocyanate groups, to be sure that side reactions are not consuming some of the isocyanate groups. This is especially true when catalysts are present. Blagonravova et al. [156] demonstrated just such a situation in the catalysed reaction of 2,4-TDI with bis(2-hydroxyethyl)adipate. The changes of concentration of NCO, OH and NH were followed with infrared. With sodium acetate as a catalyst the NCO groups disappeared at a faster rate than urethane groups were formed (probably due to trimerization of the NCO groups). With cobalt naphthenate as catalyst no side reactions were apparent.

A somewhat similar situation was observed in the reaction between nitro-substituted glycols and diisocyanates, using metal acetylacetonates as catalysts [157]. With different catalysts the rate of isocyanate homopolymer formation decreased in the order

$$Pb^{+2} > Cu^{+2} > Mn^{+3} > V^{+2} > Fe^{+3}$$

In this study ferric acetylacetonate was the best catalyst, giving a rapid rate of urethane formation, and very little homopolymerization of isocyanate.

While most kinetic studies have logically concentrated on the relatively early stages of the reactions, one paper has emphasized the rates of reactions in the latter stages of polyurethane cure [158]. A variety of techniques was used to measure the progress of cure, including infrared and chemical methods for NCO disappearance, thermal and mechanical measurements on the polymer. Essentially complete disappearance of isocyanate groups was observed in the reaction between diisocyanates and trifunctional polyhydroxy compounds, as shown in Fig. 14. With much higher functionality in the reactants, however, even the use of catalysts plus heat curing for four hours at 100°C did not succeed in driving the reactions beyond about 95% completion.

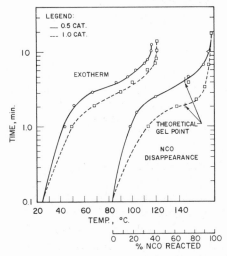

Fig. 14. Degree of reaction by titration of NCO (TP 440-2,4 TDI foam) (ref. 157).

In addition to the information presented thus far, several publications report the kinetics of polyurethane formation. These add little to the over-all understanding of the reactions, but do present specific data on certain combinations of reactants. These are outlined in Table 16.

TABLE 16
Additional kinetic studies of polyurethane formation[a]

Isocyanate	Polyol	Catalyst(s)	Reference
1 MDI	Polyester	Metal acetyl-acetonates	159
2 TDI	PPG 2000	Amines, metals	160
3 p-PDI	Polyethylene adipate, PEG		161
4 p-PDI	PPG		162
5 Nitro-diisocyanates	Nitrodiols	Metal acetyl-acetonates	157
6 80/20-TDI	Polyethers	$Bn_2Sn(laurate)_2$	163
7 TDI	PPG	Amines	164
8 HDI	DEG	Bu_3SnCl	165
9 HDI	DEG	Tin cpds.	166
10 HDI, TDI	Cl-contg. ether diols	Amine, $ZnCl_2$	167
11 HDI, TDI	Polyformal and TEG		168
12 TDI	DEG	Cu compounds	169
13 MDI, TDI	EG	Vinyl pyridine copolymers	170
14 TDI	Polyether triols		171
15 HDI	EG		172

[a] Abbreviations: TDI, tolylene diisocyanate; p-PDI, p-phenylene diisocyanate; HDI, hexamethylene diisocyanate; MDI, methylene bis(4-phenyl isocyanate); PPG, polypropylene glycol of indicated molecular weight; DEG, diethylene glycol; TEG, triethylene glycol; EG, ethylene glycol; PEG, polyethylene glycol.

7.2 POLYURETHANE FORMATION FROM BIS-CHLOROFORMATES

Polyurethanes have frequently been prepared in the laboratory from diamines and bis-chloroformates, using interfacial or solution techniques. The processes have not been widely used in industry, however. The general considerations of interfacial and solution processes described in Section 3 apply to polyurethane preparations by these methods. Specific methods have been described by Wittbecker and Katz [173], Kwolek and Morgan [174] and Lyman [118].

Kinetic measurements of the reaction between chloroformates and amines are difficult because of the great speed of the reactions. Rate coefficients have been estimated [118] to be about 10^2 to 10^4 l mole^{-1} sec^{-1}. Reaction times have been of the order of five minutes at 0—5°C in benzene-water systems [173], and 10—30 minutes at 60°C in order to get high molecular weights when using piperazine [174].

8. Polyurea and polyurea—urethane formation

Polyureas may be prepared by a variety of reactions which are closely analogous to those leading to polyurethanes (see Section 7). The most frequently used method is the reaction of a diisocyanate with a diamine

$$R(NCO)_2 + R'(NH_2)_2 \rightarrow \left[-NHCNH-R-NHCNH-R'- \right]_n$$
$$\underset{O}{\overset{\|}{}} \qquad \underset{O}{\overset{\|}{}}$$

In the preparation of block copolymers where one block is a polyurea segment, the reaction of a diisocyanate with water is often utilized, viz.

$$R(NCO)_2 + H_2O \rightarrow \left[OCN-R-NH\overset{\overset{\displaystyle O}{\|}}{C}OH \right] \xrightarrow{-CO_2}$$

$$[OCN-R-NH_2] \rightarrow \left[-NHCNH-R- \right]_n$$
$$\underset{O}{\overset{\|}{}}$$

This reaction is often employed in the production of flexible "urethane" foams, which are frequently block copolymers of polyether or polyester segments joined to polyurea segments. (The polyester or polyether segments terminate in urethane segments resulting from reaction of polyether or polyester hydroxyl end groups with the isocyanate.)

Other preparative methods for polyureas include the reaction of diamines with phosgene or bis(carbamoyl chlorides), and of bis(urethanes) with diamines.

The reactions of isocyanates with amines and with water have been reviewed in detail [116], as have the preparations of many polymers containing polyurea segments, such as water-blown polyurea-urethane foams and water- or diamine-cured polyurea-urethane elastomers [175].

8.1 THE REACTION OF ISOCYANATES WITH AMINES

The reactions of monoisocyanates with amines or water follow the general principles outlined in Section 4.1, and are closely analogous to the reactions with alcohols (see Section 7).

The rate of isocyanate/amine reaction is greatly affected by the basicity of the amine, with aliphatic amines reacting much faster than aromatic types [176] when differences in steric hindrance are not great. In addition, Naegeli et al. [177] showed that electron-withdrawing substituents on the phenyl ring of aniline reduced its reactivity, whereas electron-donating substituents increased it (negative Hammett rho effect). The opposite was true of substituents on the ring of phenyl isocyanate. Steric hindrance greatly retards the reaction, e.g., by *ortho* substituents on either the isocyanate or the amine [178, 179]. Data are shown in Table 17.

TABLE 17
Half-lives of the reaction between ArNCO and ArNH$_2$
(Each reactant 0.1N in dioxane; at 31°C)

Isocyanate	Amine	Dissociation constant of amine	Reaction half-life (min)
Phenyl	Aniline	4.6×10^{-10}	43
Phenyl	*o*-Toluidine	3.3×10^{-10}	60
Phenyl	*p*-Toluidine	20.0×10^{-10}	5
Phenyl	*o*-Chloroaniline	0.023×10^{-10}	(less than 32% in 1200 min)
o-Tolyl	Aniline		202
o-Tolyl	*o*-Toluidine		> 1000
p-Tolyl	Aniline		54
p-Tolyl	*p*-Toluidine		25—30

The very slow rate of reaction between an *ortho*-substituted isocyanate and an *ortho*-substituted amine in the cure of polyurea-urethanes based on tolylene diisocyanate has been considered to be of significant importance in the final cure of such polymers [180].

Naegeli et al. [177] reported mild catalysis by tertiary amines and carboxylic acids, but not by water, inorganic acids, salts or bases. In contrast, Craven [179] found that the typical tertiary amines and acids had little catalytic effect in the systems he studied. Certain substituted ureas appeared to catalyse the reaction to a greater extent than many tertiary amines. Ten mole % of butyric acid reduced the half-life of the reaction between phenyl isocyanate and *o*-toluidine by 57%, and 10 mole % of N-phenyl-N'-*o*-tolylurea reduced it by 38%. Arnold et al. [178] and

Craven [179] reported that the rate of reaction could be expressed as

$$\text{rate} = k[\text{isocyanate}][\text{amine}]^n$$

with the product of the reaction being catalytic in some cases. Where the product (urea) is a strong catalyst, n in the above equation approaches 2.0; when the product is a very weak catalyst n approaches 1.0. The results of the reactions between phenyl isocyanate and three amines are shown in Table 18. (Values of k apply only to the initial reaction, and are not true rate coefficient values.) The dependence of the rate on amine concentration (n) was shown to decrease as the reaction progressed.

TABLE 18
Reaction between isocyanates and amines
(Each reactant 0.1N in dioxane; 31°C)

Isocyanate	Amine	Catalytic action of product	Order of initial reaction $(n)^a$	k^a
Phenyl	o-Toluidine	Strong	1.85	0.24
Phenyl	Aniline	Weak	1.7	0.58
Phenyl	p-Toluidine	None	0.95	1.2
p-Tolyl	p-Toluidine	Weak	1.7	1.6

^{a} k and n from the equation: rate $= k[\text{isocyanate}][\text{amine}]^n$.

It was also shown that adding 0.025 M N-phenyl-N'-o-tolyl urea to the solution of phenyl isocyanate and o-toluidine catalysed the reaction so that the initial rate was as great as the ultimate rate in the absence of added urea. The reaction sequence proposed as a suitable explanation of these observations [178, 179] was

$$\text{Ar}'\text{NCO} + \text{ArNH}_2 \underset{k_2}{\overset{k_1}{\rightleftharpoons}} \begin{bmatrix} \text{Ar}'\text{NCO} \\ \uparrow \\ \text{ArNH}_2 \end{bmatrix}$$

$$k_3 \diagup \text{ArNH}_2 \qquad \text{urea} \diagdown k_4$$

$$\text{Ar}'\text{NHCONHAr} + \text{ArNH}_2 \qquad 2\,\text{Ar}'\text{NHCONHAr}$$

This sequence would account for catalysis by the amine reactant itself as well as catalysis by the product of the reaction, the substituted urea.

Product catalysis in the reaction between phenyl isocyanate and o-toluidine was also reported by Ozaki and Hoshino [181].

A mechanism similar to that of Arnold et al. [178] was described by Baker and Bailey [182] who also observed an additional mild catalysis by triethylamine and a weaker catalyst, pyridine; dimethylaniline was not catalytic. Sauer and Kasparian [183] also found a third-order reaction when phenyl isocyanate and aniline were present in about equal concentration. However, the order was said to be fourth when the isocyanate was initially in considerable excess.

A recent paper by Cooper et al. [155] reported a fairly strong catalytic effect of an unidentified zinc compound on the reaction of an aromatic isocyanate with 3,3'-dichlorobenzidine. Axelrood et al. [184] found diethylene triamine and stannous octoate to be powerful catalysts for the reaction of phenyl isocyanate and an aromatic diamine, whereas dibutyltin dilaurate and cobalt naphthenate had only a mild effect in their system.

8.2 REACTIONS OF ISOCYANATES WITH WATER

The reaction of isocyanates with water is a complex one and may involve several possible mechanisms. Naegeli et al. [185] proposed the following possibilities

$$RNCO + H_2O \longrightarrow RNHCOOH$$

$$RNHCOOH \xrightarrow[I]{RNCO} RNH\overset{O}{\overset{\|}{C}}O\overset{O}{\overset{\|}{C}}NHR \xrightarrow[I]{-CO_2} RNHCONHR$$

$$RNHCOOH \xrightarrow[II]{CO_2} RNH_2 \xrightarrow[II]{RNCO} RNHCONHR$$

$$RNH_2 \longrightarrow RNHCOO^- RNH_3^+ \xrightarrow[III]{-CO_2} RNHCONHR$$

RNHCOOH
III

and experience confirms the reasonableness of the proposal. Consecutive reaction of isocyanate with the product, the disubstituted urea, is also a probability, of course.

Reaction sequence I should be favoured under conditions when the carbamic acid is relatively stable but is fairly reactive toward the isocyanate. Sequence II would be favoured when the carbamic acid decomposes rapidly and the amine thus liberated reacts quickly with the isocyanate. This sequence is the one most often given in simplified discussions of the isocyanate/water reaction. Sequence III might increase in significance when the carbamic acid is not stable but the amine and the isocyanate react only at a slow rate, for example when steric hindrance is pronounced.

The effect of substituents on an aromatic isocyanate in its reaction with excess water was studied by Naegeli et al. [185]. The products of the reaction were analysed for amine and disubstituted urea. The substituent (X) on XArNCO could obviously affect each step of each of the three reaction sequences proposed. To illustrate, it could affect each of the steps of sequence II separately

$$XArNCO + H_2O \xrightarrow{k_1} XArNH_2 + CO_2$$

$$XArNCO + XArNH_2 \xrightarrow{k_2} XArNHCONHArX$$

If k_1 were much larger than k_2, then with an excess of water one would expect a high yield of amine, $XArNH_2$. On the other hand, if k_2 were much larger than k_1, a high yield of the substituted urea should be obtained.

With X = H, 4-CH_3O, 4-CH_3 and 3-CH_3O, a high yield of urea was obtained in every case, as would be expected from the effect of these substituents on an aromatic amine, with regard to rate of reaction with an isocyanate. When X = 2-NO_2 and 2,4-$(NO_2)_2$, a high yield of amine was obtained. This behaviour was the result in part of reduction of base strength of the amine by the negative substituents, so that k_2 was small. It may also be assumed that these negative substituents resulted in increased values for k_1. The net effect was that k_1 was much greater than k_2 for these reactions.

The effect of substituents on the isocyanate/water reaction is thus in agreement with other isocyanate reactions with active hydrogen compounds. Negative substituents promote the addition of water to the isocyanate, but those same groups on the resulting amine retard the addition of that amine to the isocyanate. In extreme cases, e.g., 2,4-dinitrophenyl isocyanate, the water may add to the isocyanate so fast and the resulting amine may add so slowly that all isocyanate is consumed by water, none by amine, and a high yield of amine is obtained. Electron-donating substituents on the isocyanate may retard the addition of water, but the same groups on the resulting amine promote the addition of that amine so that only the disubstituted urea is obtained.

The kinetics of the isocyanate/water reaction appear to be similar to those of the isocyanate/amine and isocyanate/alcohol reactions. Morton et al. [186, 187] showed that the rate depended on the water concentration, as indicated in Table 19. From these data it was shown that a plot of $(H_2O)/k$ versus (H_2O) gave a straight line, in agreement with the mechanism:

$$\text{RNCO} + \text{H}_2\text{O} \underset{k_2}{\overset{k_1}{\rightleftharpoons}} [\text{complex}] \xrightarrow[k_3]{(\text{H}_2\text{O})} \text{RNHCONHR} + \text{H}_2\text{O} + \text{CO}_2$$

From the data in Table 19 the activation energy for the phenyl isocyanate/water reaction was calculated to be 11.0 kcal mole^{-1}.

TABLE 19
The reaction of phenyl isocyanate with water
(0.5M isocyanate in dioxane)

Water/isocyanate mole ratio	Temperature (°C)	$k \times 10^4$ (1 mole^{-1} sec^{-1})
4:1	25	1.42
2:1	25	0.77
1:1	25	0.41
1:1	35	0.73
1:1	50	1.53

The reaction between o-tolyl isocyanate and water in dioxane at 80°C with triethylamine catalyst was studied by Shkapenko et al. [188]. Supporting the results of Morton et al., the rate of isocyanate disappearance was found to be dependent on the NCO : H$_2$O ratio.

Catalysis of the isocyanate/water reaction by tertiary amines is well known, especially from research in the foam industry. Based on the rate of carbon dioxide evolution, the rate of the catalysed reaction is generally increased as the base strength of the tertiary amine catalyst is increased [189]. As in the case of catalysis of the isocyanate/hydroxyl reaction by triethylene diamine [142, 143], this amine catalyses the isocyanate/water reaction much more vigorously than would be predicted from its base strength in water (Table 20) [189]. Again this strong effect may be due to the absence of steric hindrance, or due to its relatively higher base strength in non-aqueous systems [121].

TABLE 20
Reaction of phenyl isocyanate with water at 23°C
(0.072M isocyanate, 0.036M water, 0.0014M catalyst)

Catalyst	pk_b	$k \times 10^4$ (1 mole^{-1} sec^{-1})	
		in benzene	in dioxane
None		0.64	0.016
Triethylamine	3.36		0.66
Triethylenediamine	5.40	6.0	1.8
N,N'-Dimethylpiperazine	5.71		0.65
N-Ethylmorpholine	6.49		0.33

Certain organometallic compounds have been found to be strong catalysts for the isocyanate/water reaction in dilute solution and relatively high catalyst concentrations [145]. In dioxane solution, at 70°C, with 0.25M phenyl isocyanate, 0.125M water and catalyst concentration at or extrapolated to 0.025M, the relative results shown in Table 21 were obtained. Somewhat in contrast to these results, Wolfe [147] found that stannous octoate and dibutyltin dilaurate were milder catalysts than was triethylamine.

TABLE 21
Relative catalyst activity for the iso-
cyanate/water reaction
(dioxane at 70°C)

Catalyst	Relative rate
None	1.1
Triethylamine	47
Triethylenediamine	380
Tributyltin acetate	14,000
Dibutyltin diacetate	100,000

8.3 POLYUREA AND POLYUREA—URETHANE FORMATION

Careful studies of polyurea formations are generally quite difficult. Low-molecular-weight disubstituted ureas have such low solubility in most solvents that precipitation causes experimental difficulties; the problem is even worse with many polyureas. In addition the rather limited thermal stability of polyureas at temperatures of about 200°C has detracted from what otherwise might have been a significant interest in these polymers for fibres. The major commercial interest has not been in pure polyureas, but rather in polymers containing polyurea blocks, as in water-blown polyurethane foams and in polyurea—urethane elastomers. The complexity of these commercial systems, which are nearly always crosslinked, has made kinetic studies difficult.

A group of papers by Fedotova et al. [190—194] has provided some rate data on polyurea formation. In a study of the reaction between HDI (hexamethylene diisocyanate) and several aromatic diamines at 0.2 mole l^{-1} concentration of each reactant, at 20—80°C, the rate was said to be nearly second order [190]. Beyond 50% reaction the rate coefficients became smaller, although the rate was said to be still second order. When using HDI or TDI, with 4,4'-diaminodiphenylmethane, o-toluidine or o-dianisidine, the rate coefficients were in the range of 6.5 x 10^{-4} to 0.13 l mole^{-1} sec^{-1}, depending on the isocyanate and amine [191]. Activation energies varied from 4.1 to 6.7 kcal mole^{-1}. Additional studies using several aromatic diamines in methyl ethyl ketone solvent at 25°C, with a calorimetric technique confirmed the catalytic

568

effect of aliphatic tertiary amines and several tin compounds [192]. Some of the problems inherent in the study of these systems were reported later, when it was recognized that the reaction between HDI and 4,4'-diaminodiphenyl sulphone at moderate temperatures gave low conversion, with products containing no more than three monomer units. When the polymerization was conducted in the solid state at 150°C, crosslinking occurred [193]. The progress of the reaction between dodecamethylene diamine and TDI in chlorobenzene at 20°C was followed by changes in the specific viscosity of the polymer, and reaction with several other diisocyanates was also reported, with no rate data given [194].

An unusually valuable study of the kinetics of polyurea—urethane formation from systems normally used to prepare water-blown "polyurethane" foams was described by Hartley et al. [195]. Some typical systems were found to consist of a single phase initially, while certain others were complicated by the existence of two phases, with phase inversion occurring. Studies included measurements of the rate of heat evolution, gas evolution and viscosity increase, as well as analysis for the presence of reaction products after intervals of time.

Viscosity measurements at constant temperature were made using a conventional Shirley—Ferranti cone-and-plate viscometer. In some cases complications arose due to phase inversion or to the fracture of low-molecular-weight polymer. One case was a typical "one-shot" flexible polyether foam system. On the other hand, a polyether prepolymer prepared from 2000 M.W. poly(oxypropylene)glycol and 80:20-TDI, reacted with water, formed a homogeneous system initially, and was suitable for study with this technique. The rate of viscosity increase at several temperatures is shown in Fig. 15.

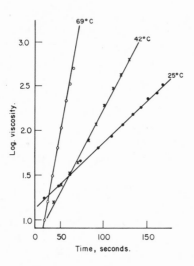

Fig. 15. Isothermal viscosity rise. Polyether—TDI prepolymer foam (ref. 195).

One-shot polyether foams were studied, using a variety of catalysts. The formula contained 100 parts by weight of poly(oxypropylene)triol of 3000 M.W., 38 parts of 80:20-TDI, 2.9 of water, 0.3 of 4-dimethyl-aminopyridine, 0.5 of N,N-dimethylbenzylamine, varying amounts of metal catalysts, and 0.1 part of X-520 siloxaneoxyalkylene copolymer. All of the gas was evolved from these systems within 60 sec after mixing. Viscosity measurements were not satisfactory due to fracture of the polymeric phase. Analysis of the reaction mixture at the end of 55 sec reaction time indicated the relative rate of formation of various products, as indicated in Table 22. The importance of selecting the proper catalyst to avoid undesirable side reactions is readily apparent. The results shown in Table 22 indicate that both tin catalysts promote the isocyanate/water reaction more than the isocyanate/hydroxyl reaction in the system studied. This is unusual, since other reports, though often of dilute solution studies, have shown the tin catalysts to promote the isocyanate/ hydroxyl reaction more [145, 147, 196].

The reaction between 65/35-TDI (65% 2,4-isomer, 35% 2,6-isomer) and a poly(oxypropylene)glycol, in the presence of small amounts of water, was studied by Aleksandrova and Lakosina [196]. The reaction rates were followed by disappearance of NCO, representing the combination of the two reactions, and by the rate of CO_2 evolution as a measure of the NCO/water reaction. Tertiary amines were found to catalyse the reaction with water about the same as that with the glycol. Dibutyltin dilaurate catalysed the TDI/glycol reaction about 1.5 times as strongly as the TDI/water reaction.

Aleksandrova et al. [197] have continued to report progress in studying the reactions of diisocyanates in systems simulating real foam systems.

TABLE 22

Comparison of the rate of formation of various chemical species in one-shot polyether foams during the first 55 sec. of reaction

Catalyst	Comparative rate of formation of				
	Urea	Biuret	Urethane	Allophanate	Isocyanurate
Stannous octoate 0.4 part	Very fast	Slow	Fast	None	None
Dibutyltin dilaurate 0.1 part	Very fast	Fast	Slow	None	None
Zinc dinonyldi-thiophosphate 0.1 part	Very fast	Very fast	Very slow	None	None
Lead octoate 0.1 part	Very fast	None	Very fast	None	Fast

They found the heat of reaction for 80:20 TDI and water to be 38.3 ± 1.74 kcal mole^{-1}, and that for 80:20 TDI and a poly(oxypropylene)triol to be 42.5 ± 0.76 kcal mole^{-1} (with two equivalents of NCO per mole in each case). Both separate reactions gave data which fit second-order kinetics up to 50% reaction. Preliminary attempts at calculating kinetic coefficients for both reactions during polyurea—urethane foam formation were given. It is believed that further efforts along this line will be needed to clarify the kinetics satisfactorily, however.

9. Miscellaneous reactions

An ever-increasing number of polycondensation reactions is being studied for preparative purposes. Some kinetic data are available on a few of them. In some cases many steps are involved, making reliable kinetic analysis difficult. In many other cases we may expect more data in the future, as the importance of the reactions becomes clearer. The following examples are not all-inclusive, but do illustrate the range of recent studies of polycondensation systems.

Examples of very well-known reactions which involve so many steps that kinetic analysis is difficult are the condensations of formaldehyde with compounds such as phenol, cresol, urea and melamine. In the reaction with phenol in alkaline medium the formation of methylol derivatives

is relatively fast, compared to the coupling of aromatic rings

The coupling reaction is the faster one in acid medium, however. This difference is utilized to prepare low-molecular-weight phenolic prepolymers, using an alkaline catalyst. These prepolymers are moderately stable at low temperatures, and may be crosslinked by acidification.

Ryabukhin [198] has proposed functional and kinetic relations for the condensation of phenol and formaldehyde in both acid and alkaline media. In alkaline systems the concentration of reacted phenol (x) at any time was found to be in agreement with the equation

$$x = a - [(na - y)n^{-1}a^{(1-n)/n}]^n$$

where a = original concentration of phenol, n = functionality of the phenol (3 in alkaline medium) and y = concentration of reacted formaldehyde. In acid medium, where $n = 1$, the data fit the equation

$$x = a - [(na - 2y)n^{-1}a^{(1-n)/n}]^n$$

In alkaline media the rate coefficients for the formation of mono-, di and trimethylolphenols had relative values of 1.0/0.75/0.25, respectively.

Schiemann and Hartman [199] studied the reactions involving o-cresol

They found the values for the first four rate coefficients to be $(1 \, \text{mole}^{-1} \, \text{sec}^{-1})$

$$k_1 = 12.6 \cdot 10^{-5} \quad k_3 = 18.0 \cdot 10^{-5}$$
$$k_2 = 11.4 \cdot 10^{-5} \quad k_4 = 8.4 \cdot 10^{-5}$$

Other recent references to phenol—formaldehyde condensations include those of Yeddanapalli et al. [200, 201]. The complications of melamine/formaldehyde and urea/formaldehyde reactions kinetics are analogous, and have been examined in two recent papers [202, 203].

Some data have been reported for another reaction of formaldehyde, the Prins reaction, with α-methylstyrene. With water or 40:60 water—dioxane mixture as solvent, the reaction was reported to be first order kinetically in each reactant. A direct relation between the log of the rate coefficient and the acidity function H_0 of the mineral acid catalyst was found [204].

The kinetics of aromatic polyether formation by the reaction of bisphenol salts with 4,4'-dichlorodiphenyl sulphone was studied by Schulze and Baron [205]. The potassium salts of three bisphenols were used, with dimethyl sulphoxide solvent. The rate was considerably faster in the early stages of the reaction, when the diphenoxide was present in higher concentrations than phenoxide chain ends. The faster reaction was assumed to be

$$\sim\!\!\text{Ar}'\text{Cl} + {}^-\text{O}\!-\!\text{Ar}\!-\!\text{O}^- \xrightarrow{k_1} \sim\!\!\text{Ar}'\text{OArO}^-$$

and the slower reaction in later stages to be

$$\sim\!\!\text{Ar}'\text{Cl} + {}^-\text{O}\!-\!\text{Ar}\!\sim \xrightarrow{k_2} \sim\!\!\text{ArOAr}'\!\sim$$

Values for k_1 and k_2 and activation energies for each of the three reactant combinations are shown in Table 23. The steric effect of the methyl substituents in the tetramethylbiphenol is readily apparent. Other differences in rate coefficients are accounted for by electronic effects.

The activation energy for the condensation of the sodium salt of bisphenol A with bis[4-(chloromethyl)phenyl]ether has been reported [206] to be 21.2 kcal mole^{-1}. At 185.5°C the second-order rate coefficient was 0.00023 mmole g^{-1} sec^{-1}.

The preparation of polyphenylene sulphides from p-halothiophenols also involves a difference in the reactivity of end groups on monomers, compared to those on the ends of polymer chains ($k_1 \neq k_2$), the processes being

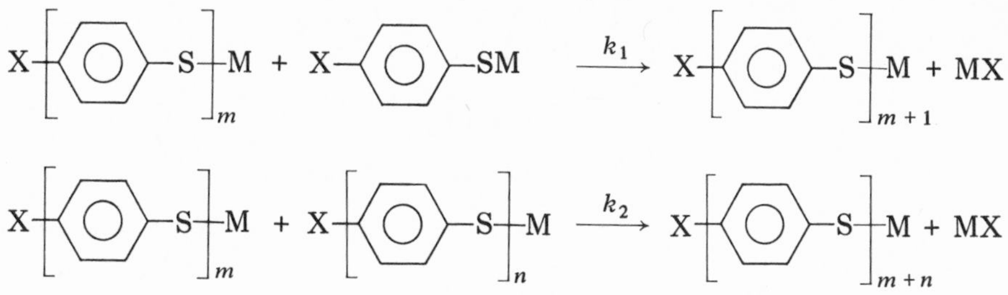

$$X\left[\!\!\left\langle\bigcirc\right\rangle\!\!-S\right]_m\!\!M + X\!\!-\!\!\left\langle\bigcirc\right\rangle\!\!-SM \xrightarrow{k_1} X\left[\!\!\left\langle\bigcirc\right\rangle\!\!-S\right]_{m+1}\!\!M + MX$$

$$X\left[\!\!\left\langle\bigcirc\right\rangle\!\!-S\right]_m\!\!M + X\!\!-\!\!\left[\!\left\langle\bigcirc\right\rangle\!\!-S\right]_n\!\!M \xrightarrow{k_2} X\left[\!\!\left\langle\bigcirc\right\rangle\!\!-S\right]_{m+n}\!\!M + MX$$

TABLE 23

Reaction rate coefficients for bisphenols which react with 4,4′-dichlorodiphenylsulphone in DMSO

Bisphenol	Temperature, (°C)	k_1	k_2	E_1	E_2
		(l equiv^{-1} sec^{-1})		(kcal equiv^{-1})	
A	80.0	0.0047	0.0014	~13	20.3
	79.8	0.0032	0.0010		
	80.0	0.0052	0.0015		
	90.1	0.0072	0.0028		
	89.5	0.0090	0.0029		
	100.0	0.0125	0.0058		
	110.0	a	0.013		
	120.0	a	0.024		
4,4′-Biphenol	80.0	0.0072	0.00037	~14	19
	120.0	0.057	0.0057		
2,2,2′,2′-Tetra- methylbiphenol	80.0	0.0040	0.00003	~12	—
	120.0	0.023	0.0014		

a Could not be determined from data.

Lenz et al. [207] reported studies with model compounds which were in agreement with this expectation. They also reported the over-all second order rate coefficients shown in Table 24.

TABLE 24

Over-all rate coefficients for the polymerization of alkali p-halothiophenoxides in pyridine at 250°C

 $X\!\!-\!\!\left\langle\bigcirc\right\rangle\!\!-SM$ X	M	k (l mole^{-1} sec^{-1})
F	Na	0.00043
Cl	Na	0.00018
Br	Na	0.0022
I	Na	0.014
Br	Li	0.043
Br	K	0.0016

Rembaum et al. [208, 209] have used a displacement reaction to prepare salts of polyamines (called "ionene" polymers), viz.

$$(CH_3)_2N(CH_2)_x N(CH_3)_2 + Br(CH_2)_y Br \rightarrow \left[\begin{array}{cc} CH_3 & CH_3 \\ | & | \\ N(CH_2)_x & N(CH_2)_y \\ | & | \\ CH_3 & CH_3 \end{array} \right]_n^{2n+} + 2n\ Br^-$$

These polymers, having a very high charge density, have been of interest as polyelectrolytes. The reaction has been reported to be bimolecular [208]. In preparing the 3,4-ionene bromide ($x = 3$, $y = 4$) [209] in 4:1 dimethyl formamide:methanol at $45°C$, the rate coefficient was found to be 0.0243 l equiv^{-1} sec^{-1}. The activation energy was 12.3 kcal mole^{-1}.

Low-molecular-weight polymers have been prepared by the condensation of glycols with phosphonic acid dichlorides, $RP(O)Cl_2$. The reactions have been reported to be second order, and are relatively slow [210, 211].

Friedel—Crafts reactions with halomethyl aromatic compounds have been used to prepare several types of polymers. The reaction is usually unsatisfactory because of the formation of either low-molecular-weight or crosslinked polymers. Rate studies are complicated by the multifunctional nature of the reactants, and often by limited solubility of the products. The kinetics of the first two steps of the reaction between benzene and p-bis-chloromethylbenzene with $SnCl_4$ catalysts were investigated by Grassie and Meldrum [212]. The activation energy was found to be about 10 kcal mole^{-1}.

Oxidative coupling of phenols to polyaryl ethers is now a well-established commercial process. The reactions are complex, and anomalous kinetic features have been encountered. These have been discussed recently by Vollmert and Saatweber [213].

The interaction of beryllium acetylacetonate with bis(acetoacetyl) compounds leads to coordination polymers. With bis(acetoacetyl)phenyl ether in dimethylformamide the reaction was found to be reversible [214], viz.

$$\left[\begin{array}{cc} O & OH \\ \| & | \\ CH_3CCH{=}C \end{array} \!\!-\!\!R \right]_2 + \left[\begin{array}{c} CH_3 \\ \diagdown C{-}O \\ \diagup \\ CH \\ \diagdown C{=}O \\ H_3C \end{array} \!\!\!Be \right]_2 \rightleftharpoons$$

$$\left[\begin{array}{c} CH_3 \\ C=O \cdots \quad O-C \quad CH_3 \\ CH \quad Be \quad CH \\ R-C-O \quad O=C \end{array}\right]_x \quad + \quad 2x \ CH_3COCH_2COCH_3$$

At 55—85°C the reaction followed second-order kinetics, with an activation energy of 25 kcal mole^{-1}. Exchange reactions between polymer chains occurred readily.

REFERENCES

1 R. W. Lenz, *Organic Chemistry of Synthetic High Polymers*, Interscience, New York, 1967.
2 P. J. Flory, Condensation Polymerization and Constitution of Condensation Polymers, in R. E. Burk and O. Grummitt, Eds., *High Molecular Weight Organic Compounds*, Frontiers in Chemistry, Vol. VI, Interscience, New York, 1949, pp. 211—283.
3 P. J. Flory, *Principles of Polymer Chemistry*, Cornell Univ. Press, New York, 1953.
4 R. H. Kienle, P. A. van der Meulen and F. E. Petke, *J. Amer. Chem. Soc.*, *61* (1939) 2258, 2268.
5 L. B. Sokolov, *Synthesis of Polymers by Polycondensation*, Izdatel'stvo Khimiya, Moscow, 1966; translated by J. Schmorak, Israel Program for Scientific Translations, Ltd., 1968.
6 P. W. Morgan, *Condensation Polymers: By Interfacial and Solution Methods*, Interscience, New York, 1965.
7 P. J. Flory, *J. Amer. Chem. Soc.*, *61* (1939) 3334.
8 S. Ephraim, A. E. Woodward and R. B. Mesrobian, *J. Amer. Chem. Soc.*, *80* (1958) 1326.
9 E. S. Gould, *Mechanism and Structure in Organic Chemistry*, Holt, New York, 1959, pp. 183—184.
10 K. A. Andrianov, D. Ya. Zhinkin and A. G. Kuznetzov, *Zh. Obshch. Khim.*, *29* (1959) 1504; L. B. Sokolov, *Synthesis of Polymers by Polycondensation*, Izdatel'stvo Khimiya, Moscow, 1966; translated by J. Schmorak, Israel Program for Scientific Translations, Ltd., 1968, p. 97.
11 L. B. Sokolov, *Synthesis of Polymers by Polycondensation*, Izdatel'stvo Khimiya, Moscow, 1966; translated by J. Schmorak, Israel Program for Scientific Translations, Ltd., 1968, p. 163.
12 P. W. Morgan and S. L. Kwolek, *J. Polym. Sci.*, *40* (1959) 300; Part A-1(4) (1963) 1147.
13 E. S. Gould, *Mechanism and Structure in Organic Chemistry*, Holt, New York, 1959, pp. 314—334.
14 L. P. Hammett, *Physical Organic Chemistry*, McGraw-Hill, New York, 1940.
15 J. D. Roberts and M. C. Caserio, *Basic Principles of Organic Chemistry*, Benjamin, New York, 1965, pp. 53Q—533.
16 H. H. Jaffe, *Chem. Rev.*, *53* (1953) 191.
17 M. S. Newman, *J. Amer. Chem. Soc.*, *72* (1950) 4783.
18 J. W. Baker, M. M. Davies and J. Gaunt, *J. Chem. Soc.*, (1949) 24.
19 J. W. Baker and J. B. Holdsworth, *J. Chem. Soc.*, (1947) 713.
20 J. W. Baker and J. Gaunt, *J. Chem. Soc.*, (1949) 9, 19, 27.

576

21 D. E. Koshland, *Chem. Eng. News*, (July 6, 1970) 54.
22 F. H. Brock, *J. Org. Chem.*, *24* (1959) 1802.
23 E. S. Gould, *Mechanism and Structure in Organic Chemistry*, Holt, New York, 1959, pp. 251—305.
24 J. March, *Advanced Organic Chemistry: Reactions, Mechanisms and Structure*, McGraw-Hill, New York, 1968, pp. 251—302; see also L. P. Hammett, *Physical Organic Chemistry*, McGraw-Hill, New York, 1940, pp. 131—156; J. D. Roberts and M. C. Caserio, *Basic Principles of Organic Chemistry*, Benjamin, New York, 1965, pp. 287—297.
25 J. D. Roberts and M. C. Caserio, *Basic Principles of Organic Chemistry*, Benjamin, New York, 1965, pp. 388—9, 664.
26 L. P. Hammett, *Physical Organic Chemistry*, McGraw-Hill, New York, 1940, pp. 154, 190.
27 J. Berzelius, *Rapp. Ann.*, *26* (1847), 260.
28 A. V. Lorenzo, *Amer. Chem. Phys.*, *67* (1863) 293.
29 W. H. Carothers, *J. Amer. Chem. Soc.*, *51* (1929) 2548, 2560.
30 D. A. S. Ravens and I. M. Ward, *Trans. Faraday Soc.*, *57* (1961) 150.
31 W. Griehl, *Faserforsch. Textiltech.*, *7* (1956) 464.
32 V. V. Korshak and S. V. Vinogradova, *Polyesters*, Pergamon Press, London (1965).
33 C. K. Ingold, *Structure and Mechanism in Organic Chemistry*, Cornell University Press, Ithaca, New York, 1953.
34 G. Reinisch, H. Zimmermann and G. Rafler, *Eur. Polym. J.*, *6* (1970) 205; *Faserforsch. Textiltech.*, *20* (1969) 225.
35 S. D. Hamann, D. H. Solomon and J. D. Swift, *J. Macromol. Sci., Chem.*, *A2* (1968) 153.
36 C. H. DePuy and R. W. King, *Chem. Rev.*, *60* (1960) 399.
37 R. Taylor, G. G. Smith and W. H. Wetzel, *J. Amer. Chem. Soc.*, *84* (1962) 4817.
38 I. Goodings in *Thermal Degradation of Polymers*, Soc. of Chemical Industry Monograph No. 13, London, 1961, p. 211.
39 L. H. Buxbaum, *Angew. Chem.*, *80* (1968) 225.
40 H. Schnell, *Chemistry and Physics of Polycarbonates*, Interscience, New York, 1964.
41 S. G. Hovenkamp and J. P. Munting, *J. Polym. Sci., Part A1*, *8* (1970) 679.
42 F. Dobinson, unpublished data.
43 I. Goodman and B. F. Nesbitt, *Polymer*, *1* (1960) 384; *J. Polym. Sci.*, *48* (1960) 423.
44 H. Zahn and B. Seidl, *Makromol. Chem.*, *23* (1957) 31.
45 L. H. Peebles, Jr., M. W. Huffman and C. T. Ablett, *J. Polym. Sci., Part A1*, *7* (1969) 479.
46 W. Sweeny, *J. Appl. Polym. Sci.*, *5* (1961) 15.
47 C. N. Hinshelwood and A. Legard, *J. Chem. Soc.*, (1935) 587; R. H. Fairclough and C. N. Hinshelwood, *J. Chem. Soc.*, (1939) 593; A. C. Rolfe and C. N. Hinshelwood, *Trans. Faraday Soc.*, *30* (1934) 935.
48 P. J. Flory, *Chem. Rev.*, *39* (1946) 137.
49 M. Davies, *Research (London)*, *2* (1949) 544.
50 A. Tang and K. Yao, *J. Polym. Sci.*, *35* (1959) 219.
51 C. H. Huang, Y. Simono and T. Onizuka, *Kobunshi Kagaku*, *23* (1966) 1408.
51a J. F. Kemkes, *J. Polym. Sci., Part C*, *22* (1969) 713.
52 F. Mares, V. Bazant and K. Krupicka, *Collect. Czech. Chem. Commun.*, *34* (1969) 2208.
53 I. Vansco-Szmercsanyi, K. Maros-Greger and E. Makay-Bodi, *Eur. Polym. J.*, *5* (1969) 155.
54 M. T. Pope and R. J. P. Williams, *J. Chem. Soc.*, (1959) 3579.

55 C. E. H. Bawn and M. B. Huglin, *Polymer, 3* (1962) 257.
56 S. G. Hovenkamp, I.U.P.A.C. International Symposium on Macromolecular Chemistry, Budapest, 1969, Preprints, 1 (1969) 113.
57 H. Zimmermann, *Faserforsch. Textiltech., 11* (1962) 481.
58 K. Yoda, K. Kimoto and T. Toda, *Kogyo Kagaku Zasshi, 67* (1964) 909.
59 K. Yoda, *Makromol. Chem., 136* (1970) 311.
60 W. Gordy and W. J. O. Thomas, *J. Chem. Phys., 24* (1956) 439.
61 C. M. Fontana, *J. Polym. Sci., Part A1, 6* (1968) 2343.
62 C. M. Fontana in P. H. Plesch, Ed., *The Chemistry of Cationic Polymerization,* Pergamon Press, New York, 1963, Chapter 5.
63 W. Griehl and G. Schnock, *Faserforsch. Textiltech., 8* (1957) 408; *J. Polym. Sci., 30* (1958) 413.
64 L. H. Peebles, Jr. and W. S. Wagner, *J. Amer. Chem. Soc., 63* (1959) 1206.
65 G. Challa, *Rec. Trav. Chim., Pays-Bas., 79* (1960) 90.
66 S. G. Hovenkamp, *J. Polym. Sci., Part A1, 7* (1969) 3428.
67 E. Turska and A. M. Wrobel, *Polymer, 11* (1970) 415.
68 G. Challa, *Makromol. Chem., 38* (1960) 105.
69 G. Challa, *Makromol. Chem., 38* (1960) 123.
70 G. Challa, *Makromol. Chem., 38* (1960) 138.
71 R. W. Stevenson and H. R. Nettleton, *J. Polym. Sci., Part A1, 6* (1968) 889.
72 R. W. Stevenson, *J. Polym. Sci., Part A1, 7* (1969) 395.
73 P. Cefelin and J. Malek, *Collect. Czech. Chem. Commun., 34* (1969) 419.
74 G. A. Campbell, E. F. Elton and E. G. Bobalek, *J. Appl. Polym. Sci., 14* (1970) 1025.
75 P. J. Flory, *Principles of Polymer Chemistry,* Cornell Univ. Press, New York, 1953, p. 347.
76 P. J. Flory and F. S. Leutner, U.S. Patents Nos. 2,589,688 and 2,623,034.
77 S. G. Entelis, G. P. Kondrat'eva and N. M. Chirkov, *Vysokomol. Soedin., 3*[7] (1961) 1044; *3*[8] (1961) 1170.
78 W. H. Eareckson, *J. Polym. Sci., 40* (1959) 399.
79 A. Dems, J. Pieniazek and E. Turska, *Bull. Acad. Pol. Sci., Ser. Sci. Chim., 13* (1965) 177; *Chem. Abstr., 63* (1965) 849.
80 D. A. Holmer, O. A. Pickett, Jr. and J. H. Saunders, *J. Polym. Sci., Part A1,* (1972) 1547.
81 P. F. van Velden, G. M. van der Want, D. Heikens, C. A. Kruissink, P. H. Hermanns and A. J. Staverman, *Rec. Trav. Chim., Pays-Bas, 74* (1955) 1376.
82 J. C. Sheehan and G. P. Hess, *J. Amer. Chem. Soc., 77* (1955) 1067.
83 H. A. Staab, *Angew. Chem. Int. Ed. Engl., 1* (1962) 351.
84 B. G. Achhammer, F. W. Reinhart and G. M. Kline, *J. Appl. Chem., 1* (1951) 301.
85 S. Straus and L. Wall, *J. Res. Nat. Bur. Stand., 60* (1958) 39.
86 P. H. Hermanns, *J. Appl. Chem., 5* (1955) 493.
87 G. Champetier and R. Vergoz, *Rec. Trav. Chim., Pays-Bas, 69* (1950) 85.
88 P. J. Flory, U.S. Patent 2,244,192 (1941).
89 P. J. Flory, *Principles of Polymer Chemistry,* Cornell Univ. Press, New York, 1953, p. 83.
90 V. V. Korshak and T. M. Frunze, *Synthetic Heterochain Polyamides,* Daniel Davey & Co., New York, 1964.
91 J. Charles, J. Cologne and G. Descotes, *C.R. Acad. Sci., 256* (1963) 3107.
92 B. A. Zhubanov, S. R. Rafikov, L. V. Pavletenko and L. B. Rukhina, *Izv. Akad. Nauk. Kaz. SSR, Ser. Khim., 17*[4] (1967) 69; *Chem. Abstr., 69* (1968) 10763.
93 J. Cologne and E. Fichet, *Bull. Soc. Chim. Fr.,* (1955) 412.
94a N. Ogata, *Makromol. Chem., 43* (1961) 117.
94b V. V. Korshak, T. M. Frunze and L. Yi-nan, *Vysokomol. Soedin., 2* (1960) 984.

578

95 R. D. Chapman, unpublished data.
96 V. V. Korshak and T. M. Frunze, *Synthetic Heterochain Polyamides*, Daniel Davey & Co., New York, 1964, p. 124.
97 A. V. Volokhina, M. P. Bogdanov and G. I. Kudryavtsev, *Vysokomol. Soedin.*, *2* (1960) 92.
98 G. N. Chelnokova, S. R. Rafikov and V. V. Korshak, *Dokl. Akad. Nauk. SSSR*, *64* (1949) 353.
99 S. J. Allen and J. G. N. Drewitt, U.S. Patent 2,558,031 (1951).
100 C. F. Horn, B. T. Freure, H. Vineyard and H. J. Decker, *J. Appl. Polym. Sci.*, 7 (1963) 887.
101 J. March, *Advanced Organic Chemistry: Reactions, Mechanisms and Structure*, McGraw-Hill, New York, 1968, p. 338.
102 V. V. Korshak and T. M. Frunze, *Synthetic Heterochain Polyamides*, Daniel Davey & Co., New York, 1964, p. 86.
103 T. M. Bogert and R. R. Renshaw, *J. Amer. Chem. Soc.*, *30* (1908) 1140.
104 C. E. Sroog, A. L. Endrey, S. V. Abramo, C. E. Berr, W. M. Edwards and K. L. Olivier, *J. Polym. Sci.*, *Part A3*, (1965) 1373.
105 L. A. Laius, M. I. Bessonov, Y. V. Kallistova, N. A. Adrova and F. S. Florinskii, *Vysokomol. Soedin.; A9* (1967) 2185.
106 P. W. Morgan, *Condensation Polymers: By Interfacial and Solution Methods*, Interscience, New York, 1965, p. 137.
107 L. B. Sokolov and L. V. Turetskii, *Vysokomol. Soedin.*, *2* (1960) 711.
108 J. Preston, *J. Polym. Sci.*, *Part A1*, *4* (1966) 529.
109 F. Dobinson and J. Preston, *J. Polym. Sci.*, *Part A1*, *4* (1966) 2093.
110 W. B. Black and J. Preston, *Fiber-Forming Aromatic Polyamides* in *Man-Made Fibers, Science and Technology*, Interscience, New York, 1968, Vol. 2, p. 297.
111 H. S. Morgan and F. Dobinson, unpublished data.
112 L. B. Sokolov, *Synthesis of Polymers by Polycondensation*, Izdatel'stvo Khimiya, Moscow, 1966; translated by J. Schmorak, Israel Program for Scientific Translations, Ltd., 1968, pp. 26, 109.
113 S. Porejko, L. Makaruk and M. Kepka, *Polimery*, *12*[10] (1967) 465; *Chem. Abstr.*, *69* (1968) 3211.
114 A. Cannepin, G. Champetier and A. Parisot, *J. Polym. Sci.*, *8* (1952) 35.
115 W. R. Sorenson and T. W. Campbell, *Preparative Methods of Polymer Chemistry*, 2nd Ed., Interscience, New York, 1968, p. 101.
116 J. H. Saunders and K. C. Frisch, *Polyurethanes, Chemistry and Technology. I. Chemistry*, Interscience, New York, 1962.
117 A. Farkas and G. A. Mills, Catalytic Effects in Isocyanate Reactions, in D. D. Eley, P. W. Selwood and P. B. Weisz, Eds., *Advances in Catalysis*, *13* (1962) 393; Academic Press, New York.
118 D. J. Lyman, *Rev. in Macromol. Chem.*, *1* (1966) 191.
119 P. C. Johnson, in J. M. Buist and H. Gudgeon, Eds., *Advances in Polyurethane Technology*, Elsevier, Essex, 1968, Chapter 1.
120 J. H. Saunders and K. C. Frisch, *Polyurethanes, Chemistry and Technology. I. Chemistry*, Interscience, New York, 1962, especially pp. 211—215; and *Polyurethanes, Chemistry and Technology. II. Technology*, Interscience, New York, 1964, especially pp. 8—43.
121 A. Farkas and P. F. Strohm, *Ind. Eng. Chem. Fundam.*, *4* (1965) 32.
122 M. E. Bailey, V. Kirss and R. G. Spaunburgh, *Ind. Eng. Chem.*, *48* (1956) 794.
123 M. Kaplan, *J. Chem. Eng. Data*, *6* (1961) 272.
124 J. Burkus and C. F. Eckert, *J. Amer. Chem. Soc.*, *80* (1958) 5948.
125 J. L. O'Brien and A. S. Pagano, paper presented at the Delaware Regional Meeting of the American Chemical Society, February 5, 1958.
126 M. Sato, *J. Amer. Chem. Soc.*, *82* (1960) 3893.

579

127 H. F. Brock, *J. Phys. Chem.*, 65 (1961) 1638.

128 A. DiGiacomo, *J. Phys. Chem.*, 65 (1961) 696.

129 I. C. Kogon, *J. Org. Chem.*, 24 (1959) 438.

130 R. E. Cunningham and T. G. Mastin, *J. Org. Chem.*, 24 (1959) 1585.

131 L. C. Case, *J. Chem. Eng. Data*, 5 (1960) 347.

132 L. C. Case, *J. Appl. Polym. Sci.*, 8 (1964) 533.

133 L. C. Case and K. W. Li, *J. Appl. Polym. Sci.*, 8 (1964) 935.

134 L. L. Ferstandig and R. A. Scherrer, *J. Amer. Chem. Soc.*, 81 (1959) 4838.

135 L. Mandasescu, A. Petrus and I. Matei, *Rev. Roum. Chim.*, 9(8—9) (1964) 491; *Chem. Abstr.*, 62 (1965) 16006.

136 O. Bayer, *Angew. Chem.*, A59 (1947) 257.

137 E. C. Lovering and K. J. Laidler, *Can. J. Chem.*, 40 (1962) 26.

138 M. Pestermer and D. Lauerer, *Angew. Chem.*, 72 (1960) 612.

139 H. Pracejus and G. Ochme, *Z. Chem.*, 4 (1964) 426; *Chem. Abstr.*, 62 (1965) 13081.

140 Z. Brzozowski, *Rocz. Chem.*, 38 (1964) 1279; *Chem. Abstr.*, 61 (1964) 16047.

141 A. E. Oberth and R. S. Bruenner, *J. Phys. Chem.*, 72 (1968) 845.

142 A. Farkas and K. G. Flynn, *J. Amer. Chem. Soc.*, 82 (1960) 642.

143 A. Farkas, G. A. Mills, W. E. Erner and J. B. Maerker, *Ind. Eng. Chem.*, 51 (1959) 1299.

144 J. J. Tazuma and H. K. Latourette, paper presented at the American Chemical Society Meeting, Atlantic City, September, 1956.

145 F. Hostettler and E. F. Cox, *Ind. Eng. Chem.*, 52 (1960) 609.

146 J. W. Britain and P. G. Gemeinhardt, *J. Appl. Polym. Sci.*, 4 (1960) 207.

147 H. W. Wolfe, Jr., Foam Bulletin, *Catalyst Activity in One-Shot Urethane Foam*, E. I. DuPont deNemours and Co., March 16, 1960.

148 F. G. Willeboordse, F. E. Critchfield and R. L. Meeker, *J. Cell. Plast.*, 1 (1965) 76.

149 S. L. Reegen and K. C. Frisch, *J. Polym. Sci., Part A1*, 8 (1970) 2883.

150 A. J. Bloodworth and A. G. Davies, *Proc. Chem. Soc., London*, (1963) 264.

151 H. L. Heiss, F. P. Combs, P. G. Gemeinhardt, J. H. Saunders and E. E. Hardy, *Ind. Eng. Chem.*, 51 (1959) 929.

152 C. D. Nolen, Belg. Pat. 645,558 (Mobay Chem. Co.); *Chem. Abstr.*, 63 (1965) 10145.

153 M. E. Bailey, C. E. McGinn and R. G. Spaunburgh, paper presented at the American Chemical Society Meeting, Atlantic City, September 1956.

154 K. L. Hoy, R. A. Martin and R. H. Peterson, 152nd National American Chemical Society Meeting, September 1966, Org. Coatings and Plastics Div. preprints, p. 55.

155 W. Cooper, R. W. Pearson and S. Darke, *The Industrial Chemist*, 36 (1960) 121.

156 A. A. Blagonravova, I. A. Pronina, A. V. Uvarov, G. V. Rudnaya and S. M. Aref'eva, *Lakokrasoch. Mater. Ikh Primen.*, (1965) (4) 1; *Chem. Abstr.*, 64 (1966) 6761.

157 J. R. Fisher, *Tetrahedron Suppl.*, 1, 19 (1963) 97.

158 W. C. Darr, P. G. Gemeinhardt and J. H. Saunders, *J. Cell. Plast.*, 2 (1966) 266.

159 L. B. Wiesfeld, *J. Appl. Polym. Sci.*, 5 (1961) 424.

160 S. Ronssin, *Ind. Plastiques Mod. (Paris)*, 13(1) (1961) 13; *Chem. Abstr.*, 55 (1961) 15384.

161 T. Yokoyama and T. Iwaso, *Kogyo Kagaku Zasshi*, 63 (1960) 1835; *Chem. Abstr.*, 52 (1962) 2397e.

162 T. Tanaka, T. Yokoyama and T. Iwasa, *Kogyo Kagaku Zasshi*, 66 (1963) 158; *Chem. Abstr.*, 59 (1963) 5346.

163 H. G. Wissman, L. Rand and K. C. Frisch, *J. Appl. Polym. Sci.*, 8 (1964) 2971.

164 B. A. Fomenko, V. A. Orlov and O. G. Tarakanov, *Plast. Massy*, 10 (1964) 47; *Chem. Abstr.*, 62 (1965) 10603.

580

165 T. E. Lipatova, L. A. Bakalo, A. L. Sirotinskaya and O. P. Syutkina, *Vysokomol. Soedin. Ser. A, 10*(4) (1968) 859; *Chem. Abstr., 69* (1968) 19559k.

166 T. E. Lipatova, L. A. Bakalo and R. A. Loktionova, *Polym. Sci. USSR, 10* (1968) 1799.

167 L. M. Koncharenko, L. A. Bakalo and K. A. Kornev, *Sin. Fiz.-Khim. Polim.*, (1967) 98; *Chem. Abstr., 69* (1968) 97209k.

168 E. N. Novgorodov and A. P. Khardin, *Khim. Tekhnol.*, (1968) 82; *Chem. Abstr., 72* (1970) 32328r.

169 T. E. Lipatova and G. S. Shapoval, *Sin. Fiz.-Khim. Polim.*, (1968) (5) 133; *Chem. Abstr., 70* (1969) 4916w.

170 V. Pshezhetskii, I. Massukh and V. A. Kabanov, *J. Polym. Sci., Part C, 22* (1967) 309; *Vysokomol. Soedin. Ser. B, 9*(11) (1967) 834; *Chem. Abstr., 68* (1968) 30151w.

171 V. M. Zolotarev, *Zh. Prikl. Spektrosk.* 7(5) (1967) 743; *Chem. Abstr., 68* (1968) 50408p; see also T. E. Lipatova and S. A. Zubko, *Dokl. Akad. Nauk SSSR, 184*(4) (1969) 877; *Chem. Abstr., 70* (1969) 107054t.

172 N. A. Lipatnikov et al., *Sin. Fiz.-Khim. Polim.*, (1968) (5) 96; *Chem. Abstr., 70* (1969) 4695y.

173 E. L. Wittbecker and M. Katz, *J. Polym. Sci., 40* (1959) 367.

174 S. L. Kwolek and P. W. Morgan, *J. Polym. Sci., Part A2, 6* (1964) 2693.

175 J. H. Saunders and K. C. Frisch, *Polyurethanes, Chemistry and Technology. II. Technology*, Interscience, New York, 1964.

176 T. L. Davis and F. Ebersole, *J. Amer. Chem. Soc., 56* (1934) 885.

177 C. Naegeli, A. Tyabji and L. Conrad, *Helv. Chim. Acta, 21* (1938) 1127.

178 R. G. Arnold, J. A. Nelson and J. J. Verbanc, *Chem. Revs., 57* (1957) 47.

179 R. L. Craven, paper presented at the American Chemical Society Meeting, Atlantic City, September 1956.

180 J. H. Saunders, *Rubber Chem. Technol., 33* (1960) 1293.

181 S. Ozaki and T. Hoshino, *Nippon Kagaku Zasshi, 80* (1959) 664; *Chem. Abstr., 55* (1961) 4398a.

182 J. W. Baker and D. N. Bailey, *J. Chem. Soc.*, (1957) 4649, 4652, 4663.

183 K. Sauer and M. H. Kasparian, *J. Org. Chem., 26* (1961) 3498.

184 S. L. Axelrood, C. W. Hamilton and K. C. Frisch, paper presented at the American Chemical Society Meeting, April 1961.

185 C. Naegeli, A. Tyabji, L. Conrad and F. Litwan, *Helv. Chim. Acta, 21* (1938) 1100.

186 M. Morton and M. A. Diesz, paper presented at the American Chemical Society Meeting, Atlantic City, September 1956.

187 M. Morton, M. A. Deisz and M. Ohta, *Degradation Studies on Condensation Polymers*, U.S. Dept. of Commerce Report PB131795, March 31, 1957.

188 G. Shkapenko, G. J. Gmitter and E. E. Gruber, *Ind. Eng. Chem., 52* (1960) 605.

189 B. G. Alzner and K. C. Frisch, *Ind. Eng. Chem., 51* (1959) 715.

190 N. I. Skripchenko, O. Y. Fedotova and I. P. Losev, *Tr. Mosk. Khim.-Tekhnol. Inst., 42* (1963) 126; *Chem. Abstr., 62* (1965) 7603.

191 O. Y. Fedotova, I. Selleshi and I. P. Losev, *Tr. Mosk. Khim.-Technol. Inst., 42* (1963) 120; *Chem. Abstr., 62* (1965) 6366.

192 O. Y. Fedotova, A. G. Grozdov and I. A. Rusinovskaya, *Vysokomol. Soedin, 7*(12) (1965) 2028; *Chem. Abstr., 64* (1966) 9544.

193 O. Y. Fedotova, A. G. Grozdov and G. S. Kolesnikov, *Vysokomol. Soedin, 9*(6) (1967) 459; *Chem. Abstr., 67* (1967) 100427.

194 H. S. Kolesnikov, O. Y. Fedotova and T. Z. Tin, *Polymer Sci., USSR, 10* (1968) 3129.

195 F. D. Hartley, M. M. Cross and F. W. Lord, in J. M. Buist and H. Gudgeon, Eds., *Advances in Polyurethane Technology*, Elsevier, Essex, 1968, pp. 127—140.

196 Y. V. Aleksandrova and T. A. Lakosina, *Plast. Massy*, 7 (1965) 15; *Chem. Abstr.*, *63* (1965) 11305.

197 Y. V. Aleksandrova, Y. V. Sharikov and O. G. Tarakanov, *Polym. Sci. USSR*, *11* (1970) 2786.

198 A. G. Ryabukhin, *Vysokomol. Soedin. Ser. A11*(11) (1969) 2562.

199 G. Schiemann and E. Hartman, *Makromol. Chem.*, *63* (1963) 174.

200 K. C. Eapen and L. M. Yeddanapalli, *Makromol. Chem.*, *119* (1968) 4.

201 D. J. Francis and L. M. Yeddanapalli, *Makromol. Chem.*, *125* (1969) 119.

202 K. Sato, *Bull. Chem. Soc. Jap.*, *41* (1968) 7.

203 J. W. Aldersley and M. Gordon, *J. Polym. Sci. Part C*, *16* (1967) 4567.

204 J. Gaillard, M. Hellin and F. Coussemant, *Bull. Soc. Chim. Fr.*, (1967) (9) 3360; *Chem. Abstr.*, *68* (1968) 28975.

205 S. R. Schulze and A. L. Baron, *Addition and Condensation Polymerization Processes*, Advances in Chemistry Series, *91* (1969) 692.

206 M. Tokarzewska, *Polimery*, *12*(11) (1967) 512; *Chem. Abstr.*, *69* (1968) 10760.

207 R. W. Lenz, C. E. Handlovits and H. A. Smith, *J. Polym. Sci.*, *58* (1962) 351.

208 A. Rembaum, W. Baumgartner and A. Eisenberg, *J. Polym. Sci.*, *B*, *6* (1968) 159.

209 A. Rembaum, H. Rile and R. Somoano, *Polym. Lett.*, *8* (1970) 457.

210 M. F. Sorokin and I. Manoviciu, *Mater. Plast.*, *3*(4) (1966) 191; *Chem. Abstr.*, *66* (1967) 2892.

211 L. N. Mashlyskovski, B. I. Ionin and I. S. Okhrimenko, *Khim. Org. Soedin. Fosfora, Akad. Nauk SSSR, Otd. Obshch. Tekh. Khim.*, (1967) 244; *Chem. Abstr.*, *68* (1968) 114705.

212 N. Grassie and I. G. Meldrum, *Eur. Polym. J.*, *5*(1) (1969) 195; *Chem. Abstr.*, *70* (1969) 97225.

213 B. Vollmert and D. Saatweber, *Angew. Makromol. Chem.* (1968) (2) 114.

214 S. V. Vinogradova, V. V. Korshak and M. G. Vinogradov, *J. Polym. Sci., Part C*, *16* (1967) 2565.

Chapter 8

The Polymerization of *N*-Carboxy-α-Amino Acid Anhydrides

C. H. BAMFORD and H. BLOCK

Introduction

The monomers discussed in this Chapter have the general structure shown in (I) and are variously termed oxazolid-2,5-diones, *N*-carboxy-α-

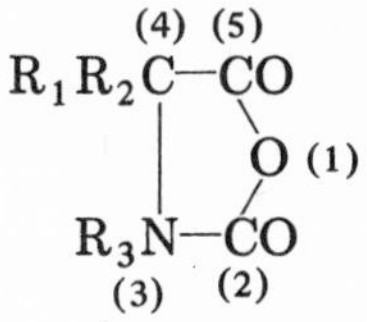

$$
\begin{array}{c}
\overset{(4)\ (5)}{R_1R_2C\!-\!CO} \\
| \qquad \diagdown \\
\qquad\quad O\ (1) \\
| \qquad \diagup \\
\underset{(3)\quad (2)}{R_3N\!-\!CO}
\end{array}
$$

(I)

amino acid anhydrides (NCAs) and Leuchs anhydrides (after their discoverer) [1]. The first of these is the systematic nomenclature and is used in Chemical Abstracts, but we shall use the second term, since this has the advantage of indicating the parent α-amino acid from which the monomer is derived. The polymerization reactions we shall discuss all lead to polymers of the parent α-amino acid. In many instances conditions can be chosen so that high polymers are formed. This aspect of the chemistry of NCAs endows them with special importance, since it provides a relatively simple way of synthesizing optically pure poly-α-amino acids which are otherwise accessible only with difficulty. These materials are of interest for a number of reasons. They have structural analogies with proteins and have played an important part in the elucidation of chain conformations in the latter; further their well-defined conformations have made them valuable in the study of polymer behaviour. Finally, possible commercial applications are of current interest. A number of reviews of these topics are available [2—6].

The preparation of a poly-α-amino acid involves first, synthesis of the monomer, and secondly its polymerization. We are mainly concerned here with the latter aspect. Methods of monomer synthesis commonly follow one of two alternative routes viz. reaction of the α-amino acid with phosgene (1) or cyclization of an *N*-alkoxy carbonyl derivative of the α-amino acid by treating with $SOCl_2$, PCl_5 or similar reagents (2).

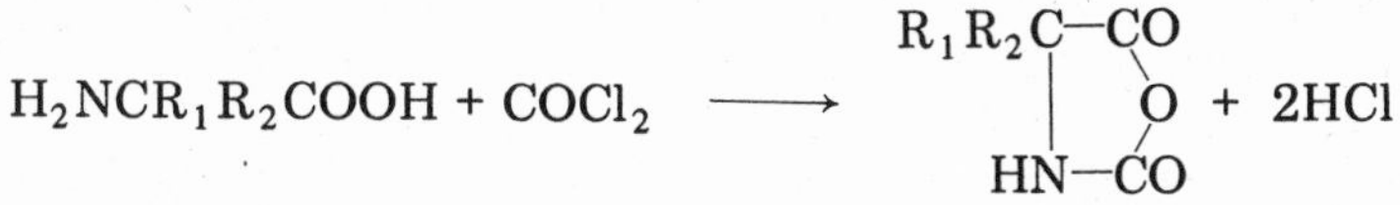

$$
H_2NCR_1R_2COOH + COCl_2 \longrightarrow
\begin{array}{c}
R_1R_2C\!-\!CO \\
| \qquad \diagdown \\
\qquad\quad O \\
| \qquad \diagup \\
HN\!-\!CO
\end{array}
+ 2HCl \qquad (1)
$$

References pp. 634—637

$$R_4O \cdot CONR_3CR_1R_2COOH \xrightarrow{PCl_5} (R_4O \cdot CONR_3CR_1R_2COCl) \rightarrow$$

$$(I) + R_4Cl \qquad (2)$$

Complications arise in both cases when reactive groups such as $-OH$, $-NH_2$, $-COOH$, $-SH$ are present in R_1, R_2 (or R_3). Such groups must be suitably masked before cyclization. If the parent poly-α-amino acid is ultimately required, the ease of subsequent removal of the protecting group is an important consideration. Details of suitable procedures have recently been reviewed [6] and will not be discussed further here.

Polymerization follows the over-all stoichiometry (3) and is most

$$n\begin{pmatrix} R_1R_2C{-}CO \\ | \quad\;\; O \\ R_3N{-}CO \end{pmatrix} \longrightarrow +R_3NCR_1R_2CO\overline{)_n} + nCO_2 \qquad (3)$$

conveniently effected by the addition of a base, but it can also be achieved in other ways, for example, thermally or by the presence of anhydrous hydrogen fluoride [6, 7]. The nature of the terminal groups of the polymer chains is determined, as will be discussed below, both by the nature of the initiator and the conditions of polymerization.

The evolution of carbon dioxide during polymerization has been used to follow the course of reaction [7—12]; other techniques of monomer estimation include sodium methoxide titration [13], spectrophotometry [14] making use of the absorbance of the NCA carbonyls at 1860 or 1790 cm^{-1} and colorimetry [15] based on the derived hydroxamate—ferric ion complex (514 nm).

For the successful synthesis of poly-α-amino acids of high molecular weight an understanding of the mechanism of polymerization is essential. We devote the remainder of this chapter to a detailed discussion of the kinetics and mechanisms.

General principles

Inspection of structure (I) will show that there are a number of alternative points for nucleophilic attack on the NCA molecule. The carbonyl groups at positions 2 and 5 can and do, in suitable circumstances, react with bases; however, as will be apparent later, only the latter type of process can sustain a propagating chain leading to a polymer. If only the 5-CO were involved in chain growth, the mechanism of polymerization would be simple. For example, with primary base initiation, reaction would proceed exclusively as shown in (4).

$$\begin{matrix} R_1R_2C-CO \\ | \quad\quad O \\ R_3N-CO \end{matrix} + R_4NH_2 \longrightarrow \begin{matrix} R_1R_2C-CONHR_4 \\ | \\ R_3NH + CO_2 \end{matrix} \quad (a)$$

$$\text{(I)} \qquad\qquad\qquad\qquad \text{(II)} \qquad\qquad\qquad\qquad (4)$$

$$\text{(I) + (II)} \rightarrow \begin{matrix} R_1R_2C-CONR_3CR_1R_2CONHR_4 \\ | \\ R_3NH + CO_2 \end{matrix} \quad (b)$$

etc. Such a simple situation is rarely encountered in practice. Attack by base on 2-CO leading to a ureido derivative (III) (reaction (5)) constitutes a termination reaction, since it removes propagating base groups. Even though (4) is superficially simple, each addition may involve multiple

$$\sim\!\sim NHR_3 + \begin{matrix} R_1R_2C-CO \\ | \quad\quad O \\ R_3N-CO \end{matrix} \longrightarrow \begin{matrix} R_1R_2C-COOH \\ | \\ R_3N-CONR_3\!\sim\!\sim \end{matrix} \quad (5)$$

$$\text{(III)}$$

stages, so that the over-all kinetics are complex. This is the case in the amine-initiated polymerization of sarcosine NCA (I : $R_1 = R_2 = H$, $R_3 = CH_3$).

Monomers with $R_3 = H$ are found to present a third possibility for nucleophilic attack, the NCA acting as a weak Brönsted acid, as in (6), in which B is a base [18—24]. Anions such as (IV), although normally present

$$\begin{matrix} R_1R_2C-CO \\ | \quad\quad O \\ HN-CO \end{matrix} + B \rightleftharpoons \begin{matrix} R_1R_2C-CO \\ | \quad\quad O \\ \underline{N}-CO \end{matrix} + BH^+ \quad (6)$$

$$\text{(IV)}$$

in very low concentration, become kinetically significant whenever they participate in other reactions which result in their consumption, since the proton exchanges in (6) occur very rapidly and maintain the equilibrium. This situation, which is encountered even when B is a weak base, has not been sufficiently appreciated.

Finally, it has been suggested [23] that ionization at C4 may occur if $R_1 = H$ and $R_3 \neq H$ when the initiating base is very strong.

References pp. 634—637

The reactions we have been considering form the skeleton of the actual mechanisms of polymerizations which are encountered in practice. However, these rarely occur in isolation so that over-all mechanism is often very complex. For clarity we shall first consider the consequences of some simple idealized reaction schemes.

1. Addition reactions of amines

The reactions shown in (4a) and (4b) constitute the initiation and propagation reactions, respectively. The rate of each step depends on the strength of the attacking base, increasing as the latter increases. In practice, after a small number of additions the propagating peptide base will have a strength effectively independent of chain-length, and the rate of propagation therefore becomes independent of chain-length. In kinetic studies it is common to employ preformed oligomers[*] as initiators for this reason. In the absence of termination reactions mechanism (4) implies constancy of base concentration, and the rate equation is therefore

$$-d[M]/dt = k[M][I] \tag{7}$$

where M is the monomer (NCA) and I the initiating oligomer. The reaction then proceeds until all the monomer is consumed. From (7) it follows that in this simple case [M] is given as a function of time by the relation

$$\ln([M]/[M]_0) = k[I]t \tag{8}$$

in which $[M]_0$ is the initial concentration of NCA.

The reactions shown in (4) proceed through the intermediate formation of substituted carbamic acid, e.g.

$$R_1R_2C\text{—CO}\underset{R_3N\text{—CO}}{\overset{}{\Big|}}O + R_4NH_2 \longrightarrow \underset{R_3NCOOH}{R_1R_2C\text{—CONHR}_4} \quad \text{(a)}$$

$$\downarrow$$

$$\underset{R_3NH}{R_1R_2C\text{—CONHR}_4} + CO_2 \quad \text{(b)} \tag{9}$$

Frequently this is not of kinetic significance (see, however, Section 5) on account of the rapid decomposition of carbamic acid. Naturally stabiliza-

---

[*] These may be prepared by the reaction of the NCA with a primary or secondary base of which the strength is sufficiently high to ensure that no unreacted base remains. In these circumstances isolation and purification of the oligomer is not necessary. The absence of termination can be checked by estimation of basic end-groups.

tion of the acid as a salt suppresses the polymerization; this forms the basis of some techniques of stepwise peptide synthesis [25] (Section 7).

Nucleophilic attack of base on 2-CO as in (5) occurs relatively infrequently and constitutes a possible termination step. Since it results in the destruction of amine it has been detected by a diminution in the free base concentration. However, observation of such a decrease is not necessarily a diagnostic indication of mechanism (5) since it may also arise from other reactions (Section 7).

Table 1 reproduces values of second-order rate coefficients based on eqns. (7) and (8) given by Katchalski and Sela [3]. It should be emphasized that these data may not truly reflect the simple mechanism discussed above for reasons outlined in Section 4.

Since the process represented by reactions (4a) and (4b) is a simple addition polymerization it leads to a distribution of molecular weights of the Poisson type [2]. At any stage of the reaction the concentration of polymer molecules containing n α-amino acid units $[P_n]$ is given by

$$\frac{[P_n]}{[I]_0} = \left(\frac{\Delta M}{[I]_0}\right)^{n-p} \frac{e^{-\Delta M/[I]_0}}{(n-p)!}, \quad n \geqslant p. \tag{10}$$

In (10) ΔM represents the conversion i.e. $[M]_0 - [M]$, and $[I]_0$ is the concentration of initiating base added. In deducing this relation it has been assumed that initiation is effected by a homogeneous polymer with degree of polymerization p and that termination is absent. The final weight fraction of n-mer is

$$w_n = \frac{n}{p + [M]_0/[I]_0} \left(\frac{[M]_0}{[I]_0}\right)^{n-p} \frac{e^{-[M]_0/[I]_0}}{(n-p)!}, \quad n \geqslant p. \tag{11}$$

For present purposes, the most important feature of the Poisson distribution is its extreme sharpness. This may be illustrated by considering the ratio $\bar{P}_w/\bar{P}_n$ ($\bar{P}_w$ being the weight-average degree of polymerization), which is equal to $1 + 1/\bar{P}_n^2$. For moderate values of $\bar{P}_n$ the ratio is therefore relatively close to unity (the value holding for strictly homogeneous polymers).

Few experimental data are available for establishing the existence of Poisson distributions in poly α-amino acids prepared in this way. Fessler and Ogston [26] showed from sedimentation measurements that polysarcosine samples had sharp distributions which could be Poissonian, and Pope et al. [27], employing column fractionation, demonstrated a Poisson

TABLE 1
Polymerization rates of N-carboxy-α-amino acid anhydrides (NCAs) in solution

NCA	Solvent	Initiator	Temp. (°C)	$10^2 k$ (mole^{-1} l sec^{-1})
DL-Alanine	Dioxan	n-Hexylamine	35	4.3
DL-α-Amino-n-butyric acid	Dioxan	n-Hexylamine	35	1.7
Glycine	Dioxan	n-Hexylamine	35	34
Glycine	NN-Dimethyl-formamide	Diethylamine	25	60
γ-Benzyl-L-glutamate	Dioxan	n-Hexylamine	34	4.4
γ-Benzyl-L-glutamate	Dioxan	n-Hexylamine	25	3.2 (0.6)
γ-Benzyl-L-glutamate	Dioxan	Diethylamine	25	5.0
γ-Benzyl-L-glutamate	Benzene	n-Hexylamine	25	31 (6.2)
γ-Benzyl-L-glutamate	Benzene	Diethylamine	25	41.6
γ-Benzyl-L-glutamate	Nitrobenzene	n-Hexylamine	25	11 (2.1)
γ-Benzyl-L-glutamate	Nitrobenzene	Diethylamine	25	7
γ-Benzyl-L-glutamate	Chloroform	n-Hexylamine	25	30 (4.5)
γ-Benzyl-L-glutamate	N,N-Dimethyl-formamide	Diethylamine	25	16
L-Leucine	Dioxan	Diethylamine	25	1.2
L-Leucine	Benzene	Diethylamine	25	33
DL-Leucine	Dioxan	n-Hexylamine	35	0.6

DL-Leucine	Nitrobenzene	Preformed polymer	25	4.2
DL-Leucine	o-Nitroanisole	Preformed polymer	25	2.4
ε-Benzyloxycarbonyl-L-lysine	Dioxan	n-Hexylamine	35	4.3
ε-Benzyloxycarbonyl-L-lysine	Benzene	Diethylamine	25	3.3
ε-Benzyloxycarbonyl-L-lysine	N,N-Dimethylformamide	Diethylamine	25	16.6
ε-Benzyloxycarbonyl-L-lysine	N,N-Dimethylformamide	Diethylamine	25	5.8
ε-Benzyloxycarbonyl-DL-lysine	Dioxan	n-Hexylamine	35	1.2
DL-Phenylalanine	Benzene	Diethylamine	25	16.6
DL-Phenylalanine	Nitrobenzene	Preformed polymer	25	1.6
DL-Phenylalanine	Nitrobenzene	Preformed polymer	45	3.5
DL-Phenylalanine	N,N-Dimethylformamide	Diethylamine	25	5.8
L-Proline	N,N-Dimethylformamide	Diethylamine	25	75
Sarcosine	Nitrobenzene	Polysarcosine	25	25
Sarcosine	Acetophenone	Polysarcosine	25	26
DL-Valine	Dioxan	n-Hexylamine	35	0.2

distribution of molecular weights in polysarcosine prepared by initiation with preformed polymer.

As we have already stated, if the initiating base or the oligomers formed during reaction have reactivities differing from that of the "long-chain" propagating species, complex kinetics may be expected. A relatively simple situation which includes some of these complications is shown in (12).

$$
\begin{aligned}
M + I &\rightarrow P_2 + CO_2 & k' \\
M + P_2 &\rightarrow P_3 + CO_2 & k \\
&\cdot \cdot \cdot \cdot \cdot \cdot \cdot \cdot \cdot \\
M + P_n &\rightarrow P_{n+1} + CO_2 & k
\end{aligned}
\tag{12}
$$

Here the initiating step only has a rate coefficient different from that of the succeeding reactions. The rate of decomposition of NCA is given by

$$
-\frac{d[M]}{dt} = [M]\left(k'[I] + k \sum_{n=2}^{\infty} [P_n]\right)
\tag{13}
$$

and the rate of consumption of initiator by

$$
-\frac{d[I]}{dt} = k'[M][I]
\tag{14}
$$

Since $\sum_{2}^{\infty} [P_n] = [I]_{\infty} - [I]$ we find from (13) and (14) that

$$
[M]_0 - [M] = (1 - \alpha)([I]_0 - [I]) + \alpha[I]_0 \ln \frac{[I]_0}{[I]}
\tag{15}
$$

in which $\alpha = k/k'$. By elimination of [M] from (14) and (15) we obtain a differential equation relating [I] and t which contains the constants $[M]_0$, $[I]_0$ and α. If these latter are known, the equation may be integrated numerically to give the values of [I], and hence by (15) the values of [M] at various times. Ballard and Bamford [10] have treated the polymerization of DL-phenylalanine NCA initiated by DL-phenylalanine dimethylamide, in this way. In this system the initiator is a relatively strong base and the reaction decelerates in the early stages and finally follows a first-order course. (We shall show later (Section 4) that this behaviour may be explained in terms of a mechanism different from that of scheme (12)). When the initiating base is relatively weak, as is the case for example with p-chloraniline, the opposite type of behaviour is encountered, the reaction accelerating in the initial stages, so that the conversion-time curve has a sigmoid shape [28].

Ballard and Bamford [10] calculated the molecular weight distribution expected for the kinetic scheme (12) and showed that

$$\frac{[P_n]}{[I]_0} = (-1)^n \frac{\alpha^{n-2}}{(1-\alpha)^{n-1}} \left(\frac{[I]}{[I]_0}\right)^\alpha \left\{1 + \sum_1^{n-2} \frac{\beta^s}{s!} - \left(\frac{[I]}{[I]_0}\right)^{1-\alpha}\right\}, \quad n \geqslant 3$$

(16)

where $\beta = (1-\alpha)\ln[I]/[I]_0$. For a given conversion [I] may be calculated from [M] with the aid of (15) and substitution of this value into (16) then allows the distribution to be calculated. It is often convenient to express the truncated exponential in (16) in terms of an incomplete gamma function; the equation then becomes

$$\frac{[P_n]}{[I]_0} = (-1)^n \frac{\alpha^{n-2}}{(1-\alpha)^{n-1}} \left(\frac{[I]}{[I]_0}\right)^\alpha \left\{e^\beta S - \frac{[I]}{[I]_0}^{1-\alpha}\right\},$$

(17)

where

$$S = 1 - \frac{1}{(n-2)!} \int_0^\beta e^{-x} x^{n-2} \, dx = 1 - \frac{\Gamma(n-1,\beta)}{(n-2)!}$$

In the early stages of reaction the distribution (16) differs considerably from the Poisson form, although with increasing conversion the difference decreases. If the final value of $\bar{P}_n$ is large the ultimate distribution may be effectively Poissonian.

Coombes and Katchalski [29] have considered a slightly more complex version of this mechanism in which a second propagation coefficient operates above a critical degree of polymerization. Katchalski et al. [30] calculated the molecular weight distribution obtained in a system following scheme (12) but also including a bimolecular termination step. Various authors have analysed more complex systems in which the initiator is a polymeric species. Thus Gold [31] has shown that initiation by a poly α-amino acid with a Poisson distribution leads to a polymeric product with an over-all Poisson distribution, and Katchalski et al. [32] demonstrated that in multichain polymers synthesized from polyfunctional initiators Poisson distributions also arise.

2. Polymerization involving ionization of the NCA

We have already indicated (eqn. (6)) that, by virtue of the weakly acidic character of NCAs unsubstituted at the 3 position, anions of type (IV) may readily be formed. These can themselves act as strong nucleophiles in a manner analogous to primary and secondary bases in that they may add to the 5-CO of another monomer molecule. Thus reaction (18) constitutes the first step in the so-called aprotic- (or strong-base-) initiated polymerization.

$$
\begin{array}{c}
\mathrm{R_1R_2C-CO} \\
\quad | \qquad \diagdown \\
\quad | \qquad\; \mathrm{O} \\
\mathrm{HN-CO}
\end{array}
\;+\;
\begin{array}{c}
\mathrm{R_1R_2C-CO} \\
\quad | \qquad \diagdown \\
\quad | \qquad\; \mathrm{O} \\
\underline{\mathrm{N}}-\mathrm{CO}
\end{array}
\xrightarrow{\;(a)\;}
$$

$$(IV)$$

$$
\begin{array}{c}
\mathrm{R_1R_2C-CO} \\
\quad | \qquad \diagdown \\
\quad | \qquad\; \mathrm{O} \\
\mathrm{N-CO} \\
\quad | \\
\mathrm{R_1R_2C-CO} \\
\quad | \\
\mathrm{HN-CO\bar{O}}
\end{array}
\;\;
\begin{array}{c}
(b) \\
\mathrm{BH^+} \\
\xrightarrow{\qquad} \\
\text{or NCA}
\end{array}
\;\;
\begin{array}{c}
\mathrm{R_1R_2C-CO} \\
\quad | \qquad \diagdown \\
\quad | \qquad\; \mathrm{O} \\
\mathrm{N-CO} \\
\quad | \\
\mathrm{R_1R_2C-CO} \\
\quad | \\
\mathrm{NH_2 + CO_2 + B\;or\;(IV)}
\end{array}
\qquad (18)
$$

$$(V) \qquad\qquad\qquad (VI)$$

The product anion (V) decarboxylates (18b) on acquiring a proton from an acid species present (e.g. the conjugate acid $\mathrm{BH^+}$ of the initiating base product in (6) or an NCA molecule). Formation of species such as (VI) diversifies the routes to polymerization. Thus (VI) is a primary base and can react by the mechanism already described with any oxazolidone-2,5-dione system present, including monomer (I) or a molecule of its own kind. These processes are indicated in (19a) and (19b), respectively.

$$
\begin{array}{c}
\mathrm{R_1R_2C-CO} \\
\quad | \qquad \diagdown \\
\quad | \qquad\; \mathrm{O} \\
\mathrm{N-CO} \\
\quad | \\
\mathrm{R_1R_2C-CO} \\
\quad | \\
\mathrm{NH_2}
\end{array}
\;+\;
\begin{array}{c}
\mathrm{R_1R_2C-CO} \\
\quad | \qquad \diagdown \\
\quad | \qquad\; \mathrm{O} \\
\mathrm{HN-CO}
\end{array}
\longrightarrow
\begin{array}{c}
\mathrm{R_1R_2C-CO} \\
\quad | \qquad \diagdown \\
\quad | \qquad\; \mathrm{O} \\
\mathrm{N-CO} \\
\quad | \\
\mathrm{R_1R_2C-CO} \\
\quad | \\
\mathrm{NH} \\
\quad | \\
\mathrm{R_1R_2C-CO} \\
\quad | \\
\mathrm{NH_2 + CO_2}
\end{array}
$$

$$(VI) \qquad\qquad\qquad\qquad (VII)$$

(19a)

$$
\begin{array}{c}
\underset{|}{R_1R_2C-CO} \\
\quad\quad O \\
\underset{|}{N-CO} \\
\underset{|}{R_1R_2C-CO} \\
NH_2
\end{array}
\;+\;
\begin{array}{c}
\underset{|}{R_1R_2C-CO} \\
\quad\quad O \\
\underset{|}{N-CO} \\
\underset{|}{R_1R_2C-CO} \\
NH_2
\end{array}
\;\longrightarrow\;
\begin{array}{c}
\underset{|}{R_1R_2C-CO} \\
\quad\quad O \\
\underset{|}{N-CO} \\
\left(\underset{|}{R_1R_2C-CO} \atop NH\right)_2 \\
\underset{|}{R_1R_2C-CO} \\
NH_2 + CO_2
\end{array}
\quad (19b)
$$

(VI)   (VI)   (VIII)

Further, the anion (IV) may react with the terminal ring of (VI), (VII), (VIII) and similar species, leading to growth from the end carrying the ring as shown in (20).

$$
\begin{array}{c}
\underset{|}{R_1R_2C-CO} \\
\quad\quad O \\
\underset{|}{N-CO} \\
\cdots CO
\end{array}
\;+\;
\begin{array}{c}
\underset{|}{R_1R_2C-CO} \\
\quad\quad O \\
\underset{|}{\overset{}{N}-CO}
\end{array}
\;(a)\;\longrightarrow
$$

(VI) (VII), (VIII), etc.   (IV)

$$
\begin{array}{c}
\underset{|}{R_1R_2C-CO} \\
\quad\quad O \\
\underset{|}{N-CO} \\
\underset{|}{R_1R_2C-CO} \\
\underset{|}{N-CO\overset{-}{O}} \\
\cdots CO
\end{array}
\;\xrightarrow[(b)]{BH^+ \text{ or } NCA}\;
\begin{array}{c}
\underset{|}{R_1R_2C-CO} \\
\quad\quad O \\
\underset{|}{N-CO} \\
\underset{|}{R_1R_2C-CO} \\
\underset{|}{NH + B \text{ or } (IV) + CO_2} \\
\cdots CO
\end{array}
\quad (20)
$$

The mechanism of eqn. (20) is closely similar to that encountered in the anionic polymerization of ϵ-caprolactam [33] and highlights the lactam character of the NCA molecule.

In general, the mechanisms which have been discussed form the basis of most polymerizations of NCAs. Clearly, they often may lead to very complex kinetic behaviour, particularly since primary and secondary amines can initiate either by direct nucleophilic attack or by proton abstraction. The elucidation of the various chemical steps, their consequences and their relative importance in specific instances have been based on extensive chemical and kinetic data which we present below. Alternative mechanisms for the polymerization of NCAs have been proposed, but have in the main been rejected because they are not consistent with some of the experimental observations. The reader is referred to the extensive literature for a description of alternative mechanisms [2—6, 34—36]. Certain individual NCAs, on account of their chemical structure, diverge from the general pattern of behaviour in polymerization; such cases are described in Section 7.

Further complexity is introduced into the mechanism of NCA polymerization by physical factors arising from the structure of the growing chain. These important effects are discussed in Section 8.

3. Experimental evidence relating to the polymerization of NCAs

According to the mechanisms outlined in Sections 1 and 2, substitution in the 3-position should preclude polymerization by the ionization route. This is indeed found to be so; for example, sarcosine and proline NCAs ((IX), (X), respectively) cannot be polymerized by tertiary amines when

$$
\begin{array}{cc}
CH_2\!-\!CO & \quad H_2 \\
\quad\quad\;\; O & \quad C\!-\!CH\!-\!CO \\
CH_3N\!-\!-\!CO & CH_2\;\;|\quad\quad O \\
& \quad C\!-\!N\!-\!CO \\
& \quad H_2 \\
(IX) & (X)
\end{array}
$$

species capable of readily donating protons (particularly water) are rigorously excluded [18—20, 37, 38(b)] (Fig. 1). This crucial matter has, in the past, been the subject of considerable controversy, but most recent work confirms the view expressed above, provided that the initiator is not a very strong base. Some extremely strong bases such as sodium triphenylmethyl appear to initiate the polymerization of proline NCA, but under these conditions the monomer may ionize at the 4-position [23]. This aspect, apparently peculiar to proline NCA, is discussed further in Section 5. Naturally, primary and secondary amines initiate the polymerization of sarcosine and proline NCAs, and other bases such as tertiary amines and chloride ion in suitable solvents do so in the presence of proton donors [18]. Particularly significant as a proton donor is 3-methyl hydantoin (XI) because of its structural similarity to the NCA ring system.

$$
\begin{array}{c}
CH_2\!-\!CO \\
\quad\quad\;\; NMe \\
HN\!-\!-\!CO \\
(XI)
\end{array}
$$

It has been established that the polymerization of sarcosine NCA can be initiated by tertiary bases [18] and chloride ion [37] when (XI) is

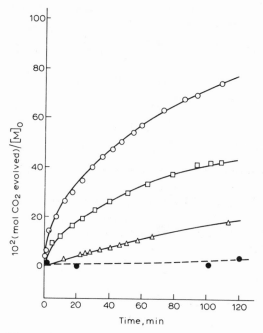

Fig. 1. Conversion—time curves for the polymerization of γ-ethyl-L-glutamate NCA (0.2 mole l^{-1}) in N,N-dimethylformamide by pyridine and homologues (2.0 mole l^{-1}) at 25°C. ○ 2,6-lutidine, □ α-picoline, △ pyridine. The broken curve shows a comparative experiment with sarcosine NCA (0.2 mole l^{-1}) and 2,6-lutidine (2.0 mole l^{-1}). Bamford and Block [19].

present. Of even greater significance is the behaviour of 3-phenyl oxazolid-2,5-dione (XII), which, on account of the weakness of the derived substi-

$$
\begin{array}{c}
\text{CH}_2\text{—CO} \\
| \qquad\quad \text{O} \\
\text{PhN——CO} \\
\text{(XII)}
\end{array}
$$

tuted aniline as a base, is not able to propagate significantly after ring-opening (cf. Section 7). It will not react at all with tertiary bases or chloride ion in the absence of (XI), but, in the presence of the latter, (XII) participates in reaction (21) giving (XIII) which has been isolated and characterized [18].

$$
\begin{array}{ccc}
\text{CH}_2\text{—CO} & \overset{B}{\rightleftharpoons} & \text{CH}_2\text{—CO} \\
| \qquad\ \text{NMe} & & | \qquad\ \text{NMe} + \text{BH}^+ \\
\text{HN——CO} & & \text{N——CO} \\
\text{(XI)} & &
\end{array} \qquad (21)
$$

(reaction (21) continued overleaf)

References pp. 634—637

596

$$(21)$$

The similarity between this reaction and that of eqn. (18) provides supporting evidence for the validity of the latter.

More direct evidence for mechanism (20a) may be adduced from the observation that hydantoin 3-acetic acid (XIV) can be isolated from the reaction between glycine NCA and lithium chloride in N,N-dimethylformamide solution (eqn. (22)) [37, 39].

$$(22)$$

As might be anticipated, dilution of the reaction mixture increases the rate of the intramolecular condensation process relative to the polymerization, thereby increasing the yield of (XIV). By contrast, cyclization of (VI) through the 5-CO is sterically impossible so that diketopiperazines are not formed. This steric effect persists as the length of the peptide chain increases during polymerization (reactions (19) and (20)) until, in the case of glycine, the growing species ((XV) eqn. (23)) contains five glycyl residues in addition to the ring. At this stage cyclization can occur (with glycine NCA) to give the cyclic hexapeptide (XVI) (reaction (23)) [37, 39a].

$$(23)$$

The absence of cyclization at an earlier stage is responsible for the unique property of these monomers in providing an easy and effective route to high polymers by aprotic base initiation. This contrasts with other synthetic routes which are mainly employed in preparing sequential polypeptides [40]. The behaviour of glycine in forming a cyclic hexapeptide is not found with most other NCAs since the development of secondary chain structures in the "tail" of the intermediate precludes cyclization.

The occurrence of the ionization mechanism (eqns. (6) and (18)) in preference to the alternative addition mechanism (eqn. (24)) has been demonstrated by a study of the initiating activities of the series of bases pyridine, α-picoline, 2,6-lutidine [21]. Mechanism (24) is analogous to the normal tertiary base-catalysed acylation mechanism encountered with anhydrides such as acetic anhydride [41]. Steric factors determine the

$$
\begin{array}{c}
\text{R}_1\text{R}_2\text{C--CO} \\
\quad | \quad \backslash \\
\quad\quad \text{O} \quad + \quad \overset{}{\underset{\text{R}_3}{\bigcirc}}\underset{\text{N}}{\quad}\text{R}_4 \quad \longrightarrow \\
\text{HN--CO}
\end{array}
\qquad
\begin{array}{c}
\overset{}{\underset{\text{R}_3}{\bigcirc}}\underset{\text{N}}{\quad}\text{R}_4 \\
\text{R}_1\text{R}_2\text{C--}\overset{+}{\text{C}}\text{--O}^- \\
\quad | \quad\quad \text{O} \\
\quad\quad \text{HN--CO} \\
\text{(XVI)}
\end{array}
$$

$$
\text{(XVI)} \; + \;
\begin{array}{c}
\text{R}_1\text{R}_2\text{C--CO} \\
\quad | \quad \backslash \\
\quad\quad \text{O} \quad \longrightarrow \quad \text{(V)} \; + \; \overset{}{\underset{\text{R}_3}{\bigcirc}}\underset{\underset{\text{H}}{\overset{+}{\text{N}}}}{\quad}\text{R}_4 \\
\text{HN--CO}
\end{array}
\tag{24}
$$

order of activity of these catalysts, acting as Lewis bases, as:

pyridine ($\text{R}_3 = \text{R}_4 = \text{H}$) $>$ α-picoline ($\text{R}_3 = \text{H}$, $\text{R}_4 = \text{Me}$) $>$ 2,6-lutidine ($\text{R}_3 = \text{R}_4 = \text{Me}$)

for mechanism (24); this order is reversed for mechanisms (6) and (18), in which the bases ionize the NCA by a Brönsted process. In practice the latter order of activity is found, confirming the predominance of (6) and (18) in NCA polymerizations (see Fig. 1).

A number of alternative mechanisms which have been proposed for the polymerization of NCAs catalysed by strong bases involve intermediate steps such as the direct addition of the initiator to the 5-CO of the ring, followed by ring opening and subsequent polymerization via the carbamate ion acting as an adding base [34, 35]. We do not favour these suggestions in view of the evidence already presented. Further, they are in conflict with observations of Goodman and co-workers [23, 24] who showed that labelled initiators such as $^{14}CH_3ONa$ and $PhCH_2NH^{14}COONa$ do not normally introduce ^{14}C into the growing chains. The conclusion that these initiators have a purely catalytic function is inescapable.

It has on occasion been argued [3, 4] that the ionization mechanism is unlikely because the conjugate base of the NCA (IV) is likely to be present in very low concentration, (since lactams are weak acids and tertiary bases are weak bases) and could therefore only account for low rates for reactions of type (18) and (20). Although species (IV) is undoubtedly present in low concentration the conclusion that it is likely to be kinetically insignificant is false. The weakness of an NCA as an acid implies that the conjugate base (IV) is strong; since steps (18a) and (20a) have rates dependent both on the concentration of (IV) and the rate coefficient for addition of (IV) to the 5-CO, the kinetically significant parameter is the product of the concentration and the rate coefficient. This product may clearly have a significant magnitude.

The ionization mechanism for the polymerization of NCAs initiated by tertiary amines and other aprotic initiators such as chloride ion or carboxylate anions [38(a)] involves several types of growth process. After the initial ionization (reaction (6)) and formation of the smallest bifunctional intermediate (VI) (reaction (18)) polymer growth may proceed by (i) the mechanism of eqns. (20a, b) involving addition to the ring in (VI), and (ii) the protic mechanisms shown in (19) which involve addition of primary or secondary amines to 5-CO groups. Further, the total base concentration will, at least in the early stages of reaction, increase by virtue of the processes shown in (18). It must also be realized that the ionization (6) can involve any base in the system, protic or aprotic, so that species of types (VI), (VII), (VIII) and homologues can all contribute to the build-up in concentration of (IV). (This fact is significant not only when tertiary bases are used as initiators but also, as elaborated below, with primary and secondary base initiation.) A detailed kinetic analysis taking all these processes into account is too complex to be analytically useful. Thus it is not surprising that the unravelling of the mechanisms encountered in NCA polymerization has proceeded mainly on the basis of chemical evidence and has relied only to a relatively minor extent on purely kinetic observations.

The relative kinetic importance of the reactions we have outlined would be expected to change throughout the course of a polymerization. This is illustrated by the following considerations. The reactivity of the 5-CO of an oxazolidine-2,5-dione system depends markedly on whether the 3-position is unsubstituted or acylated (as in (VI), (VII), (VIII), etc.): electron-withdrawal by the acyl substituent activates the 5-CO to nucleophilic attack. Thus the rate coefficient of (18a) is less than that of (20a) and the rate coefficient of (19a) is less than that of (19b). In all but the very earliest and latest stages of polymerization reaction (20a) appears to predominate and gives rise to pseudo first-order kinetics in [NCA] until the concentration of NCA has fallen to such a low value that (19b) becomes significant. This situation is reached near the end of the polymerization; the coupling of bifunctional intermediates (19b) is then evidenced by a dramatic increase in the molecular weight of the products [42].

The molecular weights and molecular weight distributions of poly-α-amino acids resulting from an aprotic polymerization naturally reflect the polymerization mechanism. Generally, polymers of high molecular weight are obtained, with distributions intermediate between the Poisson and "most probable" types (Section 4) [43, 44]. It has been argued that the ionization mechanism should lead to a "most probable" distribution and indeed this would be so if reaction (20a) were absent. This latter process in isolation could, in suitable circumstances, lead to narrow distributions. The necessary condition is that the rate of initiation (18a) should be much lower than that of the propagation (20a). We have seen above that this is to be expected. In practice, polymers are generally isolated before the coupling process (19b) is complete, so that the observation of intermediate distributions is not surprising.

So far, little mention has been made of possible termination reactions in polymerizations initiated by aprotic bases. The formation of hydantoin 3-acetic acid (XIV) and cyclohexaglycine (XVI) during the polymerization of glycine NCA has been experimentally established (p. 596) but there do not appear to be any reports of related products from other NCAs. Termination reactions specific to certain monomers have been recorded and are referred to below; they all involve interactions between the side chain R_1 with the propagating amine. General termination processes such as amidation of 2-CO giving ureido derivatives (reaction (5)) are not generally observed with aprotic initiation. This is perhaps not surprising in view of the more facile addition reactions at 5-CO, particularly reaction (20a). Steric shielding of 2-CO in (VI), (VII), (VIII) etc. also reduces the probability of reaction at this group, except for the formation of hydantoin derivatives (cf. reaction (22)) which is encouraged by the favourable ring size. Experimentally, it is not easy to establish a small extent of ureido-formation in a high polymeric material. It seems clear that, since polymers of high molecular weight are readily formed, termination processes are of relatively minor importance.

4. Polymerization of NCAs unsubstituted at the 3-position initiated by primary and secondary amines.

To achieve a full appreciation of the reactions of NCAs with primary and secondary amines we must realize that there is no a priori reason for excluding the reaction paths discussed for tertiary-base initiation, provided that the 3-position of the NCA ring is unsubstituted. Thus both

600

nucleophilic addition to the 5-CO and proton abstraction from the 3-position may occur; the relative rates of these disparate reactions depend upon the chemical nature of the reactants and particularly upon the strength and stereochemistry of the amine. For example, the polymerization of γ-benzyl-L-glutamate NCA occurs more rapidly when initiated by di-isopropylamine than when initiated by n-hexylamine, both initiators being at a concentration equal to 1/15 of the initial monomer concentration (Fig. 2) [22]. This result cannot be attributed to a greater reactivity of di-isopropylamine compared to n-hexylamine in attacking the 5-CO group of the NCA according to eqn. (4) because the enhanced reaction rate persists well beyond the conversion at which the initiator would be completely consumed. It would therefore appear that the alternative ionization route (reactions (6) and (18)) must also contribute; this conclusion is strongly supported by the observation that, with di-isopropylamine as initiator, the weight-average degree of polymerization of the poly-γ-benzyl-L-glutamate formed at the end of the reaction is 150, approximately, rather than the value near 15 which would be expected from mechanism (4) above [22]. Even with hexylamine as initiator the latter value is exceeded ($\bar{P}_w$ = 60, approximately). The relative ineffectiveness of di-isopropylamine in reacting with the 5-CO of the NCA ring, shown by the low rate of reaction with sarcosine NCA (Fig.

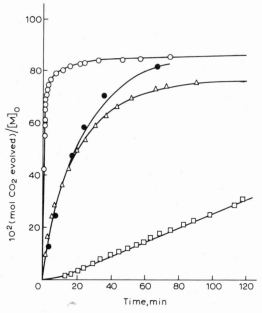

Fig. 2. Conversion—time curves for polymerization of NCAs (0.224 mole l^{-1}) initiated by amines at 25°C in N,N-dimethylformamide solution. ○ γ-ethyl-L-glutamate NCA/di-isopropylamine, ● sarcosine NCA/n-hexylamine, △ γ-ethyl-L-glutamate NCA/n-hexylamine, □ sarcosine NCA/di-isopropylamine. Bamford and Block [22].

2), is a consequence of steric restraint by the bulky isopropyl groups; such steric factors are much less important when proton abstraction from the 3-position is involved. There is some indication that reaction at the 5-CO position may involve an intermediate of the types (XVII) and (XVIII),

$$(I) + > NH \rightleftharpoons \underset{(XVII)}{\overset{-NH+}{\underset{R_3N-CO}{\overset{|}{\underset{O}{\overset{|}{R_1R_2C-C-O^-}}}}}} \rightleftharpoons \underset{(XVIII)}{\overset{-N}{\underset{R_3N-CO}{\overset{|}{\underset{O}{\overset{|}{R_1R_2C-C-OH}}}}}} \qquad (25)$$

since reaction (25) appears to have kinetic significance in the polymerization of sarcosine NCA ($R_1 = R_2 = H$, $R_3 = CH_3$; see Section 5). Such a pre-equilibrium would be influenced both by the base strength and the steric properties of the initiating amine.

The duality in the mechanisms of initiation and propagation described above is reflected in the molecular weights and molecular distributions of the polymeric products. Thus, Stewart and Stahmann [43] found by column chromatography double-peaked distributions in poly-L-lysine samples prepared from ε-benzyloxycarbonyl-L-lysine NCA by initiation with ammonia, the samples having both low and high molecular weight

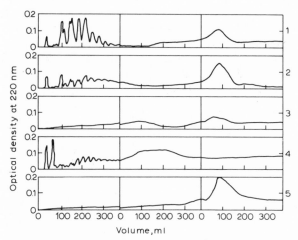

Fig. 3. Chromatograms of five poly-L-lysine samples eluted with three continuous convex gradients. Preparative details:

Curve	1	2	3	4	5
Initiator	NH_3	NH_3	NH_3	$LiOCH_3$	$NaOCH_3$
$[M]_0/[I]_0$	6	12	50	6	320

Stewart and Stahmann [43].

602

components (Fig. 3). Sodium methoxide as initiator gave only a high molecular weight component, although samples prepared by initiation with lithium methoxide showed the presence of some low molecular weight material (Fig. 3). Similar results have been reported for polymers (poly-L-lysine [45] and poly-γ-benzyl-L-glutamate [46]) prepared by initiation with n-butylamine. The dual mechanism described above is consistent with these observations.

5. Polymerization of NCA's carrying a substituent at the 3-position

The presence of a substituent in the 3-position of the NCA precludes equilibrium (6) and hence prevents the occurrence of reaction (18). It follows that, on the basis of the mechanisms described above, such NCA's should not polymerize unless a protic base (for example a primary or secondary amine) or other source of protons (for example, 3-methyl hydantoin) is present. If it could be established that polymerization does proceed with an aprotic base in aprotic media then some other mechanism of polymerization must be operative. This matter has been of central importance in discussions of various mechanisms of polymerization which have been advanced (Section 3). Experimentally, it is not easy to obtain definitive evidence because of the high sensitivity of NCA's to protonic impurities (such as water and alcohols) in the presence of bases. It has been shown [18, 19, 38a] that proline NCA (X) and sarcosine NCA ($I : R_1 = R_2 = H, R_3 = CH_3$) do not polymerize in the presence of tertiary bases under strictly aprotic conditions. With alkoxides, realization of such conditions is difficult, but it would appear that, at least with proline NCA, such strong bases can bring about ionization of the methine hydrogen and hence initiate polymerization as shown in (26). Evidence for this mechanism is provided by the observation that while sodium methoxide enriched

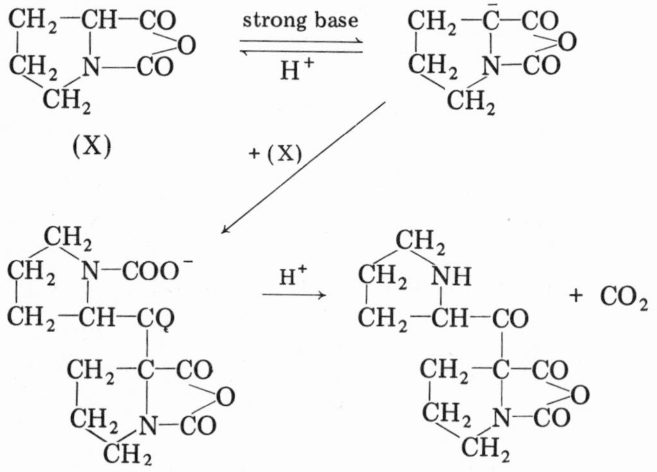

(26)

with $^{14}CH_3ONa$ initiates the polymerization of proline NCA the resulting polymer does not contain ^{14}C [23]. Investigations with radioactive tracers, together with the failure of tertiary bases to initiate the polymerization of 3-substituted NCAs under aprotic conditions, indicate that neither a general anhydride exchange mechanism [34] (27) nor propagation through imide ions (28) can provide routes to polymer formation.

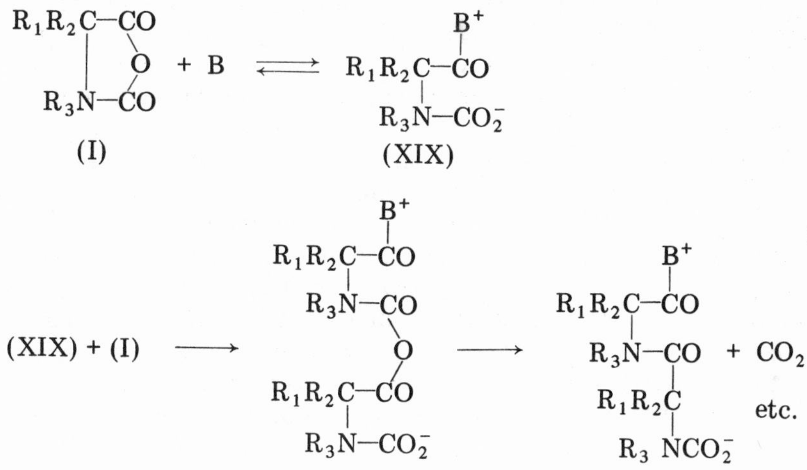

$$
\begin{array}{c}
R_1R_2C-CO \\
\quad \diagdown \\
\quad\quad O \;+\; B \;\rightleftharpoons\; R_1R_2C-CO \\
R_3N-CO \\
\qquad\qquad\qquad\qquad\qquad R_3N-CO_2^- \\
(I) \qquad\qquad\qquad\qquad (XIX)
\end{array}
$$

$$
(XIX) + (I) \longrightarrow
\begin{array}{c}
\overset{B^+}{|} \\
R_1R_2C-CO \\
R_3N-CO \\
\qquad\diagdown \\
\qquad\quad O \\
R_1R_2C-CO \\
R_3N-CO_2^-
\end{array}
\longrightarrow
\begin{array}{c}
\overset{B^+}{|} \\
R_1R_2C-CO \\
R_3N-CO \;+\; CO_2 \\
R_1R_2C \\
R_3\;NCO_2^- \quad \text{etc.}
\end{array}
\qquad (27)
$$

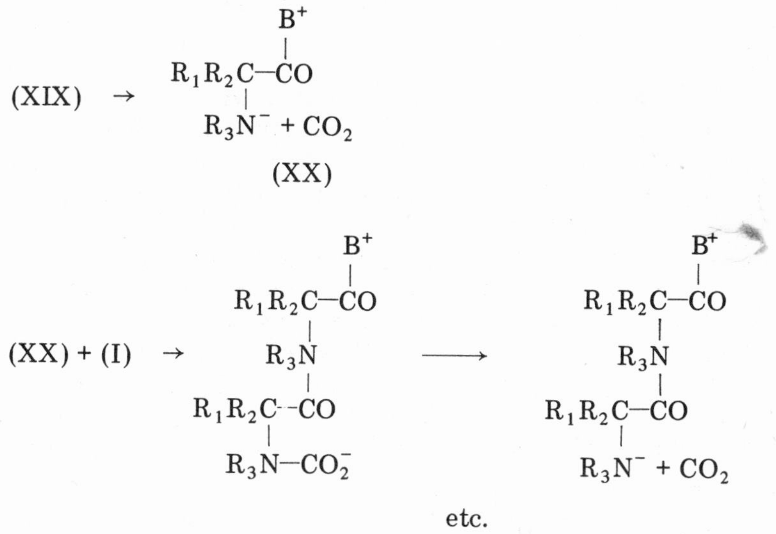

$$
(XIX) \;\rightarrow\;
\begin{array}{c}
\overset{B^+}{|} \\
R_1R_2C-CO \\
R_3N^- + CO_2 \\
(XX)
\end{array}
$$

$$
(XX) + (I) \;\rightarrow\;
\begin{array}{c}
\overset{B^+}{|} \\
R_1R_2C-CO \\
R_3N \\
R_1R_2C--CO \\
R_3N-CO_2^-
\end{array}
\longrightarrow
\begin{array}{c}
\overset{B^+}{|} \\
R_1R_2C-CO \\
R_3N \\
R_1R_2C-CO \\
R_3N^- + CO_2
\end{array}
\qquad (28)
$$

etc.

If B is methoxide ion enriched with ^{14}C, the latter should be incorporated in the polymer according to (27) or (28). Further, carbamate anions such as $Ph^{14}CH_2CO_2^-$ should initiate polymerization when $R_3 \neq H$ and lead to incorporation of ^{14}C. None of these results is found in practice. See also refs. 38 and 80.

604

On the basis of the preceding discussion it might be assumed that the polymerization of 3-substituted NCAs initiated by primary or secondary amines would be mechanistically simple and follow the route (4) with kinetics expressed by eqns. (7) and (8). This is not generally the case. For example, with sarcosine NCA ((I) : $R_1 = R_2 = H$, $R_3 = CH_3$), one of the earliest studied monomers, the kinetic behaviour was shown to be complex [9], although the over-all stoichiometry obeyed equation (4) and studies of molecular weight distribution indicated values of $\bar{M}_w/\bar{M}_n$ consistent with the expected Poisson distribution (Section 1). Progress in understanding the mechanism of the polymerization was made possible by the observation [10] that the CO_2 evolved catalyses the polymerization (the order in $[CO_2]$ being complex); thus a desirable prerequisite for kinetic experimentation on this system is a constant (or zero) partial pressure of CO_2. It was observed that this polymerization has an activation energy significantly lower than that expected for a simple acylation reaction; under certain conditions of CO_2 pressure the over-all activation energy assumes negative values [10]. Further, the order in propagating base is greater than unity, and therefore suggests base catalysis. All these observations are consistent with catalysis by the carbamic acid arising from interaction of the propagating base with CO_2:

$$\text{\textasciitilde\textasciitilde}COCH_2NHCH_3 + CO_2 \rightleftharpoons \text{\textasciitilde\textasciitilde}COCH_2N(CH_3)CO_2H \qquad (29)$$

Since the forward reaction in (29) is exothermic, the equilibrium is displaced to the left by increase in temperature; this factor accounts in part for the anomalous temperature coefficient of reaction rate mentioned above. The apparent catalysis by propagating base is also explicable as acid catalysis since the carbamic acid is stoichiometrically derived from the base by reaction (29). That true base catalysis is not operative has been shown by the observation that addition of tertiary bases does not affect the reaction rate [17]. Further, the polymerization is catalysed by other weak acids such as hydrocinnamic [17] and α-picolinic acids [10, 17], which, if present in sufficient concentration under conditions of low CO_2 pressure, reduce the order in initiating base to unity. Thus, under such conditions, with hydrocinnamic acid (HX) as catalyst the simple kinetic form (30) is achieved.

$$\text{rate} = k[M][HX][I] \qquad (30)$$

The mechanism of the complex propagation reaction is now thought to be that illustrated in eqns. (31)—(34).

$$\text{\textbackslash\textbackslash\textbackslash COCH}_2\text{NHCH}_3 + \begin{array}{c} \text{CH}_2-\text{CO} \\ | \quad\quad \backslash \\ \quad\quad\quad\text{O} \\ | \quad\quad / \\ \text{CH}_3\text{N}---\text{CO} \end{array} \underset{}{\overset{K}{\rightleftharpoons}} \begin{array}{c} \text{\textasciitilde COCH}_2\text{NCH}_3 \\ | \\ \text{CH}_2-\text{C}-\text{OH} \\ | \quad\quad \backslash \\ \quad\quad\quad\text{O} \\ | \quad\quad / \\ \text{CH}_3\text{N}---\text{CO} \end{array} \tag{31}$$

(XXI)

$$(\text{XXI}) \xrightarrow{k_1} \begin{array}{c} \text{\textasciitilde COCH}_2\text{NCH}_3 \\ | \\ \text{CH}_2-\text{C}=\text{O} \\ | \\ \text{CH}_3\text{N}-\text{CO}_2\text{H} \end{array} \tag{32}$$

(XXII) (HX)

$$(\text{XXII}) \underset{k_{-2}}{\overset{k_2}{\rightleftharpoons}} \begin{array}{c} \text{\textasciitilde COCH}_2\text{NCH}_3 \\ | \\ \text{CH}_2-\text{C}=\text{O} \\ | \\ \text{CH}_3\text{NH} + \text{CO}_2 \end{array} \tag{33}$$

$$\begin{array}{c} \text{\textasciitilde COCH}_2\text{NCH}_3 \\ | \\ \text{CH}_2-\text{C}-\text{O}-\text{H} \\ | \quad\quad \text{O} \\ \text{CH}_3\text{N}\cdots\text{CO} \\ | \\ \text{X}-\text{H} \ (\text{XXI}) \end{array} \xrightarrow{k_3} \begin{array}{c} \text{\textasciitilde COCH}_2\text{NCH}_3 \\ | \\ \text{CH}_2-\text{C}=\text{O} \\ | \\ \text{CH}_3\text{NH} + \text{H}^+\text{X}^- + \text{CO}_2 \end{array} \tag{34}$$

The pre-equilibrium (31) (equilibrium constant K) leading to the adduct (XXI) is, in part, responsible for the observed negative temperature coefficient, since K will decrease with increasing temperature. Reaction (32) is the ring-opening reaction of (XXI) leading to a carbamic acid (XXII) while (33) represents the reversible decarboxylation of (XXII). In the absence of added acid catalysts reaction (34) is the decomposition of (XXI) catalysed by the carbamic acids HX. The kinetic behaviour of the reaction at constant non-zero CO_2 pressure in the absence of added acids is complex on account of the catalysis by the carbamic acid species (XXII) (HX). Even at the lowest achievable partial pressures of CO_2 the finite rate of decomposition of (XXII) (reaction (33)) affects the rate of polymerization, that is the rate coefficient k_2 is involved in the kinetics [17]. It is reasonable to assume a stationary state for carbamic acid concentration

$$\frac{d[\text{HX}]}{dt} = k_1[\text{XXI}] - k_2[\text{HX}] + k_{-2}[\text{I}][\text{CO}_2] = 0 \tag{35}$$

where [I] represents the concentration of free base

$$[I] = [I]_0 - [HX], \tag{36}$$

$[I]_0$ being the concentration of initiating base added. The rate of evolution of CO_2, which is equal to the rate of consumption of monomer, is

$$\frac{d[CO_2]}{dt} = -\frac{d[M]}{dt} = k_2[HX] - k_{-2}[I][CO_2] + k_3[HX][XXI] \tag{37}$$

which, with the aid of (35) becomes

$$-\frac{d[M]}{dt} = (k_1 + k_3[HX])[XXI]. \tag{38}$$

If the equilibrium in (31) is established and lies well over to the left we have $[XXI] = K[M][I]$, hence it follows from (38) that

$$-\frac{d[M]}{dt} = K[M][I](k_1 + k_3[HX]). \tag{39}$$

In order to obtain [M] as a function of time, the values of [HX] and [I] given by eqns. (35) and (36) are inserted into (39) and the resulting expression integrated. It is thus found that

$$[M] - [M]_0 + \frac{\lambda^2}{\mu} \ln \frac{[M]}{[M]_0} + \left(2\lambda - \mu - \frac{\lambda^2}{\mu}\right) \ln \frac{[M] + \mu}{[M]_0 + \mu}$$

$$= -\frac{k_2}{k_1}(k_1 + k_3[I]_0)[I]_0 t \tag{40}$$

in which

$$\lambda = \frac{k_2}{Kk_1}\left(1 + \frac{k_{-2}}{k_2}[CO_2]\right)$$

$$\mu = \left\{k_1 + \frac{k_{-2}}{k_2}(k_1 + k_3[I]_0)[CO_2]\right\}\left\{\frac{Kk_1}{k_2}(k_1 + k_3[I]_0)\right\}^{-1} \tag{41}$$

The complex expression (40) has been shown adequately to reproduce the observed kinetic behaviour for a range of values of $[I]_0$ and $[CO_2]$. The numerical values of the kinetic parameters at 25°C are given in (42) [17].

$$Kk_1 = 1.32 \, \text{mole}^{-1} \, l \, \text{min}^{-1}$$

$$k_2 = 1.40 \, \text{min}^{-1}$$

$$Kk_3 = 800 \, \text{mole}^{-2} \, l^2 \, \text{min}^{-1}$$

$$k_{-2}/k_2 = 4.45 \times 10^{-3} \, (\text{cm Hg})^{-1}$$

(42)

The above discussion refers to catalysis by carbamic acid in the absence of added acids. However, results obtained with hydrocinnamic acid are understandable in terms of the same mechanism. At high concentration of

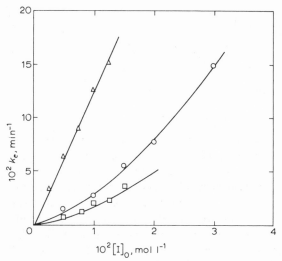

Fig. 4. Dependence of first-order rate coefficient k_e for the polymerization of sarcosine NCA in N,N-dimethylformamide on concentration of preformed polymer $[I]_0$ in the absence and presence of hydrocinnamic acid. "Zero" CO_2 pressure and $25°C$. Hydrocinnamic acid concentrations: □ Zero, ○ 2.5×10^{-3} mole l^{-1}, △ 2.5×10^{-2} mole l^{-1}. The lowest curve (points □) is calculated from an approximate form of (40) with parameters given in (42). Bamford et al. [17].

acid, catalysis is very rapid compared to catalysis by carbamic acid and the reaction consists essentially of the forward reaction in (31) and the rapid catalysed decomposition of the product (XXI). The rate is therefore first-order in [I], (cf. (30)), provided that the propagating base is not significantly neutralized by the acid. At low hydrocinnamic acid concentrations, carbamic acid catalysis makes a major contribution, so that the reaction shows an over-all order in $[I]_0$ exceeding unity for reasons already described (Fig. 4).

6. Special aspects of initiation

There have been a number of studies [47—52] of the polymerization of NCAs by zinc and aluminium alkyls, particularly with a view to achieving stereo-selective polymerization (Section 8). Initiation with di-n-butyl zinc is attributed to initial attack of the alkyl at the 3-position of the NCA, to give compounds (XXIII) or (XXIV)

$$
\text{ZnBu}_2 \; + \;
\begin{array}{c}
\text{CH}_3\text{CH—CO} \\
| \quad\quad \diagdown \\
\quad\quad\quad \text{O} \\
| \quad\quad \diagup \\
\text{HN—CO}
\end{array}
\quad \longrightarrow \quad
\begin{array}{c}
\text{CH}_3\text{CH—CO} \\
| \quad\quad \diagdown \\
\quad\quad\quad \text{O} \\
| \quad\quad \diagup \\
\text{BuZnN—CO}
\end{array}
\; + \; \text{BuH}
$$

(XXIII)

$$\Big\downarrow \text{NCA} \tag{43}$$

$$
\begin{array}{c}
\text{OC—CHCH}_3 \quad\quad \text{CH}_3\text{CH—CO} \\
\diagup \quad\quad | \quad\quad\quad\quad\quad | \quad\quad \diagdown \\
\text{O} \quad\quad\quad\quad\quad\quad\quad\quad\quad\quad\quad \text{O} \\
\diagdown \quad\quad\quad\quad\quad\quad\quad\quad\quad\quad \diagup \\
\text{OC—N \quad——Zn——\quad N—CO}
\end{array}
$$

(XXIV)

The nature of the reactions occurring in these systems was inferred from the yields of butane under various conditions, from spectroscopic (IR, NMR) studies and from work with model compounds, especially n-butyl-zinc oxazolidonate (prepared by reaction of ZnBu_2 and oxazolid-2-one) and n-butylzinc N,N-diethylcarbamate. Compound (XXIII) is considered to be converted readily into (XXIV). Subsequent polymerization is stated to start from the coupling of intermediates of the type (XXIV); the product is a zinc carbamate derivative which reacts with the NCA with ultimate formation of a mixed anhydride which then decarboxylates. For details the reader is referred to the original paper [50]. As pointed out by the authors, initiation shows some resemblance to the strong-base type of initiation already discussed in that an "activated NCA" (IV) plays a central role in both (if the Zn—N bond is incipiently ionic). However, the important propagation process in strong-base initiation represented by eqn. (20) is replaced in the systems under consideration by the coupling of zinc-activated NCA species.

In aluminium alkyl-initiated polymerizations it has been proposed [48] that initial attack by the alkyl at the 3- or 5-position of the NCA ring can occur:

3-position:
$$
\begin{array}{c}
\text{CH}_3\text{CH—CO} \\
| \quad\quad \diagdown \\
\quad\quad\quad \text{O} \\
| \quad\quad \diagup \\
\text{HN—CO}
\end{array}
\; + \; \text{AlR}_3 \quad \longrightarrow \quad
\begin{array}{c}
\text{CH}_3\text{CH—CO} \\
| \quad\quad \diagdown \\
\quad\quad\quad \text{O} \\
| \quad\quad \diagup \\
\text{N—CO} \\
| \\
\text{Al} \\
\diagup \diagdown \\
\text{R} \quad \text{R}
\end{array}
\; + \; \text{RH} \tag{44}
$$

(XXV)

5-position:

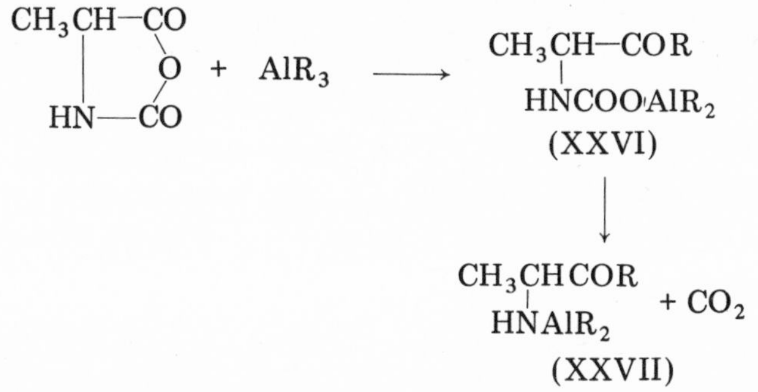

$$CH_3CH-CO$$
$$| \quad \backslash O + AlR_3 \longrightarrow$$
$$HN-CO$$

$$CH_3CH-COR$$
$$|$$
$$HNCOOAlR_2$$

(XXVI)

$$\downarrow$$

$$CH_3CHCOR$$
$$| \quad + CO_2$$
$$HNAlR_2$$

(XXVII)

$$(45)$$

Subsequent coupling between (XXV) and (XXVI) followed by decarboxylation is postulated:

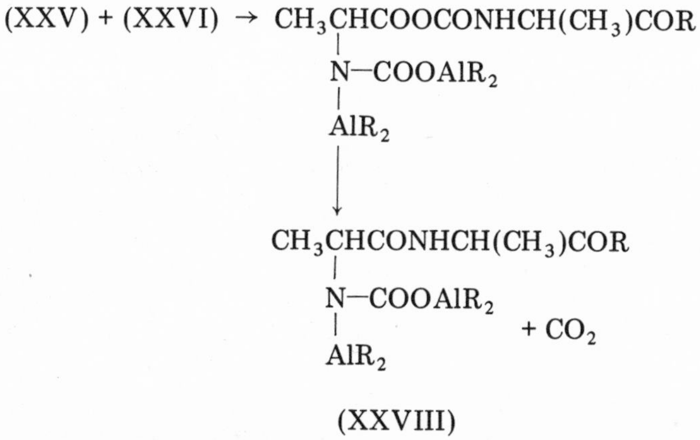

$$(XXV) + (XXVI) \rightarrow CH_3CHCOOCONHCH(CH_3)COR$$
$$|$$
$$N-COOAlR_2$$
$$|$$
$$AlR_2$$

$$\downarrow$$

$$CH_3CHCONHCH(CH_3)COR$$
$$|$$
$$N-COOAlR_2$$
$$| \qquad + CO_2$$
$$AlR_2$$

(XXVIII)

$$(46)$$

(XXVIII) may also arise by coupling of (XXV) and (XXVII). These reactions provide routes to polymer formation. Recently Makino et al. [52] have further studied these systems, with particular reference to the mechanism and stereochemistry (Section 8) of polymerization; these workers demonstrated that the presence of a hydrogen atom at position 3 of the NCA ring is a necessary requirement for polymerization, thus indicating the similarity to the strong-base mechanism.

A novel method of polymerization of NCAs has been described by Kopple and Katz [7], who reported polymerization following solution of the monomer in anhydrous hydrogen fluoride. The proposed over-all reaction is

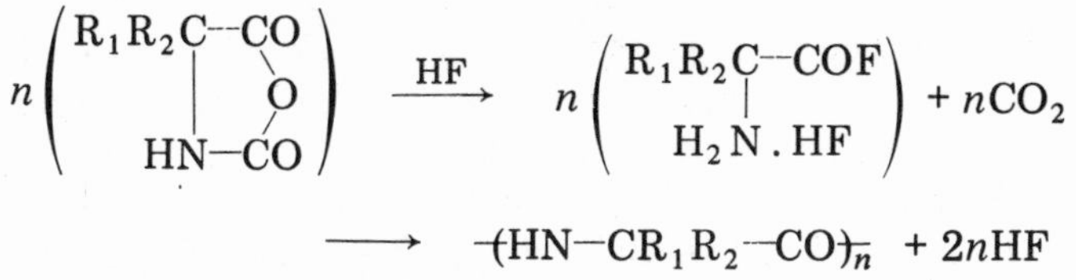

$$n \begin{pmatrix} R_1R_2C-CO \\ | \quad \backslash O \\ HN-CO \end{pmatrix} \xrightarrow{HF} n \begin{pmatrix} R_1R_2C-COF \\ | \\ H_2N.HF \end{pmatrix} + nCO_2 \qquad (47)$$

$$\longrightarrow \quad -(HN-CR_1R_2-CO)_n + 2nHF$$

but side-reactions must occur because CO is evolved and iso-valeraldehyde-2,4-dinitrophenylhydrazone can be isolated after suitable treatment of the reaction mixture. The by-products are explicable in terms of the following reaction proposed by Kopple et al. [8]

$$R_1R_2C-C\overset{\displaystyle O}{\underset{\displaystyle F}{\diagdown}} \qquad \longrightarrow \qquad R_1R_2C = NH_2^+ + CO \qquad (48)$$

$$H_2\overset{+}{N}-H \cdots (F-H)_n$$

This type of acid-catalysed polymerization does not produce high polymers.

7. Effects of substituent groups

As would be anticipated, the nature of substituent groups has a profound influence on the reactivity and polymerizability of an NCA. The rates of reaction between NCA and base molecules are affected by stereochemical and electronic factors; further, in suitable circumstances, interactions between propagating species and side-chain groups may occur.

Ballard and Bamford [53] have discussed the stereochemical aspects of the reaction between NCAs and primary and secondary bases. These workers measured initial rates of reaction (corresponding to conversions of 5% or less) between a number of NCAs and three bases (glycine dimethylamide and derivatives of it, $NHR_4CH_2CONMe_2$); values of the second-order rate coefficient k for reaction (4a) are presented in Table 2. These three bases are not very different in strength and changes in k arising from this cause should not exceed a factor of 2. From the table it is seen that with R_1 = Me and R_2 = R_3 = H the decrease in k accompanying a change in R_4 from H to iPr amounts to a factor of 10 approximately. This is in a direction opposite to that expected from changes in base strength and indicates the occurrence of steric interference when R_4 = iPr; reference has already been made to this type of behaviour (Section 3). When R_1 = R_3 = Me, we see that if R_4 = H or Me the rates are somewhat lower than those mentioned above (R_1 = Me, R_2 = R_3 = H), but only by a factor less than 4; however, if R_4 = iPr the rates in the two cases differ by a factor of 5000. This cannot be an electronic effect, but it must arise from steric interference of the isopropyl group of the base and the methyl group on the nitrogen atom. Interference between bulky substituents in the base and on the nitrogen atom is also evident from the results (Table 2) with R_1 = R_2 = H and R_3 = Me or iPr. These findings are consistent with a transition state in which the nitrogen atom of the attacking base is located between C5 and the nitrogen in the ring, with its centre above the plane of the ring. Ballard and Bamford give reasons for

TABLE 2

Reactions between N-carboxy-anhydrides and bases, $NHR_4CH_2CONMe_2$ in nitrobenzene at 15°C

N-Carboxy-anhydride			Base	k	Rates[a] of addition on two sides of ring:	
R_1	R_2	R_3	R_4	$(mole^{-1}\,l\,sec^{-1})$	same side as R_1	side opposite to R_1
Me	H	H	H	1.13 ± 0.10	m	n
			Me	0.48 ± 0.02	s	n
			iPr	0.10 ± 0.01	vs	m
H	H	Me	H	0.82 ± 0.03	n	
			Me	0.57 ± 0.03	n	
			iPr	7.4×10^{-4}	s	
Me	H	Me	H	0.45 ± 0.02	m	m
			Me	0.16 ± 0.03	s	m
			iPr	2×10^{-5}	vs	vs
Ph . CH$_2$	H	H	H	1.0 ± 0.2	m	n
			Me	0.40 ± 0.07	s	n
			iPr	0.30 ± 0.02	vs	m
H	H	iPr	H	0.83 ± 0.03	n	
			Me	0.48 ± 0.04	n	
			iPr	$<10^{-7}$	vs	
Me	Me	H	H	0.29 ± 0.05	m	
			Me	8.4×10^{-4}	s	
			iPr	1×10^{-6} approx.	vs	
N-Carboxy-L-proline			H	2.68 ± 0.08	vs	f
			Me	2.00 ± 0.07	vs	f
			iPr	0.12 ± 0.02	vs	m

[a] f = fast, n = normal, m = medium, s = slow, vs = very slow.

believing that the base approaches C5 along a path which is slightly inclined to the plane of the ring and passes more or less directly above (or below) the ring nitrogen.

A methyl substituent at C4 gives rise to steric interference in the transition state which is particularly severe with secondary bases. This is illustrated by the behaviour of the NCA of α-aminoisobutyric acid (R_1 = R_2 = Me, R_3 = H) which has a very low reactivity to the secondary bases (R_4 = Me or iPr), but a moderate reactivity towards the primary base (R_4 = H). In general, if addition on the side of the ring remote from R_1 were unaffected by R_1, the maximum change in rate coefficient attributable to a bulky substituent on C4 would amount to a factor of 2. However, it is likely that if R_1 and R_3 are both alkyl then for steric reasons their centres will lie on opposite sides of the mean plane of the ring, so that addition to C5 will be obstructed on both sides. Thus introduction of a bulky group on C4 could lead to a reduction in k exceeding a factor of 2 if $R_3 \neq H$, particularly when the attacking base is secondary. These conclusions are borne out in the results in Table 2.

L-proline NCA is comparatively reactive. In part this may arise from strain in the molecule. However, the fused rings confer a saucer-like shape on the molecule and attack at C5 from the convex side should be relatively free from obstruction by substituents on the ring nitrogen and C4. This is probably the explanation of the comparatively fast reaction of L-proline NCA with N-isopropylglycine dimethylamide (R_4 = iPr), of which the rate coefficient is approximately 6000 times that of the corresponding reaction with N-methyl-DL-alanine NCA (R_1 = R_3 = Me, R_2 = H).

The rates of addition of the bases on the two sides of the rings of various NCA's are summarized qualitatively in the right-hand portion of Table 2.

It may be noted that in the work described no account has been taken of the possible participation of the aprotic base type of initiation; we do not believe that this invalidates the major conclusions.

The ability of an NCA to polymerize may naturally be low for electronic, as well as for steric, reasons. Thus a substituent at the 3-position which, after ring opening, makes the resulting base extremely weak, may preclude propagation, so that the NCA is virtually non-polymerizable. This is the case with N-phenyl glycine NCA [18, 54] and other N-aryl NCA's [54], N-acetylglycine NCA [55] and N-triphenyl-methyl-glycine and -alanine NCAs [56]. The reactions occurring when these derivatives are treated with primary or secondary bases are represented by eqn. (4a) with R_3 = Ph, Ar, CH_3CO and Ph_3C.

Polymerization of certain NCAs is complicated by the intervention of side-reactions involving groups attached to C4. The first examples reported concern the NCAs of the γ-esters of glutamic acid; in polymerizations initiated by primary or secondary bases, propagating base groups are lost by the cyclization reaction (49) with formation of a pyrrolidone derivative (XXIX) which is unreactive in polymerization [57].

$$\begin{array}{c} \text{~~CH-NH}_2 \\ | \\ \text{CH}_2 \quad \text{CO-OR} \\ \backslash \; / \\ \text{CH}_2 \end{array} \longrightarrow \begin{array}{c} \text{~~CH-NH} \\ | \qquad \backslash \\ \text{CH}_2 \quad \text{CO} + \text{ROH} \\ \backslash \; / \\ \text{C} \\ \text{H}_2 \end{array} \qquad (49)$$

(XXIX)

Shalitin [58] has reported a marked reduction in rate during the polymerization of O-acetylhydroxyproline NCA and attributed this to the O→N acetyl shift (amidation) shown in (50).

$$\begin{array}{c} \text{H}_2 \\ \text{C} \\ \text{~~CO-CH} \quad \text{CHOCOCH}_3 \\ \text{HN---CH}_2 \end{array} \longrightarrow \begin{array}{c} \text{H}_2 \\ \text{C} \\ \text{~~CO-CH} \quad \text{CHOH} \\ \text{CH}_3\text{CON---CH}_2 \end{array} \qquad (50)$$

Reactions (49) and (50) are termination processes specific to the particular monomers mentioned and their occurrence reduces the average molecular weights and broadens the molecular weight distributions of the products.

It has been reported by Bailey [25] that use of a high concentration of aliphatic tertiary base at low temperatures enables the controlled addition of single amino acid residues to peptide amine end-groups to be achieved by reaction with an NCA. The success of this procedure depends on the quantitative formation of a trialkylammonium carbamate (XXX) by interaction of the carbamic acid intermediate in reaction (9a) with the tertiary base; the over-all reaction is shown in (51)

$$\text{~~NH}_2 + \begin{array}{c} \text{R}_1\text{R}_2\text{C-CO} \\ | \qquad \backslash \\ | \qquad \text{O} \\ \text{R}_3\text{N-CO} \end{array} \xrightarrow{\text{N(R}_4)_3} \begin{array}{c} \text{~~NHCO-CR}_1\text{R}_2 \\ | \\ \text{R}_3\text{NCOO}^- \; \overset{+}{\text{N}}(\text{R}_4)_3\text{H} \end{array} \qquad (51)$$

(XXX)

Unlike the free carbamic acid, (XXX) does not decarboxylate readily and is therefore unable to enter into propagation reactions at low temperatures.

8. Stereoisomeric effects

Lundberg and Doty [12] first reported that the initiation of the polymerization of racemic monomer by preformed, optically pure, homopolymer is selective in the sense that the enantiomorph with stereo-chemistry identical to that of the main chain is preferentially initiated. Thus a homopolymer consisting of L-residues initiates polymerization of the NCA of the L-amino acid more effectively than that of the D isomer. Further, these workers found that the rate of primary-amine-initiated polymerization of mixtures of D- and L-γ-benzylglutamate NCAs varied with enantiomorphic feed composition, being a minimum for the racemic mixture (Fig. 5). Subsequently, similar observations have been made with other types of initiator e.g. sodium methoxide and tertiary bases [35]. Lundberg and Doty [12] noticed that the rate of polymerization in racemic mixtures is lower than that calculated on the basis of absolute preference of one type of chain for its own enantiomorph (Fig. 5). This

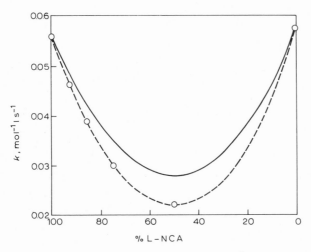

Fig. 5. Propagation coefficients (k) at 25°C for the polymerization of γ-benzyl-glutamate NCA in N,N-dimethylformamide as a function of D/L composition $[M]_0/[I]_0 = 20$. The solid line is calculated on the assumption that the L—NCA does not react with polymers ending in D-residues (absolute preference). Initiation by n-hexylamine. Lundberg and Doty [12].

implies that not only is a "wrong" addition relatively difficult, but also that subsequent propagation after such an addition is inhibited regardless of the stereochemistry of the NCA reacting. In support of this view it was found that initiating polymers composed of 15 units of γ-benzyl-L-glutamate and one terminal unit of the D-enantiomorph (statistical values) react comparatively slowly with both D- and L-γ-benzylglutamate NCA's, the addition being 6—8 times slower than the simple L—L reaction [19].

Lundberg and Doty [12] suggested that this specificity is connected with the existence of α-helical conformations. It may be envisaged as the result of adsorption of NCA on the terminal residues of the helix by hydrogen-bonding as depicted in Fig. 6 due to Weingarten [59]. Evidence

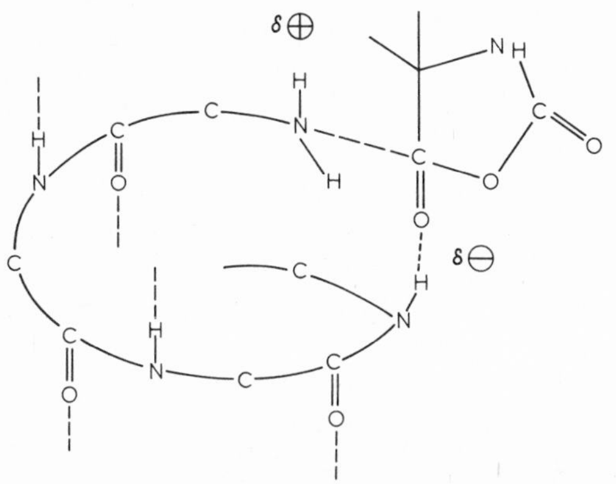

Fig. 6. Schematic representation of NCA molecule adsorbed on terminal residues of α-helix. Weingarten [59].

that an effect of this kind may be significant is forthcoming. The optical specificity is much greater when the initiating polymer is a 15-mer (some α-helix) than when it is a 3-mer (no α-helix) [60]. It is also removed by the addition of lithium perchlorate which, by preferential hydrogen-bonding, reduces the number of bound NCA molecules [60]. Models show that the approach to the transition state for reaction between the terminal base group of a helical polymer and an NCA molecule is more facile if the residues of the helix and the monomer have the same optical configuration. This is not surprising. According to the considerations advanced by Huggins [61], the side-chain of the wrong residue makes bad contacts with the body of the helix; however, these can be relieved by rotation of the residue about the C'—αC bond, but, as a result, sterically favourable contact for reaction becomes less likely.

Direct evidence for the adsorption of both enantiomorphs of γ-benzyl-glutamate NCA on to the L-polymer in dioxan was obtained by Williams et al. [62] using vapour pressure osmometry to monitor the free-monomer concentration. The results are shown in Table 3. It is of interest that the D-monomer is adsorbed preferentially by an L-polymer (K_{LD}/K_{LL} = 1.4). The authors do not state that the amine end-group of the polymer was deactivated in this work; if not, complications which might arise from in situ polymerization appear to have been disregarded.

It has been known for some time that NCAs of racemic α-amino acids can be converted to optically active poly α-amino acids with the aid of optically active alcohols or combinations of such alcohols with metal alkyls as initiators. Such processes have been termed "asymmetric-selective" or "stereo-selective". Some of these investigations will be described subsequently. Bührer and Elias [63] have recently made a comprehensive investigation of the kinetics of polymerization of DL-leucine NCA initiated by a series of optically active primary and

TABLE 3

Adsorption of γ-benzyl-glutamate NCA on to poly-γ-benzyl-L-glutamate ($\bar{M}_n$ = 4000) at 30°C in dioxan (Concentrations in mole l^{-1})

	L-NCA	D-NCA
10^3 [NCA] added	4.0	2.0
10^3 free [NCA]	3.31	1.56
10^3 base molar polymer concentration	28.1	28.1
K l(base mole)$^{-1}$	7.5	10.3

secondary amines in a variety of solvents at temperatures between 10 and 55°C. These authors discussed their results in terms of the ideal copolymerization equation

$$\frac{d[L]}{d[D]} = R \frac{[L]}{[D]} \tag{52}$$

where [L] and [D] are the molar concentrations of L- and D-monomers and R is a constant. R was calculated from measurements of the fractional conversion of the monomer and the optical yield of the polymer, p = (base molar rotation $[M]_D^{25}$ of polymer)/(base molar rotation of pure L-polymer). Results are presented in Table 4.

The authors discuss the various mechanisms which can lead to a constant value of R in (52). They conclude: "1. There are many reasonable models giving equations with constant R in the case of catalyst-controlled polymerization mechanisms. 2. Under certain conditions, an end-controlled mechanism also leads to constant R. 3. Helical growth can be included in some of the models without changing the mathematical expressions." It should be noted that control by helical growth is kinetically a special case of end-control.

In view of the work of Williams et al. [62], Bührer and Elias [63] assume that adsorption equilibria exist prior to the incorporation of the adsorbed NCA into the chain. There are four such equilibria:

$$
\begin{aligned}
&P_L^* + L \rightleftarrows P_L^* \cdot L; \quad K_{LL} = [P_L^* \cdot L]/([P_L^*][L]) \\
&P_L^* + D \rightleftarrows P_L^* \cdot D; \quad K_{LD} = [P_L^* \cdot D]/([P_L^*][D]) \\
&P_D^* + L \rightleftarrows P_D^* \cdot L; \quad K_{DL} = [P_D^* \cdot L]/([P_D^*][L]) \\
&P_D^* + D \rightleftarrows P_D^* \cdot D; \quad K_{DD} = [P_D^* \cdot D]/([P_D^*][D])
\end{aligned}
\tag{53}
$$

In (53) P_L^* and P_D^* represent chains with active L and D ends in the case of an end-controlled polymerization such as the one we are considering (or, in the case of the enantiomorphic catalyst site model, catalyst sites preferring L- and D-monomers, respectively), $P_L^* \cdot L$, $P_L^* \cdot D$ etc. are the

TABLE 4
Polymerization of DL-leucine N-carboxy anhydride with 2 mole-% initiator in 3% dioxane solution at 25°C

Initiator	% Conversion	$[M]_D^{25}$, polymer[a]	% Opt. yield $10^2 p$	$\frac{\eta_{sp}}{c}$ (cm^3 g^{-1})	R	$\bar{R}$	Polymer enriched in
(+)—BA	26.9	+1.3	<1	14.2[b]				
(+)—BA	33.4	+0.7	<1	16.2[b]				
(+)—BA	59.8	+0.5	<1	19.7[b]		≈1	(D)
(+)—BA	78.8	+0.5	<1	21.7[b]				
(−)—OA	13.1	+0.6	<1	8.4[c]				
(−)—OA	18.9	0	0	9.5[c]				
(−)—OA	34.7	0	0	11.3[b]		≈1	(D)
(−)—OA	47.4	0	0	13.4[b]				
(+)—PEA	18.4	+14.8	11.2	13.2[b]	1.28		
(+)—PEA	21.0	+14.9	11.2	10.8[b]	1.28	1.26 ±0.02	D
(+)—PEA	45.9	+10.3	7.7	13.4[b]	1.24		
(+)—PEA	50.2	+9.7	7.3	14.2[b]	1.24		
(+)—NEA	20.4	+1.4	≈1	16.3[b]			
(+)—NEA	47.5	+1.5	≈1	17.0[b]		≈1	(D)
(+)—NEA	55.9	+0.9	≈1	19.2[b]			
(+)—NEA	68.0	+0.7	≈1	20.9[b]			
(+)—MPA	25.1	+2.4	1.8	13.6[b]	1.04		
(+)—MPA	28.5	+2.0	1.5	15.4[b]	1.04	1.04 ±0.01	D
(+)—MPA	39.9	+2.1	1.6	16.2[b]	1.03		
(+)—MPA	58.6	+1.9	1.4	19.3[b]	1.04		
L-(+)—AME	34.8	−7.9	6.0	15.0[b]	1.16		
L-(+)—AME	52.2	−6.4	4.8	20.2[b]	1.15	1.15 ±0.01	L
L-(+)—AME	55.5	−6.3	4.7	19.8[b]	1.15		
L-(+)—AME	56.1	−5.9	4.4	20.4[b]	1.15		

618

TABLE 4—*continued*

Initiator	% Conversion	$[M]_D^{25}$, polymer[a]	% Opt. yield $10^2 p$	$\dfrac{\eta_{sp}}{c}$ (cm^3 g^{-1})	R	$\bar{R}$	Polymer enriched in
L–(+)–VME	20.8	−4.4	3.3	11.9^b	1.08		
L–(+)–VME	31.8	−4.3	3.2	13.1^b	1.08	1.09	
L–(+)–VME	66.6	−3.4	2.5	19.4^b	1.10	±0.02	L
L–(+)–VME	73.6	−3.3	2.5	18.3^b	1.11		
L–(+)–LME	14.3	−4.1	3.1	9.8^c	1.07		
L–(+)–LME	15.6	−4.0	3.0	9.9^c	1.06	1.07	
L–(+)–LME	39.3	−3.6	2.7	12.2^b	1.07	±0.01	L
L–(+)–LME	65.2	−3.3	2.5	13.2^b	1.09		
L–(+)–LEE	23.9	−7.1	5.3	17.9^c	1.13		
L–(+)–LEE	30.1	−7.8	5.8	16.1^d	1.15	1.14	
L–(+)–LEE	45.6	−6.1	4.6	16.7^b	1.14	±0.01	L
L–(+)–LEE	62.6	−5.0	3.8	17.9^b	1.14		
L–(+)–LPE	67.3	−5.1	3.8	18.5^b	1.15		
L–(+)–LPE	74.0	−4.2	3.2	18.8^b	1.14	1.15	
L–(+)–LPE	78.9	−4.1	3.1	18.8^b	1.16	±0.01	L
L–(+)–LPE	94.3	−1.6	1.2	16.3^b	1.15		
L–(+)–LBE	13.4	−4.4	3.3	7.5^c	1.08		
L–(+)–LBE	26.1	−3.9	3.0	10.8^c	1.08	1.07	
L–(+)–LBE	30.1	−4.0	3.0	10.8^c	1.07	±0.01	L
L–(+)–LBE	31.8	−3.3	2.5	11.4^c	1.06		

L—(+)—PAME	6.9	+6.1	4.6	9.6^d	1.10	1.07	D
L—(+)—PAME	15.8	+4.2	3.2	9.4^c	1.07		
L—(+)—PAME	30.8	+2.8	2.1	11.5^c	1.06	±0.03	
L—(+)—PAME	37.4	+2.0	1.5	11.8^c	1.04		
L—(+)—MME	35.1	−4.4	3.3	17.4^b	1.07	1.09	L
L—(+)—MME	42.6	−4.0	3.0	18.5^b	1.08		
L—(+)—MME	85.0	−2.1	1.6	20.6^b	1.10	±0.02	
L—(+)—MME	85.7	−2.5	1.9	22.0^b	1.12		
L—(−)—PME	15.4	−18.2	13.6	9.0^b	1.34	1.30	L
L—(−)—PME	19.6	−16.8	12.6	9.1^b	1.32		
L—(−)—PME	23.0	−14.2	10.6	9.9^b	1.27	±0.03	
L—(−)—PME	24.0	−14.3	10.8	9.7^b	1.28		
(+)—DMPEA	8.1	+1.7	1	11.8^c		≈1	(D)
(+)—DMPEA	34.4	+1.0	1	15.4^b			
(+)—DMPEA	43.9	+0.7	≈1	17.5^b			
(+)—DMPEA	87.3	+0.9	1	20.5^b			

^{a} $[M]_D^{25}$ of poly-L-leucine in trifluoroacetic acid is −133.3.
^{b} $c = 2 \cdot 10^{-2}$ g cm^{-3}; ^{c} $c = 1.5 \cdot 10^{-2}$ g cm^{-3}; ^{d} $c = 1.2 \cdot 10^{-2}$ g cm^{-3}.
BA = 2-butylamine, AO = 2-octylamine, PEA = α-phenylethylamine,
NEA = α-(1-naphthyl)-ethylamine, MPA = α-methylphenethylamine,
AME = alanine methyl ester, VME = valine methyl ester,
LME = leucine methyl ester, LEE = leucine ethyl ester,
LPE = leucine propyl ester, LBE = leucine t-butyl ester,
PAME = phenylalanine methyl ester, MME = methionine methyl ester,
PME = proline methyl ester, DMPEA = N,α-dimethyl-β-phenylethylamine.

corresponding adsorption complexes and K_{LL}, K_{LD} ... the respective equilibrium constants for adsorption. The rate-determining steps will be the addition reactions of the adsorbed monomers to the chain ends, so that the rate equations will be:

$$
\begin{aligned}
V_{LL} &= k_{LL} \, [P_L^* \cdot L] = k_{LL} K_{LL} \, [P_L^*][L] \\
V_{LD} &= k_{LD} \, [P_L^* \cdot D] = k_{LD} K_{LD} \, [P_L^*][D] \\
V_{DL} &= k_{DL} \, [P_D^* \cdot L] = k_{DL} K_{DL} \, [P_D^*][L] \\
V_{DD} &= k_{DD} \, [P_D^* \cdot D] = k_{DD} K_{DD} \, [P_D^*][D] .
\end{aligned} \tag{54}
$$

The relative consumptions of L- and D-monomers are given by

$$
\frac{d[L]}{d[D]} = \frac{k_{DL} K_{DL} + k_{LL} K_{LL} \, [P_L^*]/[P_D^*]}{k_{DD} K_{DD} + k_{LD} K_{LD} \, [P_L^*]/[P_D^*]} \equiv R \, \frac{[L]}{[D]} \tag{55}
$$

According to (55), constancy of R implies constancy of $[P_L^*]/[P_D^*]$. This condition is always fulfilled for an enantiomorphic catalyst of which the different sites remain in constant concentration. In an end-controlled mechanism, a ratio $[P_L^*]/[P_D^*] \neq 1$ can be achieved by use of an optically active initiator; clearly if propagation proceeds only by homopropagation (i.e. $V_{LD} = V_{DL} = 0$) then $[P_L^*]/[P_D^*]$ remains constant in a living system. Bührer and Elias [63] point out that the condition $[P_L^*]/[P_D^*] =$ constant may be fulfilled in two ways: either both $[P_L^*]$ and $[P_D^*]$ are individually constant or both concentrations vary with time in the same manner. The latter is unlikely to hold since it is not consistent with a living system in which the total concentration of active centres is constant, i.e. in which $d([P_L^*] + [P_D^*])/dt = 0$. The conclusion is therefore drawn that homopropagation is dominant i.e. that there is a high degree of selectivity by a growing chain for its own antipode. A very small amount of heteropropagation is not excluded. The individual chains thus consist exclusively of L- or D-units, or at least of long blocks. According to Bührer and Elias, therefore, the reaction may be described as an "active species controlled asymmetric-stereoselective polymerization". The inequality of $[P_L^*]$ and $[P_D^*]$, and hence the asymmetry, arises from preferential reaction of the optically active initiator with one of the NCA antipodes.

The authors also show that the experimental data of Falcetta [64], on the polymerization of DL-alanine NCA initiated by optically active α-phenylethylamine in tetrahydrofuran, can be interpreted in terms of eqn. (52) with constant R (= 1.65 ± 0.08 at $25°C$).

In the writers' opinion it is significant that the investigations referred to earlier, in which it was found that the presence of one "wrong" residue at the chain-end reduces the rate of addition of both L- and D-monomers, appear to cast some doubt on the justification for the a priori use of the ideal copolymerization eqn. (52). Penultimate residues would seem to influence propagating activity and the process of chain growth would more correctly be described by Markoff kinetics, taking into account at least the effect of penultimate groups [65].

There have been several reports that, during the primary-base initiated polymerization of single enantiomorphs of NCAs, marked increases in rate set in rather sharply when the growing chains reach a size of 7—12 units [12, 14, 59, 66]. Some of the experimental findings were a subject of controversy [60, 67, 68]. It was suggested [12, 59, 66] that, since at this stage a change in conformation from random coil to α-helix takes place [69], the higher rate is connected with the adoption of the α-helical form. Thus Lundberg and Doty [12] proposed that the acceleration is connected with a change in the entropy of activation resulting from "an increased accommodation provided for the anhydride molecule at the end of a helix". However, subsequent work has tended to discount the importance of this effect. The original observations of Doty, Blout and co-workers [12, 14, 66] were made on γ-benzyl-L-glutamate NCA in dioxan solution. This system, like many others, is now thought to be complicated by polymer aggregation and subsequent work has confirmed that this phenomenon leads to complex kinetic behaviour. Ballard and Bamford [70] first showed that the onset of heterogeneity during the polymerization of sarcosine NCA in benzene was accompanied by an increase in the rate of reaction (Fig. 7). The formation of an α-helix in this system is excluded; it

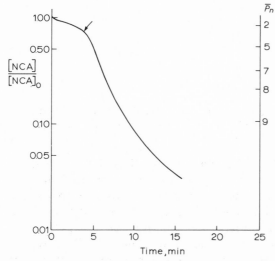

Fig. 7. Polymerization of sarcosine NCA in benzene at 25°C initiated by n-hexylamine. $[NCA]_0 = 0.100$ mole l^{-1}, $[n\text{-hexylamine}]_0 = 10^{-2}$ mole l^{-1}. Axis on right: $\bar{P}_n$ = mean size of growing polymer chains. The arrow marks the point at which separation of polymer begins. Ballard and Bamford [70].

was suggested that the aggregated phase is relatively rich in monomer and that the local increase in concentration gives rise to the observed increase in rate. Subsequently Ballard et al. [60] suggested that a similar phenomenon is operative in the system γ-benzyl-L-glutamate NCA/dioxan. These authors also remarked that the two-stage conversion—time curves reported for other systems such as L-leucine NCA/nitrobenzene [12], DL-alanine NCA/dioxan [59] and glycine NCA/dioxan [59] may be explained in terms of the development of heterogeneity.

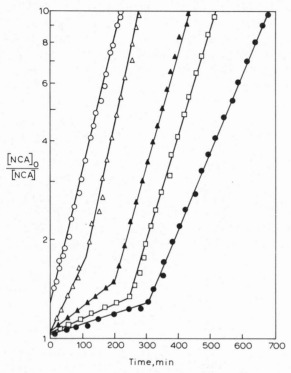

Fig. 8. First-order plot for the polymerization of γ-benzyl-L-glutamate NCA in dioxan at 25°C initiated by n-hexylamine. $[NCA]_0 = 0.10$ mole l^{-1}. Values of $[NCA]_0/[I]_0$: ○ 5.45; △ 11; ▲ 44.5; □ 65.5; ● 100.9. Williams et al. [62].

More recently Williams et al. [62] have reinvestigated the γ-benzyl-L-glutamate NCA/dioxan system, making extensive rate and adsorption measurements. Figure 8 presents some of their results which clearly show the "transition points" in the conversion—time curves. The values of $\bar{P}_n$ at the transition points are plotted in Fig. 9 as a function of $[NCA]_0/[I]_0$; it can be seen that the transition does not occur at some fixed $\bar{P}_n$ as would be expected if a change in the secondary polymer structure were involved. The results were rationalized by assuming the formation of polymer aggregates coupled with the adsorption of NCA on to growing polymer prior to reaction.

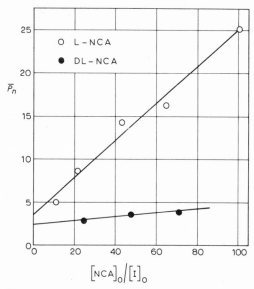

Fig. 9. Average degree of polymerization P_n at the transition point between the two linear portions of the first-order plots as a function of $[NCA]_0/[I]_0$ for the polymerization of ○ γ-benzyl-L-glutamate NCA and ● γ-benzyl DL-glutamate NCA. Dioxan solution at 25°C, initiation by n-hexylamine. Williams et al. [62].

Williams et al. [62] also studied the copolymerization of D- and L-enantiomorphs in dioxan solution and obtained transition points similar to those shown for the L-NCA in Fig. 8. However, in these systems the transition occurred at lower $\bar{P}_n$ (Fig. 9), a result which is contrary to that expected if a coil-helix transition were responsible for the change in rate. Kinetic analysis was carried out in terms of simultaneous polymerization in both phases following prior adsorption of monomer. (The general procedure was similar to that of Bührer and Elias [63] (p. 615 et seq.) and was used by them). It was found that only in the associated phase is selectivity essentially complete with no cross-over reactions. The analysis leads to an estimate of $K_{LD}/K_{LL} = 1.6 \pm 0.6$, in reasonable agreement with the value determined directly (p. 615).

In general agreement with the views outlined is the observation that, in solvents in which aggregation is insignificant no marked changes in rate are found during polymerization. Thus Lundberg and Doty [12] reported a single rate coefficient for the polymerization of γ-benzyl-L-glutamate NCA in N,N-dimethylformamide. Further, polymers isolated from these reactions were shown to have relatively narrow molecular weight distributions, $\bar{P}_w/\bar{P}_n$ approaching that expected for a Poisson distribution. This contrasts with the results obtained for polymerizations showing a rate transition, which give rise to broader distributions [12].

Ballard et al. [60] and Bamford and Block [19] have demonstrated smaller changes in the rate of homopolymerization of γ-benzyl-

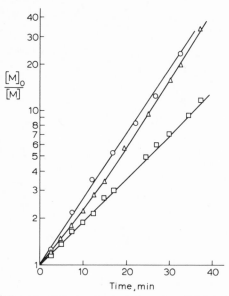

Fig. 10. Effect of chain-length of initiator $\bar{P}_n$ on the polymerization of γ-ethyl-L-glutamate NCA. Initiation by pre-formed poly-γ-ethyl-L-glutamate (1.5×10^{-2} mole l^{-1}) in N,N-dimethylformamide at $15°C$. $\bigcirc$ $\bar{P}_n = 25$, $[M]_0/[I]_0 = 5$, $\triangle$ $\bar{P}_n = 5$, $[M]_0/[I]_0 = 25$, $\square$ $\bar{P}_n = 5$, $[M]_0/[I]_0 = 5$. Bamford and Block [19].

L-glutamate- and γ-ethyl-L-glutamate-NCAs when the reactions are initiated by preformed L polymers of differing $\bar{P}_n$ in N,N-dimethylformamide solution. Figure 10 shows the influence of the chain-length of the initiator ($\bar{P}_n$) on the rate of polymerization and demonstrates that an increase of 50%, approximately, occurs as $\bar{P}_n$ is increased from 5 to 25. Further, under conditions such that there is an increase in mean chain-length from 5 to 30 during reaction, there is a steady increase in rate coefficient. In the absence of stereoselective influences and the known tendency of polypeptide chains to adopt α-helical conformations in the range of $\bar{P}_n$ between 8—12, these findings support a real but small enhancement in rate attributable to the α-helical conformer. Such rate increases arise from adsorption of the monomer by the α-helix, which is also partly responsible for stereoselective polymerization (p. 614 et seq.). As in the latter case, the rate increases are removed by the addition of $LiClO_4$ which competes with the NCA for adsorption on polymer sites [60].

Asymmetric selectivity in copolymerization of the enantiomorphs of alanine NCA initiated by organoaluminium derivatives has also been reported [47, 49, 51]. The authors state that polyalanine obtained from an L-rich monomer feed with the aid of $AlEt_3$ is, in the initial stages of reaction, enriched in the D isomer, but that with extensive conversion the major enantiomorph in the monomer feed is preferentially selected. This surprising result was not observed when di-isobutyl aluminium

N,N-diethylamide or methanol were used as initiators. Makino et al. [51] state "the steric control may be considered to take place by the steric cooperation between the ultimate unit [of the growing chain] and the approaching NCA through the coordination of an aluminium atom". Model compounds have been investigated in an effort to elucidate the stereochemical consequences of some of the proposed reactions [52].

9. The chain effect

In 1956 Ballard and Bamford [71] reported that the polymerization of several NCAs (e.g. DL phenylalanine NCA) with preformed polysarcosine as initiator shows unusual features. Thus the initial rate of reaction is a function of the chain length of the initiating polymer $\bar{P}_n$, the reaction rate increasing with $\bar{P}_n$ and approaching a limit when $\bar{P}_n = 20$, approximately (Fig. 11). The enhanced rate is high, and may be twenty times as great as that observed with sarcosine dimethylamide ($\bar{P}_n = 1$ effectively) as initiator. This phenomenon, which is so dependent on the chain-length of the polymeric initiator, was termed the "chain effect". As the reaction proceeds there is a marked reduction in rate coefficient which is consequent upon the formation of the block copolymer. The addition of a polymer of the α-amino acid whose NCA is being polymerized also reduces the rate, as does addition of acetylated polysarcosine. For the occurrence of the chain-effect it is necessary for the initiating centre (base) to be attached to the polysarcosine chain; thus a mixture of N-acetyl-polysarcosine with an equivalent of sarcosine dimethylamide gives only the "normal" rate of polymerization.

Intrinsic viscosity measurements on polymers isolated after the chain-effect reactions showed that the materials were true block copolymers, their viscosities being identical with those of copolymers prepared by

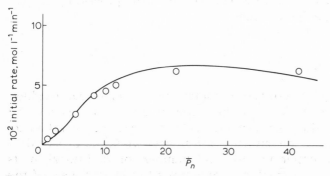

Fig. 11. Polymerization of DL-phenylalanine NCA initiated by polysarcosine dimethylamide with different degrees of polymerization $\bar{P}_n$. Nitrobenzene solution, 15°C. $[M]_0 = 0.1$ mole l^{-1}, $[I]_0 = 5.4 \times 10^{-3}$ mole^{-1}. The curve has been calculated from eqns. (58) and (59). Ballard and Bamford [71].

polymerizing sarcosine NCA with the corresponding preformed poly-α-amino acid derivatives as initiators.

The most satisfactory explanation of these findings is that the NCA is adsorbed on the polysarcosine chain with the result that it is brought into closer proximity to the initiating base group, which therefore encounters a higher local concentration of monomer. This view is consistent with observed dependence of rate on $\bar{P}_n$ since the accessibility of an adsorbed monomer for reaction is influenced by the flexibility of the polysarcosine chain and the position of the adsorption site along the chain. The newly-formed polymer, which is probably more rigid on account of the development of secondary (e.g. α-helical) structure, progressively removes the base site from the environment of the flexible polysarcosine chains with its adsorbed monomer, so that ultimately the chain-effect enhancement is lost. Alternatively, hydrogen-bonding between the growing polymer and the polysarcosine chains could lead to relatively rigid structures of low activity. Further, the effect of addition of acylated polysarcosine is easily understood, since it adsorbs monomer and so competes with the active polysarcosine.

The kinetic scheme employed by Ballard and Bamford [71] in the analysis of their experimental data resembles the Michaelis—Menten mechanism descriptive of enzymic reactions. Equation (56) represents adsorption of the monomer by the polysarcosine chain, S_i being a sarcosine residue in position i, counting the residues along the chain from the terminal base group, and E_i the adsorbed NCA molecule on this site.

$$M + S_i \underset{\longleftarrow}{\overset{K_i}{\longrightarrow}} E_i \tag{56}$$

Molecule E_i is decomposed by the base group at the end of the same chain with liberation of carbon dioxide, presumably according to eqn. (4), so that we may represent this (intra-chain) stage of the process by (57) in which X_n is a growing polymer chain

$$E_i \, (+X_n) \xrightarrow{k_i} CO_2 + S_i \, (+X_{n+1}) \tag{57}$$

(arising from NCA polymerization) containing n amino-acid residues. Differences in behaviour between the initial polysarcosine chains and those to which a few additional α-amino acid units have been attached are ignored in this simple treatment; such differences would be expected to become increasingly important as the number of additional units increases, so that this treatment can only apply to the early part of the reaction.

If it is assumed that all the sites are equivalent as far as adsorption is concerned i.e. $K_i = K$ for all i, $[E] = \Sigma_i \, [E_i]$ is given by

$$[E] = \frac{1}{2K} \left[1 + K[S] + K[M]_0 - \{(1 + K[S] + K[M]_0)^2 \right.$$
$$\left. - 4K^2 [M]_0 [S]\}^{1/2} \right] \tag{58}$$

[S] being the total concentration of sarcosine residues. The total rate of reaction is then

$$\frac{d[CO_2]}{dt} = [E] \sum_1^n k_i/n \equiv \bar{k}_n [E] \tag{59}$$

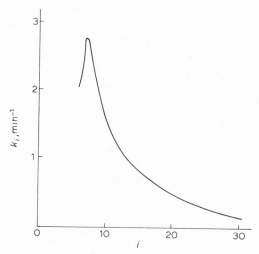

Fig. 12. Polymerization of DL-phenylalanine NCA initiated by polysarcosine dimethylamide. Nitrobenzene solution at 15°C. Velocity coefficient k_i (eqn. (57)) as a function of position of the reaction site i in the polysarcosine chain. Ballard and Bamford [71].

$\bar{k}_n$ being the mean value of k_i for an n-mer. Ballard and Bamford chose values of the two independent parameters K, $\bar{k}_n$ in (58), (59) to give the most satisfactory agreement with their data. From the dependence of $\bar{k}_n$ upon n they deduced the variation of k_i with i shown in Fig. 12, which shows a sharp maximum in k_i at $i = 7$. As the authors point out, this value for i is indicative of limited flexibility of the polysarcosine chain.

Direct evidence for the adsorption of NCAs on polysarcosine has been obtained [72] from infrared observations of the N—H stretching frequency of the NCA at 3437 cm^{-1} (DL-phenylalanine NCA) and the C=O stretching vibration of polysarcosine (of which the terminal base group had been deactivated) near 1680 cm^{-1}. These investigations revealed the existence of hydrogen bonding between the monomer and main-chain carbonyls (XXXI)

References pp. 634—637

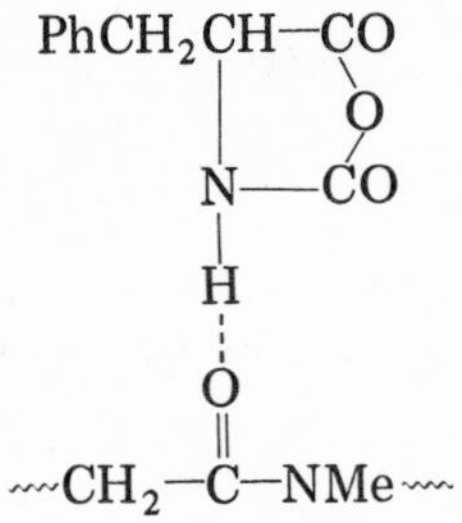

628

$$PhCH_2CH{-}CO$$

(XXXI)

and allowed the equilibrium constant for adsorption to be estimated as 3.5 ± 0.3 mole^{-1} l in methylene chloride solution at $29°C$. In nitrobenzene solution it was impossible to derive precise values of K owing to interference by solvent bands but an approximate value of 5 mole^{-1} l was estimated. Bamford and Price [72] concluded that these values of K for both solvent systems were compatible with the kinetic data.

More recently Ballard [73] has reported that poly-L-proline does not show the chain-effect with DL-phenylalanine NCA. He has also studied a number of block copolymers of L-proline and sarcosine as initiators with results shown in Table 5. The complex behaviour observed is discussed by Ballard in terms of availability to peptide sites for NCA adsorption. It is concluded that poly-L-proline does not adsorb the NCA. The existence of a chain effect with the initiators $(Sarc)_{10}(Pro)_{10}X$ and $(Sarc)_5(Pro)_{10}X$ requires explanation; it may be connected with the occurrence of conformations of the proline chain in which the terminal base group X is not far removed from the sarcosine "tail".

A series of N-substituted polyglycines including polysarcosine has been studied by Imanishi et al. [74, 75], the new substituents being ethyl,

TABLE 5
Polymerization of DL-phenylalanine NCA. Initiation by poly-L-proline and block copolymers of L-proline and sarcosine. (Nitrobenzene solution at $25°C$. $[NCA]_0 = 0.1$ mole l^{-1})

Initiating polymer	$10^3 [I]_0$, mole l^{-1}	10^2 Initial rate of polymerization, mole l^{-1} min^{-1}
$(Pro)_2{-}X$	7.3	0.75
$(Pro)_{10}{-}X$	7.3	0.75
$(Sarc)_2{-}X$	5.4	1.0
$(Sarc)_{10}{-}X$	5.4	9.2
$(Pro)_{10}(Sarc)_{10}{-}X$	7.0	6.2
$(Sarc)_{10}(Pro)_{10}{-}X$	7.0	7.2
$(Pro)_{10}(Sarc)_5{-}X$	8.0	4.6
$(Sarc)_5(Pro)_{10}{-}X$	8.0	4.5

X = terminal base group.

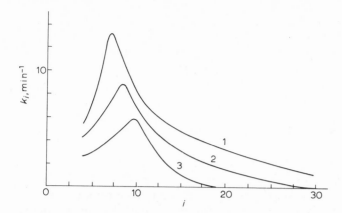

Fig. 13. Chain-effect polymerization of DL-phenylalanine NCA in nitrobenzene solution at 25°C with different polymeric initiators. Rate coefficient k_i (eqn. (57)) as a function of position of the reaction site i in the polymer chain. Initiators: 1 Polysarcosine diethylamide, 2 Poly(N-ethylglycine)diethylamide, 3 Poly(N-n-propylglycine)diethylamide. Sisido et al. [75].

n-propyl, n-butyl, isobutyl and benzyl, together with poly-N-methyl-DL-alanine. All but the last two show some measure of chain-effect, but the magnitude of the effect decreases as the bulkiness of the N-alkyl group increases. This is accompanied by an increase in the i-value of the maximum rate coefficient (see Fig. 13) [75]. It is clear that two factors may influence the magnitude of the chain-effect, namely the inherent chain stiffness of the macromolecular catalyst and the extent of monomer adsorption on to the catalyst. Both factors would be influenced by the bulkiness of substituents in the catalyst chain. The studies of Imanishi et al. [74, 75] indicate that, for the series of N-substituted polyglycines (N-methyl, N-ethyl, N-propyl) investigated, the equilibrium constants for adsorption vary little from the value of 3.3 mole^{-1} l in methylene chloride at "ambient temperatures" with chain-lengths of 30, approximately. With poly-N-methyl alanine the equilibrium constant is lower — 1.3 mole^{-1} l. The magnitude of the chain-effect was found to follow the order polysarcosine > poly-N-ethylglycine > poly-N-propylglycine and is effectively zero with poly-N-methylalanine. For these three catalysts chain-stiffness is the origin of the decrease in activity with increasing substituent size; with poly-N-methylalanine a similar cause operated since this polymer is thought to have low flexibility in solution. Poly-N-benzyl glycine has a terminal group which is a very weak base and is consequently relatively unreactive; it is reported to show no chain-effect although low solubility of the polymer makes study difficult.

A phenomenon related to the chain-effect has been observed [76] when polymeric catalysts containing basic units in the chain are employed to initiate the polymerization of DL-β-phenylalanine-, 4-nitro-DL-β-phenylalanine- and 2,4-dinitro-DL-β-phenylalanine-NCAs. The enhancement of rate in these systems arises from adsorption of the monomers, which hydrogen-bond to the polymer phase. The catalysts studied were poly-2-vinylpyridine, poly-N-vinyl-2-ethyl imidazole, the copolymer of 2-vinylpyridine with styrene and the copolymers of N-vinyl pyrolidine and N,N-diethyl acrylamide with N-vinyl-2-ethyl imidazole. A correlation between rate of reaction and base strength and concentration was observed.

An interesting recent application of the chain effect is its use as a means of influencing copolymer structure [77]. This is possible since the chain effect arises from a selective process confined to NCAs with R_3 = H. In the polysarcosine-initiated copolymerization of the NCAs of γ-ethyl-L-glutamate and sarcosine, selection of the former monomer by the polysarcosine template induces a rapid polymerization to give an initial block-like polymer of sarcosine and γ-ethyl-L-glutamate, on which a third block, rich in sarcosine, subsequently grows by normal solution polymerization. The presence of a (helical) block of γ-ethyl-L-glutamate residues in this polymer was established by optical rotatory dispersion measurements. A copolymer of the same composition prepared by conventional initiation under conditions such that the chain effect is largely inoperative showed no such block character.

10. Copolymerization

From the earlier discussion it will be amply apparent that block copolymers may be synthesized from NCAs by the use of a preformed poly-α-amino acid to initiate the polymerization of the NCA of a different α-amino acid. The kinetics of such a reaction would be expected to be determined by features already discussed; in general, the rate coefficient for initiation may differ from that of propagation (Section 1) and phenomena attributable to stereoselectivity, polymer aggregation and monomer adsorption (Sections 8 and 9) may arise.

In (Section 8) we have considered kinetic phenomena encountered during the copolymerization of L- and D-enantiomorphs of a single type of NCA. Copolymerization of mixtures of NCAs of different α-amino acids to give "random" copolymers is readily realizable [2], one of the earliest examples reported being the synthesis of a DL-phenylalanine L-leucine copolymer by Woodward and Schramm [78]. However, there are few reports on the kinetics of random copolymerization. Shalitin and Katchalski [79] studied the copolymerization of the NCAs of γ-benzyl L-glutamate (A) and ε,N-carbobenzoxy L-lysine (B) initiated by diethylamine in N,N-dimethylformamide at 25°C and obtained the interesting result that the over-all rate of reaction (measured by the CO_2 evolution) is equal to the sum of the rates of reaction of the individual monomers under similar conditions. The copolymerization is represented schematically in (60)

$$
\begin{array}{lll}
\text{~~}A + B & \xrightarrow{\ k_{AA}\ } & \text{~~}AA \\
\text{~~}A + B & \xrightarrow{\ k_{AB}\ } & \text{~~}AB \\
\text{~~}B + A & \xrightarrow{\ k_{BA}\ } & \text{~~}BA \\
\text{~~}B + B & \xrightarrow{\ k_{BB}\ } & \text{~~}BB
\end{array}
\tag{60}
$$

where $\text{~~}A$ represents a propagating chain with a terminal A unit and so forth. Shalitin and Katchalski [79] concluded from their findings that $k_{AA} = k_{BA} \equiv k_A$ and $k_{AB} = k_{BB} \equiv k_B$, so that the propagation is characterized by two different rate coefficients instead of the normal four. Thus the rate of propagation is determined mainly by the nature of the NCA rather than the character of the base group; this simplification stems from the fact that the strengths of the peptide base groups concerned are not very different. The composition of the copolymer formed in the initial stages should therefore be given by the ideal copolymerization eqn. (61) (analogous to (52))

$$
\left(\frac{[A]}{[B]} \right)_{polymer} = R\, \frac{[A]_0}{[B]_0}
\tag{61}
$$

in which $R = k_A / k_B$. This conclusion was verified experimentally, R having the value 2.5. Since the monomers are consumed at different rates, in general a drift in composition of the monomer mixture as the reaction proceeds with a corresponding change in copolymer composition would be expected. Analysis of the copolymers at different stages of reaction confirmed this prediction, the initial and final copolymers being enriched in glutamate and lysine residues, respectively. The effective over-all rate coefficient for the copolymerization falls with increasing conversion for similar reasons.

The following pairs of NCAs behaved similarly when copolymerized under the same conditions [79]: L-phenylalanine + ϵ,N-carbo-benzoxy-L-lysine, L-phenylalanine + γ-benzyl-L-glutamate, glycine + ϵ,N-carbobenzoxy-L-lysine, ϵ,N-carbobenzoxy-L-lysine + DL-alanine and γ-benzyl-L-glutamate + DL-alanine. On the other hand, copolymerization of γ-benzyl-L-glutamate and DL-alanine NCAs behaved in a different manner and showed retardation in the later stages.

Shalitin and Katchalski [79] derived analytical expressions for the molecular weight and compositional distributions of the copolymers. Poissonian distributions were calculated for the over-all molecular weights and for the concentrations of molecules containing a specified number of A units regardless of the number of B present.

Use of N,N-dimethylformamide as solvent in copolymerizations avoids complications arising from aggregation which have been discussed in

Section 8; it will be recalled that γ-benzyl-L-glutamate NCA polymerizes in a simple first-order manner in this medium [12]. Copolymerization in less polar media would be expected to be much more complex, but no relevant kinetic data are available.

11. Further kinetic considerations

In spite of the considerable effort which has been devoted to investigation of the polymerization of NCAs little information is available which permits calculation of the enthalpy and entropy of activation of the rate-determining processes. While this is an unusual situation, it is perhaps not surprising for reasons already discussed. We now consider the available data.

Ballard and Bamford [80] reported values of the propagation coefficient k (eqn. (7)) for primary-base initiation of the polymerization of two NCA's at different temperatures. These are presented in Table 6. It is clear that the energies of activation for the polymerization of DL-leucine and DL-phenylalanine NCAs are quite small and the frequency factors (A) are lower than the normal value (10^{11}—10^{12} mole^{-1} l sec^{-1}) by several orders of magnitude. On the other hand, with γ-benzyl-L-glutamate NCA in nitrobenzene solution, the activation energy is much higher and the frequency factor is somewhat bigger than the normal value. For the polymerization of γ-benzyl-L-glutamate NCA in N,N-dimethylformamide Shalitin [58] reported an activation energy of 5.5 kcal mole^{-1} and a frequency factor of 1.9×10^3 mole^{-1} l sec^{-1} while Lundberg and Doty [12] gave the activation energy as 6.6 kcal mole^{-1}; it would therefore appear that there is a very large medium effect or a mechanistic change with this NCA.

According to eqn. (25), we should expect that the transition state in the reaction of a primary or secondary base with an NCA would be more polar than the reactants. In a polar solvent, therefore, orientation of solvent molecules around the transition complex would lead to strongly negative entropies of activation and low frequency factors. Ballard and Bamford [80] proposed that the propagating amino groups in the γ-benzyl-L-glutamate NCA polymerization partake in intramolecular complex formation with ester groups, with formation of structures which do not differ greatly in polarity from the transition state in propagation. The result of this would be to increase the activation energy and raise the frequency factor to about the normal value. According to Ballard and Bamford [80] the polymerization of this NCA is relatively insensitive to the polarity of the medium, the rate coefficient being reduced only by a factor of three, approximately, when the solvent is changed from nitrobenzene to dioxan (Table 6). This polymerization appears to be

TABLE 6
Values of $10^2 k$ (mole^{-1} l sec^{-1}) (eqn. (7)), activation energies (E) and frequency factors (A); initiation by preformed polymers

NCA	DL-leucine		DL-phenylalanine	γ-benzyl-L-glutamate	
Solvent	Nitrobenzene	o-Nitroanisole	Nitrobenzene	Nitrobenzene	Dioxan
Temperature (°C)					
25		2.38	1.56		
25.2	4.17			7.04	2.17
45	8.72	5.83	3.50		
E (kcal mole^{-1})	6.95	8.5	7.6	20.2	
A (mole^{-1} l sec^{-1})	5.0×10^3	3.8×10^4	6.0×10^3	3.9×10^{13}	

the only instance in which the activation parameters of a single enantiomorph are known and the need for further experiments is evident.

The rate of the aprotic polymerization of NCAs, involving an ionization mechanism, is very sensitive to the nature of the medium. For example, it is stated [80] that while DL-phenylalanine NCA dissolved in pure tri-n-butylamine (dielectric constant =3, approximately) reacts only slowly, a rapid reaction ensues on adding N,N-dimethylformamide (dielectric constant = 38) or nitrobenzene (dielectric constant = 36). Not only the over-all rate of polymerization, but also the relative rates of competing processes are influenced by the solvent. Ballard et al. [39] reported that the yield of hydantoin 3-acetic acid (XIV) obtained from glycine NCA on tertiary base initiation varies from solvent to solvent. Although the yield is generally higher in more polar media such as N,N-dimethylformamide and acetonitrile, there is not a particularly close correlation with dielectric constant and it is clear that properties of the medium other than dielectric constant are important. Thus in nitrobenzene solution the yield of hydantoin 3-acetic acid is only 13% of that in N,N-dimethylformamide under similar conditions.

In Sections 8 and 9 we have seen that a number of diverse phenomena may play important roles in the polymerization of NCAs. These include adsorption of the NCA on to the polymer chains, aggregation of polymer chains and conformational effects. All these could influence the apparent activation parameters $\Delta H^{\neq}$ and $\Delta S^{\neq}$ and each would in general be affected by the nature of the solvent. For example, if adsorption is important, the relevant kinetic parameter is the product of an equilibrium constant and a rate coefficient, so that the over-all temperature coefficient does not depend solely on the activation energy of the propagation step. Again, the rate of chain-effect polymerization, in which hydrogen-bonding of NCA to polymer is involved, is markedly affected by the nature of the solvent [71], being much lower in N,N-dimethylformamide than in nitrobenzene. In view of these complexities and those of a purely chemical nature resulting from the structure of the NCA ring, the comparative paucity of kinetic data of the conventional kind is not surprising.

REFERENCES

1 H. Leuchs, *Chem. Ber.*, *39* (1906) 857.
2 C. H. Bamford, A. Elliott and W. E. Hanby, *Synthetic Polypeptides*, Academic Press, New York, 1956.
3 E. Katchalski and M. Sela, *Advan. Protein Chem.*, *13* (1958) 243.
4 G. D. Fasman, Y. Shalitin and N. Tooney, in N. H. Bikales (Ed.), *N-Carboxyanhydrides, Encyclopedia of Polymer Science and Technology*, Interscience Publ. *2* (1965) 837; M. Szwarc, The Kinetics and Mechanism of *N*-Carboxy-α-amino acid Anhydride (NCA) polymerization to Poly-amino acids, *Fortschr. Hochpolymer,*

Forsch., *4* (1965) 1; G. D. Fassman (Ed.), Poly-α-amino acids: protein models for conformational studies, *Biological Macromolecules*, Arnold (London) and M. Dekker (N.Y.) 1 (1967).

5 Y. Shalitin, in K. C. Frisch and S. L. Reegen (Eds.), *Kinetics and Mechanism of Polymerization, Ring Opening Polymerization*, M. Dekker, 2 (1969) 421.

6 C. H. Bamford and H. Block, in K. C. Frisch (Ed.), *Carboanhydrides in Cyclic Monomers, High Polymers*, Wiley Interscience, *26* (1972) 687.

7 K. D. Kopple and J. J. Katz, *J. Amer. Chem. Soc.*, *78* (1956) 6199.

8 K. D. Kopple, L. A. Quarterman and J. J. Katz, *J. Org. Chem.*, *27* (1962) 1062.

9 S. G. Waley and J. Watson, *Proc. Roy. Soc. Ser. A, 199* (1949) 499.

10 D. G. H. Ballard and C. H. Bamford, *Proc. Roy. Soc. Ser. A, 223* (1954) 495.

11 N. Noma and T. Tsuchida, *Kobunshi Kagaku, 6* (1949) 109; J. W. Breitenbach and K. Allinger, *Monatsh. Chem.*, *84* (1953) 1103; M. Idelson and E. R. Blout, *J. Amer. Chem. Soc.*, *79* (1957) 3948; A. Patchornik and Y. Shalitin, *Bull. Res. Council Israel, 5A* (1956) 300; *Anal. Chem.*, *33* (1961) 1887; H. Block, *J. Sci. Instrum.*, *41* (1964) 370.

12 R. D. Lundberg and P. Doty, *J. Amer. Chem. Soc.*, *79* (1957) 3961.

13 A. Berger, M. Sela and E. Katchalski, *Anal. Chem.*, *25* (1953) 1554.

14 M. Idelson and E. R. Blout, *J. Amer. Chem. Soc.*, *79* (1957) 3948.

15 T. K. Miwa and M. A. Stahmann, in M. A. Stahmann (Ed.), *Polyamino Acids, Polypeptides and Proteins*, Wisconsin Univ. Press, Madison, 1962, 81.

16 M. Sela and A. Berger, *J. Amer. Chem. Soc.*, *77* (1955) 1893.

17 C. H. Bamford, H. Block, D. Mason and A. W. Openshaw, Symposium on the Chemistry of Polymerization Processes, Society of Chemical Industry, Monograph No. 20 (1965) 304.

18 C. H. Bamford, H. Block and A. C. P. Pugh, *J. Chem. Soc.*, (1961) 2057.

19 C. H. Bamford and H. Block, in M. A. Stahmann (Ed.), *Polyamino acids, Polypeptides and Proteins*, University of Wisconsin Press, 1962, 65.

20 D. G. H. Ballard and C. H. Bamford, *J. Chem. Soc.*, (1956) 381.

21 C. H. Bamford and H. Block, *J. Chem. Soc.*, (1961) 4989.

22 C. H. Bamford and H. Block, *J. Chem. Soc.*, (1961) 4992.

23 M. Goodman and U. Arnon, *Biopolymers, 1* (1963) 500; *J. Amer. Chem. Soc.*, *86* (1964) 3384.

24 M. Goodman and J. Hutchison, *J. Amer. Chem. Soc.*, *87* (1965) 3524; *88* (1966) 3627.

25 F. Wessely, *Z. Physiol. Chem.*, *146* (1925) 72; J. L. Bailey, *J. Chem. Soc.*, (1950) 3461; W. Langenbeck and P. Kresse, *J. Prakt. Chem.* [4], *2* (1955) 261; R. G. Denkewalter, D. F. Veber, F. W. Holly and R. Hirschmann, *J. Amer. Chem. Soc.*, *91* (1969) 502; R. G. Strachan, W. J. Paleveda Jr., R. F. Nutt, R. A. Vitali, D. F. Veber, M. J. Dickinson, V. Garsky, J. E. Deak, E. Walton, S. R. Jenkins, F. W. Holly and R. Hirschmann, *J. Amer. Chem. Soc.*, *91* (1969) 503; S. R. Jenkins, R. F. Nutt, R. S. Dewey, D. F. Veber, F. W. Holly, W. J. Paleveda, Jr., T. Lanza, Jr., R. G. Strachan, E. F. Schoenewaldt, H. Barkemeyer, M. J. Dickinson, J. Sondey, R. Hirschmann and E. Walton, *J. Amer. Chem. Soc.*, *91* (1969) 504; D. F. Veber, S. L. Varga, J. D. Milkowski, H. Joshua, J. B. Conn, R. Hirschmann and R. G. Denkewalter, *J. Amer. Chem. Soc.*, *91* (1969) 506; R. Hirschmann, R. F. Nutt, D. F. Veber, R. A. Vitali, S. L. Varga, T. A. Jacob, F. W. Holly and R. G. Denkewalter, *J. Amer. Chem. Soc.*, *91* (1969) 507.

26 J. H. Fessler and A. G. Ogston, *Trans. Faraday Soc.*, *223A* (1954) 495.

27 M. T. Pope, T. J. Weakley and R. J. P. Williams, *J. Chem. Soc.*, (1959) 3442.

28 J. W. Breitenbach and K. Allinger, *Monatsch. Chem.*, *84* (1953) 1103.

29 J. D. Coombes and E. Katchalski, *J. Amer. Chem. Soc.*, *82* (1960) 5280.

30 E. Katchalski, Y. Shalitin and M. Gehatia, *J. Amer. Chem. Soc.*, *77* (1955) 1925.

31 L. Gold, *J. Chem. Phys.*, *21* (1953) 1190.

636

32 E. Katchalski, M. Gehatia and M. Sela, *J. Amer. Chem. Soc.*, *77* (1955) 6175.

33 O. Wichterle, *Forschr. Hochpoly. Forsch.*, *2* (1961) 578.

34 T. Wieland, *Angew. Chem.*, *63* (1951) 7; *66* (1954) 507.

35 M. Idelson and E. R. Blout, *J. Amer. Chem. Soc.*, *80* (1958) 2387.

36 J. Kopple, *J. Amer. Chem. Soc.*, *79* (1957) 6442.

37 D. G. H. Ballard, C. H. Bamford and F. J. Weymouth, *Proc. Roy. Soc., Ser. A, 227* (1955) 155.

38 (a) D. G. H. Ballard and C. H. Bamford, *Chem. Soc. Special Publ. No. 2* (1955) 25; (b) C. H. Bamford, *Chem. Soc. Special Publ. No. 2* (1955) 45; K. Schlögl, *Chem. Soc. Special Publ. No. 2* (1955) 47.

39 D. G. H. Ballard, C. H. Bamford and F. J. Weymouth, *Nature, 174* (1954) 173.

39a C. H. Bamford and F. J. Weymouth, *J. Amer. Chem. Soc.*, *77* (1955) 6368.

40 J. Noguchi and T. Hayakawa, *J. Amer. Chem. Soc.*, *76* (1954) 2846; J. Noguchi, T. Hayakawa, T. Motoi, J. Suzuki and M. Ebata, *J. Chem. Soc., Japan, 76* (1955) 646, 648, 651; J. Noguchi, S. Ishino, T. Hayakawa and M. Iida, *J. Chem. Soc., Japan, 76* (1955) 457; F. J. Weymouth, *Chem. Ind., Brit. Inds. Fair Rev.*, (1956) R34; De Los F. De Tar and N. F. Estrin, *Tetrahedron Lett.* (1966) 5985; B. J. Johnson, *J. Chem. Soc. (C)*, (1967) 2638; A. M. Tamburro, A. Scattuvin and F. Marchiori, *Gazz. Chim. Ital.*, *98* (1968) 638.

41 V. Gold and E. G. Jefferson, *J. Chem. Soc.*, (1953) 1409, 1416.

42 E. Peggion, E. Scoffone, A. Conani and A. Portolan, *Biopolymers, 4* (1966) 695.

43 J. W. Stewart and M. A. Stahmann, in M. A. Stahmann (Ed.), *Polyamino acids, Polypeptides and Proteins*, University of Wisconsin Press, Madison (1962) 95.

44 E. Scoffone, E. Peggion, A. Cosani and M. Terbojevich, *Biopolymers, 3* (1965) 535; J. T. Yang, *J. Amer. Chem. Soc.*, *80* (1958) 1783, 5139.

45 H. A. Sober, in M. A. Stahmann (Ed.), *Polyamino acids, Polypeptides and Proteins*, University of Wisconsin Press, Madison (1962) 105.

46 E. R. Blout and J. Asadourian, *J. Amer. Chem. Soc.*, *78* (1956) 955.

47 T. Tsuruta, S. Inoue and K. Matsuura, *Makromol. Chem.*, *63* (1963) 219; K. Matsuura, S. Inoue and T. Tsuruta, *Makromol. Chem.*, *80* (1964) 149.

48 K. Matsuura, S. Inoue and T. Tsuruta, *Makromol. Chem.*, *103* (1967) 140.

49 M. Yoneyama, S. Inoue and T. Tsuruta, *Makromol. Chem.*, *107* (1967) 241.

50 T. Makino, S. Inoue and T. Tsuruta, *Makromol. Chem.*, *131* (1970) 147.

51 T. Makino, S. Inoue and T. Tsuruta, *Makromol. Chem.*, *150* (1971) 137.

52 T. Makino, S. Inoue and T. Tsuruta, *Makromol. Chem.*, *162* (1972) 235.

53 D. G. H. Ballard and C. H. Bamford, *J. Chem. Soc.*, (1958) 355.

54 F. Fuchs, *Chem. Ber.*, *55* (1922) 2943; H. Leuchs and W. Manasse, *Chem. Ber., 40* (1907) 3235; F. Wessely, *Z. Physiol. Chem.*, *146* (1925) 72.

55 E. Dyer, F. L. McCarthy, R. L. Johnson and E. V. Nagle, *J. Org. Chem.*, *22* (1957) 78.

56 H. Block and M. E. Cox, in G. T. Young (Ed.), *Peptides. Proceedings of the Fifth European Symposium*, Pergamon Press (1963) 83.

57 W. E. Hanby, S. G. Waley and J. Watson, *J. Chem. Soc.*, (1950) 3239; A. J. Hubert, R. Buijle and B. Hargitay, *Nature, 182* (1958) 259.

58 Y. Shalitin, Ph.D. Thesis, Hebrew University, Jerusalem, Israel (1958).

59 H. Weingarten, *J. Amer. Chem. Soc.*, *80* (1958) 352.

60 D. G. H. Ballard, C. H. Bamford and A. Elliott, *Makromol. Chem.*, *35* (1960) 222.

61 M. L. Huggins, *J. Amer. Chem. Soc.*, *74* (1952) 3963.

62 F. D. Williams, M. Eshaque and R. D. Brown, *Biopolymers, 10* (1971) 753.

63 H. G. Bührer and H. G. Elias, *Makromol. Chem.*, *169* (1973) 145.

64 J. J. Falcetta, Ph.D. Thesis, Polytechnic Inst. of Brooklyn (1969).

65 G. E. Ham (Ed.), *Copolymerization*, Vol. 18 of *High Polymers*, Interscience (1964).

66 P. Doty and R. D. Lundberg, *J. Amer. Chem. Soc.*, *78* (1956) 4810.

67 D. G. H. Ballard and C. H. Bamford, *J. Amer. Chem. Soc.*, *79* (1957) 2336.
68 R. D. Lundberg and P. Doty, *J. Amer. Chem. Soc.*, *79* (1957) 2338.
69 J. C. Mitchell, A. E. Woodward and P. Doty, *J. Amer. Chem. Soc.*, *79* (1957) 3955.
70 D. G. H. Ballard and C. H. Bamford, *J. Chem. Soc.*, (1959) 1039.
71 D. G. H. Ballard and C. H. Bamford, *Proc. Roy. Soc. Ser. A*, *236* (1956) 384.
72 C. H. Bamford and R. C. Price, *Trans. Faraday Soc.*, *61* (1965) 2208.
73 D. G. H. Ballard, *Biopolymers*, 2 (1964) 463.
74 Y. Imanishi, M. Sisido, T. Higashimura and S. Okamura, IUPAC Symposium on Macromolecular Chemistry Abstracts, Budapest (1969) 19.
75 M. Sisido, Y. Imanishi and S. Okamura, *Biopolymers*, *7* (1969) 937; *9* (1970) 791; *Polymer J.*, *1* (1970) 198.
76 K. Suguoki, Y. Imanishi, T. Higashimura and S. Okamura, *Biopolymers*, *7* (1969) 917, 925.
77 C. H. Bamford, H. Block and Y. Imanishi, *Biopolymers*, *4* (1966) 1067.
78 R. B. Woodward and C. H. Schramm, *J. Amer. Chem. Soc.*, *69* (1947) 1551.
79 Y. Shalitin and E. Katchalski, *J. Amer. Chem. Soc.*, *82* (1960) 1630.
80 C. H. Bamford, A. Elliott and W. E. Hanby, *Synthetic Polypeptides*, Academic Press, New York, 1956, Chapter 3.

Index

A

acetaldehyde, from preparation of polyethylene terephthalate, 502

—, pentane solution of, 332, 333

—, polymerization of, 368–371

acetic acid, effect on polymerization of formaldehyde, 352, 354

—, — methylmethacrylate, 42, 43

—, reaction + Et stearate, 518

acetic anhydride, effect on polymerization of chloral, 374

—, — formaldehyde, 349, 366

—, reaction + HClO₄, 70

acetone, PhCOCl + PhNH₂ in, 537

—, polymerization of formaldehyde in, 122, 337, 347–349

acetonitrile, CH₃OH + PhNCO in, 482, 485

—, polycondensation in, 520

—, polymerization of methylmethacrylate in, 229

—, — trioxane in, 314

acetophenone, polymerization of NCAs in, 589

acetylacetone, and epoxide polymerization, 267

acetylcaprolactam, effect on lactam polymerization, 402, 407, 421, 425, 429, 438, 439

acetyl hexafluoroantimonate, catalysis of tetrahydrofuran polymerization by, 291, 292, 299

acetyl perchlorate, catalysis of formaldehyde polymerization by, 344, 346, 347

acetyl piperidone, and lactam polymerization, 429

acrylonitrile, anionic copolymerization of, 56–58

—, — polymerization of, 2, 40, 41

—, electrochemical copolymerization of, 299

—, Ziegler—Natta polymerization of, 146, 230, 243

activated complex, *see* transition state

activation energy, of anionic polymerization of styrene and dienes, 11, 20, 31, 35, 39, 45, 47

—, of Ar diamines + isocyanates, 567

—, of esterification and polyesterification, 507, 514, 515, 519

—, of initiation in cationic polymerization, 92, 98, 99, 118

—, — Ziegler—Natta polymerization, 169–172

—, of ion-pair formation, 36

—, of isocyanates + ROH, 544, 545, 548

—, of Li alkyl dimer dissociation, 12

—, of PhCOCl + PhNH₂, 537

—, of PRNCO + H₂O, 566

—, of polyamidation, 529, 530, 556, 557

—, of polycondensation, 572–575

—, of polyethylene terephthalate decomposition, 501

—, of polymerization of butadiene, 212, 216, 217, 220, 222

—, — butene, 224

—, — C₂H₄, 193, 195–197, 199, 200, 202

—, — C₃H₆, 204

—, — CH₃CHO, 371

—, — CH₂O, 338, 347, 352, 356, 361, 362

—, — cyclic oxides, 325

—, — dioxolane, 304

—, — epoxides, 261, 262, 264

—, — hexene-1, 224

—, — lactams, 426, 436, 451, 452, 461, 464

—, — methylmethacrylate, 229, 230

—, — 4-methylpentene-1, 225

—, — NCAs, 604, 632, 633

—, — oxepane, 302

—, — oxetane, 276, 279, 280, 282, 325

—, — styrene, 209–211

—, — tetrahydrofuran, 292, 300, 325

—, — N-vinylcarbazole, 87

—, of propagation in cationic polymerization, 92, 98, 99, 118

—, — Ziegler—Natta polymerization, 180

—, of transfer in Ziegler—Natta polymerization, 184

—, of Ziegler—Natta copolymerization, 235, 243

activity coefficient, and ion-pair dissociation, 77

—, and polymerization of lactams, 384, 456, 457

acylium salts, initiation of tetrahydrofuran polymerization by, 108, 291, 292, 299

adipic acid, decarboxylation of, 500, 525

—, polycondensation with, 495, 497, 504, 505, 507, 508, 521 et seq.

alanine methyl ester, initiation of NCA polymerization by, 617

O

P

654